Eder, Schlögl (Eds.)
Nanocarbon-Inorganic Hybrids

Also of Interest

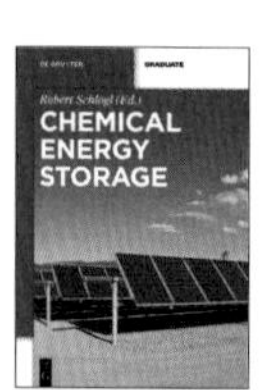

Chemical Energy Storage
Robert Schlögl (Ed.), 2012
ISBN 978-3-11-026407-4, e-ISBN 978-3-11-026632-0

Functional Materials
For Energy, Sustainable Development and Biomedical Sciences
Mario Leclerc, Robert Gauvin (Eds.), 2014
ISBN 978-3-11-030781-8, e-ISBN 978-3-11-030782-5

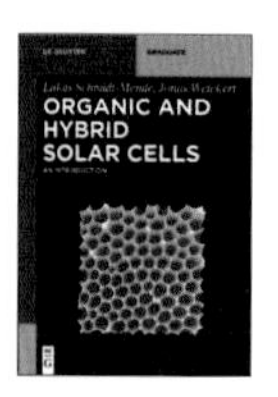

Organic and Hybrid Solar Cells
An Introduction
Lukas Schmidt-Mende, Jonas Weickert, 2015
ISBN 978-3-11-028318-1, e-ISBN 978-3-11-028320-4

Complex Oxides
Materials Physics, Synthesis, Characterization and Applications
Jan Seidel, 2015
ISBN 978-3-11-030424-4, e-ISBN 978-3-11-030425-1

Green
The International Journal of Sustainable Energy Conversion and Storage
Robert Schlögl (Editor-in-Chief)
ISSN 1869-8778

Nanocarbon-Inorganic Hybrids

Next Generation Composites for Sustainable Energy Applications

Edited by
Dominik Eder and Robert Schlögl

DE GRUYTER

Editors
Prof. Dominik Eder
Westfälische Wilhelms University
Department of Physical Chemistry
Correnstr. 28/30
48149 Münster
Germany

Prof. Robert Schlögl
Fritz Haber Institute of the Max Planck Society
Department of Inorganic Chemistry
Faradayweg 4–6
14195 Berlin
Germany

ISBN 978-3-11-026971-0
e-ISBN 978-3-11-026986-4
Set-ISBN 978-3-11-026987-1

Library of Congress Cataloging-in-Publication Data
A CIP catalog record for this book has been applied for at the Library of Congress.

Bibliographic information published by the Deutsche Nationalbibliothek
The Deutsche Nationalbibliothek lists this publication in the Deutsche Nationalbibliografie; detailed bibliographic data are available in the Internet at http://dnb.dnb.de.

Cover image: Simone Brandt/gettyimages/creative
Typesetting: PTP-Berlin, Protago-TEX-Production GmbH, Berlin
Printing and binding: Hubert & Co. GmbH & Co. KG, Göttingen
♾Printed on acid-free paper
Printed in Germany

www.degruyter.com

MIX
Papier aus verantwortungsvollen Quellen
FSC
www.fsc.org
FSC® C016439

Preface

Today's world is facing major challenges that directly affect our modern life. The demand for energy is expected to double by the year 2050. As a consequence of increased energy utilization the need is growing to protect our environment from the adverse effects of pollution and the destruction of natural habitats. To ensure sufficient supply of clean water and air and to nourish the growing population without conflicts and poverty are additional global challenges. In this arena, the transformation of the fossil energy system into a sustainable operation and the technical increase of energy efficiency are key objectives of chemical sciences with their ability to create novel fuels, materials and processes of molecular transformations. Using the energy of sunlight to split water into hydrogen as a clean energy source and storing energy in batteries and supercapacitors are two popular examples of energy science. Both challenges critically involve the availability of novel carbon materials. Carbon is the most versatile chemical element for designing molecules and materials. It enables us to address a wide range of functional characteristics by varying the assembly of only one type of atoms interacting with each other in essentially only two binding modes, i.e. sp^2/sp^3 hybridization.

The unlimited number of combinations of the two basic bonding motives allows the realization of molecular and supra-molecular properties limited only by our imagination. Hetero-atomic additions to the carbon backbone give additional chemical and structural diversity that needs exploitation in interface-controlled material science applications. The interplay between combining building blocks of carbon and decorating the products with hetero-elements forms the basis of a knowledge-based carbon material science. A critical strategy of material design is to combine carbon with other materials with diverging properties into spatial arrangements that create synergistic functions. Such materials, known as composites, were developed for carbon-based systems with the maturation of polymer science and are implemented today in numerous products ranging from materials for packaging, dental and medical use, energy production and storage, to structural materials for lightweight applications.

Nanocarbon composites are multiphase materials, in which a nanostructured *filler* (*i.e.* particles, whiskers, fibers, nanotubes or lamellae) is dispersed in an organic (*i.e.* polymer) or inorganic (*i.e.* carbon, ceramic or metal) *matrix*. In the last few years, carbon in the form of nanotubes (CNTs) in addition to nanostructured fibers or as graphene has attracted wide interest as a filler for nanocomposites. Typical application profiles are a high electrical conductivity in transparent conducting polymers or a remarkable fracture toughness reinforcing ceramics such as hydroxyapatite for bone replacement. Nanocomposites have a considerable impact on large-scale industrial applications of lightweight structural materials in aerospace and e-mobility, of electrically conducting plastics for electronic applications or as packaging materials with reduced gas permeability for foodstuff and air-sensitive goods.

Nanocarbon hybrids are a new class of composite materials in which the carbon nanostructures are compounded with thin layers of metals, semiconductors, inorganic glasses or ceramics. The carbon component gives ready access for gases and fluids to a large fraction of the inner surface area of the inorganic compound. In addition, a large interface between two materials with different bulk properties allows for the design of materials with interface-dominated properties. Their special appeal arises from charge and energy transfer through this interface, giving rise to transport properties different from the linear combination of the respective bulk properties. Although still at an early stage of research, such hybrid materials have demonstrated potential in applications concerning energy conversion and environmental protection. These include improved sensitivities in bio/chemical sensors, increased energy densities in batteries and larger capacities in supercapacitors, higher currents in field emission devices, more efficient charge separation and thus superior activities in photocatalysts and improved efficiencies in photovoltaics.

This book is dedicated exclusively to the family of nanocarbon hybrids covering a multidisciplinary research field that combines materials chemistry and physics with nanotechnology and applied energy sciences. It provides both introductory material on fundamental principles as well as reviews of the current research. Therefore, this book should be helpful for Master and PhD students wishing to become familiar with a modern field of knowledge-driven material science as well as for senior researchers and industrial staff scientists who explore the frontiers of knowledge.

The **first part of this book** introduces the concept of nanocarbons as building blocks. It establishes a scientific foundation for their subsequent use in hybrids and composites. *Chapter 1* provides a concise introduction into the world of carbon nanotubes (CNTs), explaining their unique structural characteristics, synthesis routes and key characterization techniques. It summarizes the profile of exceptional properties of CNT. *Chapter 2* concentrates on the synthesis and characterization of graphene-based materials, including single-layer and few-layer graphene as well as graphene oxide and its chemically/thermally reduced counterpart. The chapter further demonstrates that the dispersion of nanocarbons remains a key challenge for their implementation into hybrids. Chapters 3 and 4 are dedicated to the post-synthesis processing of nanocarbons. In particular, *Chapter 3* focuses on the chemical functionalization of CNTs, providing examples for a whole range of covalent and noncovalent functionalization routes. *Chapter 4* offers a comprehensive review on doping and filling of CNTs and the effect of defects on the hybridization of CNTs with polymers.

The **second part of this book**, comprising Chapters 5 to 10, is dedicated to the synthesis of nanocarbon hybrids and composites. *Chapter 5* begins by identifying the general synthesis routes towards nanocarbon hybrids, which can be categorized into *ex situ* (*i.e.* "building block") and *in situ* approaches, and comparing their advantages and disadvantages on the basis of some of the most intriguing recent results. In general, the *ex situ* route is a two-step process in which the inorganic compound is synthesized first, taking advantage of the existing wealth of knowledge in synthesizing nanomaterials (*i.e.* structure-property relationship). In a second step, the inorganic compound is linked to the nanocarbon surface via covalent, noncovalent or electrostatic interactions. In the *in situ* approach the inorganic or polymeric compound is grown on the (modified) nanocarbon surface from molecular precursors via (electro)chemical, vapor-based or physical deposition techniques, exploiting the stabilizing effects of nanocarbon as templates and as local heat sinks.

The examples discussed in *Chapter 5* cover a wide range of synthetic aspects, yet concentrate on hybrids involving CNTs, while *Chapter 6* summarizes recent developments on the hybridization of graphene-based materials. *Chapter 7* on the other hand introduces sustainable carbon materials made from hydrothermal carbonization (HTC) as promising candidates for hybrid materials. *Chapter 8* then combines nanocarbons with polymers and documents that engineering the interfaces is a challenge that is equally important in the synthesis of nanocarbon hybrids and of carbon composites. The book section is concluded by *Chapters 9 and 10*, which are dedicated to specific examples of hybrids. Chapter 9 describes hybrids whose components are all carbon based, such as CNTs hybridized with graphene, while Chapter 10 discusses the incorporation of graphene oxide into metal-organic framework structures (MOFs).

The **third part of the book** highlights the potential of nanocarbon hybrids for various important applications, particularly concerning environmental and sustainable energy applications. These include electrode materials in batteries and electrochemical capacitors (*Chapters 11 and 12*), sensors and emitters in field emission devices (*Chapter 13*), electrocatalysts in fuel cells (*Chapter 14*), supports for heterogeneous catalysts (*Chapter 15*), next-generation photocatalysts (*Chapter 16*), as well as active compounds in electrochromic and photovoltaic applications (*Chapters 17 and 18*). All these chapters discuss the benefits of nanocarbon hybrids in the respective application, identify major challenges and critically review the present state of research with the most intriguing recent developments. Finally, *Chapter 19* elaborates on the importance of defects and edge atoms in graphene-based hybrid materials.

This book illustrates that nanocarbon hybrid materials are an exciting new class of multi-purpose composites with great potential to become the next-generation energy materials. The synergistic effects in nanocarbon hybrids are manifold and it is clear that a detailed fundamental understanding of their origins will be essential to exploit the options given by combining classes of materials with diverging properties.

We foremost express our deepest thanks to our colleagues who spent considerable time and effort in writing the chapters in this book. We hope that this book will be useful to those interested in the subject of nanocarbon hybrids from many different perspectives and that it will establish a sound foundation for future research. We further would like to thank Julia Lauterbach and Karin Sora of De Gruyter Publishers for their tireless support and guidance. It is a particular pleasure to acknowledge the students of the Münster group for their invaluable help in proof-reading.

June 2014 — Dominik Eder and Robert Schlögl

Contents

Preface —— v

Contributing authors —— xvii

Part I: **Nanocarbon building blocks**

Paul Gebhardt and Dominik Eder

1 A short introduction on carbon nanotubes —— 3

1.1 Introduction —— **3**
1.2 Structural aspects —— **4**
1.2.1 Chirality —— **4**
1.2.2 Defects —— **5**
1.2.3 Doping —— **6**
1.3 Properties of CNTs —— **7**
1.3.1 Mechanical properties —— **7**
1.3.2 Electronic properties —— **8**
1.3.3 Thermal properties —— **9**
1.4 Characterization —— **10**
1.5 Synthesis —— **11**
1.5.1 Laser ablation —— **12**
1.5.2 Arc discharge —— **12**
1.5.3 Molten salt route / electrolytic process —— **13**
1.5.4 Chemical vapor deposition (CVD) —— **13**
1.6 Post-synthesis treatments —— **14**
1.6.1 Purification —— **14**
1.6.2 Separation of metallic and semiconducting CNTs —— **15**
1.6.3 Functionalization —— **16**
1.6.4 Assembly —— **18**
1.7 Summary —— **18**

Keith Paton

2 Synthesis, characterisation and properties of graphene —— 25

2.1 Introduction —— **25**
2.2 Properties —— **25**
2.3 Synthesis —— **26**
2.3.1 Micromechanical cleavage —— **26**
2.3.2 Liquid phase exfoliation —— **27**
2.3.3 Precipitation from metals/CVD —— **30**
2.3.4 Epitaxial growth from SiC —— **31**
2.4 Characterization —— **32**

Michele Melchionna and Maurizio Prato
3 Functionalization of carbon nanotubes — 43
3.1 Introduction — **43**
3.2 Functionalization. Why? — **44**
3.3 Types of functionalization — **46**
3.3.1 Covalent functionalization — **46**
3.3.2 Noncovalent functionalization — **54**
3.4 Functionalization with metals — **61**
3.5 Summary — **65**

S.M. Vega-Diaz, F. Tristán López, A. Morelos-Gómez, R. Cruz-Silva, and M. Terrones
4 The importance of defects and dopants within carbon nanomaterials during the fabrication of polymer composites — 71
4.1 Introduction — **71**
4.1.1 Carbon nanostructures and their properties — **72**
4.1.2 Doped carbon nanostructures — **74**
4.1.3 Defects in carbon nanostructures — **76**
4.1.4 Functionalization of carbon nanostructures for nanocomposites — **79**
4.2 Incorporation of nanocarbons into polymer composites and hybrids — **83**
4.2.1 Types of polymer composites — **83**
4.2.2 Synthesis approaches — **86**
4.3 Properties — **89**
4.3.1 Mechanical properties — **89**
4.3.2 Thermal properties — **93**
4.3.3 Electrical properties — **95**
4.3.4 Optical properties — **97**
4.3.5 Biocompatibility — **98**
4.3.6 Biodegradation — **99**
4.3.7 Permeability — **102**
4.4 Summary — **104**

Part II: Synthesis and characterisation of hybrids

Cameron J. Shearer and Dominik Eder
5 Synthesis strategies of nanocarbon hybrids — 125
5.1 Introduction — **125**
5.2 *Ex situ* approaches — **127**
5.2.1 Covalent interactions — **127**
5.2.2 Noncovalent interactions — **129**
5.3 *In situ* approaches — **134**
5.3.1 *In situ* polymerization — **135**

5.3.2 Inorganic hybridization from metal salts — 137
5.3.3 Electrochemical processes — 142
5.3.4 Sol–gel processes — 146
5.3.5 Gas phase deposition — 148
5.4 Other nanocarbons — 152
5.5 Comparison of synthesis techniques — 153
5.6 Summary — 154

C.N.R. Rao, H.S.S. Ramakrishna Matte, and Urmimala Maitra
6 Graphene and its hybrids with inorganic nanoparticles, polymers and other materials — 171
6.1 Introduction — 171
6.2 Synthesis — 172
6.3 Nanocarbon (graphene/C_{60}/SWNT) hybrids — 175
6.4 Graphene-polymer composites — 178
6.5 Functionalization of graphene and related aspects — 182
6.6 Graphene-inorganic nanoparticle hybrids — 185
6.7 Graphene hybrids with SnO_2, MoS_2 and WS_2 as anodes in batteries — 189
6.8 Graphene-MOF hybrids — 192
6.9 Summary — 195

Markus Antonietti, Li Zhao, and Maria-Magdalena Titirici
7 Sustainable carbon hybrid materials made by hydrothermal carbonization and their use in energy applications — 201
7.1 Introduction — 201
7.2 Hydrothermal synthesis of carbonaceous materials — 202
7.2.1 From pure carbohydrates — 202
7.2.2 From complex biomass — 209
7.2.3 Energy applications of hydrothermal carbons and their hybrids — 210
7.3 Summary — 221

Juan J. Vilatela
8 Nanocarbon-based composites — 227
8.1 Introduction — 227
8.2 Integration routes: From filler to other more complex structures — 228
8.2.1 Filler route — 229
8.2.2 Evaluation of reinforcement — 230
8.2.3 Other properties — 232
8.3 Hierarchical route — 235
8.3.1 Structure and improvement in properties — 236
8.3.2 Other properties — 238

8.4 Fiber route — 240
8.4.1 Different assembly routes — 241
8.4.2 Assembly properties and structure — 243
8.4.3 Assembly composites — 245
8.4.4 Other properties of nanocarbon assemblies — 248
8.5 Summary — 248

Robert Schlögl
9 Carbon-Carbon Composites — 255
9.1 Introduction — 255
9.2 Typology of C3 materials — 256
9.3 Synthesis — 259
9.4 Identification of the structural features of C3 material — 264
9.5 Surface chemistry — 266
9.6 Summary — 268

Teresa J. Bandosz
10 Graphite oxide-MOF hybrid materials — 273
10.1 Introduction — 273
10.2 Building blocks — 274
10.2.1 Graphite oxide — 274
10.2.2 Metal Organic Frameworks: MOF-5, HKUST-1 and MIL-100(Fe) — 275
10.3 Building the hybrid materials: Surface texture and chemistry — 276
10.4 MOF-Graphite oxides composites as adsorbents of toxic gases — 281
10.4.1 Ammonia — 282
10.4.2 Nitrogen dioxide — 284
10.4.3 Hydrogen sulfide — 286
10.5 Beyond the MOF-Graphite oxides composites — 288
10.6 Summary — 289

Part III: **Applications of nanocarbon hybrids**

Dang Sheng Su
11 Batteries/Supercapacitors: Hybrids with CNTs — 297
11.1 Introduction — 297
11.2 Application of hybrids with CNTs for batteries — 298
11.2.1 Lithium ion battery — 298
11.2.2 Lithium sulfur battery — 307
11.2.3 Lithium air battery — 308
11.3 Application of hybrids with CNTs in supercapacitor — 310
11.3.1 CNT-based carbon hybrid for supercapacitors — 311
11.3.2 CNT-based inorganic hybrid for supercapacitors — 313
11.4 Summary — 314

Zhong-Shuai Wu, Xinliang Feng, and Klaus Müllen
12 Graphene-metal oxide hybrids for lithium ion batteries and electrochemical capacitors — 319
12.1 Introduction — 319
12.2 Graphene for LIBs and ECs — 320
12.3 Graphene-metal oxide hybrids in LIBs and ECs — 321
12.3.1 Typical structural models of graphene-metal oxide hybrids — 321
12.3.2 Anchored model — 323
12.3.3 Encapsulated model — 327
12.3.4 Sandwich-like model — 330
12.3.5 Layered model — 332
12.3.6 Mixed models — 335
12.4 Summary — 336

John Robertson
13 Nanocarbons for field emission devices — 341
13.1 Introduction — 341
13.2 Carbon nanotubes – general considerations — 343
13.2.1 Field emission from nanocarbons — 346
13.2.2 Emission from nanowalls and CNTs walls — 346
13.3 Applications — 347
13.3.1 Field emission electron guns for electron microscopes — 347
13.3.2 Displays — 348
13.3.3 Microtriodes and E-beam lithography — 349
13.3.4 Microwave power amplifiers — 351
13.3.5 Ionization gauges — 352
13.3.6 Pulsed X-ray sources and tomography — 352
13.4 Summary — 353

Panagiotis Trogadas and Peter Strasser
14 Carbon, carbon hybrids and composites for polymer electrolyte fuel cells — 357
14.1 Introduction — 357
14.2 Carbon as electrode and electrocatalyst — 357
14.2.1 Structure and properties — 357
14.2.2 Electrochemical properties — 360
14.2.3 Applications — 362
14.3 Carbon, carbon hybrids and carbon composites in PEFCs — 368
14.3.1 Carbon as structural component in PEFCs — 368
14.3.2 Carbon as PEFC catalyst support — 369
14.3.3 Carbon hybrids and composites as ORR electrocatalysts — 379
14.4 Summary — 385

Benjamin Frank
15 Nanocarbon materials for heterogeneous catalysis — 393
15.1 Introduction — **393**
15.2 Relevant properties of nanocarbons — **394**
15.2.1 Textural properties and macroscopic shaping — **394**
15.2.2 Surface chemistry and functionalization — **397**
15.2.3 Confinement effect — **400**
15.3 Nanocarbon-based catalysts — **401**
15.3.1 Dehydrogenation of Hydrocarbons — **402**
15.3.2 Dehydrogenations of alcohols — **407**
15.3.3 Other reactions — **410**
15.4 Nanocarbon as catalyst support — **412**
15.4.1 Catalyst preparation strategies — **412**
15.4.2 Applications in heterogeneous catalysis — **416**
15.5 Summary — **422**

Gabriele Centi and Siglinda Perathoner
16 Advanced photocatalytic materials by nanocarbon hybrid materials — 429
16.1 Introduction — **429**
16.1.1 Hybrid *vs.* composite nanomaterials — **430**
16.1.2 Use of nanocarbon hybrid materials in photoreactions — **432**
16.2 Nanocarbon characteristics — **433**
16.2.1 The role of defects — **435**
16.2.2 Modification of nanocarbons — **437**
16.2.3 New aspects — **437**
16.2.4 Nanocarbon quantum dots — **438**
16.3 Mechanisms of nanocarbon promotion in photoactivated processes — **440**
16.4 Advantages of nanocarbon-semiconductor hybrid materials — **443**
16.5 Nanocarbon-semiconductor hybrid materials for sustainable energy — **447**
16.6 Summary — **448**

Jiangtao Di, Zhigang Zhao, and Qingwen Li
17 Electrochromic and photovoltaic applications of nanocarbon hybrids — 455
17.1 Introduction — **455**
17.2 Nanocarbon Hybrids for electrochromic materials and devices — **456**
17.2.1 Intrinsic electrochromism of nanocarbons — **456**
17.2.2 Synthesis and electrochromic properties of nanocarbon–metal oxide hybrids — **457**
17.2.3 Electrochromic properties of nanocarbon–polymer hybrids — **459**

17.3 Nanocarbon hybrids for photovoltaic applications — **461**
17.3.1 Working mechanisms of PECs and OPVs — **461**
17.3.2 Nanocarbon hybrids for PECs — **462**
17.3.3 Nanocarbon hybrids for OPVs — **468**
17.4 Summary — **469**

Rubén D. Costa and Dirk M. Guldi
18 Carbon nanomaterials as integrative components in dye-sensitized solar cells — 475
18.1 Today's dye-sensitized solar cells. Definition and potential — **475**
18.2 Major challenges in improving the performance of DSSCs — **477**
18.3 Carbon nanomaterials as integrative materials in semiconducting electrodes — **479**
18.3.1 Interlayers made out of carbon nanomaterials — **479**
18.3.2 Implementation of carbon nanomaterials into electrode networks — **480**
18.4 Carbon nanomaterials for solid-state electrolytes — **484**
18.4.1 Fullerene-based solid-state electrolytes — **484**
18.4.2 CNTs-based solid-state electrolytes — **485**
18.4.3 Graphene-based solid-state electrolytes — **487**
18.5 Versatility of carbon nanomaterials-based hybrids as novel type of dyes — **488**
18.5.1 Fullerene-based dyes — **488**
18.5.2 Graphene-based dyes — **490**
18.6 Photoelectrodes prepared by nanographene hybrids — **492**
18.6.1 Preparation of photoelectrodes by using noncovalently functionalized graphene — **492**
18.6.2 Preparation of photoelectrodes by preparing nanographene-based building blocks *via* electrostatic interactions — **494**
18.7 Summary — **496**

Ljubisa R. Radovic
19 Importance of edge atoms — 503
19.1 Introduction — **503**
19.2 External edges — **505**
19.3 Internal edges — **515**
19.4 Edge reconstruction — **519**
19.5 Summary — **522**

Index — 527

Contributing authors

Markus Antonietti
Max Planck Institute of Colloids and Interfaces
Potsdam, Germany
e-mail: pape@mpikg.mpg.de
Chapter 7

Teresa J. Bandosz
CUNY Energy Institute
The City College of New York
New York, NY, USA
e-mail: Tbandosz@ccny.cuny.edu
Chapter 10

Gabriele Centi
Department of Electronic Engineering,
Chemistry and Industrial Engineering
University of Messina
Messina, Italy
e-mail: centi@unime.it
Chapter 16

Rubén D. Costa
Department of Chemistry and Pharmacy
Friedrich–Alexander–University
Erlangen, Germany
e-mail: ruben.costa@fau.de
Chapter 18

Rudolfo Cruz-Silva
Research Center for Exotic Nanocarbons (JST)
Shinshu University
Wakasato, Nagano, Japan
e-mail: rcruzsilva@shinshu-u.ac.jp
Chapter 4

Jiangtao Di
Suzhou Institute of Nanotech and Nanobionics
Suzhou, China
e-mail: jiangtao.di@utdallas.edu
Chapter 17

Xinliang Feng
Max Planck Institute for Polymer Research
Mainz, Germany
e-mail: feng@mpip-mainz.mpg.de
Chapter 12

Benjamin Frank
Department of Inorganic Chemistry
Fritz Haber Institute of the Max Planck Society
Berlin, Germany
e-mail: benjamin.frank@mail.de
Chapter 15

Paul Gebhardt
Department of Physical Chemistry
Westfälische Wilhelms-University
Münster, Germany
e-mail: paul.gebhardt@uni-muenster.de
Chapter 1

Dirk M. Guldi
Department of Chemistry and Pharmacy
Friedrich–Alexander–University
Erlangen, Germany
e-mail: guldi@chemie.uni-erlangen.de
Chapter 18

Aarón Morelos-Gómez
Faculty of Engineering
Shinshu University
Wakasato, Nagano, Japan
e-mail: amorelos@shinshu-u.ac.jp
Chapter 4

Qingwen Li
Suzhou Institute of Nanotech & Nanobionics
Suzhou, China
e-mail: qwli2007@sinano.ac.cn
Chapter 17

F. Tristán López
Research Center for Exotic Nanocarbons (JST)
Shinshu University
Wakasato, Nagano, Japan
e-mail: ftristan@shinshu-u.ac.jp
Chapter 4

Urmimala Maitra
New Chemistry Unit and International Centre for Materials Science
Jawaharlal Nehru Centre for Advanced Scientific Research (JNCASR)
Bangalore, India
urmi@jncasr.ac.in
Chapter 6

H.S.S. Ramakrishna Matte
New Chemistry Unit and International Centre for Materials Science
Jawaharlal Nehru Centre for Advanced Scientific Research (JNCASR)
Bangalore, India
krishnamatte@gmail.com
Chapter 6

Michele Melchionna
INSTM, Unit of Trieste
Department of Chemistry and Pharmacy
University of Trieste
Trieste, Italy
e-mail: melchionnam@units.it
Chapter 3

Klaus Müllen
Max Planck Institute for Polymer Research
Mainz, Germany
e-mail: muellen@mpip-mainz.mpg.de
Chapter 12

Keith Paton
The Naughton Institute
CRANN, Trinity College Dublin
Dublin, Ireland
e-mail: patonk@tcd.ie
Chapter 2

Siglinda Perathoner
Department of Chemical Engineering Industrial Chemistry and Materials Science
University of Messina
Messina, Italy
e-mail: perathon@unime.it
Chapter 16

Maurizio Prato
Department of Chemistry and Pharmacy
University of Trieste
Trieste, Italy
e-mail: prato@units.it
Chapter 3

Ljubisa R.Radovic
The Pennsylvania State University
University Park, PA, USA
and
Department of Chemical Engineering
University of Concepión
Concepión,Chile
e-mail: ljrradovic@gmail.com
Chapter 19

C.N.R. Rao
CSIR Centre for Excellence in Chemistry
New Chemistry Unit and International Centre for Materials Science
Jawaharlal Nehru Centre for Advanced Scientific Research (JNCASR)
Bangalore, India
e-mail: cnrrao@jncasr.ac.in
Chapter 6

John Robertson
Department of Engineering
University of Cambridge
Cambridge, UK
e-mail: jr@eng.cam.ac.uk
Chapter 13

Cameron J. Shearer
Institute of Physical Chemistry
Westfälische Wilhelms-University Münster
Münster, Germany
e-mail: cshea_01@wwu.de
Chapter 5

Dang Sheng Su
Department of Inorganic Chemistry
Fritz Haber Institute of the Max Planck Society
Berlin, Germany
e-mail: dangsheng@fhi-berlin.mpg.de
Chapter 11

Peter Strasser
Department of Chemistry
Chemical Engineering division
Technical University Berlin
Berlin, Germany
e-mail: pstrasser@tu-berlin.de
Chapter 14

Mauricio Terrones
Department of Physics
Eberly College of Science
Pennsylvania State University
University Park, PA, USA
and
Research Center for Exotic Nanocarbons (JST)
Shinshu University
Wakasato, Nagano, Japan
e-mail: mutt11@psu.edu
Chapter 4

Maria-Magdalena Titirici
School of Engineering and Materials Science
Queen Mary University of London
Mile End, Eng, E406
m.m.titirici@qmul.ac.uk
Chapter 7

Panagiotis Trogadas
Department of Chemistry
Chemical Engineering division
Technical University Berlin
Berlin, Germany
e-mail: p.trogadas@ucl.ac.uk
Chapter 14

Sofia M. Vega-Diaz
Research Center for Exotic Nanocarbons (JST)
Shinshu University
Wakasato, Nagano, Japan
e-mail: sofia.mvega@gmail.com
Chapter 4

Juan José Vilatela
IMDEA-Materials
E.T.S. de Ingenieros de Caminos
Madrid, Spain
e-mail: juanjose.vilatela@imdea.org
Chapter 8

Zhong-Shuai Wu
Max Planck Institute for Polymer Research
Mainz, Germany
e-mail: wuzs@mpip-mainz.mpg.de
Chapter 12

Li Zhao
Institute of Coal Chemistry
Chinese Academy of Sciences
Taiyuan 030001
China
Chapter 7

Zhigang Zhao
Suzhou Institute of Nanotech and Nanobionics
Suzhou, China
e-mail: zgzhao2011@sinano.ac.cn
Chapter 17

Part I: **Nanocarbon building blocks**

Paul Gebhardt and Dominik Eder

1 A short introduction on carbon nanotubes

1.1 Introduction

Carbon nanotubes (CNTs) constitute a nanostructured carbon material that consists of rolled up layers of sp^2 hybridized carbon atoms forming a honeycomb lattice. After diamond, graphite and fullerenes, the one-dimensional tubular structure of CNTs is considered the 4th allotrope of carbon (graphene is the 5th).

For a long time the discovery of CNTs was attributed to the S. Iijima, who investigated soot formation during the production of fullerenes via arc discharge and observed the presence of "microfibrils" made of concentric carbon layers. Later he identified these microfibrils as possessing a honeycomb structure and being capped with fullerenes [1]. However, although this seminal paper in *Nature* in 1991 undoubtedly initiated a research field that has become tremendously popular, it was not the first publication on CNTs. In fact, electron micrographs of CNTs and carbon fibers have been published throughout the decades since as early as the 1950s, usually while being studied as by-products, impurities or catalyst poison in heterogeneous catalysis [2, 3]. It is now widely accepted that the first report on CNTs was published in 1952 by Radushkevich and Lukyanovich [4], who investigated the hydrogenation of CO over Fe catalysts under reaction conditions today deemed as suitable for CNT production. They documented the formation of "unusual carbon structures" by transmission electron microscopy (TEM). In contrast to these multi-walled CNTs (MWCNTs), the first report on single-walled carbon nanotubes (SWCNTs) was published in 1993 by Kiang *et al.* [5] and Iijima *et al.* [6] simultaneously (in the same issue of *Nature* on consecutive pages).

The remarkable interest in CNTs arose mainly because of their unique properties (see Section 1.3) that commend them for a wide range of applications involving both *science fiction*, *i.e.* space elevator [7], bullet-proof shirts [8], artificial muscles [9], and *real life applications*, *i.e.* field emission sources [10], Li-ion batteries [11], electrochemical storage devices [12], molecular sensors [13], hydrogen storage [14] and enhancing plant growth [15]. The incorporation of CNTs into organic or ceramic matrices (*i.e.* nanocomposites) and coating them with functional materials (*i.e.* hybrids) has increased their applicability and so stimulated further interest.

This chapter provides a concise summary of the most important concepts and characteristics of CNTs including structural aspects (*i.e.* chirality, defects, doping), properties (*i.e.* mechanical, electronic, thermal), synthesis and characterization techniques and post-processing strategies (*i.e.* purification, separation, functionalization), and is thus intended as an introduction for newcomers.

1.2 Structural aspects

1.2.1 Chirality

The structure of CNTs can be understood as sheets of graphene (*i.e.* monolayers of sp^2 hybridized carbon, see Chapter 2) rolled-up into concentric cylinders. This results in the saturation of part of the dangling bonds of graphene and thus in a decrease of potential energy, which counterbalances strain energy induced by curvature and thus stabilizes the CNTs. Further stabilization can be achieved by saturating the dangling bonds at the tips of the tubes so that in most cases CNTs are terminated by fullerene caps. Consequently, the smallest stable fullerene, *i.e.* C60, which is ~ 0.7 nm in diameter, thus determines the diameter of the smallest CNT. The fullerene caps can be opened by chemical and heat treatment, as described in Section 1.5.

CNTs may consist of just one layer (*i.e.* single-walled carbon nanotubes, SWCNTs), two layers (DWCNTs) or many layers (MWCNTs) and per definition exhibit diameters in the range of $0.7 < d < 2$ nm, $1 < d < 3$ nm, and $1.4 < d < 150$ nm, respectively. The length of CNTs depends on the synthesis technique used (Section 1.1.4) and can vary from a few microns to a current world record of a few cm [16]. This amounts to aspect ratios (*i.e.* length/diameter) of up to 10^7, which are considerably larger than those of high-performance polyethylene (PE, Dyneema). The aspect ratio is a crucial parameter, since it affects, for example, the electrical and mechanical properties of CNT-containing nanocomposites.

The structure of SWCNTs is characterized by the concept of chirality, which essentially describes the way the graphene layer is wrapped and is represented by a pair of indices (n, m). The integers n and m denote the number of unit vectors (a_1, a_2) along the two directions in the hexagonal crystal lattice of graphene that result in the chiral vector C_n (Fig. 1.1):

$$C_n = na_1 + ma_2.$$

If $m = 0$, the nanotubes are called "*zigzag*" nanotubes, if $n = m$, the nanotubes are defined as "*armchair*" nanotubes, and all other orientations are called "*chiral*". The deviation of C_n from a_1 is expressed by the inclination angle θ and ranges from 0° ("armchair") to 30° ("zigzag") [17].

The lattice vector and thus (n, m) can be also used to calculate the tube diameter following

$$d = \frac{a\sqrt{m^2 + mn + n^2}}{\pi}$$

where $a = 1.42 \times \sqrt{3}$ Å $= 0.246$ nm corresponds to the lattice constant in the graphite sheet (C-C distance for sp^2 hybridized carbon: 1.42 Å).

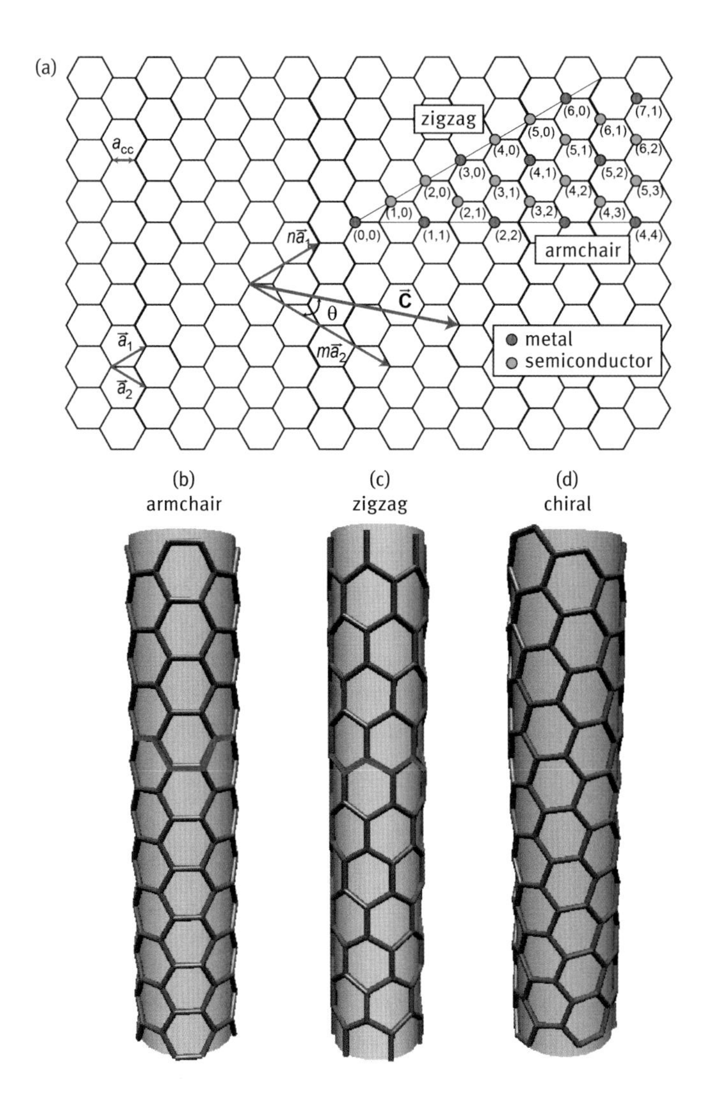

Fig. 1.1: (a) Schematic of unrolled SWCNT showing chiral vector C_n and the effect of m and n on the electronic properties of SWCNTs. (b, c, d) The direction of the chiral vector affects the appearance of the nanotube showing (b) (4,4) armchair, (c) (6,0) zigzag and (d) (5,3) exemplary chiral shape. With kind permission from [18].

1.2.2 Defects

Depending on the synthesis procedure (see Section 1.4) and purification methods (Section 1.6.1), the structure of synthesized carbon nanotubes may include a range of defects (see Chapter 4).

"**Topological defects**" describe the presence of rings other than hexagons, *i.e.* pentagons (n_5) and heptagons (n_7), which result in "kinks" and "elbows" in the usually planar hexagonal carbon layer. Strictly speaking, the fullerene caps are also topological defects as pentagons are essential for a spherical carbon structure. Pentagons and heptagons further accumulate near other defect sites in CNTs [19]. However, no ring sizes other than pentagons and heptagons have yet been observed [20]. Common occurrences of topological defects are pentagon-heptagon pairs directly connected to each other, which are also called Stone–Wales defects. This particular defect structure is more energetically favorable than other combinations [21]. $n_5 - n_7$ pairs can also be the center for intramolecular and intermolecular junctions, connecting two parts of one CNT with different helicity or two different CNTs [22]. The density of Stone–Wales defects is typically small due to the high activation barrier for the bond rotation (*i.e.* few eV). Stone–Wales defects further affect the absorption and charge transfer characteristics and increase strain energy, which leads to enhanced reactivity (*i.e.* for nucleophilic attack). A related group of defects are dislocations and vacancies. Vacancies can be induced by electron irradiation (*e.g.* by electron microscopy) and are also centers of enhanced chemical reactivity. On the other hand, vacancies can be annihilated in the presence of neighboring pentagon-heptagon pairs via an atomic exchange mechanism [23].

Although a honeycomb lattice theoretically consists of sp^2 atoms, the carbon's ability to represent intermediate states of hybridization leads to another kind of defect to counterbalance the strain energy induced by high curvature. This so-called **rehybridization** results in a higher π-character of the C-C bonds [24]. Furthermore, local sp^3 hybridization can be induced though chemical treatment, such as after thermal elimination of functional groups.

In general, most of these defects considerably affect the electronic, mechanical and chemical properties of CNTs [25]. For instance, the presence of just one carbon vacancy in the outer (*i.e.* load-bearing) carbon layer reduces the CNT's tensile strength by 30 % [26]. Defects are also scattering centers that limit the ballistic transport of electrons. However, structural defects on the outer surface of CNTs can also be beneficial when it comes to hybridizing them with metal oxides or other materials. Defects can serve as reaction centers for functionalization and as nucleation sites for crystal growth, influencing both crystallization (*i.e.* crystal structure, crystal defects) and growth (*i.e.* size, morphology) of the coating material [27–29].

1.2.3 Doping

Another set of defects is created though the introduction of heteroatoms into CNTs, providing an attractive tool for adjusting their electronic characteristics. Heteroatoms can be incorporated in different ways. An early approach [30] was **intercalation** between the walls of MWCNTs. This can be achieved *ex situ* [31–33], *i.e.* processing ready-made MWCNTs (*e.g.* annealing in ammonia), or *in situ*, *i.e.* during CNT synthesis [30].

In the latter case the experimental procedure depends on the synthesis method employed (see Section 1.5).

Alternatively, carbon atoms in the CNT lattice can be substituted for light elements such as nitrogen or boron (“**on-wall doping**”) [34], which again can be achieved *ex situ* or *in situ* (*i.e.* using nitrogen-containing carbon precursors in a CVD process). *Ex situ* substitutional doping requires the removal of a carbon atom and is thus energetically challenging.

The idea of nitrogen doping of carbon materials has been investigated since the 1960s [35]. In graphitic materials like CNTs, nitrogen can be bonded in two ways [36]: In the graphite-like bonding, only one carbon atom is replaced by a nitrogen atom, being covalently bonded to the three neighboring carbon atoms. The other possibility is the additional existence of a stabilizing defect next to the nitrogen atom, called pyridine type N.

In general, differences in chemical bonding and electron configuration between carbon atoms and dopants mandate the deviation from the geometric and electronic equilibrium structure of the aromatic layers in CNTs. As a consequence, topological defects such as Stone–Wales defects are formed with increased probability [37].

In addition, differences in electron density between carbon and the dopant result in considerably altered electronic properties, because the dopant can either act as an electron donor (*e.g.* nitrogen) or acceptor (*e.g.* boron). In the case of graphite-like doping, these electrons are injected into the πsystem of the graphitic structure, which leads to the desired n-type doping [38]. Pyridine-type N doping may produce n- or p-type behavior, depending on the doping level and the number of carbon vacancies [36].

1.3 Properties of CNTs

CNTs have received tremendous interest from both fundamental and applied research due to their unique physical properties [39]. These properties stem from their aromatic nature as well as the quasi 1D tubular geometry that renders many properties anisotropic. Table 1.1 summarizes the most important properties of CNTs.

1.3.1 Mechanical properties

CNTs are among the world’s strongest materials. The mechanical characteristics of CNTs stand out due to their very high stiffness and tensile strength. The Young’s modulus, *i.e.* a measure of stiffness and thus of how much a given material deforms upon application of mechanical stress, is about 1 TPa, which is comparable to in-plane graphite [40].

Tensile strength is the maximum mechanical stress that can be applied to a material without breaking it. Values have been calculated that reach up to 800 GPa for SWCNTs [41, 42]. Experimental data are typically significantly lower. Measuring the mechanical properties of individual CNTs constitutes a considerable challenge. The most reliable yet tedious approach is to identify an individual tube on a substrate and connect each end to an AFM tip. Still, the data are often not comparable due to the presence of a random number of structural defects (*i.e.* vacancies), which dramatically affect the strength of the CNT. Pugno *et al.* calculated that the presence of just one defect in the outer wall can reduce the tensile strength of a CNT by one third [7]. Therefore, the experimental data for single MWCNTs range from 15 to 270 GPa [43, 44]. In spite of these large differences, CNTs have the highest tensile strength of all known materials and even the "weakest" of the investigated CNTs are ~ 30 times stronger than steel [43].

1.3.2 Electronic properties

The electronic properties of CNTs are of particular importance for hybrid materials and strongly depend on the structure of the CNTs. Theoretical [45–48] and experimental results [49] reveal that SWCNTs are either metallic or semiconducting depending on diameter and chirality, while MWCNTs are generally metallic (due to band alignment upon interlayer interactions).

Flat single-layer graphene is a zero band-gap semiconductor [50], in which every direction for electron transport is possible. However, when the graphene sheet is rolled up to form a SWCNT, the number of allowed states is limited by quantum confinement in the radial direction [17], *i.e.* the movement of electrons is confined by the periodic boundary condition [51]:

$$C \cdot k = 2\pi q$$

in which C is the circumference of the CNT, k is the wave vector and q is an integer. As mentioned above, the orientation of the wave vector k depends on the (n, m) chirality on the tube. In the case of $|n - m| = 3q$, k passes through a Kpoint in the Brillouin Zone and the CNT is metallic, while for $|n - m| \neq 3q$, no Kpoint will be passed by k and the band structure is characterized by a finite band gap, *i.e.* the CNT is semiconducting (Fig. 1.2). The band gap decreases with increasing diameter of the CNT. However, very small CNTs might show exceptional behavior due to curvature effects [52] which result in a deviation from this classification.

In principle, metallic CNTs are ballistic conductors (*i.e.* electrons move without scattering), which means that the resistance is independent of the nanotube length. In MWCNTs, the conductance is defined by the outermost layer [53]. In general, the ballistic nature of CNTs is affected by the presence of structural defects [54]. Still, experimental values for electric current density are exceptionally high, *i.e.* up to 4 ·

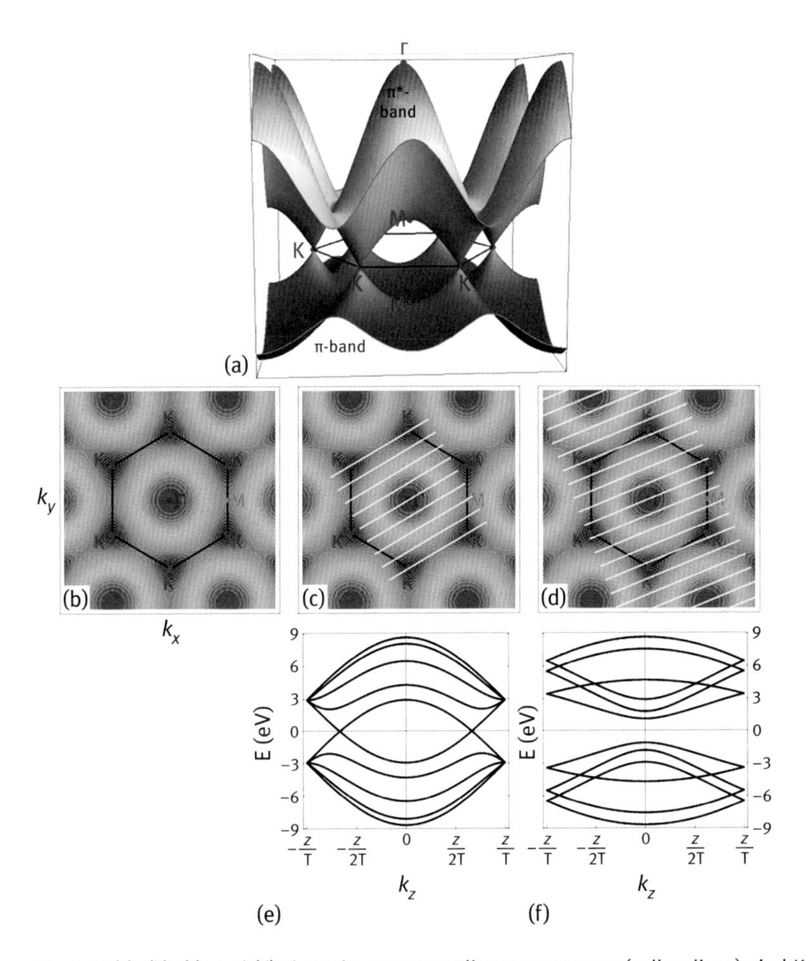

Fig. 1.2: (a), (b), (c), and (d) show the corresponding wave vectors (yellow lines) ajnd K-points (red dots) for structures in reciprocal space, as calculated by the tight binding method, belonging to (a) and (b) graphene, (c) metallic (4,4) SWCNT, (d) semiconducting (5,3) SWCNT, respectively. The blue dots represent energy maxima and the pink points represent saddle (M) points. (e) and (f) represent electronic band structures: (e) shows a (4,4) metallic nanotube where wave vectors cross a K-point and (f) shows a (5,3) semiconducting nanotube where no wave vectors cross a K-point. With kind permission from [18].

10^9 A/cm^2, which is more than 1000 times greater than in metals such as copper or aluminum [55] and is only outmatched by graphene [56].

1.3.3 Thermal properties

The thermal conductivity of individual SWCNTs was calculated to be as large as 6600 Wm^{-1}K^{-1} in axial direction, but only 1.52 Wm^{-1}K^{-1} perpendicular to its axis

[57]. Although the experimental values are significantly lower (*e.g.* 3500 $Wm^{-1}K^{-1}$ [58]), thermal conductivity is still roughly ten times that of copper, which is generally considered a material with good thermal conduction.

CNTs also possess a high thermal stability when treated in inert atmosphere (*e.g.* 2800 °C). In air, however, SWCNTs and MWCNTs will oxidize, albeit at comparatively high temperatures ranging between ~ 650 and 750 °C, respectively [59]. The presence of defects can to some extent affect the oxidation of CNTs, slightly reducing the oxidation temperature. Furthermore, the oxidation temperature of CNTs can be reduced considerably by the presence of metal oxide particles on the surface of CNTs (such as in nanocarbon-inorganic hybrids), in some cases down to 320 °C [60]. This catalytic activity of metal oxides correlates with their reducibility, *i.e.* the feasibility to create an oxygen vacancy, which suggests that the catalytic oxidation of CNTs proceeds via lattice oxygen at the metal oxide-CNT interface, presumably according to the Mars–van Krevelen mechanism.

Tab. 1.1: Properties of CNTs, from [39] with kind permission from ACS Publications.

Property	SWCNTs	MWCNTs	Graphite
Specific gravity	0.8 g/cm^3	< 1.8 g/cm^3	2.26 g/cm^3
Young's modulus	~ 1.4 TPa	~ 0.3–1 TPa	1 TPa (in plane)
Strength	50–500 GPa	15–60 GPa	
Resistivity	5–50 μΩ cm		50 μΩ cm (in plane)
Thermal conductivity	3000 $Wm^{-1}K^{-1}$		3000 $Wm^{-1}K^{-1}$ (in plane) 6 $Wm^{-1}K^{-1}$ (in plane)
Thermal expansion	Negligible		$-1 \times 10^{-6} K^{-1}$ (in plane) $29 \times 10^{-6} K^{-1}$ (c-axis)
Thermal stability	600–800 °C in air 2800 °C in vacuum		

1.4 Characterization

This section provides brief insights on some of the most important characterization techniques used for CNTs and other nanocarbons in addition to microscopy-related (*i.e.* SEM, TEM, AFM, STM) and diffraction (*i.e.* X-ray, electron) techniques.

Raman spectroscopy is one of the most powerful techniques for the characterization of nanocarbons. It is also a convenient technique because it involves almost no sample preparation and leaves the material unharmed. There are four characteristic bands for CNTs: The band at ~ 200 cm^{-1} is called radial breathing mode (RBM). It depends on the curvature and can be used to calculate the diameter of SWCNTs [61]. The relatively broad D-band at 1340 cm^{-1} is assigned to sp^2-related defects and disorder in the graphitic structure of the material. The tangential C-C stretching mode is located at ~ 1560 cm^{-1} (G-band). The second order mode of the D-band can be observed (G′-band, also known as 2D-band) at 2450–2650 cm^{-1} [62]. The quantity of defects is often evaluated by comparing the I_D/I_G ratio of the D- and G-bands [63]. Information on the chirality of SWCNTs can be obtained by correlating the RBM and the anti-Stokes Raman lines [63].

Information obtained: Diameter of SWCNTs, defects/crystallinity of CNTs, number of layers (if few).

Photoluminescence spectroscopy is used to analyze the electronic properties of semiconducting CNTs [64]. The emission wavelength is particularly sensitive to the tube diameter [65] and chemical defects [66]. However, a more dedicated sample preparation is required in order to eliminate van der Waals and charge transfer interactions between bundled CNTs. This can be done via ultrasonication or treatment of the bundles with surfactants that separate individual CNTs and suppress interactions between them [67].

Information obtained: Chirality of semiconducting SWCNTs (and thus diameter), band gap, bundling.

X-ray photoelectron spectroscopy (XPS) and infrared spectroscopy (FTIR) are used to gather information about the surface properties of CNTs, including functionalization, doping and structural modifications. In a study on SWCNTs, the C1s consisted of signals for sp^2 (2843.0 eV), sp^3 (285.0 eV) and oxygen bonded carbon (288.5 eV). Compared to highly-ordered pyrolytic graphite (HOPG) [68], the C1s peak of CNTs is significantly broadened because of defects and C-H bonds on the surface.

Computational simulations [68] suggest the possibility of identifying 7–8 **infrared**-active vibrational modes that depend on the chiral structure of CNTs. This was supported by experimental infrared spectroscopy [69] of SWCNTs, where features around 1598 and 874 cm^{-1} were found that could be linked to the calculated results.

Information obtained: sp^2/sp^3 level (*i.e.* defect), functional groups, dopants, chirality.

1.5 Synthesis

The topic of different synthesis techniques is well covered in several text books [70, 71] and review articles [72–74]. Thus, only a very brief introduction on the most important approaches is given here.

1.5.1 Laser ablation

Among the first techniques, applied primarily for SWCNTs, is the vaporization of a mixture of graphite target through ablation using a pulsed [75] or continuous [76] laser. In order to obtain SWCNTs, the graphite target needs to be modified with a metal catalyst (*e.g.* Co, Ni), the process needs to be carried out in an inert atmosphere (argon/nitrogen) and under reduced pressure (*e.g.* 650 mbar) and can be optimized by increasing the temperature to about 1200 °C [77]. Interestingly, helium is not used because its low molecular weight results in too high a cooling rate, which negatively affects nanotube formation [78].

Scott and co-workers [79] proposed a mechanism for CNT formation that involves the initial formation of a carbon cloud consisting of short carbon entities (C_1–C_3) and vaporized catalyst. Upon cooling, the condensation of the catalyst hinders the aggregation of carbon blocks into larger clusters, while supporting the anisotropic addition of carbon units into SWCNTs instead. The deposited SWCNTs are then collected from the cooled chamber walls after the synthesis.

In general, this synthesis approach features a good control over the diameter by adjusting the pressure in the reaction chamber and produces SWCNTs with few structural defects. Disadvantages are the high cost and poor scalability of this method due to the usage of a laser.

1.5.2 Arc discharge

Ablation of graphite through arc discharge constitutes another method that has become popular due to its simplicity. It involves applying an electric field (DC) between two graphite rods in an argon-filled reaction chamber. This stimulates an arc discharge between the graphite electrodes that creates very high local temperatures (~ 4000 °C). As a result, carbon is vaporized from the anode and re-deposited on the cathode, forming MWCNTs [80]. During the process, the anode is moved in order to provide a constant distance between the electrodes (*i.e.* distance of 1 cm). Although this technique is primarily used to produce MWCNTs, SWCNTs can also be obtained in high yields using a catalyst on the cathode or anode (*e.g.* Co [5] or Ni-Y [81]).

The advantages of this method include the simple experimental setup as well as the possibility to produce CNTs in relatively large quantities. However, it typically produces relatively short CNTs with a wide range of diameters as well as low purity, often producing fullerenes, graphite sheets or amorphous carbon as side products [82].

1.5.3 Molten salt route / electrolytic process

One synthesis approach that does not rely on CNT formation from the gas phase is molten salt synthesis. The reactor consists of a vertically oriented quartz tube that contains two graphite electrodes (*i.e.* anode is also the crucible) and is filled with ionic salts (*e.g.* LiCl or LiBr). An external furnace keeps the temperature at around 600 °C, which leads to the melting of the salt. Upon applying an electric field the ions penetrate and exfoliate the graphite cathode, producing graphene-type sheets that wrap up into CNTs on the cathode surface. Subsequently, the reactor is allowed to cool down, washed with water, and nanocarbon materials are extracted with toluene [83]. This process typically yields 20–30 % MWCNTs of low purity.

This process is very simple and requires low temperatures. However, the MWCNTs predominantly contain a large number of structural defects as well as amorphous carbon impurities. Furthermore, a significant part of the salt remains encapsulated within the CNTs.

1.5.4 Chemical vapor deposition (CVD)

Chemical vapor deposition (CVD) is versatile and currently the most popular method for the synthesis of SWCNTs and MWCNTs as well as other nanocarbons. CVD requires (1) a carbon source (*e.g.* methane, acetylene, toluene, ethanol etc.), (2) a catalyst (*e.g.* Fe, Co, Ni and various alloys), and (3) a suitable energy source (*e.g.* heat, plasma, laser). In the simplest case, *i.e.* the thermal CVD process, the carbon source is injected into a tube reactor along with an inert or reducing feed gas and decomposed at temperatures between 550–1200 °C. The resulting small carbon units then diffuse into the catalyst nanoparticles, which were either pre-deposited on a suitable substrate (*e.g.* silica, alumina) or continuously injected along with the carbon source (*e.g.* ferrocene-toluene solutions [84]). The key requirement for a suitable catalyst is a low solubility of carbon in the metal [71]. When the carbon concentration reaches saturation, carbon atoms segregate to the surface and precipitate in the thermodynamically favored hexagonal structure, still attached to the metal surface via 3d-π interactions [85]. According to the *yarmulke* mechanism [86], the driving force for this precipitation is the compensation of the high surface energy of the catalyst nanoparticles by the very low surface energy of this graphene-type sheet. These layers can act as nucleation sites for the growth of CNTs, which grow as long as carbon atoms are provided via diffusion through the bulk catalyst. In most cases the metal catalyst is pre-deposited onto a substrate such as silica or alumina. Depending on the strength of interaction between the catalyst particles and the support, the CNTs grow either via a base-growth model (*i.e.* metal particles remain attached to the substrate) or a tip-growth mechanism (*i.e.* the particles are lifted off the surface upon CNT growth and remain encapsulated near the CNTs' tips). In general, MWCNTs require the growth of multiple layers on the catalyst particle and favor larger catalyst particles. In contrast, SWCNTs are primarily produced on small catalyst particles, where the formation of additional layers is energetically restricted by their curvature [87]. The *yarmulke* mechanism further implies that all layers grow simultaneously.

The pre-deposition of catalyst particles enables the growth of vertically aligned CNTs [88] on substrates with positional control of the growth, such as is required for electrical devices for example [89]. In the modification of "continuously feeding the catalyst", variation of reaction time provides excellent control of the CNT length. Finally, it enables the continuous production of hierarchical nanocarbon assemblies, such as CNT fibers [84], which are macroscopic materials that harness the nanoscopic properties of CNTs and offer great potential for applications (*i.e.* rope, cables, and components in textiles...). In contrast, the CNTs often contain more defects than those produced *e.g.* via laser ablation, while part of the metal catalyst generally remains within the CNT sample, typically encapsulated within the interior walls, and may require post-synthesis purification (see next section). Still, the CVD process is a versatile and scalable technique that renders it the preferred method both for scientific as well as commercial production of high quality CNTs in large yields [90].

1.6 Post-synthesis treatments

1.6.1 Purification

Depending on the synthesis technique, CNTs may contain various impurities, including fullerenes and irregular carbon structures on the surface (*i.e.* amorphous) as well as residual salts and metal catalysts often encapsulated within a carbon shell.

Metal catalyst particles that are not encapsulated or covered by a carbon shell can easily be removed by washing the CNTs in mineral acids such as HCl. If the metal particles are covered by an "amorphous" carbon shell, this acid treatment can be assisted by microwave heating [91]. In contrast, metal particles that are encapsulated within closed CNTs require harsher conditions, such as ultrasonication/heating under reflux in strongly oxidizing acids (*e.g.* HNO_3/H_2SO_4), which first open the tubes and then dissolve the metal residues. It further renders the CNTs hydrophilic by incorporating oxygen functionalities, such as carboxylic, epoxy and hydroxyl groups (see Section 1.6.3). This property enables better dispersion in aqueous solutions as well as subsequent hybridization with inorganic compounds into composites and especially hybrid materials. However, apart from being a tedious and nasty chemical process, oxidative acid treatment does not provide sufficient control over the type, number and location of such functional groups (and thus the inorganic coating).

A more gentle approach would be the oxidative treatment in humid air at elevated temperatures [92]. This method selectively removes the amorphous carbon, while keeping the corrosion of the carbon surface to a minimum. The degree of purification/functionalization is determined by the vapor pressure of water as well as the temperature. In the latter case, the applied temperature must remain well below that of CNT oxidation (*i.e.* 600–700 °C), which strongly depends on the presence of catalytic impurities. For example, the presence of reducible metal oxides attached to the CNT's surface can reduce the oxidation temperature down to values as low as 330 °C [60].

A more sensitive approach to purification is the usage of non-oxidizing heat treatments [39, 93]. The amorphous carbon coating can be easily removed via annealing at ~ 1000 °C in flowing inert atmosphere. The tips of CNTs can be opened and much of the metal residues removed upon raising the temperature to about 1600 °C, depending on the melting point of the metal particles. At even higher temperatures (2000 °C, Ar), graphitization of the CNTs may even lead to an annealing of structural defects [39].

Tab. 1.2: Summary of typical purification techniques for CNTs. [a] Treatment can remove metal catalyst residues. [b] Carbon residues (*e.g.* amorphous or organic aromatic debris). [c] Purification introduces covalently bonded functional groups. [d] Only if not covered with carbon or encapsulated within CNT. [e] Only amorphous carbon around metal particles. From [39] with kind permission from ACS Publications.

	Metals[a]	**Carbon**[b]	**Functional**[c]	**Comments**
Heating in air/O_2		X	X	opens tips
Heating in wet O_2		X		
HCl_{conc}	X[d]			assisted with microwave or magnetic field
HNO_3/H_2SO_4	X	X	X	open tips, shortens CNTs
Microwave		X[e]		aided by acids (*e.g.*, HCl)
Ar @ 2000 °C	X	X		anneals crystal structure

1.6.2 Separation of metallic and semiconducting CNTs

As mentioned above, the electronic properties of SWCNTs depend on their chirality and may be semiconducting or metallic. There is still no satisfying way to produce just one sort of SWCNTs, which would require the exact control of catalyst particle size at elevated temperature. Hence, the separation of semiconducting from metallic SWCNTs is of paramount importance for their application in, for example, electric devices, field emission and photovoltaics etc.

Common separation methods can be divided into chemical and physical routes. **Chemical approaches** rely on the interaction of the surface of different CNT types with surfactant molecules. Early work has shown that octadecylamine [94] and agarose gel [95] adsorb preferably on semiconducting SWCNTs, while diazonium reagents [96] and DNA [97, 98] show preference with metallic tubes. The assemblies with adsorbed molecular species are considerably larger and heavier than the individual CNTs and can subsequently be separated via standard separation methods like ion exchange, chromatography or microfiltration. With these techniques CNTs can be separated depending on length [99], diameter [97] and chirality [100].

The **physical approach** uses alternating current (ac-) dielectrophoresis to separate metallic and semiconducting SWCNTs in a single step without the need for chemical modifications [101]. The difference in dielectric constant between the two types of SWCNTs results in an opposite movement along an electric field gradient between two electrodes. This leads to the deposition of metallic nanotubes on the microelectrode array, while semiconducting CNTs remain in the solution and are flushed out of the system. Drawbacks of this separation technique are the formation of mixed bundles of CNTs due to insufficient dispersion and difficulties in up-scaling the process [102].

1.6.3 Functionalization

The surface properties of CNTs are paramount for their hybridization with other components. The formation of large bundles due to van der Waals interactions between hydrophobic CNT walls further limits the accessibility of individual tubes. Functionalization of CNTs can enhance their dispersion in aqueous solvent mixtures and provide a means for tailoring the interfacial interactions in hybrid and composite materials. Functionalization techniques can be divided in covalent and non-covalent routes, which will be described in greater detail in Chapter 3.

Covalent functionalization requires the formation of a chemical bond with a surface carbon, which means the disruption of the graphitic structure and thus harsh reaction conditions. The simplest way is the aforementioned ultrasonication in oxidizing acids under elevated temperatures that results in carboxyl- and other oxygenated functionalities [104, 105]. It is very difficult to control the type, amount and exact location of these groups, which typically accumulate near defects and CNT tips due to enhanced reactivity of these sites [106, 107]. However, a great variety of covalent functionalization strategies have been developed over the last few years (see Chapter 3) that provide a wide range of functional groups. SWCNTs have been the focus of research because their high curvature induces enhanced reactivity that allows sidewall functionalization (Fig. 1.3). In contrast, the low reactivity of MWCNTs, *i.e.* essentially as inert as graphite, requires the presence of structural defects for covalent functionalization (Fig. 1.3). Covalent functionalization offers a very stable modification that can withstand high pressure and temperature, convection and sonication and has therefore been realized with aryl diazonium salts [108], azomethine ylides [109], nitrenes [110], and organic radicals [111]. Furthermore, fluorine functionalization [112] has been carried out which allows subsequent interactions with alkyllithium [113] and Grignard [114] reagents or terminal diamines [115]. However, introducing an additional covalent bond induces a change in hybridization towards sp^3 hybridized carbon atoms. Since the outer CNT layer is responsible for a good part of the conductance properties of CNTs

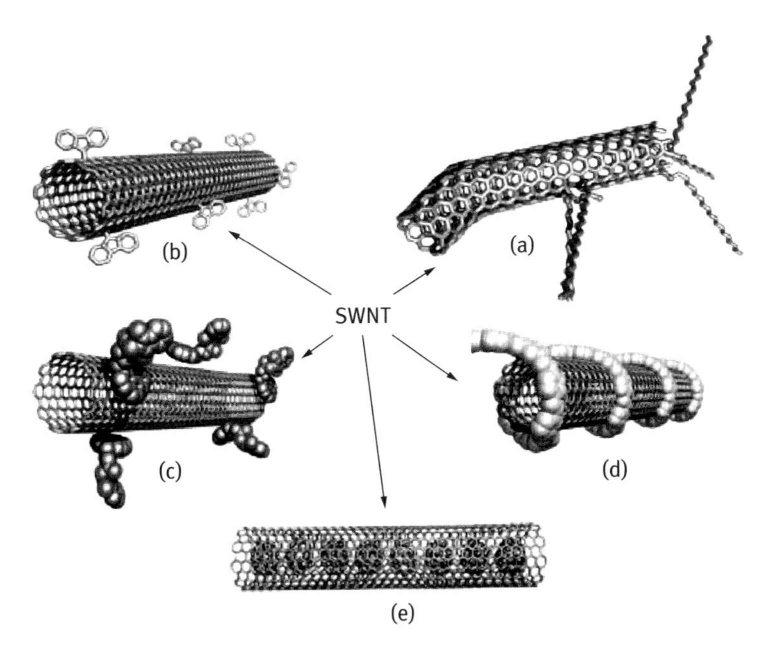

Fig. 1.3: Functionalization pathways for SWNTs: (a) defect-group functionalization, (b) covalent sidewall functionalization, (c) noncovalent exohedral functionalization with surfactants, (d) noncovalent exohedral functionalization with polymers, and (e) endohedral functionalization with, for example, C_{60}. For methods (b)–(e), the tubes are drawn in idealized fashion, but defects are found in real situations. From [103] with kind permission of Wiley.

as well as for their tensile strength (compare Section 1.3.2), covalent functionalization has a considerable, typically negative impact on their properties.

The alternative **noncovalent functionalization** does not rely on chemical bonds but on weaker Coulomb, van der Waals or π-π interactions to connect CNTs to surface-active molecules such as surfactants, aromatics, biomolecules (*e.g.* DNA), polyelectrolytes and polymers. In most cases, this approach is used to improve the dispersion properties of CNTs [116], for example via charge repulsion between micelles of sodium dodecylsulfate [65] adsorbed on the CNT surface or a large solvation shell formed by neutral molecule (*e.g.* polyvinylpyrrolidone) [117] around the CNTs.

One good example of noncovalent functionalization for subsequent hybridization is the use of benzyl alcohol (BA) [118]. π-π interactions between the aromatic ring of BA and the CNT sidewalls result in a good dispersibility in ethanol. Furthermore, BA offers a well-ordered and well-distributed functionalization [119] of hydroxyl groups on the sidewalls of the CNTs that can be used to hybridize the material with a large number of metal oxides using conventional chemical methods [60].

1.6.4 Assembly

The macroscopic design of CNT materials is an important factor regarding possible applications. Due to their anisotropy (see Section 1.3), the respective orientation and ordered structure can strongly affect the macroscopic properties of CNTs. Numerous attempts have been made in this direction, including different kinds of aligned CNTs [120]; CNT carpets [121], films [122, 123] and fibers [124]. Of these, this chapter will only focus on the latter.

In contrast to early post-synthesis attempts of fiber formation [125, 126], a more advanced route produces CNT fibers directly from a CVD oven [84]. It uses a CVD setup with an oxygen-containing carbon source and a catalyst mixture of ferrocene and thiophene. At temperatures around 1150 °C the forming aerogel can be drawn by winding it onto a rotating rod, collecting a continuous flow of fiber onto a spindle. That way, fibers can be wound at speeds up to 100 m/min with strengths beyond 2 GPa/SG, which make them some of the world's strongest fibers [124].

1.7 Summary

Carbon nanotubes comprise a very promising material for various applications and especially as an active component in composites and hybrids as will be documented in the other chapters of this book. Harnessing these nanoscopic assets in a macroscopic material would maximize CNTs' potential and applicability. The choice of synthesis technique and purification method, which define size, type, properties, quality and purity of CNTs as well as their processability, is crucial for their implementation into composites and hybrids.

For example, the exceptional electronic properties of CNTs are most distinct in metallic nanotubes, having a very high conductivity. In a realistic application as photosensitizer, however, a semiconductor is required, which implies further demands on the chiral structure of the CNTs (see Section 1.3.2). The use of a hypothetical CNT-inorganic hybrid with a semiconductor coating on the CNT surface in a dye-sensitized solar cell (DSSC) [127] also implies the need for closed caps. Otherwise, electrolyte could leak inside the nanotube channel, leading to a direct CNT electrolyte interface, and thus, to a short circuit. As a consequence, the need for closed tips reduces the options for surface functionalization, because a harsh acid treatment would also lead to an opening at the tips. One answer to this problem could be the use of CNTs with a defect-rich surface, defects being a possible nucleation point and reactive site for possible coatings. The trade-off for this, however, may be increase electrical resistance caused by the larger amount of defect sites in the CNT walls.

It is still a great challenge to obtain control of the different characteristics and the specific properties of CNTs that restrict their use in commercial real-life applications. However, considering the ongoing progress in research and development on CNTs and other nanocarbons as well as their hybrids and composites, combined with the steadily increasing industrial production of CNTs, the future of CNTs appears bright and promising.

Bibliography

[1] Iijima, S. *Nature* **1991**, *354*, 56.
[2] Monthioux, M.; Kuznetsov, V. L. *Carbon* **2006**, *44*, 1621.
[3] Oberlin, A.; Endo, M.; Koyama, T. *Journal of Crystal Growth* **1976**, *32*, 335.
[4] Radushkevich, L. V.; Lukyanovich, V. M. *VM Zurn Fisic Chim* **1952**, 88.
[5] Bethune, D. S.; Klang, C. H.; de Vries, M. S.; Gorman, G.; Savoy, R.; Vazquez, J.; Beyers, R. *Nature* **1993**, *363*, 605.
[6] Iijima, S.; Ichihashi, T. *Nature* **1993**, *363*, 603.
[7] Pugno, N. M. *J. Phys.: Condens. Matter* **2006**, *18*, S1971.
[8] Mylvaganam, K.; Zhang, L. C. *Appl. Phys. Lett.* **2006**, *89*, 123127.
[9] Aliev, A. E.; Oh, J.; Kozlov, M. E.; Kuznetsov, A. A.; Fang, S.; Fonseca, A. F.; Ovalle, R.; Lima, M. D.; Haque, M. H.; Gartstein, Y. N.; Zhang, M.; Zakhidov, A. A.; Baughman, R. H. *Science* **2009**, *323*, 1575.
[10] Yue, G. Z.; Qiu, Q.; Gao, B.; Cheng, Y.; Zhang, J.; Shimoda, H.; Chang, S.; Lu, J. P.; Zhou, O. *Appl. Phys. Lett.* **2002**, *81*, 355.
[11] Zhong, D. Y.; Zhang, G. Y.; Liu, S.; Wang, E. G.; Wang, Q.; Li, H.; Huang, X. J. *Appl. Phys. Lett.* **2001**, *79*, 3500.
[12] Frackowiak, E.; Béguin, F. *Carbon* **2002**, *40*, 1775.
[13] Wang, J. *Electrocatalysis* **2005**, *17*, 7.
[14] Darkrim, F. L.; Malbrunot, P.; Tartaglia, G. P. *International Journal of Hydrogen Energy* **2002**, *27*, 193.
[15] Khodakovskaya, M.; Dervishi, E.; Mahmood, M.; Xu, Y.; Li, Z.; Watanabe, F.; Biris, A. S. *ACS Nano* **2009**, *3*, 3221.
[16] Wen, Q.; Zhang, R.; Qian, W.; Wang, Y.; Tan, P.; Nie, J.; Wei, F. *Chemistry of Materials* **2010**, *22*, 1294.
[17] Terrones, M. *Ann. Rev. Mat. Res.* **2003**, *33*, 419.
[18] Shearer, C. J., Flinders University, 2011.
[19] Suenaga, K.; Wakabayashi, H.; Koshino, M.; Sato, Y.; Urita, K.; Iijima, S. *Nat Nano* **2007**, *2*, 358.
[20] Ebbesen, T. W.; Takada, T. *Carbon* **1995**, *33*, 973.
[21] Jeong, B. W.; Ihm, J.; Lee, G.-D. *Phys. Rev. B* **2008**, *78*, 165403.
[22] Charlier, J. C. *Accounts of Chemical Research* **2002**, *35*, 1063.
[23] Lee, G.-D.; Wang, C.-Z.; Yoon, E.; Hwang, N.-M.; Ho, K.-M. *Appl. Phys. Lett.* **2010**, *97*, 093106.
[24] Kane, C. L.; Mele, E. J. *Phys. Rev. Lett.* **1997**, *78*, 1932.
[25] Blase, X.; Benedict, L. X.; Shirley, E. L.; Louie, S. G. *Phys. Rev. Lett.* **1994**, *72*, 1878.
[26] Pugno, N. M. *J. Phys.: Condens. Matter* **2008**, *20*, 474205.
[27] Eder, D.; Windle, A. H. *J. Mater. Chem.* **2008**, *18*, 2036.
[28] Krissanasaeranee, M.; Wongkasemjit, S.; Cheetham, A. K.; Eder, D. *Chemical Physics Letters* **2010**, *496*, 133.
[29] Ren, Z.; Kim, E.; Pattinson, S. W.; Subrahmanyam, K. S.; Rao, C. N. R.; Cheetham, A. K.; Eder, D. *Chem. Sci.* **2011**, *3*, 209.

[30] Zhou, O.; Fleming, R. M.; Murphy, D. W.; Chen, C. H.; Haddon, R. C.; Ramirez, A. P.; Glarum, S. H. *Science* **1994**, *263*, 1744.
[31] Duclaux, L. *Carbon* **2002**, *40*, 1751.
[32] Hamwi, A.; Alvergnat, H.; Bonnamy, S.; Béguin, F. *Carbon* **1997**, *35*, 723.
[33] Maurin, G.; Bousquet, C.; Henn, F.; Bernier, P.; Almairac, R.; Simon, B. *Chemical Physics Letters* **1999**, *312*, 14.
[34] Ayala, P.; Arenal, R.; Rümmeli, M.; Rubio, A.; Pichler, T. *Carbon* **2010**, *48*, 575.
[35] Marchand, A.; Zanchetta, J. V. *Carbon* **1966**, *3*, 483.
[36] Terrones, M.; Jorio, A.; Endo, M.; Rao, A. M.; Kim, Y. A.; Hayashi, T.; Terrones, H.; Charlier, J. C.; Dresselhaus, G.; Dresselhaus, M. S. *Materials Today* **2004**, *7*, 30.
[37] Golberg, D.; Bando, Y.; Bourgeois, L.; Kurashima, K.; Sato, T. *Carbon* **2000**, *38*, 2017.
[38] Krstić, V.; Rikken, G. L. J. A.; Bernier, P.; Roth, S.; Glerup, M. *EPL [Europhysics Letters]* **2007**, *77*, 37001.
[39] Eder, D. *Chem. Rev.* **2010**, *110*, 1348.
[40] Robertson, D. H.; Brenner, D. W.; Mintmire, J. W. *Phys. Rev. B* **1992**, *45*, 12592.
[41] Ruoff, R. S.; Lorents, D. C. *Carbon* **1995**, *33*, 925.
[42] Yakobson, B. I.; Brabec, C. J.; Bernholc, J. *Phys. Rev. Lett.* **1996**, *76*, 2511.
[43] Yu, M.-F.; Lourie, O.; Dyer, M. J.; Moloni, K.; Kelly, T. F.; Ruoff, R. S. *Science* **2000**, *287*, 637.
[44] Demczyk, B. G.; Wang, Y. M.; Cumings, J.; Hetman, M.; Han, W.; Zettl, A.; Ritchie, R. O. *Materials Science and Engineering: A* **2002**, *334*, 173.
[45] Hamada, N.; Sawada, S.-i.; Oshiyama, A. *Phys. Rev. Lett.* **1992**, *68*, 1579.
[46] Mintmire, J. W.; Dunlap, B. I.; White, C. T. *Phys. Rev. Lett.* **1992**, *68*, 631.
[47] Saito, R.; Fujita, M.; Dresselhaus, G.; Dresselhaus, M. S. *Appl. Phys. Lett.* **1992**, *60*, 2204.
[48] Saito, R.; Fujita, M.; Dresselhaus, G.; Dresselhaus, M. S. *Phys. Rev. B* **1992**, *46*, 1804.
[49] Odom, T. W.; Huang, J.-L.; Kim, P.; Lieber, C. M. *Nature* **1998**, *391*, 62.
[50] Wallace, P. R. *Phys. Rev.* **1947**, *71*, 622.
[51] Joselevich, E. *ChemPhysChem* **2004**, *5*, 619.
[52] Lu, X.; Chen, Z. *Chem. Rev.* **2005**, *105*, 3643.
[53] Delaney, P.; Di Ventra, M.; Pantelides, S. T. *Appl. Phys. Lett.* **1999**, *75*, 3787.
[54] Wei, B. Q.; Vajtai, R.; Ajayan, P. M. *Appl. Phys. Lett.* **2001**, *79*, 1172.
[55] Hong, S.; Myung, S. *Nat Nano* **2007**, *2*, 207.
[56] Akturk, A.; Goldsman, N. *J. Appl. Phys.* **2008**, *103*, 053702.
[57] Berber, S.; Kwon, Y.-K.; Tománek, D. *Phys. Rev. Lett.* **2000**, *84*, 4613.
[58] Pop, E.; Mann, D.; Wang, Q.; Goodson, K.; Dai, H. *Nano Lett.* **2006**, *6*, 96.
[59] Thostenson, E. T.; Li, C.; Chou, T.-W. *Composites Science and Technology* **2005**, *65*, 491.
[60] Aksel, S.; Eder, D. *J. Mater. Chem.* **2010**, *20*, 9149.
[61] Rols, S.; Righi, A.; Alvarez, L.; Anglaret, E.; Almairac, R.; Journet, C.; Bernier, P.; Sauvajol, J. L.; Benito, A. M.; Maser, W. K.; Muñoz, E.; Martinez, M. T.; de laFuente, G. F.; Girard, A.; Ameline, J. C. *Eur. Phys. J. B* **2000**, *18*, 201.
[62] Mamedov, A. A.; Kotov, N. A.; Prato, M.; Guldi, D. M.; Wicksted, J. P.; Hirsch, A. *Nat Mater* **2002**, *1*, 190.
[63] Dresselhaus, M. S.; Jorio, A.; Hofmann, M.; Dresselhaus, G.; Saito, R. *Nano Lett.* **2010**, *10*, 751.
[64] Belin, T.; Epron, F. *Materials Science and Engineering: B* **2005**, *119*, 105.
[65] O'Connell, M. J.; Bachilo, S. M.; Huffman, C. B.; Moore, V. C.; Strano, M. S.; Haroz, E. H.; Rialon, K. L.; Boul, P. J.; Noon, W. H.; Kittrell, C.; Ma, J.; Hauge, R. H.; Weisman, R. B.; Smalley, R. E. *Science* **2002**, *297*, 593.
[66] Lauret, J. S.; Voisin, C.; Cassabois, G.; Roussignol, P.; Delalande, C.; Filoramo, A.; Capes, L.; Valentin, E.; Jost, O. *Physica E* **2004**, *21*, 1057.
[67] Lefebvre, J.; Fraser, J. M.; Homma, Y.; Finnie, P. *Appl Phys A* **2004**, *78*, 1107.

[68] Pham-Huu, C.; Keller, N.; Roddatis, V. V.; Mestl, G.; Schlogl, R.; Ledoux, M. J. *Physical Chemistry Chemical Physics* **2002**, *4*, 514.

[69] Kuhlmann, U.; Jantoljak, H.; Pfänder, N.; Bernier, P.; Journet, C.; Thomsen, C. *Chemical Physics Letters* **1998**, *294*, 237.

[70] Harris, P. J. F. *Carbon Nanotube Science: synthesis, properties and applications*; Cambridge University Press: Cambridge, UK, 2009.

[71] Dresselhaus, M. S.; Dresselhaus, G.; Avouris, P. *Carbon nanotubes: synthesis, structure, properties, and applications*; Springer: Berlin; New York, 2001.

[72] Baddour, C. E.; Briens, C. *Int. J. Chem. React. Eng.* **2005**, *3*, 404.

[73] Prasek, J.; Drbohlavova, J.; Chomoucka, J.; Hubalek, J.; Jasek, O.; Adam, V.; Kizek, R. *J. Mater. Chem.* **2011**, *21* (40), 15872.

[74] See, C. H.; Harris, A. T. *Ind. Eng. Chem. Res.* **2007**, *46*, 997.

[75] Eklund, P. C.; Pradhan, B. K.; Kim, U. J.; Xiong, Q.; Fischer, J. E.; Friedman, A. D.; Holloway, B. C.; Jordan, K.; Smith, M. W. *Nano Lett.* **2002**, *2*, 561.

[76] Bolshakov, A. P.; Uglov, S. A.; Saveliev, A. V.; Konov, V. I.; Gorbunov, A. A.; Pompe, W.; Graff, A. *Diamond and Related Materials* **2002**, *11*, 927.

[77] Thess, A.; Lee, R.; Nikolaev, P.; Dai, H.; Petit, P.; Robert, J.; Xu, C.; Lee, Y. H.; Kim, S. G.; Rinzler, A. G.; Colbert, D. T.; Scuseria, G. E.; Tománek, D.; Fischer, J. E.; Smalley, R. E. *Science* **1996**, *273*, 483.

[78] Muñoz, E.; Maser, W. K.; Benito, A. M.; Martıìnez, M. T.; de la Fuente, G. F.; Maniette, Y.; Righi, A.; Anglaret, E.; Sauvajol, J. L. *Carbon* **2000**, *38*, 1445.

[79] Scott, L. T.; Jackson, E. A.; Zhang, Q.; Steinberg, B. D.; Bancu, M.; Li, B. *J. Am. Chem. Soc.* **2012**, *134*, 107.

[80] Gamaly, E. G.; Ebbesen, T. W. *Phys. Rev. B* **1995**, *52*, 2083.

[81] Journet, C.; Maser, W. K.; Bernier, P.; Loiseau, A.; de la Chapelle, M. L.; Lefrant, S.; Deniard, P.; Lee, R.; Fischer, J. E. *Nature* **1997**, *388*, 756.

[82] Popov, V. N. *Mat. Sci. Eng.: R.* **2004**, *43*, 61.

[83] Hsu, W. K.; Terrones, M.; Hare, J. P.; Terrones, H.; Kroto, H. W.; Walton, D. R. M. *Chemical Physics Letters* **1996**, *262*, 161.

[84] Li, Y.-L.; Kinloch, I. A.; Windle, A. H. *Science* **2004**, *304*, 276.

[85] Charlier, J.-C.; Iijima, S. In *Carbon Nanotubes*; Dresselhaus, M. S., Dresselhaus, G., Avouris, P., Eds.; Springer: Berlin, Heidelberg, 2001, p 55.

[86] Dai, H.; Rinzler, A. G.; Nikolaev, P.; Thess, A.; Colbert, D. T.; Smalley, R. E. *Chemical Physics Letters* **1996**, *260*, 471.

[87] Dupuis, A.-C. *Progress in Materials Science* **2005**, *50*, 929.

[88] Constantopoulos, K. T.; Shearer, C. J.; Ellis, A. V.; Voelcker, N. H.; Shapter, J. G. *Adv. Mat.* **2010**, *22*, 557.

[89] Fan, S.; Liang, W.; Dang, H.; Franklin, N.; Tombler, T.; Chapline, M.; Dai, H. *Physica E* **2000**, *8*, 179.

[90] Donaldson, K.; Aitken, R.; Tran, L.; Stone, V.; Duffin, R.; Forrest, G.; Alexander, A. *Toxicol. Sci.* **2006**, *92*, 5.

[91] Park, T.-J.; Banerjee, S.; Hemraj-Benny, T.; Wong, S. S. *J. Mater. Chem.* **2006**, *16*, 141.

[92] Salzmann, C. G.; Llewellyn, S. A.; Tobias, G.; Ward, M. A. H.; Huh, Y.; Green, M. L. H. *Adv. Mat.* **2007**, *19*, 883.

[93] Andrews, R.; Jacques, D.; Qian, D.; Dickey, E. C. *Carbon* **2001**, *39*, 1681.

[94] Chattopadhyay, D.; Galeska, I.; Papadimitrakopoulos, F. *J. Am. Chem. Soc.* **2003**, *125*, 3370.

[95] Tanaka, T.; Jin, H.; Miyata, Y.; Fujii, S.; Suga, H.; Naitoh, Y.; Minari, T.; Miyadera, T.; Tsukagoshi, K.; Kataura, H. *Nano Lett.* **2009**, *9*, 1497.

[96] Strano, M. S.; Dyke, C. A.; Usrey, M. L.; Barone, P. W.; Allen, M. J.; Shan, H.; Kittrell, C.; Hauge, R. H.; Tour, J. M.; Smalley, R. E. *Science* **2003**, *301*, 1519.

[97] Arnold, M. S.; Stupp, S. I.; Hersam, M. C. *Nano Lett.* **2005**, *5*, 713.

[98] Zheng, M.; Jagota, A.; Strano, M. S.; Santos, A. P.; Barone, P.; Chou, S. G.; Diner, B. A.; Dresselhaus, M. S.; McLean, R. S.; Onoa, G. B.; Samsonidze, G. G.; Semke, E. D.; Usrey, M.; Walls, D. J. *Science* **2003**, *302*, 1545.

[99] Feng, Q.-P.; Xie, X.-M.; Liu, Y.-T.; Gao, Y.-F.; Wang, X.-H.; Ye, X.-Y. *Carbon* **2007**, *45*, 2311.

[100] Arnold, M. S.; Green, A. A.; Hulvat, J. F.; Stupp, S. I.; Hersam, M. C. *Nat Nano* **2006**, *1*, 60.

[101] Krupke, R.; Hennrich, F.; Löhneysen, H. v.; Kappes, M. M. *Science* **2003**, *301*, 344.

[102] Krupke, R.; Hennrich, F. *Adv. Eng. Mat.* **2005**, *7*, 111.

[103] Hirsch, A. *Angewandte Chemie International Edition* **2002**, *41*, 1853.

[104] Liu, J.; Rinzler, A. G.; Dai, H.; Hafner, J. H.; Bradley, R. K.; Boul, P. J.; Lu, A.; Iverson, T.; Shelimov, K.; Huffman, C. B.; Rodriguez-Macias, F.; Shon, Y.-S.; Lee, T. R.; Colbert, D. T.; Smalley, R. E. *Science* **1998**, *280*, 1253.

[105] Rinzler, A. G.; Liu, J.; Dai, H.; Nikolaev, P.; Huffman, C. B.; Rodríguez-Macías, F. J.; Boul, P. J.; Lu, A. H.; Heymann, D.; Colbert, D. T.; Lee, R. S.; Fischer, J. E.; Rao, A. M.; Eklund, P. C.; Smalley, R. E. *Appl Phys A* **1998**, *67*, 29.

[106] Hamon, M. A.; Hu, H.; Bhowmik, P.; Niyogi, S.; Zhao, B.; Itkis, M. E.; Haddon, R. C. *Chemical Physics Letters* **2001**, *347*, 8.

[107] Peng, H.; Alemany, L. B.; Margrave, J. L.; Khabashesku, V. N. *J. Am. Chem. Soc.* **2003**, *125*, 15174.

[108] Bahr, J. L.; Yang, J.; Kosynkin, D. V.; Bronikowski, M. J.; Smalley, R. E.; Tour, J. M. *J. Am. Chem. Soc.* **2001**, *123*, 6536.

[109] Georgakilas, V.; Kordatos, K.; Prato, M.; Guldi, D. M.; Holzinger, M.; Hirsch, A. *J. Am. Chem. Soc.* **2002**, *124*, 760.

[110] Holzinger, M.; Vostrowsky, O.; Hirsch, A.; Hennrich, F.; Kappes, M.; Weiss, R.; Jellen, F. *Angewandte Chemie International Edition* **2001**, *40*, 4002.

[111] Ying, Y.; Saini, R. K.; Liang, F.; Sadana, A. K.; Billups, W. E. *Organic Letters* **2003**, *5*, 1471.

[112] Mickelson, E. T.; Huffman, C. B.; Rinzler, A. G.; Smalley, R. E.; Hauge, R. H.; Margrave, J. L. *Chemical Physics Letters* **1998**, *296*, 188.

[113] Saini, R. K.; Chiang, I. W.; Peng, H.; Smalley, R. E.; Billups, W. E.; Hauge, R. H.; Margrave, J. L. *J. Am. Chem. Soc.* **2003**, *125*, 3617.

[114] Khabashesku, V. N.; Billups, W. E.; Margrave, J. L. *Accounts of Chemical Research* **2002**, *35*, 1087.

[115] Stevens, J. L.; Huang, A. Y.; Peng, H.; Chiang, I. W.; Khabashesku, V. N.; Margrave, J. L. *Nano Lett.* **2003**, *3*, 331.

[116] Vaisman, L.; Wagner, H. D.; Marom, G. *Advances in Colloid and Interface Science* **2006**, *128–130*, 37.

[117] Moore, V. C.; Strano, M. S.; Haroz, E. H.; Hauge, R. H.; Smalley, R. E.; Schmidt, J.; Talmon, Y. *Nano Lett.* **2003**, *3*, 1379.

[118] Eder, D.; Windle, A. H. *Adv. Mat.* **2008**, *20*, 1787.

[119] Cooke, D. J.; Eder, D.; Elliott, J. A. *The Journal of Physical Chemistry C* **2010**, *114*, 2462.

[120] Yan, Y.; Chan-Park, M. B.; Zhang, Q. *Small* **2007**, *3*, 24.

[121] Chattopadhyay, D.; Galeska, I.; Papadimitrakopoulos, F. *J. Am. Chem. Soc.* **2001**, *123*, 9451.

[122] Wu, Z.; Chen, Z.; Du, X.; Logan, J. M.; Sippel, J.; Nikolou, M.; Kamaras, K.; Reynolds, J. R.; Tanner, D. B.; Hebard, A. F.; Rinzler, A. G. *Science* **2004**, *305*, 1273.

[123] Prasad, D.; Zhiling, L.; Satish, N.; Barrera, E. V. *Nanotechnology* **2004**, *15*, 379.

[124] Koziol, K.; Vilatela, J.; Moisala, A.; Motta, M.; Cunniff, P.; Sennett, M.; Windle, A. *Science* **2007**, *318*, 1892.

[125] Davis, V. A.; Ericson, L. M.; Parra-Vasquez, A. N. G.; Fan, H.; Wang, Y.; Prieto, V.; Longoria, J. A.; Ramesh, S.; Saini, R. K.; Kittrell, C.; Billups, W. E.; Adams, W. W.; Hauge, R. H.; Smalley, R. E.; Pasquali, M. *Macromolecules* **2003**, *37*, 154.
[126] Gommans, H. H.; Alldredge, J. W.; Tashiro, H.; Park, J.; Magnuson, J.; Rinzler, A. G. *J. Appl. Phys.* **2000**, *88*, 2509.
[127] Grätzel, M. *Journal of Photochemistry and Photobiology C: Photochemistry Reviews* **2003**, 4, 145.

Keith Paton

2 Synthesis, characterisation and properties of graphene

2.1 Introduction

As a result of its ultra-thin, two-dimensional structure, coupled with its remarkable combination of physical properties, graphene has become one of the most studied nanomaterials of recent times. Combining high mechanical strength and stiffness [1] with electrical conductivity surpassing that of copper, as well as low optical absorption [2] has led to a wide range of possible applications for graphene. These include polymer composite reinforcement [3–6], solar cells [7, 8], transparent conducting films [9, 10], membranes [11, 12] and batteries [13] amongst many others.

2.2 Properties

Graphene has been ascribed a wide range of superlative properties, such as the thinnest, lightest, strongest material known to man. Being only one atomic layer thick, the claim for the thinnest material is perhaps obvious, and this also contributes to its high level of optical transparency. Single-layer graphene absorbs 2.3 % of the incident light, a value that can be defined by fundamental constants, $\pi a \approx 2.3\,\%$ and is almost invariant across the visible range [14]. The electrical properties have been extensively probed [15] with electron mobility of up to $230{,}000\,\text{cm}^2 \cdot V^{-1} \cdot s^{-1}$ at a carrier density of $2 \times 10^{11}\,\text{cm}^{-3}$ for suspended and annealed samples [16]. For samples on a substrate, the mobility is typically an order of magnitude lower due to scattering from the substrate phonons [17]. It should be noted however that these values were obtained at low temperature (5 K) with the highest room temperature mobility having been measured using high-κ dielectric solvents, as $\sim 7 \times 10^4\,\text{cm}^2 \cdot V^{-1} \cdot s^{-1}$ [18].

The combination of low optical absorbance and high electrical conductivity has attracted a lot of interest for transparent conductor applications. When coupled with its flexibility, it is widely seen as a possible replacement for indium-doped tin oxide (ITO), which has a sheet resistance of $\sim 100\,\Omega/\text{cm}$ at 90 % transparency. By growing graphene on copper foils, sheet resistances of $\sim 125\,\Omega/\text{cm}$ at 97.4 % transparency have been achieved [19]. This has been improved by combining four layers with doping of the graphene, giving resistance of $\sim 30\,\Omega/\text{cm}$ at 90 % transparency, all done on 30-inch roll-to-roll production scale.

Indentation testing of suspended graphene flakes has yielded a stiffness of ~ 1 TPa for pristine, monolayer graphene, with a strength of ~ 130 GPa and strain to break of ~ 25 % [1]. This gives a stiffness to strength ratio of ~ 8, close to the value of 9 predicted by Griffith [20]. These values are in close agreement with those of bulk graphite (in-plane) and carbon nanotubes. As for carbon nanotubes, graphene has been incorporated into polymer matrices to provide an enhancement in mechanical properties [21]. For example, incorporating graphene into a polyurethane matrix allowed the properties of the resulting composite to be tuned between thermoplastic-like behavior (low stiffness, low strength, high strain to break) and thermoset-like behavior (high stiffness, high strength but low strain to break) [22]. Indeed, by careful control of the graphene loading, a composite with high stiffness, high strength as well as high strain to break could be produced.

The thermal conductivity of suspended graphene has been calculated by measuring the frequency shift of the G-band in the Raman spectrum with varying laser power. These measurements yielded a value for thermal conductivity of 4840–5300 $W \cdot m^{-1} \cdot K^{-1}$ [23], better than that of SWCNTs, with the exception of crystalline ropes of nanotubes, which gave values up to 5800 $W \cdot m^{-1} \cdot K^{-1}$ [24]. Even when deposited on a substrate, the measured thermal conductivity is $\sim$ 600 $W \cdot m^{-1} \cdot K^{-1}$ [25], higher than in commonly used heat dissipation materials such as copper and silver.

2.3 Synthesis

2.3.1 Micromechanical cleavage

Micromechanical cleavage (Fig. 2.1(a)) is a technique that has been used by crystallographers for many years to obtain samples of well-defined crystal faces. In 1999 it was suggested that this method could be used to obtain few, or even single atomic layers of graphite, *i.e.* graphene [26]. In 2005, Geim *et al.* published results showing successful micromechanical cleavage of single layer graphene [27]. This technique involves using sticky-tape to repeatedly peel layers of the crystal from the bulk, eventually reaching single atomic layer thicknesses. As well as graphene, the authors demonstrated exfoliation of several layered materials including MoS_2, $NbSe_2$ and BN. The identification of monolayers of these materials was enabled by optical microscopy when deposited on a suitable substrate.

This method is capable of producing large flakes, of high quality, offering the highest mobilities of any production method [28–30], with room temperature values of up to 20 000 $cm^2 \cdot V^{-1} \cdot s^{-1}$ reported [29]. Due to its low throughput it is not a feasible option for large-scale production, but it remains a widely used method for producing good quality graphene for research applications.

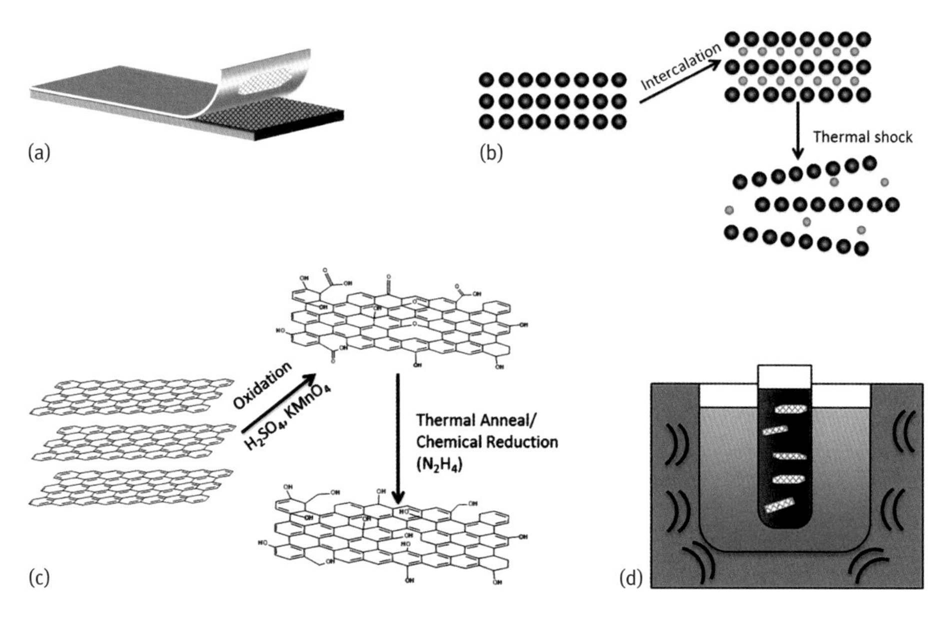

Fig. 2.1: "Top-down" synthesis methods. (a) Micromechanical cleavage; (b) ion intercalation; (c) graphite oxide; (d) liquid-phase exfoliation.

2.3.2 Liquid phase exfoliation

In contrast, exfoliation of graphene in liquid environments offers a route to large-scale production, from simple starting materials. There are various approaches that have been developed to enable effective exfoliation of graphene in liquids.

2.3.2.1 Intercalation

While the interlayer van der Waals bonding is weak in comparison to the in-plane bonds, it nonetheless presents a barrier to exfoliation of graphene from the crystal. One approach to overcoming this barrier is to first weaken these bonds by intercalation (Fig. 2.1(b)). The resulting graphite intercalation compounds (GIC) have a long history [31, 32], and have been studied and used for their own properties (including high conductivity, superconductivity and energy storage) rather than as a route to two-dimensional materials. This approach forces small molecules or ions in-between the layers of carbon atoms in pristine graphite, increasing the interlayer spacing and weakening the van der Waals forces between them. A wide range of compounds have been used as intercalants, including metal fluorides, metal chlorides and alkali metals. The degree of intercalation is described by a "stage" number, indicating the number of carbon layers between intercalation layers, *i.e.* stage 1 compounds have intercalant between every carbon layer, stage 2 compounds have intercalates between every second carbon layer. The introduction of these intercalating compounds increases the interlayer spacing by as much as three times [33], and so they offer a good starting point for graphene production [33–36]. With the layer spacing thus increased, and the bonding correspondingly weakened, it is then relatively easy to separate the layers. While this is often described as being spontaneous, in all cases, at least some level of mechanical agitation or mixing is applied [34, 37, 38].

While GICs allow high yield production of graphene [34–36], they have the drawback of often requiring controlled atmospheres for their preparation and storage, as they tend to oxidize in air [32–36], although graphene intercalated with $FeCl_3$ has been reported to be stable in air for up to one year [39]. This lack of stability, coupled with the additional processing steps involved in their preparation has limited the use of this approach to date as a graphene production method.

2.3.2.2 Graphite oxide

An alternative approach to weakening the interlayer bonding in graphite is to first oxidize the graphite, which was first investigated by Brodie in 1859 [40]. The process involving reaction with potassium chloride (KCl_3) was modified by Staudenmaier to use concentrated sulfuric acid and adding the KCl_3 stepwise during the process, and TEM images published by Ruess and Vogt in 1948 showed single sheets of graphene oxide. These reactions however yield chlorine dioxide (ClO_2) gas, which can explosively decompose into chlorine and oxygen. A safer and quicker method was developed by Hummers that avoided these explosive by-products, using sulfuric acid, sodium nitrate and potassium permanganate [41].

The oxidation reaction (Fig. 2.1(c)) introduces functional groups to both the basal plane (hydroxyl and epoxide) and flake edges (lactone, phenol and quinone) [42–45]. These groups weaken the van der Waals forces between the layers, as well as making the resulting flakes hydrophilic, allowing them to be dispersed easily in water [42, 46] and simple organic solvents [42, 47]. This exfoliation is achieved through either sonication [42, 48], stirring [49] or thermal expansion [50]. While this approach can yield large flakes [51], the functional groups strongly affect the properties of the graphene, in particular the electrical conductivity [10, 52]. The graphene oxide can be chemically [45, 46, 53–55] or thermally [10, 50, 56] reduced, in order to remove the functional groups, and a range of chemistries has been developed to achieve this. The use of alkali solutions has been used since the 1960s [45], and hydrazine [53], hydrides [47, 55], hydroquinone [55] and p-phenylene [54] have all been used. A mechanism involving UV-assisted photocatalysis has also been demonstrated, wherein GO is reduced as it accepts electrons from irradiated TiO_2 nanoparticles [57]. The reduction has also be shown to be possible using a camera flash [58], and focused sunlight [59]. However, none of these methods is able to completely return GO to pristine graphene.

Although the electrical properties of GO are significantly inferior to pristine graphene, the presence of the functional groups on the flakes can be a considerable advantage when considering composites [6].

2.3.2.3 Pristine graphite

Although the weakening of the intersheet bonding though graphite intercalation or graphite oxide allows easier subsequent exfoliation, it is not necessary. Graphene has been successfully exfoliated from pristine graphite without any prior pre-treatments through ultrasonication in solvents [60–63] and aqueous solutions [63–67] (Fig. 2.1(d)). Exfoliation occurs due to hydrodynamic forces associated with cavitation (formation and collapse of bubbles in liquids) during the sonication process [68–70]. These cavitation events can create local temperatures of 5000 K and pressures of several hundred atmospheres, sufficient to overcome the intersheet bonds.

Graphene is most effectively dispersed in solvents in which the interfacial energy between the graphene and the liquid is minimized [71]. If the difference in surface energies is too high (or too low), the graphene flakes will re-aggregate and sediment out of the dispersion. By measuring the concentration of graphene dispersed in a wide range of solvents with varying surface energies, it was found that those with surface energy of ~ 70 mJ/m^2 are most effective [72, 73]. This is in close agreement with contact angle measurements for the surface energy of graphite (~ 62 mJ/m^2). Many solvents with surface energy close to this value (*e.g.* NMP, DMF, GBL, CHP) have drawbacks with regards to toxicity [74, 75], and all have high boiling points (> 150 °C), making removal of solvent difficult. Exfoliation in low boiling point solvents, which also lack the issues of toxicity, has been demonstrated [76], although the concentrations achieved are lower than in surface energy matched solvents.

Stabilization in water (surface tension ~ 72 mN/m) can be achieved through the use of surfactants [77] such as SDBS [78] or sodium cholate [79], or long chain polymers such as polyvinyl alcohol. While this allows a non-toxic and easily removed solvent to be used, the presence of the surfactant or polymer molecules can be detrimental to the subsequent applications [80].

The yield of these sonication approaches is low (~ 1 wt% graphene) and the unexfoliated graphite material needs to be separated from the graphene dispersion. Typically this is achieved through centrifugation, which allows a selection of flake sizes by controlling the applied centrifugal force [81]. Furthermore, by recycling the sediment back through the process with fresh solvent, the overall yield can be increased. The use of density gradient centrifugation can also be used to isolate monolayers of graphene [79, 82], where the density of the flake is modified by the surfactant coating, allowing separation. Rate-zonal separation has also been used to produced graphene dispersions with a narrow range of lateral flake sizes [83].

2.3.3 Precipitation from metals/CVD

As well as the "top-down" methods described above, it is also possible to produce graphene with "bottom-up" approaches (Fig. 2.2). Growth of graphite on transition metal surfaces was first described in the 1940s [84], with details of the mechanism described in the 1970s [85]. This synthetic graphite is formed by precipitation of carbon from Co, Ni or Pt annealed at high temperature, with single layer graphene formed as islands on the surface (Fig. 2.2(a)). A similar mechanism leads to the formation of "Kish" graphite from molten iron during steelmaking as the supersaturated iron cools [86, 87]. The solubility of carbon in most metals is only a few wt% and the best graphite growth is obtained from metals that do not form carbides, such as Ni, Cu, Pt and Au [48, 88]. Metals such as Ti, Hf and Zr are not suitable as they form stable carbides [89, 90], as well as having a large lattice mismatch with graphite.
Various techniques have been used to deposit the carbon onto the metal surface, including physical vapor deposition, chemical vapor deposition and flash evaporation. When using a pure carbon source, PVD or flash evaporation is used to deposit the carbon on the surface. The substrate is then heated to allow the carbon to diffuse into the metal as a solid solution. Upon cooling this carbon precipitates out as graphitic layers. The thickness of the graphite film obtained is controlled by the carbon solubility, metal thickness and cooling rate [91]. Careful control of these parameters allows single layer graphene to be grown over large areas [92].

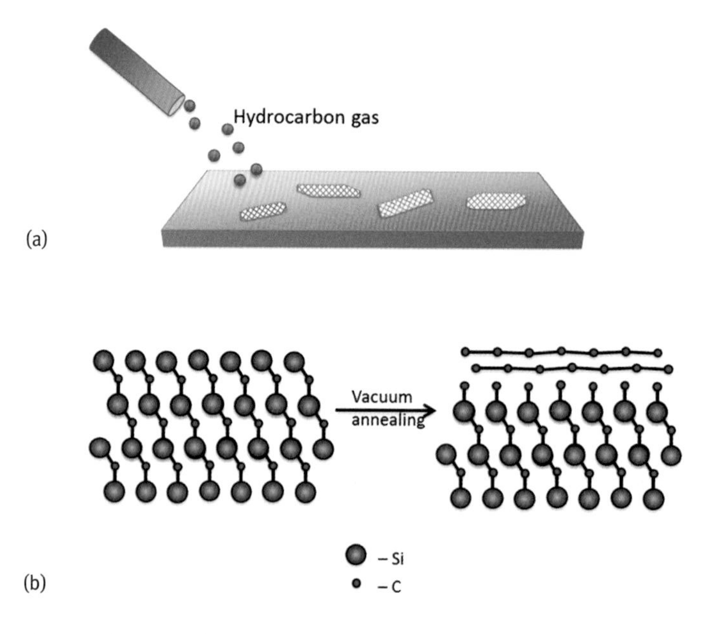

Fig. 2.2: "Bottom-up" synthesis methods. (a) Catalytic vapor deposition; (b) epitaxial growth from SiC.

Chemical vapor deposition is widely used for growing thin films, and has been successfully used to grow graphene, most notably on copper [92]. While it was initially thought that growth on copper was self-limiting to monolayer graphene [88], recent work has shown this not to be the case [93]. Indeed the parameter space for successful growth of good quality graphene on Cu is somewhat small, as a result of low catalytic activity of Cu and relatively low sublimation temperatures. An alternative material is nickel, which has a better catalytic activity, allowing lower growth temperatures, as well as theoretically ideal lattice matching on the {111} crystal faces [94]. For both metals, the growth is controlled by a balance between the flux of carbon onto the surface of the catalyst and the diffusion into the bulk [95]. Growth of graphene layers only occurs on the catalyst surface, and so the second layer grows underneath the initial monolayer, and is proposed to only occur after complete coverage of the monolayer. By control of this carbon flux balance, the carbon concentration in the bulk of the catalyst can be limited, giving very little precipitation upon cooling [95].

Due to the growth mechanism, the resulting graphene film is polycrystalline, with the domain sizes governed by the nucleation rate of the initial islands. This is turn can be controlled by the temperature and pressure in the reaction chamber, with single crystal domains of up to 0.5mm observed. The use of bi-metallic catalyst films has also been demonstrated to reduce the nucleation density of graphene, giving larger final domain sizes [96].

One drawback of both the previous techniques is that most applications most suited to graphene produced in these methods require it to be on an insulating substrate. CVD has been attempted on various insulating substrates, including sapphire [97], quartz, ZrO_2 [98], MgO [99] and Si_3N_4 [100] with some limited success, but much more work is required in this area. Graphene has also been deposited on quartz by spray pyrolysis of sodium ethoxide [101] although characterization of this material has shown it to be highly defective in nature.

2.3.4 Epitaxial growth from SiC

One solution to the problem of an insulating substrate is epitaxial growth from SiC, which goes back to 1896, when graphite was produced for lubrication applications [102]. Silicon is thermally desorbed from the surface of a SiC crystal, and the resulting carbon layer undergoes a surface reconstruction to form graphene [103, 104] (Fig. 2.2(b)). Initially this graphene layer is covalently bonded to the SiC underneath, but can be detached by hydrogen intercalation [105–107]. Furthermore, the process is not self-limiting [104], and further graphene layers will be formed beneath this initial one, with the bottom layer always bonded to the underlying SiC.

Control of the annealing atmosphere [108–110] and choice of SiC face from which Si is desorbed [110, 111] offer some control over the domain sizes. For example, graphene growth is much faster on the C-face than on the Si-face [110, 111], while annealing at near atmospheric pressures yields domains almost 1000 times larger than under UHV [105, 109].

A major drawback to this method is the cost of SiC wafers, which are upwards of 15 times the price of Si wafers. Graphene has been grown from a thin SiC layer deposited on sapphire, but the properties were markedly inferior compared to graphene produced from bulk SiC [112].

2.4 Characterization

The characterization of graphene often involves several techniques in conjunction in order to build up a complete picture of the material. The techniques typically include electron microscopy, Raman spectroscopy, X-ray photo-emission spectroscopy (XPS), Fourier-transform infrared spectroscopy (FTIR) and thermal-gravimetric analysis (TGA).

Electron microscopy can most obviously provide information on the lateral dimensions of the graphene produced. Transmission electron microscopy can also be used to measure the number of atomic layers in a graphene sample. One of the most straightforward methods to assess this is by simply counting the edges visible in transmission electron micrographs. While this is often used, it has a limitation of missing any overlapping edges, and as a result tends to underestimate the number of layers. A more reliable method to confirm monolayer graphene is electron diffraction, comparing the relative intensity between the $\{\bar{2}110\} - \{10\bar{1}0\}$ diffraction spots [113, 114] (Fig. 2.3). For monolayer graphene, the inner, $\{10\bar{1}0\}$ spots have a higher intensity, whereas for bi-layer or more the outer, $\{\bar{2}110\}$ spots have a higher intensity. The effect is more clearly seen if the diffraction pattern is recorded as a function of tilt angle. This approach however can only be used to distinguish between monolayer and multilayer. A more quantitative approach is to use electron energy loss spectroscopy (EELS) to determine the number of layers. The thickness of the graphene layer can be shown to be given by $t = \lambda \ln(I_l/I_0)$ where λ is the mean free path for energy loss, I_0 is the intensity of the filtered image (using the zero-loss peak) and I_l is the total intensity of the (unfiltered) image [115]. By measuring the spectrum from a folded flake, where the number of layers can be confirmed by electron diffraction, a value of λ can be obtained, allowing the number of layers to be determined. It has also been shown that high resolution imaging using an annular dark-field detector in scanning-TEM allows not only structure information such as grain boundaries [116] but also identification of the nature of substitutional defects [117].

While electron microscopy is an invaluable tool for characterization of graphene, it is a costly and often time-consuming technique. In contrast, Raman spectroscopy is a quick and easy method to obtain a wide range of properties of graphene, including flake size, layer number, defect density, and doping levels amongst others [118].

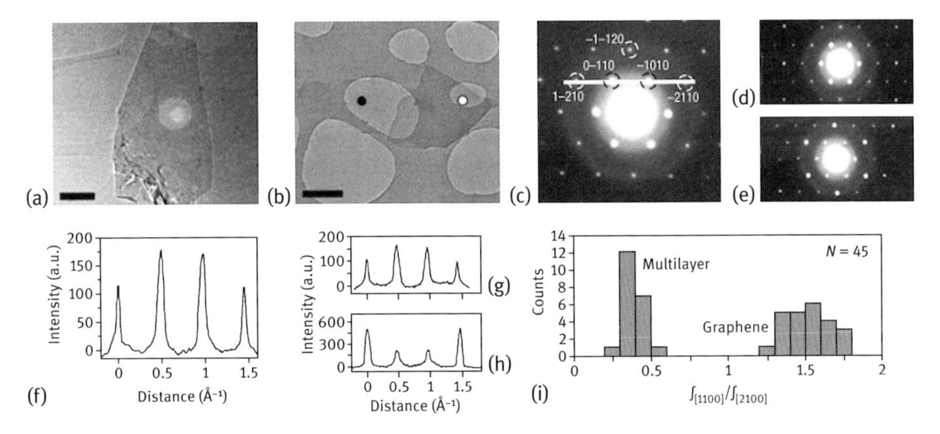

Fig. 2.3: Evidence of monolayer graphene from TEM [72]. (a) and (b) High-resolution TEM images of solution-cast monolayer (a) and bilayer (b) graphene (scale bar 500 nm); (c) electron diffraction pattern of the sheet in (a), with the peaks labeled by Miller-Bravais indices; (d) and (e) electron diffraction patterns taken from the positions of the black (d) and white spots (e), respectively, of the sheet shown in (b), using the same labels as in (c). The graphene is clearly one layer thick in (d) and two layers thick in (e); (f)–(h) Diffracted intensity taken along the $1\bar{2}10$ to $\bar{2}110$ axis for the patterns shown in (c)–(e), respectively; (i) Histogram of the ratios of the intensity of the 1100 and 2110 diffraction peaks for all the diffraction patterns collected. A ratio > 1 is a signature of graphene.

A typical Raman spectrum for graphene consists of three main peaks of interest, the D-peak at ~ 1350 cm^{-1}, the G-peak at ~ 1580 cm^{-1} and the 2D peak (sometimes called G′) at ~ 2700 cm^{-1} [119, 120] (Fig. 2.4). While the G-peak, the result of in-plane phonon mode, is present in all graphitic carbon materials, the D-peak, an in-plane “breathing mode” of the 6-membered rings, only becomes Raman active in the presence of defects. The ratio of these two peaks is commonly used as a measure of the defect density of graphite materials, as it is for carbon nanotubes. However, it should be noted that as well as vacancies and sp^3-type defects, the D-peak is also seen at the edges of graphene sheets [121]. As the intensity of the G-band is proportional to the area sampled ($I(G) \propto L_a^2$) whereas the intensity of the D-band is proportional to the total length of the edge ($I(D) \propto L_a$), the ratio $I(D)/I(G) \propto 1/L_a$, gives an estimate of the flake size or domain size [122, 123]. Therefore, as the size of graphene flakes decreases, the D/G intensity ratio will increase even in the absence of basal plane defects [124]. While this holds for samples with low basal plane defects, it needs to be adjusted for samples with higher point defect densities. For a sample with an average defect separation of L_D probed by a laser with spot size L_L, there will be, on average, $(L_L/L_D)^2$ defects probed, and so $I(D) \propto (L_L/L_D)^2$. The G-band intensity will be proportional to the area probed ($I(G) \propto L_L^2$) and so $I(D)/I(G) \propto 1/L_D^2$ [120]. Due to the origin of the D-band, this picture fails at very high defect density, as the material tends towards amorphous carbon, at which point the D-band intensity begins to fall with increasing defect density [125].

The number of layers, in contrast, can be obtained by examination of the shape of the 2D (also known as the G′) peak, which is a second order process of the D-peak, and is always Raman active. While for single layer graphene, this comprises a single peak, with intensity up to 4 times the intensity of the G-peak, 2-layer graphene is fitted by 4 peaks (due to splitting of the electronic band structure [119]). In a similar way, the peak shape varies with increasing layer number until by 5 layers it is almost indistinguishable from that of bulk graphite, with two peaks with intensity ~ 1/4 and 1/2 of the G-peak.

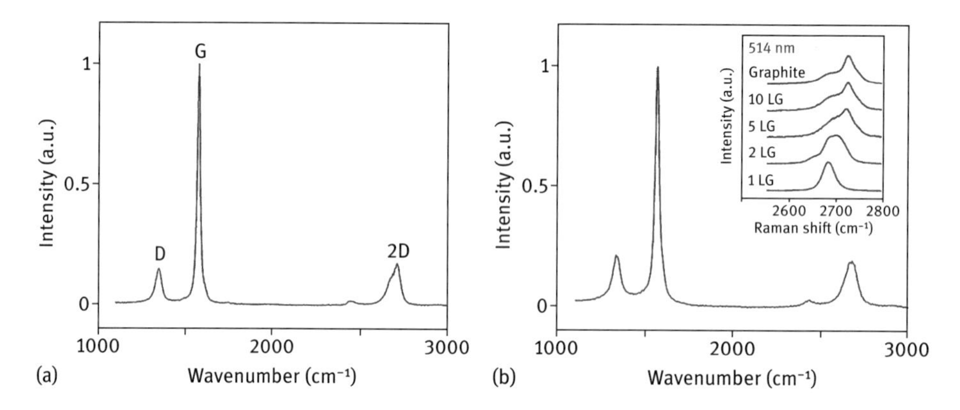

Fig. 2.4: Typical Raman spectra of: (a) graphite, with peaks labeled as discussed in the text; (b) graphene (liquid-phase exfoliated). Inset: Evolution of 2D-band with increasing layer numbers [120].

Further characterization of graphene routinely uses AFM to obtain both the number of layers and lateral sizes of flakes [126]. The presence of residual surfactant on the surface of the flakes needs to be taken into account in order to obtain an accurate measure of layer number [127]. X-ray photo-emission spectroscopy (XPS) is also widely used to characterize any functional groups and defects [128–130], as well as to indicate the presence of residual solvent molecules [129]. Standard chemical analysis techniques are also widely used to identify the nature of any functional groups, as well as the degree of functionalization. Elemental analysis provides a quantitative measure of the C/O/N ratio in the material, though this needs to be combined with a Karl–Fischer titration, due to the hygroscopic nature of functionalized graphene [131]. Fourier-transform infrared (FTIR) spectroscopy is also widely used to characterize the nature of the functional groups [58, 132].

Bibliography

[1] C. Lee, X. Wei, J.W. Kysar, J. Hone, Measurement of the elastic properties and intrinsic strength of monolayer graphene, Science, 321 (2008) 385–388.

[2] R.R. Nair, P. Blake, A.N. Grigorenko, K.S. Novoselov, T.J. Booth, T. Stauber, et al., Fine structure constant defines visual transparency of graphene, Science, 320 (2008) 1308.

[3] D. Cai, M. Song, A simple route to enhance the interface between graphite oxide nanoplatelets and a semi-crystalline polymer for stress transfer, Nanotechnology, 20 (2009) 315708.

[4] H. Kim, C.W. Macosko, Processing-property relationships of polycarbonate/graphene composites, Polymer, 50 (2009) 3797–3809.

[5] J. Liang, Y. Huang, L. Zhang, Y. Wang, Y. Ma, T. Guo, et al., Molecular-level dispersion of graphene into poly(vinyl alcohol) and effective reinforcement of their nanocomposites, Advanced Functional Materials, 19 (2009) 2297–2302.

[6] S. Stankovich, D.A. Dikin, G.H.B. Dommett, K.M. Kohlhaas, E.J. Zimney, E.A. Stach, et al., Graphene-based composite materials, Nature, 442 (2006) 282–286.

[7] X. Wang, L. Zhi, K. Müllen, Transparent, conductive graphene electrodes for dye-sensitized solar cells, Nano Letters, 8 (2008) 323–327.

[8] X. Wang, L. Zhi, N. Tsao, Z. Tomović, J. Li, K. Müllen, Transparent carbon films as electrodes in organic solar cells, Angewandte Chemie (International Ed. in English), 47 (2008) 2990–2992.

[9] G. Eda, G. Fanchini, M. Chhowalla, Large-area ultrathin films of reduced graphene oxide as a transparent and flexible electronic material, Nature Nanotechnology, 3 (2008) 270–274.

[10] C. Mattevi, G. Eda, S. Agnoli, S. Miller, K.A. Mkhoyan, O. Celik, et al., Evolution of electrical, chemical, and structural properties of transparent and conducting chemically derived graphene thin films, Advanced Functional Materials, 19 (2009) 2577–2583.

[11] K. Chung, C.-H. Lee, G.-C. Yi, Transferable GaN layers grown on ZnO-coated graphene layers for optoelectronic devices, Science (New York, N.Y.) 330 (2010) 655–657.

[12] T.J. Booth, P. Blake, R.R. Nair, D. Jiang, E.W. Hill, U. Bangert, et al., Macroscopic graphene membranes and their extraordinary stiffness, Nano Letters, 8 (2008) 2442–2446.

[13] M. Pumera, Electrochemistry of graphene: New horizons for sensing and energy storage, Chemical Record, 9 (2009) 211–223.

[14] R.R. Nair, P. Blake, A.N. Grigorenko, K.S. Novoselov, T.J. Booth, T. Stauber, et al., Fine structure constant defines visual transparency of graphene, Science, 320 (2008) 1308.

[15] A.K. Geim, K.S. Novoselov, The rise of graphene, Nature Materials, 6 (2007) 183–191.

[16] K.I. Bolotin, K.J. Sikes, Z. Jiang, M. Klima, G. Fudenberg, J. Hone, et al., Ultrahigh electron mobility in suspended graphene, Solid State Communications, 146 (2008) 351–355.

[17] J.-H. Chen, C. Jang, S. Xiao, M. Ishigami, M.S. Fuhrer, Intrinsic and extrinsic performance limits of graphene devices on SiO2, Nature Nanotechnology, 3 (2008) 206–209.

[18] F. Chen, J. Xia, D.K. Ferry, N. Tao, Dielectric screening enhanced performance in graphene FET, Nano Letters, 9 (2009) 2571–2574.

[19] S. Bae, H. Kim, Y. Lee, X. Xu, J.-S. Park, Y. Zheng, et al., Roll-to-roll production of 30-inch graphene films for transparent electrodes, Nature Nanotechnology, 5 (2010) 574–578.

[20] A.A. Griffith, The phenomena of rupture and flow in solids, Philosophical Transactions of the Royal Society A: Mathematical, Physical and Engineering Sciences, 221 (1921) 163–198.

[21] T. Ramanathan, A.A. Abdala, S. Stankovich, D.A. Dikin, M. Herrera-Alonso, R.D. Piner, et al., Functionalized graphene sheets for polymer nanocomposites, Nature Nanotechnology, 3 (2008) 327–331.

[22] U. Khan, P. May, A. O'Neill, J.N. Coleman, Development of stiff, strong, yet tough composites by the addition of solvent exfoliated graphene to polyurethane, Carbon, 48 (2010) 4035–4041.
[23] A.A. Balandin, S. Ghosh, W. Bao, I. Calizo, D. Teweldebrhan, F. Miao, et al., Superior thermal conductivity of single-layer graphene, Nano Letters, 8 (2008) 902–907.
[24] J. Hone, M. Whitney, C. Piskoti, A. Zettl, Thermal conductivity of single-walled carbon nanotubes, Physical Review B, 59 (1999) R2514–R2516.
[25] J.H. Seol, I. Jo, A.L. Moore, L. Lindsay, Z.H. Aitken, M.T. Pettes, et al., Two-dimensional phonon transport in supported graphene, Science (New York, N.Y.), 328 (2010) 213–216.
[26] X. Lu, M. Yu, H. Huang, R.S. Ruoff, Tailoring graphite with the goal of achieving single sheets, Nanotechnology, 10 (1999) 269–272.
[27] K.S. Novoselov, D. Jiang, F. Schedin, T.J. Booth, V. V Khotkevich, S. V Morozov, et al., Two-dimensional atomic crystals, Proceedings of the National Academy of Sciences of the United States of America, 102 (2005) 10451–10453.
[28] A.S. Mayorov, D.C. Elias, I.S. Mukhin, S. V Morozov, L.A. Ponomarenko, K.S. Novoselov, et al., How close can one approach the Dirac point in graphene experimentally?, Nano Letters, 12 (2012) 4629–4634.
[29] Z.H. Ni, L.A. Ponomarenko, R.R. Nair, R. Yang, S. Anissimova, I. V Grigorieva, et al., On resonant scatterers as a factor limiting carrier mobility in graphene, Nano Letters, 10 (2010) 3868–3872.
[30] D.C. Elias, R. V. Gorbachev, a. S. Mayorov, S. V. Morozov, A.A. Zhukov, P. Blake, et al., Dirac cones reshaped by interaction effects in suspended graphene, Nature Physics, 7 (2011) 701–704.
[31] M. Inagaki, Applications of graphite intercalation compounds, Journal of Materials Research, 4 (1989) 1560–1568.
[32] M.S. Dresselhaus, G. Dresselhaus, Intercalation compounds of graphite, Advances in Physics, 51 (2002) 1–186.
[33] A. Lerf, H. He, M. Forster, J. Klinowski, Structure of graphite oxide revisited, Journal of Physical Chemistry B, 5647 (1998) 4477–4482.
[34] C. Vallés, C. Drummond, H. Saadaoui, C. a Furtado, M. He, O. Roubeau, et al., Solutions of negatively charged graphene sheets and ribbons, Journal of the American Chemical Society, 130 (2008) 15802–15804.
[35] P.K. Ang, S. Wang, Q. Bao, J.T.L. Thong, K.P. Loh, High-throughput synthesis of graphene by intercalation-exfoliation of graphite oxide and study of ionic screening in graphene transistor, ACS Nano, 3 (2009) 3587–3594.
[36] A. Catheline, C. Vallés, C. Drummond, L. Ortolani, V. Morandi, M. Marcaccio, et al., Graphene solutions, Chemical Communications (Cambridge, England), 47 (2011) 5470–5472.
[37] E.M. Milner, N.T. Skipper, C.A. Howard, M.S.P. Shaffer, D.J. Buckley, K.A. Rahnejat, et al., Structure and morphology of charged graphene platelets in solution by small-angle neutron scattering, Journal of the American Chemical Society, 134 (2012) 8302–8305.
[38] A. Catheline, C. Vallés, C. Drummond, L. Ortolani, V. Morandi, M. Marcaccio, et al., Graphene solutions, Chemical Communications (Cambridge, England), 47 (2011) 5470–5472.
[39] I. Khrapach, F. Withers, T.H. Bointon, D.K. Polyushkin, W.L. Barnes, S. Russo, et al., Novel highly conductive and transparent graphene-based conductors, Advanced Materials, 24 (2012) 2844–2849.
[40] M.B.-C. Brodie, Sur le poids atomique du graphite, Annalen Der Chemie und Parmacie, 114 (1860) 466–472.
[41] J. Hummers, William S, R.E. Offerman, Preparation of graphitic oxide, Journal of the American Chemical Society, 80 (1957) 1339.

[42] S. Stankovich, D.A. Dikin, R.D. Piner, K. a. Kohlhaas, A. Kleinhammes, Y. Jia, et al., Synthesis of graphene-based nanosheets via chemical reduction of exfoliated graphite oxide, Carbon, 45 (2007) 1558–1565.

[43] A. Lerf, H. He, M. Forster, J. Klinowski, Structure of graphite oxide revisited, Journal of Ph., 5647 (1998) 4477–4482.

[44] W. Cai, R.D. Piner, F.J. Stadermann, S. Park, M. a Shaibat, Y. Ishii, et al., Synthesis and solid-state NMR structural characterization of 13C-labeled graphite oxide, Science (New York, N.Y.), 321 (2008) 1815–1817.

[45] H.P. Boehm, A. Clauss, G. Fischer, U. Hofmann, Surface Properties of Extremely Thin Crystals, in: Proceedings of the Fifth Conference on Carbon, (1962) 73–80.

[46] X. Li, X. Wang, L. Zhang, S. Lee, H. Dai, Chemically derived, ultrasmooth graphene nanoribbon semiconductors, Science (New York, N.Y.), 319 (2008) 1229–1232.

[47] Y. Si, E.T. Samulski, Synthesis of water soluble graphene, Nano Letters, 8 (2008) 1679–1682.

[48] X. Li, W. Cai, J. An, S. Kim, J. Nah, D. Yang, et al., Large-area synthesis of high-quality and uniform graphene films on copper foils, Science (New York, N.Y.), 324 (2009) 1312–1314.

[49] J.R. Lomeda, C.D. Doyle, D. V Kosynkin, W.-F. Hwang, J.M. Tour, Diazonium functionalization of surfactant-wrapped chemically converted graphene sheets, Journal of the American Chemical Society, 130 (2008) 16201–16206.

[50] H.C. Schniepp, J.-L. Li, M.J. McAllister, H. Sai, M. Herrera-Alonso, D.H. Adamson, et al., Functionalized single graphene sheets derived from splitting graphite oxide, The Journal of Physical Chemistry, B. 110 (2006) 8535–8539.

[51] C.-Y. Su, Y. Xu, W. Zhang, J. Zhao, X. Tang, C.-H. Tsai, et al., Electrical and spectroscopic characterizations of ultra-large reduced graphene oxide monolayers, Chemistry of Materials, 21 (2009) 5674–5680.

[52] H.A. Becerril, J. Mao, Z. Liu, R.M. Stoltenberg, Z. Bao, Y. Chen, Evaluation of solution-processed reduced graphene oxide films as transparent conductors, ACS Nano, 2 (2008) 463–470.

[53] C. Gómez-Navarro, R.T. Weitz, A.M. Bittner, M. Scolari, A. Mews, M. Burghard, et al., Electronic transport properties of individual chemically reduced graphene oxide sheets, Nano Letters, 7 (2007) 3499–3503.

[54] Y. Chen, X. Zhang, P. Yu, Y. Ma, Stable dispersions of graphene and highly conducting graphene films: a new approach to creating colloids of graphene monolayers, Chemical Communications (Cambridge, England), (2009) 4527–4529.

[55] A.B. Bourlinos, D. Gournis, D. Petridis, T. Szabó, A. Szeri, I. Dékány, Graphite oxide: chemical reduction to graphite and surface modification with primary aliphatic amines and amino acids, Langmuir, 19 (2003) 6050–6055.

[56] H. Wang, J.T. Robinson, X. Li, H. Dai, Solvothermal reduction of chemically exfoliated graphene sheets, Journal of the American Chemical Society, 131 (2009) 9910–9911.

[57] G. Williams, B. Seger, P. V Kamat, TiO2-graphene nanocomposites. UV-assisted photocatalytic reduction of graphene oxide, ACS Nano, 2 (2008) 1487–1491.

[58] L.J. Cote, R. Cruz-Silva, J. Huang, Flash reduction and patterning of graphite oxide and its polymer composite, Journal of the American Chemical Society, 131 (2009) 11027–11032.

[59] V. Eswaraiah, S.S. Jyothirmayee Aravind, S. Ramaprabhu, Top down method for synthesis of highly conducting graphene by exfoliation of graphite oxide using focused solar radiation, Journal of Materials Chemistry, 21 (2011) 6800.

[60] F. Torrisi, T. Hasan, W. Wu, Z. Sun, A. Lombardo, T.S. Kulmala, et al., Inkjet-printed graphene electronics, ACS Nano, 6 (2012) 2992–3006.

[61] P. Blake, P.D. Brimicombe, R.R. Nair, T.J. Booth, D. Jiang, F. Schedin, et al., Graphene-based liquid crystal device, Nano Letters, 8 (2008) 1704–1708.

[62] Y. Hernandez, V. Nicolosi, M. Lotya, F.M. Blighe, Z. Sun, S. De, et al., High-yield production of graphene by liquid-phase exfoliation of graphite, Nature Nanotechnology, 3 (2008) 563–568.
[63] T. Hasan, F. Torrisi, Z. Sun, D. Popa, V. Nicolosi, G. Privitera, et al., Solution-phase exfoliation of graphite for ultrafast photonics, Physica Status Solidi (B), 247 (2010) 2953–2957.
[64] M. Lotya, Y. Hernandez, P.J. King, R.J. Smith, V. Nicolosi, L.S. Karlsson, et al., Liquid phase production of graphene by exfoliation of graphite in surfactant/water solutions, Journal of the American Chemical Society, 131 (2009) 3611–3620.
[65] A.A. Green, M.C. Hersam, Solution phase production of graphene with controlled thickness via density differentiation, Nano Letters, 9 (2009) 4031–4036.
[66] O.M. Maragó, F. Bonaccorso, R. Saija, G. Privitera, P.G. Gucciardi, M.A. Iatì, et al., Brownian motion of graphene, ACS Nano, 4 (2010) 7515–7523.
[67] R.J. Smith, M. Lotya, J.N. Coleman, The importance of repulsive potential barriers for the dispersion of graphene using surfactants, New Journal of Physics, 12 (2010) 125008.
[68] J. Lindley, T.J. Mason, Sonochemistry. Part 2 – Synthetic applications, Chemical Society Reviews, 16 (1987) 275–311.
[69] J.P. Lorimer, T.J. Mason, Sonochemistry. Part 1 – The physical aspects, Chemical Society Reviews, 16 (1987) 239–274.
[70] K.S. Suslick, Sonochemistry, Science (New York, N.Y.), 247 (1990) 1439–1445.
[71] J.N. Israelashvili, Intermolecular and Surface Forces, Third Edition, Elsevier Inc., 2011.
[72] Y. Hernandez, V. Nicolosi, M. Lotya, F.M. Blighe, Z. Sun, S. De, et al., High-yield production of graphene by liquid-phase exfoliation of graphite, Nature Nanotechnology, 3 (2008) 563–568.
[73] Y. Hernandez, M. Lotya, D. Rickard, S.D. Bergin, J.N. Coleman, Measurement of multicomponent solubility parameters for graphene facilitates solvent discovery, Langmuir, 26 (2010) 3208–1323.
[74] H.M. Solomon, B.A. Burgess, G.L. Kennedy, R.E. Staples, 1-Methyl-2-Pyrrolidone (NMP): reproductive and developmental toxcicity study by inhalation in the rat, Drug and Chemical Toxicology, 18 (1995) 271–293.
[75] G.L. Kennedy, H. Sherman, Acute and subchronic toxicity of dimethylformamide and dimethylacetamide following various routes of administration, Drug and Chemical Toxicology, 9 (1986) 147–170.
[76] A. O'Neill, U. Khan, P.N. Nirmalraj, J. Boland, J.N. Coleman, Graphene Dispersion and exfoliation in low boiling point solvents, The Journal of Physical Chemistry C, 115 (2011) 5422–5428.
[77] R.J. Smith, M. Lotya, J.N. Coleman, The importance of repulsive potential barriers for the dispersion of graphene using surfactants, New Journal of Physics, 12 (2010) 125008.
[78] M. Lotya, Y. Hernandez, P.J. King, R.J. Smith, V. Nicolosi, L.S. Karlsson, et al., Liquid phase production of graphene by exfoliation of graphite in surfactant/water solutions, Journal of the American Chemical Society, 131 (2009) 3611–3620.
[79] A.A. Green, M.C. Hersam, Solution phase production of graphene with controlled thickness via density differentiation, Nano Letters, 9 (2009) 4031–4036.
[80] P.N. Nirmalraj, T. Lutz, S. Kumar, G.S. Duesberg, J.J. Boland, Nanoscale mapping of electrical resistivity and connectivity in graphene strips and networks, Nano Letters. 11 (2011) 16–22.
[81] U. Khan, A. O'Neill, H. Porwal, P. May, K. Nawaz, J.N. Coleman, Size selection of dispersed, exfoliated graphene flakes by controlled centrifugation, Carbon, 50 (2011) 470–475.
[82] A.A. Green, M.C. Hersam, Emerging methods for producing monodisperse graphene dispersions, The Journal of Physical Chemistry Letters, 1 (2010) 544–549.
[83] X. Sun, D. Luo, J. Liu, D.G. Evans, Monodisperse chemically modified graphene obtained by density gradient ultracentrifugal rate separation, ACS Nano, 4 (2010) 3381–3389.

[84] H. Lipson, a. R. Stokes, The structure of graphite, Proceedings of the Royal Society A: Mathematical, Physical and Engineering Sciences, 181 (1942) 101–105.

[85] F.J. Derbyshire, A.E.B. Presland, D.L. Trimm, Graphite formation by the dissolution-precipitation of carbon in cobalt, nickel and iron, Carbon, 13 (1975) 111–113.

[86] P.L. Walker, G. Imperial, Structure of Graphites: graphitic character of kish, Nature, 180 (1957) 1185–1185.

[87] S.M. Winder, D. Liu, J.W. Bender, Synthesis and characterization of compound-curved graphite, Carbon, 44 (2006) 3037–3042.

[88] X. Li, W. Cai, L. Colombo, R.S. Ruoff, Evolution of graphene growth on Ni and Cu by carbon isotope labeling, Nano Letters, 9 (2009) 4268–4272.

[89] A. Fernández Guillermet, Analysis of thermochemical properties and phase stability in the zirconium-carbon system, Journal of Alloys and Compounds, 217 (1995) 69–89.

[90] L. Kaufman, Coupled phase diagrams and thermochemical data for transition metal binary systems-VI, CALPHAD, 3 (1979) 45–76.

[91] J.C. Shelton, H.R. Patil, J.M. Blakely, Equilibrium segregation of carbon to a nickel (111) surface: A surface phase transition, Surface Science, 43 (1974) 493–520.

[92] X. Li, W. Cai, J. An, S. Kim, J. Nah, D. Yang, et al., Large-area synthesis of high-quality and uniform graphene films on copper foils, Science, 324 (2009) 1312–1314.

[93] P.R. Kidambi, C. Ducati, B. Dlubak, D. Gardiner, R.S. Weatherup, M.-B. Martin, et al., The parameter space of graphene chemical vapor deposition on polycrystalline Cu, The Journal of Physical Chemistry C, 116 (2012) 22492–22501.

[94] Y. Gamo, A. Nagashima, M. Wakabayashi, M. Terai, C. Oshima, Atomic structure of monolayer graphite formed on Ni(111), Surface Science, 374 (1997) 61–64.

[95] R.S. Weatherup, B. Dlubak, S. Hofmann, Kinetic control of catalytic CVD for high-quality graphene at low temperatures, ACS Nano, 6 (2012) 9996–10003.

[96] R.S. Weatherup, B.C. Bayer, R. Blume, C. Ducati, C. Baehtz, R. Schlögl, et al., In situ characterization of alloy catalysts for low-temperature graphene growth, Nano Letters, 11 (2011) 4154–4160.

[97] M.A. Fanton, J.A. Robinson, C. Puls, Y. Liu, M.J. Hollander, B.E. Weiland, et al., Characterization of graphene films and transistors grown on sapphire by metal-free chemical vapor deposition, ACS Nano, 5 (2011) 8062–8069.

[98] A. Scott, A. Dianat, F. Börrnert, A. Bachmatiuk, S. Zhang, J.H. Warner, et al., The catalytic potential of high-κ dielectrics for graphene formation, Applied Physics Letters, 98 (2011) 073110.

[99] G. Ning, Z. Fan, G. Wang, J. Gao, W. Qian, F. Wei, Gram-scale synthesis of nanomesh graphene with high surface area and its application in supercapacitor electrodes, Chemical Communications, 47 (2011) 5976.

[100] J. Sun, N. Lindvall, M.T. Cole, K.B.K. Teo, A. Yurgens, Large-area uniform graphene-like thin films grown by chemical vapor deposition directly on silicon nitride, Applied Physics Letters, 98 (2011) 252107.

[101] C.R. Herron, K.S. Coleman, R.S. Edwards, B.G. Mendis, Simple and scalable route for the “bottom-up” synthesis of few-layer graphene platelets and thin films, Journal of Materials Chemistry, 21 (2011) 3378.

[102] E.F. Acheson, Article of Caborundum and Process of the Manufacture Thereof, U.S. Patent US615648, 1898.

[103] C. Berger, Z. Song, T. Li, X. Li, A.Y. Ogbazghi, R. Feng, et al., Ultrathin epitaxial graphite?: 2D electron gas properties and a route toward, Journal of Physical Chemistry B, 108 (2004) 19912–19916.

[104] K. V Emtsev, A. Bostwick, K. Horn, J. Jobst, G.L. Kellogg, L. Ley, et al., Towards wafer-size graphene layers by atmospheric pressure graphitization of silicon carbide, Nature Materials, 8 (2009) 203–207.

[105] J. Hass, F. Varchon, J. Millán-Otoya, M. Sprinkle, N. Sharma, W. de Heer, et al., Why multilayer graphene on 4H-SiC(0001¯) behaves like a single sheet of graphene, Physical Review Letters, 100 (2008) 125504.

[106] C. Riedl, C. Coletti, T. Iwasaki, A.A. Zakharov, U. Starke, Quasi-free-standing epitaxial graphene on SiC obtained by hydrogen intercalation, Physical Review Letters. 103 (2009) 246804.

[107] S. Goler, C. Coletti, V. Piazza, P. Pingue, F. Colangelo, V. Pellegrini, et al., Revealing the atomic structure of the buffer layer between SiC(0001) and epitaxial graphene, Carbon, 51 (2013) 249–254.

[108] R. Tromp, J. Hannon, Thermodynamics and kinetics of graphene growth on SiC(0001), Physical Review Letters, 102 (2009) 106104.

[109] K. V Emtsev, A. Bostwick, K. Horn, J. Jobst, G.L. Kellogg, L. Ley, et al., Towards wafer-size graphene layers by atmospheric pressure graphitization of silicon carbide, Nature Materials, 8 (2009) 203–207.

[110] W.A. de Heer, C. Berger, M. Ruan, M. Sprinkle, X. Li, Y. Hu, et al., Large area and structured epitaxial graphene produced by confinement controlled sublimation of silicon carbide, Proceedings of the National Academy of Sciences of the United States of America, 108 (2011) 16900–16905.

[111] W.A. de Heer, The development of epitaxial graphene for 21st century electronics, arXiv Preprint arXiv:1012.1644. (2010).

[112] T.J. McArdle, J.O. Chu, Y. Zhu, Z. Liu, M. Krishnan, C.M. Breslin, et al., Multilayer epitaxial graphene formed by pyrolysis of polycrystalline silicon-carbide grown on c-plane sapphire substrates, Applied Physics Letters, 98 (2011) 132108.

[113] J.C. Meyer, A.K. Geim, M.I. Katsnelson, K.S. Novoselov, T.J. Booth, S. Roth, The structure of suspended graphene sheets, Nature, 446 (2007) 60–63.

[114] J.C. Meyer, A.K. Geim, M.I. Katsnelson, K.S. Novoselov, D. Obergfell, S. Roth, et al., On the roughness of single- and bi-layer graphene membranes, Solid State Communications, 143 (2007) 101–109.

[115] D.B. Williams, C.B. Carter, Transmission Electron Microscopy: A Textbook for Materials Science, 1996.

[116] P.Y. Huang, C.S. Ruiz-Vargas, A.M. van der Zande, W.S. Whitney, M.P. Levendorf, J.W. Kevek, et al., Grains and grain boundaries in single-layer graphene atomic patchwork quilts, Nature. 469 (2011) 389–392.

[117] O.L. Krivanek, M.F. Chisholm, V. Nicolosi, T.J. Pennycook, G.J. Corbin, N. Dellby, et al., Atom-by-atom structural and chemical analysis by annular dark-field electron microscopy, Nature, 464 (2010) 571–574.

[118] A.C. Ferrari, D.M. Basko, Raman spectroscopy as a versatile tool for studying the properties of graphene, Nature Nanotechnology. 8 (2013) 235–246.

[119] A.C. Ferrari, J.C. Meyer, V. Scardaci, C. Casiraghi, M. Lazzeri, F. Mauri, et al., Raman Spectrum of graphene and graphene layers, Physical Review Letters, 97 (2006) 187401.

[120] A.C. Ferrari, D.M. Basko, Raman spectroscopy as a versatile tool for studying the properties of graphene, Nature Nanotechnology, 8 (2013) 235–246.

[121] C. Casiraghi, A. Hartschuh, H. Qian, S. Piscanec, C. Georgi, A. Fasoli, et al., Raman spectroscopy of graphene edges, Nano Letters, 9 (2008) 1433–1441.

[122] F. Tuinstra, Raman spectrum of graphite, The Journal of Chemical Physics, 53 (1970) 1126.

[123] D.S. Knight, W.B. White, Characterization of diamond films by Raman spectroscopy, Journal of Materials Research, 4 (2011) 385–393.
[124] U. Khan, A. O'Neill, M. Lotya, S. De, J.N. Coleman, High-concentration solvent exfoliation of graphene, Small (Weinheim an Der Bergstrasse, Germany), 6 (2010) 864–871.
[125] A.C. Ferrari, J. Robertson, Interpretation of Raman spectra of disordered and amorphous carbon, Physical Review B, 61 (2000) 95–107.
[126] C. Soldano, A. Mahmood, E. Dujardin, Production, properties and potential of graphene, Carbon, 48 (2010) 2127–2150.
[127] M. Lotya, P.J. King, U. Khan, S. De, J.N. Coleman, High-concentration, surfactant- stabilized graphene dispersions, ACS Nano, 4 (2010) 3155–3162.
[128] P.K. Ang, S. Wang, Q. Bao, J.T.L. Thong, K.P. Loh, High-throughput synthesis of graphene by intercalation-exfoliation of graphite oxide and study of ionic screening in graphene transistor, ACS Nano, 3 (2009) 3587–3594.
[129] J.N. Coleman, Liquid exfoliation of defect-free graphene, Accounts of Chemical Research, 46 (1) (2012) 14–22.
[130] C. Vallés, C. Drummond, H. Saadaoui, C.A. Furtado, M. He, O. Roubeau, et al., Solutions of negatively charged graphene sheets and ribbons, Journal of the American Chemical Society, 130 (2008) 15802–15804.
[131] S. Stankovich, R.D. Piner, X. Chen, N. Wu, S.T. Nguyen, R.S. Ruoff, Stable aqueous dispersions of graphitic nanoplatelets via the reduction of exfoliated graphite oxide in the presence of poly(sodium 4-styrenesulfonate), Journal of Materials Chemistry, 16 (2006) 155.
[132] V. Eswaraiah, S.S. Jyothirmayee Aravind, S. Ramaprabhu, Top down method for synthesis of highly conducting graphene by exfoliation of graphite oxide using focused solar radiation, Journal of Materials Chemistry, 21 (2011) 6800.

Michele Melchionna and Maurizio Prato

3 Functionalization of carbon nanotubes

3.1 Introduction

Carbon nanotubes (CNTs) represent a very symbolic material since the advent of the so-called “nanotechnology revolution”. From a purely fictional point of view, CNTs have enjoyed great popularity in many sci-fi novels and TV shows, with documentaries emphasizing their leading role in the science of the new millennium. Such popularity has found even better fortune amongst the scientific community, with carbon nanotubes being the focus of massive research efforts, resulting in the proliferation of articles, reviews and textbooks describing their fascinating properties and potential applications.

A retrospective of the great promises posed by CNTs indicates that these materials have indeed fulfilled many of the expectations, but the space for new discoveries and breakthroughs is still vast. The state-of-the-art shows CNTs as versatile components in a range of very diverse applications: their unique structural and electronic properties, which are dependent on their size/surface ratio, are for example exploited in organic photovoltaic systems [1] and integrated circuits [2], where they can enhance energy conversion and storage performances; in heterogeneous catalysis [3], where they act as useful alternative supports due to their inertness, mechanical and thermal stability as well as tunable topography; in biosensors [4], due to their distinct optical/spectroscopic properties; and in nanomedicine [5], where they can act as vehicles for new drug delivery strategies. Regardless of the type of application, however, carbon nanotubes are seldom used in their pristine state. It has become an almost inevitable prerequisite for the tubes to be subject to *functionalization* prior to their utilization. The reasons why functionalization of nanotubes is desirable are multifaceted, and the purpose of this chapter is to demonstrate the significance of functionalization of carbon nanotubes, and give the reader a flavor of the general methods of functionalization, the most common reactions involved and the effects that such modifications have on CNTs’ properties.

Multifunctionalization, that is the attachment of more than one functional group to the carbon nanotube leading to integration of the chemically modified CNTs into more complex systems, will not be treated herein. Despite this being an important feature for many applications, the intricacy of the topic goes beyond the purpose of the present chapter.

3.2 Functionalization. Why?

Functionalization is one of the most investigated topics of research on carbon nanotubes. There is a series of practical reasons why introducing functional groups is an advantageous event, and many of them relate to the structure and properties of CNTs. Carbon nanotubes can be conceptually represented as rolled up sheets of graphene into a tubular structure [6]. They are defined as single-walled carbon nanotubes (SWCNTs) when composed of a single rolled-up graphene sheet, or as multi-walled carbon nanotubes (MWCNTs) when composed of multiple concentric rolled-up layers (Fig. 3.1).

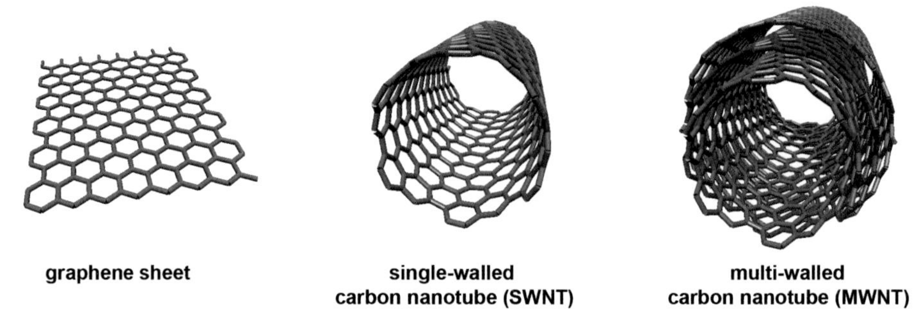

Fig. 3.1: Examples of typical carbon nanostructures.

One obvious but key difference between graphene and CNTs is the curvature introduced on the honeycomb planar arrangement. The loss of graphite-like planarity results in a distortion from pure sp^2 hybridization of the carbon atoms of the benzene ring motif. Despite this distortion, van der Waals forces are still exerted between individual tubes. Such forces, although intrinsically weak, become significant (order of 1000eV) due to the large surface area [7] resulting in aggregation of the tubes into bundles (Fig. 3.2). Formation of bundles dramatically reduces the solubility of CNTs in most organic or aqueous solvents, which is a serious limitation for processes requiring solution phase chemistry to homogeneously disperse the nanotubes within host materials.

Chemical modification, through addition of functional groups that disrupt the sp^2 pattern or isolation of individual tubes by noncovalent wrapping of various compounds, comes to our aid as one of the simplest and most effective ways to debundle the aggregates, thus improving solubility and processability.

Another complication in CNT applicability arises from the way the graphene sheet is rolled up to create the cylindrical structures, which is usually called “helicity”. Depending on the angle of the wrapping, three different structures (different helicities) can result: (1) armchair, (2) chiral or (3) zigzag (Fig. 3.3). Such structures exhibit differ-

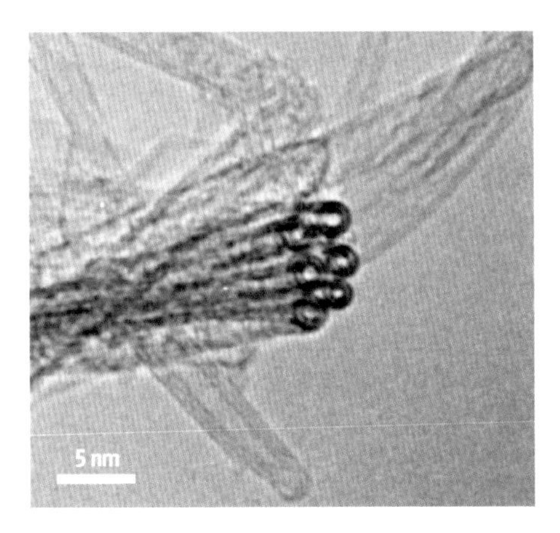

Fig. 3.2: Electron microscopy image of a carbon nanotubes bundle [8]. Adapted with permission from [8], © 2005, American Chemical Society.

ent electronic properties as a consequence of the different π-π* band gaps [9]. It is possible to mathematically predict electronic behaviors of the three different structures on the basis of specific geometrical parameters defined along the graphene plane [9]. It turns out that armchair tubes are always metallic, whereas zigzag and chiral tubes can be either metallic or semiconductors.

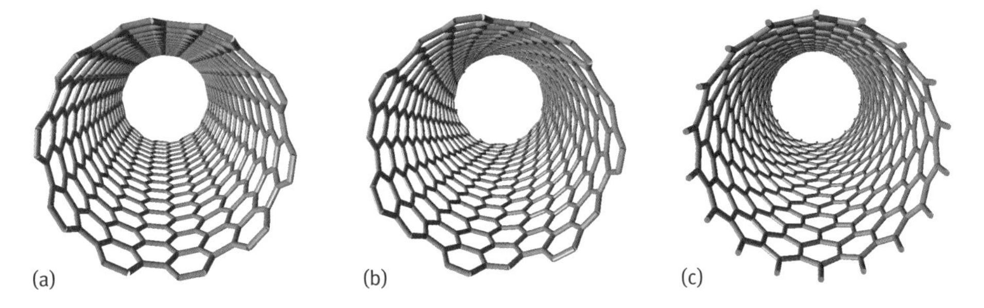

Fig. 3.3: (a) Metallic "armchair" tube. (b) Semiconducting "chiral" tube. (c) Conducting "zigzag" tube.

For many applications, such as optoelectronics or nanoscale electronic and sensing devices, only one type of electronic behavior is desired. Unfortunately, all different synthetic methodologies available to-date always afford mixtures of the above structures, with a selective synthesis of single structures yet to be accomplished. It therefore becomes imperative to conceive effective separation protocols. One approach is to exploit the stronger interaction that certain molecules have with one of the electronic structures. For example, affinity of octadecylamine (ODA) towards the semiconducting type leads to additional stability to the physisorbed ODA on the tube sidewalls, leading to selective precipitation of the metallic counterparts after increasing dispersion concentration [11]. In contrast, under certain conditions diazonium reagents react preferentially with metallic SWCNTs, with selectivity dictated by the electron population near the Fermi level and are thus able to stabilize the transition state that leads to bond formation [12].

Functionalization is also an exceptional way to bind the tubes to other organic or inorganic compounds, providing access to more complex systems that merge the properties of the two components to create materials with new properties for diverse applications.

Finally, it is worth mentioning that functionalization of CNTs generally results in a lowered toxicity of the tubes, this being of paramount importance in nanomedicine applications.

3.3 Types of functionalization

The vast majority of functionalization methods of carbon nanotubes belong to two broad categories: (a) covalent and (b) noncovalent functionalization of the external CNT surface. The former is achieved by covalent attachment of functional groups to the C-C double bond of the π-conjugated framework. The latter is based on the adsorption through van der Waals type bonds of various functional entities.

Functionalization of carbon nanotubes with metals can be achieved by different techniques exploiting either the covalent or the noncovalent approach. This topic, which is important for many applications, will be briefly discussed in a separate section after the description of the two methods.

3.3.1 Covalent functionalization

Covalent functionalization of CNTs exploits the curvature of the honeycomb arrangement of conjugated benzene rings. The curvature introduces a strain in the planar sp^2 hybridized hexagonal motif, concomitant with a π-orbital misalignment, resulting in a higher reactivity as compared to graphene. Synthetic modification leads to a change of configuration of the carbon atoms from sp^2 to sp^3, which locally relieves such strain. The same situation is found in fullerenes, although the spherical nature of the latter systems is synonymous of a more pronounced curvature and therefore much higher reactivity.

When covalent functionalization of CNTs is considered, it is necessary to provide a slightly more detailed structural description of a carbon nanotube. The perfection of the hexagonal network is only an idealized portrait of the nanotube structure. In reality, carbon nanotubes present several irregularities in their crystallographic periodicity. Such irregularities are referred to as *defects* [13]. Defects can be essentially divided into four groups: (1) a *topological* defect is observed when a nonhexagonal ring is introduced to break the benzene pattern, (2) *rehybridization* occurs when a carbon atom changes from an sp^2 to a sp^3 configuration, (3) *incomplete bonding defects* are associated with the presence of vacancies or dislocation and (4) *doping* defines a situation in which other elements are introduced (Fig. 3.4).

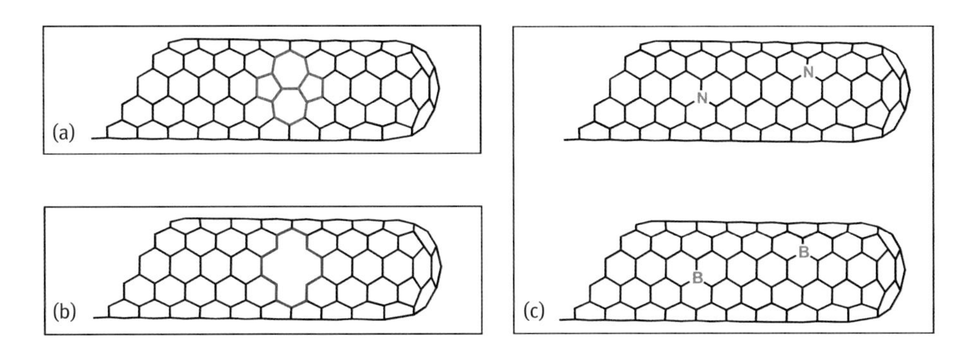

Fig. 3.4: Typical defects occurring in CNTs: (a) topological, (b) vacancy, (c) doping (examples with nitrogen and boron).

The existence of these defects is fundamental to the covalent modification of the tube, as they can augment the local curvature, acting as catalyzing sites for reaction of neighboring atoms.

One more structural differentiation that can influence functionalization is the tip *vs*. sidewall region. The tips are the two ending parts of the tube; in these two regions, the curvature is increased and the shape more resembles that of a hemisphere, with reactivity expected to be similar to that of fullerenes. In contrast, sidewalls present reduced pyramidalization angles and therefore different behaviors towards functionalization. Reactions involving the use of harsh conditions can result in a fracture of the tubes, enabling production of shorter tubes with open tips, where the aromatic pattern is interrupted and carbon atoms are more reactive.

Opening and shortening of the tubes is a sought-after event for certain applications and can be induced *ad hoc* by different techniques such as ball milling, mechanical cutting, electrochemically and chemically. The open tips can be closed again if required [14].

The diameter of the nanotube is an additional important parameter, with smaller tubes presenting enhanced curvature and consequently enhanced reactivity. One last aspect affecting reactivity is the helicity of the carbon nanotubes. In metallic CNTs, the aromaticity is slightly lower than in the semiconducting types, rendering the former more susceptible to functionalization.

Covalent functionalization may also be subdivided into two groups, depending on whether the modification is carried out on previously oxidized CNTs or not. In the former case, the various functionalities react with the carboxyl and carbonyl groups introduced on the tube during oxidation protocols. The latter case involves direct reaction with the tube's aromatic framework; this requires use of highly reactive species as it involves a change in hybridization of the carbon atoms from sp^2 (trigonal planar geometry) to sp^3 (tetrahedral) with loss of aromaticity. For the sidewall region, this is a relatively high energy process, while for the cap region it is more favorable due to the more dramatic curvature.

3.3.1.1 Acid oxidation

It was discussed earlier how a major hurdle for the manipulation of CNTs is their insolubility in most organic and aqueous solvents, and how introducing functional groups is a simple and powerful tool to improve solubility. However, the first functionalization step is expected to be the most problematic due to the primary utilization of as yet nonfunctionalized pristine nanotubes.

A typical preliminary step to derivatize carbon nanotubes turns out to be also a valid method of purification. The tubes are subject to strong oxidative conditions, typically by treatment with strong concentrated acids such as HNO_3 or H_2SO_4 together with heating or sonication. This procedure has the dual utility of removing metallic traces deriving from CNT preparation and of creating structural defects along with the formation of functional groups, generally carboxyl, carbonyl and hydroxyl groups [15].

Because of the aggressive conditions, the tubes are also shortened and opened, and the oxygen functionalities predominantly anchor to the tips of the tubes.

Generation of carbonyl and carboxyl groups can also be achieved by plasma etching, which is a particularly useful technique because when applied in an inert atmosphere (*e.g.* N_2) affords the introduction of basic functionalities [16].

3.3.1.2 Amidation and esterification

The carboxylic functionalities inserted onto the tubes can be used as platforms to obtain further transformations (Fig. 3.5). A commonly utilized route is the reaction of carboxylic groups with thionyl chloride or oxalyl chloride to prepare the corresponding acyl chlorides, which are useful intermediates for amidation or esterification reactions. Amides can also be prepared directly from the acids by means of standard solution chemistry conditions, using carbodiimide derivatives in the presence of the selected amine.

The appropriate choice of amine led to the synthesis by Haddon and co-workers of a series of shortened nanotubes exhibiting high solubility and dispersibility that are suitable for a range of different applications [17]. In this regard, the assembling of water-soluble graft co-polymers based on the linking *via* amide bond of poly(aminobenzene sulfonic acid) (PABS) and polyethylene glycol (PEG) to SWCNTs was particularly ingenious [18].

Amidation of carbon nanotubes comes across as a reliable and relatively easy way to access materials that combine CNTs' properties with those of other interesting sys-

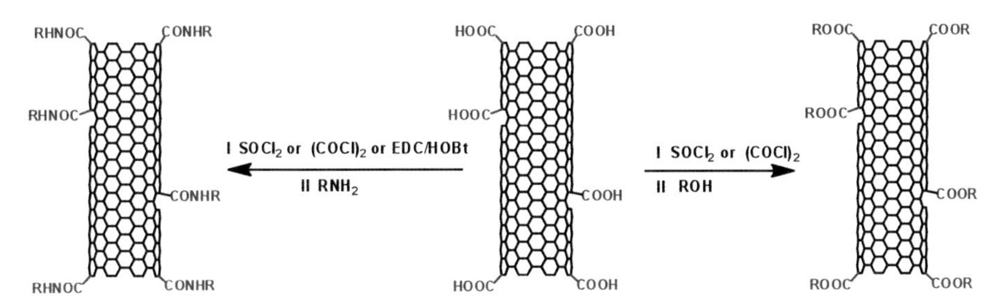

Fig. 3.5: Oxidized carbon nanotubes undergoing amidation and esterification reactions.

tems. Examples include linkages to DNA, proteins, enzymes and antibodies [19] as well as the intriguing SWCNT-fullerene hybrid, which has been the subject of studies related to electronic communication between the ground states of the two moieties [20].

Esterification constitutes a valuable alternative to the amidation strategy. As with amidation, the formation of the ester bond is performed following a first reaction step with acyl chloride. The ester bond has been extensively utilized to attach many organic and inorganic moieties. Porphyrins are a classic example of substrates covalently bound *via* esterification strategies: their photoinduced electron transfer to the nanotube has been studied for applications in molecular electronics and photovoltaic devices (Fig. 3.6) [21].

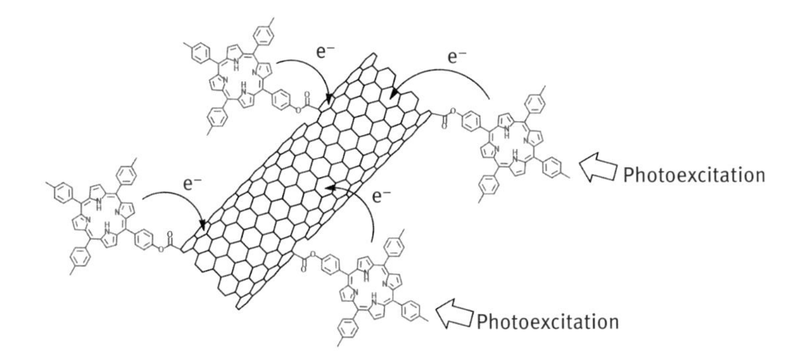

Fig. 3.6: Photoinduced electron transfer from porphyrins linked through ester bond to the carbon nanotube sidewalls. Adapted with permission from [21], © 2005, American Chemical Society.

3.3.1.3 Halogenation

Halogenation constitutes an important reaction as the fluorine atoms can be further substituted by other functionalities (Fig 3.7) [22]. Substitution with alkyl groups using Grignard or organolithium reagents is one of the most important examples, as alkyl groups considerably enhance solubility of CNTs in organic solvents [23].

The fluorination of carbon nanotubes was first reported by Margrave and is traditionally performed by using elemental fluorine in the presence of small amounts of HF which serves as catalyst [24]. The loading turns out to be very high, with one fluorine atom every two carbon atoms.

The reverse reaction to recover the unsubstituted nanotubes is possible by means of anhydrous hydrazine in *i*PrOH [25].

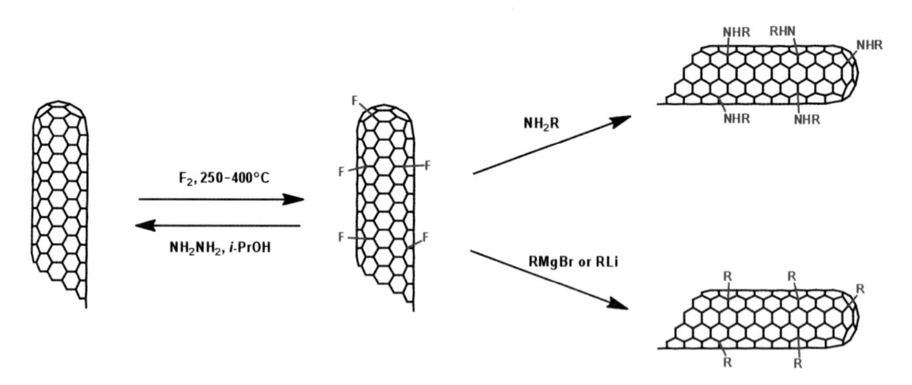

Fig. 3.7: Fluorination and further substitution of pristine carbon nanotubes.

Similarly, chlorination, bromination and iodination are achieved by high temperature and high pressure reactions with $SOCl_2$, Br_2 and I_2 respectively. The type of halogen does not seem to significantly affect the fluorescence emission, but rather influences the fluorescence lifetimes [26].

3.3.1.4 Cycloadditions

Cycloadditions are one class of reactions that provide a wide variety of differently functionalized nanotubes (Fig. 3.8). As for halogenations, cycloadditions are generally carried out directly on the nonoxidized pristine CNTs, which requires the use of highly reactive species or/and harsh conditions. The most common cycloaddition reactions include the addition of carbenes, nitrenes, 1,3-dipolar cycloadditions and Diels–Alder reactions.

Carbenes

Carbenes were among the first reactants to be explored in addition reactions. Phenyl (bromodichloromethyl)-mercury has been used to introduce dichlorocarbene, which produced large optical changes at Fermi level transitions [27]. It was recently reported that under certain conditions dichlorocarbene is also able to effect a transition from

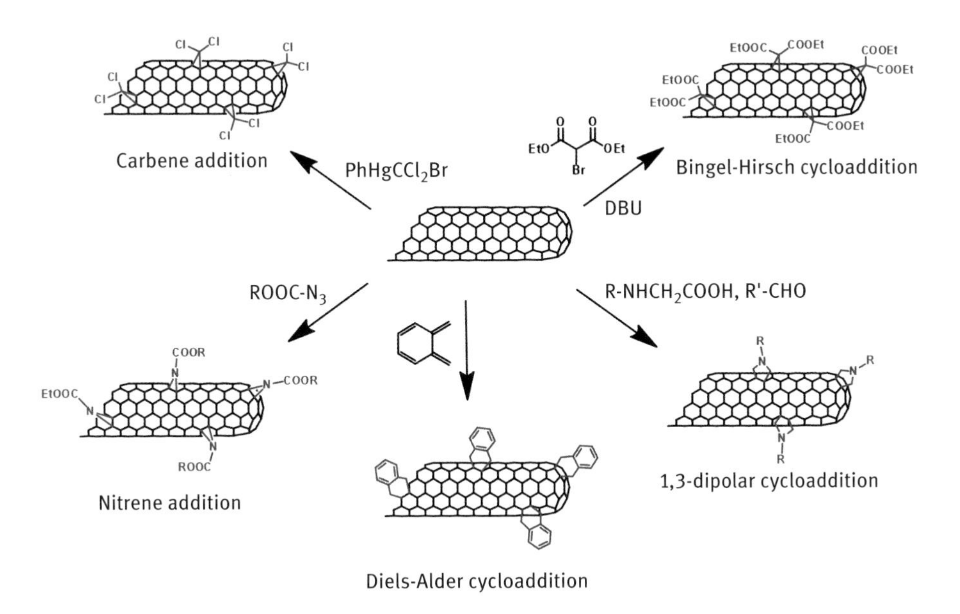

Fig. 3.8: Schematic representation of the most common types of cycloaddition.

semiconducting to metallic type. This transition was reversible upon thermal annealing [151].

The addition of a particularly useful reagent, bearing two ester functionalities, is obtained by means of the Bingel–Hirsch cyclopropanation [29]. Although the mechanism may not be of carbene-type, the result is formally equivalent to a carbene addition. The presence of the ester groups renders this derivative susceptible for further modifications.

Nitrenes

Addition of nitrenes was first reported by Hirsch and co-workers who described the addition of alkyl azidoformate at 160 °C, which proceeds *via* N_2 extrusion to yield the alkoxycarbonylaziridino-SWCNT [30]. The product could be isolated by exploiting its solubility in DMSO, with precipitation of insoluble impurities. This strategy can be extended to a range of more complex substituents, including longer alkyl chains, crown ethers, aromatic groups, dendrimers and oligoethylene glycol units.

Nitrene cycloaddition was also used by Yinghuai and co-workers to append C_2B_{10} carborane cages to SWCNTs [31]. One of the derivatives showed very interesting potential in medicinal applications as it was found that boron atoms had higher concentrations in tumor cells than in blood or other organs when administered to mice. This finding could inspire further research on the use of these systems as nanovehicles for boron delivery into the tumor cells, an important step in cancer treatment by boron neutron capture therapy.

1,3-dipolar cycloadditions

A typical 1,3-dipolar cycloaddition involves the use of azomethine ylides as the reactive species. Ylides are formed *in situ* by thermal condensation of α-amino acids and aldehydes, which then react to form the pyrrolidine-CNT system [32]. The R substituent on the 5-membered heterocycle attached to the SWCNT can be varied by the selection of the amino acid and the aldehyde, giving access to a range of different functional groups on the nanotube sidewalls.

The choice of the R group leads to materials exhibiting properties which can find applications in diverse fields: long alkyl chains help solubilization in organic solvents, whereas triethylene glycols have been used as spacers to attach water-solubilizing moieties, or even electron-donor systems such as ferrocene, which may find applications in photovoltaics, or as bio- and electrochemical sensors [33].

Diels–Alder

One problem associated with $4\pi + 2\pi$ cyclizations of carbon nanotubes is the reversibility of the process, and for this reason Diels–Alder reactions have been a less used synthetic route. One of the most representative examples of Diels–Alder functionalization was reported by Langa and co-workers, who performed MW-assisted addition of *o*-quinodimethane on pentanol-ester functionalized SWCNTs [34].

More recently, the Diels–Alder cycloaddition was used as a single-step strategy for attaching diene-capped polymers onto CNTs sidewalls through "grafting to" methods (see Section 3.3.1.6 for more details). Examples include attachment of furfuryl-functionalized polystyrene [35] and cyclopentadienyl-functionalized polymethyl methacrylate [36].

3.3.1.5 Radical additions

This class of reactions has received attention from numerous research groups and a wide range of radical additions to carbon nanotubes has been reported (Fig. 3.9). The highly reactive radical species are generated by different procedures, typically by means of photolysis, electrolysis, decomposition of organic peroxides and thermal oxidation/reduction by inorganic salts involving single electron transfers.

Photolysis of the carbon-iodine bond in iodo-perfluoroalkyl compounds has been employed to enable the formation of the corresponding pefluoroalkyl radical [37]. Different perfluorinated alkyl chains can thus be covalently bound to the CNT. Interestingly, the authors reported that no change in solubility of the nanotubes was noted after the perfluoro-alkylation.

Thermal decomposition of peroxides generates a variety of radical groups which can covalently react with CNTs. This strategy has been employed in particular by Peng and co-workers who functionalized sidewalls with benzoyl, lauroyl and carboxyalkyl derivatives [38].

Diazonium salts are another useful source of free radicals, and the formation of the reactive species can be achieved by reductive electrolysis or direct treatment with diazonium tetrafluoroborate salts [39]. By this route, several aryl derivatives could be introduced onto the nanotube sidewalls [40]. Aryl groups bearing halogen or alkyne functionalities are particularly interesting as they can be further reacted in Pd-catalyzed coupling reactions (Suzuki, Heck) or in "click chemistry" reactions to create products with great potential in materials science [41].

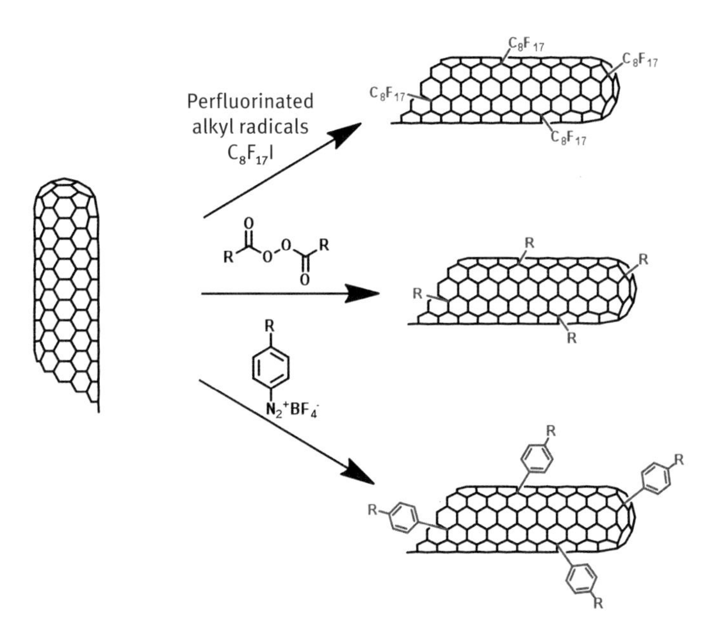

Fig. 3.9: Functionalization of pristine carbon nanotubes by radical addition.

One final example worth mentioning is the reductive alkylation/arylation with lithium and alkyl/aryl halides in liquid ammonia. This is a two-step process in which negatively charged nanotubes are formed *via* electron transfer from the metal. This step is relatively easy and fast due to the CNTs "electron sink" properties, and it enables exfoliation of the tubes through electrostatic repulsion; in the second stage, the alkyl/aryl halides react with the charged tubes to form a radical anion which can dissociate into the alkyl radical and the halide anion, with the former species undergoing addition to the CNT sidewalls [42].

3.3.1.6 Grafting polymers

In the section discussing Diels–Alder cycloadditions, it was shown how this reaction can be exploited as a way to link polymeric chains to the nanotube sidewalls. Attachment of polymers to carbon nanotubes is an important possibility for the chemistry of nanotubes as even low degrees of derivatization considerably enhance their solubility.

The Diels–Alder pathway illustrated above belongs to what is called the "grafting to" method, which consists of attaching a defined polymer chain through transformation of an appropriate end group.

There are a number of other reactions that can be used: cycloaddition of an azide group allowed binding of polystyrene [43], while a radical coupling was exploited to graft polymers prepared by a nitroxide-mediated radical polymerization [121]. Other end groups could be used: it was shown above how amide bonds were utilized to attach water-soluble polymers; amino or hydroxyl moieties are other conventional groups.

Other reactions employed in "grafting to" methodologies rely on the prior formation of carbanionic polymers. For example, strong bases or organometallic reagents such as sodium hydride or butyl lithium, can afford the polymeric anion which then reacts with the tube [45].

An alternative approach for covalent polymer attachment is called the "grafting from" method. This method is based on binding the monomer precursor to the nanotube and generating chain growth directly from the tube. By these means, several types of polymers could be grafted, such as polystyrene-sulfonate [46], polyvinylpyridine [47], polystyrene [48] and many others.

There are various strategies to carry out this type of *in situ* grafting; the use of an initiating agent may be required (generally to form radicals or anions), but other methodologies have been developed not requiring such agents, for example in the ultrasonic induced emulsion polymerization of acrylates [49].

3.3.2 Noncovalent functionalization

The second type of functionalization of carbon nanotubes is based on noncovalent interactions, such as CH-π, π-π stacking, van der Waals and electrostatic forces.

As for the covalent type, modification of CNTs by this approach has as its first goal to lead to debundling of the tubes, thus increasing their solubility and facilitating their manipulation. However, while the covalent method destroys the extended aromatic framework, noncovalent interactions preserve the original regular carbon network. This is important in those applications requiring use of the nanotubes without alteration of their electronic and optical properties, a process that normally occurs when the aromatic periodicity is disrupted.

It is possible to divide the noncovalent functionalization according to the type of molecules used, thus four categories are found: (1) surfactants, (2) polymers, (3) biomolecules, and (4) small aromatic molecules. A fifth special type of noncovalent type of functionalization that will also be briefly described is endohedral filling.

3.3.2.1 Surfactants

One characteristic feature of surfactants is their amphiphilic nature. These molecules present two moieties: the hydrophobic moiety (usually a hydrocarbon chain) interacts with the nanotube sidewalls, while the hydrophilic part, called polar head group, is generally charged or has zwitterionic character. It has the double function of helping solubility in aqueous solvents and of providing additional stabilization towards tubes aggregation by coulombic charge repulsion.

Surfactants are therefore effective solubilizers that can exfoliate CNTs by physical adsorption, which occurs at interfaces, allowing self-assembling into supramolecular structures [50].

The general procedure for dispersion involves use of sonication, which causes an "unzipping" mechanism of dispersion as proposed by Smalley and co-workers [51]. Typical surfactants used for this purpose are sodium dodecylsulfate (SDS) or sodium dodecylbenzene sulfonate (SDBS). The latter shows a stronger interaction as a result of the presence of the aromatic ring. It also appears that longer and more branched hydrocarbon chains interact more efficiently [52].

Other commonly used surfactants are cetyltrimethylammonium bromide (CTAB), Brij, Tween, Triton X and Siloxane polyether copolymer (PSPEO) [53].

One challenge posed by the use of surfactants is that their efficiency critically depends on the concentration of both surfactants and nanotubes. There is a "sweet spot" of concentration ratios for the preparation of stably dispersed nanotubes. Too low or too high concentrations lead, by different processes, to aggregation, with only very low levels of debundled tubes. Knowledge of this critical range of concentrations still usually relies on conducting time-consuming trials, despite Vigolo *et al.* having demonstrated that a domain of homogeneously dispersed CNTs can be defined in principle [54].

3.3.2.2 Polymers

The combination of polymers and carbon nanotubes is a widely investigated field. The same solubility issues discussed above when surfactants were treated are also addressed here, with polymers being a valid strategy for dispersion of CNTs. Conjugated polymers can homogeneously be wrapped around the tubes, as demonstrated by atomic force microscopy, and efficiently isolate them [55]. The wrapping occurs with formation of supramolecular complexes and can be helical or nonhelical (through π-stacking) depending on the flexibility of the polymer backbone (Fig. 3.10).

By these means, polymers are utilized to create new composite materials, which combine the good mechanical, electronic and optical properties of the two systems and find applications in solar cells, light emitting devices, sensors etc. The prepara-

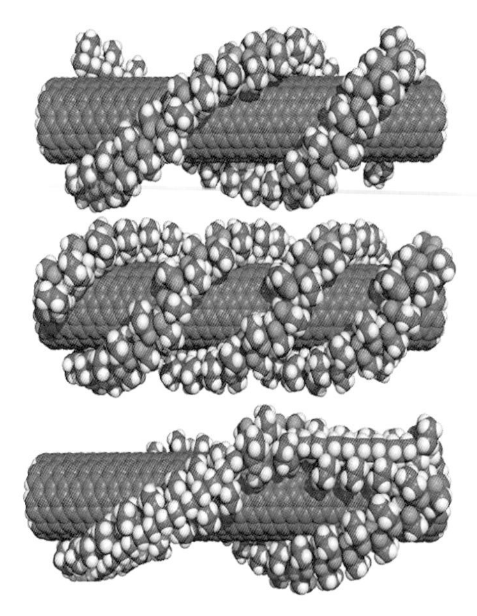

Fig. 3.10: Model representation of polymer wrapping around carbon nanotubes. Reprinted from J. M. O'Connell *et al.*, Reversible water-solubilization of single-walled carbon nanotubes by polymer wrapping, Chemical Physics Letters, 342, 265–271, Copyright (2001), with permission from Elsevier.

tion is relatively simple, involving stirring and sonication of the tube in organic or aqueous solutions of the polymer.

Poly(phenylenevinylene) derivatives are amongst the most studied as far as carbon nanotubes are concerned. They helically envelop the CNT sidewalls resulting in formation of composites with greatly enhanced conductivity with applications in optoelectronics [56].

Polythiophenes (PTs)/CNTs composites have emerged as an intriguing system for use as photovoltaic devices and field effect transistors [57]. Swager and Bao independently reported methods for the assembling of PTs/CNTs systems and showed their great potential as transparent conductive films [58]. Another interesting application arises from the possibility to functionalize the polythiophene backbone for applications as chemical sensors [134].

Nonconjugated hydrocarbon polymers could also be combined with carbon nanotubes, with polystyrene being the most studied example. The composites are generally prepared by solution or shear mixing techniques, resulting in materials with improved mechanical properties [60].

An additional opportunity that arises from applying polymer-based materials is the preparation of water-dispersible composites, which is an essential feature for biomedical purposes, as it is possible to attach bio-active molecules to the polymer/CNTs systems and specifically deliver them to cells. In this manner, plasmid DNA, siRNA (Fig. 3.11) and several anticancer agents have been successfully bound and delivered [61]. The stratagem to generate materials with good solubility in aqueous media usually involves the presence of water-soluble polar groups (*e.g.* phosphates, protonated amines *etc.*) embedded in the polymer chain.

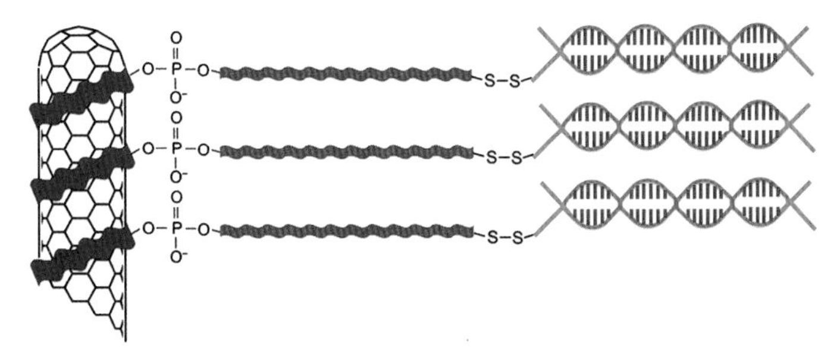

Fig. 3.11: Functionalization of SWNTs with PL-PEG2000-NH_2 (PL = phospholipid, in blue; PEG in pink) for the conjugation of thiol–siRNA (in red/orange) through disulfide linkages. Adapted with permission from [61], © 2007, Wiley-VCH.

It has also been observed that the attachment of conjugated polymer nanoparticles (CPNs) based on bromohexyl-fluorenes can be carried out to generate CPNs/CNTs composites with no significant fluorescence-quenching characteristics [62]. Quenching of light emission is normally associated with the presence of the carbon nanotube; if this can be avoided, the resulting composites may have potential applications in bioimaging and biosensing, exploiting direct fluorescent labeling of the nanotube structure.

Wrapping of conjugated polymers is not the only method of noncovalent interaction. Another approach is to provide the polymers with end groups (generally small aromatic molecules) that are able to noncovalently interact with the carbon framework *via* π-stacking. This can be exploited to create architectures where the macromolecule is free to extend and dangle out of the CNT surface, determining different properties. The most investigated anchor groups are pyrene derivatives, and a number of polymers have been attached by these means (Fig. 3.12) [63].

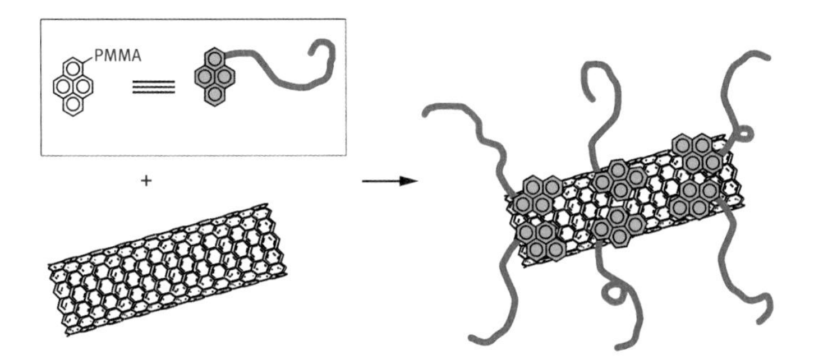

Fig. 3.12: Schematic representation of noncovalent interaction (*i.e.*, π-stacking) between the CNT surface and pyrene-PMMA polymer conjugates [62].

3.3.2.3 Biomolecules

The exploration of biomolecules as components in CNT-based systems is a rapidly flourishing research field. The combination of the CNTs' electronic and optical features with those of many biomolecules is appealing for many applications in medicine, optical device technologies and other fields.

Covalent attachment of biomolecules has been carried out by various groups, and the resulting systems proved to be useful as drug delivery carriers into infected cells. Another tantalizing prospect deriving from integration of carbon nanotubes and biomolecules is the assembly of new bioelectronic materials where the conducting and recognition properties of the two systems are merged to generate more efficient biosensors [64]. However, for this last application, the disruption caused by covalent modification is sometimes detrimental due to the reduced aromatic conjugation. In contrast, noncovalent adsorption of biomolecules occurs with no dramatic alteration of the conductivity capacity of the graphene-like framework; however, such conductivity is still remarkably sensitive to the adsorbed biomolecule, which makes these systems particularly appealing for use as electrochemical biosensors.

Several CNT/biomolecules electrochemical biosensors have thus been prepared, including a hemoglobin system for hydrogen peroxide detection [65], the myoglobin composite for nitric oxide [66] or hydrogen peroxide detection [67], the hemin conjugate for oxygen gas sensing [68] and the cholesterol esterase system for blood analysis [69].

Proteins, and particularly enzymes, are amongst the most used biomolecules and it has been demonstrated that CNTs can effectively promote electron-transfer reactions even in those glycoproteins where the redox center is located in the core of the protein. The proteins presumably interact by means of their hydrophobic regions. Amperometric glucose biosensors based on the direct electron transfer between glucose oxidase (GOx) and CNTs have also been synthesized [70]. GOx attached to CNTs have also been used to detect glucose by exploiting the GOx-catalyzed reaction of β-D-glucose to the D-glucono-1,5-lactone with a H_2O_2 co-product; this latter is involved in the partial reduction of the $[Fe(CN)_6]^{3+}$ surface, resulting in a reversible coupling of the CNT near-infrared fluorescence to the glucose concentration (Fig. 3.13) [71].

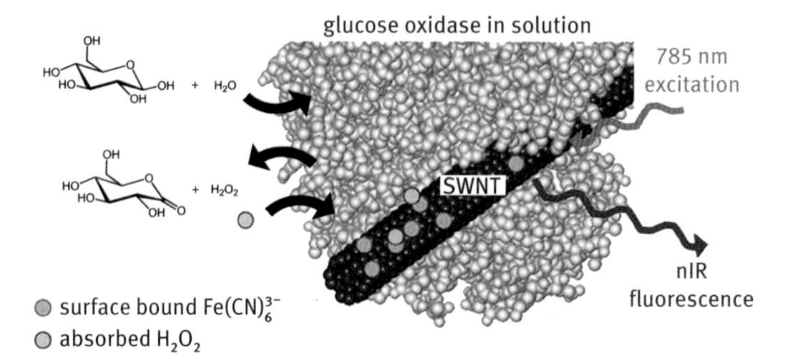

Fig. 3.13: A glucose biosensor based on CNTs (SWNT) and the enzyme glucose oxidase (in grey). Adapted by permission from Macmillan Publishers Ltd., [71], © (2005).

One of the approaches to attach a biomolecule to CNTs is to introduce a functional group that can noncovalently bind the graphene motif *via* π-π stacking. As in the case of polymers discussed previously, pyrene moieties proved to be very effective and they are in most cases the anchor group of choice [72].

It is important to mention DNA strands as one of the moieties that interact most efficiently with CNTs, yielding hybrids that exhibit excellent dispersion in aqueous media. Among the diverse applications, DNA-wrapping was employed to successfully separate metallic and semiconducting tubes [73].

3.3.2.4 Small aromatic molecules

It was already shown above how pyrene derivatives turn out to be useful anchor groups to append polymers and biomolecules. Indeed, the pyrene nucleus interacts with the nanotube sidewalls through π-π stacking and its use is mostly devoted to providing noncovalent attachment of other functional molecules. A particularly useful pyrene derivative is *N*-succinimidyl-1-pyrenebutanoate (Fig. 3.14), which is adsorbed onto the graphitic framework in DMF or MeOH solutions, and shows high resistance to desorption in aqueous media. This is an important aspect, as the succinimidyl ester group can be chosen in such a way as to be reactive towards nucleophilic substitution by primary and secondary amines, thus opening the door to decoration of CNTs with proteins such as ferritin, streptavidin, or biotinyl-3,6-dioxaoctanediamine and other biomolecules [74].

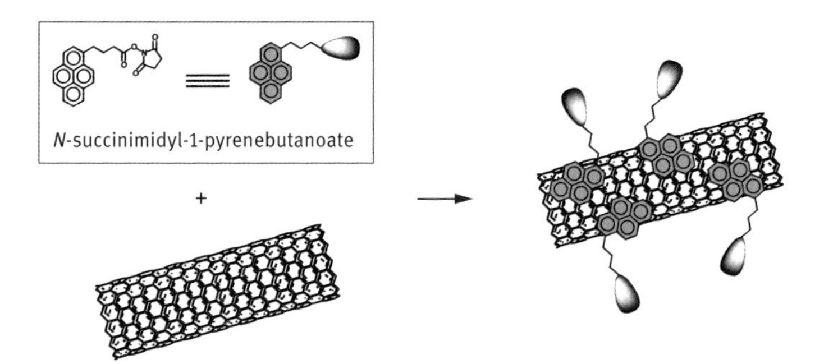

Fig. 3.14: Schematic representation of noncovalent interaction (*i.e.*, π-stacking) between the CNT surface and a pyrene derivative for protein functionalization. Adapted with permission from [74], © 2009, American Chemical Society.

High solubilization of SWCNTs in water (as high as 0.20 mg/mL^{-1}) was obtained by decorating the sidewalls with a charged pyrene derivative, 1-(trimethylammonium acetyl) pyrene. The positively charged pyrene moiety can be associated with negatively charged electron donors, for instance anionic porphyrin derivatives, to generate electron donor-acceptor nanohybrids, which could find applications as field-effect transistors [75].

Porphyrins are also able to directly interact with the nanotube sidewalls. For example, tetraphenyl porphyrin (H_2-TPP) has been reported to interact with nanotubes to form TPPs/CNTs compounds that are stable for days. Stability has been enhanced by using a micelle-assisted approach, leading to stable structures with potential applications in light harvesting devices [76].

Metal coordination to the porphyrin negatively affects the interaction with the nanotubes, with adsorption depleting according to the metal used [77].

3.3.2.5 Endohedral filling

Encapsulation of different entities inside the CNT channel stands alone as an alternative noncovalent functionalization approach. Many studies on the filling of carbon nanotubes with ions or molecules focus on how the presence of these fillers affects the physical properties of the tubes. From a different point of view, confinement of materials inside the cylindrical structure could be regarded as a way to protect such materials from the external environment, with the tubes acting as a nanoreactor or a nanotransporter. It is fascinating to envision specific reactions between molecules occurring inside the aromatic cylindrical framework, tailored by CNT characteristic parameters such as diameter, affinity towards specific molecules, *etc.*

The first ever reported molecule to undergo encapsulation was fullerene, which spontaneously and accidentally ended up in the tubes during post-processing of raw tubes prepared *via* the pulsed laser vaporization method. This could be considered a milestone in the self-assembling of a new class of nanomaterials [78].

A detailed study of the C_{60}/CNT system (named "peapod", Fig. 3.15) was conducted, highlighting how the fullerene appears to be the perfect candidate for encapsulation, as a consequence of the similar graphitic scaffold, an ideal structural match between the C_{60} spherical shape and the internal CNT channels and strong van der Waals interactions [79].

Since this pioneering work, other higher order carbon spheres (C_{70}, C_{78}, C_{80}, C_{82}, and C_{84}) [80] and their metal-derivatized analogues (Gd@C_{82} [81], Sm@C_{82} [82], Dy@C_{82} [83], Ti_2@C_{80} [84], Gd2@C_{92} [85], La@C_{82} [86], Sc_2@C_{84} [87], Ca@C_{82} [88] and Ce@C_{82} [89]) have been prepared.

Despite the strong interaction with the internal walls, the molecular barrier for motion of the fullerene derivatives is relatively weak, due to the uniformity of the polyaromatic surface [90].

Other organic molecules as well as metal ions can also be trapped. Among the organic molecules, particular relevance is given to biomolecules. It has been proved that several inserted enzymes retain their catalytic activity after encapsulation, while filling with DNA has been followed by fluorescence spectroscopy, and this could be exploited for electronic DNA sequencing and nanotechnology of gene delivery systems [91].

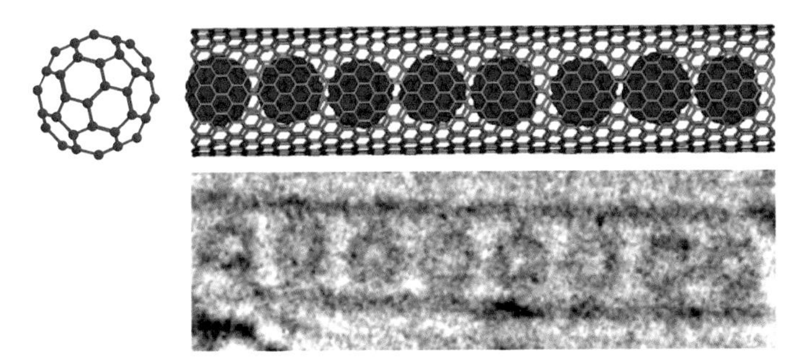

Fig. 3.15: Schematic representation (top) and microscopy image (bottom) of the "peapod" system where fullerene molecules are encapsulated inside CNTs. Adapted with permission from [90], © 2005, American Chemical Society.

As far as inorganic salts are concerned, they are normally introduced by blending their molten state with the CNTs or by sublimation. Many inorganic salts have been used with most of the transition metals and alkali/alkaline earth metals, with halides being the most typical anions [92], together with hydroxides [93]. The tubes can also be doped with individual metals, their oxides and with organometallic species such as metallocenes (Fig. 3.16) [94]. Fabrication of these materials is driven by potential applications in nanoelectronics.

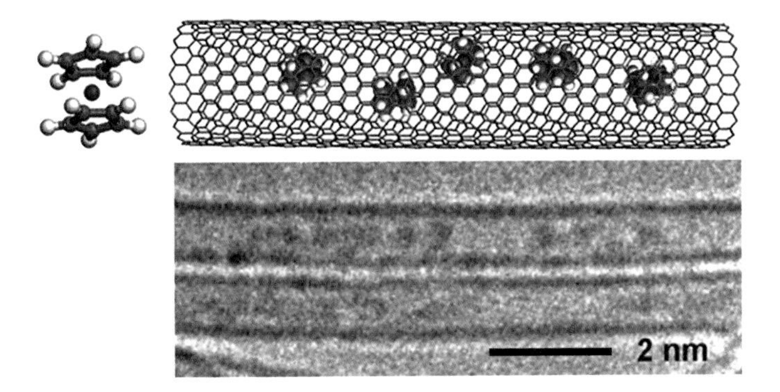

Fig. 3.16: Schematic representation (top) and microscopy image (bottom) of metallocene molecules encapsulated inside CNTs. Adapted with permission from [90], © 2005, American Chemical Society.

3.4 Functionalization with metals

Fabrication of materials combining metals and nanotubes is an attractive and intensively explored field. Particularly with a view to downscaling conventional technology processes by optimization of material combinations, it shows great potential as an alternative to develop new single-system materials.

Functionalization of CNTs with metal and metal oxide species can be achieved by different methodologies giving rise to the two distinct classes of *nanocomposites* [95] and *CNT-inorganic hybrids* [96]. The former class shows properties that add up those of the individual components, and the latter presents a merging that gives rise to materials with new properties not necessarily reminiscent of the two individual systems. Particular emphasis has been given over the last few years to inorganic nanoparticles as building blocks for CNTs/inorganic hybrids [96].

Definitions of *in situ* and *ex situ* preparation methods are frequently found in specialized articles, reviews or textbooks. *In situ* methods refer to the possibility to assemble the inorganic compounds directly on the pristine (or modified) CNTs and *ex situ* methods to binding such materials in a post-assembling step *via* some linking agent [96].

Regardless of the preparative methodology, the type of interaction between the metal species and the nanotube framework will still belong to the covalent/noncovalent type.

Covalent linkages through the carboxylic groups introduced during oxidation can anchor metal complexes (Fig. 3.17) to enable the synthesis of CNT-inorganic hybrids for applications in nanoelectronics [96].

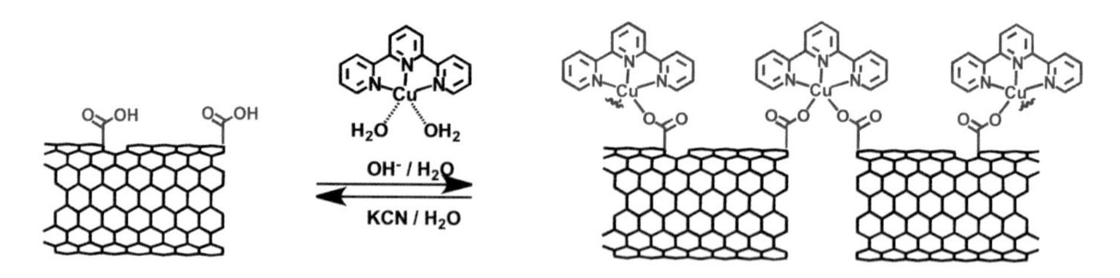

Fig. 3.17: Covalent attachment of Cu-complex moieties.

Amidation and esterification are also widely exploited. CNTs-amide bonds are found as part of linking agents employed to covalently attach nanoparticles, generally metals or semiconducting quantum dots (QDs), *via ex situ* synthesis. These linkers are frequently bifunctional, possessing an additional anchoring site that is bound to the nanoparticle side. For example, modified CdSe QDs have been attached *via* an amine-terminated amido linker (Fig. 3.18) [97], while Au nanoparticles are often attached through mercapto-terminated linkers [98].

Synthesis of SiO_2- and TiO_2-nanoparticles hybrids with MWCNTs has also been reported (Fig. 3.19); the reaction proceeds by means of phosphonic acid-modified and alkoxy silane-modified CNTs which can cap the oxides and template their assembly onto the CNT's surface [99].

Many noble metal nanoparticles (with Ru, Au, Pd, Pt, Os, Ag, Ir, Rh) have been deposited on CNTs and their catalytic activity towards a set of different reactions explored, showing the benefit of the presence of the graphitic support for catalysis.

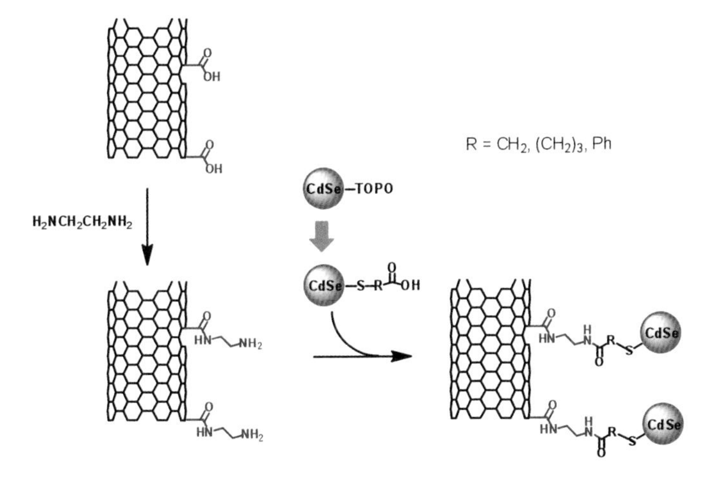

Fig. 3.18: Linkage of modified CdSe QDs through amide bond. Adapted with permission from [97], © 2002, American Chemical Society.

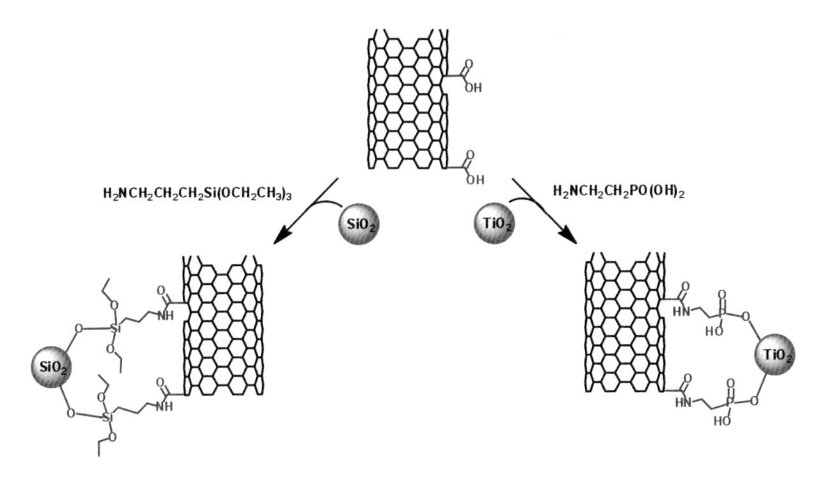

Fig. 3.19: Linkage of SiO_2 and TiO_2 nanoparticles *via* silane and phosphonic acid bonds, respectively. Adapted with permission from [99], © 2002, American Chemical Society.

Noncovalent derivatization with metal complexes or nanoparticles is again based on van der Waals, electrostatic, H-bond and π-π stacking interactions.

Pyrenes appear to be once again amongst the most useful linkers, and the attachment of metal-complexed porphyrins has already been discussed in the section on small aromatic molecules. Porphyrin rings bearing a metal center can also be directly adsorbed on the CNTs sidewalls, and they have been shown to exhibit an increased charge transfer from the metal to the tubes, presumably mediated by the aromatic moieties [100]. A similar behavior is exhibited by the phthalocyanine-based complexes [101]. Other aromatic molecules can be used for appendages, including triphenylphosphines, and benzyl derivatives. For example, TiO_2 units have been attached through adsorption of benzyl alcohol. This strategy provides a uniform coating of the tubes with the metal oxide (Fig. 3.20) [102].

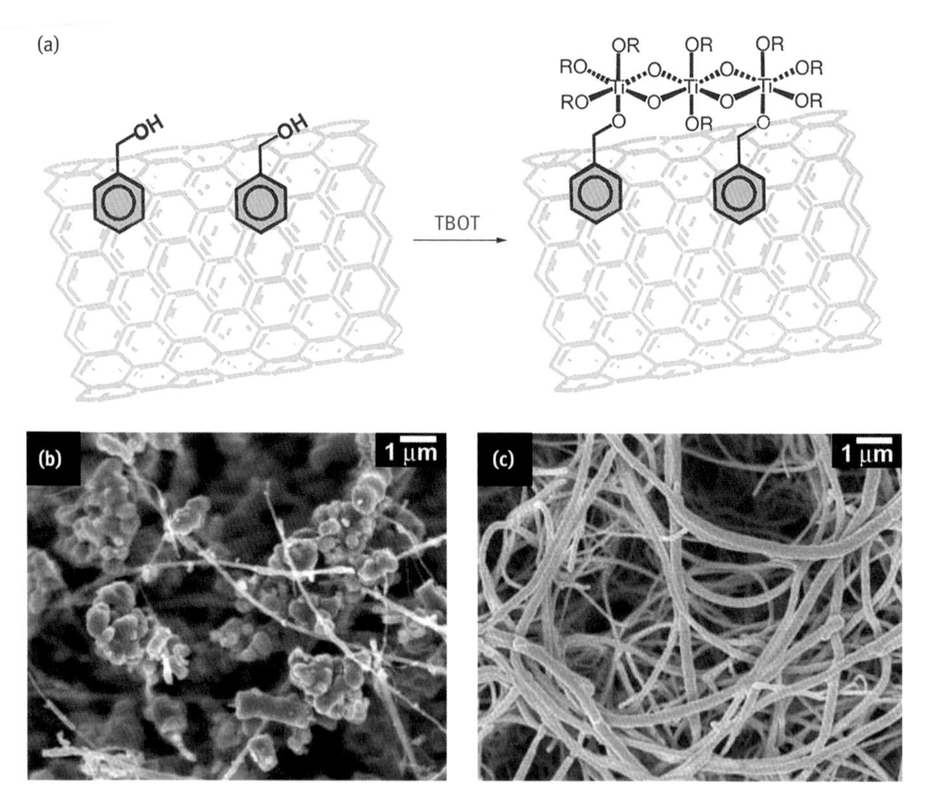

Fig. 3.20: (a) Schematic representation of noncovalent attachment (*i.e.*, π-stacking) of benzyl alcohol molecules for subsequent coordination with Titanium resulting in the coating of the tube. (b) SEM image of pristine CNTs prior to derivatization. (c) SEM image of CNTs after derivatization with benzyl alcohol and subsequent coordination with Titanium resulting in the coating of the tubes. Adapted with permission from [102], © 2008 Wiley-VCH.

The possibility of using electrostatic charge attraction has been exploited in the preparation of gold dendrimer encapsulated nanoparticles (DENs), which under appropriate conditions can be fully distributed along the surface of monodispersed MWCNTs (Fig. 3.21) [103].

More recently, electrostatic adsorption of Fe_3O_4 nanoparticles has been reported, showing superparamagnetism properties for the system [225].

In general, the protocol for electrostatic attachment of metal nanoparticles proceeds through a preliminary deposition of polyelectrolytes, which also establishes whether the tubes are positively or negatively charged: cationic poly(diallyldimethyl-

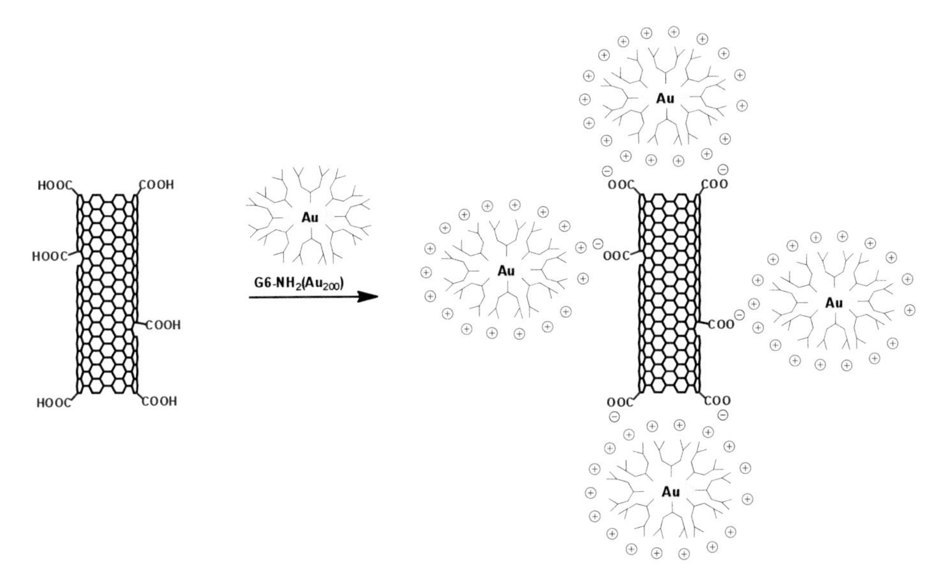

Fig. 3.21: Gold DENs' attachment on the MWCNT surface through electrostatic interactions. Adapted with permission from [103], © 2010, American Chemical Society.

ammonium chloride) (PDDA) can attract and bind negatively charged Au nanoparticles [176], while if a second layer of anionic poly(sodium 4-styrenesulfonate) (PSS) is wrapped, adsorption of positively charged SiO_2 can be accomplished [106].

These are just a few examples of CNT systems comprising metal species. The conjugation of the two classes of materials is an actively explored field, and new examples are being continuously reported.

3.5 Summary

Nanoscience is a new flourishing research area that is projecting tantalizing opportunities for applications in virtually every scientific area.

Because of their exceptional mechanical, electronic and optical properties, carbon nanotubes (CNTs) are amongst the most studied nanomaterials, particularly when used in combination with other molecular systems.

Functionalization of carbon nanotubes becomes essential for multiple reasons. Firstly, chemical modification can allow debundling and therefore solubilization of the tubes, which is an important feature for their processability. Secondly, insertion of functional groups enables attachment of more complex moieties that find applications in several fields.

There are numerous reactions and molecules that can be used to modify nanotubes; nevertheless, all reactions are based on two main types of interaction: covalent and noncovalent binding mode. Covalent modification of the surface is a stronger type of interaction, and final products are generally more stable. However, one major drawback of this type of functionalization is that it causes disruption of the conjugated aromatic motif by changing local hybridization of the carbon atoms ($sp^2 \rightarrow sp^3$). Such a modification affects the electronic and optical properties of the carbon nanotube, and is a shortcoming for those applications requiring preservation of CNTs' characteristics.

On the other hand, noncovalent adsorption leaves CNTs' electronic and optical properties almost intact, as the type of modification is through van der Waals, π-π stacking, H-bonding and electrostatic interactions, that do not determine any pyramidalization process. Such benefit is traded off by a decreased stability of CNT-functionalized systems, as a consequence of the weaker interactions.

The examples given in the present chapter are not comprehensive of the immense work carried out on carbon nanotubes, but they are just meant to give the reader some coordinates to move in this fascinating field.

New CNT-based systems deriving from more and more clever and sophisticated methodologies are proliferating every day, ensuring that the frontiers of nanotechnology are being incessantly pushed forward.

Bibliography

[1] (a) Dai, L. M.; Chang, D. W.; Baek, J. B.; Lu, W., Small 2012, 8(8), 1130–1166. (b) Cataldo, S.; Salice, P.; Menna, E.; Pignataro, B., Energy & Environmental Science (2012), 5(3), 5919–5940. (c) Ferguson, A. J.; Blackburn, J. L.; Kopidakis, N., Materials Letters 2013, 90, 115–125.

[2] (a) Liu, Z.; Jiao, L.; Yao, Y.; Xian, X.; Zhang, J., Advanced Materials (Weinheim, Germany) (2010), 22(21), 2285–2310. (b) Avouris,P., Physics Today 2009, 62(1), 34–40. (c) Rawlett, A. M.; Hopson, T. J.; Amlani, I.; Zhang, R.; Tresek, J.; Nagahara, L. A.; Tsui, R. K.; Goronkin, H., Nanotechnology (2003), 14(3), 377–384.

[3] (a) Wu, B.; Kuang, Y.; Zhang, X.; Chen, J., Nano Today 2011, 6(1), 75–90. (b) Mabena, L. F.; Sinha R. S.; Mhlanga, S. D.; Coville, N. J., Applied Nanoscience 2011, 1(2), 67–77.

[4] Chen, Z.; Zhang, X.; Yang, R.; Zhu, Z.; Chen, Y.; Tan, W., Nanoscale (2011), 3(5), 1949–1956.

[5] (a) Fabbro, C.; Ali-Boucetta, H.; Ros, T. D.;, Kostarelos, K.; Bianco, A.; Prato, M., Chem. Commun. 2012, 48, 3911–3926. (b) Fabbro, A.; Bosi, S.; Ballerini, L.; Prato, M., Carbon Nanotubes: Artificial Nanomaterials to Engineer Single Neurons and Neuronal Networks. ACS Chem. Neurosci. 2012, 3, 611–618. (c) Montellano, A.; Ros, T. D.; Bianco, A.; Prato, M., Nanoscale 2011, 3, 4035–4041. (d) Bianco, A.; Kostarelos, K.; Prato, M., Chem. Commun. 2011, 10182–10188.

[6] Iijima S. Nature 1991, 354(6348), 56–58.

[7] Coleman, J. N.; Fleming, A.; Maier, S.; O'Flaherty, S.; Minett, A. I.; Ferreira, M. S.; Hutzler, S.; Blau, W. J. J. Phys. Chem. B., 2004, 108(11), 3446–3450.

[8] Hertel, T.; Hagen, A; Talalaev, V.; Arnold, K.; Hennrich, F; Kappes, M.; Rosenthal, S.; McBride, J.; Ulbricht, H.; Flahaut, E., Nano Lett. 2005, 5 (3), 511–514.

[9] (a) Jorio, A.; Dresselhaus, M. S.; Dresselhaus, G., Carbon Nanotubes: Advanced Topics in the Synthesis, Structure, Properties and Applications, 2008 Springer-Verlag, Berlin. (b) Ajayan, P. M., Chem. Rev., 1999, 99, 1787. (c) Baughman, R. H.; Zakhidov, A. A.; de Heer, W. A., Science 2002, 297, 787.

[10] Hamada, N.; Sawada, S.; Oshiyama, A., Phys Rev Lett 1992, 68, 1579–1581.

[11] Chattopadhyay, D.; Galeska, L.; Papadimitrakopoulos, F. J. Am. Chem. Soc. 2003, 125, 3370.

[12] Strano, M. S.; Dyke, C. A.; Usrey, M. L.; Barone, P. W.; Allen, M. J.; Shan, H. W.; Kittrell, C.; Hauge, R. H.; Tour, J. M.; Smalley, R. E. Science 2003, 301, 1519.

[13] Charlier, J.-C., Accounts of Chemical Research 2002, 35(12), 1063–1069.

[14] (a) Ajayan, P. M.; Ebbesen, T. W.; Ichihashi, T.; Iijima, S.; Tanigaki, K.; Hiura, H., Nature 1993, 362(6420), 522–5. (b) Datsyuk, V.; Kalyva, M.; Papagelis, K.; Parthenios, J.; Tasis, D.; Siokou, A.; Kallitsis, I.; Galiotis, C., Carbon 2008, 46(6), 833–840. (c) Tsang, S. C.; Chen, Y. K.; Harris, P. J. F.; Green, M. L. H., Nature London 1994, 372(6502), 159–62. (d) Skowronski, J. M.; Scharff, P.; Pfander, N.; Cui, S., Advanced Materials 2003, 15(1), 55–57.

[15] Liu, J.; Rinzler, A. G.; Dai, H.; Hafner, R. K. Bradley, J. H.; Boul, P. J.; Lu, A.; Iverson, T.; Shelimov, K.; Huffman, C. B.; Rodriguez-Marcias, F.; Shon, Y.-S.; Lee, T. R.; Colbert, D. T.; Smalley, R. E., Science 1998, 280, 1253.

[16] Klein, K. L.; Melechko, A. V.; McKnight, T. E.; Retterer, S. T.; Rack, P. D.; Fowlkes, J. D.; Joy, D. C.; Simpson, M. L. J. Appl. Phys. 2008, 103, 061301.

[17] Hamon, M. A.; Chen, J.; Hu, H.; Chen, Y.; Itkis, M. E.; Rao, A. M.; Eklund, P. C.; Haddon, R. C., Adv Mater 1999 11,834–840.

[18] Zhao, B.; Hu, H.; Yu, A.; Perea, D.; Haddon, R. C., J Am. Chem. Soc. 2005, 127, 8197–8203.

[19] (a) Baker, S.; Cai, W.; Lasseter, T.; Hammers, R. J., Nano Lett. 2002, 2, 1413–1417 (b) Jiang, K.; Schadler, L.S.; Siegel, R.W.; Zhang, X.; Zhang, H.; Terrones, M. J Mater Chem 2004, 14, 37–39. (c) Wohlstadter, J. N.; Wilbur, J. L.; Sigal, G. B.; Biebuyck, H. A.; Billadeau, M. A.; Dong, L., Adv. Mater 2003, 15, 1184–1187 (d) Patolsky, F.; Weizmann, Y.; Willner, I., Angew, Chem. Int. Ed. 2004, 43, 2113–2117. (e) Sardesai, N.; Pan, S.; Rusling, J., Chem. Commun. 2009, 33, 4968–4970.

[20] (a) Delgado, J. L.; de la Cruz, P.; Urbina, A.; López Navarrete, J. T.; Casado, J.; Langa, F., Carbon 2007, 45,2250–2252. (b) Giordani, S.; Colomer, J. F.; Cattaruzza, F.; Alfonsi, J.; Meneghetti, M.; Prato, M.; Bonifazi, D., Carbon 2009, 47, 578–588.

[21] Baskaran, D.; Mays, J. W.; Zhang, X. P.; Bratcher, M. S., J. Am. Chem. Soc. 2005, 127, 6916–6917.

[22] Khabashesku, V. N.; Billups, E. W.; Margrave, J. L., Acc. Chem. Res. 2002, 35, 1087–1095.

[23] Saini, R. K.; Chiang, I. W.; Peng, H.; Smalley, R. E.; Billups, W. E.; Hauge, R. H.; Margrave, J. T.; J., Am. Chem. Soc. 2003, 125, 3617–3621.

[24] Mickelson, E. T.; Huffman, C. B.; Rinzler, A. G.; Smalley, R. E.; Hauge, R. H.; Margrave, J. L., Chem. Phys. Lett. 1998, 296, 188–194.

[25] Mickelson, E. T.; Chiang, I. W.; Zimmerman, J. L.; Boul, P.; Lozano, J.; Liu, J.; Smalley, R. E.; Hauge, R. H.; Margrave, J. L., J. Phys. Chem. B. 1999, 103, 4318–4322.

[26] Qian, Z.; Ma, J.; Zhou, J.; Lin, P.; Chen, C.; Chen, J.; Feng, H., Journal of Materials Chemistry 22, (41), 22113.

[27] Hu, H.; Zhao, B.; Hamon, M. A.; Kamaras, K.; Itkis, M. E.; Haddon, R. C., J. Am. Chem. Soc. 2003, 125, 14893–14900.

[28] Liu, C.; Zhang, Q.; Stellacci, F.; Marzari, N.; Zheng, L. X.; Zhan, X.Y., Small 2011, 7(9), 1257–1263.

[29] (a) Bingel, C., Chem. Ber. 1993, 126, 1957–1959. (b) Camps, X.; Hirsch, A., J. Chem. Soc. 1997, 1, 1595–1596.

[30] Holzinger, M,; Vostrowsky, O.; Hirsch, A.; Hennrich, F.; Kappes, M.; Weiss, R.; Jellen, F., Angew. Chem. 2001, 113, 4132–4136.

[31] Yinghuai, Z.; Peng, A.T.; Carpenter, K.; Maguire, J.A.; Hosmane, N. S.; Takagaki, M., J. Am. Chem. Soc. 2005, 127, 9875– 9880.

[32] Tasis, D.; Tagmatarchis, N.; Bianco, A.; Prato, M., Chem. Rev., 2006, 106, 1105.

[33] (a) Guldi, D. M.; Marcaccio, M.; Paolucci, D.; Paolucci, F.; Tagmatarchis, N.; Tasis, D.; Vazquez, E.; Prato, M., Angew. Chem., Int. Ed., 2003, 42, 4206. (b) Callegari, A.; Marcaccio, M.; Paolucci, D.; Paolucci, F.; Tagmatarchis, N.; Tasis, D.; Vazquez, E.; Prato, M., Chem. Commun., 2003, 2576.
[34] Delgado, J. L.; de la Cruz, P.; Langa, F.; Urbina, A.; Casado, J.; Lopez Navarrete, J. T., Chem. Commun., 2004, 1734.
[35] Bernal, M. M.; Liras, M.; Verdejo, R.; Lopez-Manchado, M. A.; Quijada-Garrido, I.; Paris, R., Polymer 2011, 52, (25), 5739–5745.
[36] Zydziak, N.; Hubner, C.; Bruns, M.; Barner-Kowollik, C., Macromolecules 2011, 44, (9), 3374–3380.
[37] Holzinger, M.; Abraham, J.; Whelan, P.; Graupner, R.; Ley, L.; Hennrich, F.; Kappes, M.; Hirsch, A., J. Am. Chem. Soc. 2003, 125, 8566–8580.
[38] Peng, H.; Alemany, L. B.; Margrave, J. L.; Khabashesku, V.N., J. Am. Chem. Soc. 2003, 125, 15174–15182.
[39] Price, B. K.; Tour, J. M., J. Am. Chem. Soc. 2006, 128, 12899–12904.
[40] Dyke, C. A.; Tour, J. M., J. Phys. Chem. A. 2004, 108, 11151–11159.
[41] (a) Campidelli, S.; Ballesteros, B.; Filoramo, A.; Diìaz, D.; de la Torre, G.; Torres, T.; Aminur Rahman, G. M.; Ehli, C.; Kiessling, D.; Werner, F.; Sgobba, V.; Guldi, D. M.; Cioffi, C.; Prato, M.; Bourgoin, J. P., J. Am. Chem. Soc. 2008,130, 11503–11509. (b) Yoo, B. K.; Myung, S.; Lee, M, Hong, S.; Chun, K.; Paik, H. J.; Kim, J.; Lim, J. K.; Joo, S. W., Mater. Lett. 2006, 60, 3224–3226. (c) Gómez-Escalonilla, M. J.; Atienzar, P.; Fierro, J. L. G.; García, H.; Langa, F., J. Mater. Chem. 2008, 18, 1592–1600.
[42] (a) Liang, F.; Sadana, A. K.; Peera, A.; Chattopadhyay, J.; Gu, Z.; Hauge, R.H.; Billups, W. E., Nano Lett 2004, 4, 1257–1260. (b) Chattopadhyay, J.; Sadana, A. K.; Liang, F.; Beach, J. M.; Xiao, Y.; Hauge, R.H.; Billups, W. E., Org Lett 2005, 7, 4067–4069.
[43] Qin, S.; Qin, D.; Ford, W. T.; Resasco, D. E.; Herrera, J. E., Macromolecules 2004, 37, 752.
[44] Liu, Y.; Yao, Z.; Adronov, A. Macromolecules 2005, 38, 1172.
[45] Wu, W.; Zhang, S.; Li, Y.; Li, J.; Liu, L.; Qin, Y.; Guo, Z. X.; Dai, L.; Ye, C.; Zhu, D., Macromolecules 2003, 36, 6286.
[46] Qin, S.; Qin, D.; Ford, W. T.; Herrera, J. T.; Resasco, D. E.; Bachilo, S. M.; Weisman, R. B., Macromolecules 2004, 37, 3965.
[47] Qin, S.; Qin, D.; Ford, W. T.; Herrera, J. T.; Resasco, D. E., Macromolecules 2004, 37, 9963.
[48] Viswanathan, G.; Chakrapani, N.; Yang, H.; Wei, B.; Chung, H.; Cho, K.; Ryu, C. Y.; Ajayan, P. M. J. Am. Chem. Soc. 2003, 125, 9258.
[49] Xia, H.; Wang, Q.; Qiu, G. Chem. Mater. 2003, 15, 3879.
[50] Britz, D. A.; Khlobystov, A. N., Chem. Soc. Rev., 2006, 35(7), 637–659.
[51] Strano, M. S.; Moore, V. C.; Miller, M. K.; Allen, M. J.; Haroz, E. H.; Kittrell, C.; Hauge, R. H.; Smalley, R. E., Journal of Nanoscience and Nanotechnology, 2003, 3(1–2), 81–86.
[52] (a) Priya, B. R.; Byrne, H. J., J. Phys. Chem. C. 2008, 112, 332–337. (b) Nish. A.; Hwang, J-Y.; Doig, J.; Nicholas, R..J., Nat. Nanotech 2007, 2, 640–646.
[53] (a) Hu, C. Y.; Liao, H. L.; Li, F. Y.; Xiang, J. H.; Li, W. K.; Duo, S. W.; Li, M. S., Mater. Lett. 2008, 62, 2585. (b) Priya, B. R.; Byrne, H. J., J. Phys. Chem. C. 2008, 112, 332. (c) White, B.; Banerjee, S.; O'Brien, S.; Turro, N. J.; Herman, I. P., J. Phys. Chem. C. 2007, 11, 13684. (d) McDonald, T. J.; Engtrakul, C.; Jones, M.; Rumbles, G.; Heben, M. J., J. Phys. Chem. B. 2006, 110, 25339. (e) Britz, D. A.; Khlobystov, A. N. Chem. Soc. Rev. 2006, 35, 637.
[54] Vigolo, B.; Poulin, P.; Lucas, M.; Launois, P.; Bernier, P., Applied Physics Letters 2002, 81, (7), 1210–1212.
[55] O'Connell, M. J.; Boul, P., Ericson, L. M.; Huffman, C.; Wang, Y.; Haroz, E.; Kuper, C.; Tour, J.; Ausman, K. D.; Smalley, R. E., Chem. Phys. Lett. 2001, 342, 265–71.

[56] Curran, S. A.; Ajayan, P. M.; Blau, W. J.; Carroll, D. L.; Coleman, J. N.; Dalton, A. B.; Davey, A. P.; Drury, A.; McCarthy, B.; Maier, S.; Strevens, A., Adv. Mater 1998, 10, 1091–1093.
[57] Yu, G.; Gao, J.; Hummelen, J. C.; Wudl, F.; Heeger, A. J., Science,1995, 270, 1789.
[58] (a) Gu, H.; Swager, T. M., Adv. Mater, 2008, 20, 4433. (b) Lee, H. W.; You, W.; Barman, S.; Hellstrom, S.; LeMieux, M. C.; Oh, J. H.; Liu, S.; Fujiwara, T.; Wang, W. M.; Chen, B.; Jin, Y. W.; Kim, J. M.; Bao, Z., Small, 2009, 5, 1019.
[59] (a) Wang, F.; Gu, H.; Swager, T. M., J. Am. Chem. Soc., 2008, 130,5392. (b) Wang, F.; Yang, Y.; Swager, T. M., Angew. Chem., Int. Ed., 2008, 47, 8394.
[60] (a) Andrews, R.; Jacques, D.; Qian, D.; Rantell, T., Acc. Chem. Res. 2002, 35, 1008. (b) Tibbetts, G. G.; McHugh, J., J. Mater. Res., 1999, 14, 2871. (c) Safadi, B.; Andrews, R.; Grulke, E. A. J., Appl. Polym. Sci. 2002, 84, 2660.
[61] Liu, Z.; Winters, M.; Holodniy, M.; Dai, H., Angew. Chem. Int. Ed. 2007. 46, 2023–2027.
[62] Baykal, B.; Ibrahimova, V.; Er, G.; Bengu, E.; Tuncel, D., Chem. Commun. 2010, 46, 6762.
[63] Meuer, S.; Braun, L.; Zentel, R., Macromol. Chem. Phys. 2009, 210, 1528–1535.
[64] (a) Davis, J. J.; Coles, R. J.; Hill, H. A. O., J. Electroanal. Chem. 1997, 440, 279. (b) Wang, J. Electroanalysis 2005, 17, 7.
[65] (a) Cai, C.; Chen, J. Anal. Biochem. 2004, 325, 285. (b) Zhao, Y. D.; Bi, Y. H.; Zhang, W. D.; Luo, Q. M. Talanta 2005, 65, 489.
[66] (a) Zhao, G. C.; Zhang, L.; Wei, X. W.; Yang, Z. S., Electrochem. Commun. 2003, 5, 825. (b) Zhang, L.; Zhao, G. C.; Wie, X. W.; Yang, Z. S., Electroanalysis 2005, 17, 630.
[67] (c) Zhao, G. C.; Zhang, L.; Wei, X. W. Anal. Biochem. 2004, 329, 160.
[68] (a) Ye, J. S.; Wen, Y.; Zhang, W. D.; Cui, H. F.; Gan, L. M.; Xu, G. Q.; Sheu, F. S. J., Electroanal. Chem. 2004, 562, 241.
[69] Li, G.; Liao, J. M.; Hu, G. Q.; Ma, N. Z.; Wu, P. J., Biosens. Bioelectron. 2005, 20, 2140.
[70] Guiseppi-Elie, A.; Lei, C.; Baughman, R. H., Nanotechnol. 2002, 13, 559–564.
[71] Barone, P. W.; Baik, S.; Heller, D. A.; Strano, M. S., Nature Mat. 2005, 4, 86–92.
[72] Chen, R. J.; Zhang, Y.; Wang, D.; Dai, H., J. Am. Chem. Soc. 2001, 123, 3838–3839.
[73] (a) Zheng, M.; Jagota, A.; Strano, M. S.; Santos, A. P.; Barone, P.; Chou, S. G.; Diner, B. A.; Dresselhaus, M. S.; McLean, R. S.; Onoa, G. B.; Samsonidze, G. G.; Semke, E. D.; Usrey, M.; Walls, D. J., Science 2003, 302, 1545. (b) Arnold, M. S.; Stupp, S. I.; Hersam, M. C. Nano Lett. 2005, 5, 713.
[74] Zhao,Y.-L.; Stoddart, J. F., Accounts Chem. Res. 2009, 42 (8), 1161–1171.
[75] (a) Guldi, D. M.; Rahman, G. M. A.; Jux, N.; Tagmatarchis, N.; Prato, M., Angew. Chem., Int. Ed. 2004, 43, 5526–5530. (b) Ehli, C.; Rahman, G. M. A.; Jux, N.; Balbinot, D.; Guldi, D. M.; Paolucci, F.;Marcaccio, M.; Paolucci, D.; Melle-Franco, M.; Zerbetto, F.; Campidelli, S.; Prato, M., J. Am. Chem. Soc. 2006, 128, 11222–11231.
[76] C. Roquelet et al., Chem. Phys. (2012), http://dx.doi.org/10.1016/j.chemphys.2012.09.004.
[77] Basiuk, E. V.; Basiuk, V. A.; Santiago, P.; Puente-Lee, I., Journal of Nanoscience and Nanotechnology 2007, 7, (4–5), 1530–1538.
[78] Smith, B. W.; Monthioux, M.; Luzzi, D. E.. Nature 1998, 396, 323–324.
[79] Ulbricht, H.; Moos, G.; Hertel, T., Phys. Rev. Lett. 2003, 90, No. 095501.
[80] (a) Hirahara, K.; Bandow, S.; Suenaga, K.; Kato, H.; Okazaki, T.; Shinohara, H.; Iijima, S., Phys. Rev. B 2001, 64, 115420. (b) Kataura, H.; Maniwa, Y.; Abe, M.; Fujiwara, A.; Kodama, T.; Kikuchi, K., Appl. Phys. A 2002, 74, 349.
[81] Suenaga, K.; Tence, M.; Mory, C.; Colliex, C.; Kato, H.; Okazaki, T.;Shinohara, H.; Hirahara, K.; Bandow, S.; Iijima, S. Science 2000, 290, 2280.
[82] Okazaki, T.; Suenaga, K.; Hirahara, K.; Bandow, S.; Iijima, S.; Shinohara, H., J. Am. Chem. Soc. 2001, 123, 9673.
[83] Chiu, P. W.; Gu, G.; Kim, G. T.; Philipp, G.; Roth, S., Appl. Phys. Lett. 2001, 79, 3845.

[84] Debarre, A.; Jaffiol, R.; Julien, C.; Richard, A.; Nutarelli, D.; Tchenio, P., Chem. Phys. Lett. 2003, 380, 6.
[85] Suenaga, K.; Taniguchi, R.; Shimada, T.; Okazaki, T.; Shinohara, H.; Iijima, S. Nano Lett. 2003, 3, 1395.
[86] Suenaga, K.; Okazaki, T.; Hirahara, K.; Bandow, S.; Kato, H.; Taninaka, A.; Shinohara, H.; Iijima, S., Appl. Phys. A 2003, 76, 445.
[87] Suenaga, K.; Okazaki, T.; Wang, C.-R.; Bandow, S.; Shinohara, H.; Iijima, S., Phys. ReV. Lett. 2003, 90, 055506.
[88] Gloter, A.; Suenaga, K.; Kataura, H.; Fujii, R.; Kodama, T.; Nishikawa, H.; Ikemoto, I.; Kikuchi, K.; Suzuki, S.; Achiba, Y.; Iijima, S., Chem. Phys. Lett. 2004, 390, 462.
[89] Khlobystov, A. N.; Porfyrakis, K.; Kanai, M.; Britz, D. A.; Ardavan, A.; Shinohara, H.; Dennis, T. J. S.; Briggs, G. A. D. Angew. Chem., Int. Ed. 2004, 43, 1386.
[90] Khlobystov, A. N.; Britz, D. A.; Briggs, G. A. D., Acc. Chem. Res. 2005, 38, 901–909.
[91] (a) Ito, T.; Sun, L.; Crooks, R. M., Chem. Commun. 2003, 1482. (b) Gao, H.; Kong, Y.; Cui, D.; Ozkan, C. S., Nano Lett. 2003, 3, 471.
[92] (a) Philp, E.; Sloan, J.; Kirkland, A. I.; Meyer, R. R.; Friedrichs, S.; Hutchison, J. L.; Green, M. L. H., Nat. Mater. 2003, 2, 788. (b) Brown, G.; Bailey, S.; Sloan, J.; Xu, C.; Friedrichs, S.; Flahaut, E.; Coleman, K. S.; Hutchison, J. L.; Dunin-Borkowski, R. E.; Green, M. L. H. Chem. Commun. 2001, 845. (c) Xu, C.; Sloan, J.; Brown, G.; Bailey, S.; Clifford Williams V.; Friedrichs, S.; Coleman, K. S.; Flahaut, E.; Hutchison, J. L.; Dunin-Borkowski, R. E.; Green, M. L. H., Chem. Commun. 2000, 2427.
[93] Thamavaranukup, N.; Hoppe, H. A.; Ruiz-Gonzalez, L.; Costa, P. M.; Sloan, J.; Kirkland, A.; Green, M. L. H., Chem. Commun. 2004, 1686.
[94] (a) Stercel, F.; Nemes, N. M.; Fischer, J. E.; Luzzi, D. E., Mater. Res. Soc. Symp. Proc. 2002, 706, 245. (b) Li, L.-J.; Khlobystov, A. N.; Wiltshire, J. G.; Briggs, G. A. D.; Nicholas, R., J. Nat. Mater. 2005, 4, 481.
[95] (a) Kickelbick, G., Hybrid Materials: Synthesis, Characterization and Application, Wiley-VCH, Weinheim, Germany, 2007. (b) Schubert, U.; Huesing, N.; Lorenz, A. Chem. Mater. 1995, 7, 2010. (c) Judenstein, P.; Sanchez, C. J. Mater. Chem. 1996, 6, 511.
[96] (a) Eder, D., 2010 Chem. Rev. 110, 1348–1385. (b) Vilatela, J. J., Eder, D. 2012 Chem. Sus. Chem. 5, 456–478.
[97] Banerjee, S.; Wong, S. S., Nano Lett. 2002, 2, 195
[98] (a) Ellis, A. V.; Vijayamohanan, K.; Goswami, R.; Chakrapani, N.; Ramanathan, L. S.; Ajayan, P. M.; Ramanath, G. Nano Lett. 2003, 3, 279. (b) Rahman, G. M.; Guldi, D. M.; Zambon, E.; Pasquato, L.; Tagmatarchis, N.; Prato, M. Small 2005, 1, 527. (c) Han, L.; Wu, W.; Kirk, F. L.; Luo, J.; Maye, M. M.; Kariuki, N. N.; Lin, Y.; Wang, C.; Zhong, C. J. Langmuir 2004, 20.
[99] Sainsbury, T.; Fitzmaurice, D., Chem. Mater. 2004, 16, 3780.
[100] (a) Murakami, H.; Nomura, T.; Nakashima, N. Chem. Phys. Lett. 2003, 378, 481. (b) Guldi, D. M.; Taieb, H.; Rahman, G. M. A.; Tagmatarchis, N.; Prato, M. AdV. Mater. 2005, 17, 871
[101] Wang, X.; Liu, Y.; Qiu, W.; Zhu, D. J. Mater. Chem. 2002, 12, 1636.
[102] a) Eder, D.; Windle, A. H., Adv. Mater. 2008, 20, 1787. b) Eder, D., Kinloch, I. A., Windle, A. H., J. Mater. Chem. 2008, 18, 2036.
[103] Herrero, M. A.; Guerra, J.; Myers, V. S.; Goìmez, M. V.; Crooks, R. M., Prato M., ACS Nano 2010, 4, 905–912.
[104] Liu, Y.; Jiang, W.; Li, S.; Li, F. S., Applied Surface Science 2009, 255, (18), 7999–8002.
[105] Jiang, K.; Eitan, A.; Schadler, L. S.; Ajayan, P. M.; Siegel, R. W.; Grobert, N.; Mayne, M.; Reyes-Reyes, M.; Terrones, H.; Terrones, M. Nano Lett. 2003, 3, 275.
[106] (a) Correa-Duarte, M. A.; Perez-Juste, J.; Sanchez-Iglesias, A.; Giersig, M. Angew. Chem., Int. Ed. 2005, 44, 4375. (b) Kim, B.; Sigmund, W. M. Langmuir 2004, 20, 8239.

S.M. Vega-Diaz, F. Tristán López, A. Morelos-Gómez, R. Cruz-Silva, and M. Terrones

4 The importance of defects and dopants within carbon nanomaterials during the fabrication of polymer composites

This chapter reviews the modification of carbon nanostructures using different approaches such as: (a) doping, addition of foreign atoms into the carbon nanostructures, (b) introduction of structural defects (c) functionalization involving covalent or noncovalent bonding with different molecules. The dispersion and the physicochemical properties are analyzed for composites prepared using carbon nanostructures. The fabrication of composite materials with carbon nanostructures has been challenging due to the difficulty of uniformly dispersing carbon nanostructures within a host matrix. Various strategies have been developed to overcome this obstacle. These involve the addition of compatibilizing agents to the carbon nanostructures (by covalent or noncovalent functionalization) and acid treatment (introduction of defects into the hexagonal network and oxygenated functional groups) amongst others. These strategies are designed specifically to improve the dispersion of carbon nanomaterials in the host matrix, but frequently the effect of the particular carbon structure or its properties are not considered, especially when the target property is biocompatibility.

4.1 Introduction

The field of carbon nanostructure research is vast and novel, and it experienced a major breakthrough after the discovery of fullerenes in 1985 [1], and their subsequent bulk synthesis in 1990 [2]. This event opened the minds of various scientists towards discovering novel carbon allotropes. Promptly, yet another allotrop of carbon was observed by Iijima [3], although it had previously been produced by M. Endo *et al.* in the 1970s by chemical vapor deposition (CVD) [4]. The most recent important advance in the quest for novel forms of carbon constitutes the isolation of graphene layers [5], which exhibit unique and exceptional electrical properties [6]. In addition, graphene nanoribbons have recently been synthesized and produced using diverse methods [7].

These novel carbon nanostructures can also be modified by (a) *doping*, that is the addition of foreign atoms into the carbon nanostructure, (b) by the *introduction of structural defects* that modify the arrangement of the carbon atoms and (c) by *functionalization* involving covalent or noncovalent bonding with other molecules. These modifications opened up new perspectives in developing novel composite materials with different matrices (ceramic, polymer and metals). For example, polymer composites containing carbon nanostructures have attracted considerable attention due to their unique mechanical and/or physicochemical properties amongst others. Unfortunately, carbon nanostructures agglomerate and entangle easily due to strong van der Waals interactions, thus leading to the formation of voids within the composites that limit their efficiency and range of possible applications. One strategy to overcome this issue is to introduce defects into the carbon nanostructures or by functionalizing these structures in order to improve the interactions between the polymer (host matrix) and the carbon nanostructures (filler). The better the interaction between the matrix and the carbon nanostructures, the lower the entanglement-aggregation of the carbon structures within the matrix during processing. Furthermore, the introduction of dopants or other types of defect can also significantly modify the pristine properties of carbon nanostructures and can lead to strong interactions of the matrix with the filler. This can result in composites displaying unprecedented electronic and optical properties, in addition to enhancing the biocompatibility of composite materials.

4.1.1 Carbon nanostructures and their properties

In this section, different nanocarbons and their chemical and physical properties are discussed (for more details see Chapters 1 and 2). Furthermore, the types of defects that can be embedded within these carbon nanostructures are explained, as well as their resulting chemical and physical properties.

Layers of sp^2 hybridized carbon may result in different structures such as graphite, which behaves as a conducting semimetal and is composed of several stacked layers, each exhibiting a hexagonal lattice. Highly crystalline graphite, known as highly oriented pyrolytic graphite (HOPG), might display a high conductivity of *ca.* 10^4 S cm^{-1}. Now, if a single layer is isolated, known as graphene (Fig. 4.1(a)), it exhibits ballistic electronic transport [5] and high thermal transport of *ca.* 5×10^3 W/mK [8], both at room temperature. When graphene is oxidized into graphene oxide (Fig. 4.1(b)), the oxygen functionalities and vacancies generated induce curvature to the structure. [9, 10] When heavily oxidized, graphene oxide exhibits a very large electrical resistivity making it almost an insulator with an electrical conductivity of lower than 1×10^{-3} S cm^{-1} [11]. This oxidized carbon can be chemically synthesized by exfoliating graphite using strong acids. It can be reduced by thermal or chemical routes; however it never recovers the same degree of crystallinity as the starting pristine graphite used for its synthesis [9, 12]. The electrical conductivity of graphene oxide can be tailored upon reduction conditions, and relatively high electrical conductivities of *ca.* 160 S cm^{-1} can be reached [11, 13]. If a layer of graphene exhibits a finite width accompanied by an elongated length, then a graphene nanoribbon is formed (Fig. 4.1(c)). Depending upon the ribbon edges (armchair, zigzag or chiral), and their width, these can behave as insulators, metals or semiconductors [14–16].

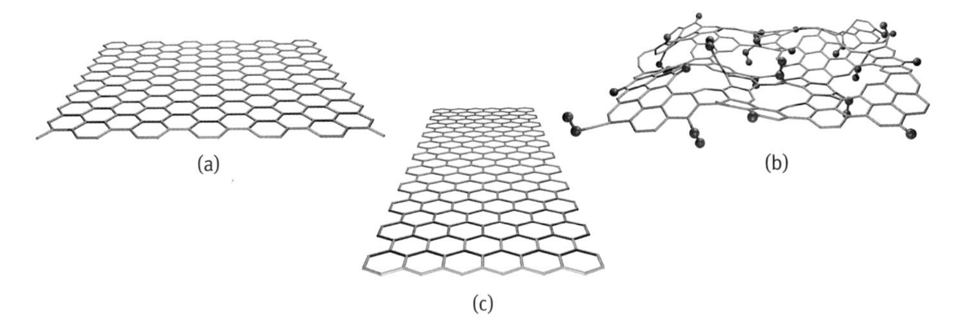

Fig. 4.1: Molecular models of (a) graphene, (b) graphene oxide and (c) graphene nanoribbon.

A graphene layer folded into a cylinder can be considered a single-walled carbon nanotube (SWCNT) (Fig. 4.2(a)). According to the angle used to fold it, the cylinder edges might be armchair, zigzag or chiral. Carbon nanotubes (CNTs) can also possess two, three or many layers, and these are called double-walled carbon nanotubes (DWCNTs) (Fig. 4.2(b)), triple-walled carbon nanotubes (TWCNTs) (Fig. 4.2(c)), and multi-walled carbon nanotubes (MWCNTs), respectively. Interestingly, individual MWCNTs may reach low resistivities ranging between 1×10^{-4} and 5×10^{-6} Ω cm [7]. SWCNTs with armchair terminations are all metallic and zigzag SWCNTs are mainly semiconductors (2/3 of all zigzag tubes); while all chiral tubes are semiconductors. The chirality of SWCNTs can also determine their stiffness. For example, zigzag SWCNTs are less stiff than armchair SWCNTs [18]. On the other hand, MWCNTs can exhibit a Young's modulus as high as 1.28 TPa despite their diameter [19].

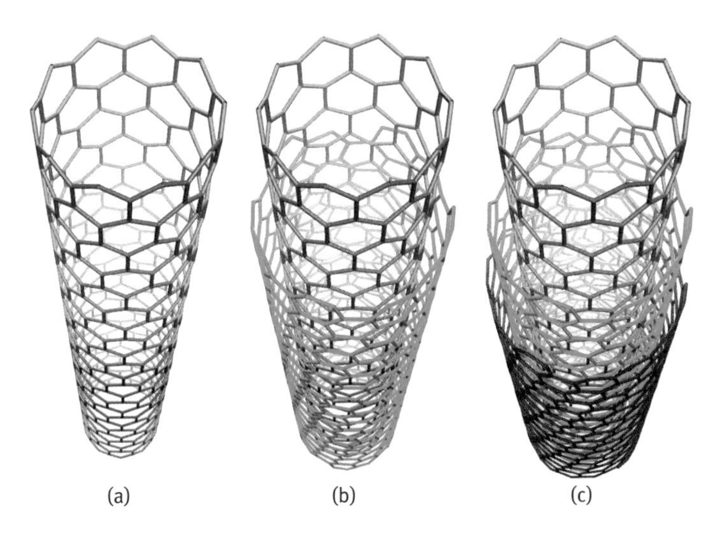

Fig. 4.2: Molecular models of (a) single-walled carbon nanotube, (b) double-walled carbon nanotube, and (c) multi-walled carbon nanotube.

A pure form of sp^3 hybridized carbon is known as diamond and this may also be synthesized at the nanoscale *via* detonation processing. Depending on their sizes, these are classified as: nanocrystalline diamond (10 nm–100 nm), ultrananocrystalline diamond (< 10 nm) and diamondoids (hydrogenated molecules, 1 nm–2 nm). Nanodiamond exhibits low electron mobility, high thermal conductivity and its transparency allows spectro-electrochemistry [20, 21]. However, ultrananocrystalline diamond exhibits poor electron mobility, poor thermal conductivity and redox activity [21, 22].

4.1.2 Doped carbon nanostructures

Carbon nanostructures can be doped by the introduction/interaction of foreign atoms. Different doping categories include (a) exohedral doping or intercalation, (b) endohedral doping or encapsulation or (c) in-plane or substitutional doping.

Exohedral doping (intercalation) occurs when foreign atoms are situated in the interstitial sites or between the graphitic layers of the material (Fig. 4.3(a)). For example, a bundle of SWCNTs consists of closely packed nanotubes, which leads to interstitial channels between the nanotubes (Fig. 4.3(a)). Typically alkali atoms can be intercalated along these channels. The intercalation of iodine in DWCNT ropes has been shown enhance the electrical properties of the tubes, reaching resistivity values of *ca.* 1.5×10^{-5} Ω cm, five times lower than pristine DWCNTs [8]. When lithium is intercalated within carbon nanotube bundles, the doped system displays excellent performance needed for Li-ion batteries with high reversible capacity of 1000 mAh g^{-1} [9]. In some cases the alkali atoms (K, Na, Li and Cs) can be intercalated within the graphitic layers, thus increasing the interlayer spacing [25–27]. In addition, sulfur oxide species can be intercalated within the graphitic layers of carbon nanostructures [28]. Exohedral doping can also take place by chemical functionalization, which modifies the chemical and physical properties of the resulting material. The electrical conductivity of graphene is decreased by oxidation [6], however its hydrophilicity is increased [10], thus making it easier to process in polar solvents for further applications including medicine and composite fabrication. The use of surfactants and polymers to wrap CNTs has been widely used to tailor their dispersibility in polar and non-polar solvents[11]. It is also possible to increase the electrical conductivity of CNT bundles by doping with acids. For example, the use of HNO_3, H_2SO_3, H_2SO_4, $NH_4S_2O_8$, and $SOCl_2$ resulted in electrical resistivities ranging from 10^{-3} to 10^{-4} Ω cm [31–35].

Endohedral doping (encapsulation) of other materials within carbon nanostructures can be carried out by nano-capillary effects or during synthesis (Fig. 4.3(b)). A great variety of halides, oxides, metals and alloys have been encapsulated within CNTs [36–41]. When transition metals are encapsulated, the entire sample can exhibit high magnetic coercivities *ca.* 0.22 T [42, 43]. The encapsulation of C_{60} molecules can also be accomplished and if the material is heat treated at high temperatures

(> 1200° C), these fullerenes start to fuse together creating another SWCNT, thus resulting in a DWCNT [44]. Recently, graphene nanoribbons have also been synthesized inside these SWCNTs. In particular, after exposure to an electron beam inside the microscope sulfur functionalized fullerenes render sulfur terminated graphene nanoribbons [45]. Carbon chains have been also be created inside carbon nanotubes by thermal treatment of C_{60} molecules trapped inside SWCNTs at temperatures above 1400 °C [46–48]. It is envisaged that semiconductor SWCNTs with encapsulated carbon chains may give rise to superconductivity and ferromagnetism due to weak coupling between both structures [49], but additional experiments should be carried out along this line

In-plane doping using elements close to carbon in the periodic table can be easily achieved (Fig. 4.3(c)). In this context, nitrogen and boron doping have been widely studied in carbon nanostructures, showing novel properties including high electrical conductivity, biocompatibility, reactivity and so on. Graphene-like structures doped with nitrogen can display different configurations, most commonly the substitution of a carbon atom within the hexagonal carbon lattice and pyridinic sites. However, other configurations include N-doping close to vacancy sites [50]. N-doping will induce positive curvature (*via* the stabilization of pentagons) within CNTs thus inducing bamboo-like morphologies. In addition, nitrogen atoms can introduce states near the Fermi level, and these are capable of converting semiconducting CNTs into metallic CNTs. In this context, N-doping can indeed increase the electrical conductivity of CNTs [51, 52], graphene [53, 54] and amorphous-like carbon [55].

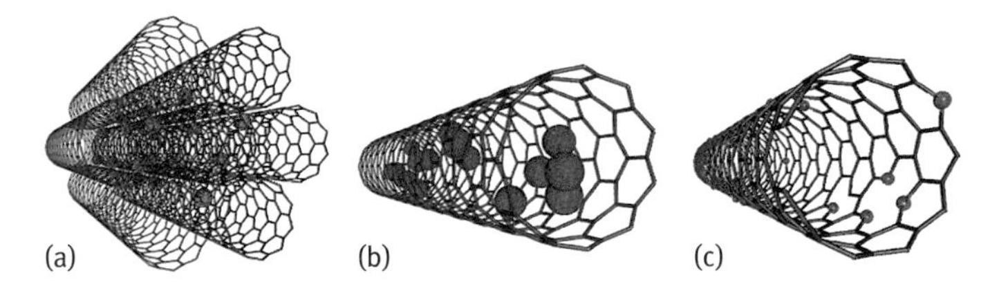

Fig. 4.3: Molecular models of (a) exohedral doping (intercalation), (b) endohedral doping (encapsulation) and (c) in-plane (substitutional) doping.

Substitutional boron within graphene-like structures is the most common way to dope sp^2 hybridized carbon [56]. The boron atom introduces hole-type carriers within the system [57]. It may decrease the resistivity in carbon nanotubes by one order of magnitude (7.4×10^{-5}–7.7×10^{-4} Ω cm) when compared to its undoped counterpart (5.33×10^{-4}–1.9×10^{-3} Ω cm). However, this change in transport also depends on the level of doping [58–60]. For example, boron doped SWCNTs can result in superconductivity with T_c = 12 K [61], and it is envisaged that heavily doped boron graphene can behave as a superconductor [62]. Sulfur doping has also been accomplished in MWCNTs, and the structures exhibit kinks caused by sulfur atoms [63]. Phosphorus doping has also been reported for MWCNTs, SWCNTs and graphene, showing n-type semiconducting behavior [64, 65]. Heterodoping can also be achieved, for example BN, PN and NS systems have been reported [66–71].

4.1.3 Defects in carbon nanostructures

Defects in carbon nanostructures can be classified into (a) structural defects, (b) topological defects, (c) high curvature and (d) non-sp^2 carbon defects. Even slight changes within the carbon nanostructure can modify the chemical and physical properties. Some defects in carbon systems results in high chemical reactivity, mainly due to the accumulation of electrons in the vicinity of the dopant. These defects can be used as anchoring sites in order to make the carbon nanostructures more compatible with ceramic or polymer matrices, thus enhancing interactions between carbon structures (filler) and the host matrices.

(a) Structural defects are those capable of significantly transforming the curvature of the carbon nanostructure. This is achieved by the introduction of pentagons or heptagons within the sp^2 hybridized lattice. Pentagons will introduce positive curvature whereas heptagons will induce negative curvature (Fig. 4.4(a)). Theoretical calculations indicate that isolated pentagons or heptagons will result in loop currents at these sites, thus creating a magnetic moment [72]. In a CNT a pentagon on one end and a heptagon on the other end can induce a bend with a 30° angle [73]. These types of defects can also be observed in graphene [74–77]. Curved sites can indeed exhibit an increase in electron density and therefore become more chemically active [78]. When fabricating composites, this reactivity can be used to enhance the interactions established between the carbon nanostructure (filler) and the host matrix.

(b) Topological defects or bond rotations do not induce significant changes in the overall curvature of the lattice. When a C-C bond is rotated 90° a Thrower–Stone–Wales (TSW)-type defect is created. This defect is also known as 5-7-7-5 pairs in which two pentagons and two heptagons are adjacent [79–82]. Recently, a wide variety of defects exhibiting vacancies, pentagons and heptagons have been observed, and theoretically predicted for graphene (Fig. 4.4(b)–(d)) [77, 83–85].

(c) High curvature. Highly curved graphene surfaces or narrow diameter nanotubes (< 1 nm) can be more reactive than flat graphene due to the distortion of π-orbitals. Loop formations with high curvature have been observed at the edges of graphene, graphene nanoribbons and graphitic nanoribbons (Fig. 4.5) [86–88]. These loop edges or bends in graphene and graphene nanoribbons reduce the strain energy [12]. When graphene is highly functionalized, folding of its edges can be modified and in some cases more than one folding can be formed [90].

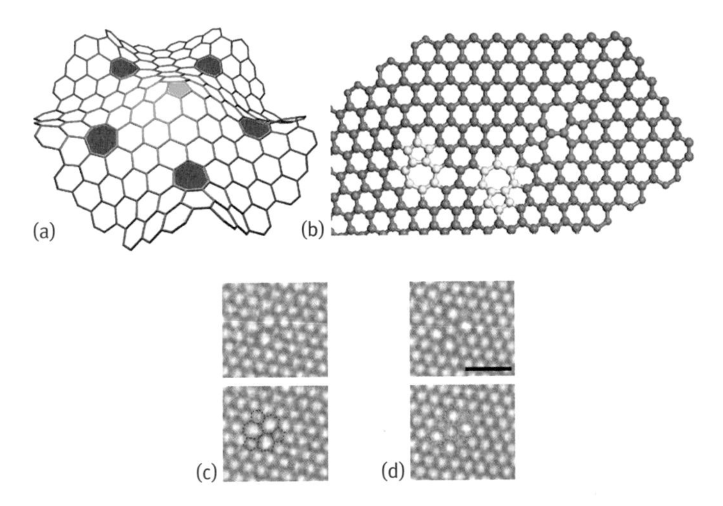

Fig. 4.4: Molecular model of (a) surface containing positive and negative curvature induced by a pentagon (red) and a heptagon (blue), respectively; (b) molecular model of 5-7 defects (yellow) and a Thrower–Stone–Wales defect (green) [82]; (c) high-resolution transmission electron microscopy (HRTEM) images of bond rotations V_2 (555-777) divacancy, and (d) V_2(5555-6-7777) divacancy within graphene; scale bar is 1 nm [75].

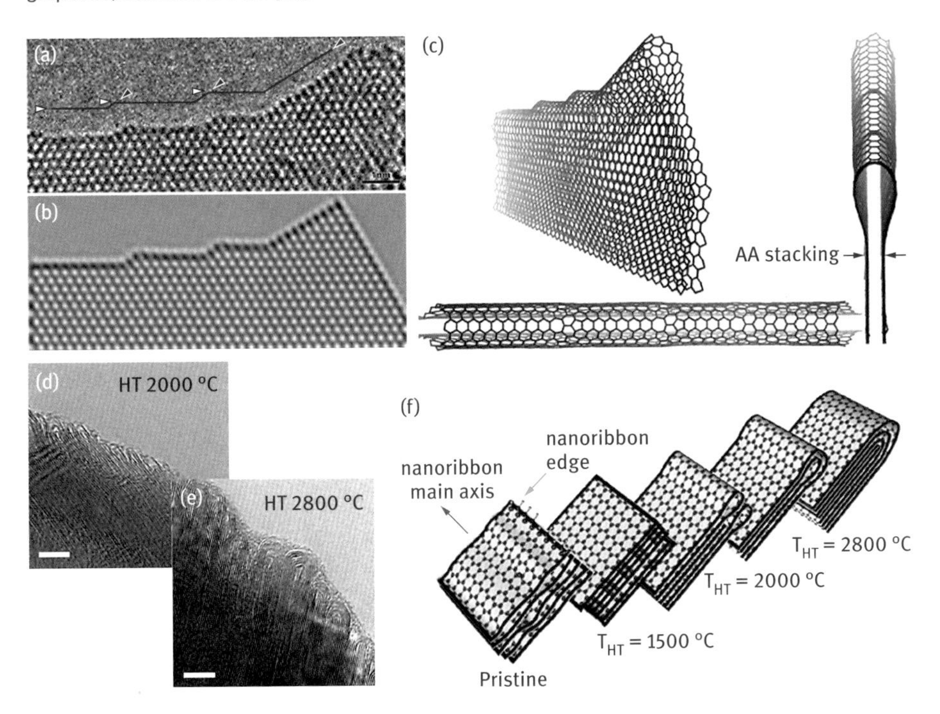

Fig. 4.5: (a) HRTEM, (b) simulated and (c) models of a folded edge of graphene, observed perpendicular to the plane [86]; (d) and (e) HRTEM images (scale bar is 5 nm) and (f) models of loop edges created in graphitic nanoribbons heat treated at different temperatures [88].

(d) Non-sp^2 carbon defects. These defects consist of carbon chains, edges, adatoms and vacancies. Electron beam irradiation on graphene or multilayers of graphene may initiate the growth of holes, and in certain cases carbon chains can be created during this process (Fig. 4.6(a), (b)) [91]. Interestingly, calculations of carbon chains coupled with zigzag graphene nanoribbon electrodes may work as a spin filter with almost 100 % spin polarization [92]. It has been observed that the edges of graphene can be two times more reactive than the flat surface [93]. When the edges exhibit pentagons or heptagons, these sites will exhibit higher chemical activity [94]. Adatoms are generally localized in the vicinity of other defects and have been observed by HRTEM (Fig. 4.6(c), (d)) [78, 95]. Vacancies in carbon nanostructures have also been widely

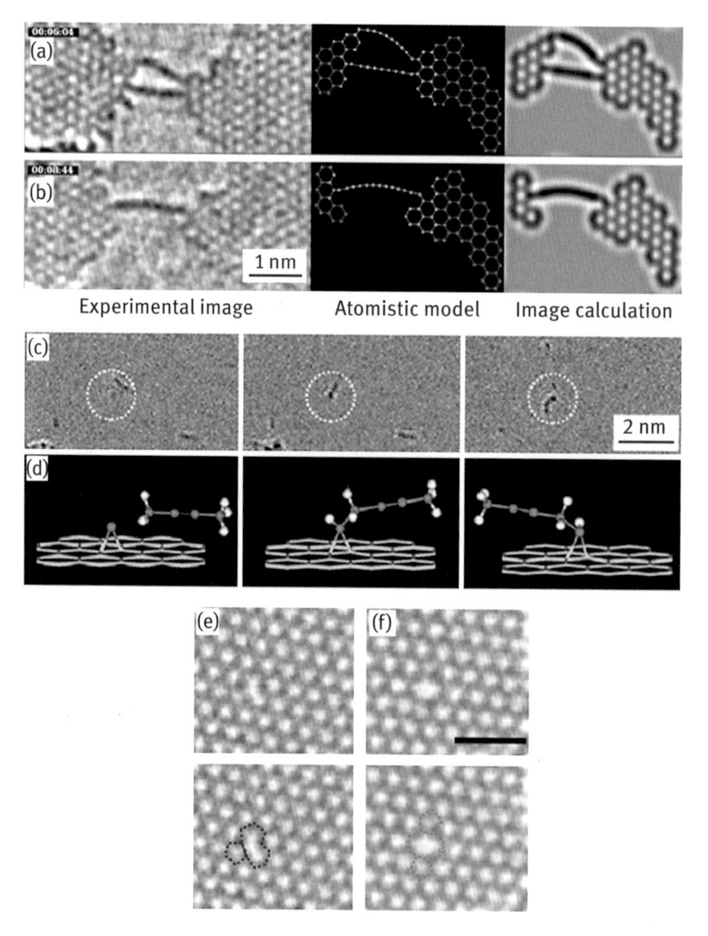

Fig. 4.6: (a) Two and (b) one carbon linear chain connecting two graphene nanoribbons [92]; (c) HRTEM and (d) molecular model of a carbon atom and an alkyne chain attached to the carbon adatom [95]; HRTEM images of (e) V_1 (5-9) single vacancy, and (f) V_2 (5-8-5) divacancy in graphene; scale bar is 1 nm [76].

studied by HRTEM (Fig. 4.6(e), (f)). They might display high chemical reactivity and magnetic moments [96, 97]. In addition, vacancies in graphene may help to maintain several layers parallel and avoid loop formation at the edges [98].

4.1.4 Functionalization of carbon nanostructures for nanocomposites

Nanocarbon structures such as fullerenes, carbon nanotubes and graphene, are characterized by their weak interphase interaction with host matrices (polymer, ceramic, metals) when fabricating composites [99, 100]. In addition to their characteristic high surface area and high chemical inertness, this fact turns these carbon nanostructures into materials that are very difficult to disperse in a given matrix. However, uniform dispersion and improved nanotube/matrix interactions are necessary to increase the mechanical, physical and chemical properties as well as biocompatibility of the composites [101, 102].

In order to overcome this drawback, there are two main approaches for the surface modification of carbon nanostructures that reoccur in the literature. The first one is covalent functionalization, mainly by chemical bonding of functional groups; and the second one is noncovalent functionalization, mainly by physical interactions with other molecules or particles. Both strategies have been used to provide different physical and chemical properties to the carbon nanostructures. Those that will be presented here are only a few examples of the modifications that can be achieved in carbon nanostructure surfaces and composite fabrication.

In the beginning, functionalization reactions were applied to fullerenes [1], later to CNTs [4, 3], and recently to graphene [5]. Although both functionalization approaches have clear differences, they share the same intrinsic objective: the creation of defects or doping within the surface of the carbon nanostructures in order to facilitate the interactions between the matrix and the filler.

After functionalization, carbon nanostructures can be further processed by solvent-assisted techniques, such as layer-by-layer assembly, spin-coating, and filtration to synthesize and prepare the hybrids or composites [103]. Another advantage of functionalization consists of avoiding the formation of agglomerates of carbon nanostructures during the composite processing. In the following sections, we will describe some aspects of both functionalization approaches, emphasizing the synthesis of composites.

4.1.4.1 Covalent functionalization of nanocarbons

When the formation of covalent bonds is established between functional groups and a surface, a covalent or chemical functionalization is reached. The main characteristic of this type of functionalization is the change in the carbon hybridization from sp^2 to sp^3 [104]. Although this covalent functionalization provides the possibility to obtain a wide variety of chemical groups in different types of carbon nanostructures (fullerenes, carbon nanotubes and graphene), there is a major drawback. Covalent functionalization favors the formation of a large number of defects within the graphitic layers, thus affecting their mechanical and transport properties. There is also a second major drawback regarding the covalent functionalization, and this consists of the use of strong acids or heavy oxidants that result in severe damage of graphene-like surfaces.

Covalent functionalization in fullerenes

The first carbon nanostructure that was covalently modified to alter surface properties was the C_{60} fullerene. After its discovery in the mid-1980s, numerous functionalization methods were developed for obtaining two important classes of derivatives, methanofullerenes and fullerene Diels–Alder adducts.

Some of these compounds were found later to have biological or material science applications [105]. Methanofullerenes, which are fullerenes with a substitutional group covalently bonded to a pair of carbons of the six-membered rings (methane bridge), have been more widely studied for the first purpose [106] and it was found that they were ideal for the preparation of water-soluble fullerenes with biological functions. Some examples of this type of functionalized fullerenes are water-soluble inhibitors of HIV protease [107, 108]; a water-soluble fullerene-peptide conjugate with chemotactic activity [109], and a fullerene-nucleotide conjugate as G-selective DNA-cleaving agent [110]. It is therefore clear that surface modifications contribute to increased biological interactions in fullerenes.

Covalent functionalization of fullerenes has also been used to obtain surface-modified fullerenes that are more compatible to polymer matrices in order to fabricate composites. In this context, four basic strategies were developed. The first one allows the fullerenes to react during the monomer polymerization, so that the fullerene can be attached to the polymer chain [111, 112]. Second, an already synthesized polymer is treated using specific conditions that allow the chemical reaction with fullerenes [113, 114]. Third, the fullerenes are chemically bonded to a monomer which is polymerized or co-polymerized to obtain the modified monomer [115, 116]. Fourth, a dendrimer can be synthesized around a fullerene which then acts as a nucleus [117, 118].

Synthesis of fullerene-polymers has been possible by taking full advantage of controlled fullerene surface modification by covalent functionalization. Examples of these materials include the preparation of fullerene containing polyurethane or polyester by polycondensation [119], the covalent grafting of C_{60} onto polydiene chains [120, 121], fullerenated polycarbonates through cycloaddition reactions [122, 123], synthesis of C_{60}-end-capped polymers using different polymers such as: polystyrene (PS) [114, 124]; polyethylene oxide (PEO) [125, 126]; poly(1-phenyl-1-butyne) (PPB) [127]: poly(methyl methacrylate) (PMMA) [128]; oligothienylenevinylenes (OTV) [129]; oligophenylenevinylene (OPV) [130]; (4-hydroxyl-2,2,6,6-tetramethyl-piperdinyloxyl)-terminated polystyrene (TEMPOL- terminated PS) [131], *etc.*

Although there have been great advances in covalent functionalization of fullerenes to obtain surface-modified fullerene derivatives or fullerene polymers, the application of these compounds in composites still remains unexplored, basically because of the low availability of these compounds [132]. However, until now, modified fullerene derivatives have been used to prepare composites with different polymers, including acrylic [133, 134] or vinyl polymers [135], polystyrene [136], polyethylene [137], and polyimide [138, 139], amongst others. These composite materials have found applications especially in the field of optoelectronics [140] in which the most important applications of the fullerene-polymer composites have been in the field of photovoltaic and optical-limiting materials [141]. The methods to covalently functionalize fullerenes and their application for composites or hybrid materials are very well established and they have set the foundations that later were applied to the covalent functionalization of other carbon nanostructures including CNTs and graphene.

Covalent functionalization in carbon nanotubes

Following the identification of CNTs [4, 3] and the optimized synthesis methods including arc discharge and catalytic pyrolysis, applications were sought for these structures. However, CNTs exhibit characteristic properties, for example, low dispersibility, that limit their processability and thus their applications in composite fabrication.

The use of CNTs in composites for optical, mechanical, electronic, biological and medical applications, *etc.*, requires the chemical modification of their surface in order to meet specific requirements depending on the application [140]. While searching for how to perform the covalent functionalization of CNTs, it was found that the tips of CNTs were more reactive than their sidewalls [142, 143].

Among covalent functionalization techniques, oxidation of CNTs is probably the method that has been explored the most. The most common oxidation reactions performed on CNTs involve two main routes: gas phase and liquid phase treatments. The first involves oxidative-gas or oxidative plasma treatments. The second uses some acids (for example HNO_3, H_2SO_4, *etc.*), sulfonitric mixture or "piranha" mixture (H_2SO_4/H_2O_2) in order to produce etching and/or cutting CNTs by introducing carboxyl (-COOH) and hydroxyl (-OH) groups at the opened ends and also on their sidewalls [144–147].

It has also been demonstrated that CNT sidewalls can be covalently fluorinated [148–150], or they can be derivatized with certain highly reactive chemicals such as dichlorocarbene [142]. In this context, Chen *et al.* applied derivatization chemistry with thionychloride and octadecylamine in order to obtain organic soluble SWCNTs and later they performed a reaction with dichlorocarbene that led to the covalent functionalization of the nanotube walls.

More recently, microwave chemistry has also been used to achieve covalent functionalization. In particular, these treatments can functionalize CNTs with sulfonated and carboxylic groups using a mixture of nitric and sulfuric acid under microwave radiation for 3 min, thus resulting in highly dispersible CNTs in ethanol and water

[151, 152]. In general, these methods are partially efficient since they induce a large amount of defects that deteriorate the sidewalls and diminish their electronic and mechanical properties [153].

There are several chemical reactions that can be used as an alternative to achieve covalent functionalization of CNTs. Two of them are amidation and/or esterification reactions. Both reactions take advantage of the carboxylic groups sitting on the sidewalls and tips of CNTs. In particular, they are converted to acyl chloride groups (-CO-Cl) *via* a reaction with thionyl (SO) or oxalyl chloride before adding an alcohol or an amine. This procedure is very versatile and allows the functionalization of CNTs with different entities such as biomolecules [154–156], polymers [157], and organic compounds [158, 159] among others.

Covalent functionalization in graphene

Although similar strategies used to functionalize carbon nanotubes or fullerenes can also be applied to graphene, it is possible to find plenty of examples in the literature dealing with the covalent functionalization of graphene, graphene oxide or reduced graphene oxide [160–163]. As mentioned above, functionalization is a key strategy able to obtain biocompatible graphene-based materials. For example, pegylated nanographene oxide sheets have been synthesized and described by Sun *et al.*, and this material showed high solubility and compatibility in biological environments [164]. Another example of covalent functionalization and enhancement of biocompatibility is the material reported by Bao *et al.*, in which chitosan is covalently anchored to graphene nanoplatelets and used as drug delivery carriers [165]. These examples illustrate how covalent functionalization of graphene can be achieved in order to obtain highly biocompatible composites.

4.1.4.2 Noncovalent functionalization of nanocarbons

One of the main characteristics of noncovalent functionalization is the fact that it does not compromise the conjugated system of the graphitic layers, and this constitutes a major advantage. In addition, this strategy allows easy tuning of interfacial properties on the carbon nanostructures mainly by π-π stacking or hydrophobic interactions. As a result of taking advantage of noncovalent functionalization, it is possible to preserve and tune the desired properties as well as to obtain a remarkably good solubility of graphene and CNTs. There are some strategies used to achieve noncovalent functionalization such as aromatic molecule functionalization (π-π stacking), polymer wrapping, surfactant, and the endohedral method.

Functionalization by π-π stacking has been widely used in CNTs [166–168], and recently applied to graphene [169–171] with remarkable results. For example, Qi *et al.* have demonstrated an effective method to prepare amphiphilic reduced graphene oxide sheets using a poly(ethylene glycol) copolymer [172]. This material might possess high stability in physiological media as well as a high degree of biocompatibility.

Wrapping is another strategy which allows the functionalization of CNTs and graphene using biopolymers or even single or double strand DNA, as well as RNA [173–176]. The direct use of biopolymers increases the biocompatibility of CNTs, thus allowing applications such as drug delivery systems and highly specific biosensors.

Summarizing, noncovalent functionalization methods can be used to prepare materials with specific biological properties because they are quick, efficient and clean. In order to increase biocompatibility of carbon nanostructures, these materials now need to be integrated into living systems and to be potentially used as tissue regeneration scaffolds, prostheses or drug deliverers.

4.2 Incorporation of nanocarbons into polymer composites and hybrids

4.2.1 Types of polymer composites

Polymer composites are materials which are fabricated by dispersing nanomaterials (filler) with dimensions ranging from 10 to 100 Å, within a matrix that can be an organic polymer. The resulting composite significantly enhances the polymer's properties [177]. In particular, polymer composites may be divided into three families (see Chapter 8):

- **Mixtures**: These materials are characterized by introducing nanoscale size materials into the polymeric matrix only by physical mixing. Here the nanomaterials are usually in a state of agglomeration. In this case, the nanomaterials are in a weight content less than 10 %.
- **Reinforced materials**: These materials possess the same principle as the mixtures but they differ in that the nanomaterial weight content into the polymer matrix is higher than 30 %.
- **Hybrids**: The main characteristic of these materials is that the nanomaterial introduced into the polymer matrix establishes a chemical bond with a thin layer of polymer.

Many different materials are good candidates to be dispersed into polymers. Table 4.1 shows some examples.

These nanomaterials may exhibit three different structures (Fig. 4.7), depending on the interactions established between the polymer and the dispersed material [178]. These structures are:

Separate phases: This structure occurs when the polymer and the dispersed material are completely immiscible.

Intercalation: This occurs when the polymer is intercalated in the dispersed material. This phenomenon results in alternated polymeric chains within the nanometric material.

Tab. 4.1: Candidate nanomaterials commonly used to produce composites [177, 178].

Chemical Nature	Examples
Pure elements	Graphite, carbon nanotubes, fullerene, nanodiamond, *etc.*
Calcogenic metals	PbS, TiS_2, MoS_2
Metal phosphates	$Zr(HPO_4)$
Clays and silicates sheets	Montmorillonite, hectonite, saponite, fluorohectonite, vermiculite, Kaolin, *etc.*
Double layer hydroxides	$M_6Al_2(OH)_{16}CO_3{\cdot}nH_2O$; M=Mg, Zn

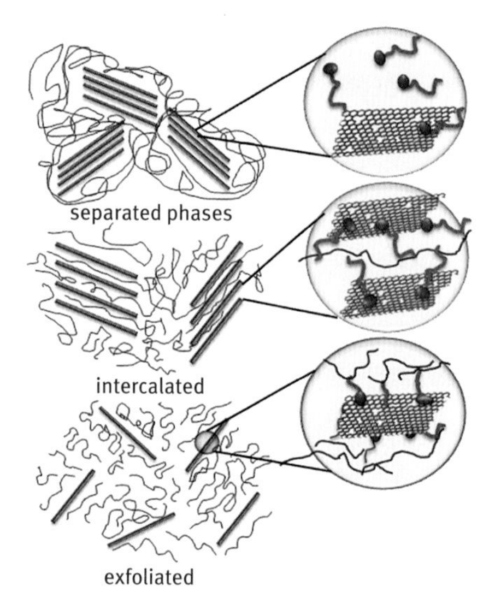

Fig. 4.7: The three idealized structures of polymer composites. This scheme explains the different structures that a polymer composite can present.

Exfoliated or delaminated: This is observed when the nanomaterial is completely dispersed in the polymer matrix, a highly desirable condition.

There are four main concepts involved in the synthesis and processing of nanocomposites: exfoliation, orientation, compatibility and re-aggregation.

4.2.1.1 Exfoliation and orientation

It is necessary to disperse the nanomaterials in the best possible manner, especially those layered structures such as graphite, graphene or clays. It is important to obtain very thin (*ca.* one nanometer) and very wide (*ca.* 500 nanometers) nanostructures dispersed in the polymer matrices to achieve optimal gas permeability and to improve their mechanical properties without affecting structural quality, using a small amount of the nanomaterial. The particle orientation also has an important effect on the properties of the nanocomposite. Nanoparticles need to be dispersed within the polymer so that are parallel to the material's surface. This condition ensures a maximum "tortuous path" for gases when migrating through the polymer and for some electronic applications the orientation of the nanomaterial facilitates the percolation and the passage of electrons through the polymer, thus improving electrical conductivity values.

4.2.1.2 Compatibility and re-aggregation

When synthesizing these composites, there is a high probability of re-aggregation of the nanomaterials. If this happens, then a true nanocomposite is unsuccessful because instead of being improved, the properties become negatively affected. Therefore, the nanomaterials need to be engineered or go through a surface modification in order to avoid re-aggregation.

4.2.1.3 Nanocarbon polymer hybrids

Hybrid polymers are composites containing one organic component (*e.g.* polymer) and another part, distinct in nature to the matrix (*e.g.* the nanomaterial), and these parts interact through chemical bonding so that the polymer layer bonded to the filler is distinct from the polymer matrix. The properties of these materials are improved by the introduction of the nanomaterial into the polymer matrix. Some of the properties that can be enhanced include the Young's modulus, resistance to flammability, low permeability to gases, and increased heat resistance.

Nanomaterial dispersion within a polymer matrix is generally difficult and requires the use of several chemical steps and specific chemical agents. When the nanomaterials are inorganic, stabilizing agents are needed. These agents contain a hydrophilic part that has the ability to attach to the inorganic nanomaterials, and another part which is attached to the polymer. The presence of non-sp^2 carbon defects, consisting of carbon chains, edges, adatoms, dopants and vacancies, might exhibit high chemical activity. In some cases, the use of compatibilizing agents helps to improve the interactions of the nanomaterial and the polymer matrix. Some of the most employed agents are listed and described in the following paragraphs.

The first chemical agents used in composite fabrication were amino acids [179] for synthesizing Nylon 6. Other agents are also used in the synthesis of nanomaterials, however, the most popular are alkylammonium ions due to their ability to exchange ions located between the layers of the clay. In addition, silanes have been widely used due to their ability to react with the hydroxyl groups on the surface and edges of the clay layers.

Amino acids: Amino acids are molecules that have an amino group ($-NH_2$) and a carboxylic acid (-COOH) group. Under acidic conditions, this molecule donates a proton from the -COOH group to the $-NH_2$ group. These conditions promote a cationic exchange between the $-NH_3^+$ cations formed in the amino acid molecules and the cations (*e.g.* Na^+, K^+, ...) interspersed between the clay layers, thus changing the hydrophilic surface of the clay. In the synthesis of Nylon 6, ω-amino acids have been used with great success because they are capable of polymerizing the ε-caprolactam [180].

Alkylammonium ions: When montmorillonite is exchanged with alkylammonium ions, it can be dispersed in a polar organic liquid, thus forming gels [181, 182]. Alkylammonium ions can also be easily inserted between the clay sheets and they constitute an alternative to the amino acids in the synthesis of nylon. Most of the alkylammonium ions used for this purpose are based on primary alkylamines placed in an acidic medium, allowing the protonation of the amine function. Its basic formula is $CH_3\,(CH_2)_n\,NH^{3+}$, where n is between 1 and 18. It should be noted that the length of the ions have a strong impact in the resulting structure of the composites [183, 184]; chain lengths greater than 8 carbon atoms favor the formation of exfoliated compounds, whereas short chains result in intercalated compounds.

4.2.2 Synthesis approaches

There are three methods for synthesizing polymers: melt state compounding, *in situ* polymerization and solvent methods [185–187, 177].

4.2.2.1 *In situ* polymerization

This technique involves the dispersion of a nanomaterial in a monomer (Fig. 4.8). This step requires a certain amount of time that depends on the polarity of the monomer molecules, the surface treatment of the nanomaterial, and the swelling temperature. For thermoplastics, the polymerization can be initiated either by the addition of an agent or by an increase in temperature. For thermosets such as epoxies or unsaturated polyesters, a curing agent or peroxide can be added in order to initiate the polymerization. Functionalized nanomaterials can improve their initial dispersion in the monomer and consequently in the composites. In the case of layered materials, such as clays or graphene, the most important step is the penetration of the monomer between the sheets, thus allowing the polymer chains to exfoliate the material. The

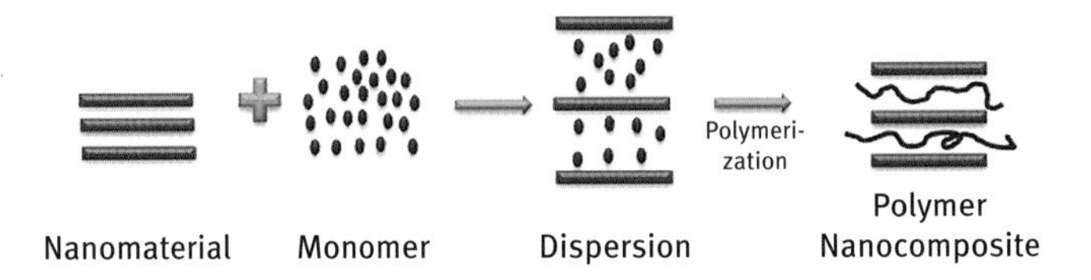

Fig. 4.8: Scheme explaining *in situ* polymerization in layered nanomaterials. The purple dots represent the monomer.

driving force of *in situ* polymerization is associated to the polarity of the monomer molecules. For polymer clay composites, during the swelling phase, the high surface energy of the clay attracts polar monomer molecules so that they diffuse between the clay layers. When the polymerization is initiated, the monomer starts to react with the agent. This reaction lowers the overall polarity of the intercalated molecules and displaces the thermodynamic equilibrium so that more polar molecules are driven between the clay layers. As this mechanism continues, the organic molecules eventually exfoliate the clay.

Furthermore, *in situ* polymerization methods enable covalent bonding between the functionalized nanomaterials and the polymer matrix using various chemical reactions. The stronger the bonding between the polymer and the nanomaterial, the better the mechanical properties of the composite material [188]. *In situ* polymerization also inhibits the possibility of migration of the nanomaterial in the polymer matrix, and prevents agglomeration of the nanomaterial during processing that can result in decreasing its mechanical properties. This technique was used initially in the synthesis of Nylon 6 with nanoclays [189]. This procedure has also been used for other polymers and additives because this method generates good dispersions and can also produce polymers chemically linked to the additive. A few examples include epoxy-clay composites [190], chitosan-g-poly(acrylic acid)/Montmorillonite [191], polylactide/clay composites [192], poly(methyl methacrylate)/MWCNTs [193], poly(*p*-phenylene benzobisoxazole)/SWCNTs [194], and graphene oxide/polyaniline [195], among others.

4.2.2.2 Solvent methods

Solvent methods are similar to *in situ* polymerizations. During the first stage, the nanomaterial is dispersed in a solvent. In some cases a surfactant can be used as a bridge between the nanomaterials and the matrix (Fig. 4.9). This reaction occurs in a liquid or gel form in the presence of a solvent such as toluene, chloroform, acetonitrile, water, acetone, *etc*. The solution of modified nanoparticles is then added to a polymeric solution under agitation (at room or elevated temperatures) in order to ensure a homogeneous dispersion of the nanomaterials in the matrix. Finally, the polymer composite

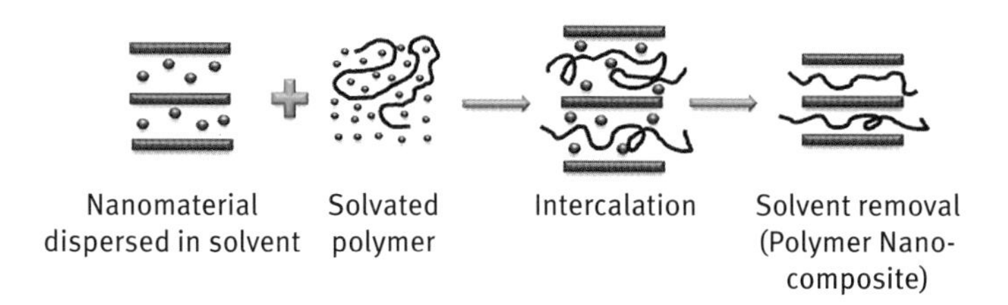

Fig. 4.9: Schematics showing the intercalation of the polymer by the "solution" approach. The blue dots represent the solvent molecules.

is recovered by precipitation or by casting. The difficulties associated with nanomaterial dispersion in a solvent by simple stirring often require the use of high-power ultrasonication in order to achieve metastable suspensions. In the case of CNTs, they are difficult to disperse because of their tendency to aggregate *via* van der Waals interactions, and different approaches such as heat treatments [196], acid treatments [197] or chemical functionalization of CNTs [198] have been used to alleviate this condition. The major advantage of solution mixing methods is that it offers the possibilities to synthesize intercalated composites based on polymers with low or even no polarity. However, it is difficult to transfer this method to industry due to problems associated with the use of large quantities of solvents. Various polymer composites have been successfully prepared by the solution mixing method including natural rubber and organophilic layered silicates [199], PS [200], polycarbonate [201], polyimides [202], and poly (methyl methacrylate) (PMMA) [203].

4.2.2.3 Melt intercalation

Melt blending uses high temperatures and high shear forces in order to disperse nanomaterials in a thermoplastic matrix (Fig. 4.10). Compatibilizers and ultrasonic tips *etc.* are often used to enhance dispersion of nanofillers on polymer matrices. The rotating screw forces the polymer resin with the nanoparticles forward into the barrel which is heated to the desired melt temperature of the molten plastic. The compression force of the screw also causes the rear end of the barrel to be at a higher temperature compared to the front part. This allows the plastic resin to melt gradually as it is pushed through the barrel. This lowers the risk of overheating and prevents degradation of the nanocomposite polymer. The melt intercalation process has become increasingly popular because of its great potential in industry. Melt blending has been reported for all the main polymers, including polyolefines (PE, PP), polyamides, polyesters (PET, PBT and others), polyurethane, polystyrene, *etc.* However, when compared to solution mixing and *in situ* polymerization, melt blending is generally less effective at dispersing nanofillers in polymers, and the technique is limited to lower concentrations of filler due to the high viscosity of the composites at high nanomaterial loadings.

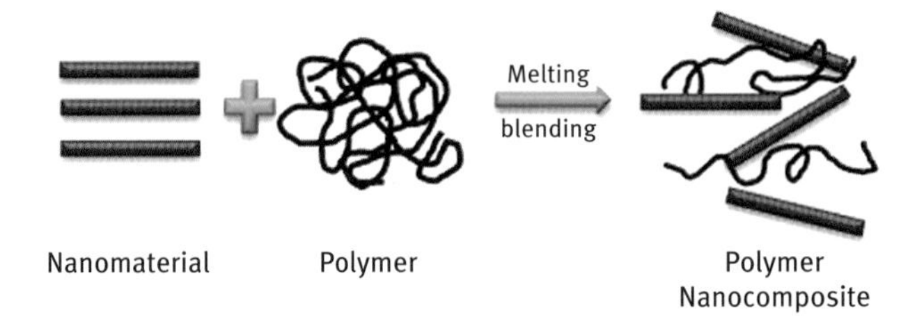

Fig. 4.10: Schematic representation of the melt intercalation process.

4.3 Properties

4.3.1 Mechanical properties

After Iijima pointed out the extraordinary strength of carbon nanotubes [204], and after their Young's modulus were measured [205], scientists and engineers have been interested in the mechanical properties of CNTs and other carbon nanomaterials. In this section, we will review the effect of defects and doping on the mechanical properties of carbon nanomaterials.

4.3.1.1 Effect of doping on carbon nanotubes

Nitrogen doping (N-doping)

Sumpter *et al.* [206] used *ab initio* calculations in conjunction with molecular dynamics to study the effect of nitrogen, phosphorus and sulfur on the mechanical properties of carbon nanostructures. N-doping seems to reduce the mechanical strength of the nanotubes, thus resulting in less robust CNTs. Figure 4.11(a) shows the results obtained from molecular dynamics simulations for the fracture strength of pristine and N-doped nanotubes. For N-doped nanotubes, there is a 50 % reduction in strain at fracture and crack propagation along the nitrogen atoms. Experimentally, nitrogen-doped CNTs contain nitrogen atoms within the lattice in addition to defects such as vacancies, functional groups and non-sp^2 hybridized carbon atoms. These defects might result in reduced mechanical properties. In this context, Cruz-Silva *et al.* [207] experimentally and theoretically studied SWCNTs doped with phosphorus (P) and phosphorus-nitrogen (PN). According to density functional calculations, substitutional P and P-N doping modified the mechanical strength, decreasing by approximately 50 % the elongation at fracture. Ganesan *et al.*[13] experimentally studied the tensile properties of MWCNTs and nitrogen-doped MWCNTs by using a microfabricated nanoindenter located inside a scanning electron microscope (SEM) chamber. The results differ from theoretical experiments because they found that pristine MWCNTs deformed and failed in a brittle fashion, whereas the nitrogen-doped MWCNTs deformed plastically prior to failure. C. Morant *et al.*[14] doped SWCNTs using low energy N_2^+ ion bombardment, and later tested the mechanical properties by measuring the indentation load using an AFM nanoindenter. The authors found that the force needed to achieve plastic deformations on the CNTs increased as the N-doping increased, *i.e.* N-doping results in hardening of the carbon nanotube. The reason might be that during ion bombardment, both the N and C atoms with sp^3 hybridization increase. The relationship established between sp^3 carbon atoms and hardness has been observed previously in many other carbon materials, especially DLC thin films.

Boron doping

Fakhrabadi *et al.* [210] studied the elasticity and failure behavior of boron doped CNTs using molecular dynamics simulations. These nanotubes were subjected to uniaxial deformation while the strain and thermal conductivity was analyzed. It was found that 4.0 at.% of boron atoms negatively affected the elongation at break. Different time frames of the simulation are shown in Fig. 4.11(b), in which the crack propagation along the B-doping atoms can be seen. Song Hai-Yang *et al.* [211] studied the effect of boron doping and boron grafting (boron adatoms) on SWCNTs' mechanical properties. They found that the moduli, tensile strength, buckling load, and buckling strain decreased after boron functionalization. In this study, particular attention was paid to the density distribution of the boron doping, and it was shown that for three different types of distribution densities, *i.e.* 6.25 %, 12.5 % and 25.0 %, the decreased magnitudes of Young's moduli of armchair (6, 6) SWCNT for boron doping (respectively boron grafts) were 10.4 %, 16.9 % and 31.2 %. These results indicate that not only the presence of dopants, but also their distribution affects the mechanical performance of carbon nanostructures.

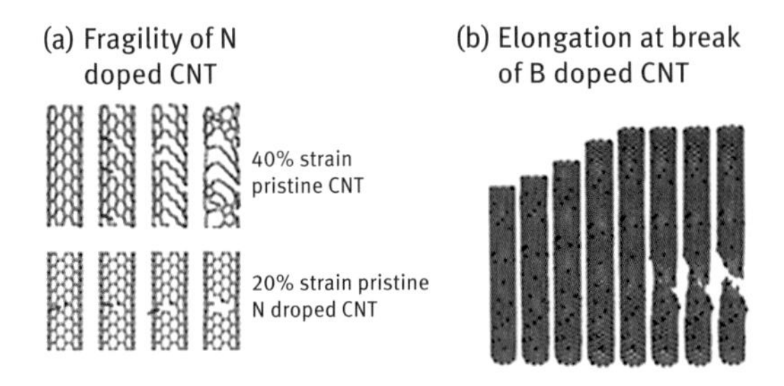

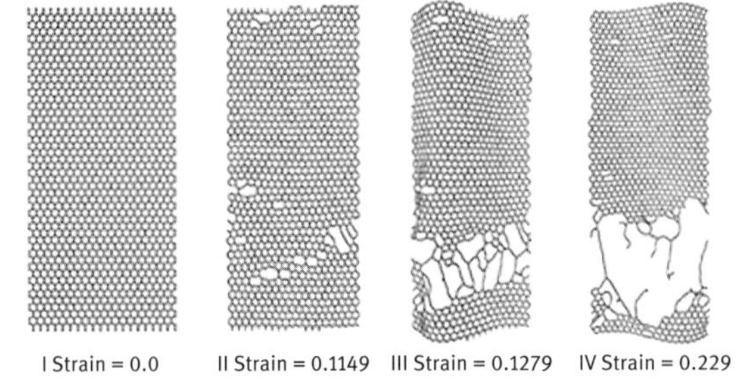

Fig. 4.11: Time frames obtained by simulation of stress-strain tests on (a) N-doped SWCNTs, (b) B-doped SWCNTs and (c) N-doped single layer graphene.

Silicon doping

Rahmandoust *et al.* [212] used finite element models of SWCNTs to calculate the mechanical properties of these doped systems, and obtained results that were in agreement with existing literature values calculated using more complex methods. Subsequently, they replaced carbon atoms with silicon atoms in zigzag and armchair SWCNTs models, and these Si doped SWCNTs showed lower modulus than the pristine ones due to introduction of Si defects.

4.3.1.2 Graphene and graphene nanoribbons

The local mechanical properties of graphene have been measured by nanoindentation resulting in the strongest material ever measured under tension [213]; the theoretical Young's modulus is close to 1.0 TPa. Mortazavi *et al.* [214] studied the effects of N-doping on the mechanical properties of single layer graphene by performing classical molecular dynamic simulations. Figure 4.11(c) depicts different time frames of a nitrogen-doped graphene sheet (*ca.* 6 %at.) at different stages of loading. It can be observed that voids originate due to bond rupture occurring between two adjacent carbon and nitrogen atoms along the loading direction. After the formation of a few voids, these voids keep growing and eventually interconnect without the formation of new voids. Finally, the coalescence of voids results in the failure of the graphene sheet. It is interesting to note that by increasing the concentration of nitrogen atoms up to 6 %, the Young's modulus of graphene was not affected, but the tensile strength decreased dramatically (35 %) after adding only 2 at.% of nitrogen atoms. It was also found that the Young's moduli of graphene are very sensitive to the presence of vacancies. In addition, boron-doped graphene simulations showed that the effects of boron atoms on the mechanical properties were less dramatic than nitrogen [215]. Boron atoms however, changed the failure behavior of single layer graphene from ductile to brittle after adding 4.0 at.% of boron atoms on the lattice. While the impact on the mechanical properties was low, the thermal conductivity was drastically affected, and with only 0.75 % of boron doping, the thermal conductivity decreased by 60 %.

Graphene nanoribbons (GNRs)

Terrones *et al.* [216] have pointed out the advantage of having graphene nanoribbons (GNRs) as reinforcement for polymer composites, although experimental measurements of the mechanical properties of graphene nanoribbons have been scarce. Nonetheless, GNRs have been widely studied using molecular dynamics simulations. Kang *et al.* [217] simulated the mechanical resonance of an armchair GNR under axial strain. The mechanical properties were found to be nonlinear, although at small strains the mechanical properties of GNRs can be estimated by classical continuum theory. These results are in agreement with previous research performed by Georgantzinos *et al.* [218] that predicted the size-dependent nonlinear mechanical properties of GNRs using a nonlinear spring-like finite-element method. Bu *et al.* [219] carried out MD simulations of the mechanical behavior of GNRs and also found that GNRs exhibit a nonlinear elastic behavior under tensile load. When the GNRs were strained above 18 %, the Young's modulus increased due to stress-stiffening. Strains at fracture as high as 30 % were found, thus reaching an intrinsic strength of 0.175 TPa. These exceptional mechanical properties are attributed to the strength and flexibility of the C-C bonds and the variations of bond angles. Interestingly, the effect of edge-states on the mechanical properties of narrow ribbons has not yet been studied.

4.3.1.3 Mechanical properties of diamond-like carbon (DLC) thin films

Not only CNTs and graphene are mechanically robust carbon nanomaterials, but also carbon-based thin films with thicknesses below 1.0 μm. Carbon thin films have a wide range of applications due to their tribological properties that depend mainly on their microstructure and composition. Graphite-based thin films are soft and usually have a low friction coefficient, whereas diamond-like carbon (DLC) films are hard, electrically insulating, optical transparent and of superior hardness. Typically, the mechanical properties of DLC films depend strongly on the bonding between carbon atoms, *i.e.* the hardness of the amorphous (a-C:H) films increases with increasing number of sp^3 C-C bonds. It has been found that silicon doping of diamond-like thin films deposited by RF magnetron sputtering increases their hardness and decreases their residual stress [220]. The reason is the presence of silicon carbide nanocrystals embedded in the diamond-like system. Conversely, nitrogen-doped DLC films deposited by PE-assisted CVD exhibit a slightly decreased hardness (*e.g.* from 18 GPa for pristine films to 16 GPa for nitrogen-doped systems containing 3.8 at.% nitrogen), but the electrical conductivity can increase 4 orders of magnitude [221]. However, nitrogen-doped DLC films result in either an increase or a decrease in the mechanical properties, depending on the microstructure of the film [222]. DLC thin films co-doped with 1.5 % of nitrogen and 10 % of boron showed higher wear resistant properties than undoped films. The hardness of silicon (1.4–1.8 %) and aluminum (0.3–0.9) doped DLC films, prepared by radio frequency magnetron sputtering, is relatively soft with hardness values below 9 GPa [223].

4.3.1.4 Mechanical properties of doped carbon nanotubes-polymer composites

Doped CNTs exhibit more reactive sites on their walls when compared to pure MWCNTs. Mild oxidation of nitrogen-doped MWCNTs (CNx MWCNTs) can introduce additional defects such as carbonyl or carboxyl that are present on their walls. In this context, Fragneaud *et al.* [224] treated CNx MWCNTs with benzoyl peroxide in order to graft aromatic groups onto their walls that were further brominated. These brominated-aryl groups were used to grow polystyrene chains by atom transfer radical polymerization. After polymerization, up to 40 wt% of polymer was grafted on the CNx MWCNTs. These PS-coated CNx nanotubes might reveal improved performance as fillers in nanocomposites due to their engineered polymer-graphene interphase. Nanocomposites containing N-doped MWCNTs and poly(bisphenol A carbonate) have been prepared by solution blending [225]. High electrical conductivity values (734 S cm^{-1}) were obtained due to the nanotube aggregation that formed bundles of tubes that allowed high conductivities.

4.3.2 Thermal properties

The addition of specific nanomaterials can result in significant changes and improvements in thermal properties of polymer composites. In the next section we review some of the principal thermal properties that are affected by introducing nanomaterials in polymer matrices in order to form composites.

4.3.2.1 Anti-flammability

The thermal properties of polymers with montmorillonite have shown excellent anti-flammability properties. The heat release rate measured by calorimetry is reduced by 63 % in a Nylon-6 silicate composite containing 2 vol% of silicate. In contrast to many commercial fire retardants, the rate of soot and CO generation does not increase during combustion. The flame-retardant behavior of these composites has been attributed to the formation of a tough char layer which may act as a mass-transport barrier, slowing the escape of volatile by-products during combustion. Similar improved char yields and flame resistance have also been observed for different polymer hosts like polycaprolactone [226], polypropylene [227] and vinyl ester layered silicate composites, ethylene-vinyl acetate copolymer (EVA) with MWCNTs [228], polyethylene (PE) with SWCNTs and MWCNTs [229], and polypropylene (PP) with MWCNTs [230]. For example, pure polymer samples exposed to an open flame continue burning after the flame is removed, until they are externally extinguished. In contrast, the composites stop burning after the flame is removed and become highly charred but maintain their initial dimensions. It is important to note that the nanodispersion of the silicate or carbon nanotubes is essential to this flame-retardant behavior. In contrast to other flame-retardant additives which tend to deteriorate in mechanical properties, the composites of silicates and carbon nanotubes show the opposite trend. It is worth mentioning that silicates are very well proven in this kind of application while in the case of carbon nanotubes there is room for further research in this field.

4.3.2.2 Thermal conductivity

The thermal conductivities of polymers depend on many factors, such as chemical constituents, bonding strength, structure type, molecular weight of side groups, molecular density distribution, type and strength of defects or structural faults, size of intermediate range order, processing conditions, temperature, *etc.* Thus, different results on thermal conductivity and crystallinity have been reported, because the crystallinity ratio of polymers and their thermal conductivity exhibit complicated dependencies. It is also worth mentioning that the crystallinity of the polymer depends greatly on the thermal history of the sample. Semicrystalline and amorphous polymers also vary considerably in their temperature dependence of thermal conductivity. At low temperature, semicrystalline polymers display a temperature dependence similar to that obtained from highly imperfect crystals, having a maximum in the temperature range near 100 K, which shifts to lower temperatures and higher thermal conductivities as the crystallinity increases [231]. Amorphous polymers display temperature dependence similar to that obtained for inorganic glasses with no maximum but a significant plateau region at low temperature range [232].

Tab. 4.2: Thermal conductivities of some materials [233].

Material	Thermal conductivity at 25°C [W/m·K]
Carbon black	6–174
Carbon nanotubes	2000–6000
Diamond	2000
Graphite	100–400
PAN based carbon fibers	8–70
Copper	483
Boron nitride nanotube	250–300
Silver	450

Carbon materials can also be used as fillers in order to increase the thermal conductivity of composites (see Table 4.2). Graphite exhibits high thermal conductivity values, in addition to low cost, and good dispersion of the platelets. However, if large concentrations of graphite are mixed with the composite, the mechanical properties of the composite can decrease significantly. For graphene sheets embedded in polymers, the thermal conductivity strongly depends on the exfoliation degree and dispersion of the sheets. Other options are carbon fibers and CNTs, however the thermal conductivity in fibers depends on the geometry of the fillers; fibers are better thermal conductors if they are vertically aligned. However, more work needs to be carried out regarding the use of CNTs (SWCNTs or MWCNTs) in the fabrication of polymer composites with high thermal properties. Other studies conducted using boron nitride nanotubes have found significant increases in the thermal conductivity values when using polymers such as PVA, PMMA, PBV, and PS [234].

4.3.2.3 Crystallization of polymers

Crystallinity is one of the most important properties of semicrystalline polymers, because it affects most of the end properties of the polymer. The areas of ordered molecular crystallographic registration are the crystals, while the folds and ties are called amorphous regions. The crystallinity within a polymer can be seen as an organization in three different levels: unit cell, lamellar structure and spherulitic structure. The organization of macromolecules has resulted in various models of crystallization due to the complexity of this phenomenon. It is worth mentioning that some nanomaterials act as nucleating agents favoring polymer crystallization and the generation of small spherulites [235]. For semicrystalline polymer composites with clay in polymorphic polymers, the sheets of clay tend to orient the polymer chains, thus favoring certain crystalline phases [236, 237]. Moreover, CNTs can act as nucleating agents, providing growing sites for the polymer chains, accelerating the overall crystal growth rate. One example of nanoconfinement effect is the formation of peculiar "shish-kebab" structures (Fig. 4.12). These structures have been observed and identified in a variety of nanostructured polymers with CNTs such as: Polyethylene/CNTs [238–242], Poly(vinylidene fluoride)(PVDF)/CNTs [243], Nylon 6,6/CNTs [241, 244], PVA/CNTs [245].

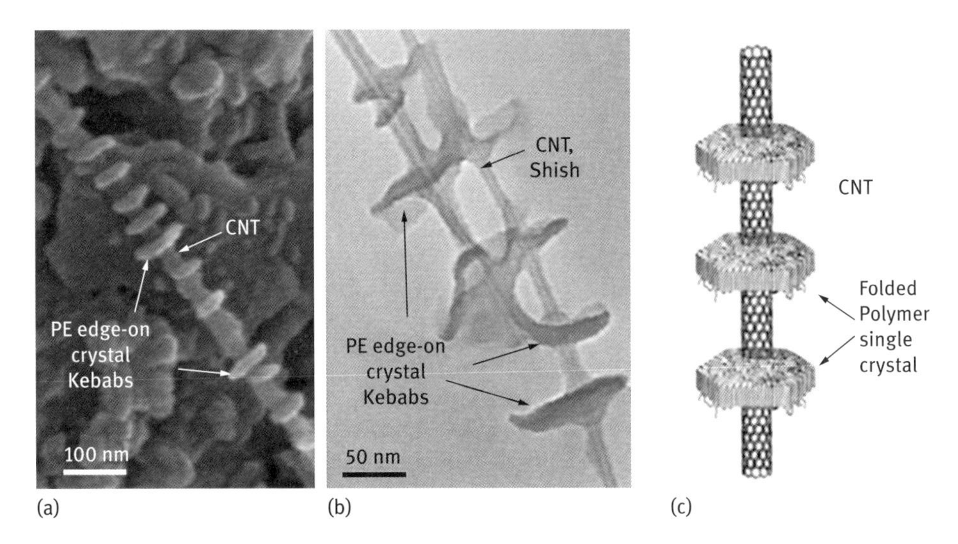

Fig. 4.12: SEM and TEM images of PE-MWCNTs "shish-kebab" structure produced by PE on MWCNTs at 103 °C, (reprinted from [15, 16] by permission of Wiley).

4.3.3 Electrical properties

It is well known that CNTs may exhibit high electrical conductivities and by adding them to other materials it is possible to increase the electrical conductivity of the entire composite. However, the addition of the adequate amount of CNTs and good dispersions are crucial for producing enhanced composites [247]. Once the content of carbon nanostructures is increasing there will be a point at which conductive paths between the nanostructures are created, known as the percolation threshold. The probability of establishing contacts between the nanostructures may increase by increasing the overall nanostructure content within the composite, thus increasing the electrical conductivity. The conductivity may follow the expression $\sigma \propto (p - p_c)^t$, where p_c is the percolation threshold (Fig. 4.13a). This is valid when $p > p_c$ and $(p - p_c)$ is small.

When the filling is higher than the percolation threshold, the conductivity increases sharply as conductive paths begin to form [248, 249].

When differentiating the electrical conductivities using alternating current (ac) and direct current (dc) against filling content, there are clear differences. At low filling ratios, the conductivity is frequency dependent, at higher frequencies, the conductivity increases following the expression $\sigma = \omega\varepsilon''\varepsilon_0$; also valid for dielectric materials in which σ is the conductivity, ε'' is the imaginary part of the dielectric constant, ω is the angular frequency, and ε_0 is the vacuum permittivity. At higher filling ratios, the conductivity does not vary at low frequencies until a critical value is reached where the conductivity values exhibit a complicated hopping mechanism (Fig. 4.13(a), (b)). When comparing ac conductivity (σ_{ac}) and dc conductivity (σ_{dc}), σ_{ac} is higher than σ_{dc} at low filling ratios, since there is very low or no percolation making σ_{dc} more difficult to overcome energy barriers. At higher filling ratios, in which percolation is possible, there might be large networks of carbon nanostructures that are not connected to the entire system, therefore σ_{ac} is more favorable [250, 251].

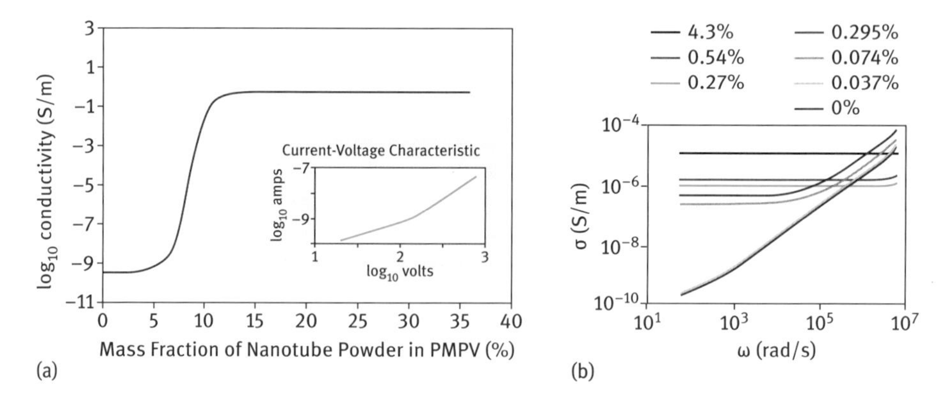

Fig. 4.13: (a) Semilogarithmic plot of conductivity versus the nanotube content (wt%) in poly(phenylene vinylene-co-2,5-dioctoxy-m-phenylene vinylene) (PMPV) [2]. (b) Frequency dependent conductivity of carbon nanotubes at different wt% in PmPV (filled symbols) and polyvinyl alcohol (PVA) (unfilled symbols) based composites [250].

Depending on the type of carbon nanostructure, the amount of mass may vary in order to achieve a percolation threshold of certain conductivity, see Table 4.3. It is evident that SWCNTs and DWCNTs do not exhibit very high conductivities; this may be due to the difficulty to achieve good dispersion within the composites. MWCNTs can be easier to functionalize, however their intrinsic conductivity is not as high as SWCNTs and DWCNTs. As mentioned previously, the reduction of graphene oxide will not recover the same properties as the initial graphite. However, due to its high surface area it might be easier to produce a homogenous composite using GO. Graphite is highly crystalline and therefore its electrical properties may be better than other car-

Tab. 4.3: Percolation threshold and conductivity for electrical transport using different types of nanocarbons in polymer composites.

Carbon structure	Percolation threshold	Electrical conductivity (S/cm)	References
MWCNT	0.0025–20 wt%	$1.53 \times 10^{-11} - 21$	[2], [17]
DWCNT	0.2–0.3 wt%	3×10^{-2}	[18]
SWCNT	3–6 wt%	1×10^{3} (at 30 wt%)	[19], [20]
Reduced graphene oxide	0.1–1 wt%	24	[21], [22]
Nanographite	0.34 vol%	500	[23]
Expanded graphite	8–10 wt%	77	[24]

bon nanostructures, but appropriate processing techniques are required, as observed for nanographite and expanded graphite [12, 13].

The length of CNTs can also influence the conductivity of the fabricated composite. With longer CNTs the composite can be more homogenous and the larger length can improve the conductivity paths [260]. The method for dispersing CNTs in composites may also greatly affect their conductivity. For example, it was found that sonication results in higher conductivity values when compared to shear mixing. However, shear mixing exhibits lower percolation thresholds. In addition, sonication cuts and partially oxidizes the CNTs, therefore the process requires a higher content of sonicated CNTs for percolation. However, the sample may also exhibit a lower amount of CNT agglomerations [261]. The CNT functionalization process may also determine the conductivity of the composite [262]. Therefore, these and other processing factors may influence the conductivity values of the fabricated composite.

4.3.4 Optical properties

It is well known that carbon nanostructures can absorb visible light, with a reflectance in the order of 10 % and in certain cases, special arrays of CNTs may exhibit the lowest value of *ca.* 5.75 % [263]. CNTs can absorb more light between 250 nm and 300 nm than the visible region (380–750 nm) due to π-π^* transition [264]. The absorption spectrum of graphene monolayers is flat and monotonous in the wavelength range between 400 and 750 nm. Here, it absorbs 2.3 % of white light, which increases with the number of layers, and its reflectance is negligible (< 0.1 %). Graphene is also very easy to transfer onto polymeric substrates, therefore graphene/polymer transfer is widely used with high transmittance (*ca.* 88 %–95 %) and high conductivity (*ca.* 1.17 kΩ/sq. sheet resistance on polyethylene terephthalate) (Fig. 4.14(a)) [265]. By varying the content of carbon nanostructures within the composite it is possible to modulate the optical band gap position [266]. In the case of CNTs, a low content of CNTs may achieve enhanced transmittances; however the electrical properties will be lower, as discussed above. If the thickness of the composite is thin enough (*ca.* 90 nm), then

80 % optical transmittance and high conductivity (*ca.* 1×10^3 S/cm) may be achieved (Fig. 4.14) [8].

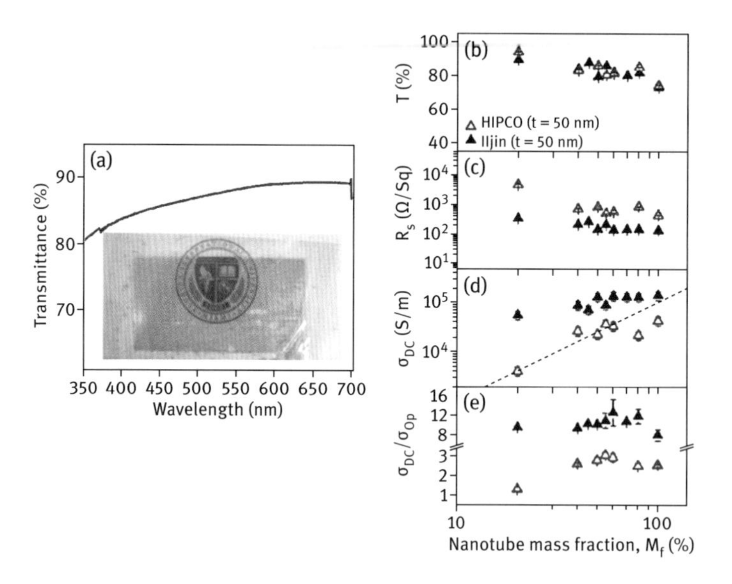

Fig. 4.14: (a) Optical transmittance of graphene on a polyethylene terephthalate (PET) flexible substrate [19]. Optical and electrical data for PEDOT:PSS-based composites with SWCNT: (b) transmittance at 550 nm, (c) sheet resistance, (d) DC conductivity and (e) ratio of DC to optical conductivity.

4.3.5 Biocompatibility

Carbon nanostructures might also possess outstanding bio-related properties. In particular, carbon nanostructures (CNTs, graphene and fullerenes) are highly versatile and they have been used in the fabrication of biosensors [267–271], to fabricate novel biomaterials [272] and as drug delivery agents for gene therapy [273, 274], and even for neuronal regeneration [275, 276].

Composites made with carbon nanostructures have demonstrated their high performance as biomaterials, basically applied in the field of tissue regeneration with excellent results. For example, P.R. Supronowicz *et al.* demonstrated that nanocomposites fabricated with polylactic acid and CNTs can be used to expose cells to electrical stimulation, thus promoting osteoblast functions that are responsible for the chemical composition of the organic and inorganic phases of bone [277]. MacDonald *et al.* prepared composites containing a collagen matrix CNTs and found that CNTs do not affect the cell viability or cell proliferation [278].

Although there are several definitions of biocompatibility, the concept named as biocompatibility is usually used to describe the ability of a material to perform a desired function without producing a negative effect on biological systems in specific situations. It also involves how they can interact with biological systems such as the human body and how these interactions determine their success in their performance [279–281]. It is clear that these interactions are a consequence of the characteristic properties displayed by the surface of the materials. Therefore, their surface functionalization (doping or defects presence) appears to be critical for the behavior of CNTs in biological systems [282, 283].

Although the biocompatibility of carbon nanostructures has been explored in the past, deep knowledge of this subject should be increased since there are contradictory studies about the biological effects of these nanostructures in living organisms. The necessity of new protocols to evaluate the biocompatibility of CNTs has arisen since it seems that current protocols cannot be applied completely to carbon nanostructures.

4.3.6 Biodegradation

4.3.6.1 Intrinsic nanocarbon biodegradation

While macroscopic carbon has a chemical reactivity low enough to be degraded significantly by biological pathways (a fact that allows the preservation of mineral carbon for millions of years or vegetal carbon for thousands of years), their nanoscopic counterparts have a considerably higher chemical activity that allows their degradation in a time span of months and even days. Allen *et al.* [284] have demonstrated that SWCNTs can be effectively degraded by horseradish peroxidase (HRP), an oxidoreductase enzyme, in approximately 12 weeks (Fig. 4.15). Even though this experiment was carried out *in vitro*, this work opened up the possibility of enzymatic degradation of SWCNTs under environmental conditions. *In vivo* studies have also shown that peroxidases can degrade PEG grafted carboxylated SWCNTs [285]. The oxidation pathway *in vivo* is a lot more complex and consequently still unclear, but by comparing different proteins it has been found that hypochlorous acid produced by myeloperoxidase is most likely the oxidizing compound responsible for the degradation of carbon nanotubes. The *in vivo* biodegradation path was also studied by injection of PEG-grafted carboxylated SWNCTs into mice. The nanotubes induced an increase and activation of neutrophils and macrophages in the local tissue, which in turn produced a high local concentration of HOCl creating favorable conditions for the SWCNTs' degradation.

The influence of oxidation defects on the enzymatic biodegradation of MWCNTs was studied with different degrees of carboxylation [286]. Since the relative amount of defects increased considerably, the degradation of N-doped MWCNTs was also studied and due to their larger amount of defects their biodegradation was faster. For instance, while some MWCNTs were still visible in the sample after 60 days of biodegradation, all N-doped MWCNTs were biodegraded after just 50 days. Indeed, previous experiments show *in vivo* that N-doped MWCNTs were less harmful and better tolerated by mouse [287] and microorganisms [288], suggesting better biocompatibility and possibly biodegradability.

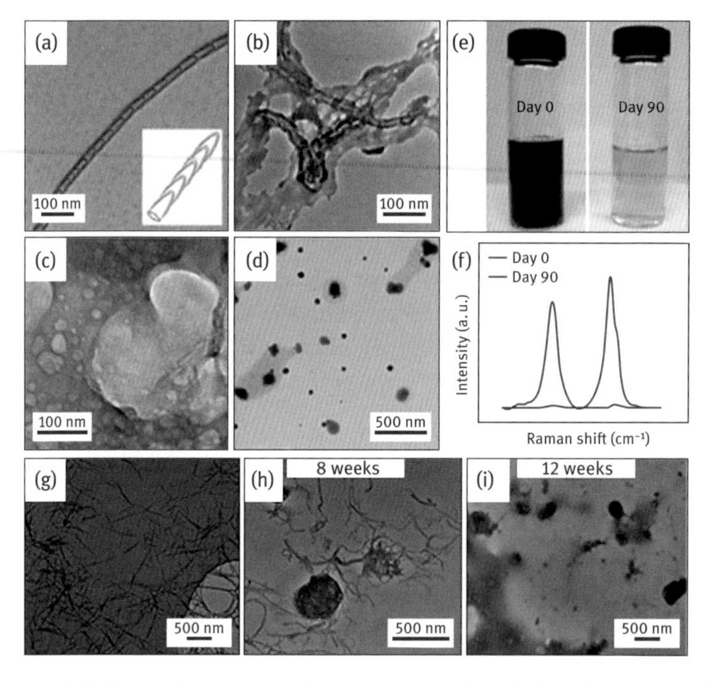

Fig. 4.15: Morphology changes during enzymatic degradation of nanotubes. TEM images of purified nitrogen-doped MWCNT (n-MWCNT) samples at the beginning of the experiment. After (b) 15 days, (c) 50 days and (d) 80 days of enzymatic degradation by HRP in aqueous medium. In (e) two vials show the difference in the n-MWCNT dispersions before (left) and after (right) enzymatic degradation. (f) Raman spectra for n-MWCNT samples before (black) and after (red) enzymatic degradation. (g) Original SWCNT dispersion, and (h) after 8 weeks and (i) 12 weeks of enzymatic degradation by HRP.

In vitro experiments of CNT degradation have shown outstanding results towards their understanding of biodegradability. Liu *et al.* [289] carried out degradation experiments of SWCNTs with different degrees of chemical functionalization in simulated phagolysosomal fluid. The morphology of the SWCNTs bundles was followed by TEM and light scattering. It was found that only carboxylated nanotubes degraded, whereas pristine, ozone treated, and aryl sulfonated SWCNTs did not degrade. These results lead to the conclusion that carboxylic functionalization of nanotubes is the method that more severely disrupts the graphitic tubes, leading to a more chemically reactive nanostructure. Russier *et al.* [290] monitored the enzymatic and nonenzymatic degradation of oxidized SWCNTs and MWCNTs by Raman spectroscopy and TEM measurements. They found that both ox-SWCNTs and ox-MWCNTs can degrade in both reaction media, the simulated PSF and the HRP solution. SWCNTs degraded considerably faster than MWCNTs, particularly when HRP and H_2O_2 were present. MWCNTs did not degrade completely, but after two months their morphology changed considerably. Not only vegetal but also mammalian peroxidases, such as *Eosinophil* peroxidase [291] have shown similar action on SWCNTs. This is a very significant result because this peroxidase is present in the lungs after inflammatory response due to adverse reaction to pollutants, and is thought to act as a natural degradation enzyme in mammalian lungs. All these works clearly show that defects in the graphitic sheets of carbon nanotubes are key factors to allow the enzymatic degradation of the samples. Recently, *in vivo* experiments have shown that intracellular myeloperoxidase can degrade SWCNTs [292] in neutrophils and macrophages cells, and more importantly, that inflammatory response was not present when biodegraded nanotubes were introduced into the lungs of mice. The complete biodegradability of N-doped MWCNTs in less than 30 days [286] is a remarkable phenomenon that should be considered when designing cellular scaffolds or biological applications of CNTs.

With the recent discovery of the graphene family of carbon materials, concerns about their biodegradability were raised. It has already been shown that enzymatic degradation of a perfect graphitic lattice might be too slow to be noticed, but after introduction of defects it can be speeded up considerably. Kotchey *et al.* [293] found that graphene oxide, a material that can be considered as a highly functionalized and defective graphene sheet, can be enzymatically degraded *in vitro* in just about 20 days. Indeed, after only 8 days of reaction with a mixture of HRP and H_2O_2, the GO sheet showed a large number of holes created by enzymatic degradation, and from this point it rapidly degraded to an over-oxidized soluble material. In contrast, reduced graphite oxide can not have been degraded, showing that defects on the graphitic sheet were crucial to enhancing their biodegradability.

4.3.6.2 Nanocarbon-polymer composites biodegradation

Nanocarbon materials are particularly interesting for the biomaterials field, not only because they can be used to improve the mechanical properties of biocompatible polymers, but also because they usually improve their biodegradability. Armentano *et al.* [294] have shown that poly(DL-lactide-co-glycolide) (PLGA)/SWCNTs composites degrade under physiological conditions (pH 7.4 and 36 °C), it depends largely on the functional groups present on CNTs. This is because PLGA polymer degrades mainly by hydrolysis, which can be promoted by the carboxyl groups present in functionalized SWCNTs. Consequently, nanocomposites prepared with SWCNTs bearing functional groups degraded considerably faster than those prepared with pristine SWCNTs. Zheng *et al.* [295] have shown that poly(epsilon-caprolactone) (PCL) covalently grafted on multi-walled carbon nanotubes (MWCNTs) can be completely degraded within 4 days by the biocatalytic action of Pseudomonas lipase. The PCL not only provided biodegradability to the MWCNTs but also processabilty, rendering these polymer-coated carbon nanotubes soluble in organic solvents. These nanotubes had a high density of PCL grafted on the surface but retained their tube-like morphology; thus these modified nanotubes can be used for biomaterials, biomedicine and artificial bones. PLLA and its copolymers are among the most interesting biocompatible polymers due to the large number of potential applications. Adeli *et al.* [296] prepared porous scaffolds of PLLA and MWCNTs by the freeze-extraction method. MWCNTs provided significant reinforcing to the PLLA matrix, improving mechanical properties and enhancing the thermal stability of the polymer. Nevertheless, biodegradation was studied *in vitro* using PBS buffer and it was found that MWCNTs decreased the rate of degradation. Liu *et al.* [297] studied nanocomposites of MWCNTs and poly(glycerol-sebacate-citrate) (PGSC) elastomer. The mechanical properties of the elastomers were improved with low MWCNTs content, due to the good dispersion and matrix-nanotube interaction. No cytotoxicity was found even at MWCNT concentrations above 1 wt%, but similar to the work of Adeli *et al.*, the degradation rate in simulated body fluid was found to decrease with the increase of MWCNT loading. The reason might be an improved interaction of the biodegradable polymer with the non-biodegradable pure carbon nanotubes. Nevertheless, functionalized SWCNTs have been found to promote biodegradation. Ray *et al.* [298] studied poly(butylene succinate), a biodegradable polymer, and its composites with MWCNTs prepared by melt blending. MWCNTs provided enhancement of the mechanical and electrical properties, improving the modulus 88 % with only 3 wt%. Not only nanocarbons, but also fullerene, carbon black and other nanocarbons have been studied as fillers in biodegradable nanocomposites. Tsuji *et al.* [299] studied the mechanical and electrical properties as well as enzymatic biodegradation of different types of nanocarbons, such as C_{60}, SWCNTs, graphitized carbon black, and carbon nanoballoons as additives for PLLA. It was found that C_{60} and SWCNTs enhanced the enzymatic degradation of PLLA, whereas the addition of carbon black and carbon nanoballoons disturbed the enzymatic degradation of PLLA. A possible explanation is that in PLLA enzymatic degradation rate depends on the diffusibility of the enzyme in the nanocarbon/PLLA interphase, and to the capability of the filler to be released by the polymer matrix. This means that fillers such as SWCNTs and C_{60} that show low interfacial adhesion result in nanocomposites with higher biodegradation rates.

4.3.7 Permeability

Among nanocarbons, graphene is certainly an interesting material for barrier applications due to its sheet-like shape. Several research studies have shown the potential application of graphene due to impermeability to gas [300] and larger molecules. The use of graphene as a protective barrier for metals against corrosion was first proposed by Ruoff's group [301] and since then other groups have also explored graphene-containing coatings against corrosion [302, 303]. Recently, not only graphene, but multilayered reduced graphite oxide coatings have been studied [304]. In addition to metals, other materials such as glass fibers can be protected with graphene coatings. For instance, glass fibers are sensitive to alkaline ion attacks, thus the use of a graphene/MWCNTs hybrid coating to protect the glass fibers in composites from environmental attack has also been studied [305]. It was found that this layer effectively inhibited the penetration of moisture/alkali ions through the coating layer, resulting in glass fiber and composite protection. Another interesting application that takes advantage of the natural barrier properties of graphene sheets is their use as an environmental protective layer, for instance, against mercury vapor, a well-known contaminant [306].

4.3.7.1 Gas barrier for polymer composites

Recently polymer containing graphene and graphene oxide (GO) have shown dramatic improvements on the barrier properties upon addition of these types of nanocarbons. Even trace concentrations of graphene can dramatically improve the barrier properties of polymer composites. Compton *et al.* [307] studied nanocomposites of crumpled graphene and polystyrene sheets and found that only 0.02 wt% results in a dramatic decrease of oxygen permeability of the composite, most likely due to the low solubility of oxygen in the graphene/polystyrene nanocomposite. Polyimide/graphite oxide nanocomposites with only 0.001 wt% of GO show an enhanced resistance to moisture, higher transmittance of light, enhanced mechanical strength, and excellent dimensional stability, simultaneously. It is interesting to point out that this "invisible" filler decreased the water vapor transmission rate from 181 g mil/(m^2 day) for pure PI to 30 g mil/(m^2 day) after adding 0.001 wt% of GO [308]. GO sheets were also uniformly dispersed in a PVA matrix. Concentrations of GO as low as 0.72 vol% improved dramatically the nanocomposite barrier properties for both oxygen and water vapor. Similarly, Kim *et al.* [309] prepared PVA/GO and PVA/RGO films and the oxygen permeability of the PVA/RGO (0.3 wt%) composite coated film was 86 times lower than that of a pure PET film. Polyethylenimine (PEI) and GO multilayered self-assembled free standing films have also shown significant reduced oxygen and carbon dioxide permeation rates [310]. PEI/GO films prepared similarly but supported on PET were used as a semitransparent oxygen barrier [311]. This bilayer PET/PEI-GO film showed excellent oxygen barrier properties, as well as high transparency and electrical conductivity. Oxygen permeability of solution prepared polynorbornene/graphene nanocomposites was reduced to one fifth of the neat polymer with only 5 wt% of graphene [312]. Free-standing graphene nanoplatelet films were studied as an oxygen barrier in polymer composites [313]. It was found that O_2 permeability of 70 wt% graphene nanoplatelets (GNP) polyetherimide paper is only 1.1 % of neat polyetherimide, despite the introduction of voids that were not interconnected. Chemically modified GO has also been studied to simultaneously reinforce while reducing permeation rates. For instance, graphene reinforced thermoplastic polyurethane showed a 10-fold increase in tensile stiffness and simultaneously a 90 % decrease in nitrogen permeation with only 3 wt% of isocyanate treated GO. [314]

4.4 Summary

After the Nylon 6 hybrid was developed in the Toyota Co. laboratory, polymer composites have experienced important developments over the past 20 years in laboratories across the world, and they continue with the development of many polymer composites with specific properties and purposes. Thus, a wide range of nanomaterial (clays, silicates, carbon nanotubes, graphene, *etc.*) and polymers have been successfully used in the synthesis of composites, and new methods of synthesis, directly applicable in industry, have been developed. The substantial improvements in mechanical and physical properties brought by polymer composites may increase their use in many different applications from food, energy, engineering, to medicine. Their excellent barrier properties combined with their good transparency make them ideal for packaging applications. Their improved thermal and mechanical properties might even extend the use of polymers in the automotive industry for under-the-hood applications. Now the use of nanomaterials has allowed to extend the application areas of engineering polymers, but more importantly, some biological and biodegradable polymers are improving their properties thus widening the area of application of these polymers. In addition, the use of polymers in medical and biological areas is increased thanks to new mechanical properties and biocompatibility of these new materials. Alternative application areas are emerging such as the development of lighter prostheses, catheters, *etc.* Besides with their antibacterial properties and greater biocompatibility than typical materials used so far, these smart composite materials are continuously being developed, opening new areas of research and application. Until now, only few commercial successes have been achieved with these new materials, but more are to come. With the current developments of new nanomaterials, a range of possibilities is opening for the fabrication of composites with unprecedented properties yet to be discovered, thus allowing the design of novel smart materials that can adapt to different needs in daily life.

Bibliography

[1] H. W. Kroto, J.R. Heath, S.C. O'Brien, R. F. Curl, R. E. Smalley, C60: Buckminsterfullerene, *Nature*, vol. 318, pp. 162–163, 1985.
[2] W. Krätschmer, L. D. Lamb, K. Fostiropoulos, D. R.Huffman, Solid C60: a new form of carbon, *Nature*, vol. 347, pp. 354–358, 1990.
[3] S. Iijima, Helical microtubules of graphitic carbon, *Nature*, vol. 354, pp. 56–58, 1991.
[4] A. Oberlin, M. Endo, T. Koyama, Filamentous growth of carbon through benzene decomposition., *Journal of Crystal Growth*, vol. 32, pp. 335–349, 1976.
[5] K. S. Novoselov, A. K. Geim, S. V. Morozov, D. Jiang, Y. Zhang, S. V. Dubonos, I. V. Grigorieva, A. A. Firsov, Electric field effect in atomically thin carbon film, *Science*, vol. 306, pp. 666–669, 2004.

[6] K. S. Novoselov, A. K. Geim, S. V. Morozov, D. Jiang, M. I. Katsnelson, I. V. Grigorieva, S. V. Dubonos, A. A. Firsov, Two-dimensional gas of massless Dirac fermions in graphene, *Nature*, vol. 438, p. 197, 2005.
[7] M. Terrones, A. R. Botello-Méndez, J. Campos-Delgado, F. López-Urías, Y. I. Vega-Cantú, F. J. Rodríguez-Macías, A. L. Elias, E. Muñoz-Sandoval, A. G. Cano-Márquez, J.-C. Charlier, H. Terrones, Graphene and graphite nanoribbons: Morphology, properties, synthesis, defects and applications., *Nano Today*, vol. 5, pp. 351–372, 2010.
[8] A. A. Balandin, S. Ghosh, W. Bao, I. Calizo, D. Teweldebrhan, F. Miao, C. N. Lau, Superior thermal conductivity of single-layer graphene, *Nano Lett.*, vol. 8, pp. 902–907, 2008.
[9] K. Erickson, R. Erni, Z. Lee, N. Alem, W. Gannett, A. Zettl, Determination of the local chemical structure of graphene oxide and reduced graphene oxide, *Adv. Mater*, vol. 22, pp. 4467–4472, 2010.
[10] K. A. Mkhoyan, A. W. Contryman, J. Silcox, D. A. Stewart, G. Eda, C. Mattevi, S. Miller, M. Chhowalla, Atomic and electronic structure of graphene-oxide, *Nano Lett.*, vol. 9, pp. 1058–1063, 2009.
[11] C. Gómez-Navarro, R. T. Weitz, A. M. Bittner, M. Scolari, A. Mews, M. Burghard, K. Kern, Electronic transport properties of individual chemically reduced graphene oxide sheets, *Nano Lett.*, vol. 7, pp. 3499–3503, 2007.
[12] C. Gómez-Navarro, J. C. Meyer, R. S. Sundaram, A. Chuvilin, S. Kurasch, M. Burghard, K. Kern, U. Kaiser, Atomic structure of reduced graphene oxide, *Nano Lett.*, vol. 10, pp. 1144–1148, 2010.
[13] I. Jung, D. A. Dikin, R. D. Piner, R. S. Ruoff, Tunable electrical conductivity of individual graphene oxide sheets reduced at low temperatures, *Nano Lett.*, vol. 8, pp. 4283–4287, 2008.
[14] K. Nakada, M. Fujita, G. Dresselhaus, M. S. Dresselhaus, Edge state in graphene ribbons: Nanometer size effect and edge shape dependence, *Phys. Rev. B*, vol. 54, pp. 17954–17961, 1996.
[15] L. Brey, H. A. Fertig, Electronic states of graphene nanoribbons studied with the Dirac equation, *Phy. Rev. B*, vol. 73, p. 235411, 2006.
[16] X. Wang, Y. Ouyang, X. Li, H. Wang, J. Guo, H. Dai, Room-temperature all-semiconducting sub-10-nm graphene nanoribbon field-effect transistors, *Phys. Rev. Lett.*, vol. 100, p. 206803, 2008.
[17] T. W. Ebbesen, H. J. Lezec, H. Hiura, J. W. Bennett, H. F. Ghaemi, T. Thio, Electrical conductivity of individual carbon nanotubes, *Nature*, vol. 382, pp. 54–56, 1996.
[18] E. W. Wong, P. E. Sheehan, C. M. Lieber, Nanobeam mechanics: elasticity, strength, and toughness of nanorods and nanotubes, *Science*, vol. 277, pp. 1971–1975, 1997.
[19] J. C. Huie, Guided molecular self-assembly; a review of recent efforts, *Smart Mater. Struct.*, vol. 12, pp. 264–271, 2003.
[20] J. Stotter, J. Zak, Z. Behier, Y. Show, G.M. Swain, Optical and electrochemical properties of optically transparent, boron-doped diamond thin films deposited on quartz, *Anal. Chem.*, vol. 74, p. 5924, 2002.
[21] M. Amanda, S. A. Schrand, C. Hens, O. A. Shenderova, Nanodiamond particles: properties and perspectives for bioapplications, *Crc. Cr.Rev. Sol. State*, vol. 34, pp. 18–74, 2009.
[22] B. Fausett, M. C. Granger, M.L. Hupert, J. Wang, G. M. Swain, D. M. Gruen, The electrochemical properties of nanocrystalline diamond thin-films deposited from C60/Argon and Methane/Nitrogen gas mixtures, *Electroanal.*, vol. 12, pp. 7–15, 2000.
[23] Y. Zhao, J. Wei, R. Vajtai, P. M. Ajayan, E. V. Barrera, Iodine doped carbon nanotube cables exceeding specific electrical conductivity of metals, *Scientific Reports* 1:83, 2011.

[24] B. Gao, A. Kleinhammes, X.P. Tang, C. Bower, L. Fleming, Y. Wu, O. Zhou, Electrochemical intercalation of single-walled carbon nanotubes with lithium, *Chem. Phys. Lett.*, vol. 307, pp. 153–157, 1999.

[25] O. Tanaike, M. Inagaki, Degradation of carbon materials by intercalation, *Carbon*, vol. 37, p. 1759–1769, 1999.

[26] W. Q. Deng, X. Xu, W. A. Goddard, New alkali doped pillared carbon materials designed to achieve practical reversible, *Phys. Rev. Lett*, vol. 92, p. 166103, 2004.

[27] L. M. Viculis, J. J. Mack, O. M. Mayer, H. T. Hahn, R. B. Kaner, Intercalation and exfoliation routes to graphite nanoplatelets, *J. Mater. Chem.*, vol. 15, p. 974–978, 2005.

[28] A. Morelos-Gómez, S. M. Vega-Díaz, V. J. González, F. Tristán-López, R. Cruz-Silva, K. Fujisawa, H. Muramatsu, T. Hayashi, X. Mi, Y. Shi, H. Sakamoto, F. Khoerunnisa, K. Kaneko, B. G. Sumpter, Y. A. Kim, V. Meunier, M. Endo, E. Muñoz-Sandoval, M. Terrones, Clean nanotube unzipping by abrupt thermal expansion of molecular nitrogen: graphene nanoribbons with atomically smooth Edges, *ACS Nano*, vol. 6, p. 2261–2272, 2012.

[29] N. I. Kovtyukhova, T. E. Mallouk, L. Pan, E. C. Dickey, Individual single-walled nanotubes and hydrogels made by oxidative exfoliation of carbon nanotube ropes, *J. Am. Chem. Soc.*, vol. 125, pp. 9761–9769, 2003.

[30] L. Vaisman, H. D. Wagner, G. Marom, The role of surfactants in dispersion of carbon nanotubes, *Adv. Colloid Interfac*, vol. 128–130, pp. 37–46, 2006.

[31] J. D. Wiggins-Camacho, K. J. Stevenson, Effect of nitrogen concentration on capacitance, density of states, electronic conductivity, and morphology of N-doped carbon nanotube electrodes, *J. Phys. Chem. C*, vol. 113, p. 19082–19090, 2009.

[32] H. Tantang, J. Y. Ong, C. L. Loh, X. Dong, P. Chen, Y. Chen, X. Hu, Lay P. Tan, L.-J. Li, Using oxidation to increase the electrical conductivity of carbon nanotube electrodes, *Carbon*, vol. 47, pp. 1867–1885, 2009.

[33] Y. Miyata, K. Yanagi, Y. Maniwa, H. Kataura, Highly stabilized conductivity of metallic single wall carbon nanotube thin films, *J. Phys. Chem. C.*, vol. 112, pp. 3591–3596, 2008.

[34] V. Skákalová, A. B. Kaiser, U. Dettlaff-Weglikowska, K. Hrnčariková, S. Roth, Effect of chemical treatment on electrical conductivity, infrared absorption, and Raman spectra of single-walled carbon nanotubes, *J. Phys. Chem. B*, vol. 109, pp. 7174–7181, 2005.

[35] U. Dettlaff-Weglikowska, V. Skákalová, R. Graupner, S. H. Jhang, B. H. Kim, H. J. Lee, L. Ley, Y. W. Park, S. Berber, D. Tománek, S. Roth, Effect of SOCl2 treatment on electrical and mechanical properties of single-wall carbon nanotube networks, *J. Am. Chem. Soc.*, vol. 127, pp. 5125–5131, 2005.

[36] S. Ijima, T. Ichihashi,, Single-shell carbon nanotubes of 1 nm diameter, *Nature*, vol. 363, p. 603, 1993.

[37] M. Monthioux, Filling single-wall carbon nanotubes, *Carbon*, vol. 40, pp. 1809–1823, 2002.

[38] D. S. Bethune, C. H. Kiang, M. S. DeVries, G. Gorman, R. Savoy, J. Vazquez, R. Beyers, Cobalt-catalyzed growth of carbon nanotubes with single-atomic-layer walls, *Nature*, vol. 363, p. 605, 1993.

[39] N. Grobert, M. Mayne, M. Terrones, J. Sloan, R.E. Dunin-Borkowski, R. Kamalakaran, T. Seeger, H. Terrones, M. Rühle, D. R. M. Walton, H.W. Kroto, J.L. Hutchison, Alloy nanowires: Invar inside carbon nanotubes, *Chem. Commun.*, vol. 5, pp. 471–472, 2001.

[40] Y. K. Chen, A. Chu, J. Cook, M. L. H. Green, P. J. F. Harris, R. Heesom, M. Humphries, J. Sloan, S. C. Tsang, J. F. C. Turner, Synthesis of carbon nanotubes containing metal oxides and metals of the d-block and f-block transition metals and related studies, *J. Mater. Chem.*, vol. 7, pp. 545–550, 1997.

[41] P. M. Ajayan, T. W. Ebbesen, Ichihashi, S. Ijima, K. Tanigaki, H. Hiura, Opening carbon nanotubes with oxygen and implications for filling, *Nature*, vol. 362, pp. 522–525, 1993.

[42] N. Grobert, W. K. Hsu, Y. Q. Zhu, J. P. Hare, H. W. Kroto, D. R. M. Walton, M. Terrones H. Terrones, Ph. Redlich, M. Rühle, R. Escudero, F. Morales, Enhanced magnetic coercivities in Fe nanowires, *Appl. Phys. Lett.*, vol. 75, pp. 3363–3365, 1999.

[43] A. Morelos-Gomez, F. Lopez-Urıas, E. Munoz-Sandoval, C. L. Dennis, R. D. Shull, H. Terrones, M. Terrones, Controlling high coercivities of ferromagnetic nanowires encapsulated in carbon nanotubes, *J. Mater. Chem.*, vol. 20, p. 5906–5914, 2010.

[44] E. Hernandez, V. Meunier, B. W. Smith, R. Rurali, H. Terrones, Buongiorno, N. Nardelli, M. Terrones, D. E. Luzzi, J. C. Charlier, Fullerene coalescence in nanopeapods: A path to novel tubular carbon, *Nano Lett.*, vol. 3, p. 1037, 2003.

[45] A. Chuvilin, E. Bichoutskaia, M. C. Gimenez-Lopez, T. W. Chamberlain, G. A. Rance, N. Kuganathan, J. Biskupek, U. Kaiser, A. N. Khlobystov, Self-assembly of a sulphur-terminated graphene nanoribbon within a single-walled carbon nanotube, *Nature Mater.*, vol. 10, p. 687–692, 2011.

[46] X. Zhao, Y. Ando, Y. Liu, M. Jinno, T. Suzuki, Carbon nanowire made of a long linear carbon chain inserted inside a multiwalled carbon nanotube, *Phys. Rev. Lett.*, vol. 90, p. 187401, 2003.

[47] J. Zhang, Y. Feng, H. Ishiwata, Y. Miyata, R. Kitaura, J. E. P. Dahl, R. M. K. Carlson, H. Shinohara, D. Tománek, Synthesis and transformation of linear adamantane assemblies inside carbon nanotubes, *ACS Nano*, vol. 6, pp. 8674–8683, 2012.

[48] C. Zhao, R. Kitaura, H. Hara, S. Irle, H. Shinohara, Growth of linear carbon chains inside thin double-wall carbon nanotubes, *J. Phys. Chem. C*, vol. 115, p. 13166–13170, 2011.

[49] J. Chen, L. Yang, H. Yang, J. Dong, Electronic and transport properties of a carbon-atom chain in the core of semiconducting carbon nanotubes, *Phys. Lett. A*, vol. 316, pp. 101–106, 2003.

[50] E. Gracia-Espino, F. López-Urías, H. Terrones, M. Terrones, Doping (10, 0)-semiconductor nanotubes with nitrogen and vacancy defects, *Mater. Express*, vol. 1, pp. 127–135, 2011.

[51] R. Czerw, M. Terrones, J.-C. Charlier, X. Blase, B. Foley, R. Kamalakaran, N. Grobert, H. Terrones, D. Tekleab, P. M. Ajayan, W. Blau, M. Ru2hle, D. L. Carroll, Identification of electron donor states in N-doped carbon nanotubes, *Nano Lett.*, vol. 1, pp. 457–460, 2001.

[52] M. Terrones, P. M. Ajayan, F. Banhart, X. Blase, D. L. Carroll, J. C. Charlier, R. Czerw, B. Foley, N. Grobert, R. Kamalakran, P. Kohler-Redlich, M. Ruhle, T. Seeger, H. Terrones, N-doping and coalescence of carbon nanotubes: synthesis and electronic properties, *Appl. Phys. A*, vol. 74, pp. 355–361, 2002.

[53] X. Li, H. Wang, J. T. Robinson, H. Sanchez, G. Diankov, H. Dai, Simultaneous Nitrogen Doping and Reduction of Graphene Oxide, *J. Am. Chem. Soc.*, vol. 131, p. 15939–15944, 2009.

[54] L. S. Panchokarla, K. S. Subrahmanyam, S. K. Saha, A. Govindaraj, H. R. Krishnamurthy, U. V. Waghmare, C. N. R. Rao, Synthesis, structure and properties of boron- and nitrogen-doped graphene, *Adv. Mater.*, vol. 21, pp. 4726–4730, 2009.

[55] N.E. Derradjia, M.L. Mahdjoubia, H. Belkhira, N. Mumumbilab, B. Angleraudb, P.Y. Tessier, Nitrogen effect on the electrical properties of CNx thin films deposited by reactive magnetron sputtering, *Thin Sol. Films*, vol. 482, pp. 258–263, 2005.

[56] W. Han, Y. Bando, K. Kurashima, T. Sato, Boron-doped carbon nanotubes prepared through a substitution reaction, *Chem. Phys. Lett.*, vol. 299, pp. 368–373, 1999.

[57] K. C. Mondal, A. M. Strydomc, R. M. Erasmus, J. M. Keartland, N. J. Coville, Physical properties of CVD boron-doped multiwalled carbon nanotubes, *Mater. Chem. Phys.*, vol. 111, pp. 386–390, 2008.

[58] Y. Miyamoto, A. Rubio, S.G. Louie, M.L. Cohen, Electronic properties of tubule forms of hexagonal BC3, *Phys. Rev. B*, vol. 50, p. 18360, 1994.

[59] W.K. Hsu, S. Firth, P. Redlich, M. Terrones, H. Terrones, Y.Q. Zhu, N. Grobert, A. Schilder, R.J.H. Clark, H.W. Kroto, D.R.M. Walton, Boron-doping effects in carbon nanotubes, *J. Mater. Chem*, vol. 10, p. 1425–1429, 2000.

[60] B. Wei, R. Spolenak, P. Kohler-Redlich, M. Rühle, E. Arzt, Electrical transport in pure and boron-doped carbon nanotubes, *Appl. Phys. Lett.*, vol. 74, p. 3149–3151, 1999.

[61] N. Murata, J. Haruyama, J. Reppert, A. M. Rao, T. Koretsune, S. Saito, M. Matsudaira, Y. Yagi, Superconductivity in thin films of boron-doped carbon nanotubes, *Phys. Rev. Lett.*, vol. 101, p. 027002, 2008.

[62] Y. A. Kim, K. Fujisawa, H. Muramatsu, T. Hayashi, M. Endo, T. Fujimori, K. Kaneko, M. Terrones, J. Behrends, A. Eckmann, C. Casiraghi, K. S. Novoselov, R. Saito, M. S. Dresselhaus, Raman spectroscopy of boron-doped single-layer graphene, *ACS Nano*, vol. 6, p. 6293–6300, 2012.

[63] J.M. Romo-Herrera, B.G. Sumpter, D.A. Cullen, H. Terrones, E. Cruz-Silva, D.J. Smith, V. Meunier, M. Terrones, An atomistic branching mechanism for carbon nanotubes: Sulfure as the triggering agent, *Angew. Chem. Int. Ed.*, vol. 47, pp. 2948–2953, 2008.

[64] M. E. H. Maia da Costa, F. H. Monteiro, A. L. Pinto, F. L. Freire, D. G. Larrude, Characterization of phosphorus-doped multiwalled carbon nanotubes, *J. Appl. Phys.*, vol. 111, p. 064315, 2012.

[65] J. Campos-Delgado, I. O. Maciel, D. A. Cullen, D. J. Smith, A. Jorio, M. A. Pimenta, H. Terrones, M. Terrones, Chemical vapor deposition synthesis of N-, P-, and Si-doped single-walled carbon nanotubes, *ACS Nano*, vol. 4, pp. 1696–1702, 2010.

[66] S. Some, J. Kim, K. Lee, A. Kulkarni, Y. Yoon, S. M. Lee, T. Kim, H. Lee, Highly air-stable phosphorus-doped n-type graphene field-effect transistors, *Adv. Mater.*, vol. 24, pp. 5481–5486, 2012.

[67] G. Larkins, Y. Vlasov, Indications of superconductivity in doped highly oriented pyrolytic graphite, *Supercond. Sci. Technol.*, vol. 24, p. 092001, 2011.

[68] J. Liu, H. Liu, Y. Zhang, Ruying Li, G. Liang, M. Gauthier, X. Sun, Synthesis and characterization of phosphorus–nitrogen doped multiwalled carbon nanotubes, *Carbon*, vol. 49, pp. 5014–5021, 2011.

[69] O. Stephan, P. M. Ajayan, C. Colliex, Ph. Redlich, J. M. Lambert, P. Bernier, P. Lefin, Doping graphitic and carbon nanotube structures with boron and nitrogen, *Sc.*, vol. 266, pp. 1683–1685, 1994.

[70] E. Cruz-Silva, D. A. Cullen, L. Gu, J. M. Romo-Herrera, E. Muñoz-Sandoval, F. López-Urías, B. G. Sumpter, V. Meunier, J.-C. Charlier,D. J. Smith, H. Terrones, M. Terrones, Heterodoped nanotubes: theory, synthesis, and characterization of phosphorus nitrogen doped multiwalled carbon nanotubes, *ACS Nano*, vol. 2, pp. 441–448, 2008.

[71] J. Liang, Y. Jiao, M. Jaroniec, S. Z. Qiao, Sulfur and nitrogen dual-doped mesoporous graphene electrocatalyst for oxygen reduction with synergistically enhanced performance, *Angew. Chem. Int. Ed.*, vol. 51, pp. 1–6, 2012.

[72] Y. Zhang, J.-P. Hu, B. A. Bernevig, X. R. Wang, X. C. Xie, W. M. Liu, Quantum blockade and loop currents in graphene with topological defects, *Phys. Rev. B*, vol. 78, p. 155413, 2008.

[73] S. Iijima, T. Ichihashi, Y. Ando, Pentagons, heptagons, and negative curvature in graphitic microtubule growth, *Nature*, vol. 356, pp. 776–778, 1992.

[74] S. G. Louie, A. Zettl, C. O. Girit, J. C. Meyer, R. Erni, M. D. Rossell, C. Kisielowski, L. Yang, C. Park, M.F. Crommie, M.L. Cohen, Graphene at the edge: stability and dynamics, *Science*, vol. 323, pp. 1705–1708, 2009.

[75] J. Kotakoski, A. V. Krasheninnikov, U. Kaiser, J. C. Meyer, From point defects in graphene to two-dimensional amorphous carbon, *Phys. Rev. Lett.*, vol. 106, p. 105505, 2011.

[76] J. C. Meyer, C. Kisielowski, R. Erni, M. D. Rossell, M. F. Crommie, A. Zettl, Direct imaging of lattice atoms and topological defects in graphene membranes, *Nano Lett.*, vol. 8, pp. 3582–3586, 2008.

[77] A. Hashimoto, K. Suenaga, A. Gloter, K. Urita, S. Iijima, Direct evidence for atomic defects in graphene layers, *Nature*, vol. 430, pp. 870–873, 2004.

[78] N. Park, M. Yoon, S. Berber, J. Ihm, E. Osawa, D. Tománek, Magnetism in all-carbon nanostructures with negative gaussian curvature, *Phys. Rev. Lett.*, vol. 91, p. 237204, 2003.

[79] P. A. Thrower, Study of defects in graphite by transmission electron microscopy, *Chem. Phys. Carbon*, vol. 5, pp. 217–319, 1969.

[80] A. Stone, D. Wales, Theoretical studies of icosahedral C60 and some related species, *Chem. Phys. Lett.*, vol. 128, pp. 501–503, 1986.

[81] M. Terrones, G. Terrones, H. Terrones, Structure, chirality, and formation of giant icosahedral fullerenes and spherical graphitic onions, *Struct. Chem.*, vol. 13, pp. 373–384, 2002.

[82] M. Terrones, H. Terrones, The role of defects in graphitic structures, *Full Sci & Tech*, vol. 4, pp. 517–533, 1996.

[83] S. G. Louie, A. Zettl, C. O. Girit, J. C. Meyer, R. Erni, M. D. Rossell, C. Kisielowski, L. Yang, C. Park, M. F. Crommie, M. L. Cohen, Graphene at the edge: stability and dynamics, *Science*, vol. 323, pp. 1705–1708, 2009.

[84] J. Lahiri, Y. Lin, P. Bozkurt, I. I. Oleynik, M. Batzill, An extended defect in graphene as a metallic wire, *Nature Nanotechnol.*, vol. 5, p. 326, 2010.

[85] H. Terrones, R. Lv, M. Terrones, M. S Dresselhaus, The role of defects and doping in 2D graphene sheets and 1D nanoribbons, *Rep. Prog. Phys.*, vol. 75, p. 062501, 2012.

[86] Z. Liu, K. Suenaga, P. J. F. Harris, S. Iijima, Open and closed edges of graphene layers, *Phys. Rev. Lett.*, vol. 102, p. 015501, 2009.

[87] M. Endo, B. J. Lee, Y. A. Kim, Y. J. Kim, H. Muramatsu, T. Yanagisawa, T. Hayashi, M. Terrones, M.S. Dresselhaus, Transitional behaviour in the transformation from active end planes to stable loops caused by annealing, *New J. Phys.*, vol. 5, p. 1211, 2003.

[88] J. Campos-Delgado, Y.A. Kim, T. Hayashi, A. Morelos-Gómez, M. Hofmann, H. Muramatsu, M. Endo, H. Terrones, R.D. Shull, M.S. Dresselhaus, M. Terrones, Thermal stability studies of CVD-grown graphene nanoribbons: Defect annealing and loop formation, *Chemical Physics Letters*, vol. 469, pp. 177–182, 2009.

[89] J. Y. Huang, F. Ding, B. I. Yakobson, P. Lud, L. Qi, J. Li, In situ observation of graphene sublimation and multi-layer edge reconstructions, *P. Natl. Acad. Sci. USA*, vol. 106, pp. 10103–10108, 2009.

[90] K. Kim, Z. Lee, B. D. Malone, K. T. Chan, B. Aleman, W. Regan, W. Gannett, M. F. Crommie, M. L. Cohen, A. Zettl, Multiply folded graphene, *Phys. Rev. B*, vol. 83, p. 245433, 2011.

[91] A. Chuvilin, J. C Meyer, G. Algara-Siller, U. Kaiser, From graphene constrictions to single carbon chains, *New J Phys.*, vol. 11, p. 083019, 2009.

[92] M. G. Zeng, L. Shen, Y. Q. Cai, Z. D. Sha, Y. P. Feng, Perfect spin-filter and spin-valve in carbon atomic chains, *Appl. Phys. Lett.*, vol. 96, p. 042104, 2010.

[93] R. Sharma, J. H. Baik, C. J. Perera, M. S. Strano, Anomalously large reactivity of single graphene layers and edges toward electron transfer chemistries, *Nano Lett.*, vol. 10, pp. 398–405, 2010.

[94] S. M. Dubois, A. Lopez-Bezanilla, A. Cresti, F. Triozon, B. Biel, J.-C. Charlier, S. Roche, Quantum transport in graphene nanoribbons: Effects of edge reconstruction and chemical reactivity, *ACS Nano*, vol. 4, pp. 1971–1976, 2010.

[95] F. Schäffel, M. Wilson, J. H. Warner, Motion of light adatoms and molecules on the surface of few-layer graphene, *ACS Nano*, vol. 5, pp. 9428–9441, 2011.

[96] I. Gerber, M. Oubenali, R. Bacsa, J. Durand, A. Gonçalves, M. F. R. Pereira, F. Jolibois, L. Perrin, R. Poteau, P. Serp, Theoretical and experimental studies on the carbon-nanotube surface oxidation by nitric acid: Interplay between functionalization and vacancy enlargement, *Chem. Eur. J.*, vol. 17, pp. 11467–11477, 2011.

[97] P. O. Lehtinen, A. S. Foster, Yuchen Ma, A. V. Krasheninnikov, R. M. Nieminen, Irradiation-induced magnetism in graphite: a density functional study, *Phys. Rev. Lett.*, vol. 93, p. 187202, 2004.
[98] X. Jia, J. Campos-Delgado, M. Terrones, V. Meunier, M. S. Dresselhaus, Graphene edges: a review of their fabrication and characterization, *Nanoscale*, vol. 3, p. 86, 2011.
[99] T. Hertel, R. E. Walkup, P. Avouris, Deformation of Carbon Nanotubes by surface van der Waals Forces, *Physical Review B*, vol. 58, pp. 13870–13873, 1998.
[100] Y. J. Dappe, M.A. Basanta, F. Flores, J. Ortega, Weak chemical interaction and van der Waals forces between graphene layers: A combined density functional and intermolecular perturbation theory approach, vol. 74, p. 205434–9, 2006.
[101] E. T. Thostenson, Z.F. Ren, T.W. Chou, Advances in the science and technology of carbon nanotubes and their composites: a review, *Composite Science and Technology*, vol. 61, pp. 1899–1912, 2001.
[102] K.A. Wepasnick, B.A. Smith, J.L. Bitter, D.H. Fairbrother, Chemical and structural characterization of carbon nanotube surfaces, *Analytical and Bioanalytical Chemistry*, vol. 396, pp. 1003–1014, 2010.
[103] T. Kuila, S. Bose, A.K. Mishra, P. Khanra, N.H. Kim, J.H. Lee, Chemical functionalization of graphene and its apllications, *Progress in Materials Science*, vol. 57, pp. 1061–1105, 2012.
[104] M. J. Park, J. K. Lee, B. S. Lee, Y. W. Lee, I. S. Choi, S. Lee, Covalent modification of multiwalled carbon nanotubes with imidazolium based ionic liquids: effect of anions on solubility, *Chemistry of Materials*, vol. 18, pp. 1546–1551, 2006.
[105] F. Diederich, C. Thilgen, Covalent fullerene chemistry, *Science*, vol. 217, pp. 317–323, 1996.
[106] F. Diederich, L. Isaacs, D. Philp, Syntheses, structures, and properties of methanofullerenes, *Chemical Society Reviews*, vol. 23, pp. 243–255, 1994.
[107] S. H. Friedman, D. L. DeCamp, R. P. Sijbesma, G. Srdanov, F. Wudl, G.L.Kenyon, Inhibition of the HIV-1 protease by fullerene derivatives: model building studies and experimental verification, *Journal American Chemical Society*, vol. 115, pp. 6506–6509, 1993.
[108] R. F.Schinazi, R. Sijbesma, G. Srdanov, C. L. Hill, F. Wudl, Synthesis and virucidal activity of a water-soluble, configurationally stable, derivatized C60 fullerene, *Antimicrobial Agents and Chemotherapy*, vol. 37, pp. 1707–1710, 1993.
[109] C. Toniolo, A. Bianco, M. Maggini, G. Scorrano, M. Prato, M. Marastoni, R. Tomatis, S. Spisani, G. Palu, E.D. Blair, A bioactive fullerene peptide, *Journal of Medicinal Chemistry*, vol. 37, pp. 4558–4562, 1994.
[110] A. S. Boutorine, M. Takasugi, C. Hélène, H. Tokuyama, H. Isobe, E. Nakamura, Fullerene-oligonucleotide conjugates-photoinduced sequence-specific DNA cleavage, *Angewandte Chemie International Edition in English*, vol. 33, pp. 2462–2465, 1994.
[111] G. A. Olah, I. Bucsi, C. Lambert, R. Aniszfeld, N. J. Trivedi, D. K. Sensharma, G. K. S. Prakash, Chlorination and bromination of fullerenes. Nucleophilic methoxylation of polychlorofullerenes and their aluminum trichloride catalyzed friedel-crafts reaction with aromatics to polyarylfullerenes, *Journal of American Chemical Society*, vol. 113, pp. 9385–9387, 1991.
[112] T. Cao, S.E. Webber, Free-radical copolymerization of fullerenes with styrene, *Macromolecules*, vol. 28, pp. 3741–3743, 1995.
[113] K. E. Geckeler, A. Hirsch, Polymer-bound C60, *Journal of the American Chemical Society*, vol. 115, pp. 3850–3851, 1993.
[114] C. Weis, C. Friedrich, R. Mmaupt, H. Frey, Fullerene-end-capped polystyrenes. Monosubstituted polymeric C60 derivatives, *Macromolecules*, vol. 28, pp. 403–405, 1995.
[115] N. Zhang, S.R. Schricker, F. Wudl, M. Prato, M. Manggini, G. Scorrano, A new C60 polymer via ring-opening metathesis polymerization, *Chemistry of Materials*, vol. 7, pp. 441–442, 1995.

[116] S. Shi, Q. Li, K. C. Khemani, F. Wudl, A polyester and polyurethane of diphenyl C61: Retention of fulleroid properties in a polymer, *Journal of the American Chemical Society*, vol. 114, pp. 10656–10657, 1992.
[117] C. J. Hawker, K. L. Wooley, J. M. J. Fréchet, Dendritic fullerenes; a new approach to polymer modification of C60, *Journal of the Chemical Society, Chemical Communications*, pp. 925–926, 1994.
[118] K. L. Wooley, C,J. Hawker, J. M. J. Fréchet, F. Wudl, G. Srdanov, S. Shi, C. Li M. Kao, Fullerene-bound dendrimers: Soluble, isolated carbon clusters, *Journal of the American Chemical Society*, vol. 115, pp. 9836–9837, 1993.
[119] S. Shi, K.C. Khemanikc, Q. C. Li, F. Wudl, A polyester and polyurethane of diphenyl-C61: retention of fulleroid properties in a polymer, *Journal of the American Chemical Society*, vol. 114, p. 10656–10657, 1992.
[120] L. Dai, A. W. H. Mau, H. J. Griesser, T. Spurling, J. W. White, Grafting of buckminsterfullerene onto polydiene: A new route to fullerene-containing polymers, *Journal of Physical Chemistry*, vol. 99, pp. 17302–17304, 1995.
[121] L. Dai, A. W. H. Mau, X. Zhang, Synthesis of fullerene- and fullerol-containing polymers, *Journal of Material Chemistry*, vol. 8, pp. 325–330, 1998.
[122] F.Y. Li, Y. L. Li, Z. X. Guo, D. B. Zhu, Y. L. Song, G. Y. Fang, Synthesis and optical limiting properties of polycarbonates containing fullerene derivative, *Journal of Physical Chemistry of Solids*, vol. 61, pp. 1101–1103, 2000.
[123] B. Z. Tang, S. M. Leung, H. Peng, N. T. Yu, K. C. Su, Direct fullerenation of polycarbonate via simple polymer reactions, *Macromolecules*, vol. 30, pp. 2848–2852, 1997.
[124] P. Zhou, G. Q. Chen, C. Z. Li, F. S. Du, Z. C. Li, F. M. Li, Synthesis of hammerlike macromolecules of C60 with well-defined polystyrene chains via atom transfer radical polymerization (ATRP) using a C60-monoadduct initiator, *Chemical Communications*, pp. 797–798, 2000.
[125] X.D. Huang, S.H. Goh, S.Y. Lee, Miscibility of C60-end-capped poly(ethylene oxide) with poly(p-vinylphenol), *Macromolecular Chemistry and Physics*, vol. 201, p. 2660–2665, 2000.
[126] X.D. Huang, S.H. Goh, Interpolymer complexes through hydrophobic interactions: C60-end-capped poly(ethylene oxide)/poly(methacrylic acid) complexes, *Macromolecules*, pp. 8894–8897, 2000.
[127] B. Z. Tang, H.Y. Xu, J. W. Y. Lam, P. P. S. Lee, K. T. Xu, Q. H. Sun, K. K. L. Cheuk, C60-containing poly(1-phenyl-1-alkynes): synthesis, light emission, and optical limiting, *Chemistry of Materials*, vol. 12, pp. 1446–1455, 2000.
[128] P. Zhou, G. Q. Chen, H. Hong, F.S. Du, Z.C. Li, F. M. Li, Synthesis of C60-endbonded polymers with designed molecular weights and narrow molecular weight distributions via atom transfer radical polymerization, *Macromolecules*, vol. 33, pp. 1948–1954, 2000.
[129] C. Martineau, P. Blanchard, D. Rondeau, J. Delaunay, J. Roncali, Synthesis and electronic properties of adducts of oligothienylenevinylenes and fullerene C60, *Advanced Materials*, vol. 14, pp. 283–287, 2002.
[130] T. Gu, D. Tsamouras, C. Melzer, V. Krasnikov, P. Gisselbrecht, M. Gross, G. Hadziioannou, J.-F. Nierengarten, Photovoltaic devices from fullerene oligophenylene ethynylene conjugates, *Chem Phys Chem*, vol. 3, pp. 124–127, 2002.
[131] C.C. Wang, J.P. He, S.K. Fu, K.J. Jiang, H.Z. Cheng, M. Wang, Synthesis and characterization of the narrow polydispersity fullerene-endcapped polystyrene, *Polymer Bulletin*, vol. 37, pp. 305–311, 1996.
[132] N. Roy, R. Senguptaa, A. K. Bhowmicka, Modifications of carbon for polymer composites and nanocomposites, *Progress in Polymer Science*, vol. 37, pp. 781–819, 2012.

[133] B. M. Ginzburg, L. A. Shibaev, V. L. Ugolkov, Effect of fullerene C60 on thermal oxidative degradation of polymethyl methacrylate prepared by radical polymerization, *Russian Journal of Applied Chemistry*, vol. 74, pp. 1329–1337, 2001.

[134] Y. N. Sazanov, M. V. Mokeev, A. V. Novoselova, V. L. Ugolkov, G. N. Fedorova, A.V. Gribanov, V. N. Zgonnik, Thermochemical reactions of polyacrylonitrile with fullerene C60, *Russian Journal of Applied Chemistry*, vol. 76, pp. 452–456, 2003.

[135] A. O. Pozdnyakov, B. M. Ginzburg, O. F. Pozdnyakov, B. P. Redkov, Desorption of fullerene C60 from a mixture with a copolymer of trifluorochloroethylene and vinylidene fluoride, *Technical Physics Letters*, vol. 2, pp. 51–53, 1997.

[136] A.O. Pozdnyakov, B.L. Baskin, O.F. Pozdnyakov, Fullerene C60 diffusion in thin layers of amorphous polymers: polystyrene and poly(amethylstyrene), *Techical Physics Letters*, vol. 30, pp. 839–842, 2004.

[137] Z. Y. Wang, L. Kuang, X. S. Meng, J. P. Gao, New route to incorporation of [60] fullerene into polymers via the benzocyclobutenone group, *Macromolecules*, vol. 31, p. 5556–5558, 1998.

[138] N.V. Kamanina, Mechanisms of optical limiting in p-conjugated organic system: fullerene doped polyimide, *Synthetic Metals*, vol. 127, pp. 121–128, 2002.

[139] G. N. Gubanova, T. K. Meleshko, V. E. Yudin, Y. A. Fadin, Y. P. Kozirev, V. I. Gofman, A. A. Mikhailov, N. N. Bogorad, A. G. Kalbin, Y. N. Panov, G. N. Fedorova, V. V. Kudryavtsev, Fullerene modified polyimide derived from 3,30,4,40-benzophenonetetracarboxylic acid and 3,30-diaminobenzophenone for casted items and its use in tribology, *Russian Journal Applied Chemistry*, vol. 76, pp. 1156–1163, 2003.

[140] L. Dai, A. W. H. Mau, Controlled synthesis and modification of carbon nanotubes and C60: Carbon nanostructures for advanced polymeric composite materials, *Advanced Materials*, vol. 13, pp. 899–913, 2001.

[141] G. Yu, J. Gao, J. C. Hummelen, F. Wudi, J. A. Heeger, Polymer photovoltaic cells: Enhanced efficiencies via a network of internal donor-acceptor heterojunctions, *Science*, vol. 270, pp. 1789–1791, 1995.

[142] J. Chen, M. A. Hamon, H. Hu, Y. Chen, A. M. Rao, P. C. Eklund, Solution properties of single wall carbon nanotubes, *Science*, vol. 282, pp. 95–98, 1998.

[143] M. A. Hamon, J. Chen, H. Hu, Y. Chen, M. E. Itkis, A. M. Rao, P. C. Eklund, R. C. Haddon, Dissolution of single walled carbon nanotubes, *Advanced Materials*, vol. 11, pp. 834–840, 1999.

[144] J. Liu, A. G. Rinzler, H. Dai, J. H. Hafner, R. K. Bradley, P. J. Boul, A. Lu, T. Iverson, K. Shelimov, C. B. Huffman, F. Rodriguez- Macias, Y.-S. Shon, T. R. Lee, D. T. Colbert, R. E. Smalley, Fullerene pipes, *Science*, vol. 280, pp. 1253–1256, 1998.

[145] J. Sloan, J. Hammer, M. Zwiefka-Sibley, M. L. H. Green, The opening and filling of single walled carbon nanotubes (SWTs), *Chemical Communications*, pp. 347–348, 1998.

[146] Y. Xing, L. Li, C. C. Chusuei, R. V Hull, Sonochemical oxidation of multiwalled carbon nanotubes, *Langmuir*, vol. 21, pp. 4185–4190, 2005.

[147] K.J. Ziegler, Z. Gu, H. Peng, E.L. Flor, R.H. Hauge, R.E. Smalley, Controlled oxidative cutting of single-walled carbon nanotubes, *Journal American Chemestry Society*, vol. 127, pp. 1541–1547, 2005.

[148] T. Nakajima, S. Kasamatsu, Y. Matsuo, Synthesis and characterization of fluorinated carbon nanotubes, *European Journal of Solid State Inorganic Chemestry*, vol. 33, pp. 831–840, 1996.

[149] E. T. Mickelson, C. B. Huffman, A. G. Rinzler, R. E Smalley, R. H. Hauge, Fluorination of single-wall carbon nanotubes, *Chemical Physics Letters*, vol. 296, pp. 188–194, 1998.

[150] E. T. Mickelson, I. W. Chiang, J. L. Zimmerman, P. J. Boul, J. Lozano, J. Liu, R.E. Smalley, R. H. Hauge, J. L. Margrave, Solvatation of fluorinated single -wall carbon nanotubes in alcohol solvents, *Journal of Physical Chemestry B.*, vol. 103, pp. 4318–4322, 1999.

[151] Y. Wang, Z. Iqbal, S. Mitra, Rapidly functionalized, water-dispersed carbon nanotubes at high concentration, *Journal of American Chemical Society*, vol. 128, pp. 95–99, 2006.

[152] Y. Chen, S. Mitra, Fast microwave-assisted purification functionalization and dispersion of multi-walled carbon nanotubes, *Journal of nanosciences and Nanotechnology*, vol. 8, pp. 5770–5775, 2008.

[153] N. Karousis, N. Tagmatarchis, Current progress on the chemical modification of carbon nanotubes, *Chemical Reviews*, vol. 110, pp. 5366–5397, 2010.

[154] P. He, M.W. Urban, Controlled phospholipid functionalization of single-walled carbon nanotubes, *Biomacromolecules*, vol. 6, pp. 2455–2457, 2005.

[155] Z. Wu, W. Feng, Y. Feng, Q. Liu, X. Xu, T. Senkino, A. Fuji, M. Ozaki, Preparation and characterization of chitosan-grafted multiwalled carbon nanotubes and their electrochemical properties, *Carbon*, vol. 45, pp. 1212–1218, 2007.

[156] X. Deng, G. Jia, H. Wang, H. Sun, X. Wang, S. Yang, T. Wang, Y. Liu, Translocation and fate of multi-walled carbon nanotubes in vivo, *Carbon*, vol. 45, pp. 1419–1424, 2007.

[157] E. B. Malarkey, R. C. Reyes, B. Zhao, R. C. Haddon, V. Parpura, Water soluble single-walled carbon nanotubes inhibit stimulated endocytosis in neurons, *Nano Letters*, vol. 8, pp. 3538–3542, 2008.

[158] M. Fagnoni, A. Profumo, D. Merli, D. Dondi, P. Mustarelli, E. Quartarone, Water-miscible liquid multiwalled carbon nanotubes, *Advanced Materials*, vol. 21, pp. 1761–1765, 2009.

[159] C. H. Andersson, H. Grennberg, Reproducibility and efficiency of carbon nanotube end-group generation and functionalization, *European Journal of Organic Chemistry*, pp. 4421–4428, 2009.

[160] M. Fang, K. G. Wang, H. B. Lu, Y. L. Yang, S. Nutt, Covalent polymer functionalization of graphene nanosheets and mechanical properties of composites, *Journal of Materials Chemestry*, vol. 19, pp. 7098–7105, 2009.

[161] H. F. Yang, C. S. Shan, F.H. Li, D.X. Han, Q. X. Zhang, L. Niu, Covalent functionalization of polydisperse chemically-converted graphene sheets with amine-terminated ionic liquid, *Chemical Communications*, vol. 26, pp. 3880–3882, 2009.

[162] H. F. Yang, F. H. Li, C. S. Shan, D. X Hang, Q.X. Zhang, L. Niu, A. Ivaska, Covalent functionalization of chemically converted graphene sheets via silane and its reinforcements, *Journal of Materials Chemestry*, vol. 19, pp. 4632–4638, 2009.

[163] J. M. Englert, C. Dotzer, G. A. Yang, M. Schmid, C. Papp, J. M. Gottfried, H. P. Steinruck, E. Spiecker, F. Hauke, A. Hirsch, Covalent bullk functionalization of graphene, *Nature Chemestry*, vol. 3, pp. 279–286, 2011.

[164] X. M. Sun, Z. Liu, K. Welshe, J. T. Robinson, A. Goodwin, S. Zaric, H. J. Dai, Nano-graphene oxide for cellular imagin and drug delivery, *Nano Research*, vol. 1, pp. 203–212, 2008.

[165] H. Q. Bao, Y. Z. Pan, Y. Ping, N. G. Sahoo, T. F. Wu, L. Li, J. Li, L. H. Gan, Chitosan-functionalized graphene oxide as a nanocarrier for drug and gene delivery, *Small*, vol. 7, pp. 1569–1578, 2011.

[166] D. A. Britz, A. N. Khlobystov, Nanocovalent interactions of molecules with single walled carbon nanotubes, *Chemical Society Reviews*, vol. 35, pp. 637–659, 2006.

[167] Z. Liu, S.M. Tabakman, Z. Chen, H. J. Dai, Preparation of carbon nanotube bioconjugates for biomedical applications, *Nature Protocols*, vol. 4, pp. 1372–1382, 2009.

[168] S. E. Campidelli, C. Klumpp, A. Bianco, D.M. Guldi, M. Prato, Functionalization of CNT: Synthesis and applications in photovoltaics and biology, *Journal of Physical Organic Chemestry*, vol. 19, pp. 531–539, 2006.

[169] H. Bai, Y. X. Xu, L. Zhao, C. Li, G. Q. Shi, Non-covalent functionalization of graphene sheets by sulfonated polyaniline, *Chemical Communications*, vol. 13, pp. 1667–1669, 2009.

[170] K. Jo, T. Lee, H. J. Choi, J. H. Park, D. J. Lee, D. W. Lee, B. S. Kim, Stable aqueous dispersion of reduced graphene nanosheets via non-covalent functionalization with conducting polymers and applications in transparents electrodes, *Langmuir*, vol. 27, pp. 2014–2018, 2011.
[171] A. Ghosh, K.V. Rao, R. Voggu, S. J. George, Non-covalent functionalization, solubilization of graphene and single-walled carbon nanotubes with aromatic donor and acceptor molecules, *Chemical Physics Letters*, vol. 488, pp. 198–201, 2010.
[172] X. Qi, K. Y. Pu, H. Li, X. Zhou, S. Wu, Q. L. Fan, B. Liu, F. Boey, W. Huang, H. Zhang, Amphiphilic graphene composites, *Angewandte Chemie International Edition*, vol. 49, pp. 9426–9429, 2010.
[173] D. Tasis, N. Tagmatarchis, A. Bianco, M. Prato, Chemistry of carbon nanotubes, *Chemical Reviews*, vol. 106, pp. 1105–1136, 2006.
[174] Y. Lin, S. Taylor, H.P. Li, K.A.S. Fernando, L.W. Qu, W. Wang, L.R. Gu, B. Zhou, Y.P. Sun, Advances toward bioapplications of carbon nanotubes, *Journal of Materials Chemistry*, vol. 14, pp. 527–541, 2004.
[175] J. Wang, Carbon-nanotubes based electrochemical biosensor: A review, *Electroanalysis*, vol. 17, pp. 7–14, 2005.
[176] A. Bianco, K. J. Kostarelos, M. Prato, Applications of carbon nanotubes in drug delivery, *Current Opinion in Chemical Biology*, vol. 9, pp. 674–679, 2005.
[177] M. Alexandre, P. Dubois, Polymer-layered silicate nanocomposites: Preparation, properties and uses of a new class of materials, *Materials Science and Engineering*, vol. 28, pp. 1–63, 2000.
[178] E. P. Giannelis, R. Krisnamoorti, E. Manias, Polymer-silicate nanocomposites: Model systems for confined polymers and polymer brushes, *Advances in Polymer Science*, vol. 138, pp. 107–147, 1999.
[179] A. Okada, A. Usuki, The chemistry of polymer-clay hybrids, *Materials Science and Engineering:*, vol. C3, pp. 109–115, 1995.
[180] A. Usuki, Y. Kojimaa, M. Kawasumia, A. Okadaa, Y. Fukushimaa, T. Kurauchia, O. Kamigaitoa, Synthesis of nylon 6-clay hybrid, *Journal of Materials Research*, vol. 8, pp. 1179–1184, 1993.
[181] A. Weiss, Organic derivatives of mica-types layer silicates, *Angew. Chem. Internat. Edit*, vol. 2, pp. 134–143, 1963.
[182] J.W. Jordan, Organophilic bentonites, *J. Phys. Colloid Chem*, vol. 53, pp. 294–306.
[183] T. Lan, T.J. Pinnavaia, Mechanism of clay tactoid exfoliation in epoxy–clay nanocomposites, *Chem. Mater.*, vol. 7, pp. 2144–2150, 1995.
[184] P. C. LeBaron, Z. Wang, T. J. Pinnavaia, Polymer-layered silicate nanocomposites: An overview, *Applied Clay Science*, vol. 15, pp. 11–29, 1999.
[185] M. G. Kanatzidis, M. Humbbard, L. M. Tonge, T. J. Marks, H. O. Marcy, C. R. Kannewurf, In situ intercalative polymerization as a route to layered conducting polymer-inorganic matrix microlaminates. polypyrrole and polythiophene in FeOCl, *Synthetic Metals*, vol. 28, pp. 89–95, 1989.
[186] R. A. Vaia, K. D. Jandt, E. J. Kramer, E. P. Giannelis, Kinetics of polymer melt intercalation, *Macromolecules*, vol. 28, pp. 8080–8085, 1995.
[187] R. A. Vaia, E. P. Giannelis, Polymer melt intercalation in organically-modified layered silicates: Model predictions and experiment, *Macromolecules*, vol. 30, pp. 8000–8009, 1997.
[188] Y. Ou, F. Yang, J. Chen, Interfacial interaction and mechanical properties of nylon 6–potassium titanate composites prepared by In-situ polymerization, *Journal of Applied Polymer Science*, vol. 64, pp. 2317–2322, 1998.
[189] Y. Kojima, A. Usuki, M. Kawasumi, A. Okada, T. Kurauchi, O. Kamigaito, Synthesis of nylon 6–clay hybrid by montmorillonite intercalated with ϵ-caprolactam, *Journal of Polymer Science Part A: Polymer Chemistry*, vol. 31, pp. 983–986, 1993.

[190] X. Kornmann, H. Lindberg, L. A. Berglund, Synthesis of epoxy-clay nanocomposites. Influence of the nature of the curing agent on structure, *Polymer*, vol. 42, pp. 4493–4499, 2001.

[191] J. Zhang, L. Wang, A. Wang, Preparation and properties of chitosan-g-poly(acrylic acid)/montmorillonite superabsorbent nanocomposite via In situ intercalative polymerization, *Ind. Eng. Chem Res.*, vol. 46, pp. 2497–2505, 2007.

[192] M. A. Paul, M.Alexandre, P. Degee, C. Calberg, R. Jerome, P. Dubois, Exfoliated polylactide/clay nanocomposites by In-situ coordination-insertion polimerization, *Macromol. Rapid. Commun.*, vol. 24, pp. 561–566, 2003.

[193] C. Velasco-Santos, A. L. Martinez-Hernandez, F. T. Fisher, R. Ruoff, V. M. Castano, Improvement of thermal and mechanical properties of carbon nanotube composites through chemical functionalization, *Chem. Mater.*, vol. 15, pp. 4470–4475, 2003.

[194] S. Kumar, T. D. Dang, F. E. Arnold, A. R. Bhattacharyya, B. G. Min, X.i Zhang, R. A. Vaia, C. Park, W. W. Adams,R. H. Hauge, R. E. Smalley, S. Ramesh, P. A. Willis, Synthesis, structure, and properties of PBO/SWNT composites, *Macromolecules*, vol. 35, pp. 9039–9043, 2002.

[195] Y. F. Huang, C. W. Lin, Facile synthesis and morphology control of graphene oxide/polyaniline nanocomposites via in-situ polymerization process, *Polymer*, vol. 53, pp. 2574–2582, 2012.

[196] T. E. Chang, A. Kisliuk, S. M. Rhodes, W. J. Brittain, A.P. Sokolov, Conductivity and mechanical properties of well-dispersed single-wall carbon nanotube/polystyrene composite, *Polymer*, vol. 47, pp. 7740–7746, 2006.

[197] J.Y. Kwon, H.D. Kim, Preparation and properties of acid-treated multiwalled carbon nanotubes/waterborne polyurethane nanocomposites, *Journal of Applied Polymer Science*, vol. 96, pp. 595–604, 2005.

[198] B. K. Zhu, S. H. Xie, Z. K. Xu, Y. Y. Xu, Preparation and properties of the polyimide/multi-walled carbon nanotubes (MWNTs) nanocomposites, *Composites Science and Technology*, vol. 66, pp. 548–554, 2006.

[199] M. A. Lopez-Manchado, B. Herrero, M. Arroyo, Organoclay–natural rubber nanocomposites synthesized by mechanical and solution mixing methods, *Polym Int*, vol. 53, pp. 766–1772, 2004.

[200] S. Stankovich, D. A. Dikin, G. H. B. Dommett, K. M. Kohlhaas, E. J. Zimney, E. A. Stach, R. D. Piner, S. B. T. Nguyen, R. S. Ruoff, Graphene-based composite materials, *Nature*, vol. 442, pp. 282–286, 2006.

[201] A. L. Higginbotham, J. R. Lomeda, A. B. Morgan, J. M. Tour, Nanocomposite, Graphite oxide flame-retardant polymer, *ACS Appl Mater Interfaces*, vol. 10, pp. 2256–2261, 2009.

[202] D. Chen, H. Zhu, T. Liu, In situ thermal preparation of polyimide nanocomposite films containing functionalized graphene sheets, *ACS ApplMater Interfaces*, vol. 2, pp. 3702–3708, 2010.

[203] T. Ramanathan, A. A. Abdala, S. Stankovich, D. A. Dikin, M. Herrera-Alonso, R. D. Piner, D. H. Adamson, H. C. Schniepp, X. Chen, R. S. Ruoff, S. T. Nguyen, I. A. Aksay, R. K. Prud'Homme, L. C. Brinson, Functionalized graphene sheets for polymer nanocomposites, *Nature Nanotechnology*, vol. 3, pp. 327–331, 2008.

[204] S. Iijima, T. Ichihashi, Single-shell carbon nanotubes of 1-nm diameter, *Nature*, vol. 363, pp. 603–605, 1993.

[205] M. J. Treacy, T. W. Ebbesen, J. M. Gibson, Exceptionally high Young's modulus observed for individual carbon nanotubes., *Nature*, vol. 381, pp. 678–680, 1996.

[206] B. G. Sumpter, J. S. Huang, V. Meunier, J. M. Romo-Herrera, E. Cruz-Silva, H.Terrones, M. Terrones, A theoretical and experimental study on manipulating the structure and properties of carbon nanotubes using substitutional dopants., *International Journal of Quantum Chemestry*, vol. 109, pp. 97–118, 2009.

[207] E. Cruz-Silva, F. López-Urías, E. Muñoz-Sandoval, B. G. Sumpter, H. Terrones, J. C. Charlier, V. Meunier, M. Terrones, Electronic transport and mechanical properties of phorous and phosphorus-nitrogen doped carbon nanotubes, *ACS Nano*, vol. 3, pp. 19131–1921, 2009.

[208] Y. Ganesan, C. Peng, Y. Lu, L. Ci, A. Srivastava, P.M. Ajayan, J. Lou, Effect of nitrogen doping on the mechanical properties of carbon nanotubes., *Acs Nano*, vol. 4, pp. 7637–7643, 2010.

[209] C. Morant, J. Andrey, P. Prieto, D. Mendiola, J.M. Sanz, E. Elizalde, XPS characterization of nitrogen-doped carbon nanotubes., *Physica Status Solidi a-Applications and Materials Science*, vol. 203, pp. 1069–1075, 2006.

[210] M. M. S. Fakhrabadi, A. Allahverdizadeh, V. Norouzifard, B. Dadashzadeh, Effects of boron doping on mechanical properties and thermal conductivities of carbon nanotubes., *Solid State Communications*, vol. 152, pp. 1973–1979, 2012.

[211] H. Y. Song, X. W. Zha, The effects of boron doping and boron grafts on the mechanical properties of single-walled carbon nanotubes., *Journal of Physics D-Applied Physics 2009, 42, 6.*, vol. 42, pp. 6 ff., 2009.

[212] M. Rahmandoust, A. Ochsner, Influence of structural imperfections and doping on the mechanical properties of single-walled carbon nanotubes., *Journal of Nano Research*, vol. 6, pp. 185–196, 2009.

[213] C. Lee, X. D. Wei, J. W. Kysar, J. Hone, Measurement of the elastic properties and intrinsic strength of monolayer graphene., *Science*, vol. 321, pp. 385–388, 2008.

[214] B. Mortazavi, S. Ahzi,V. Toniazzo,Y. Remond, Nitrogen doping and vacancy effects on the mechanical properties of graphene: A molecular dynamics study., *Physics Letters A*, vol. 376, pp. 1146–1153, 2012.

[215] B. Mortazavi, S. Ahzi, Molecular dynamics study on the thermal conductivity and mechanical properties of boron doped graphene., *Solid State Communications*, vol. 152, pp. 1503–1507, 2012.

[216] M. Terrones, O. Martin, M. Gonzalez, J. Pozuelo, B. Serrano, J. C. Cabanelas, S. M. Vega-Diaz, J. Baselga, Interphases in graphene polymer-based nanocomposites: Achievements and challenges., *Advanced Materials*, vol. 23, pp. 5302–5310, 2011.

[217] J. W. Kang, H. J. Hwang, K. S. Kim, Molecular dynamics study on vibrational properties of graphene nanoribbon resonator under tensile loading., *Computational Materials Science*, vol. 65, pp. 216–220, 2012.

[218] S. K. Georgantzinos, G. I. Giannopoulos, D. E. Katsareas, P. A. Kakavas, N. K. Anifantis, Size-dependent non-linear mechanical properties of graphene nanoribbons., *Computational Materials Science*, vol. 50, pp. 2057–2062, 2011.

[219] H. Bu, Y. F. Chen, M. Zou, H. Yi, K. D. Bi, Z.H. Ni, Atomistic simulations of mechanical properties of graphene nanoribbons., *Physics Letters A*, vol. 373, pp. 3359–3362, 2009.

[220] K. H. Er, S. G. So, The mechanical and structural properties of Si doped diamond-like carbon prepared by reactive sputtering., *Journal of Ceramic Processing Research*, vol. 12, pp. 187–190, 2011.

[221] V. Anita, T. Butuda, T. Maeda, K. Takizawa, N.Saito, O. Takai, Effect of N doping on properties of diamond-like carbon thin films produced by RF capacitively coupled chemical vapor deposition from different precursors., *Diamond and Related Materials*, vol. 13, pp. 1993–1996, 2004.

[222] H. Nakazawa, A. Sudoh, M. Suemitsu, K. Yasui, T. Itoh,T. Endoh, Y. Narita, M. Mashita, Mechanical and tribological properties of boron, nitrogen-coincorporated diamond-like carbon films prepared by reactive radio-frequency magnetron sputtering., *Diamond and Related Materials*, vol. 19, pp. 503–506, 2010.

[223] X. Q. Liu, J. Yang, J. Y. Hao, J. Y. Zheng, Q. Y. Gong, W. M. Liu, Microstructure, mechanical and tribological properties of Si and Al co-doped hydrogenated amorphous carbon films

deposited at various bias voltages., *Surface & Coatings Technology*, vol. 206, pp. 4119–4125, 2012.

[224] B. Fragneaud, K. Masenelli-Varlot, A. Gonzalez-Montiel, M. Terrones, J.Y. Cavaille, Efficient coating of N-doped carbon nanotubes with polystyrene using atomic transfer radical polymerization., *Chemical Physics Letters*, vol. 419, pp. 567–573, 2006.

[225] J. A. Talla, D. Zhang, S. A. Curran, Electrical transport measurements of highly conductive nitrogen-doped multiwalled carbon nanotubes/poly(bisphenol A carbonate) composites, *Journal of Materials Research*, vol. 22, pp. 2854–2859, 2011.

[226] E. P. Giannelis, Polymer layered silicate nanocomposites, *Adv. Mater*, vol. 8, pp. 29–35, 1996.

[227] M. LeBras., Mineral fillers in intumescent fire retardant formulations – Criteria for the choice of a natural clay filler for the ammonium polyphosphate/pentaeythritol/polypropylene system, *Fire and Materials*, vol. 20, pp. 39–49, 1996.

[228] G. Beyer, Short communication: Carbon nanotubes as flame retardants for polymers, *Fire and Materials*, vol. 26, pp. 291–293, 2002.

[229] G. Beyer, Filler blend of carbon nanotubes and organoclays with improved char as new flame retardant system for polymers and cable applications, *Fire and Materials*, vol. 29, pp. 61–69, 2005.

[230] T. Kashiwagi, E. Grulke, J. Hilding, R. Harris, W. Awad, J. Douglas, Thermal degradation and flammability properties of poly(propylene)/carbon nanotube composites, *Macromol. Rapid Commun*, vol. 23, pp. 761–765, 2002.

[231] O. Yano, H. Yamaoka, Cryogenic properties of polymers, *Prog Polym Sci*, vol. 20, pp. 585–613, 1995.

[232] W. Reese, Thermal properties of polymers at low temperatures, *Macromol Sci A*, vol. 3, pp. 1257–1295, 1969.

[233] Z. Hana, A. Fina, Thermal conductivity of carbon nanotubes and their polymer nanocomposites: A review, *Progress in Polymer Science*, vol. 36, p. 914–944, 2011.

[234] C. Zhi, Y. Bando, T. Terao, C. Tang, H. Kuwahara, D. Golberg, Towards thermoconductive, electrically insulating polymeric composites with boron nitride nanotubes as fillers, *Adv. Funct. Mater*, vol. 19, p. 1857–1862, 2009.

[235] S. M. Vega-Díaz, F.J. Medellín-Rodríguez, S. Sánchez-Valdés, B. E. Handy, J.M. Mata-Padilla,O. Dávalos-Montoya, Synthesis and morphological characterization of sodium and calcium clay nanostructured poly(E-caprolactam) [nylon 6] hybrids, *The Open Macromolecules Journal*, vol. 2, pp. 19–25, 2008.

[236] F. J. Medellin-Rodriguez, C. Burguer, B. S. Hsiao, B. Chu, R. Vaia, S. Phillips, Time-resolved shear behavior of end tethered nylon 6-clay nanocomposites followed by non-isothermal crystallization, *Polymer*, vol. 42, pp. 9015–2023, 2001.

[237] W. Zheng, X. Lu, C. L. Toh, T. H. Zheng, C. He, Effects of clay on polymorphism of polypropylene in polypropylene/clay nanocomposites, *Journal of Polymer Science: Part B: Polymer Physics*, vol. 42, p. 1810–1816, 2004.

[238] T. McNally, P. Pötschke, P. Halley, M. Murphy, D. Martin, SEJ Bell, G.P. Brennan, D. Bein, P. Lemoine, J. P. Quinn, Polyethylene multiwalled carbon nanotube composites, *Polymer*, vol. 46, p. 8222–8232, 2005.

[239] L. Zhang, T. Tao, C. Li, Formation of polymer/carbon nanotubes nano-hybrid shish–kebab via non-isothermal crystallization, *Polymer*, vol. 50, pp. 3835–3840, 2009.

[240] T. Tao, L. Zhang, J. Ma, C. Li, Production of flexible and electrically conductive polyethylene–carbon nanotube shish-kebab structures and their assembly into thin films, *Ind. Eng. Chem. Res*, vol. 51, pp. 5456–5460, 2012.

[241] C.Y. Li, L. Li, W. Cai, S. L. Kodjie, K. K. Tenneti, Nanohybrid shish-kebabs: Periodically funcionalized carbon nanotubes, *Adv. Mater*, vol. 17, pp. 1198–1202, 2005.

[242] A. R. Bhattacharyya, T. V. Sreekumar, T. Liu, S. Kumar, L. M. Ericson, Crystallization and orientation studies in polypropylene/single wall carbon nanotube composite, *Polymer*, vol. 44, p. 2373–2377, 2003.
[243] Y. W. Nam, W. N. Kim, Y. H. Cho, D. W. Chae, G. H. Kim, S. P. Hong, S. S. Hwang, S.M. Hong, Morphology and physical properties of binary blend based on PVDF and multi-walled carbon nanotube, *Macromol. Symp*, Vols. 249–250, pp. 478–484, 2007.
[244] L. Li, C. Y. Li, C. Ni, Polymer crystallization-driven, periodic patterning on carbon nanotubes, *J. AM. CHEM. SOC.*, vol. 128, pp. 1692–1699, 2006.
[245] M. L. Minus, H. G. Chae, S. Kumar, Interfacial crystallization in gel-spun poly(vinyl alchohol) single-wall carbon nanotubes composite fibers, *Macromol. Chem. Phys*, vol. 210, pp. 1799–1808, 2009.
[246] L. Li, B. Li, M. A. Hood, C. Y. Li, Carbon nanotube induced polymer crystallization: The formation of nanohybrid shish–kebabs, *Polymer*, vol. 50, p. 953–965, 2009.
[247] R. H. Schmidt, I. A. Kinloch, A. N. Burgess, A. H. Windle, The effect of aggregation on the electrical conductivity of spin-coated polymer/carbon nanotube composite films, *Langmuir*, vol. 23, pp. 5707–5712, 2007.
[248] J. N. Coleman, S. Curran, A. B. Dalton, A. P. Davey, B. McCarthy, W. Blau, and R. C. Barklie, Percolation-dominated conductivity in a conjugated-polymer-carbon-nanotube composite, *Phys. Rev. B*, vol. 58, pp. R7492–R7495, 1998.
[249] D. Aharony, A. Stauffer, Introduction to percolation theory, *Taylor and Francis, London*, 1994.
[250] B. E. Kilbride, J. N. Coleman, J. Fraysse, P. Fournet, M. Cadek, A. Drury, S. Hutzler, S. Roth, W. J. Blau, Experimental observation of scaling laws for alternating current and direct current conductivity in polymer-carbon nanotube composite thin films, *J. Appl. Phys. 92, 4024–4030 (2002)*, vol. 92, pp. 4024–4030, 2002.
[251] D. S. Mclachlan, C. Chiteme, C. Park, K. E. Wise, S. E. Lowther, P. T. Lillehei, E. J. Siochi, J. S. Harrison, AC and DC Percolative conductivity of single wall carbon nanotube polymer composites, *J. Polym. Sci. B*, vol. 43, pp. 3273–3287, 2005.
[252] J. K. W. Sandler, J. E. Kirk, I. A. Kinloch, M. S. P. Shaffer, A. H. Windle, Ultra-low electrical percolation threshold in carbon-nanotube-epoxy composites, *Polymer*, vol. 44, pp. 5893–5899, 2003.
[253] V. Tishkova, P.-I. Raynal, P. Puech, A. Lonjon, M. Le Fournier, P. Demont, E. Flahaut, W. Bacsa, Electrical conductivity and Raman imaging of double wall carbon nanotubes in a polymer matrix, *Compos. Sci. Technol.*, vol. 71, pp. 1326–1330, 2011.
[254] S. De, P. E. Lyons, S. Sorel, E. M. Doherty, P. J. King, W. J. Blau, P. N. Nirmalraj, J. J. Boland, V. Scardaci, J. Joimel, J. N. Coleman, Transparent, flexible, and highly conductive thin films based on polymer-nanotube composites, *ACS Nano*, vol. 3, pp. 714–720, 2009.
[255] R. Ramasubramaniam, J. Chen, H. Liu, Homogeneous carbon nanotube/polymer composites for electrical applications, *Appl. Phys. Lett.*, vol. 83, pp. 2928–2930, 2003.
[256] J. R. Potts, D. R. Dreyer, C. W. Bielawski, R. S. Ruoff,, Graphene-based polymer nanocomposites, *Polymer*, vol. 52, pp. 5–25, 2011.
[257] G. Eda, M. Chhowalla, Graphene-based composite thin films for electronics, *Nano Lett.*, vol. 9, pp. 814–818, 2009.
[258] X. Wu, S. Qi, J. He, G. Duan, High conductivity and low percolation threshold in polyaniline/graphite nanosheets composites, *J. Mater. Sci.*, vol. 45, pp. 483–489, 2010.
[259] W.P. Wang, Y. Liu, X.-X. Li, Y.-Z. You, Synthesis and characteristics of poly(methyl methacrylate)/expanded graphite nanocomposites, *J Appl. Polym. Sci.*, vol. 10, pp. 1427–1431, 2006.

[260] J. S. Kim, S. J. Cho, K. S. Jeong, Y. C. Choi, M. S. Jeong, Improved electrical conductivity of very long multi-walled carbon nanotube bundle/poly (methyl methacrylate) composites, *Carbon*, vol. 49, pp. 2127–2133, 2011.
[261] Y. Y. Huang, J. E. Marshall, C. Gonzalez-Lopez, E. M. Terentjev, Variation in carbon nanotube polymer composite conductivity from the effects of processing, Dispersion, Aging and Sample Size, *Mater. Express*, vol. 1, pp. 315–328, 2011.
[262] Y. J. Kim, T. S. Shin, H. D. Choi, J. H. Kwon, Y.-C. Chung, H. G, Yoon, Electrical conductivity of chemically modified multiwalled carbon nanotube/epoxy composites, *Carbon*, vol. 43, pp. 23–30, 2005.
[263] Z. P. Yang, L. Ci, J. A. Bur, S.-Y. Lin, P. M. Ajayan, Experimental observation of an extremely dark material made by a low-density nanotube array, *Nano Lett.*, vol. 8, pp. 446–451, 2008.
[264] B. Z. Tang, H. Xu, Preparation, alignment, and optical properties of soluble poly(phenylacetylene)-wrapped carbon nanotubes, *Macromolecules*, vol. 32, pp. 2569–2576, 1999.
[265] V. P. Verma, S. Das, I. Lahiri, W. Choi, Large-area graphene on polymer film for flexible and transparent anode in field emission device, *Appl. Phys. Lett.*, vol. 96, p. 203108, 2010.
[266] C. Harish, V. Sai SreeHarsha, C. Santhosh, R. Ramachandran, M. Saranya, T. Mudaliar Vanchinathan, K. Govardhan, A. Nirmala Grace, Synthesis of polyaniline/graphene nanocomposites and Its optical, electrical and electrochemical properties, *Adv. Sci. Eng. Med.*, vol. 5, pp. 140–148, 2013.
[267] J. Wang, M. Musameh, Y. H. Lin, Solubilization of carbon nanotubes by nafion toward the preparation of amperometric biosensors, *Journal of the American Chemical Society*, vol. 125, pp. 2408–2409, 2003.
[268] R. J. Chen, S. Bangsaruntip, K.A. Drouvalakis, N. W. S. Kam, M. Shim, Y. Li, W. Kim, P.J. Utz, H. Dai, Noncovalent functionalization of carbon nanotubes for highly specific electronic biosensors, *Proceeding of the National Academic of Science of the United States of America*, vol. 48, pp. 4984–4989, 2003.
[269] C. H. Lu, H. H. Yang, C.L. Zhu, X. Chen, G. N. Chen, A graphene platform for sensing biomolecules, *Angewandte Chemie International edition*, vol. 48, pp. 4785–4787, 2009.
[270] C. Shan, H. Yang, J. Song, D. Han, A. Ivaska, L. Niu, Direc electrochemistry of glucose oxidase biosensing for glucose based on graphene, *Analytical Chemistry*, vol. 81, pp. 2378–2382, 2009.
[271] Y. Shao, Jun Wang, H. Wu, J. Liu, I. A. Aksay, Y. Lina, Graphene based electrochemical sensors and biosensors: A Review, *Electroanalysis*, vol. 22, pp. 1027–1036, 2010.
[272] S. Nardecchia, M. C. Serrano, M. C. Gutiérrez, M. T. Portolés, M. L. Ferrer, F. del Monte, Osteoconductive performance of carbon nanotubes scaffolds homogeneously mineralized by flow through electrodeposition, *Advanced functional Materials*, vol. 22, pp. 4411–4420, 2012.
[273] Z. Liu, X. Sun, N. Nakayama-Ratchford, H. Dai, Supramolecular chemistry on water soluble carbon nanotubes for drung loading and delivery, *ACS Nano*, vol. 1, pp. 50–56, 2007.
[274] R. Singh, D. Pantarotto, D. McCarthy, O. Chaloin, J. Hoebeke, C. D. Partidos, J.-P. Briand, M. Prato, A. Bianco, K. Kostarelos, Binding and condensation of plasmid DNA onto functionalized carbon nanotubes: Toward the construction of nanotubes based gene delivery vectors, *Journal of American Chemical Society*, vol. 127, pp. 4388–4396, 2005.
[275] R. Singh, D. Pantarotto, D. McCarthy, O. Chaloin, J. Hoebeke, C. D. Partidos, J.-P. Briand, M. Prato, A. Bianco, K. Kostarelos, Biofunctionalized carbon nanotubes in neural regeneration: a mini-review, *Nanoscale*, vol. 5, pp. 487–497, 2013.

[276] A. Seichepine, E. Flahaut, I. Loubinoux, L. Vaysse, C. Vieu, Elucidation of the role of carbon nanotube patterns on the development of cultured neuronal cells, *Langmuir*, vol. 28, pp. 17363–17371, 2012.
[277] P. R. Supronowicz, P. M. Ajayan, K. R. Ullmann, B. P. Arulanandam, D. W. Metzger, R. Bizios, Novel current-conducting composite substrates for exposing osteoblasts to alternating current stimulation, *Journal of Biomedical Materials Research*, vol. 59, pp. 499–506, 2002.
[278] R. A. MacDonald, B.F . Laurenzi G. Viswanathan P. M. Ajayan J. P. Stegemann, Collagen–carbon nanotube composite materials as scaffolds in tissue engineering, *Journal of Biomedical Materials Research Part A*, vol. 74A, pp. 489–496, 2005.
[279] S. K. Smart, A. I. Cassady, G. Q. Lu, D. J. Martin, The biocompatibility of carbon nanotubes, *Carbon*, vol. 44, pp. 1034–1047, 2006.
[280] D. H. Zhang, M. A. Kandadai, J. Cech, S. Roth, S. A. Curran, Poly(L-lactide) (PLLA)/multiwalled carbon nanotube (MWCNT) composite: Characterization and biocompatibility evaluation, *Journal of Physical Chemistry B*, vol. 110, pp. 12910–12915, 2006.
[281] A. A. White, S. M. Best, I.A. Kinloch, Hydroxyapatite-carbon nanotube composites for biomedical applications: A review, *International Journal of Applied Ceramic Technology*, vol. 4, pp. 1–13, 2007.
[282] Z. Liu, S. Tabakman, K. Welsher, H. Dai, Carbon nanotubes in biology and medicine: In vitro and in vivo detection, imaging and drug delivery, *Nano Research*, vol. 2, pp. 85–120, 2009.
[283] B. Sitharaman, X.F. Shi, X.F. Walboomers, H.B. Liao, V. Cuijpers, L.J. Wilson, A.G. Mikos, J.A. Jansen, In vivo biocompatibility of ultra-short single-walled carbon nanotube/biodegradable polymer nanocomposites for, bone tissue engineering, *Bone*, vol. 43, pp. 362–370, 2008.
[284] B.L. Allen, P.D. Kichambare, P. Gou, I.I. Vlasova, A.A. Kapralov, N. Konduru, V.E. Kagan, A. Star, Biodegradation of single-walled carbon nanotubes through enzymatic catalysis, *Nano Letters*, vol. 8, pp. 3899–3903, 2008.
[285] I. I. Vlasova, A. V.Sokolov, A. V. Chekanov,V. A. Kostevich, V. B. Vasilyev, Myeloperoxidase-induced biodegradation of single-walled carbon nanotubes is mediated by hypochlorite., *Russian Journal of Bioorganic Chemistry*, vol. 37, pp. 453–463, 2011.
[286] Y. Zhao, B. L. Allen, A. Star, Enzymatic degradation of multiwalled carbon nanotubes, *Journal of Physical Chemistry A*, vol. 115, pp. 9536–9544, 2011.
[287] J. C. Carrero-Sanchez, A. L. Elias, R. Mancilla, G. Arrellin, H. Terrones, J. P. Laclette, M.Terrones, Biocompatibility and toxicological studies of carbon nanotubes doped with nitrogen, *Nano Letters*, vol. 6, pp. 1609–1616, 2006.
[288] A. L. Elias, J. C. Carrero-Sanchez, H. Terrones, M. Endo, J. P. Laclette, M Terrones, Viability studies of pure carbon- and nitrogen-doped nanotubes with Entamoeba histolytica: From amoebicidal to biocompatible structures, *Small*, vol. 3, pp. 1723–1729, 2007.
[289] X. Liu, R.H. Hurt, A.B. Kane, Biodurability of single-walled carbon nanotubes depends on surface functionalization, *Carbon*, vol. 48, pp. 1961–1969., 2010.
[290] J. Russier, C. Menard-Moyon, E. Venturelli, E. Gravel, G. Marcolongo, M. Meneghetti, E. Doris, A. Bianco, Oxidative biodegradation of single- and multi-walled carbon nanotubes, *Nanoscale*, vol. 3, pp. 893–896, 2011.
[291] A. A. Kapralov, W. H. Feng, F.T. Andon, B. J. Chambers, J. Klein-Seetharaman, A. Star, A. A. Shvedova, B. Fadeel, V. E. Kagan, Biodegradation of carbon nanotubes by eosinophil peroxidase, *Toxicology Letters*, vol. 211, pp. S204-S204, 2012.
[292] V. E. Kagan, N. V. Konduru, W. Feng,B. L.Allen, J.Conroy, Y.Volkov, I .I. Vlasova, N. A. Belikova, N.Yanamala, A. Kapralov, Y. Y.Tyurina, J. Shi, E. R. Kisin, A. R. Murray, J.Franks, D. Stolz, P. Gou, J. Klein-Seetharaman, B. Fadeel, A. Star, A. A. Shvedova, Carbon nanotubes degraded by neutrophil myeloperoxidase induce less pulmonary inflammation, *Nature Nanotechnology*, vol. 5, pp. 354–359, 2010.

[293] G.P. Kotchey, B.L. Allen, H. Vedala, N. Yanamala, A.A. Kapralov, Y.Y. Tyurina, J. Klein-Seetharaman, V.E. Kagan, A. Star, The enzymatic oxidation of graphene oxide, *Acs Nano*, vol. 5, pp. 2098–2108, 2011.
[294] I. Armentano, M. Dottori, D. Puglia, J.M. Kenny, Effects of carbon nanotubes (CNTs) on the processing and in-vitro degradation of poly(DL-lactide-co-glycolide)/CNT films, *Journal of Materials Science-Materials in Medicine*, vol. 19, pp. 2377–2387, 2008.
[295] H. L. Zeng, C. Gao, D.Y. Yan, Poly(epsilon-caprolactone)-functionalized carbon nanotubes and their biodegradation properties, *Advanced Functional Materials*, vol. 16, pp. 812–818, 2006.
[296] H. Adeli, S. H. S. Zein, S. H. Tan, H. M. Akil, A. L. Ahmad, Synthesis, characterization and biodegradation of novel poly(L-lactide)/multiwalled carbon nanotube porous scaffolds for tissue engineering applications., *Current Nanoscience*, vol. 7, pp. 323–333, 2011.
[297] Q. Y. Liu, J. Y. Wu, T. W. Tan, L. Q. Zhang, D. F. Chen, W. Tian, Preparation, properties and cytotoxicity evaluation of a biodegradable polyester elastomer composite., *Polymer Degradation and Stability*, vol. 94, pp. 1427–1435, 2009.
[298] S. S. Ray, S. Vaudreuil, A. Maazouz, M. Bousmina, Dispersion of multi-walled carbon nanotubes in biodegradable poly(butylene succinate) matrix., *Journal of Nanoscience and Nanotechnology*, vol. 6, pp. 2191–2195, 2006.
[299] H. Tsuji, Y. Kawashima, H. Takikawa, S. Tanaka, Poly(L-lactide)/nano-structured carbon composites: Conductivity, thermal properties, crystallization, and biodegradation., *Polymer*, vol. 48, pp. 4213–4225, 2007.
[300] J. S. Bunch, S. S. Verbridge, J. S. Alden, A.M. van der Zande, J. M. Parpia, H. G. Craighead, P. L. McEuen, Impermeable atomic membranes from graphene sheets., *Nano Letters 2008*, vol. 8, pp. 2458–2462, 2008.
[301] S. S. Chen, L. Brown, M. Levendorf, W. W. Cai, S.Y. Ju, J. Edgeworth, X.S. Li, C.W. Magnuson, A. Velamakanni, R. D. Piner, J. Y. Kang, J. Park, R.S. Ruoff, Oxidation resistance of graphene-coated Cu and Cu/Ni alloy., *Acs Nano*, vol. 5, pp. 1321–1327, 2011.
[302] C. H. Chang, T. C. Huang, C. W. Peng, T. C. Yeh, H. I. Lu, W. I. Hung, C. J. Weng,T. I. Yang, J. M. Yeh, Novel anticorrosion coatings prepared from polyaniline/graphene composites., *Carbon*, vol. 50, pp. 5044–5051, 2012.
[303] N. T. Kirkland, T. Schiller, N. Medhekar, N. Birbilis, Exploring graphene as a corrosion protection barrier., *Corrosion Science*, vol. 56, pp. 1–4, 2012.
[304] D. Kang, J. Y. Kwon, H. Cho, J. H. Sim, H. S. Hwang,C. S. Kim, Y. J. Kim, R. S. Ruoff, H. S. Shin, Oxidation resistance of iron and copper foils coated with reduced graphene oxide multilayers, *Acs Nano*, vol. 6, pp. 7763–7769, 2012.
[305] P. C. Ma, J. W. Liu, S. L. Gao, E. Mader, Development of functional glass fibres with nanocomposite coating: A comparative study., *Composites Part a-Applied Science and Manufacturing*, vol. 44, pp. 16–22, 2013.
[306] F. Guo, G. Silverberg, S. Bowers, S.P. Kim, D. Datta, V. Shenoy, R.H. Hurt, Graphene-based environmental barriers., *Environmental Science & Technology*, vol. 46, pp. 7717–7724, 2012.
[307] O. C. Compton, S. Kim, C. Pierre, J. M. Torkelson, S. T. Nguyen, Crumpled graphene nanosheets as highly effective barrier property enhancers., *Advanced Materials*, vol. 22, pp. 4759–4763, 2010.
[308] I. H. Tseng, Y. F. Liao, J. C. Chiang, M. H. Tsai, Transparent polyimide/graphene oxide nanocomposite with improved moisture barrier property., *Materials Chemistry and Physics*, vol. 136, pp. 247–253, 2012.
[309] H. M. Kim, J. K. Lee, H. S. Lee, Transparent and high gas barrier films based on poly(vinyl alcohol)/graphene oxide composites., *Thin Solid Films*, vol. 519, pp. 7766–7771, 2011.

[310] Y. H. Yang, L. Bolling, M. A. Priolo, J. C. Grunlan, Super gas barrier and selectivity of graphene oxide-polymer multilayer thin films., *Advanced materials (Deerfield Beach, Fla.)*, vol. 25, pp. 503–508, 2013.

[311] L. Yu, Y. S. Lim, J. H. Han, K. Kim, J.Y. Kim, S.Y. Choi, K. Shin, A graphene oxide oxygen barrier film deposited via a self-assembly coating method., *Synthetic Metals*, vol. 162, pp. 710–714, 2012.

[312] D. Lee, M. C. Choi, C. S. Ha, Polynorbornene dicarboximide/amine functionalized graphene hybrids for potential oxygen barrier films., *Journal of Polymer Science Part a-Polymer Chemistry*, vol. 50, pp. 1611–1621, 2012.

[313] H. Wu, L. T. Drzal, Graphene nanoplatelet paper as a light-weight composite with excellent electrical and thermal conductivity and good gas barrier properties., *Carbon*, vol. 50, pp. 1135–1145, 2012.

[314] H. Kim, Y. Miura, C. W. Macosko, Graphene/polyurethane nanocomposites for improved gas barrier and electrical conductivity., *Chemistry of Materials*, vol. 22, pp. 3441–3450, 2010.

Part II: **Synthesis and characterisation of hybrids**

Cameron J. Shearer and Dominik Eder

5 Synthesis strategies of nanocarbon hybrids

Hybridizing nanocarbons, such as carbon nanotubes (CNTs) or graphene (G), with inorganic and organic compounds is a powerful strategy toward designing next-generation functional materials. The properties of nanocarbon hybrids are highly dependent on many aspects of the hybrid such as the homogeneity, morphology, composition and structure of the coating as well as the nature of the interface. Nanocarbon hybrids can be synthesized *via* either *ex situ* or *in situ* processes. *Ex situ* processes involve the combination of pre-defined components *via* covalent or noncovalent interactions. *In situ* processes involve the synthesis of one component (usually the inorganic part) in the presence of the other (*i.e.* carbon), which is possible when a component is formed from molecular precursor units such as monomers, metal salts, metal alkoxides or other reactive intermediates. The major advantage of this route is that the nanocarbon can act as a substrate, template and "heat sink", thus considerably affecting the size, shape and structure of the inorganic compound. This has led to hybrids with uncommon or novel crystal phases and unusual geometries. A broad range of synthesis methods have been reported, each with advantages and disadvantages; this chapter aims to introduce the various concepts and general strategies and highlight some of the most intriguing recent examples.

5.1 Introduction

The downscaling of the inorganic and polymeric compounds to similar volume fractions as the nanocarbon has introduced nanocarbon hybrids as the next-generation multifunctional composite. Since the appeal of these hybrids lies in synergistic effects based on interfacial charge and energy transfer processes [1, 2], controlling the nature of the interface and maximizing the interfacial area become key challenges to designing the ideal nanocarbon hybrid. This also implies that the chemistry of the building blocks is fundamental to tailoring the interfacial and thus the material's properties. This is in contrast to nanocomposites, where the main challenges have predominantly been the assembly and dispersion of the nanocarbons in the matrix.

Nanocarbon hybrids can utilize advantageous template-assisted synthesis processes for designing novel complex materials. In this reaction design, the template can be either the nanocarbon or the hybridizing component (Fig. 5.1). This experimental design affords the fabrication of a wide range of possible hybrid material architectures including: one-dimensional structures (*e.g.* CNTs, inorganic wires, polymer fibers, biomolecules or DNA), two-dimensional structures (*e.g.* graphene, graphene oxide or thin metal/metal oxide (MO) films) or multi-dimensional networks (*e.g.* nanocarbon networks, polymer blends, metal oxide frameworks, zeolites, and membranes).

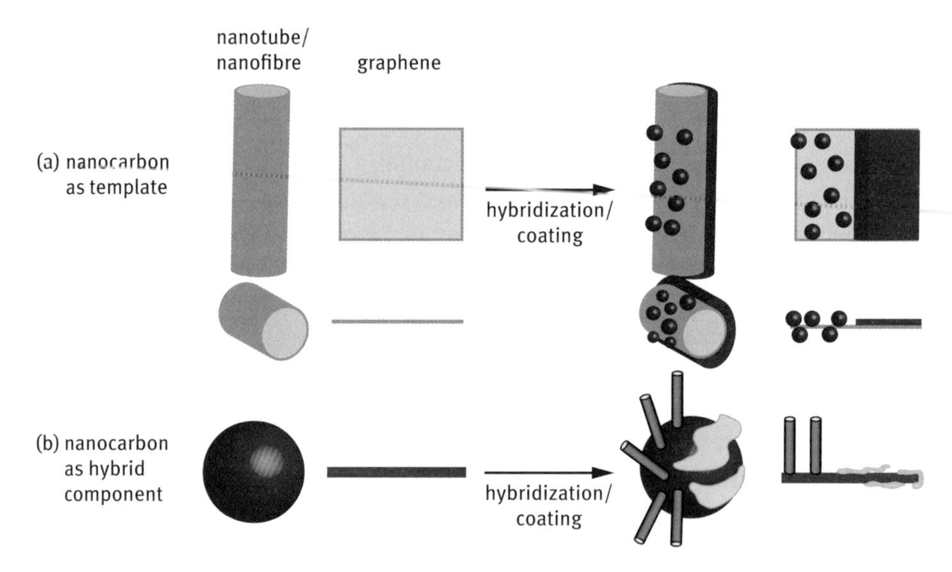

Fig. 5.1: Schematic representation of templated synthesis of nanocarbon hybrids with (a) nanocarbon as template and (b) nanocarbon as hybridizing component.

The choice of synthesis technique for nanocarbon hybrids and the extent of their synergistic function depend on the type and purity of nanocarbon and the modification of their surface chemistry (see Chapter 3). Briefly, this is because nanocarbon functionalization can disrupt the π orbital conjugation resulting in altered electrical and mechanical properties (Chapter 1). Depending on the desired application this can be beneficial, detrimental or have no observable consequence. For example, strong oxidation of graphene to graphene oxide (GO) facilitates many covalent hybridization reactions, however, the oxidation procedure reduces the nanocarbon's electrical conductivity, rendering it less effective for use in electronic applications such as transparent electrodes. Therefore, careful consideration of the type and degree of functionalization must be undertaken prior to deciding upon which hybridization route to follow. Moreover, while single-walled CNTs and graphene can be easily functionalized through covalent means, other nanocarbons, *i.e.* multi-walled CNTs or few-layer graphene, possess only little chemical reactivity towards covalent functionalization. Fortunately, numerous alternative hybridization strategies are available.

In general, the various synthesis strategies for nanocarbon hybrids can be categorized as *ex situ* and *in situ* techniques [3]. The *ex situ* ("building block") approach involves the separate synthesis of the two components prior to their hybridization. One can rely on a plethora of scientific work to ensure good control of the component's dimensions (*i.e.* size, number of layers), morphology (*i.e.* spherical nanoparticles, nanowires) and functionalization. The components are then hybridized through covalent, noncovalent or electrostatic interactions. In contrast, the *in situ* approach is a one-step process that involves the synthesis of one of the components in the presence of the other. Examples include the growth of graphene on the surface of an inorganic nanostructure as well as the deposition of the inorganic material on the surface of CNTs from molecular precursors in the form of particles, nanowires or thin films.

Recent reviews have focused on particular nanocarbon hybrids including CNT-inorganic [3], CNT-nanoparticles [4–6], graphene-inorganic [7], graphene-semiconductor [8] and nanocarbon composites and hybrids [1]. This chapter attempts to summarize all synthetic techniques for nanocarbon hybrids with emphasis placed on recent and important hybridization strategies. Carbon nanotubes and graphene will be the major focus as they are the most popular and promising nanocarbons.

5.2 *Ex situ* approaches

In this "building-block" approach, the components are synthesized separately and then hybridized *via* linking agents/methods that utilize covalent, noncovalent (van der Waals, π-π interactions, hydrogen bonding), or electrostatic interactions. The attachment of these building blocks often requires the chemical modification of at least one component to overcome the differences in surface chemistry. As a consequence deposition is often limited to the first layer. Excess nanoparticles can be removed by filtration or centrifugation.

5.2.1 Covalent interactions

As discussed earlier in this book (Chapter 3) there are many potential chemical routes toward functionalizing nanocarbons. The most popular is the oxidation of nanocarbons, particularly for CNTs by treatment in acidic solutions (*e.g.* HNO_3/H_2SO_4) or the oxidation of graphene to GO. These oxidation methods produce a range of chemical functionalities on the nanocarbons including hydroxyl, epoxy and carboxylic acid.

Carboxyl groups on the surface of nanocarbons are often used to attach amine-, hydroxyl- or mercapto-terminated species *via* amide, ester or thioester bonds (Fig. 5.2) [9, 10]. This coupling is commonly assisted by activating the carboxyl group with carbodiimide coupling agents such as dicyclohexyl carbodiimide (DCC) or 1-ethyl-3-(3-dimethylaminopropyl) carbodiimide hydrochloride (EDC) [11]. This has been used to successfully link various inorganic NPs [12, 13], quantum dots [14], DNA [15], polymers [16], peptides [17], proteins [18, 19], dendrons [20], ionic liquids [21], ferrocene [22, 23] and porphyrins [24, 25] to oxidized CNTs and graphene.

As an example, Nakayama *et al.* thermally oxidized CNTs to introduce carboxyl groups at the opened CNT tips and then covalently linked ethylenediamine as a cross-linker (*via* EDC coupling) to provide primary amino groups that formed peptide bonds with N-hydroxysuccinimide-functionalized proteins (Fig. 5.2) [26].

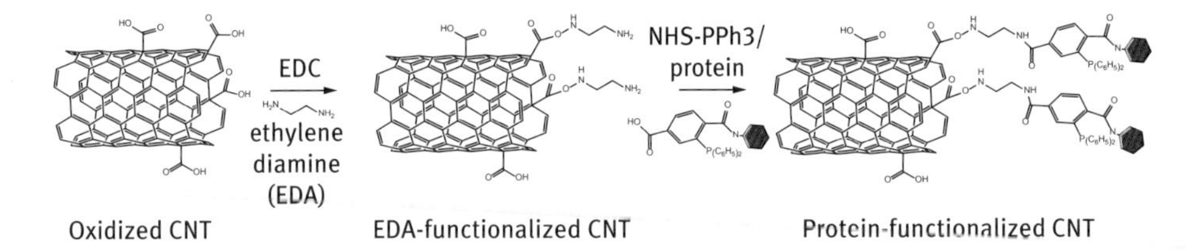

Fig. 5.2: Example of *ex situ* covalent protein hybridization of CNTs via carbodiimide assisted coupling. Redrawn from [26].

In the case of inorganic nanoparticles, covalent binding can occur between the nanocarbon and the nanoparticle capping agent. Among the various metal nanoparticles, gold has been used most frequently due to its excellent potential in biosensing and other medical applications [27]. For example, Au nanoparticles have been linked to carboxylic acid terminated CNTs using aminothiols, bifunctional thiols or thioether bonds [28]. Sfier *et al.* have demonstrated covalent immobilization of 2-aminoethanthiol capped CdSe QDs with oxidized DWCNTs and *via* DCC assisted attachment [29].

An alternative to oxidation of the nanocarbon for further chemical synthesis is diazonium/nitrene chemistry [30, 31]. This reaction is understood to occur by injection of an electron from the nanocarbon into an aryl diazonium salt which subsequently induces the formation of a reactive aryl radical, releasing N_2 and creating a nanocarbon-arene bond (Figure 5.3(a)) [32]. Such a reaction has been used by Johnson *et al.* to covalently bind prostate cancer antibodies to a CNT as a cancer biosensor [33]. Other hybrids produced through this method include CNT-Au NPs [34]. For example, Snell *et al.* produced Au hybrids with CNTs and graphene by first synthesizing nitrene modified Au NPs (Figure 5.3(b)) which were then activated *via* UV irradiation to form reactive azide intermediates which bonded with the nanocarbons without the need for any other chemical modification (Figure 5.3(c), (d)) [35]. In the case of supported single layer graphene, the extent of hybridization has been found to be highly dependent upon the underlying surface [31].

Covalent linkages can also be used to wrap graphene around or onto materials of interest. Yu *et al.* have described a general route toward preparing graphene-wrapped metal oxide nanotube and nanoparticle hybrids [36]. The synthesis process involves the deposition of a poly(allylamine) layer on the MO followed by immersion in an aqueous GO solution to covalently attach the GO to the amine terminal groups of the MO. The GO is then reduced *via* hydrazine reduction to form MO-graphene hybrids. This was completed on SnO_2, CuO and CuO nanowires and α-Fe_2O_3 nanoparticles [36].

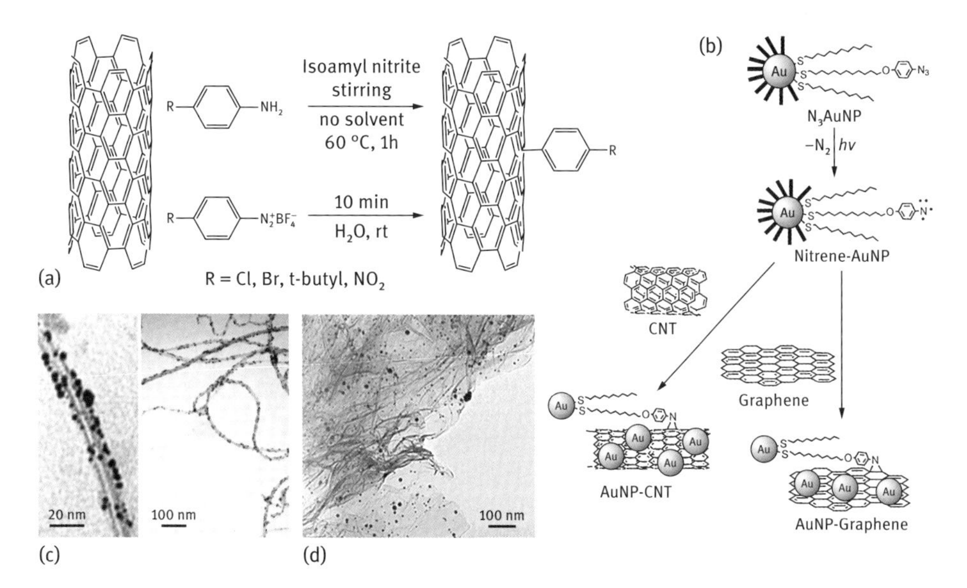

Fig. 5.3: Scheme of azide chemistry of nanocarbons. (b) Example of nitrene chemistry for the covalent attachment of Au nanoparticles to (c) CNTs and (d) graphene. (a) Redrawn from [32], (b), (c), (d) reprinted with permission from [35] (2012), © Wiley.

5.2.2 Noncovalent interactions

Noncovalent interactions such as van der Waals, hydrogen bonding, π-π stacking and electrostatic interactions have been widely used to hybridize pristine nanocarbons *via ex situ* approaches. The major advantage of this route is that the nanocarbons do not require modification prior to hybridization and their structure remains undisturbed, an important factor in many electronic applications. The strength of hybridization is weaker compared to covalent interactions but the synthetic process is generally simpler. Noncovalent attachment of small molecules to nanocarbons is often used to change the surface chemistry for subsequent *ex situ* or *in situ* hybridization.

5.2.2.1 Hydrophobic and van der Waals interactions

The majority of studies have used surfactants that wrap around nanocarbons *via* van der Waals interactions [37]. For instance, surfactants such as sodium dodecylsulfate (SDS) are commonly used to disperse CNTs in aqueous solutions [38, 39] while other surfactants, such as Pluorinc-123, are used to mechanically exfoliate graphene from graphite flakes (Fig. 5.4(a)) [40, 41]. The polar head group of the surfactant can be used to further hybridize the nanocarbon *via* a range of covalent or noncovalent interactions [42]. For example, nanoparticles of Pt [43, 44] and Pd [45] have been decorated onto SDS-wrapped MWCNTs. Similarly, Whitsitt *et al.* evaluated various surfactants for their ability to facilitate the deposition of SiO_2 NPs onto SWCNTs [46, 47]. As an example of coupling *ex situ* with *in situ* techniques, pristine MWCNTs have been wrapped (*ex situ*) in poly(sodium 4-styrenesulfonate) (PSS) prior to dispersion in ethanol along with varying ratios of $SnCl_2$ and $CuCl_2$ [48]. The surfactant wrapping led to an even dispersion of adsorbed ions which, upon solvothermal treatment (*in situ*) in the presence of thiourea, produced dispersed NPs of Cu_2S, Cu_3SnS_4, SnO_2 and Cu_2SnS_3 depending upon initial molar ratios [48].

Functionalization of inorganic nanoparticles with hydrophobic capping agents provides another hybridization route (Fig. 5.4(b)). As an example, Au clusters were first modified with a monolayer of octanethiols [49] or dodecanethiols [50] and then simply dispersed in a suspension of pristine SWCNTs in dichloromethane. Deposition of ZnO NPs onto GO has also been facilitated by the polymeric capping of the NPs with poly(vinyl pyrrolidone) [51]. This approach can be further utilized for QDs, for example oleic acid capped CdSe QDs have been found to noncovalently hybridize with SWCNTs [52]. The advantage of this approach is that both coverage and morphology of the hybrid materials can be controlled by modifying the length and functionality of the chains in the capping agents [53].

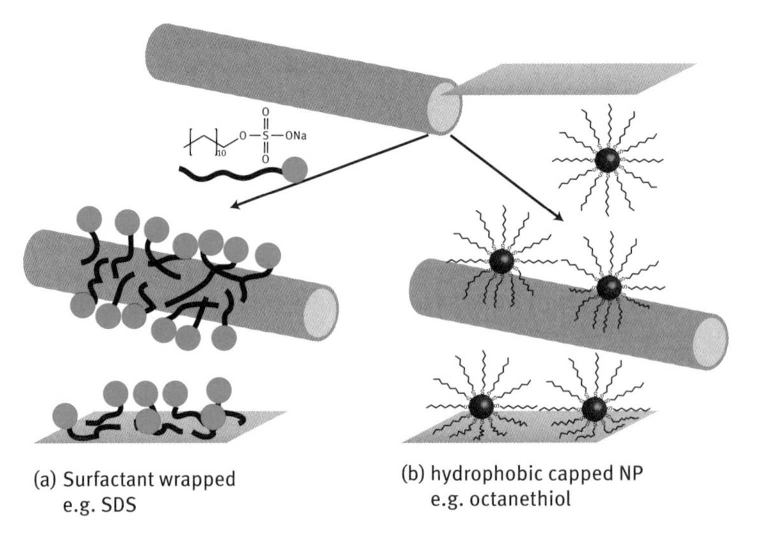

Fig. 5.4: Scheme of van der Waals interactions between (a) surfactants and (b) capped nanoparticles with nanocarbons.

5.2.2.2 π-π stacking

Another approach is based on π-π stacking, which utilizes the moderately strong interactions between delocalized π-electrons of nanocarbons and those in aromatic organic compounds, such as derivatives of pyrene [54, 55], porphyrins [56, 57], phthalocyanines [58] or combinations thereof [59], as well as peptides [60], DNA [61], benzyl alcohol [62] or triphenylphosphine [63]. These molecules are often modified with long alkyl chains that are terminated with thiol, amine, or acid groups, which can then potentially be used for further hybridization (Fig. 5.5).

The most common linker molecules are pyrene derivatives, which have been used, for example, to link Au [64, 65] and Pd [66] NPs, as well as QDs [67] to CNTs. Martin *et al.* chemically attached a π-extended tetrathiafulvalene (exTFF) group to pyrene prior to π-π hybridization on SWCNTs in order to investigate donor-acceptor interactions between the SWCNT and electron rich exTFF [68]. More recently, the same group has synthesized a complex molecule consisting of two exTFF anchors connected *via* a flexible linkage that also consists of a second generation carboxylic terminated dendron [69]. Here, π interactions between the exTFF anchors and the SWCNTs lead to hybridization while the carboxylic acid groups of the dendron (when de-protonated) lead to increased solubility in aqueous solutions [69].

Nakashima *et al.* utilized π-π stacking to hybridize SWCNTs with a porphyrin-fluorene copolymer [70]. The pyrene-like fluorene arms led to a chirality-selective interaction with semiconducting SWCNTs while the Zn center of the porphyrin was used to coordinate with capped Au NPs [70]. DNA is also a useful linking group for biological, organic and inorganic moieties. The strong interaction between the aromatic rings of graphene and the N-containing groups of DNA has been utilized by Zhang *et al.* to prepare a mixed electrode of Graphene-DNA-Au NPs with glucose oxidase for biosensing [71].

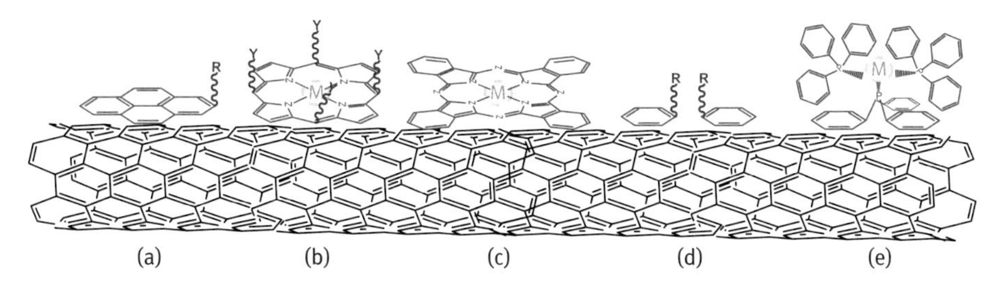

Fig. 5.5: Scheme of π-π stacking on nanocarbons (CNT shown in example) showing (a) pyrene, (b) porphyrin, (c) phthalocyanine, (d) benzyl and (e) triphenylphosphine derivatives. M, R and Y groups vary and can be used for further hybridization.

Recently, Guldi *et al.* have synthesized a porphyrin molecule with one carboxylic acid terminated arm and four hydrophobic arms. This smartly designed porphyrin molecule enabled both exfoliation of graphite to graphene and hybridization in one single step and so created a photoactive hybrid [72]. The π electrons of nanocarbons can lead to interesting CNT-graphene hybrids [73–76]. MWCNT-graphene films for use as actuators have been produced by the group of Lu [77]. GO and MWCNTs were dispersed *via* sonication prior to GO reduction by the addition of hydrazine during which π-π interactions led to the noncovalent formation of the hybrid [77].

5.2.2.3 Electrostatic interactions

Another noncovalent approach utilizes electrostatic interactions between modified nanocarbons and oppositely charged compounds. Depending upon the nature of the surface groups on both the nanocarbon and the hybridizing component, the density and strength of electrostatic interaction can be altered by changing the pH or ionic strength of solvent [78]. Positively charged hybridizing components can directly interact with oxidized CNTs or GO. For example, negatively charged GO can be wrapped around positively charged poly (diallyl dimethylammonium chloride) (PDDA) wrapped NPs such as SiO_2 [79, 80]. Another example is the PSS coating of CNTs, which has been shown to facilitate the electrostatic adsorption of cetyltrimethylammonium bromide (CTAB) capped Pt nanocubes (Fig. 5.6) [81]. This work demonstrates one of the biggest advantages of the *ex situ* hybridization route, which is the formation and attachment of metal nanoparticles with uniform size (*i.e.* 6–7 nm) and unique morphology (*i.e.* nanocubes).

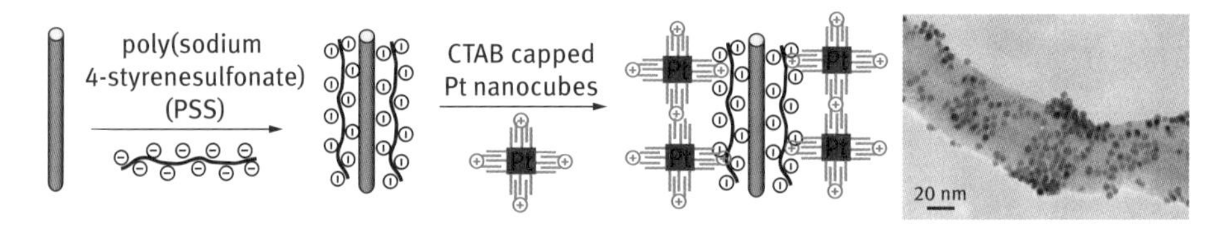

Fig. 5.6: Example of electrostatic hybridization of cetyltrimethylammonium bromide (CTAB) capped Pt nanocubes with PSS-wrapped CNTs. Reprinted with permission from [81] (2008), © Wiley.

Alternatively, oxidized CNTs have been coated with a thin film of PDDA which served as the template for negatively charged Au nanoparticles [82]. Yang *et al.* coated CNTs with a mixture of PSS and PDDA resulting in a positively charged surface which was subsequently used for the adsorption of $AuCl_4^-$ ions which, after reduction, resulted in Au NP decoration of the CNTs [83]. Without addition of polyelectrolyte, Au NPs were formed away from the CNT surface [83], Similar synthetic routes have been followed to prepare CNT-Pt [84], CNT-Fe_2O_3 [85], CNT-Al_2O_3 [86], CNT-ZrO_2 [86], CNT-TiO_2 [87], NP hybrids, as well as Graphene-$Ti_{0.87}O_2$ nanosheets [88] and GO-MnO_2 films [89].

Electrostatic interactions have recently been exploited for the synthesis of graphene-CNT hybrids. For example, poly(ethyleneimine) (PEI) coated graphene has been mixed with acid treated CNTs in a layer-by-layer method to form high surface area electrodes for supercapacitors [90]. Furthermore, Lu *et al.* prepared a supercapacitor electrode by mixing PDDA coated CNT-MnO_2 hybrid with RGO [91].

Electrostatic hybridization can also be achieved through electrophoretic deposition (Fig. 5.7(a)). Niu *et al.* electrophoretically coated an indium doped tin oxide (ITO) electrode with GO in dimthethyl formamide (DMF) [92]. The GO coated ITO was then submerged in a solution of Au NPs, a bias was applied and the negatively charged NPs coated the GO/ITO electrode. The process could be repeated a number

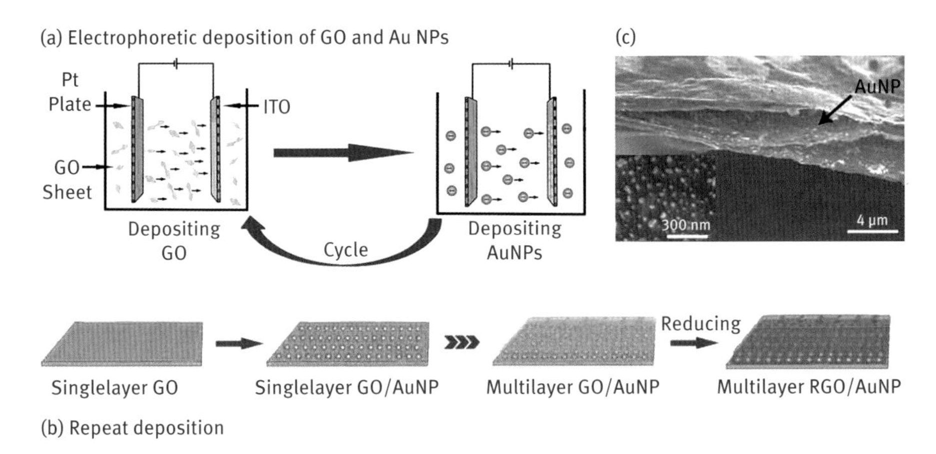

Fig. 5.7: (a), (b) Schematic of electrophoretic deposition of GO and Au NPs to form layered hybrid structure (c). Altered and reprinted with permission from [92] (2012), © Wiley.

of times to prepare a GO/Au film of desired thickness before the reduction of GO (Fig. 5.7) [92].

Ionic liquids can also be used to hybridize NPs to nanocarbons [93]. The positively charged ionic liquid 1-(3-Aminopropyl)-3-methylimidazolium bromide (NH_2-IL) was covalently attached to oxidized MWCNTs *via* DCC mediated coupling. Electrostatic interactions between NH_2-IL and $HAuCl_4$ led to precipitation of Au NPs on the MWCNTs [94]. Pt NPs were deposited in a similar fashion *via* microwave-assisted synthesis on ionic liquid (3-ethyl-1-vinylimidazolium tetrafluoroborate) functionalized MWCNTs [95]. Cui *et al.* have demonstrated the preparation of Au and Pt NPs hybridized with CNTs *via* ionic liquid assisted coating [96]. Ionic liquids were utilized to functionalize graphene flakes with Au NPs by the group of Niu [97]. GO sheets were initially coated with 3,4,9,10-perylene tetracarboxylic acid (PTCA) to negatively charge the surface which was then reacted with Au NPs that were created in the ionic liquid NH_2-IL resulting in Au NP coating of GO. The Au nanoparticles were then further hybridized with DNA for biosensing [97]. More detail on the fabrication and application of nanocarbon-ionic liquid hybrids can be found in a recent review by Tunckol *et al.* [98].

Srivastava *et al.* have recently demonstrated the multi-step synthesis of a complex hybrid, graphene-CdSe QD-Au NP-TiO_2, for photovoltaic applications, for which they have utilized many different *ex situ* routes discussed here [99]. The ionic liquid 1-Butyl-3-methyl-imidazolium trifluoromethanesulfonate was used to intercalate and electrophoretically exfoliate few-layered graphene from a graphite rod. The G-IL hybrid was then mixed with TiO_2 to form a composite and dried onto an electrode surface. The G-IL-TiO_2 electrode was immersed in a solution of mercaptopropionic acid (MPA) where covalent and noncovalent interactions bound MPA to the surface rendering it thiol functionalized. The surface was dried and transferred to a solution containing both Au NPs and trioctyl phosphine capped CdSe QDs. Electrostatic interactions with the IL, covalent reactions with the thiol moieties and van der Waals interactions between the capping agent and graphene were thus used to create a hybrid that showed very promising optoelectronic properties [99].

In summary, these few examples demonstrate the simplicity and feasibility of the *ex situ* approach, which remains the only method suitable for the hybridization of biomolecules. The main advantage is the possibility of using pre-prepared nanoparticles with controlled morphology, structure, shape and size, and therefore a good structure-property relationship. A perfect example of this is the Pt nanocube hybrids highlighted in Fig. 5.6. The downside of this route, however, is the need to chemically modify either the nanocarbon or the hybridizing component. This process is often work-intensive, and functionalization alters both the surface chemistry of the nanocarbon and also, particularly for SWCNTs and graphene, their physical properties. Furthermore, the use of pre-defined components restricts the synthesis of novel hybrid materials and thus the development of new physical properties. Furthermore, the deposition of the inorganic compounds is restricted to the first layer. Most importantly, the use of pre-defined building blocks limits the synthesis of truly novel hybrid materials with new morphologies and physical properties often found in hybrids produced *via* the *in situ* approach.

5.3 *In situ* approaches

The hybridizing component can also be formed directly on the surface of a pristine or modified nanocarbon using molecular precursors, such as organic monomers, metal salts or metal organic complexes. Depending on the desired compound, *in situ* deposition can be carried out either in solution, such as *via* direct network formation *via in situ* polymerization, chemical reduction, electro- or electroless deposition, and sol–gel processes, or from the gas phase using chemical deposition (*i.e.* CVD or ALD) or physical deposition (*i.e.* laser ablation, electron beam deposition, thermal evaporation, or sputtering).

The advantages of the *in situ* approach include an enhanced interfacial adhesion for polymers, while inorganic compound can be deposited as either amorphous, polycrystalline, or single crystalline films with good control of uniformity and thickness, depending on the chosen method and process conditions. The major difference between the *ex situ* and *in situ* approaches, however, is that in the latter the nanocarbon may act as a support to stabilize uncommon or even novel crystal phases, unusual morphologies (*i.e.* growth of nanowires or nanoplates perpendicular to the CNT surface [100]) and to prevent crystal growth during crystallization and phase-transformation processes.

The surface chemistry of the nanocarbon is required to be suitable for the chosen molecular precursor. In contrast to the *ex situ* approach, functionalization of the nanocarbon is not explicitly required, however the presence of functional groups and the use of linking agents provide attractive reaction centers that can facilitate the initial stages in hybrid synthesis.

5.3.1 *In situ* polymerization

In situ growth *via* covalent binding of a hybridizing component to a nanocarbon can be achieved in the case of polymers, dendrons and various other macromolecules which are synthesized in a stepwise manner. The *in situ* synthesis of such macromolecules potentially increases binding site density while steric effects of the nanocarbon can lead to increased variation in average polymer chain length (polydispersity) [101–103].

There are many polymerization techniques that have been successfully completed from nanocarbons. These have been covered in detail elsewhere for CNTs [104] and graphene [105]. The simplest method is the chemical grafting of a monomer followed by polymerization [106]. An interesting example is the polymerization of polyaniline (PANI) from GO. The primary amine functionality present on the aniline monomer is capable of interacting with carboxylic acid groups on the surface of GO. Xia *et al.* have followed a two phase polymerization system where GO (aqueous phase) and aniline (organic phase) interact at the phase boundary. As the polymer grows, the hybrid moves to the aqueous phase, away from the monomer, and polymerization ends producing PANI-GO hybrids [107].

Increased control of polydispersity can also be achieved using "living" polymerization reactions. "Living" polymer grafting from a nanocarbon is facilitated by an initiator molecule which is attached to the nanocarbon *via* covalent or noncovalent means. For example, Herrara *et al.* chemically attached a 2-bromopropionate initiator for atom transfer radical polymerization (ATRP) onto oxidized SWCNTs for subsequent grafting of poly(*n*-butyl methylmethacrylate) [108], poly(methyl methacrylate) [109] and poly(styrene) [110] brushes. More recently, a similar ATRP initiated synthetic process has been followed to graft poly(methylmethacrylate) from GO [111]. Figure 5.8(a) summarizes how Wang *et al.* utilized azo coupling to –OH functionalized RGO before modification with the ATRP initiator 2-bromoisobutyl for the grafting of poly(2-(ethyl(phenyl)amino)ethyl-methacrylate) [112].

An alternative to ATRP is reversible addition fragmentation chain transfer (RAFT) polymerization which is facilitated by a chain transfer agent (CTA) [113]. For example, Ellis *et al.* have demonstrated the preparation of a CNT-CTA *via* chemical linkage between oxidized CNTs and two units of a magnesium chloride dithiopropanoate salt [114]. A GO-poly(*N*-vinylcarbozole) hybrid was prepared by first using carbodiimide mediated coupling to immobilize the CTA agent *S*-1-dodecyl-*S*′-(α,α'-dimethyl-α''-acetic acid) trithiocarbonate on GO (Fig. 5.8(b)). The hybrid was found to switch between a high and low conductivity state with applied potential, leading to potential applications in memory devices [115].

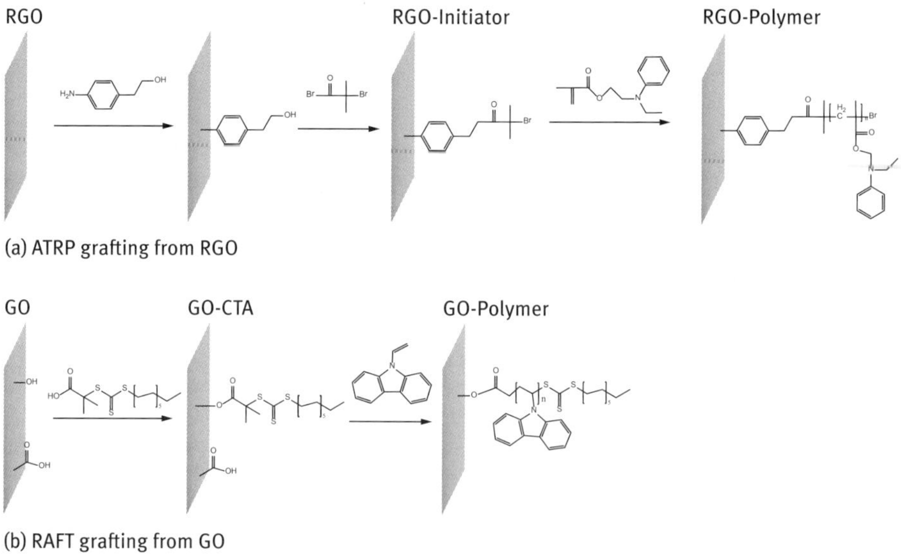

RGO RGO-Initiator RGO-Polymer

(a) ATRP grafting from RGO

GO GO-CTA GO-Polymer

(b) RAFT grafting from GO

Fig. 5.8: Examples of polymers grafted from nanocarbons. (a) An ATRP initiator covalently attached to RGO via nitrene and carbodiimide chemistry was used for the growth of poly(2-(ethyl (phenyl)amino)ethyl-methacrylate). (b) A RAFT chain transfer agent is covalently attached to GO prior to polymerization of vinylcarbozole.

Dendrimers are branched macromolecules that can be synthesized *via* divergent growth originating on a nanocarbon [116]. An early example of poly(amidoamine) dendron hybridized MWCNTS was produced by He *et al.* who oxidized MWCNTs, converted the carboxylic acid groups to acyl chlorides and then attached ethylene diamine to the acyl chloride groups leaving the MWCNTs -NH_2 terminated [117]. The amino groups could subsequently be conjugated to two methyl acrylate groups *via* Michael addition which in turn could be amino terminated *via* further reaction with ethylene diamine [117]. This process could then be repeated to produce PAMAM dendrons of desired generation (controlled by the number of reaction cycles). The dendrons could be used in applications such as photovoltaics [118, 119] or the increased functionality can be used for various hybridization reactions [120, 121].

Campidelli *et al.* have synthesized interesting linear and hyperbranched porphyrin polymers from CNTs *via* copper-catalyzed alkyne-azide cycloaddition (CuAAC) [122]. Zinc porphyrin monomers containing an azide group and one or three alkyne groups were synthesized and chemically bound to alkyne functionalized SWCNTs *via* CuAAC. Depending upon the number of alkyne functionalities either linear (single alkyne) or dendrimer-like (triple alkyne) porphyrin polymers were produced (Fig. 5.9) [122].

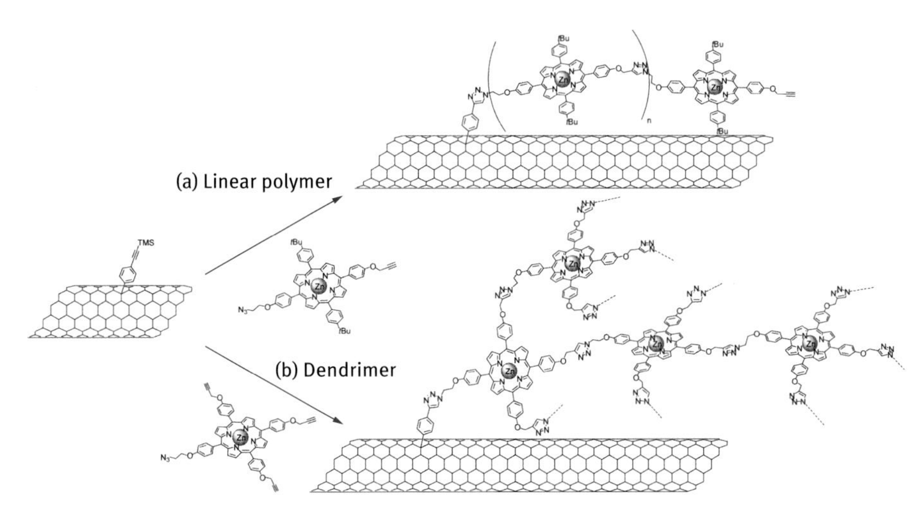

Fig. 5.9: Covalent grafting of (a) linear polymer and (b) dendrimer-like hyperbranched polymer of azo-functional porphyrin groups from alkyne functionalized CNTs. Redrawn from [122].

5.3.2 Inorganic hybridization from metal salts

Inorganic-nanocarbon hybrids can be produced from the simplest metallic building blocks: metal salts. These are readily available, and are generally highly soluble and are able to form metal NPs. Control of nanocarbon-metal complex interaction enables effective control over nucleation and growth [123, 124]. Most research has been conducted on the deposition of noble metals and alloys such as Pd [124–126], Pt [125, 127, 128], Au [125, 129, 130], Ag [127, 131, 132], various bimetallic clusters [133], as well as metal oxides such as TiO_2, [134], and Fe_2O_3 [135, 136] as they are the metals of choice for applications like heterogeneous catalysis and electrocatalysis [137–139], supercapacitors [140], gas sensors [83, 141, 142], and biosensors [71, 97, 143].

In general, metal nanoparticles are obtained *via* reduction of metal complexes, such as metal chlorides, by chemical agents (chemical reduction), or by electrons (electrodeposition). Hybrids of metal oxides are obtained by oxidation, network formation or precipitation of precursors such as metal nitrates and acetates [144].

5.3.2.1 Wet chemical processes

These techniques involve reactions in which the reduction of the precursor is carried out with liquid agents which can be aided with heat, plasma, light, ultrasound, microwave, or supercritical CO_2 [45, 145–149]. Common reducing agents include $NaBH_4$ [42], ethylene glycol [84], thiourea [48], sodium citrate [150], and formic acid [107].

Lin *et al.* oxidized MWCNTs and followed by mixing with poly(allylamine hydrochloride) (PAH) [84]. A higher coverage of PAH was obtained when KNO_3 was added to increase the ionic strength of the solution. A solution containing H_2PtCl_6

and ethylene glycol was added and the pH was increased to 12. The high pH increased the positive charge on the PAH-MWCNTs leading to electrostatic adsorption of $PtCl_6^{2-}$ ions while the ethylene glycol chemically reduced the $PtCl_6^{2-}$ resulting in a well-dispersed coating of NPs on the surface (Fig. 5.10) [84].

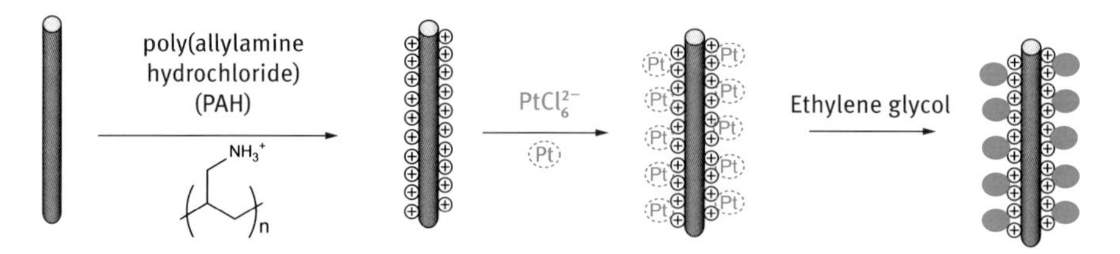

Fig. 5.10: Synthesis strategy and TEM image of CNT-Pt NP hybrid via chemical reduction of Pt ions on PAH functionalized CNTs.

Graphene oxide is required to be reduced to RGO for a majority of applications. This opens the opportunity for both GO and hybrid precursor to be reduced concurrently. $NaBH_4$ is a popular reducing agent for such purposes as it is capable of reducing both NPs and GO to form hybrid materials of RGO-NPs under relatively mild conditions. As an example, Fe-Ni NPs have been produced on PEI wrapped GO. An aqueous solution of PEI wrapped GO was prepared and solutions of $FeSO_4$ and $NiCl_2$ were added and dispersed prior to chemical reduction *via* addition of $NaBH_4$ [151]. Further examples include RGO-Rh/Ni NPs [152], RGO-Pt [153, 154], RGO-Pd [153, 155], RGO-Au [156], RGO-Ag [157], and RGO-ferrocene [23]. Alternatives to $NaBH_4$ include hydrazine hydrate for RGO-SnO_2 [158], RGO-MoS_2 [159], $NaHSO_3$ for RGO-Fe_3O_4 [160], ethylene glycol for RGO-Cu_2O [161], urea for RGO-BiOBr [162], NH_3BH_3 for RGO-CuCO [163], and benzyl alcohol for RGO-SnO_2 [164]. RGO has been shown to spontaneously reduce Au^{3+} to from NPs without the need for reducing agent [165].

Figure 5.11 summarizes a recent publication by Cai *et al.* where magnetic Fe_3O_4 NPs were wrapped with polydopamine (PDA) followed by addition of GO which was reduced onto the NPs by the PDA [166]. The now RGO-wrapped Fe_3O_4-PDA hybrids were then further functionalized by the addition of $HAuCl_4$ which was chemically reduced on the RGO by PDA thus producing Fe_3O_4-PDA-RGO-Au hybrids which exhibited both magnetic and catalytic properties [166].

The simultaneous reduction of both GO and metal precursor enables a simple one-step synthetic route toward nanocarbon hybrids based on electrostatic interactions with, adversely, reduced control of the level of GO reduction. This is not a significant problem in hybrid formation but it will significantly affect further application of the hybrid, particularly in electronic applications.
The reduction of metal salts can also produce nanocarbon-QD hybrids [167]. CNT-$Cd_{0.8}Zn_{0.2}S$ QD hybrids have been obtained by first adsorbing Cd^{2+} and Zn^{2+} ions onto oxidized MWCNTs followed by reduction *via* addition of Na_2S to precipitate the QDs

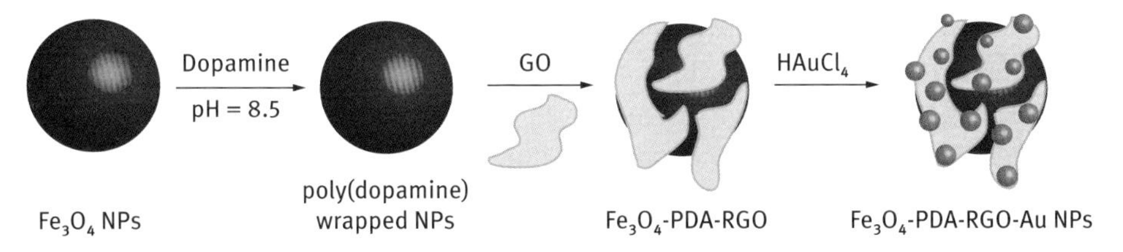

Fig. 5.11: Schematic of the stepwise preparation of Fe3O4-PDA-RGO-Au NP hybrid via ex situ electrostatic assembly and in situ Au NP reduction.

onto the MWCNT surface [168]. The authors have also reported the same strategy for preparation of RGO-$Cd_{0.8}Zn_{0.2}S$ QD hybrids [169]. This technique has been repeated by others to produce nanocarbon-QD hybrids including CNT-CdS [170, 171], Graphene-CdS [171, 172] and also graphene-CoS [173]. GO has been shown to act as a capping agent to reduce ZnO QD size during growth *via* Zn-O-C interactions. The resultant ZnO-GO QDs exhibited white light emission due to extra emission bands occurring from the constrained GO [174].

Oxidation of metallic precursors can lead to metal oxide NP or thin film formation on nanocarbons. For example, GO-Fe_3O_4 hybrids have been prepared as medical imaging contrast agents by Ajayan *et al.* [175]. In this work the authors mixed GO with $FeSO_4$ and $FeCl_3$ prior to addition of ammonia solution to increase the pH to 10 at which point precipitation of NPs on the GO occurred [175]. The precipitation reaction is a popular choice to prepare nanocarbon-metal oxide hybrids, particularly from metal nitrate precursors, and has recently been applied to CNT-CoO [176], RGO-Fe_3O_4 [177], CNT-MoO_2 [178], RGO-bimetallic clusters [179] and many others [180]. Alternative to adding base, PDDA wrapping of GO has been shown to assist in both metal precursor immobilization and hydrolysis of SnO_2 and Fe_2O_3 NPs [181].

In the case of MnO_4^-, it is known to reduce on the surface of CNTs and graphene spontaneously to produce MnO_2 NPs [182, 183]. Solvothermal assisted precipitation of metal oxides can occur in milder solutions [184]. For example, mixed metal oxide NPs of $CoFe_2O_4$ have been deposited on GO from metallic salt precursors *via* the addition of ethanolamine followed by incubation at 180 °C in a sealed vessel [185]. Mixing GO with Cd^{2+} in DMSO followed by solvothermal treatment has been shown to both reduce GO to RGO and coat with CdS QDs [186].

Selective deposition of NPs inside wide diameter CNTs is also possible [187], as demonstrated initially by Tessonier *et al.* [188]. A concentrated aqueous solution of metal salt is prepared and dropped slowly onto a relatively large amount (10 g) of wide diameter, opened MWCNTs with vigorous stirring. The slurry was then allowed to dry at room temperature during which capillary forces draw the metal ions into the MWCNTs interior prior to thermal reduction to produce metal NPs inside the MWCNTs [188].

A very important solvothermal method involves the use of supercritical CO_2 as an antisolvent that reduces the solvent strength, resulting in the precipitation of the oxide due to high saturation. For instance, using metal nitrates or halides, this method has been applied to deposit CeO_2, Al_2O_3, La_2O_3 [189], and Fe_2O_3 [190] onto CNTs where it has been found to increase NP dispersion. In the case of graphene, supercritical CO_2 has been shown to increase GO dispersion during solvothermal treatment leading to increased dispersion of Pt NPs [191].

Structured hybrids can also be grown from nanocarbons. For example, ZnO nanowires have been synthesized on a GO film on a flexible substrate [192]. This was carried out by first spin coating a GO film onto a poly(dimethyl siloxane, PDMS) substrate prior to ZnO seeding *via* spin coating of zinc acetate. Nanowire growth was then completed by immersing the seeded GO substrate into a solution containing zinc nitrate precursor and hexamethylenetetramine as structure directing agent [192].

Liquid-phase chemical reduction is suitable for the formation of metal and metal oxide NPs on nanocarbons. Careful consideration is required in designing the nanocarbon-precursor interaction and choosing the reduction/oxidation method. The synthetic process is often quite time consuming and a number of filtering/washing steps are often required. As discussed, the concurrent liquid phase reduction of GO and precursor is a simple, efficient way to produce a hybrid but the lack of control of GO reduction may affect further applications.

5.3.2.2 Photoreduction

A common alternative to chemical reduction is the reduction of adsorbed metallic precursors *via* UV light irradiation. This synthesis does not require hazardous chemical agents and has, for example, recently been applied to prepare hybrids of Pt [193], Pd [45], Au [194], Ag [194], ZnO [51] with CNTs and graphene. Light energy has also been shown to partly reduce GO in various solvents [51, 194]. A mild mixed reduction method has been used by Matthews *et al.* to prepare a RGO-Pt NP hybrid [195]. An ethanolic solution of GO and H_2PtCl_6 was prepared and thoroughly mixed prior to reduction by heating to 80 °C and illuminating with a 100 W incandescent light bulb (Fig. 5.12(a)). The combination of ethanol, heat and illumination resulted in an RGO-Pt hybrid with well-dispersed NPs of small size within 18 h. The obtained hybrids were superior to those obtained *via* hydrazine reduction in terms of NP size and distribution (Fig. 5.12(b), (c)). Without ethanol, no reduction of Pt or GO was observed while light and heat were found to significantly increase the reaction rate [195].

5.3.2.3 Gas phase reduction

Nanocarbon hybridization from adsorbed inorganic precursor materials can also be completed by gas phase reduction or oxidation. The general procedure involves first mixing the nanocarbon with the hybrid precursor (*e.g.* $PdCl_2$), filtering and drying the mixture and finally heat treating within a reducing [196] or an oxidizing [197] atmosphere. An interesting example is the work of Li *et al.* who prepared a film consisting of

Fig. 5.12: (a) Schematic for the preparation of RGO-Pt hybrid *via* reduction of Pt ions on GO with comparison of SEM images between (b) hydrazine (chemical), (Pt NPs 20–200 nm) and (c) photoreduction (Pt NPs 3 nm) of Pt ions. Altered and reproduced with permission from [195], © (2012) American Chemical Society.

alternating GO sheets and Ag^+ ions on a quartz substrate [198]. The layered hybrid was dried and both GO and Ag^+ were reduced by heating the sample to 600 °C in an Ar/H_2 atmosphere (5 : 1). The layered hybrid was found to exhibit a greater transparency and conductivity than the equivalent RGO layers [198].

The temperature required for reduction can be reduced by exploiting plasma processes. For example, Gogotsi *et al.* compared thermal reduction in H_2, H_2 plasma reduction and chemical reduction *via* ethylene glycol for the preparation of an RGO-Pd hybrid [146]. Plasma reduction was found to produce Pd nanoparticles with smaller size, size dispersion and higher dispersity than the other methods. The authors claim that the faster, simultaneous reduction of both GO and Pd^{2+} resulted in the superior hybrid [146, 199]. A generalized approach for synthesis of GO-inorganic hybrids has recently been presented by Chen *et al.* [200]. In this work, GO and precursor were mixed in water to create an even suspension. The mixed suspensions were then nebulized by an ultrasonic nebulizer to form aerosol particles which were carried through a tube furnace at 600–700 °C (see Fig. 5.13). Ar was the carrier gas for oxides, Ar/H_2 was the carrier gas for metallic NPs. The GO/precursor aerosol dried and crumpled during heat treatment and the precursor was crystallized to nanoparticles creating hybrids of crumpled RGO with Mn_3O_4, SnO_2, Ag and Pt [200].

5.3.2.4 Electroless deposition

Electroless deposition is the reduction of a metal ion without an external injection of electrons. This occurs when the redox potential is higher than that of the substrate or

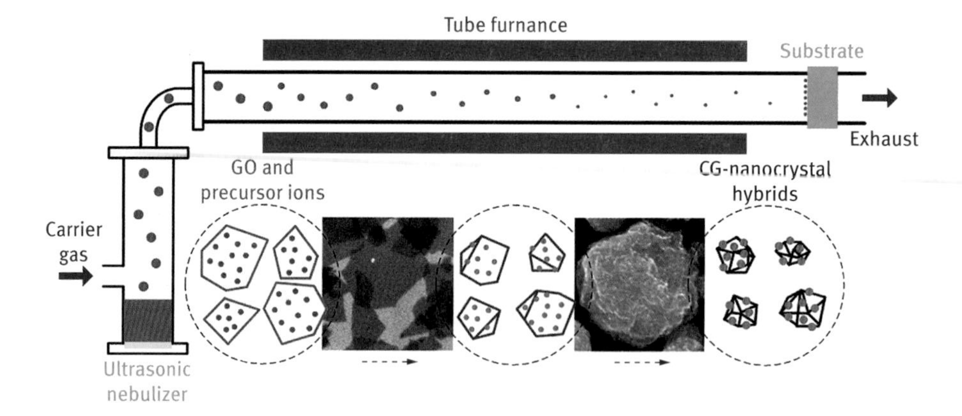

Fig. 5.13: Strategy for 1-pot synthesis of crumpled RGO-metal/metal oxide hybrid *via* nebulization of aqueous GO/precursor solution prior to compression of GO and NP formation in tube furnace at 600–700 °C. Reproduced with permission from [200], © (2012) American Chemical Society.

reducing agent. SWCNTs have a redox potential of +0.5 V (*vs*. SHE), therefore ions with a higher redox potential such as Au ($AuCl_4^-$/Au, +1.002 V) and Pt ($PtCl_4^{2-}$/Pt, +0.775 V) can reduce spontaneously onto SWCNTs [201]. Metal ions with a lower potential than the nanocarbon can be reduced by utilizing substrate-enhanced electroless deposition (SEED, Fig. 5.14(a)). This is achieved by making a thin nanocarbon film on a metallic substrate with a low redox potential *e.g.* Cu (Cu^{2+}/Cu, 0.34 V) or Zn (Zn^{2+}/Zn, -0.76 V) which supply the electrons for ion reduction onto the nanocarbon. The SEED process has been applied to prepare many CNT-NP hybrids including Au, Pt, Pd, Cu and Ag [202].

CVD growth of graphene can be completed on metal foils of Cu or Ni. Recently, pristine graphene grown *via* CVD on Cu foil has been hybridized with NPs of Au [143] (Fig. 5.14(b)) and Pd [203] using the SEED process. Solution based electroless deposition of various metals is also possible. For example, oxidized MWCNTs were first sensitized with Sn^{2+}, followed by activation with Ag^{+} and finally Cu^{2+} was reduced onto the surface using an aldehyde as the reducing agent [204].

5.3.3 Electrochemical processes

Electrochemical deposition has two main advantages over chemical reduction. First, it is much faster with most deposition completed within five minutes. Second, the size of the metal nanoparticles and their coverage on the nanocarbon can be controlled by the concentration of the metal salt and various electrochemical deposition parameters, including nucleation potential and deposition time [124, 127, 205].

Figure 5.15 depicts the general procedure of electrochemical nanocarbon hybrid formation which involves applying a potential between a nanocarbon covered

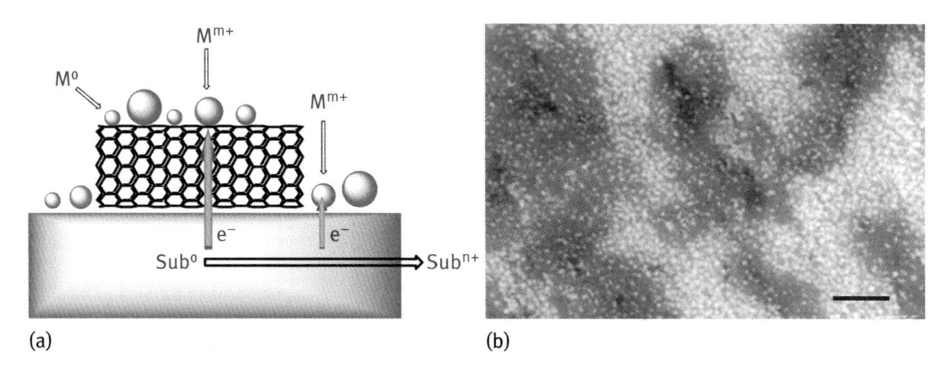

Fig. 5.14: (a) Schematic of SEED process of metal NP deposition on metal surface supported nanoparticle. (b) Example of Au NP deposition on CVD grown graphene on Cu foil, scale 1 μm. Reproduced with permission from [143], © (2012) Elsevier.

electrode and solution containing hybrid precursor (monomer, metal ion, *etc.*). The potential used depends upon the precursor and the time of deposition will increase hybrid size from NP to thin film to coating. The nanocarbon does not necessarily need to be functionalized [125]. However, similar to the chemical reduction route, a greater dispersion and smaller size distribution can be achieved by tailoring the nanocarbon surface chemistry to interact with the hybrid precursor [124]. Mild oxidation of nanocarbons can be achieved by a number of electrochemical means prior to hybridization [205].

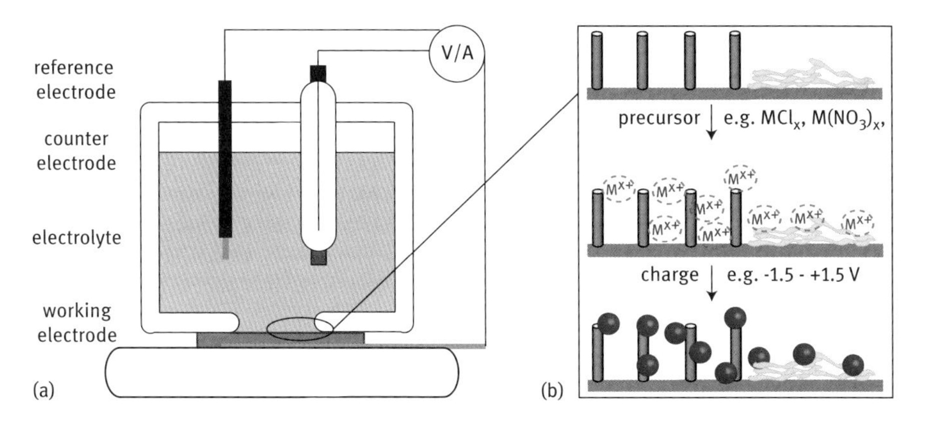

Fig. 5.15: (a) Schematic of a standard three-electrode cell with (b) general procedure for the electrochemical formation of NPs on nanocarbon electrodes.

Nanoparticles of various metals and metal oxides have been prepared on nanocarbons including Pt [139], Pd [206, 207], Ag [208, 209], Au [210], MnO_x [211], TiO_x [212], NiO_x [213], CoO_x [213], RuO_2 [214] and various bimetallic clusters [215]. For example, a GO-chitosan composite was electrostatically adsorbed to a cystamine modified gold electrode prior to the electrochemical reduction of $AgNO_3$ [132]. Chen *et al.* prepared a graphene electrode by filling a micropipette tip with a pristine CVD grown graphene layer that had been transferred onto a PMMA sheet [216]. The PMMA was dissolved in acetone and upon drying the graphene sheets assembled into a porous architecture. Silver paint was used to electrically contact the porous graphene and it was used as the working electrode and dipped into a solution of $Co(NO_3)_2$ for the electrochemical deposition of Co_3O_4 NPs [216].

The previous examples used the three-electrode electrochemical system. An alternative was utilized by Ajayan *et al.* to prepare Ag NP coated SWCNTs [217]. An electrode was fabricated consisting of SWCNTs attached to a Ti cathode and a silver contact pad as a sacrificial anode (Fig. 5.16(a)). The electrode was submerged in an aqueous solution and a potential was applied resulting in oxidation of Ag metal to Ag^{2+} ions which then subsequently deposited onto the SWCNT cathode. Although experimentally complicated, silver NPs, wires and patterns were controllably deposited on the SWCNTs (Fig. 5.16(a), (b)) [217].

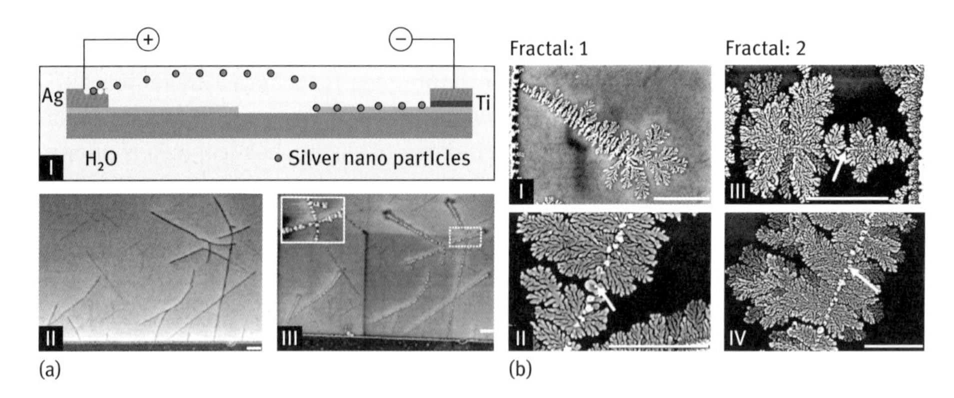

Fig. 5.16: (a) (i) Schematic of the electrochemical coating of SWCNT connected cathode from a sacrificial silver cathode with SEM images (ii) before and (iii) after Ag NP decoration. (b) Examples of fractal patterns possible with continued deposition, arrows indicate original position of SWCNT. Scale bar in all SEM images represent 1 µm. Altered and reproduced with permission from [217], © (2011) American Chemical Society.

Nanocarbon electrodes can also be prepared by utilizing the π electrons of nanocarbons and glassy carbon (GC) electrodes. For example, Cai *et al.* simply dropped an aqueous solution of RGO onto a cleaned GC electrode and allowed the solvent to dry. The π-π interactions were suitable for the subsequent deposition of Pt-Au nanostructures *via* electrochemical reduction of varying ratios of $PtCl_6^{2-}$ and $AuCl_4^-$ [133].

Polymer-nanocarbon hybrids are also possible *via* electrochemical depositions, particularly with conducting polymers [218]. Similar to the inorganic materials described above, conducting polymer films can be grown from a monomer solution onto a nanocarbon covered electrode by applying a bias [219, 220]. Thin films covering the outermost nanocarbons are produced following this method. A more uniform coating is possible by increasing the interaction between monomer and nanocarbon prior to electrodeposition [221]. Holzinger *et al.* have both covalently and noncovalently linked pyrrole-like molecules to oxidized and prisine CNTs (Fig. 5.17) [222]. In the first case, 11-1 (pyrrolyl) undecanol was covalently linked to oxidized CNTs *via* carbodiimide assisted coupling. In the second method, a pyrrole conjugated pyrene derivative was synthesized prior to noncovalent π-π stacking of the modified pyrene onto pristine CNTs. The pyrrole derivative-CNT hybrids were then electropolymerized to produce an interconnected network of ppy-CNTs [222].

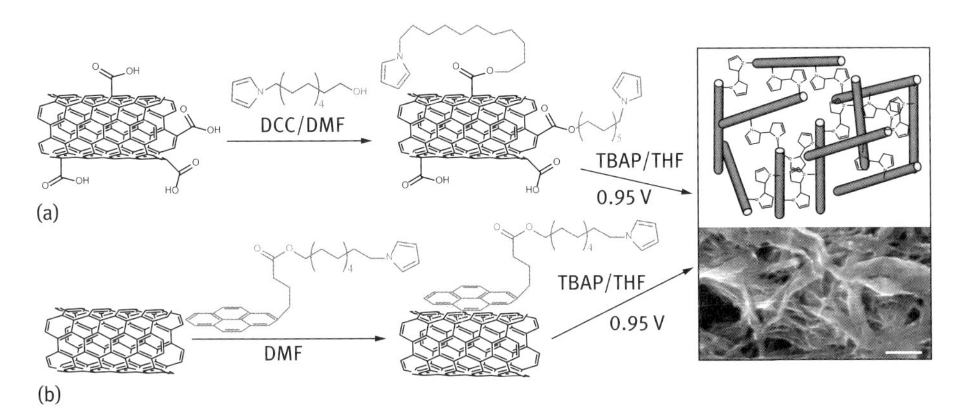

Fig. 5.17: Schematic of electropolymerization of CNT-polypyrrole hybrid via (a) ester linkage and (b) noncovalent pyrene linkage of pyrrole monomer with schematic of hybrid and SEM. Scale 200 nm. Reproduced with permission from [222], © (2008) Elsevier.

Nanocarbons can also be deposited onto surfaces *via* electrochemistry, such as electrophoretic deposition described earlier. A method for one-step electrochemical layer-by-layer deposition of GO and PANI has been reported by Chen *et al.* [199]. A solution of GO and aniline was prepared and deposited onto a working electrode *via* cyclic voltammetry. GO was reduced on the surface when a potential of approx. −1 V (*vs.* SCE) was applied compared to the polymerization of aniline which occurred at approx. 0.7 V (*vs.* SCE). Repeated continuous scans between -1.4 to 9 V (*vs.* SCE) resulted in layer by layer deposition [199]. A slightly modified method has been reported by Li *et al.* who demonstrated a general method for electrochemical RGO hybridization by first reducing GO onto glassy carbon, copper, Ni foam, or graphene paper to form a porous RGO coating [223]. The porous RGO coated electrode could then be transferred to another electrolyte solution for electrochemical deposition, PANI hybridization was shown as an example [223].

The electrochemical processes detailed here often require the nanocarbons to be deposited onto an electrode, which is advantageous for applications such as supercapacitors and batteries but detrimental for applications where nanocarbons in powder form are required *e.g.* heterogenous catalysis.

5.3.4 Sol–gel processes

The sol–gel process is a versatile, solution-based process for producing various nanoparticle, thin-film coatings, fibers, or aerogels and involves the transition of a liquid, colloidal “sol” into a solid “gel” phase. Typical starting materials for the preparation of the sol include metal salts or metal organic compounds, such as metal alkoxides, which undergo a series of hydrolysis and condensation reactions to form a colloidal or polymeric sol. Upon aging, the sol forms a wet inorganic continuous network with oxo (M–O–M) or hydroxo (M–OH–M) bridges, creating a gel.

The major drawbacks to standard sol–gel synthesis include slow growth rate and the typically amorphous product, rather than defined crystals, which requires crystallization and post annealing steps. Growth rate and crystallization of the fabricated hybrid can be improved *via* solvothermal, reflux [224], sonication, and microwave [225] treatment. However, the air oxidation of CNTs (600 °C) and graphene (450 °C) may still be lower than MO crystallization temperature. Moreover, it has been shown that the MO coatings on CNTs can drastically affect their thermal oxidation, particularly with easily reducible metal oxides (*e.g.* TiO_2 = 520 °C, Bi_2O_3 = 330 °C) [180]. It appears that metal oxides can catalyze the oxidation of CNTs *via* a Mars van Krevelen mechanism, limiting the maximum temperature of their synthesis as well as applications (*i.e.* catalysis, fuel cells).

Sol–gel growth of NPs and thin films onto nanocarbons can be achieved when there is an interaction between the nanocarbon and the precursor. The most popular are the oxygen containing groups of oxidized CNTs and GO [226, 227]. Fig. 5.18 summarizes a general strategy to prepare metal NP containing metal oxide shells around oxidized MWCNTs *via* a sol–gel route developed by the group of Fornasiero [228]. A solution of metal alkoxide is prepared in THF prior to slow addition of mercaptoundecanoic capped metal NPs followed by ultrasonic mixing. Oxidized MWCNTs in THF/DMF were then added and sonicated for 15 min prior to further addition of water to complete hydrolysis. Following this method core/shell structures of MWCNTs covered with TiO_2, ZrO_2 and CeO_2 were prepared with embedded NPs of Pt and Pd [228]. The sol–gel approach can also be used for the grafting of polymers from nanocarbons such as a recent example of solvothermal assisted Poly (N, N’–dimethylacrylamine) growth from GO [229]. The addition of urea has been shown to improve the coverage of TiO_2 NPs on oxidized CNTs [230].

Alternatively, benzyl and pyrene derivatives offer a noncovalent approach to alter nanocarbon surface chemistry and have been studied extensively [62]. For example,

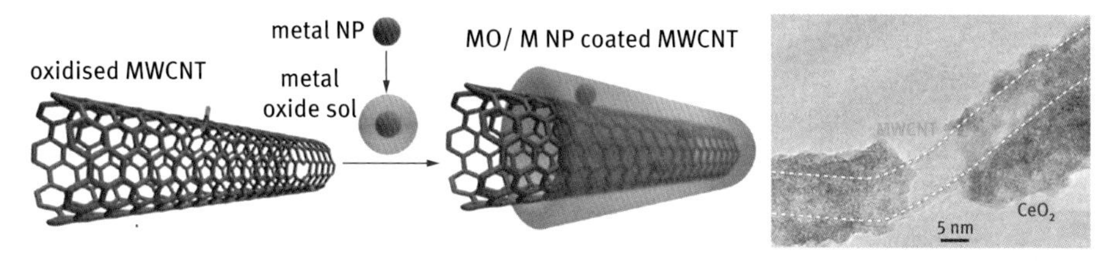

Fig. 5.18: Schematic and TEM image of reaction scheme to prepare metal nanoparticles encapsulated within metal oxide coating on oxidized MWCNTs. Metal NPs are added to developing metal alkoxide sol followed by addition of oxidized MWCNTs and water for hydrolysis. Adapted with permission from [228], © (2012) American Chemical Society.

hybrids of both CNTs and graphene with titanium silicate (TS-1), an industrially important photoactive zeolite, *via* microwave-assisted sol-gel synthesis have been prepared with benzyl alcohol as linking agent (Fig. 5.19) [225, 231]. In both cases the benzyl alcohol assisted sol–gel method produced a uniform, thin coating of the complex metal oxide TS-1 zeolite. Interestingly, the TS-1 particle size was approximately 20–30 times smaller when the CNTs were present indicating a heat-sink effect of the CNTs which facilitates the formation of TS-1 NPs with unique size and shape [231, 232]. The heat sink effect was more evident in the graphene (produced by arc discharge) hybrids where with increasing graphene concentration the TS-1 particles changed from large hexagonal prisms (0 wt%) to ellipsoidal particles (1 wt%), to rectangular plates (1–5 wt%) to spherical nanoparticles (10 nm, 5–20 wt%) [225].

Fig. 5.19: (a), (b) SEM and (c)–(f) TEM images of (a) pure TS-1, (b) TS-1 hybrid with CNTs (20 wt%), and (c)–(f) TS-1 hybrids with varying amounts of graphene showing (c) 1 wt%, (d) 5 wt%, (e) 10 wt% and (f) 20 wt%. (b) Reproduced with permission from [231], © (2010) Elsevier. (c)–(f) Adapted from [225], © (2012) Royal Society of Chemistry.

Another interesting hybrid has been prepared by Jaroniec *et al.* who utilized both precipitation and sol–gel strategies to synthesize RGO-MoS_2-TiO_2 [233]. First GO and Mo precursors were mixed along with thiorea as reducing agent prior to solvothermal treatment at 210 °C for 24 h to produce an RGO-MoS_2 hybrid. The MoS_2 was found to exist as small, thin sheets on the GO, similar to previous studies [159]. This hybrid then dispersed in an ethanol/water solution (1 : 2 v/v) and thoroughly sonicated prior to the dropwise addition of ($Ti(OC_4H_9)_4$) under stirring. Stirring was continued for 2 h during which sol–gel growth of TiO_2 was visualized by a solution color change prior to transferring to an autoclave for 10 h of solvothermal treatment at 180 °C. The obtained RGO-MoS_2-TiO_2 hybrid was shown to have a greater photocatalytic hydrogen generation than the TiO_2 alone and a composite of the materials [233].

To ensure metal oxide growth occurs only on the nanocarbon, careful consideration of the hydrolysis conditions are required. The ideal conditions vary between precursors, nanocarbon, linking mechanism and growth process to such an extent that optimization is required [232].

5.3.5 Gas phase deposition

Increased control of film composition, structure and size can be achieved by limiting the rate of reaction. This is possible using gas phase deposition where the amount of reactant is relatively low. Gas phase deposition loosely covers any hybridization strategy where at least one of the hybrid components is in the gas phase. This includes chemical vapor deposition (CVD), physical vapor deposition (PVD) and atomic layer deposition (ALD) as well as various plasma, sputtering and evaporation processes.

Gas phase approaches have the advantage that the nanocarbons do not need to be filtered or washed after hybridization making them ideal for nanocarbons produced on substrates such as CVD grown graphene films or CNT forests which tend to lose their structure upon immersion and/or drying. Consequently they are not ideal for chemically modified GO or CNTs.

5.3.5.1 Physical vapor deposition

Perhaps the simplest of the gas phase techniques, PVD requires high vacuum and a target material. The target material is vaporized *via* the input of energy (*e.g. via* heat for evaporation [234], or electron beam [235], or pulsed laser [236] or plasma sputtering [237]) where it is then used to coat a nanocarbon sample. Further modification of the target material is possible by simultaneously vaporizing multiple samples or controlling the atmosphere during deposition (*e.g.* oxygen to produce oxides) [238].

Recent examples include the thermal evaporation of Pd NPs onto graphene [239]. Here, graphene produced *via* CVD was transferred to a flexible substrate prior to Pd deposition. Thermal evaporation of Pd (nominally 1–10 nm) produced Pd NPs which were shown to be an effective hydrogen gas sensor [239]. Increased control of film thickness is obtained by using e-beam evaporation. For example, a 2 nm film of PbS has been coated onto CVD produced graphene *via* e-beam evaporation of a PbS target [240]. The PbS film was found to consist of a number nanoparticles which were used to study photoinduced charge transfer between graphene and the PbS [240]. Gruverman *et al.* have reported the uniform coating of a MWCNT forest with a ferroelectric oxide *via* pulsed laser deposition [241]. In this work, plasma enhanced CVD was used to create a vertically aligned array of MWCNTs on a silicon substrate. The ferroelectric coating was then produced by irradiating a $Pb(Zr_{0.52}Ti_{0.48})O_3$ target with a 1–2 J cm^{-2} laser at 5–7 Hz at 500–550 °C in a low oxygen atmosphere (70–80 mTorr). Care had to be taken here as the CNTs will be removed by oxidation at temperatures higher than 600 °C. The ferroelectric coated CNT forest has promise as an electronic memory storage device [241].

5.3.5.2 Chemical vapor deposition

The high energy of PVD processes can be damaging to substrate, reducing performance. Chemical vapor deposition is carried out using particles with less energy, and typically at lower pressure. The coating material can be from a vaporized target but also carried by a gas stream. Unused precursor materials are simply exhausted from the coating zone. CVD is also a common method for the production of CNTs and graphene (Chapers 1, 2). Therefore the growth of nanocarbons from interesting hybrid materials is also possible with CVD.

CVD growth can be a remarkably simple process. For example, carbon fibers made from entwined CNTs have been coated with WO_3 simply by first hanging over a crucible containing $W(OC_6H_5)_6$ precursor and heating to 400 °C for 10 min [242]. Similarly, ZnO nanorods have been grown from an RGO electrode by placing downstream from a high purity zinc source, heating to 550 °C in an argon carrier stream before addition of a small fraction of oxygen as a reaction gas [243]. Figure 5.20 shows that ZnO nanowires were grown vertically from the surface with both size and density of ZnO nanorods increasing with growth time.

Kim and co-workers compared ZnO thin films grown on graphene sheets *via* PVD and mist pyrolysis CVD (MPCVD) [244]. In this experiment graphene films grown *via* CVD were exposed to (1) sputter coating from a ZnO target and (2) a zinc acetate precursor ultrasonically atomized and carried into the reaction zone at 160 °C by a nitrogen stream. In both cases a thin ZnO film was obtained. However, SEM analysis showed many more cracks in the PVD prepared film. In addition, the resistivity of the PVD film increased 5 fold compared to minimal increase for the MPCVD case. Raman spectroscopy also indicated many more defects in the graphene lattice after PVD exposure, indicating – along with the resistivity measurements – that the higher energy ZnO particles damaged the graphene structure while ZnO growth *via* MPCVD had no detrimental effect [244].

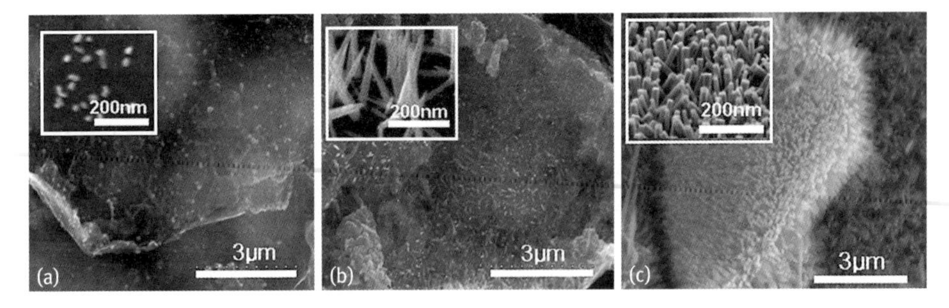

Fig. 5.20: SEM images of ZnO nanowires grown on graphene *via* CVD from a solid Zn precursor at increasing growth times from (a) 5 min to (b) 10 min and (c) 15 min, insets show higher magnification SEM images of the same surfaces. Reprinted with permission from [243], © (2012) Elsevier.

CVD processes can also be used to grow polymers from nanocarbons. An important example is the coating of a CNT forest with a thin layer of poly(tetrafluorethlylene) (PTFE) *via* hot filament CVD to produce a superhydrophobic substrate [245]. Here, a vertically aligned MWCNT forest was prepared and placed in a CVD reaction chamber. Hexafluorpropylene oxide gas was then thermally decomposed to form the reactive radical difluorocarbene (CF_2) and flowed over the CNT substrate along with a small amount of initiator where direct polymerization of PTFE onto the CNTs occurred [245].

Neocleus *et al.* have for the first time coated free-standing, hierarchical CNT fibers and forests [246] with TiO_2 *via* a simple CVD method by supporting the CNT structures over small PTFE beakers filled with precursors within an autoclave [247]. Upon thermal treatment, sublimation of the TiO_2 precursors resulted in deposition of TiO_2 onto the CNT substrates. Importantly, the addition of benzyl alcohol enabled the penetration of TiO_2 particles into the interior of the fiber, thus coating the individual CNT bundles [247].

Another route is the *in situ* growth of two types of nanocarbons. For example, nickel is often used to catalyze the growth of both CNTs and graphene *via* CVD [248]. Recently, nickel foams have been used as a substrate for the CVD growth of porous graphene films [249]. Dong *et al.* have further hybridized these films with Ni NPs *via* immersion in a solution containing $NiCl_2$ precursor and ethylene glycol as reducing agent [250]. The prepared Ni NPs were then used for the CVD growth of CNTs from the porous graphene substrate creating a conductive, porous CNT-graphene hybrid suitable for electrochemical sensing [250].

5.3.5.3 Atomic layer deposition

Atomic layer deposition is a high vacuum process where small amounts of the precursors are leaked into the system sequentially with intermittent evacuations. The ALD enables the conformal deposition of atomically thin layers with precise thickness control at low temperatures without the typical aggregate formation in the gas-phase.

The quality of the thin film depends on preferential interactions between precursor and coating substrate. However, the initial layer is clearly the most important and, in the case of nanocarbons, the surface chemistry must be tailored. For most ALD precursors, hydrophilic surface groups enhance deposition, which can be achieved by functionalizing the nanocarbon prior to placement in the ALD reaction chamber or by treating the sample within the chamber with reactive plasma. Among the many *in situ* hybridization techniques, ALD provides best control of thin film thickness.

Recent nanocarbon treatment methods investigated to enhance ALD deposition include chemical oxidation in mixed acid for CNT-TiO_x [251], mixing H_2O with precursor for CNT-TiO_2 [252], RGO-TiO_2 [253], CNT-ZnO [254, 255], graphene-SnO_2 [256] and CNT-V_2O_5 [257]. The need for CNT functionalization for the ALD coating of Pt has been studied in detail by Dameron *et al.* [258]. It was found that without functionalization, 100 ALD cycles of precursor (Trimethyl(methylcyclopentyldienyl) platinum) were required for a noticeable Pt deposition. Pt growth occurred in the form of randomly dispersed NPs indicating that growth most likely evolved from defect sites along the CNT wall. To improve deposition uniformity several functionalization strategies were investigated: (i) trimethylaluminium (TMA) pretreatment, (ii) Ar plasma, (iii) O_2 plasma, and (iv) Ar/O_2 plasma. It was found that Ar plasma treatment reduced Pt deposition, most likely a result of removal of amorphous carbon without increasing surface oxygen groups. TMA treatment resulted in a much higher Pt deposition but the coating was not uniform, presumably due to the incomplete initial functionalization of TMA or partial TMA removal under high vacuum. O_2 plasma resulted in the highest, most uniform coverage of Pt which is attributed to the even partial oxidation of the CNTs [258]. Partial CNT oxidation *via* ozone treatment has recently been investigated for the ALD growth of NiO NPs. In this work, NiO NPs were formed by cycling O_3 and bis(cyclopentadienyl) nickel precursor to create well-dispersed NiO NPs after 400 cycles [259].

Oxidative-based nanocarbon functionalization for ALD pretreatment has been shown to be quite effective in producing thin films. However, such oxidative treatments reduce electronic properties rendering them unsuitable for many applications. One alternative is coating with organic monolayers. For example, Dai *et al.* have monitored the influence of a monolayer of perylene tetracarboxylic acid (PTCA) on the ALD deposition of Al_2O_3 on CVD grown graphene (Fig. 5.21) [260]. Without PTCA, Al_2O_3 deposition occurs only on the edges of the graphene and in defect sites where dangling bonds interact with the ALD precursor. A monolayer of PTCA was formed on the graphene *via* immersion for 30 min where π-π interactions between graphene and PTCA resulted in a monolayer coating after rinsing. ALD growth of Al_2O_3 on the PTCA modified graphene led to a thin, even Al_2O_3 film deposition [260]. A similar organic monolayer, perylene tetracarboxylic dianhydride, has been evaporated onto CVD grown graphene to improve the ALD film growth and adhesion of Al_2O_3 and HfO_2 [261].

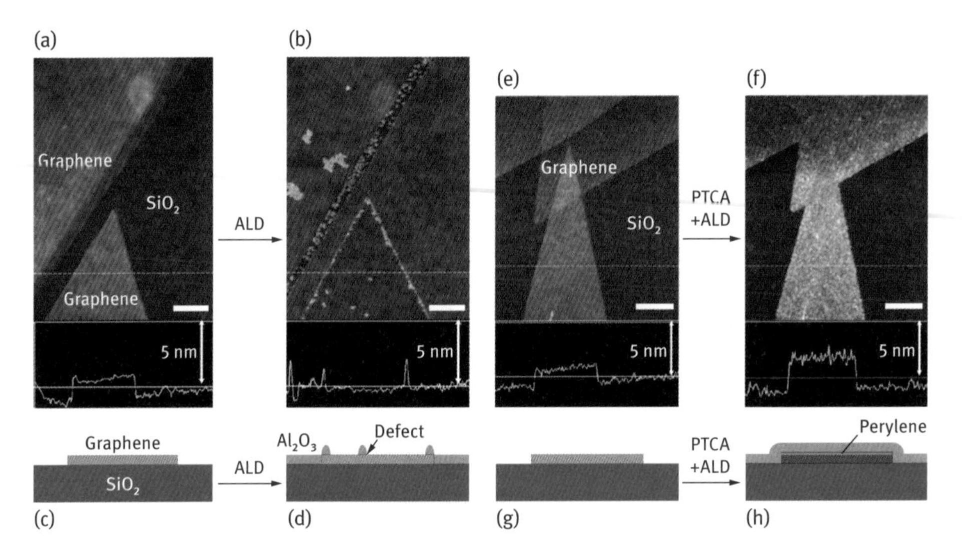

Fig. 5.21: ALD deposition of Al_2O_3 on graphene showing (a), (b), (e), (f) AFM and (c), (d), (g), (h) schematics of the process. (a)–(d) Without a hydrophilic linker deposition occurs only at graphene edges and defect sites. (e)–(h) With the use of the organic linker perylene tetracarboxylic dianhydride (PTCA) a uniform deposition is obtained. Reprinted with permission from [260], © (2008) American Chemical Society.

The highlighted examples of gas phase hybridization clearly show that they are well suited for the deposition of metals and metal oxides. They are most practical when the nanocarbons are either surface bound or grown from a surface, such as CVD grown graphene and CNT forests. Improved control of layer thickness can be obtained, in comparison to wet chemical approaches, but the synthesis strategies generally require sophisticated equipment not often present in a laboratory.

5.4 Other nanocarbons

Nanocarbons other than CNTs and graphene often exhibit similar surface chemistry and can be hybridized in a similar fashion. For example, single-walled carbon nanohorns (SWCNHs) have been oxidized *via* heat treatment in air atmosphere followed by immersion in a solution containing H_2PtCl_6. The Pt ions adsorbed to the oxidized SWCNHs and were then reduced *via* addition of sodium citrate to form Pt NPs [150].

A mesoporous carbon-CNT hybrid has been prepared using a mesoporous silica template [262]. The template was mechanically mixed with (a) phthalocyanine (Pc) or (b) Ni-phthalocyanine (Ni-Pc) followed by heating to 900 °C in Ar. Upon cooling, the silica template was removed *via* washing with hydrofluoric acid. SEM images showed that when Ni-Pc was incorporated, medusa like CNTs had grown from the sides of the mesoporous carbon [262].

Graphitic carbon nitride (g-C_3N_4) is an important semiconducting nanocarbon with a suitable band gap for solar energy conversion [263]. The low chemical reactivity of g-C_3N_4 make *ex situ* hybridization impossible while *in situ* synthesis of a hybrid component is also rarely reported. Classical composites *via* ultrasonic mixing [264] or deposition [265] of pre-prepared g-C_3N_4 and the second component are common. Alternatively, composites consisting of g-C_3N_4 can be prepared *via* the synthesis of g-C_3N_4 from its precursors in the presence of a second component (*e.g.* metal/metal oxide NP, polymer, CNT). The rigid polymeric structure of g-C_3N_4 then simply encapsulates the second component without any specific surface interaction. For example, a g-C_3N_4 composite was prepared *via* mixing MWCNTs with cyanamide, a common g-C_3N_4 precursor, *via* ultrasonication. The mixture was then dried and heated to 550 °C with nitrogen flow where g-C_3N_4 was formed, producing the composite material [266]. RGO-g-C_3N_4 composites have been prepared following a similar method [267]. G-C_3N_4 composites with PANI have been prepared *via* the opposite case where aniline was polymerized in the presence of the pre-prepared g-C_3N_4 [268].

5.5 Comparison of synthesis techniques

The advantages and disadvantages of each nanocarbon hybridization strategy discussed here are summarized in Table 5.1. The main advantage of the *ex situ* approach is that structure, size and morphology of the hybridizing component are pre-defined. This is a vital aspect when considering the hybridization of uniquely shaped moieties such as nanocubes or biomolecules. However, *ex situ* linking generally requires nanocarbon modification and consequently hybrid density is low.

The *in situ* wet chemical approach requires less nanocarbon modification, especially for electrodeposition, and can produce thin, uniform, multilayer films. This is the method of choice for nanocarbon-polymer hybrids as the increased interfacial area reduces problems of nanocarbon insolubility and subsequent aggregation. Gas phase deposition offers the greatest control of thin film thickness but is suitable almost exclusively to the deposition of metals and metal oxides.

Wet chemical approaches are easier to replicate as they generally do not require dedicated equipment, although occasionally hazardous materials are used. Conversely, gas phase hybridization requires dedicated facilities which are not available to all researchers. Gas phase and electrochemical deposition require the nanocarbon to be surface bound so are best used when the nanocarbon is prepared in this fashion.

Another important consideration involves the hybridization of porous carbon with hierarchical 3D architectures, such as fibers or arrays. Wet chemical techniques are often useless as the mandatory solvent removal/drying typically results in the at least partial collapse of the nanocarbon pore structure. Gas phase deposition is a better choice of method, although it requires the nanocarbon to be surface bound in the desired geometry (*i.e.* CNT forest or patterned single layer graphene).

Various *in situ* methods require post-treatments to induce crystallization or phase transformation. In some cases, however, the required temperatures for these treatments may be close to or even higher than those for the thermal oxidation of the nanocarbons themselves. Furthermore, it has been shown that the oxidation temperature of CNTs can be drastically reduced to as low as 330 °C, when coated with reducible metal oxides such as Bi_2O_3 [180]. This process is thought to occur *via* a Mars van Krevelen mechanism and may severely reduce the potential operating temperatures for various applications such as fuel cells.

The greatest advantage of *in situ* methods over *ex situ* processes is the benefit of using the nanocarbon as a substrate, template and "heat sink" for stabilizing metastable phases and small particle sizes and creating hybrids with unusual morphologies [232]. This enables the synthesis of new hybrid materials that may offer new properties and unknown potential for future research and application.

Tab. 5.1: Comparison of nanocarbon hybridization strategies discussed within this chapter.

	ex situ	***in situ* wet chemical**	***in situ* gas phase**
nanocarbon state	powder, suspended in solution	powder, suspended in solution OR (for electrochemical) adhered to electrode	adhered to substrate
morphology	pre-defined	coating, NPs, crystals possible	coating, NPs
structure	pre-defined	often amorphous	deposition rate dependent
size control	pre-defined	reaction conditions dependent	high, reaction time dependent
coating uniformity	linking mechanism dependent maximum monolayer	reaction condition and nanocarbon surface modification dependent	often conformal
interface	requires nanocarbon surface modification	pristine or modified nanocarbon	pristine nanocarbons are suitable

5.6 Summary

This chapter demonstrates the huge variety of synthesis techniques available for the preparation of nanocarbon hybrids, which can be categorized into *ex situ* and *in situ* approaches.

Even today, most researchers choose to oxidize nanocarbons (*i.e.* acid treatment) prior to hybridization in order to introduce functional groups and so render them hydrophilic and thus more accessible for dispersion and subsequent hybridization. However, such pretreatment does not allow sufficient control over the type, number and location of the functional groups, often limiting the quality and uniformity of the coating. Furthermore, it damages the structure of the nanocarbon considerably and thus negatively affects their electronic and mechanical properties. We therefore strongly recommend to consider noncovalent modification instead, which are often simpler, more versatile, less hazardous (*i.e.* compared with strong oxidizing acids) and – despite offering weaker interfacial interactions – typically enable more homogeneous coatings.

In all cases, the optimization of the desired hybrid structure, *i.e.* most suited for the selected application, depends on many choices, such as the type and quality of the nanocarbon (*i.e.* graphene, CNT *etc.*), and the type, structure and morphology of the active component for hybridization (*i.e.* metal oxide, biomolecule, polymer, metal nanoparticle). The extent and nature of the interface is also a crucial parameter, especially in view of fundamental studies on interfacial charge and energy transfer processes as well as for optimizing the hybrid's performance for the desired application. Interface engineering, such as adjusting the crystallization conditions on the nanocarbon surface or the deposition of an interlayer consisting of organic linker molecules, conducting polymers or dielectric inorganic barrier layers, will become an attractive focus in future research on nanocarbon hybrids.

As subsequent chapters will document, the type, structure and quality of the nanocarbon have a considerable impact on the final performance of the nanocarbon hybrid. Currently, most publications on the synthesis of nanocarbon hybrids focus on GO, which is both easy to prepare and simple to hybridize. However, the mechanical and electrical properties of GO (and also RGO) are often inferior to their pristine counterparts and in fact closer to those of activated carbon. Hence, we recommend always synthesizing and comparing various types of nanocarbons with different features and functionalizations.

Finally, the reproducibility of the nanocarbons and their hybrids is of paramount importance when implementing them into commercial devices. This will require the definition of key characteristics and the development of standard synthesis methodology that will also enable better comparison of results between research groups.

Nomenclature

Abbreviation	Definition
ALD	atomic layer deposition
ATRP	atom transfer radical polymerization
CNT	carbon nanotube
CTA	chain transfer agent
CTAB	cetyltrimethylammonium bromide

CuAAC	copper catalysed alkyne-azide cycloaddition
CVD	chemical vapor deposition
DCC	dicylohexyl carbodiimide
DMF	dimethyl formamide
DWCNT	double walled carbon nanotube
EDC	1-ethyl-3-(3-dimethylaminopropyl) carbodiimide hydrochloride
G	graphene
G	graphene oxide
GC	glassy carbon
g-C3N4	graphitic carbon nitride
IL	ionic liquid
ITO	indium doped tin oxide
MO	metal oxide
MPA	mercaptopropionic acid
MPCVD	mist pyrolysis CVD
MWCNT	multi walled carbon nanotube
NP	nanoparticle
PAH	poly(allylamine hydrochloride)
PANI	poly(aniline)
PDA	poly(dopamine)
PDDA	poly(diallyl dimethylammonium chloride)
PDMS	poly(dimethyl sulfoxide)
PEI	poly(ethyeleneimine)
PSS	poly(sodium 4-styrenesulfonate)
PTCA	perylene tetracarboxlic acid
PTFE	poly(tetrafluroethylene)
PVD	physical vapor deposition
QD	quantum dot
RAFT	reversible addition-fragmentation chain transfer
SDS	sodium dodecyl sufate
SEED	substrate enhanced electroless deposition
SWCNH	single-walled carbon nanohorn
SWCNT	single-walled carbon nanotube
TMA	trimethyl aluminum

Bibliography

[1] Vilatela, J.J. and D. Eder, *Nanocarbon composites and hybrids in sustainability: A Review.* ChemSusChem, 2012. **5**(3): p. 456–478.

[2] Shearer, C.J., A. Cherevan, and D. Eder, *Application and Future Challenges of Functional Nanocarbon Hybrids*. Adv. Mat., 2014. **26**(15): p. 2295–2318.

[3] Eder, D., *Carbon nanotube–inorganic hybrids*. Chemical Reviews, 2010. **110**(3): p. 1348–1385.

[4] Georgakilas, V., et al., *Decorating carbon nanotubes with metal or semiconductor nanoparticles.* Journal of Materials Chemistry, 2007. **17**(26): p. 2679–2694.

[5] Wildgoose, G.G., C.E. Banks, and R.G. Compton, *Metal nanopartictes and related materials supported on carbon nanotubes: Methods and applications.* Small, 2006. **2**(2): p. 182–193.

[6] Li, X., et al., *Noncovalent assembly of carbon nanotube-inorganic hybrids.* Journal of Materials Chemistry, 2011. **21**(21): p. 7527–7547.

[7] Bai, S. and X. Shen, *Graphene-inorganic nanocomposites.* RSC Advances, 2012. **2**(1): p. 64–98.

[8] Xiang, Q., J. Yu, and M. Jaroniec, *Graphene-based semiconductor photocatalysts.* Chemical Society Reviews, 2012. **41**(2): p. 782–796.

[9] Neises, B. and W. Steglich, *Simple method for the esterification of carboxylic acids.* Angewandte Chemie International Edition in English, 1978. **17**(7): p. 522–524.

[10] Neises, B. and W. Steglich, *Esterification of carboxyic acids with dicyclohexylcarbodiimide/4-dimethylaminopyridine: tert-butyl ethyl fumarate.* Organic Syntheses, 1990. **7**: p. 93–94.

[11] Valeur, E. and M. Bradley, *Amide bond formation: beyond the myth of coupling reagents.* Chemical Society Reviews, 2009. **38**(2): p. 606–631.

[12] He, F., et al., *The attachment of Fe3O4 nanoparticles to graphene oxide by covalent bonding.* Carbon, 2010. **48**(11): p. 3139–3144.

[13] Yang, C., et al., *Conjugates of graphene oxide covalently linked ligands and gold nanoparticles to construct silver ion graphene paste electrode.* Talanta, 2012. **97**(0): p. 406–413.

[14] Banerjee, S. and S.S. Wong, *Synthesis and Characterization Carbon Nanotube-Nanocrystal Heterostructures.* Nano Letters, 2002. **2**(3): p. 195–200.

[15] Daniel, S., et al., *A review of DNA functionalized/grafted carbon nanotubes and their characterization.* Sensors and Actuators B: Chemical, 2007. **122**(2): p. 672–682.

[16] Hu, H., et al., *Microwave-assisted covalent modification of graphene nanosheets with chitosan and its electrorheological characteristics.* Applied Surface Science, 2011. **257**(7): p. 2637–2642.

[17] Flavel, B., M. Nambiar, and J. Shapter, *Electrochemical Detection of Copper Using a Gly-Gly-His Modified Carbon Nanotube Biosensor.* Silicon, 2011. **3**(4): p. 163–171.

[18] Sharma, P., et al., *Bio-functionalized graphene–graphene oxide nanocomposite based electrochemical immunosensing.* Biosensors and Bioelectronics, 2013. **39**(0): p. 99–105.

[19] Gooding, J.J., et al., *Protein Electrochemistry Using Aligned Carbon Nanotube Arrays.* Journal of the American Chemical Society, 2003. **125**(30): p. 9006–9007.

[20] Palacin, T., et al., *Efficient Functionalization of Carbon Nanotubes with Porphyrin Dendrons via Click Chemistry.* Journal of the American Chemical Society, 2009. **131**(42): p. 15394–15402.

[21] Han, L., et al., *Ionic liquids grafted on carbon nanotubes as highly efficient heterogeneous catalysts for the synthesis of cyclic carbonates.* Applied Catalysis A: General, 2012. **429–430**(0): p. 67–72.

[22] Yu, J., et al., *Electron transfer through [small alpha]-peptides attached to vertically aligned carbon nanotube arrays: a mechanistic transition.* Chemical Communications, 2012. **48**(8): p. 1132–1134.

[23] Fan, L., et al., *Ferrocene functionalized graphene: preparation, characterization and efficient electron transfer toward sensors of H2O2.* Journal of Materials Chemistry, 2012. **22**(13): p. 6165–6170.

[24] Xu, Y., et al., *A graphene hybrid material covalently functionalized with porphyrin: Synthesis and optical limiting property.* Advanced Materials, 2009. **21**(12): p. 1275–1279.

[25] Zhao, H., et al., *Synthesis, characterization, and photophysical properties of covalent-linked ferrocene–porphyrin–single-walled carbon nanotube triad hybrid.* Carbon, 2012. **50**(13): p. 4894–4902.

[26] Maruyama, H., et al., *Covalent attachment of a specific site of a protein molecule on a carbon nanotube tip.* Journal of Applied Physics, 2012. **111**(7): p. 074701.

[27] Daniel, M.-C. and D. Astruc, *Gold nanoparticles: Assembly, supramolecular chemistry, quantum-size-related properties, and applications toward biology, catalysis, and nanotechnology.* Chemical Reviews, 2003. **104**(1): p. 293–346.

[28] Zanella, R., et al., *Deposition of gold nanoparticles onto thiol-functionalized multiwalled carbon nanotubes.* Journal of Physical Chemistry B, 2005. **109**(34): p. 16290–16295.

[29] Peng, X., et al., *Efficient charge separation in multidimensional nanohybrids.* Nano Letters, 2011. **11**(11): p. 4562–4568.

[30] Peng, X. and S.S. Wong, *Functional covalent chemistry of carbon nanotube surfaces.* Advanced Materials, 2009. **21**(6): p. 625–642.

[31] Wang, Q.H., et al., *Understanding and controlling the substrate effect on graphene electron-transfer chemistry via reactivity imprint lithography.* Nature Chemistry, 2012. **4**(9): p. 724–732.

[32] Dyke, C.A., et al., *Diazonium-based functionalization of carbon nanotubes: XPS and GC-MS analysis and mechanistic implications.* Synlett, 2004. **2004**: p. 155–160.

[33] Lerner, M.B., et al., *Hybrids of a genetically engineered antibody and a carbon nanotube transistor for detection of prostate cancer biomarkers.* ACS Nano, 2012. **6**(6): p. 5143–5149.

[34] Ismaili, H., F.o. Lagugneì-Labarthet, and M.S. Workentin, *Covalently assembled gold nanoparticle-carbon nanotube hybrids via a photoinitiated carbene addition reaction.* Chemistry of Materials, 2011. **23**(6): p. 1519–1525.

[35] Snell, K.E., H. Ismaili, and M.S. Workentin, *Photoactivated nitrene chemistry to prepare gold nanoparticle hybrids with carbonaceous materials.* ChemPhysChem, 2012. **13**(13): p. 3185–3193.

[36] Zhou, W., et al., *A general strategy toward graphene@metal oxide core-shell nanostructures for high-performance lithium storage.* Energy & Environmental Science, 2011. **4**(12): p. 4954–4961.

[37] Moore, V.C., et al., *Individually suspended single-walled carbon nanotubes in various surfactants.* Nano Letters, 2003. **3**(10): p. 1379–1382.

[38] Islam, M.F., et al., *High weight fraction surfactant solubilization of single-wall carbon nanotubes in water.* Nano Letters, 2003. **3**(2): p. 269–273.

[39] Blanch, A.J., C.E. Lenehan, and J.S. Quinton, *Optimizing surfactant concentrations for dispersion of single-walled carbon nanotubes in aqueous solution.* The Journal of Physical Chemistry B, 2010. **114**(30): p. 9805–9811.

[40] Fernández-Merino, M.J., et al., *Investigating the influence of surfactants on the stabilization of aqueous reduced graphene oxide dispersions and the characteristics of their composite films.* Carbon, 2012. **50**(9): p. 3184–3194.

[41] Guardia, L., et al., *High-throughput production of pristine graphene in an aqueous dispersion assisted by non-ionic surfactants.* Carbon, 2011. **49**(5): p. 1653–1662.

[42] Jiao, J., et al., *Decorating multi-walled carbon nanotubes with Au nanoparticles by amphiphilic ionic liquid self-assembly.* Colloids and Surfaces A: Physicochemical and Engineering Aspects, 2012. **408**: p. 1–7.

[43] Lee, C.-L., et al., *Preparation of Pt nanoparticles on carbon nanotubes and graphite nanofibers via self-regulated reduction of surfactants and their application as electrochemical catalyst.* Electrochemistry Communications, 2005. **7**(4): p. 453–458.

[44] Fang, B., et al., *High Pt loading on functionalized multiwall carbon nanotubes as a highly efficient cathode electrocatalyst for proton exchange membrane fuel cells.* Journal of Materials Chemistry, 2011. **21**(22): p. 8066–8073.

[45] Tan, Z., H. Abe, and S. Ohara, *Ordered deposition of Pd nanoparticles on sodium dodecyl sulfate-functionalized single-walled carbon nanotubes.* Journal of Materials Chemistry, 2011. **21**(32): p. 12008–12014.

[46] Whitsitt, E.A., et al., *LPD silica coating of individual single walled carbon nanotubes.* Journal of Materials Chemistry, 2005. **15**(44): p. 4678–4687.

[47] Whitsitt, E.A. and A.R. Barron, *Silica Coated Single Walled Carbon Nanotubes.* Nano Letters, 2003. **3**(6): p. 775–778.

[48] Wu, H., et al., *Solvothermal synthesis and optical limiting properties of carbon nanotube-based hybrids containing ternary chalcogenides.* Carbon, 2012. **50**(13): p. 4847–4855.

[49] Ellis, A.V., et al., *Hydrophobic anchoring of monolayer-protected gold nanoclusters to carbon nanotubes.* Nano Letters, 2003. **3**(3): p. 279–282.

[50] Rahman, G.M.A., et al., *Dispersable carbon nanotube/gold nanohybrids: evidence for strong electronic interactions.* Small, 2005. **1**(5): p. 527–530.

[51] Wang, J., et al., *Reduced graphene oxide/ZnO Composite: Reusable adsorbent for pollutant management.* ACS Applied Materials & Interfaces, 2012. **4**(6): p. 3084–3090.

[52] Shi, Z., et al., *Free-standing single-walled carbon nanotube-CdSe quantum dots hybrid ultrathin films for flexible optoelectronic conversion devices.* Nanoscale, 2012. **4**(15): p. 4515–4521.

[53] Sainsbury, T. and D. Fitzmaurice, *Carbon-nanotube-templated and pseudorotaxane-formation-driven gold nanowire self-assembly.* Chemistry of Materials, 2004. **16**(11): p. 2174–2179.

[54] Guo, C.X., et al., *RGD-Peptide functionalized graphene biomimetic live-cell sensor for real-time detection of nitric oxide molecules.* ACS Nano, 2012. **6**(8): p. 6944–6951.

[55] Yang, D.Q., B. Hennequin, and E. Sacher, *XPS Demonstration of π-π interaction between benzyl mercaptan and multiwalled carbon nanotubes and their use in the adhesion of Pt nanoparticles.* Chemistry of Materials, 2006. **18**(21): p. 5033–5038.

[56] Zhu, M., et al., *Enhanced photocatalytic hydrogen evolution performance based on Ru-trisdicarboxybipyridine-reduced graphene oxide hybrid.* Journal of Materials Chemistry, 2012. **22**(45): p. 23773–23779.

[57] Roquelet, C., et al., *Π-Stacking functionalization of carbon nanotubes through micelle swelling.* ChemPhysChem, 2010. **11**(8): p. 1667–1672.

[58] Wang, X., et al., *Immobilization of tetra-tert-butylphthalocyanines on carbon nanotubes: a first step towards the development of new nanomaterials.* Journal of Materials Chemistry, 2002. **12**(6): p. 1636–1639.

[59] D'Souza, F., et al., *Self-assembled single-walled carbon nanotube:Zinc–porphyrin hybrids through ammonium ion–crown ether interaction: Construction and electron transfer.* Chemistry – A European Journal, 2007. **13**(29): p. 8277–8284.

[60] Wei, G., et al., *Protein-promoted synthesis of Pt nanoparticles on carbon nanotubes for electrocatalytic nanohybrids with enhanced glucose sensing.* The Journal of Physical Chemistry C, 2011. **115**(23): p. 11453–11460.

[61] Lv, W., et al., *Graphene-DNA hybrids: self-assembly and electrochemical detection performance.* Journal of Materials Chemistry, 2010. **20**(32): p. 6668–6673.

[62] Eder, D. and A.H. Windle, *Carbon–inorganic hybrid materials: The carbon-nanotube/TiO2 interface.* Advanced Materials, 2008. **20**(9): p. 1787–1793.

[63] Mu, Y., et al., *Controllable Pt nanoparticle deposition on carbon nanotubes as an anode catalyst for direct methanol fuel cells.* The Journal of Physical Chemistry B, 2005. **109**(47): p. 22212–22216.

[64] Liu, L., et al., *Self-assembly of gold nanoparticles to carbon nanotubes using a thiol-terminated pyrene as interlinker.* Chemical Physics Letters, 2003. **367**(5–6): p. 747–752.

[65] Ding, M., et al., *Welding of gold nanoparticles on graphitic templates for chemical sensing.* Journal of the American Chemical Society, 2012. **134**(7): p. 3472–3479.

[66] Li, H., et al., *Palladium nanoparticles decorated carbon nanotubes: facile synthesis and their applications as highly efficient catalysts for the reduction of 4-nitrophenol.* Green Chemistry, 2012. **14**(3): p. 586–591.

[67] Li, C., et al., *Decoration of multiwall nanotubes with cadmium sulfide nanoparticles.* Carbon, 2006. **44**(10): p. 2021–2026.

[68] Herranz, M.Á., et al., *Spectroscopic characterization of photolytically generated radical ion pairs in single-wall carbon nanotubes bearing surface-immobilized tetrathiafulvalenes.* Journal of the American Chemical Society, 2007. **130**(1): p. 66–73.

[69] Romero-Nieto, C., et al., *Tetrathiafulvalene-based nanotweezers – noncovalent binding of carbon nanotubes in aqueous media with charge transfer implications.* Journal of the American Chemical Society, 2012. **134**(22): p. 9183–9192.

[70] Ozawa, H., et al., *Supramolecular hybrid of gold nanoparticles and semiconducting single-walled carbon nanotubes wrapped by a porphyrin–fluorene copolymer.* Journal of the American Chemical Society, 2011. **133**(37): p. 14771–14777.

[71] Zheng, J., et al., *DNA as a linker for biocatalytic deposition of Au nanoparticles on graphene and its application in glucose detection.* Journal of Materials Chemistry, 2011. **21**(34): p. 12873–12879.

[72] Malig, J., et al., *Direct exfoliation of graphite with a porphyrin – creating functionalizable nanographene hybrids.* Chemical Communications, 2012. **48**(70): p. 8745–8747.

[73] Zhang, D., et al., *Enhanced capacitive deionization performance of graphene/carbon nanotube composites.* Journal of Materials Chemistry, 2012. **22**(29): p. 14696–14704.

[74] Vinayan, B.P., et al., *Synthesis of graphene-multiwalled carbon nanotubes hybrid nanostructure by strengthened electrostatic interaction and its lithium ion battery application.* Journal of Materials Chemistry, 2012. **22**(19): p. 9949–9956.

[75] Chen, S., et al., *Chemical-free synthesis of graphene–carbon nanotube hybrid materials for reversible lithium storage in lithium-ion batteries.* Carbon, 2012. **50**(12): p. 4557–4565.

[76] Mani, V., B. Devadas, and S.-M. Chen, *Direct electrochemistry of glucose oxidase at electrochemically reduced graphene oxide-multiwalled carbon nanotubes hybrid material modified electrode for glucose biosensor.* Biosensors and Bioelectronics, 2012. **41**: p. 309–315.

[77] Lu, L., et al., *Highly stable air working bimorph actuator based on a graphene nanosheet/carbon nanotube hybrid electrode.* Advanced Materials, 2012. **24**(31): p. 4317–4321.

[78] Rance, G.A. and A.N. Khlobystov, *Nanoparticle-nanotube electrostatic interactions in solution: the effect of pH and ionic strength.* Physical Chemistry Chemical Physics, 2010. **12**(36): p. 10775–10780.

[79] Zhou, X.S., et al., *Self-assembled nanocomposite of silicon nanoparticles encapsulated in graphene through electrostatic attraction for lithium-ion batteries.* Advanced Energy Materials, 2012. **2**(9): p. 1086–1090.

[80] Sreejith, S., X. Ma, and Y. Zhao, *Graphene oxide wrapping on squaraine-loaded mesoporous silica nanoparticles for bioimaging.* Journal of the American Chemical Society, 2012. **134**(42): p. 17346–17349.

[81] Yang, W., et al., *Carbon nanotubes decorated with Pt nNanocubes by a noncovalent functionalization method and their role in oxygen reduction.* Advanced Materials, 2008. **20**(13): p. 2579–2587.
[82] Jiang, K., et al., *Selective attachment of gold nanoparticles to nitrogen-doped carbon nanotubes.* Nano Letters, 2003. **3**(3): p. 275–277.
[83] Du, N., et al., *Homogeneous coating of Au and SnO2 nanocrystals on carbon nanotubes via layer-by-layer assembly: a new ternary hybrid for a room-temperature CO gas sensor.* Chemical Communications, 2008(46): p. 6182–6184.
[84] Zhang, S., et al., *Carbon nanotubes decorated with Pt nanoparticles via electrostatic self-assembly: a highly active oxygen reduction electrocatalyst.* Journal of Materials Chemistry, 2010. **20**(14): p. 2826–2830.
[85] Stoffelbach, F., et al., *An easy and economically viable route for the decoration of carbon nanotubes by magnetite nanoparticles, and their orientation in a magnetic field.* Chemical Communications, 2005(36): p. 4532–4533.
[86] Sun, J. and L. Gao, *Attachment of inorganic nanoparticles onto carbon nanotubes.* Journal of Electroceramics, 2006. **17**(1): p. 91–94.
[87] Sun, J., et al., *Single-walled carbon nanotubes coated with titania nanoparticles.* Carbon, 2004. **42**(4): p. 895–899.
[88] Sun, P., et al., *The formation of graphene–titania hybrid films and their resistance change under ultraviolet irradiation.* Carbon, 2012. **50**(12): p. 4518–4523.
[89] Li, Z., et al., *Electrostatic layer-by-layer self-assembly multilayer films based on graphene and manganese dioxide sheets as novel electrode materials for supercapacitors.* Journal of Materials Chemistry, 2011. **21**(10): p. 3397–3403.
[90] Yu, D. and L. Dai, *Self-assembled graphene/carbon nanotube hybrid films for supercapacitors.* The Journal of Physical Chemistry Letters, 2009. **1**(2): p. 467–470.
[91] Lei, Z., F. Shi, and L. Lu, *Incorporation of MnO2-coated carbon nanotubes between graphene sheets as supercapacitor electrode.* ACS Applied Materials & Interfaces, 2012. **4**(2): p. 1058–1064.
[92] Niu, Z., et al., *Electrophoretic build-up of alternately multilayered films and micropatterns based on graphene sheets and nanoparticles and their applications in flexible supercapacitors.* Small, 2012. **8**(20): p. 3201–3208.
[93] Guo, S., S. Dong, and E. Wang, *Constructing carbon nanotube/Pt nanoparticle hybrids using an imidazolium-salt-based ionic liquid as a linker.* Advanced Materials, 2010. **22**(11): p. 1269–1272.
[94] Wang, Z., et al., *The synthesis of ionic-liquid-functionalized multiwalled carbon nanotubes decorated with highly dispersed Au nanoparticles and their use in oxygen reduction by electrocatalysis.* Carbon, 2008. **46**(13): p. 1687–1692.
[95] Wu, B., et al., *Functionalization of carbon nanotubes by an ionic-liquid polymer: Dispersion of Pt and PtRu nanoparticles on carbon nanotubes and their electrocatalytic oxidation of methanol.* Angewandte Chemie International Edition, 2009. **48**(26): p. 4751–4754.
[96] Zhang, H. and H. Cui, *Synthesis and characterization of functionalized ionic liquid-stabilized metal (gold and platinum) nanoparticles and metal nanoparticle/carbon nanotube hybrids.* Langmuir, 2009. **25**(5): p. 2604–2612.
[97] Hu, Y., et al., *Green-synthesized gold nanoparticles decorated graphene sheets for label-free electrochemical impedance DNA hybridization biosensing.* Biosensors and Bioelectronics, 2011. **26**(11): p. 4355–4361.
[98] Tunckol, M., J. Durand, and P. Serp, *Carbon nanomaterial–ionic liquid hybrids.* Carbon, 2012. **50**(12): p. 4303–4334.

[99] Narayanan, R., M. Deepa, and A.K. Srivastava, *Nanoscale connectivity in a TiO2/CdSe quantum dots/functionalized graphene oxide nanosheets/Au nanoparticles composite for enhanced photoelectrochemical solar cell performance.* Physical Chemistry Chemical Physics, 2012. **14**(2): p. 767–778.

[100] Li, X., et al., *Atomic layer deposition of ZnO on multi-walled carbon nanotubes and its use for synthesis of CNT-ZnO heterostructures.* Nanoscale Research Letters, 2010. **5**(11): p. 1836–1840.

[101] Turgman-Cohen, S. and J. Genzer, *Computer simulation of controlled radical polymerization: Effect of chain confinement due to initiator grafting density and solvent quality in "grafting from" Method.* Macromolecules, 2010. **43**(22): p. 9567–9577.

[102] Minko, S., *Grafting on Solid Surfaces: "Grafting to" and "Grafting from" Methods*, in *Polymer Surfaces and Interfaces*, M. Stamm, Editor. 2008, Springer Berlin Heidelberg. p. 215–234.

[103] Yao, Z., et al., *Polymerization from the surface of single-walled carbon nanotubes – Preparation and characterization of nanocomposites.* Journal of the American Chemical Society, 2003. **125**(51): p. 16015–16024.

[104] Spitalsky, Z., et al., *Carbon nanotube–polymer composites: Chemistry, processing, mechanical and electrical properties.* Progress in Polymer Science, 2010. **35**(3): p. 357–401.

[105] Kuilla, T., et al., *Recent advances in graphene based polymer composites.* Progress in Polymer Science, 2010. **35**(11): p. 1350–1375.

[106] Kumar, N.A., et al., *Electrochemical supercapacitors based on a novel graphene/conjugated polymer composite system.* Journal of Materials Chemistry, 2012. **22**(24): p. 12268–12274.

[107] Qiu, J.-D., et al., *Controllable deposition of a platinum nanoparticle ensemble on a polyaniline/graphene hybrid as a novel electrode material for electrochemical sensing.* Chemistry – A European Journal, 2012. **18**(25): p. 7950–7959.

[108] Qin, S., et al., *Polymer brushes on single-walled carbon nanotubes by atom transfer radical polymerization of n-butyl methacrylate.* Journal of the American Chemical Society, 2003. **126**(1): p. 170–176.

[109] Kong, H., C. Gao, and D. Yan, *Controlled functionalization of multiwalled carbon nanotubes by in situ atom transfer radical polymerization.* Journal of the American Chemical Society, 2003. **126**(2): p. 412–413.

[110] Qin, S., et al., *Functionalization of single-walled carbon nanotubes with polystyrene via grafting to and grafting from methods.* Macromolecules, 2004. **37**(3): p. 752–757.

[111] Goncalves, G., et al., *Graphene oxide modified with PMMA via ATRP as a reinforcement filler.* Journal of Materials Chemistry, 2010. **20**(44): p. 9927–9934.

[112] Wang, D., et al., *Graphene functionalized with azo polymer brushes: Surface-initiated polymerization and photoresponsive properties.* Advanced Materials, 2011. **23**(9): p. 1122–1125.

[113] Beckert, F., et al., *Sulfur-functionalized graphenes as macro-chain-transfer and RAFT agents for producing graphene polymer brushes and polystyrene nanocomposites.* Macromolecules, 2012. **45**(17): p. 7083–7090.

[114] Ellis, A.V., M.R. Waterland, and J. Quinton, *Water-soluble carbon nanotube chain-transfer agents (CNT-CTAs).* Chemistry Letters, 2007. **36**(9): p. 1172–1173.

[115] Zhang, B., et al., *Growing poly(N-vinylcarbazole) from the surface of graphene oxide via RAFT polymerization.* Journal of Polymer Science Part A: Polymer Chemistry, 2011. **49**(9): p. 2043–2050.

[116] Newkome, G.R. and C.D. Shreiner, *Poly(amidoamine), polypropylenimine, and related dendrimers and dendrons possessing different* $1 \rightarrow 2$ *branching motifs: An overview of the divergent procedures.* Polymer, 2008. **49**(1): p. 1–173.

[117] Bifeng, P., et al., *Growth of multi-amine terminated poly(amidoamine) dendrimers on the surface of carbon nanotubes*. Nanotechnology, 2006. **17**(10): p. 2483.
[118] Bissett, M.A., et al., *Dendron growth from vertically aligned single-walled carbon nanotube thin layer arrays for photovoltaic devices*. Physical Chemistry Chemical Physics, 2011. **13**(13): p. 6059–6064.
[119] Campidelli, S., et al., *Dendrimer-functionalized single-wall carbon nanotubes: Synthesis, characterization, and photoinduced electron transfer*. Journal of the American Chemical Society, 2006. **128**(38): p. 12544–12552.
[120] Bissett, M.A., et al., *Dye functionalisation of PAMAM-type dendrons grown from vertically aligned single-walled carbon nanotube arrays for light harvesting antennae*. Journal of Materials Chemistry, 2011. **21**(46): p. 18597–18604.
[121] Vögtle, F., et al., *Functional dendrimers*. Progress in Polymer Science, 2000. **25**(7): p. 987–1041.
[122] Hijazi, I., et al., *Formation of Linear and Hyperbranched Porphyrin Polymers onto Carbon Nanotubes via CuAAC "Grafting from" Approach*. Journal of Materials Chemistry, 2012. **22**(39): p. 20936.
[123] Walsh, F.C. and M.E. Herron, *Electrocrystallization and electrochemical control of crystal growth: fundamental considerations and electrodeposition of metals*. Journal of Physics D: Applied Physics, 1991. **24**(2): p. 217.
[124] Guo, D.-j. and H.-l. Li, *Electrochemical synthesis of Pd nanoparticles on functional MWNT surfaces*. Electrochemistry Communications, 2004. **6**(10): p. 999–1003.
[125] Quinn, B.M., C. Dekker, and S.G. Lemay, *Electrodeposition of Noble Metal Nanoparticles on Carbon Nanotubes*. Journal of the American Chemical Society, 2005. **127**(17): p. 6146–6147.
[126] Hsieh, C.-T., Y.-Y. Liu, and A.K. Roy, *Pulse electrodeposited Pd nanoclusters on graphene-based electrodes for proton exchange membrane fuel cells*. Electrochimica Acta, 2012. **64**(0): p. 205–210.
[127] Day, T.M., et al., *Electrochemical templating of metal nanoparticles and nanowires on single-walled carbon nanotube networks*. Journal of the American Chemical Society, 2005. **127**(30): p. 10639–10647.
[128] Hsieh, C.-T., et al., *Fabrication of flower-like platinum clusters onto graphene sheets by pulse electrochemical deposition*. Electrochimica Acta, 2012. **64**(0): p. 177–182.
[129] Hu, Y., et al., *Graphene–gold nanostructure composites fabricated by electrodeposition and their electrocatalytic activity toward the oxygen reduction and glucose oxidation*. Electrochimica Acta, 2010. **56**(1): p. 491–500.
[130] Rout, C.S., et al., *Au nanoparticles on graphitic petal arrays for surface-enhanced Raman spectroscopy*. Applied Physics Letters, 2010. **97**(13): p. 133108–133103.
[131] Rout, C.S., A. Kumar, and T.S. Fisher, *Carbon nanowalls amplify the surface-enhanced Raman scattering from Ag nanoparticles*. Nanotechnology, 2011. **22**(39): p. 395704.
[132] Wang, L., et al., *A novel hydrogen peroxide sensor based on Ag nanoparticles electrodeposited on chitosan-graphene oxide/cysteamine-modified gold electrode*. Journal of Solid State Electrochemistry, 2012. **16**(4): p. 1693–1700.
[133] Hu, Y., et al., *Bimetallic Pt-Au nanocatalysts electrochemically deposited on graphene and their electrocatalytic characteristics towards oxygen reduction and methanol oxidation*. Physical Chemistry Chemical Physics, 2011. **13**(9): p. 4083–4094.
[134] Khalilian, M., Y. Abdi, and E. Arzi, *Formation of well-packed TiO 2 nanoparticles on multiwall carbon nanotubes using CVD method to fabricate high sensitive gas sensors*. Journal of Nanoparticle Research, 2011. **13**(10): p. 5257–5264.

[135] Huang, H., et al., *Fabrication of new magnetic nanoparticles (Fe3O4) grafted multiwall carbon nanotubes and heterocyclic compound modified electrode for electrochemical sensor.* Electroanalysis, 2010. **22**(4): p. 433–438.

[136] Irantzu, L., et al., *Carbon nanotube surface modification with polyelectrolyte brushes endowed with quantum dots and metal oxide nanoparticles through in situ synthesis.* Nanotechnology, 2010. **21**(5): p. 055605.

[137] Léger, J.M., *Preparation and activity of mono- or bi-metallic nanoparticles for electrocatalytic reactions.* Electrochimica Acta, 2005. **50**(15): p. 3123–3129.

[138] Jamil, E., et al., *Electrodeposition of gold thin films with controlled morphologies and their applications in electrocatalysis and SERS.* Nanotechnology, 2012. **23**(25): p. 255705.

[139] Maiyalagan, T., et al., *Electrodeposited Pt on three-dimensional interconnected graphene as a free-standing electrode for fuel cell application.* Journal of Materials Chemistry, 2012. **22**(12): p. 5286–5290.

[140] Dai, L., et al., *Carbon nanomaterials for advanced energy conversion and storage.* Small, 2012. **8**(8): p. 1130–1166.

[141] Zhang, H.L., et al., *Vapour sensing using surface functionalized gold nanoparticles.* Nanotechnology, 2002. **13**(3): p. 439.

[142] Kong, J., M.G. Chapline, and H. Dai, *Functionalized carbon nanotubes for molecular hydrogen sensors.* Advanced Materials, 2001. **13**(18): p. 1384–1386.

[143] Gutés, A., C. Carraro, and R. Maboudian, *Single-layer CVD-grown graphene decorated with metal nanoparticles as a promising biosensing platform.* Biosensors and Bioelectronics, 2012. **33**(1): p. 56–59.

[144] Veličković, Z., et al., *Adsorption of arsenate on iron(III) oxide coated ethylenediamine functionalized multiwall carbon nanotubes.* Chemical Engineering Journal, 2012. **181–182**(0): p. 174–181.

[145] Hu, X., et al., *A general route to prepare one- and three-dimensional carbon nanotube/metal nanoparticle composite nanostructures.* Langmuir, 2007. **23**(11): p. 6352–6357.

[146] Xu, W., et al., *Low-temperature plasma-assisted preparation of graphene supported palladium nanoparticles with high hydrodesulfurization activity.* Journal of Materials Chemistry, 2012. **22**(29): p. 14363–14368.

[147] Xing, Y., *Synthesis and electrochemical characterization of uniformly-dispersed high loading Pt nanoparticles on sonochemically-treated carbon nanotubes.* The Journal of Physical Chemistry B, 2004. **108**(50): p. 19255–19259.

[148] Siamaki, A.R., et al., *Microwave-assisted synthesis of palladium nanoparticles supported on graphene: A highly active and recyclable catalyst for carbon–carbon cross-coupling reactions.* Journal of Catalysis, 2011. **279**(1): p. 1–11.

[149] Lin, Y., et al., *Platinum/carbon nanotube nanocomposite synthesized in supercritical fluid as electrocatalysts for low-temperature fuel cells.* The Journal of Physical Chemistry B, 2005. **109**(30): p. 14410–14415.

[150] Liu, Y., et al., *Metal-assisted hydrogen storage on Pt-decorated single-walled carbon nanohorns.* Carbon, 2012. **50**(13): p. 4953–4964.

[151] Zhou, X., et al., *Deposition of Fe-Ni nanoparticles on polyethyleneimine-decorated graphene oxide and application in catalytic dehydrogenation of ammonia borane.* Journal of Materials Chemistry, 2012. **22**(27): p. 13506–13516.

[152] Wang, J., et al., *Rhodium-nickel nanoparticles grown on graphene as highly efficient catalyst for complete decomposition of hydrous hydrazine at room temperature for chemical hydrogen storage.* Energy & Environmental Science, 2012. **5**(5): p. 6885–6888.

[153] Jeon, S., D. Kim, and M. Ahmed, *Different length linkages of graphene modified with metal nanoparticles for oxygen reduction in acidic media*. Journal of Materials Chemistry, 2012. **22**(32): p. 16353–16360.

[154] Qiu, J.-D., et al., *Controllable deposition of platinum nanoparticles on graphene as an electrocatalyst for direct methanol fuel cells*. The Journal of Physical Chemistry C, 2011. **115**(31): p. 15639–15645.

[155] Yang, J., et al., *An effective strategy for small-sized and highly-dispersed palladium nanoparticles supported on graphene with excellent performance for formic acid oxidation*. Journal of Materials Chemistry, 2011. **21**(10): p. 3384–3390.

[156] Fan, G.-Q., et al., *Plasmonic-enhanced polymer solar cells incorporating solution-processable Au nanoparticle-adhered graphene oxide*. Journal of Materials Chemistry, 2012. **22**(31): p. 15614–15619.

[157] Kim, J.D., et al., *Preparation of reusable Ag-decorated graphene oxide catalysts for decarboxylative cycloaddition*. Journal of Materials Chemistry, 2012. **22**(38): p. 20665–20670.

[158] Chen, C.-M., et al., *Chemically derived graphene-metal oxide hybrids as electrodes for electrochemical energy storage: pre-graphenization or post-graphenization?* Journal of Materials Chemistry, 2012. **22**(28): p. 13947–13955.

[159] Li, Y., et al., *MoS2 nanoparticles grown on graphene: an advanced catalyst for the hydrogen evolution reaction*. Journal of the American Chemical Society, 2011. **133**(19): p. 7296–7299.

[160] Chen, W., et al., *Self-assembly and embedding of nanoparticles by in situ reduced graphene for preparation of a 3D graphene/nanoparticle aerogel*. Advanced Materials, 2011. **23**(47): p. 5679–5683.

[161] Zhang, Y., et al., *Green and controlled synthesis of Cu2O-graphene hierarchical nanohybrids as high-performance anode materials for lithium-ion batteries via an ultrasound assisted approach*. Dalton Transactions, 2012. **41**(15): p. 4316–4319.

[162] Tu, X., et al., *One-pot synthesis, characterization, and enhanced photocatalytic activity of a BiOBr–graphene composite*. Chemistry – A European Journal, 2012. **18**(45): p. 14359–14366.

[163] Yan, J.-M., et al., *Rapid and energy-efficient synthesis of a graphene-CuCo hybrid as a high performance catalyst*. Journal of Materials Chemistry, 2012. **22**(22): p. 10990–10993.

[164] Baek, S., et al., *A one-pot microwave-assisted non-aqueous sol-gel approach to metal oxide/graphene nanocomposites for Li-ion batteries*. RSC Advances, 2011. **1**(9): p. 1687–1690.

[165] Kong, B.-S., J. Geng, and H.-T. Jung, *Layer-by-layer assembly of graphene and gold nanoparticles by vacuum filtration and spontaneous reduction of gold ions*. Chemical Communications, 2009(16): p. 2174–2176.

[166] Zeng, T., et al., *A novel Fe3O4-graphene-Au multifunctional nanocomposite: Green synthesis and catalytic application*. Journal of Materials Chemistry, 2012. **22**(35): p. 18658–18663.

[167] Zhang, N., et al., *Assembly of CdS nanoparticles on the two-dimensional graphene scaffold as visible-light-driven photocatalyst for selective organic transformation under ambient conditions*. The Journal of Physical Chemistry C, 2011. **115**(47): p. 23501–23511.

[168] Liu, X., et al., *Preparation of multiwalled carbon nanotubes/Cd0.8Zn0.2S nanocomposite and its photocatalytic hydrogen production under visible-light*. International Journal of Hydrogen Energy, 2012. **37**(2): p. 1375–1384.

[169] Zhang, J., et al., *Noble metal-free reduced graphene oxide-ZnxCd1–xS nanocomposite with enhanced solar photocatalytic h2-production performance*. Nano Letters, 2012. **12**(9): p. 4584–4589.

[170] Wang, X.-F., et al., *Signal-on electrochemiluminescence biosensors based on CdS–carbon nanotube nanocomposite for the sensitive detection of choline and acetylcholine*. Advanced Functional Materials, 2009. **19**(9): p. 1444–1450.

[171] Ye, A., et al., *CdS-graphene and CdS-CNT nanocomposites as visible-light photocatalysts for hydrogen evolution and organic dye degradation.* Catalysis Science & Technology, 2012. **2**(5): p. 969–978.

[172] Lv, X.-J., et al., *Hydrogen evolution from water using semiconductor nanoparticle/graphene composite photocatalysts without noble metals.* Journal of Materials Chemistry, 2012. **22**(4): p. 1539–1546.

[173] Das, S., et al., *Synthesis of graphene-CoS electro-catalytic electrodes for dye sensitized solar cells.* Carbon, 2012. **50**(13): p. 4815–4821.

[174] Son, D.I., et al., *Emissive ZnO-graphene quantum dots for white-light-emitting diodes.* Nature Nanotechnology, 2012. **7**(7): p. 465–471.

[175] Narayanan, T.N., et al., *Hybrid 2D nanomaterials as dual-mode contrast agents in cellular imaging.* Advanced Materials, 2012. **24**(22): p. 2992–8.

[176] Liang, Y., et al., *Oxygen reduction electrocatalyst based on strongly coupled cobalt oxide nanocrystals and carbon nanotubes.* Journal of the American Chemical Society, 2012. **134**(38): p. 15849–15857.

[177] Shi, W., et al., *Achieving high specific charge capacitances in Fe3O4/reduced graphene oxide nanocomposites.* Journal of Materials Chemistry, 2011. **21**(10): p. 3422–3427.

[178] Song, J., et al., *Synergistic effect of molybdenum nitride and carbon nanotubes on electrocatalysis for dye-sensitized solar cells.* Journal of Materials Chemistry, 2012. **22**(38): p. 20580–20585.

[179] Wang, P., et al., *A one-pot method for the preparation of graphene–Bi2MoO6 hybrid photocatalysts that are responsive to visible-light and have excellent photocatalytic activity in the degradation of organic pollutants.* Carbon, 2012. **50**(14): p. 5256–5264.

[180] Aksel, S. and D. Eder, *Catalytic effect of metal oxides on the oxidation resistance in carbon nanotube-inorganic hybrids.* Journal of Materials Chemistry, 2010. **20**(41).

[181] Su, Y., et al., *Two-dimensional carbon-coated graphene/metal oxide hybrids for enhanced lithium storage.* ACS Nano, 2012. **6**(9): p. 8349–8356.

[182] Raney, J.R., et al., *In situ synthesis of metal oxides in carbon nanotube arrays and mechanical properties of the resulting structures.* Carbon, 2012. **50**(12): p. 4432–4440.

[183] Dong, X., et al., *Synthesis of a MnO2–graphene foam hybrid with controlled MnO2 particle shape and its use as a supercapacitor electrode.* Carbon, 2012. **50**(13): p. 4865–4870.

[184] Zhan, Y., et al., *Preparation, characterization and electromagnetic properties of carbon nanotubes/Fe3O4 inorganic hybrid material.* Applied Surface Science, 2011. **257**(9): p. 4524–4528.

[185] Xie, J., et al., *Self-assembly of CoFe2O4/graphene sandwich by a controllable and general route: towards high-performance anode for Li-ion batteries.* Journal of Materials Chemistry, 2012. **22**(37): p. 19738–19743.

[186] Li, Q., et al., *Highly efficient visible-light-driven photocatalytic hydrogen production of CdS-cluster-decorated graphene nanosheets.* Journal of the American Chemical Society, 2011. **133**(28): p. 10878–10884.

[187] Pan, X. and X. Bao, *The effects of confinement inside carbon nanotubes on catalysis.* Accounts of Chemical Research, 2011. **44**(8): p. 553–562.

[188] Tessonnier, J.-P., et al., *Pd nanoparticles introduced inside multi-walled carbon nanotubes for selective hydrogenation of cinnamaldehyde into hydrocinnamaldehyde.* Applied Catalysis A: General, 2005. **288**(1–2): p. 203–210.

[189] Sun, Z., et al., *Coating carbon nanotubes with metal oxides in a supercritical carbon dioxide–ethanol solution.* Carbon, 2007. **45**(13): p. 2589–2596.

[190] Sun, Z., et al., *A highly efficient chemical sensor material for H2S: α-Fe2O3 nanotubes fabricated using carbon nanotube templates*. Advanced Materials, 2005. **17**(24): p. 2993–2997.
[191] Wu, C.-H., et al., *Unique Pd/graphene nanocomposites constructed using supercritical fluid for superior electrochemical sensing performance*. Journal of Materials Chemistry, 2012. **22**(40): p. 21466–21471.
[192] Hwang, J.O., et al., *Vertical ZnO nanowires/graphene hybrids for transparent and flexible field emission*. Journal of Materials Chemistry, 2011. **21**(10): p. 3432–3437.
[193] Li, Z., et al., *Triphenylamine-functionalized graphene decorated with Pt nanoparticles and its application in photocatalytic hydrogen production*. International Journal of Hydrogen Energy, 2012. **37**(6): p. 4880–4888.
[194] Pasricha, R., et al., *Directed nanoparticle reduction on graphene*. Materials Today, 2012. **15**(3): p. 118–125.
[195] Tjoa, V., et al., *Facile photochemical synthesis of graphene-Pt nanoparticle composite for counter electrode in dye sensitized solar cell*. ACS Applied Materials & Interfaces, 2012.
[196] Sheng, W., et al., *Synthesis, activity and durability of Pt nanoparticles supported on multi-walled carbon nanotubes for oxygen reduction*. Journal of The Electrochemical Society, 2011. **158**(11): p. B1398–B1404.
[197] Zhou, X., et al., *Facile synthesis of nanospindle-like Cu2O/straight multi-walled carbon nanotube hybrid nanostructures and their application in enzyme-free glucose sensing*. Sensors and Actuators B: Chemical, 2012. **168**: p. 1–7.
[198] Zhou, Y., et al., *Electrostatic self-assembly of graphene–silver multilayer films and their transmittance and electronic conductivity*. Carbon, 2012. **50**(12): p. 4343–4350.
[199] Tang, Y., et al., *One-step electrodeposition to layer-by-layer graphene–conducting-polymer hybrid films*. Macromolecular Rapid Communications, 2012. **33**(20): p. 1780–1786.
[200] Mao, S., et al., *A general approach to one-pot fabrication of crumpled graphene-based nanohybrids for energy applications*. ACS Nano, 2012. **6**(8): p. 7505–7513.
[201] Choi, H.C., et al., *Spontaneous reduction of metal ions on the sidewalls of carbon nanotubes*. Journal of the American Chemical Society, 2002. **124**(31): p. 9058–9059.
[202] Qu, L. and L. Dai, *Substrate-enhanced electroless deposition of metal nanoparticles on carbon nanotubes*. Journal of the American Chemical Society, 2005. **127**(31): p. 10806–10807.
[203] Zhao, H., et al., *Fabrication of a palladium nanoparticle/graphene nanosheet hybrid via sacrifice of a copper template and its application in catalytic oxidation of formic acid*. Chemical Communications, 2011. **47**(7): p. 2014–2016.
[204] Peng, Y. and Q. Chen, *Fabrication of copper/multi-walled carbon nanotube hybrid nanowires using electroless copper deposition activated with silver nitrate*. Journal of The Electrochemical Society, 2012. **159**(2): p. D72–D76.
[205] Guo, D.-J. and H.-L. Li, *High dispersion and electrocatalytic properties of Pt nanoparticles on SWNT bundles*. Journal of Electroanalytical Chemistry, 2004. **573**(1): p. 197–202.
[206] Mubeen, S., et al., *Palladium nanoparticles decorated single-walled carbon nanotube hydrogen sensor*. The Journal of Physical Chemistry C, 2007. **111**(17): p. 6321–6327.
[207] Sundaram, R.S., et al., *Electrochemical modification of graphene*. Advanced Materials, 2008. **20**(16): p. 3050–3053.
[208] Chen, Y.-C., et al., *Silver-decorated carbon nanotube networks as SERS substrates*. Journal of Raman Spectroscopy, 2011. **42**(6): p. 1255–1262.
[209] Yu, A., et al., *Silver nanoparticle–carbon nanotube hybrid films: Preparation and electrochemical sensing*. Electrochimica Acta, 2012. **74**(0): p. 111–116.

[210] Shahrokhian, S. and S. Rastgar, *Electrochemical deposition of gold nanoparticles on carbon nanotube coated glassy carbon electrode for the improved sensing of tinidazole.* Electrochimica Acta, 2012. **78**(0): p. 422–429.

[211] Yu, G., et al., *Enhancing the supercapacitor performance of graphene/MnO2 nanostructured electrodes by conductive wrapping.* Nano Letters, 2011. **11**(10): p. 4438–4442.

[212] Frank, O., et al., *Structural properties and electrochemical behavior of CNT-TiO2 nanocrystal heterostructures.* physica status solidi (b), 2007. **244**(11): p. 4040–4045.

[213] Kuan-Xin, H., et al., *Electrodeposition of nickel and cobalt mixed oxide/carbon nanotube thin films and their charge storage properties.* Journal of The Electrochemical Society, 2006. **153**(8): p. A1568–A1574.

[214] Kim, I.-H., et al., *Synthesis and characterization of electrochemically prepared ruthenium oxide on carbon nanotube film substrate for supercapacitor applications.* Journal of The Electrochemical Society, 2005. **152**(11): p. A2170–A2178.

[215] He, Z., et al., *Electrodeposition of Pt–Ru nanoparticles on carbon nanotubes and their electrocatalytic properties for methanol electrooxidation.* Diamond and Related Materials, 2004. **13**(10): p. 1764–1770.

[216] Wang, X., et al., *A graphene-cobalt oxide based needle electrode for non-enzymatic glucose detection in micro-droplets.* Chemical Communications, 2012. **48**(52): p. 6490–6492.

[217] Sahoo, S., et al., *Controlled assembly of Ag nanoparticles and carbon nanotube hybrid structures for biosensing.* Journal of the American Chemical Society, 2011. **133**(11): p. 4005–4009.

[218] Heeger, A.J., *Semiconducting polymers: the Third Generation.* Chemical Society Reviews, 2010. **39**(7): p. 2354–2371.

[219] Bozlar, M., F. Miomandre, and J. Bai, *Electrochemical synthesis and characterization of carbon nanotube/modified polypyrrole hybrids using a cavity microelectrode.* Carbon, 2009. **47**(1): p. 80–84.

[220] Wang, D.-W., et al., *Fabrication of graphene/polyaniline composite paper via in situ anodic electropolymerization for high-performance flexible electrode.* ACS Nano, 2009. **3**(7): p. 1745–1752.

[221] Santhosh, P., A. Gopalan, and K.-P. Lee, *Gold nanoparticles dispersed polyaniline grafted multiwall carbon nanotubes as newer electrocatalysts: Preparation and performances for methanol oxidation.* Journal of Catalysis, 2006. **238**(1): p. 177–185.

[222] Cosnier, S. and M. Holzinger, *Design of carbon nanotube-polymer frameworks by electropolymerization of SWCNT-pyrrole derivatives.* Electrochimica Acta, 2008. **53**(11): p. 3948–3954.

[223] Chen, K., et al., *Three-dimensional porous graphene-based composite materials: electrochemical synthesis and application.* Journal of Materials Chemistry, 2012. **22**(39): p. 20968–20976.

[224] Gebeyehu, N., et al., *Ultrathin TiO2-coated MWCNTs with excellent conductivity and smsi nature as pt catalyst support for oxygen reduction reaction in PEMFCs.* Journal of Materials Chemistry, 2012.

[225] Ren, Z., et al., *Hybridizing photoactive zeolites with graphene: a powerful strategy towards superior photocatalytic properties.* Chemical Science, 2012. **3**(1): p. 209–216.

[226] Shakir, I., et al., *MoO3-MWCNTs nanocomposites photocatalyst with control of light-harvesting under visible and natural sunlight irradiation.* Journal of Materials Chemistry, 2012.

[227] Nethravathi, C., et al., *Hydrothermal synthesis of a monoclinic VO2 nanotube–graphene hybrid for use as cathode material in lithium ion batteries.* Carbon, 2012. **50**(13): p. 4839–4846.

[228] Cargnello, M., et al., *Multiwalled carbon nanotubes drive the activity of metal@oxide core–shell catalysts in modular nanocomposites.* Journal of the American Chemical Society, 2012. **134**(28): p. 11760–11766.
[229] Yang, B., et al., *Embedding graphene nanoparticles into poly (N, N'-dimethylacrylamine) to prepare transparent nanocomposite films with high refractive index.* Journal of Materials Chemistry, 2012. **22**(39): p. 21218–21224.
[230] Takenaka, S., et al., *Coverage of carbon nanotubes with titania nanoparticles for the preparation of active titania-based photocatalysts.* Applied Catalysis B: Environmental, 2012. **125**(0): p. 358–366.
[231] Krissanasaeranee, M., et al., *Complex carbon nanotube-inorganic hybrid materials as next-generation photocatalysts.* Chemical Physics Letters, 2010. **496**(1–3): p. 133–138.
[232] Eder, D. and A.H. Windle, *Morphology control of CNT-TiO2 hybrid materials and rutile nanotubes.* Journal of Materials Chemistry, 2008. **18**(17): p. 2036–2043.
[233] Xiang, Q., J. Yu, and M. Jaroniec, *Synergetic effect of MoS2 and graphene as cocatalysts for enhanced photocatalytic H2 production activity of TiO2 nanoparticles.* Journal of the American Chemical Society, 2012. **134**(15): p. 6575–6578.
[234] Yu, K., et al., *Significant improvement of field emission by depositing zinc oxide nanostructures on screen-printed carbon nanotube films.* Applied Physics Letters, 2006. **88**(15): p. 153123.
[235] Jiang, L. and L. Gao, *Fabrication and characterization of carbon nanotube–titanium nitride composites with enhanced electrical and electrochemical properties.* Journal of the American Ceramic Society, 2006. **89**(1): p. 156–161.
[236] Gwon, H., et al., *Flexible energy storage devices based on graphene paper.* Energy & Environmental Science, 2011. **4**(4): p. 1277–1283.
[237] Zhu, Y., et al., *Multiwalled carbon nanotubes beaded with ZnO Nanoparticles for ultrafast nonlinear optical switching.* Advanced Materials, 2006. **18**(5): p. 587–592.
[238] Kim, H. and W. Sigmund, *Zinc oxide nanowires on carbon nanotubes.* Applied Physics Letters, 2002. **81**(11): p. 2085–2087.
[239] Chung, M.G., et al., *Flexible hydrogen sensors using graphene with palladium nanoparticle decoration.* Sensors and Actuators B: Chemical, 2012. **169**(0): p. 387–392.
[240] Zhang, D., et al., *Understanding charge transfer at PbS-decorated graphene surfaces toward a tunable photosensor.* Advanced Materials, 2012. **24**(20): p. 2715–2720.
[241] Ashok, K., et al., *Ferroelectric–carbon nanotube memory devices.* Nanotechnology, 2012. **23**(16): p. 165702.
[242] Yao, Z., et al., *Aligned coaxial tungsten oxide-carbon nanotube sheet: a flexible and gradient electrochromic film.* Chemical Communications, 2012. **48**(66): p. 8252–8254.
[243] Wu, C., et al., *Improving the field emission of graphene by depositing zinc oxide nanorods on its surface.* Carbon, 2012. **50**(10): p. 3622–3626.
[244] Shin, K.-S., et al., *High quality graphene-semiconducting oxide heterostructure for inverted organic photovoltaics.* Journal of Materials Chemistry, 2012. **22**(26): p. 13032–13038.
[245] Lau, K.K.S., et al., *Superhydrophobic carbon nanotube forests.* Nano Letters, 2003. **3**(12): p. 1701–1705.
[246] Li, Y.-L., I.A. Kinloch, and A.H. Windle, *Direct Spinning of Carbon Nanotube Fibers from Chemical Vapor Deposition Synthesis.* Science, 2004. **304**(5668): p. 276–278.
[247] Neocleus, S., et al., *Hierarchical carbon nanotube-inorganic hybrid structures involving CNT arrays and CNT fibers.* Functional Materials Letters, 2011. **4**(1): p. 83–89.
[248] Mattevi, C., H. Kim, and M. Chhowalla, *A review of chemical vapour deposition of graphene on copper.* Journal of Materials Chemistry, 2011. **21**(10).
[249] Chen, Z., et al., *Three-dimensional flexible and conductive interconnected graphene networks grown by chemical vapour deposition.* Nature Materials, 2011. **10**(6): p. 424–428.

[250] Dong, X., et al., *Synthesis of graphene-carbon nanotube hybrid foam and its use as a novel three-dimensional electrode for electrochemical sensing.* Journal of Materials Chemistry, 2012. **22**(33): p. 17044–17048.
[251] Jin, S.H., et al., *Conformal coating of titanium suboxide on carbon nanotube networks by atomic layer deposition for inverted organic photovoltaic cells.* Carbon, 2012. **50**(12): p. 4483–4488.
[252] Hsu, C.-Y., et al., *Supersensitive, ultrafast, and broad-band light-harvesting scheme employing carbon nanotube/TiO2 core–shell nanowire geometry.* ACS Nano, 2012. **6**(8): p. 6687–6692.
[253] Xiangbo, M., et al., *Controllable synthesis of graphene-based titanium dioxide nanocomposites by atomic layer deposition.* Nanotechnology, 2011. **22**(16): p. 165602.
[254] Wen, H.-C., et al., *ZnO-coated carbon nanotubes: an enhanced and red-shifted emission band at UV-VIS wavelength.* Journal of Materials Chemistry, 2012. **22**(27): p. 13747–13750.
[255] Hu, C.J., et al., *ZnO-coated carbon nanotubes: flexible piezoelectric generators.* Advanced Materials, 2011. **23**(26): p. 2941–2945.
[256] Li, X., et al., *Tin oxide with controlled morphology and crystallinity by atomic layer deposition onto graphene nanosheets for enhanced lithium storage.* Advanced Functional Materials, 2012. **22**(8): p. 1647–1654.
[257] Chen, X., et al., *MWCNT/V2O5 core/shell sponge for high areal capacity and power density Li-ion cathodes.* ACS Nano, 2012. **6**(9): p. 7948–7955.
[258] Dameron, A.A., et al., *Aligned carbon nanotube array functionalization for enhanced atomic layer deposition of platinum electrocatalysts.* Applied Surface Science, 2012. **258**(13): p. 5212–5221.
[259] Tong, X., et al., *Enhanced catalytic activity for methanol electro-oxidation of uniformly dispersed nickel oxide nanoparticles – carbon nanotube hybrid materials.* Small, 2012.
[260] Wang, X., S.M. Tabakman, and H. Dai, *Atomic layer deposition of metal oxides on pristine and functionalized graphene.* Journal of the American Chemical Society, 2008. **130**(26): p. 8152–8153.
[261] Alaboson, J.M.P., et al., *Seeding atomic layer deposition of high-k dielectrics on epitaxial graphene with organic self-assembled monolayers.* ACS Nano, 2011. **5**(6): p. 5223–5232.
[262] Jo, Y., et al., *Highly interconnected ordered mesoporous carbon-carbon nanotube nanocomposites: Pt-free, highly efficient, and durable counter electrodes for dye-sensitized solar cells.* Chemical Communications, 2012. **48**(65): p. 8057–8059.
[263] Wang, X., S. Blechert, and M. Antonietti, *Polymeric graphitic carbon nitride for heterogeneous photocatalysis.* ACS Catalysis, 2012. **2**(8): p. 1596–1606.
[264] Wang, Y., et al., *Enhancement of photocatalytic activity of Bi2WO6 hybridized with graphite-like C3N4.* Journal of Materials Chemistry, 2012. **22**(23): p. 11568–11573.
[265] Peng, F., et al., *A carbon nitride/TiO2 nanotube arrays heterojunction visible-light photocatlyst: synthesis, characterization, and photoelectrochemical properties.* Journal of Materials Chemistry, 2012.
[266] Ge, L. and C. Han, *Synthesis of MWNTs/g-C3N4 composite photocatalysts with efficient visible light photocatalytic hydrogen evolution activity.* Applied Catalysis B: Environmental, 2012. **117–118**(0): p. 268–274.
[267] Xiang, Q., J. Yu, and M. Jaroniec, *Preparation and enhanced visible-light photocatalytic H2-production activity of graphene/C3N4 composites.* The Journal of Physical Chemistry C, 2011. **115**(15): p. 7355–7363.
[268] Ge, L., C. Han, and J. Liu, *In situ synthesis and enhanced visible light photocatalytic activities of novel PANI-g-C3N4 composite photocatalysts.* Journal of Materials Chemistry, 2012. **22**(23): p. 11843–11850.

C.N.R. Rao, H.S.S. Ramakrishna Matte, and Urmimala Maitra

6 Graphene and its hybrids with inorganic nanoparticles, polymers and other materials

Graphene has emerged as one of today's most exciting materials with many potential applications. Prompted by the results obtained with graphene, there have been investigations of graphene-like materials formed by inorganic layered compounds such as MoS_2 and BN. In this chapter, we briefly present the synthesis and relevant aspects of graphene and inorganic analogues of graphene. Equally importantly, we discuss the synthesis and properties of hybrids of graphene with other nanocarbons, polymers, graphene-like MoS_2 and other inorganic nanoparticles. Some of the hybrids exhibit interesting properties and are promising as useful materials. We end the chapter with a brief discussion of hybrids of graphene with metal organic frameworks.

6.1 Introduction

Graphene, a single-layer of sp^2 carbon atoms tightly packed in a honeycomb lattice, has generated great interest in the scientific community in the last few years owing to its extraordinary properties and their potential applications [1–2]. It exhibits an ambipolar electric field effect along with ballistic conduction of charge carriers, high carrier mobility, and quantum Hall effect at room temperature [3]. Other novel aspects of graphene include high transparency towards visible light, high elasticity, high thermal conductivity, unusual magnetic properties, high surface area and remarkable gas storage [4]. Though graphene generally refers to a single layer of sp^2 carbon atoms, there have been many important investigations on bi- and few-layered graphenes as well. Interest in the synthesis and chemical modification of graphite-related materials dates back to 1840 [5], but real progress has been made only since 2004. There has been much progress in graphene research focusing on the synthesis, characterization, properties and applications. In this chapter, we present an overview of different synthesizing routes to prepare single- and few-layer graphenes as well as inorganic graphene analogues. We then examine the synthesis of the hybrids of nanocarbons (C_{60}/SWNT/graphene) and their applications. Subsequently, we present the mechanical and electrical properties of graphene hybrids, graphene-nanoparticle hybrids, graphene-MoS_2/WS_2/SnO_2 hybrids. Finally, we take a glimpse at some of the recent work on graphene-metal organic framework (MOF) hybrids.

6.2 Synthesis

Graphene. In the process of micromechanical cleavage, Novoselov *et al.* [1] pressed patterned HOPG square mesas on a photo resist spun over a glass substrate and repeatedly peeled off layers using scotch tape. This method was later simplified just to peel off one or a few sheets of graphene using scotch tape and to directly deposit the sheets on SiO_2/Si substrate. The result is graphene of high quality suitable for various studies, but production is limited by low throughput and low yield. Production of graphene in bulk quantities is essential for practical applications and there has been much progress in the synthetic aspects of graphene, resulting in the generation of high quality single- and few-layer graphenes.

Ultrathin epitaxial graphene composed of one to three graphene layers is prepared by the thermal decomposition of the (0001) surface of 6H-SiC [6]. Single- to few-layer graphenes have been obtained by decomposing SiC coated on a thin Ni film [7]. Chemical vapor deposition (CVD) is another important approach to produce large-area and high-quality graphene films. This is carried out on transition metal substrates at high temperatures [8]. The CVD process has been carried out using methane, benzene and other hydrocarbons such as ethylene, acetylene and LPG using nickel and cobalt as catalysts [9]. The growth of graphene films depends on the carbon source and the reaction parameters. CVD was carried out on a nickel foil by passing methane (60–70 sccm) or ethylene (4–8 sccm) along with a high flow of hydrogen of about 500 sccm at 1000 °C for 5–10 min. When using benzene as the hydrocarbon source, benzene vapor diluted with argon and hydrogen was decomposed at 1000 °C for 5 min. On a cobalt foil, acetylene (4 sccm) and methane (65 sccm) were decomposed at 800 and 1000 °C, respectively. Figure 6.1 shows field-emission scanning electron microscope (FESEM), transmission electron microscope (TEM) images as well as the Raman spectrum of graphene sheets obtained by CVD on a nickel foil. These graphene samples show G-band at 1580 cm^{-1} and 2 D-band around 2670 cm^{-1} with a narrow line width of 30–40 cm^{-1}. The narrow line width and relatively high intensity of the 2D-band confirm these Raman spectra correspond to graphenes having 1–2 layers [8]. A radio frequency plasma enhanced chemical vapor deposition (PECVD) system has also been used to grow graphene films on a variety of substrates [10].

Few-layer graphenes are synthesized by the thermal exfoliation of graphite oxide at high temperatures in an inert atmosphere [11]. In this process, dried graphite oxide is subjected to a sudden thermal shock in a tube-furnace at 1050 °C. Graphene prepared using this procedure is termed exfoliated graphene ("EG"). Recently, microwave radiation was also employed to give a thermal shock in order to carry out exfoliation process [12]. Enoki and co-workers [13] prepared nanographite by thermal conversion of nanodiamond in an inert atmosphere ("DG"). This procedure has been examined in detail. Nanodiamond particles with size 4–6 nm were annealed in a graphite furnace at different temperatures (1650, 1850, 2050 and 2200 °C) for 1 h in a helium atmosphere [9] These samples are designated as DG-1650, DG-1850, DG-2050 and DG-2200, respec-

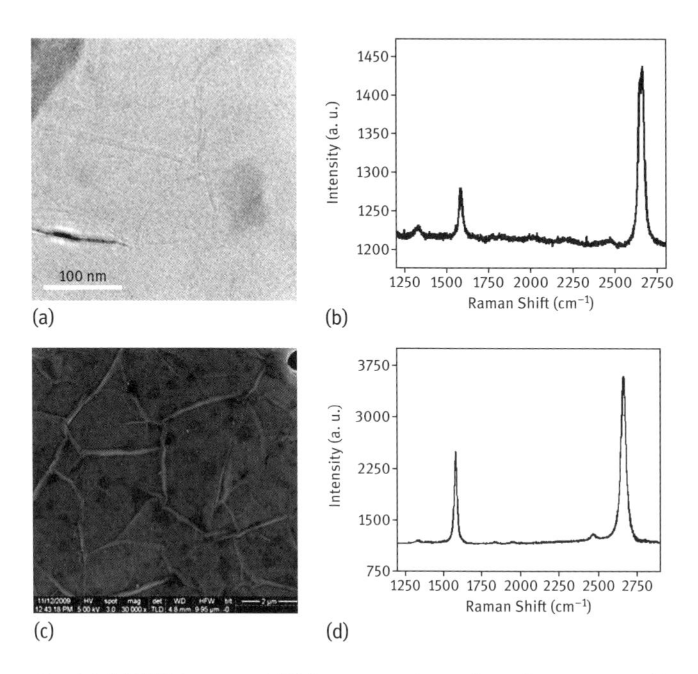

Fig. 6.1: (a) TEM image and (b) Raman spectrum of graphene prepared by the thermal decomposition of methane (70 sccm), (c) FESEM image and (d) Raman spectrum of graphene obtained by benzene (argon passed through benzene with flow rate of 200 sccm) at 1000 °C on a nickel sheet (from [8]).

tively. A slight increase in the number of layers is found with the increase in annealing temperatures.

Synthesis of graphene by the arc evaporation of graphite has been reported [14]. This procedure yields graphene ("HG") sheets with 2–3 layers having flake size of 100–200 nm (Fig. 6.2(a), (b)). This makes use of the knowledge that the presence of H_2 during the arc discharge process terminates the dangling carbon bonds with hydrogen and prevents the formation of closed structures. In a typical experiment, a graphite rod (6 mm in diameter and 50 mm long) was used as the anode and another graphite rod (13 mm in diameter and 60 mm in length) was used as the cathode. The conditions that are favorable for obtaining graphene in the inner walls are the high current (above 100 A), the high voltage (> 50 V), and the high pressure of hydrogen (above 200 torr). This method has been conveniently employed to dope graphene with boron and nitrogen [15]. To prepare boron- and nitrogen-doped graphene (BG and NG), arc discharge is carried out in the presence of H_2 + (diborane or boron) and H_2 + (pyridine or ammonia) respectively. Later, arc discharge of graphite in an air atmosphere was reported to synthesize graphene nanosheets of ~100–200 nm wide predominantly with two layers [16].

Ultrasonic exfoliation of graphite oxide and subsequent chemical reduction has been used to prepare reduced graphene oxide (rGO) [17]. This is a viable method for

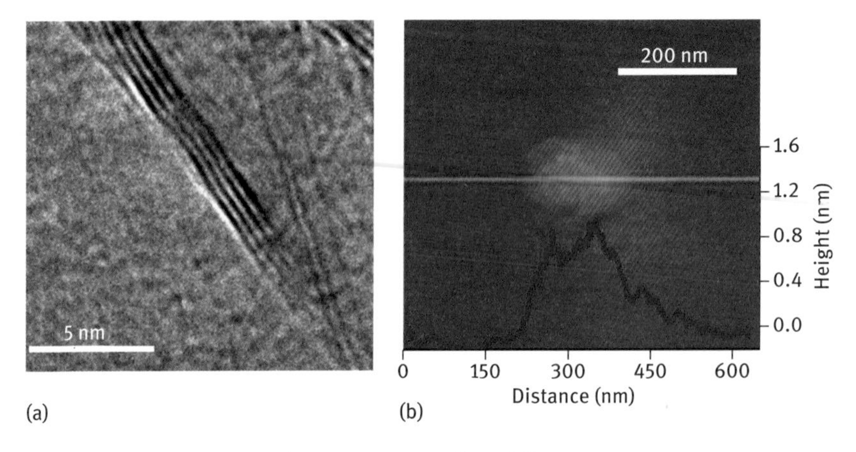

Fig. 6.2: (a) TEM image and (b) AFM image of HG (from [2]).

large-scale production. Different reducing agents have been employed in this process. Hydrothermal and microwave routes are used to reduce graphene oxide in water with hydrazine hydrate and ethylene glycol as reducing agents [9]. In this procedure, a homogeneous mixture of 25 ml of graphene oxide (GO) solution (1mg/1ml) and 2 ml of ethylene glycol or 1 ml of hydrazine hydrate is taken in a 50 mL polytetrafluoroethylene (PTFE) lined bomb. In the case of micro-wave treatment, the sealed autoclave is subjected to microwave irradiation (power 600 W, 200 °C) for 10 minutes. In the hydrothermal process, the autoclave is kept in an oven at 170 °C for 24 h under autogenous pressure and allowed to cool to room temperature gradually. Ding *et al.* [18] reduced the GO using UV irradiation to obtain single- to few-layered graphene sheets without the use of any photocatalyst. Nanosecond laser pulses of KrF eximer laser (335 nm and 532 nm) were shown to effectively reduce dispersions of GO to thermally and chemically stable graphene [19]. High quality rGO can be prepared by irradiating GO with sunlight, ultraviolet light and KrF excimer laser [20–21]. Both sunlight and ultraviolet light reduce GO well after prolonged irradiation, but laser irradiation produces graphene with negligible oxygen functionalities within a short time. Laser irradiation of graphene oxide appears to be an efficient procedure for large-scale synthesis of graphene.

Graphene analogues. The outstanding properties exhibited by graphene have aroused interest in the inorganic layered analogues such as metal chalcogenides (MX_2, M = Mo,W; X = S, Se), boron nitride (BN) and carboboronitrides (BCN). Most of these materials can be prepared as single- or few-layers structures by ultrasonication or mechanical exfoliation. Graphene analogues of MoS_2, WS_2, $MoSe_2$ and WSe_2 have been prepared by different chemical methods [22–23]. Method 1: Few-layer MoS_2, WS_2, $MoSe_2$ and WSe_2 were obtained by intercalation of their bulk counterparts with lithium followed by sonication in water. Method 2: Molybdic acid/tungstic acid was ground with excess of thiourea/selenourea (1 : 48) and heated at 773 K for 3 h under nitrogen atmosphere. Method 3: Few-layered MoS_2 and $MoSe_2$ were produced by the reaction of KSCN/Se with H_2MoO_4 under hydrothermal conditions at 180 °C. Method 4: Few-layer graphene-like sheets of MoS_2 and $MoSe_2$ have been prepared by the microwave method. In this procedure, ethylene glycol or water is used as a solvent with the same reactants used in the case of hydrothermal method. Method 5 deals with the laser irradiation of bulk layered metal dichalcogenides in dimethylformamide to produce inorganic graphenes. The products obtained from the above reactions are characterized by employing TEM, atomic force microscopy (AFM), X-ray diffraction (XRD) and Raman spectroscopy.

XRD patterns of graphene analogues do not exhibit the (002) reflection, in contrast to the bulk counterparts indicating that the materials contain few layers (Fig. 6.3(a)). The Raman spectrum of few-layered MoS_2 shows bands at 406.5 and 381.2 cm^{-1} due to the A_{1g} and E_{2g} modes. The softening of the A_{1g} mode and an increase in full-width half maximum (FWHM) are seen in the graphene analogues. Similar results are also obtained in few-layered WS_2, $MoSe_2$ and WSe_2. The broadening of the Raman bands is considered to be due to the phonon confinement. In Fig. 6.3(b), we show a high-resolution TEM image of single-layer $MoSe_2$ recorded with an FEI TITAN^{3TM} aberration-corrected microscope.

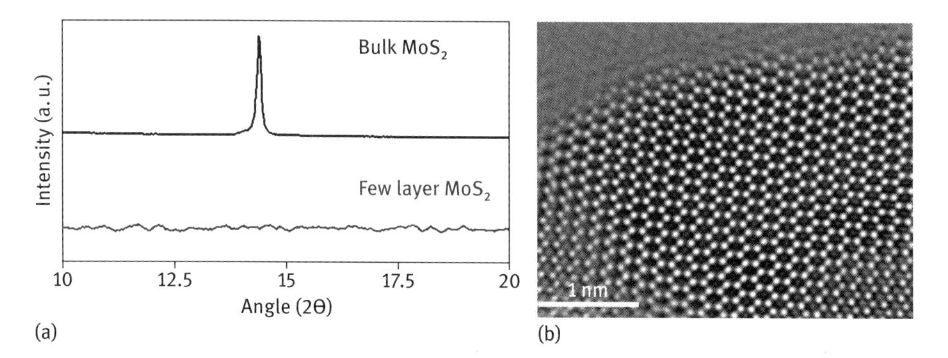

Fig. 6.3: XRD patterns of (a) bulk MoS_2 and few-layer MoS_2 (b) High resolution TEM image of $MoSe_2$ (from [2]).

6.3 Nanocarbon (graphene/C_{60}/SWNT) hybrids

All-carbon or carbon-rich hybrid materials consisting of fullerene-CNTs, fullerene-graphene and CNTs-graphene have gained a lot of attention and significant interest in the last 5–6 years [24–25]. Reduced graphene oxide (rGO)-wrapped C_{60} wires were prepared *via* liquid-liquid interfacial precipitation method and it was found that π-π interaction was the driving force for the assembly of rGO sheets and C_{60} wires. The assembly of nanocarbons and the electron transfer at interfaces induces interesting properties like exhibiting p-type behavior in the hybrid where rGO shows ambipolar and C_{60} has n-type characteristics. Photovoltaic applications have been carried out on these hybrids and have shown relatively low efficiency [26]. Ropes and bundles of carbon nanotubes have been formed along with the graphene by the reduction of GO admixed with the nanotubes [27]. Nanotube films containing small amounts of graphene are reported to be transparent conductors [28]. Layer-by-layer assembly is employed for the formation of nanofilms of reduced graphene oxide with multi-walled carbon nanotubes [29–30]. It has been reported that a flexible field-emission device has been fabricated using single-walled carbon nanotube (SWNT) network films as the conducting electrodes and thin multi-walled CNT/ tetraethylorthosilicate (TEOS) hybrid films as the emitters. P-type doping with gold ions and passivation with TEOS made the SWNT network film highly conductive and as such a good alternative to ITO electrodes [31]. A mechanically flexible, transparent thin film transistor has been made using graphene as a conducting electrode and single-walled carbon nanotubes (SWNTs) as a semiconducting channel. The resulting devices exhibited a mobility of

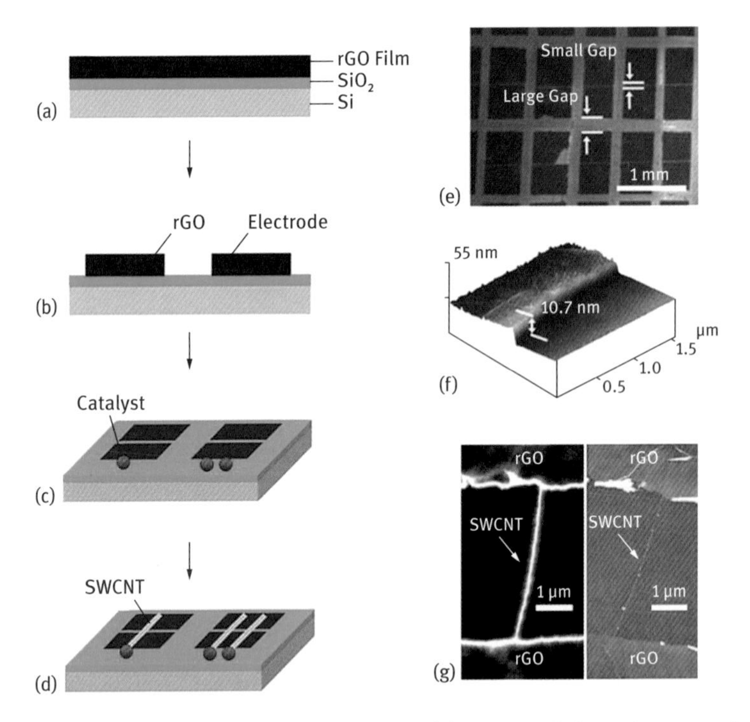

Fig. 6.4: (a) Preparation of rGO film on SiO_2. (b) Creation of electrode patterns in rGO films. (c) Deposition of catalyst in selected positions on the substrate. (d) Controlled growth of CNTs to bridge rGO electrodes by CVD. (e) SEM image of patterned rGO electrodes generated by our "scratching" method. (f) 3D-AFM image of the edge of a typical rGO electrode. The measured thickness of rGO film is 10.7 nm. (g) SEM (left) and AFM (right) images of a fabricated all-carbon device, i.e. an individual CNT-rGO device (from [3]).

~ 2 $cm^2\ V^{-1}\ s^{-1}$, On/Off ratio of ~ 10^2, with the transmittance of ~ 81 % [32]. A simple method has been developed to successfully fabricate a transistor of directly grown CNTs (active material) with patterned rGO electrodes with mobility of 748 cm^2/Vs [33]. Figure 6.4 shows a schematic illustration of the experimental procedures for generation of a CNT-based electronic device with patterned rGO as electrodes. Solar cells made up of hybrids containing C_{60}/SWNTs/rGO as the active layer and an additional evaporated C_{60} as blocking layer yield an efficiency of 0.21 %, which was increased to 0.85 % by replacing C_{60} with C_{70} [25].

Hybrids of graphene with C_{60} and SWNTs prepared at the interface are of interest because of their possible superior and synergistic properties. Hybrid films of graphene-C_{60} and graphene-SWNT hybrids have been prepared using liquid-liquid interface. The hybrid film mainly consists of C_{60} films and flower-like structures. Since C_{60} is on the surface of graphene, it is difficult to differentiate the two carbons from

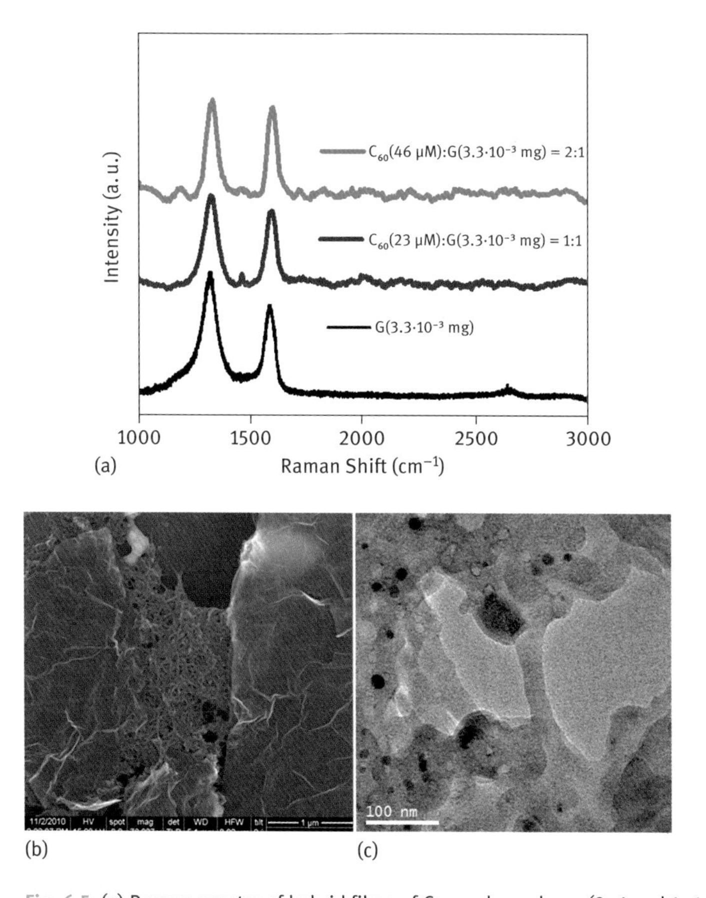

Fig. 6.5: (a) Raman spectra of hybrid films of C_{60} and graphene (2 : 1 and 1 : 1 represent the increase in concentration of C_{60} with the same weight of graphene). (b) and (c) are the FESEM and TEM images of the hybrid films of graphene with SWNT after time 12 hours after assembly (from [35]).

microscopic images. For this purpose, we have carried out spectroscopic measurements. In Fig. 6.5(a) we show the Raman G-band of a few-layer graphene film along with the bands in the hybrids of graphene with C_{60} at different concentrations. Pristine graphene shows the G-band at 1590 cm^{-1} while hybrids containing C_{60} exhibit stiffening of the G-band. The G-band occurs at 1596 cm^{-1} when the concentration of C_{60} is 23 μM and shifts to 1601 cm^{-1} when the C_{60} concentration is 46 μM. These results suggest the occurrence of charge transfer interaction between C_{60} and few-layer graphene, similar to that found between graphene and electron-acceptor molecules like TCNE and nitrobenzene [34]. UV-visible absorption spectra of the hybrid films have been examined.

Hybrids films of SWNT and graphene have been prepared with the concentration of SWNTs at 3.3 μg in 10 ml and of few-layer graphene at 3.3 μg in 10 ml. Figure 6.5(b) shows that the graphene flakes are interconnected by the SWNTs like the bridges. This interconnection leads to the formation of continuous films in the millimeter range. This type of assembly enhances the transport properties and can possibly be used in transparent conducting electrodes. The SWNTs are completely embedded in the graphene flakes by making a uniform hybrid. These hybrid films using TEM have been characterized and Fig. 6.5(b) and (c) show the TEM images with signatures of both graphene and SWNTs supporting the FESEM observations [35].

6.4 Graphene-polymer composites

Polymer nanocomposites have gained importance due to their unique properties and numerous potential applications in automotive, aerospace, construction and electrical industries. Traditionally, polymer nanocomposites are prepared with naturally available or synthetically made clays. However, such polymer-clay nanocomposites suffer from low thermal and electrical conductivities. Graphene with its high surface area, high aspect ratio, good thermal and electrical conductivity, transparency and flexibility and most of all low cost synthesis has turned out to be a good nanofiller for polymer nanocomposites [36–40]. It is well understood that the performance of a polymer composite not only depends of the properties of the filler but also on the processing method of the composite and the content of the filler. In the case of graphene-polymer composites too, the properties of the composite are highly dependent on the fabrication process which ultimately determines the dispersibility of the filler in the matrix and the bonding between the filler and the matrix. Conventional techniques of synthesis include solution mixing [41–45], blending of melt [46–47], and *in situ* synthesis of the polymer with graphene [48–49]. Graphene can be easily stabilized in solutions by ultrasonication or functionalization allowing homogenous dispersion and full intercalation of the filler into the polymer matrices [50–53]. Often functionalization of graphene is carried out to increase its dispersibility and thus better mixing. Long chain amine and amide functionalized graphene form homogenous dispersion in organic solvents and can be used to form uniform Graphene-PMMA composites by solution mixing [54–55]. Functional groups on graphene can also be used to form strong interactions with the polymer matrices containing polar groups, for example GO and rGO can be used to form strongly interacting composites with PVA [41, 45]. A new route to covalently bonded polymer-graphene nanocomposites involves ODA functionalized graphite oxide was reacted with methacrylol chloride to incorporate polymerizable -C=C- functionality on the graphene surface which was subsequently employed in *in situ* polymerization of methacrylate [56]. Zhou *et al.* [57] used polymerized ionic liquids to prepare graphene-polyaniline composite.

Mechanical properties. Like most nanofiller-polymer composites, graphene-polymer composites also show improved thermal, electrical, mechanical properties. An added advantage of using graphene is that its large surface area and huge aspect ratio give rise to great enhancements in mechanical properties compared to other inorganic-polymer nanocomposites. Pristine graphene is known to have very high breaking strength and modulus of 42 Nm^{-1} and 1 T.Pa respectively [58]. Graphene-polymer nanocomposites in general show very high strength and modulus with much lower nanofiller concentration compared to traditional polymer nanocomposites [44, 52, 54, 59–60]. With as little as 0.6 wt% of functionalized graphene (FG) in PVA or PMMA the elastic modulus and strength increased by 70 % and 45 % respectively [54]. Dynamic scanning calorimetry (DSC) studies on the composites reveal that even with only 0.6 wt% of FG in the matrix the crystallinity of the composite increased significantly. This, along with high surface area, surface roughness, homogenous dispersion and intimate mixing of the filler with matrix give rise to such enhanced mechanical properties. Cheng *et al.* [59] have observed that tensile strength and Young's modulus of the PVA composites containing 1 wt% of GO increase by almost 88 % and 150 % as compared to PVA. The elongation break was also increased by 22 % [59], Possible interactions like hydrogen bonding between the functional groups on GO with the polar PVA matrix could be the probable reason behind such strong mechanical properties.

Ramanathan *et al.* [44] compared enhancement in mechanical properties of composites with two types of graphene – expanded graphite (EGr) and functionalized graphene sheets (FGS) – and observed better modulus and strength in FGS-PMMA nanocomposites (as shown in Fig. 6.6(a)). An examination of the fracture surface of both the composites under SEM reveals an extraordinary difference in the interfacial interactions between the polymer matrix and graphene in the two systems. EGr protrudes cleanly from the fracture surface indicating a weak interfacial bond, while FGS are thickly coated with the polymer indicating strong interactions between the filler and the matrix. Functionalized graphene thus interacts well with polymer and possibly all along the 2D-sheet. Other nanocarbons like 0D nanodiamond (ND) and 1D single-walled carbon nanotubes (SWNT) also show enhanced mechanical properties when added to polymers [61–63]. Surprisingly, an extraordinary synergy is observed

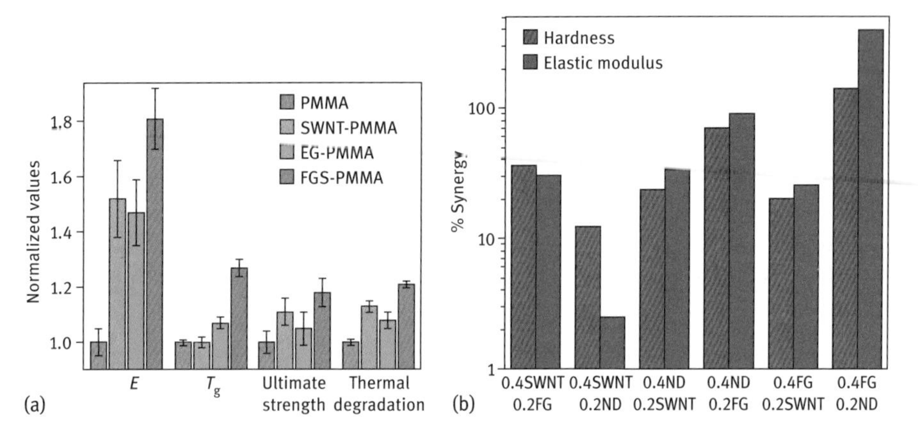

Fig. 6.6: (a) A comparison of mechanical and thermal properties of PMMA hybrids with SWNT, expanded graphite (EGr) and functionalized graphene sheets (FGS). (b) Percentage synergy in hardness and modulus of the hybrids with binary nanocarbon fillers (from [44]).

in mechanical properties when 2D-graphene is mixed with 0D-ND and 1D-SWNT. The synergistic increase implies that these composites showed higher values for strength and modulus as compared to those of the individual nanofiller-polymer composites added together [45]. Figure 6.6(b) shows the percentage synergy observed for different binary combinations of nanocarbons (E= elastic modulus and H= hardness). The synergistic effect was dramatic in the PVA composites containing ND and FG, showing 4- and 1.5-fold increases in modulus and strength, respectively in PVA-0.4FG-0.2ND composite. PVA-0.4ND-0.2FG composite on the other hand showed slightly less synergy in E and H values of about 92 % and 71 % respectively. However, synergy was not apparent in the case of SWNT+ND composites, probably because PVA-SWNT alone gives rise to large values of E and H. DSC studies revealed a variation in the percent crystallinity of around 2 % in the binary composites as compared to the single composite, suggesting that increasing crystallinity is not the cause of the observed synergy.

Electrical and thermal properties. Doping of a polymer with carbon nanotubes to form a composite was shown for the first time to increase the conductivity by ten orders of magnitude [64]. Carbon nanotubes are claimed to form conduction paths in the polymer leading to a percolation. Polymer nanocomposites containing graphene show enhanced electrical properties. PANI-nanographite composites have been fabricated with conductivities as high as 33S/cm with only 1.5 wt% of graphene loading [65]. Nanoscale dispersion of graphite sheets serves as a conducting network in-between the polymer matrix even at very low loading. The π-conjugated structure of the quinoid rings of PANI can also interact with the aromatic graphene sheets increasing conductivity [66]. Conductivity of non-conducting polymers can also be increased to a great extent by addition of graphene as nanofiller [36, 39]. Volume electrical conductivity of the PMMA has been increased to 60 S/cm with 8 wt% loading of graphite nanoplates [55]. Conductivity of bio-based polyesters was increased to 0.33 S/cm with only 1.06 wt% of graphene loading [67]. Composites of reduced graphene oxide (rGO), arc discharge graphene (HG) and acid functionalized thermally exfoliated graphene (EG) have been prepared with PMMA and PVA and their electrical properties compared. The conductivity increases with the increase in graphene content in the matrix. Figure 6.7(a) shows the variation of electrical conductivity of PVA-EG composite with increasing graphene content. There seems to be a direct relationship of conductivity to the dispersibility and the interaction of graphene filler with the polymer matrix. Thus, although conductivity of pristine graphene samples follows the order rGO > HG > EG, PVA-EG, composite shows the highest values of conductivity, possibly because the acid functionalized EG disperses well in water and interacts better with the polar PVA. Following the same reasoning, PMMA-rGO exhibits higher conductivity as compared to PMMA-HG. Dielectric properties of these composites also follow a similar trend. They increase with increasing filler content (inset in Fig. 6.7(a)) and are about 10 times higher in PVA-EG composites compared to PMMA-rGO.

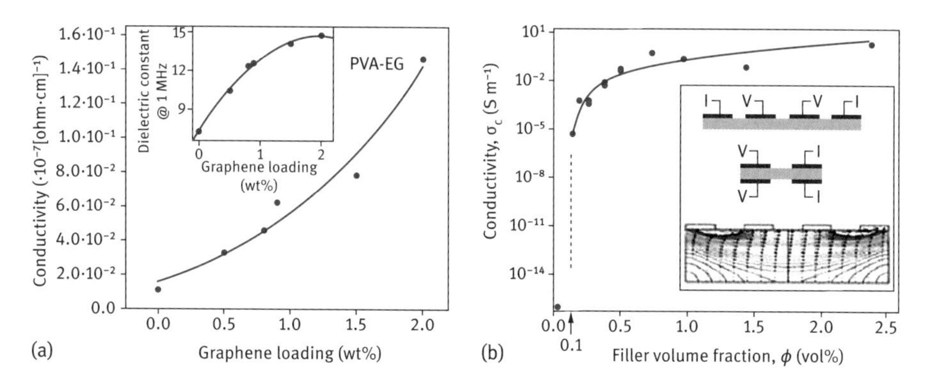

Fig. 6.7: (a) The variation of electrical conductivity of PVA-EG hybrid with increasing graphene content. Inset shows the dependence of dielectric constant for the hybrid. (b) The variation of conductivity of the polystyrene-graphene hybrid with filler content. Inset shows the four probe setup for in-plane and transverse measurements and the computed distributions of the current density for in-plane condition (reference [8]).

Hu *et al.* showed a decrease in electrical resistivity of PVA by four orders of magnitude with a percolation threshold of 6 wt% [68], while biodegradable polylactide-graphene nanocomposites were prepared with a percolation threshold as low as 3–5 wt% [46]. For polystyrene-graphene composites, percolation occurred at only 0.1 % of graphene filler, a value three times lower than those for other 2D-filler [69]. Figure 6.7(b) shows the variation of conductivity of the polystyrene-graphene composite with filler content. A sharp increase in conductivity occurs at 0.1 % (the percolation threshold) followed by a saturation. The inset shows the four probe set up for in-plane and transverse measurements, respectively. The in-plane resistivity was found to be 10 times lower than the transverse resistivity. The bottom figure in the inset shows the computed distributions of the current density shown by contour lines in a specimen in in-plane condition. High dispersibility and high surface area of graphene are presumed to be the reason for observance of such a low threshold for percolation.

Graphene nanofillers also contribute high thermal conductivity to polymer composites. On addition of graphene as a nanofiller, the composites also exhibit enhanced glass transition temperature. For example, even with 0.5 wt% graphene nanosheets, the T_g increases from 119 °C for neat PMMA to 131 °C for the composite [56]. With PVA, an improvement of thermal stability of 45 °C and an increase in the glass transition temperature of 14 °C was obtained [70]. Thermal conductivity enhancement of 300 % was observed in epoxy composites with a conductivity of 6.44 W/mK with only 25 vol% loading while conventional fillers require ~ 70 vol% loading to achieve such high values [71]. The thermal degradation temperature of composites can also be increased on addition of graphene as filler [46]. An unprecedented shift in glass transition temperature of over 40 °C was observed for poly(acrylonitrile) with 1 wt% of graphene and nearly 30 °C for 0.5 wt% of graphene for PMMA [44]. Functionalized graphene sheets as nanofiller showed better thermal properties as compared to expanded graphite (as seen in Fig. 6.7(a)) because of their better dispersibility and greater interaction with the matrix.

Graphene-PANI hybrids show capacitance values as large as 480F/g at current density of 0.1A/g [49]. With functionalized graphene, the capacitance rises to 525 F/g at current density of 0.3 A/g [72]. By *in situ* electrochemical polymerization on ITO nanorods, PANI-graphene hybrids with a highest specific capacitance of 878.57 F g^{-1} and with the charge loading of 500 mC at a current density of 1 A g^{-1} were observed [68]. Bio-sensing of glucose [50], detection of heavy metal ions, piezoresistive [53] and shape memory polymers are some of the other applications of graphene-polymer hybrids that have attracted attention in the recent years.

6.5 Functionalization of graphene and related aspects

It often becomes necessary to prepare dispersions of graphene in organic or aqueous media [73–74]. For this purpose, different approaches have been successfully employed for few-layer graphene. The two main approaches for obtaining this type of graphene are covalent functionalization or by means of noncovalent interactions. There has been some recent effort to carry out covalent and noncovalent functionalization of graphene with aromatic molecules, which help to exfoliate and stabilize the individual graphene sheets and to modify their electronic properties [75–84].

Noncovalent functionalization of graphene is important, as it does not affect the electronic structure and planarity of this 2D material. Stable aqueous dispersions of polymer-coated graphitic nanoplatelets can be prepared through an exfoliation and *in situ* reduction of graphite oxide in the presence of poly(sodium 4-styrenesulfonate) [85]. GO is functionalized with water-soluble pyrenebutyric acid in the presence of a base, followed by the reduction using hydrazine. The flexible graphene film has 7 orders of magnitude more conductivity than that of the GO precursor [86]. Wang *et al.* have reported the noncovalent functionalization of graphene with carboxylate-terminated perylene molecules. Uniform ultrathin atomic layer deposition coating on hybrid was achieved over a large area which is useful for fabricating various devices [87]. Noncovalent functionalization of graphene nanoplatelets with single-stranded DNA (ssDNA) has increased the solubility of graphene to as high as 2.5 mg/l in water [88]. Xu *et al.* have reported a GO/DNA hydrogel with high mechanical strength, excellent environmental stability, high dye-adsorption capacity, and self-healing function [89]. Graphene functionalized with polyaniline can be used as a high-performance flexible electrode. The gravimetric and volumetric capacitances of the hybrid reach 233 F g^1 and 135 F cm^{-3} respectively, more than those of graphene paper (147 Fg^{-1} and 64 F cm^{-3}) [90].

Noncovalent functionalization and solubilization of graphene has been carried out by using different surfactants and aromatic molecules. By employing a variety of surfactants, in particular, Igepal CO-890 [polyoxyethylene (40) nonylphenyl ether] (IGP), solubilization of few-layer graphene in water has been accomplished. By using 1-pyrenebutanoic acid succinimidyl ester, PYBS, one large aromatic system, solubilization of graphene has been induced in dimethylformamide through π–π interactions [80]. Interaction of the potassium salt of coronene tetracarboxylic acid (CS) with few-layer graphene causes exfoliation and selectively solubilizing single-and double-layer graphenes in water through molecular charge transfer interaction. Noncovalent functionalization and solubilization of graphene in water is achieved by employing CS to yield monolayer graphene-CS hybrids (see Fig. 6.8) [75–84]. Recently, interaction of electron-donor and -acceptor molecules with graphenes has been exploited to modify the electronic properties of graphene through ground-state charge transfer [91–95].

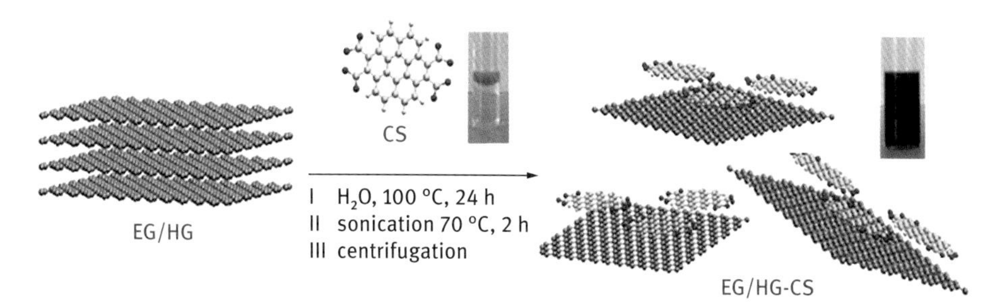

Fig. 6.8: Illustration of the exfoliation of few-layer graphene with CS to yield monolayer graphene-CS hybrids (from [75]).

Haddon and co-workers [96] functionalized graphene with long hydrocarbon chains. Oxidative treatment of microcrystalline graphite with nitric and sulfuric acid produced oxidized graphite, which contains –OH and –COOH groups and is readily soluble in water. Treatment with $SOCl_2$ converts these acidic functionalities to –COCl groups. Further treatment with octadecylamine (ODA) provides a way to introduce long hydrocarbon chains as graphene–ODA is soluble in polar solvents. This method of functionalization has been studied with different graphene samples [11, 80]. Wet chemical exfoliation of graphite in association with an *in situ* covalent functionalization of intermediately generated graphene is accomplished by coupling reductive graphite activation with oxidative arylation by using organic aryldiazonium [97]. If one can attach another functional molecule to graphene, hybrid materials are obtained. Hybrid materials of graphene have the advantage of multifunctionality, comprising properties of both graphene and the functionalizing agent [74, 83, 98–99]. Organic solution-processable functionalized-graphene hybrids with porphyrins have been synthesized and studied for their photophysical and optical-limiting properties [98]. Covalent functionalization of graphene with porphyrin and fullerene enhances the nonlinear optical (NLO) performance in the nanosecond regime [99]. Aryl radical functionalization of epitaxial graphene opens the band gap of graphene and also shows disorder-induced magnetism in the graphene sheet with antiferromagnetic regions mixed with superparamagnetic and ferromagnetic clusters [73]. Amine-terminated oligothiophenes can be covalently attached to GO nanoplatelets through the amide bond. This hybrid shows fluorescence quenching, suggesting electron or energy transfer between GO and oligothiophene and it can be used for optical limiting property [100]. Melucci *et al.* covalently attached optically active silane-terminated oligothiophene groups to GO under microwave conditions, making the hybrids soluble in water or apolar organic solvents [101]. Amine-terminated PEG can be grafted onto GO nanoplatelets through amide bond formation. It is soluble in several aqueous biological solutions such as serum or cell medium [102]. Ferrocene has been covalently attached to GO at room temperature on solid phase alumina. The ferrocene/GO hybrid shows interesting magnetic properties, and the hybrid is found to be more magnetic than the pristine graphene [103].

A novel hybrid has been obtained by covalently integrating electron-rich oligo(phenylenevinylene) amine (**OPV-amine**) into graphene. Detailed microscopic studies of **EG-OPV** generated from DMSO solution show the characteristic morphology of graphene. TEM images reveal the presence of mostly bi-and multilayered sheets of the hybrids along with the stacked graphene sheets, with an interlayer spacing varying from 2–2.4 nm, which is ~ 7 times higher than the usual layer separation in few-layer graphene (Fig. 6.9(a)). AFM studies on the **EG-OPV** hybrid showed the formation of graphene sheets having lateral dimensions extended up to micrometers (Fig. 6.9(b)). The **EG-OPV** hybrid also reveals a stepwise increase in the height profile, as can be seen in Fig. 6.9(c). The height profile shown in Fig. 6.9(d) corresponds to ~ 2.64 nm, which may be due to a single-layer of the **EG-OPV** hybrid [104].

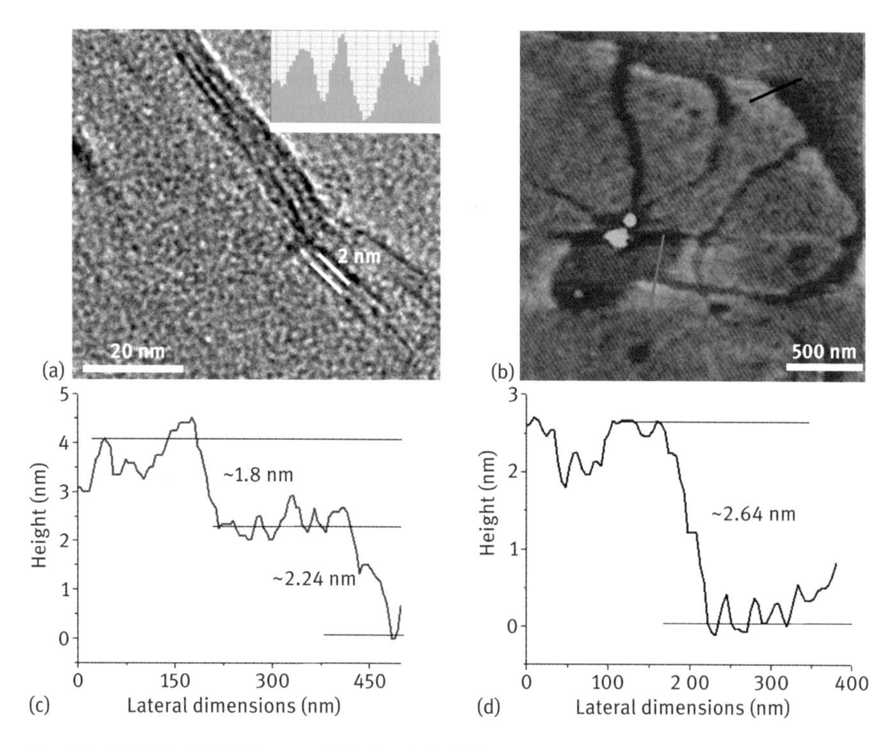

Fig. 6.9: (a) TEM, (b) AFM image of EG-OPV hybrid, height profiles of black and blue lines are shown in (c) and (d) respectively (from [104]).

6.6 Graphene-inorganic nanoparticle hybrids

Metal and semiconducting nanoparticles have been of prime interest for the past two decades because of their unique electronic, optical, magnetic and catalytic properties [30]. These properties differ from their bulk counterparts and depend on their size and shape. Nanocarbons have been used as support material for the dispersion and stabilization of metal nanoparticles due to their large chemically active surface and stability at high temperatures [105]. Decoration of inorganic nanoparticles changes the electronic structure of the parent materials through charge transfer and other interactions [106]. The combination of these two materials may lead to a successful integration of their properties in new hybrid materials that may be useful in catalysis, nanoelectronics, optics and nanobiotechnology [105–106].

Carbon nanotubes have been decorated with metal nanoparticles of Au, Ag, Pt, Pd and Co employing different procedures such as the click reaction [107], microwave treatment [106], electroless plating [108] and laser irradiation of mixtures of the individual components [109]. Pt-carbon nanotube hybrids have been used as catalysts for the conversion of nitrobenzene to aniline [110]. Tessonier *et al.* report that MWNTs covered with Pd nanoparticles on the interior walls can be employed for the selective hydrogenation of cinnamaldehyde to hydrocinnamaldehyde [111]. Pd-MWNT hybrids obtained by the deposition of Pd nanoparticles on MWCNTs in supercritical CO_2 have been used for the hydrogenation of olefins, including the conversion of stilbene into 1,2-diphenylethane [112–113]. Pd nanoparticles (NPs) supported on CNTs have been useful for sensing of hydrogen with a detection limit of around 400 ppm [114].

Gold nanoparticles have been decorated on GO nanoplatelets by the direct reduction of $AuCl_4$ using $NaBH_4$ in a GO THF suspension [115]. A similar protocol has been employed for the synthesis of Pt NPs decorated with rGO and deposited as a film on glassy carbon electrodes in a proton exchange membrane fuel cell. The rGO/Pt hybrid delivers a maximum power output of 161 mW/cm^2 compared with 96 mW/cm^2 for Pt NPs without the support [116]. Hybrids of positively charged gold nanoparticles (GNPs) and pyrene functionalized graphene (PFG) show strong electrocatalytic activity and high electrochemical stability [117]. On decoration with platinum on graphene, there is a drastic increase in capacitance value which is due to the high surface area of the hybrid by arresting the aggregation of graphene sheets [118]. Au films deposited on single-layer graphene are used as surface enhanced Raman scattering (SERS) substrates [119]. Silver-decorated graphene oxide (Ag–GO) exhibits a superior antibacterial activity towards Escherichia coli (E. coli) showing a synergistic effect [120]. Palladium nanoparticle-graphene hybrids can be used as an efficient catalyst for the Suzuki reaction [121]. Three-dimensional Pt-on-Pd bimetallic nanodendrites supported on graphene nanosheets are found to be electrocatalysts for methanol oxidation [122]. Dong *et al.* have reported the synthesis of Pt and Pt–Ru nanoparticles on the surfaces of GO nanosheets by ethyleneglycol reduction. They have studied the effect of GO as a catalyst support on the electrocatalytic activity of Pt and Pt–Ru nanoparticles for both methanol and ethanol oxidation [123].

Graphene has been decorated with metal nanoparticles such as Au, Ag, Pt, Pd and Co employing different chemical methods [124]. The influence of metal nanoparticles on the electronic structure of graphene has been examined by microscopic and spectroscopic techniques along with the first-principles calculations. There is stiffening in the position of the G-band, and the relative intensity of the 2D-band to the G-band decreases (Fig. 6.10(a)). The shifts in the G- band show meaningful trends with the ionization energies of the metals as well as the charge transfer energies. Figure 6.10(b) shows the frequency shifts of the G-band of EG against the ionization energy (IE) of the metal. Interestingly, the magnitude of the band shifts generally decreases with increasing ionization energy of the metal [125].

Magnetic properties of graphene hybrids with nanoparticles of ZnO, TiO_2, Fe_3O_4, $CoFe_2O_4$, and Ni have been studied. Higher values for saturation magnetization are obtained compared to those of the individual components and for their mechanical mixtures. Raman studies show significant shifts in the G-band of graphene due to charge transfer interaction between ZnO, TiO_2 with graphene where ZnO acting as an electron donor while TiO_2 as an acceptor. Hybrids of both Fe_3O_4 and $CoFe_2O_4$ with

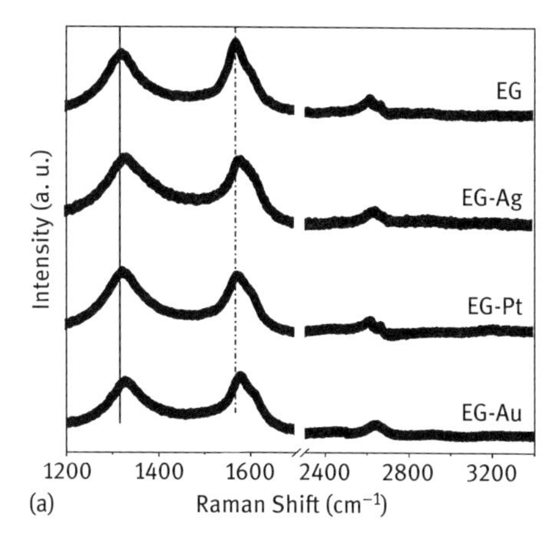

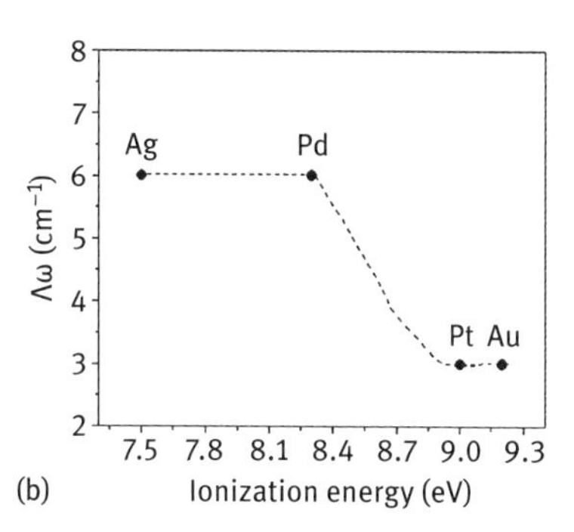

Fig. 6.10: (a) Raman spectra of EG, EG-Ag, EG-Pt and EG-Au. (b) Variation in the position of the G-band with the ionization energy of the metal. The dotted curve is to guide the eye (from [125]).

graphene showed softening of the G-band revealing a similar charge transfer interaction [126].

Graphene hybrids with semiconducting nanoparticles have much better photocatalytic properties than their pristine counterparts. TiO_2-graphene based hybrids have been prepared for efficient dye degradation. P25-graphene hybrid photocatalyst prepared under hydrothermal conditions using P25 and graphene oxide, had a higher performance for methyl blue degradation compared to pure P25 or P25/CNT hybrid. The improved performance was probably due to the higher efficiency in photo-excited electron transfer between dyes and graphene [127]. Photodegradation of methylene blue (MB) and rhodamine B (RB) have been studied with graphene (pure graphene as well as boron- and nitrogen-doped graphenes) -TiO_2 hybrids. MB, which is a good electron donor, has been found to interact strongly with electron-deficient boron-doped graphene. On the other hand, RB – which is not such a good electron donor – was found to interact strongly with electron-rich nitrogen-doped graphene causing a faster degradation of the dye (Fig. 6.11) [128]. In addition to photodegradation, the TiO_2 /graphene hybrids are also found to exhibit good performance in water splitting to generate hydrogen [129–130]. Zhang and co-workers have prepared the rGO/TiO_2 hybrid by the hydrolysis of tetra-butyl titanate in a dispersion of rGO followed by thermal annealing at 450 °C. When the GO content in the hybrid was 5 wt%, the amount of hydrogen evolution was enhanced to 17.2 μmol within 2 h with a rate of 8.6 μmol h^{-1}, which is 1.9 times higher than that of P25 (4.5 μmol h^{-1}) [129]. Liu *et al.* have found that TiO_2-graphene hybrids prepared *in situ* show an enhanced lithium insertion and extraction kinetics compared with bare TiO_2 especially at high charge/discharge rates [131].

Quantum dots (QDs) on various matrices has been extensively studied for their promising optoelectronic applications. To enhance the photocurrent generated by

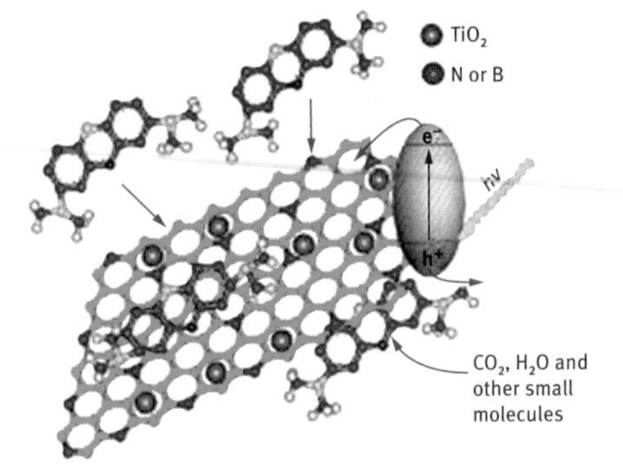

Fig. 6.11: Schematic showing the mechanism for the photo degradation of methylene blue and Rhodamine B (from [128]).

these semiconductor-matrix systems, it is essential to retard the recombination of electron-hole pairs. Carbon nanotubes and conductive polymers decorated with semiconductor nanoparticles have been found to possess this desirable property [132–133]. In this context, graphene is an ideal two-dimensional matrix [134]. CdSe nanoparticles can be attached to reduced graphene oxide (rGO) by adding rGO to the reaction mixture during the process of synthesizing CdSe nanoparticles [135]. Kim *et al.* have electrochemically deposited CdSe nanocrystal thin films over graphene [136]. A dispersion of graphite oxide in an aqueous solution of Cd^{2+}/Zn^{2+} on treatment with H_2S results in the deposition of metal sulfide nanoparticles over graphene sheets [137–138]. Fluorescent graphene-CdSe hybrids can be prepared by interaction of graphene with CdSe nanoparticles in the presence of long chain surfactants [139]. Interaction between graphene oxide and CdSe has been studied using life-time measurements. The rate constants for energy and electron transfer for the CdSe–GO hybrids were found to be 5.5×10^8 and 6.7×10^8 s^{-1} respectively. Devices fabricated from CdSe–graphene hybrids in solar cells display an improved photocurrent response of ~150 % over those with bare CdSe [140].

The liquid-liquid interface is conveniently used to prepare GO-CdSe nanohybrids. Figure 6.12(a) shows an FESEM image of a rGO–CdSe hybrid film extending over several micrometers. The image shows that CdSe nanoparticles are uniformly decorated over rGO. Figure 6.12(b) gives the EDX analysis corresponding to the FESEM image in Fig. 6.12(a). Figure 6.12(c) and (d) show the TEM and HREM images of the rGO-CdSe hybrid films. The HREM image indicates the presence of lattice fringes of the CdSe nanoparticles. The inset in Fig. 6.12(d) shows the electron diffraction pattern of the rGO-CdSe hybrid, which confirms the single crystalline nature of CdSe nanoparticles over rGO. Raman studies have confirmed the presence of charge transfer interactions between CdSe and graphene [35].

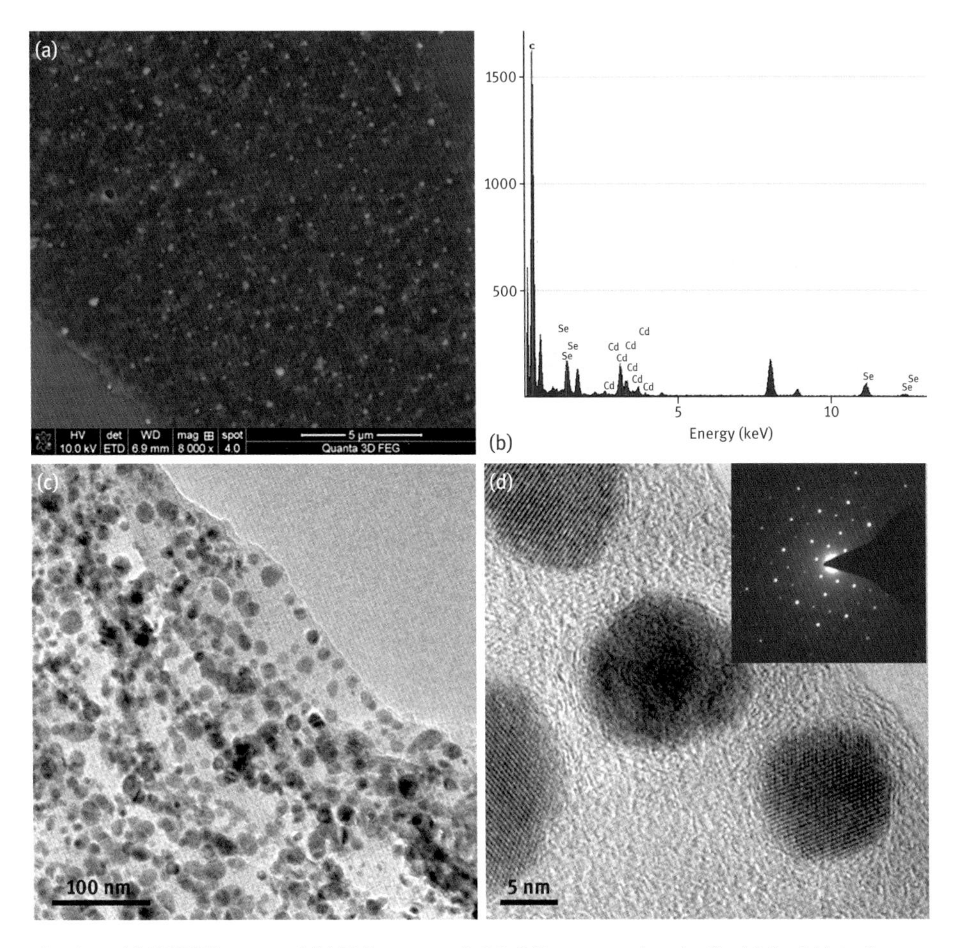

Fig. 6.12: (a) FESEM image and (b) EDX pattern of rGO-CdSe prepared at the liquid-liquid interface. (c) Low and (d) high magnification TEM images of rGO-CdSe (from [35]).

6.7 Graphene hybrids with SnO_2, MoS_2 and WS_2 as anodes in batteries

With increasing demand for energy, lithium-ion batteries have emerged as potential candidates for energy storage and portability. This has initiated an intensive search for Li-ion battery based materials with high power density, high energy, low cost, electrochemical stability, light weight and fewer environmental concerns. However, rechargeable batteries have not yet replaced primary batteries as the dominant form of energy storage because of their high cost for a limited cycling life, slow recharging, unsatisfactory energy density and low power capability. The commercially available anode material for lithium-ion batteries is graphite with a relatively small capacity (372 mAh/g) [141]. Electrochemical applications of graphene have been studied recently because of graphene's large surface area and high chemical tolerance [134]. Taking this as an advantage, it has been used as a matrix for improving the electrochemical performance of various nanomaterials including metal and metal oxides [142–143].

SnO_2 nanoparticles decorated on graphene have been examined for their electrochemical properties. These hybrids exhibit a high capacity of 810 mA h g^{-1} for the first cycle and reach 570 mA h^{-1} after 30 cycles. Bare SnO_2 electrodes show a capacity of 550 mA h g^{-1} for the first cycle and this value drops to 60 mA h g^{-1} after 15 cycles. It is believed that the existence of graphene can confine the volume change of the electrode in the charge/discharge process and the porous structure of the electrode can further buffer its volume change [144]. Wang *et al.* have prepared an SnO_2/graphene hybrid by reducing graphene oxide in the presence of $SnCl_2$ at 120 °C [145]. The hybrid shows a discharge capacity of 1420 mA h g^{-1} for the first cycle and a reversible capacity of 765 mA h g^{-1}. Aksay *et al.* have reported the synthesis of well-defined nanostructured SnO_2 on graphene sheets. The hybrid has the capacity of 625 mA h g^{-1} after 10 cycles, without any significant fading even after 100 cycles [146].

The SnO_2-rGO hybrid is synthesized by a two-step solution-phase hydrothermal process and has been investigated for battery applications. The first discharge and charge capacities for SnO_2, SnO_2-rGO are 1484 and 813 $mAhg^{-1}$ and 1451 and 1163 $mAhg^{-1}$ respectively. SnO_2-rGO showed a smaller irreversible capacity loss of 288 $mAhg^{-1}$ compared with 671 $mAhg^{-1}$ for SnO_2 (see Fig. 6.13). The corresponding coulombic efficiency was 55 % for SnO_2 and 80 % for SnO_2-rGO. The lower irreversible capacity loss and higher coulombic efficiency in the case of SnO_2-rGO is quite remarkable and is better than any other SnO_2 material reported in the literature [147]. After 90 cycles, the discharge capacity for SnO_2 -rGO hybrids (460 $mAhg^{-1}$) is more than that of bare SnO_2 (220 $mAhg^{-1}$) [148].

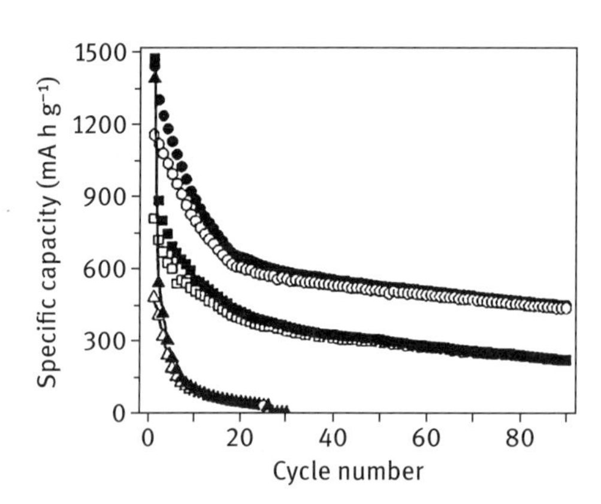

Fig. 6.13: Cycle number *vs*. capacity of SnO_2 (■, □) SnO_2-rGO (•, ○), and solid SnO_2 nanoparticles (▲, △); size: 5–10 nm. Filled symbols represent discharge, empty symbols represent charge (from [148]).

Intensive research on the hybrids containing graphene and graphene-like MoS_2 has revealed significant improvement in both the cyclability and rate capabilities. If graphene-like MoS_2 is uniformly dispersed in a graphene matrix, it inhibits aggregation and increases conductivity leading to enhanced electrochemical properties. *In situ* reduction of MoS_2 nanoflakes on graphene nanosheets led to formation of MoS_2/graphene hybrids that exhibit a specific reversible capacity of 1290 mAh/g after 50 cycles at current density of 100mA/g [149]. Das *et al.* reported the synthesis of hybrids of MoS_2 with amorphous carbon *via* a hydrothermal method with storage capacity of 800 mAh/g at a current of 400 mA/g [150]. Nanohybrids of single-layer MoS_2, graphene and amorphous carbon, synthesized using a facile solution-phase method and annealed in H_2/N_2 atmosphere at 800 °C, were used for battery applications and the hybrids not only had a reversible capacity (900–1100 mAh/g) but also high cyclic stability at 100 mA/g [151]. Graphene-like MoS_2/amorphous-carbon hybrids prepared by hydrothermal route using sodium molybdate, sulfocarbamide and glucose followed by annealing at 800 °C for 2 h in a stream of 10 % H_2/N_2 atmosphere had a reversible capacity of 912 mAh/g after 100 cycles [152]. MoS_2/Graphene hybrids prepared *via* L-cysteine-assisted solution phase method followed by annealing in a H_2/N_2 atmosphere at 800 °C for 2 h and were tested for the Li-ion batteries and found to have a specific capacity of ~1100 mAh/g at a current density of 100 mA/g with no capacity fading even after 100 cycles [153].

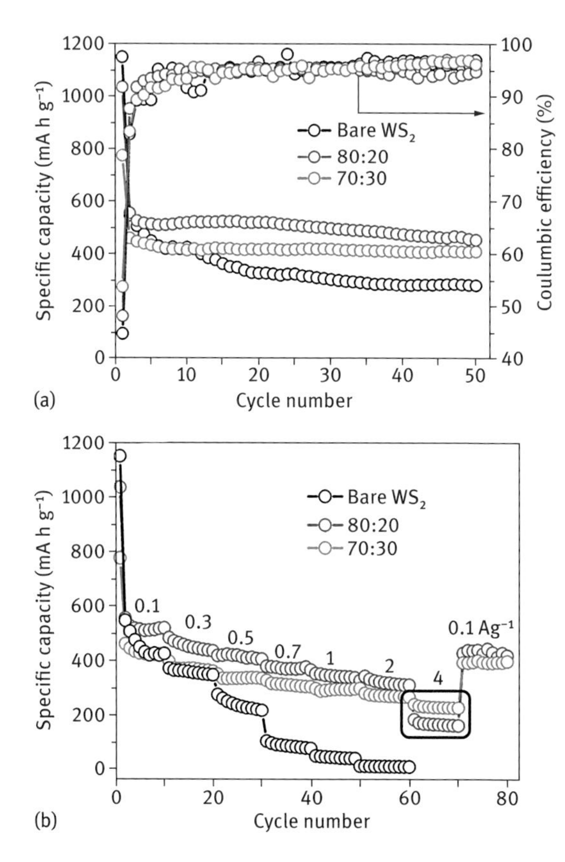

Fig. 6.14: (a) The galvanostatic cycling performance and coulombic efficiencies of WS_2, 80 : 20 and 70 : 30 at a current density of 100 mA g^{-1}. (b) The specific capacities of bare WS_2, 80 : 20 and 70 : 30 at various discharge currents of 0.1, 0.3, 0.5, 0.7, 1, 2, and 4 Ag^{-1} (from [154]).

Graphene-WS_2 hybrids synthesized in a facile approach by using sodium borohydride as the reduced agent along with the dispersed GO and few-layer WS_2 in water under reflux conditions, have been characterized with microscopic and spectroscopic techniques. The hybrids display high specific capacity and most importantly exceptional stability compared to the bare WS_2. Figure 6.14(a) shows the discharge specific capacities and coulombic efficiencies of WS_2 and WS_2-rGO at a current density of 100 mA g^{-1} (voltage range = (0.01- 3) V) for the first 50 cycles. In the first cycle, the bare WS_2, and 80 : 20 (graphene:WS_2) electrodes show a discharge capacity of approximately 1149 mAh g^{-1} and 1034 mAh g^{-1} respectively. After 50 cycles, the discharge capacity of the bare WS_2 electrode decreased significantly and was equal to 278 mAh g^{-1} compared to the discharge capacity of 80 : 20 electrodes (which was as high as 451 and with 82 % efficiency with respect to the second cycle), indicating excellent electrochemical performance. Reduced graphene oxide plays an important role in enhancing the rate capability of e WS_2-rGO hybrids (see Fig. 6.14(b)) [154].

6.8 Graphene-MOF hybrids

Metal-organic-frameworks (MOFs) have attracted great interest due to their high and tunable porosity, unique adsorptive and unusual ion exchange properties [155]. Graphene-MOF hybrids combine the positive attributes of carbonaceous graphene, like high reactive surface area, with those of MOFs. A graphene-MOF hybrid was prepared by Petit and co-workers [156] to demonstrate the combination of the high dispersive interactions of graphene with the high porosity of MOFs to obtain enhanced porosity nature with well-developed and flexible functionality of the pores. MOF-5 was chosen for its small pore size and the hybrids containing 5 %, 10 % and 20 % of GO were prepared utilizing the epoxy and carboxyl groups in GO to bind to the zinc oxide clusters. The hybrid with 10 wt% GO resulted in the most porous material with about 12 % increase in the porosity parameters compared to that of the hypothetical physical mixture of the components. Hybrids of HKUST-1, a Cu based MOF with graphene layers, resulted in enhanced ammonia adsorption at room temperature in both dry and moist conditions [157]. The hybrid was water stable and exhibited two successive changes of color of the adsorbent during breakthrough tests confirming reactive adsorption. Adsorption capacities of the hybrid were higher than those of the calculated adsorption capacity of physical mixtures, suggesting a synergistic effect.

Jahan *et al.* [158] used graphene as a structure-directing agent to form extended structures uniquely different from the original MOF. Benzoic acid functionalization of graphene (BFG) nanosheets was employed to enhance the density of reactive functional groups on the graphene's surface. Mixing the BFG with precursors used for synthesizing MOF-5 gave rise to MOF nanowires with a diameter directly dependent on the lateral dimensions of the basal plane of the BFG. Intercalation of graphene in the MOF imparts new electrical properties to the otherwise insulating MOF. Azobenzoic acid-functionalized rGO has been incorporated into Zn^{2+}-based MOF to form MOF-A-GO hydrogels [159]. These hydrogels exhibit a gradual decrease in fluorescence intensity upon addition of trinitrotoluene. Charge-transfer interactions between electron deficient aromatic rings of TNT and the electron-rich aromatic groups of the MOF-A-GO resulted in fluorescence quenching. The original ligand showed much lower fluorescence intensity compared to the MOF-A-GO and hence lesser sensitivity to quenching by TNT. Similarly, Zn-based MOF-5-GO hybrids with microporous structure and a small proportion of mesopores have been found to exhibit a high H_2S uptake. The porosity directly depended on the amount of GO loaded and could also be tuned by introduction of glucose during the synthesis, to observe very high values of H_2S uptake [160].

Pyridine-functionalized graphene has been used as a building block to form graphene-metalloporphyrin MOF with iron-porphyrin (see Fig. 6.15) [161]. Addition of pyridine-functionalized graphene changed the crystallization process of iron-

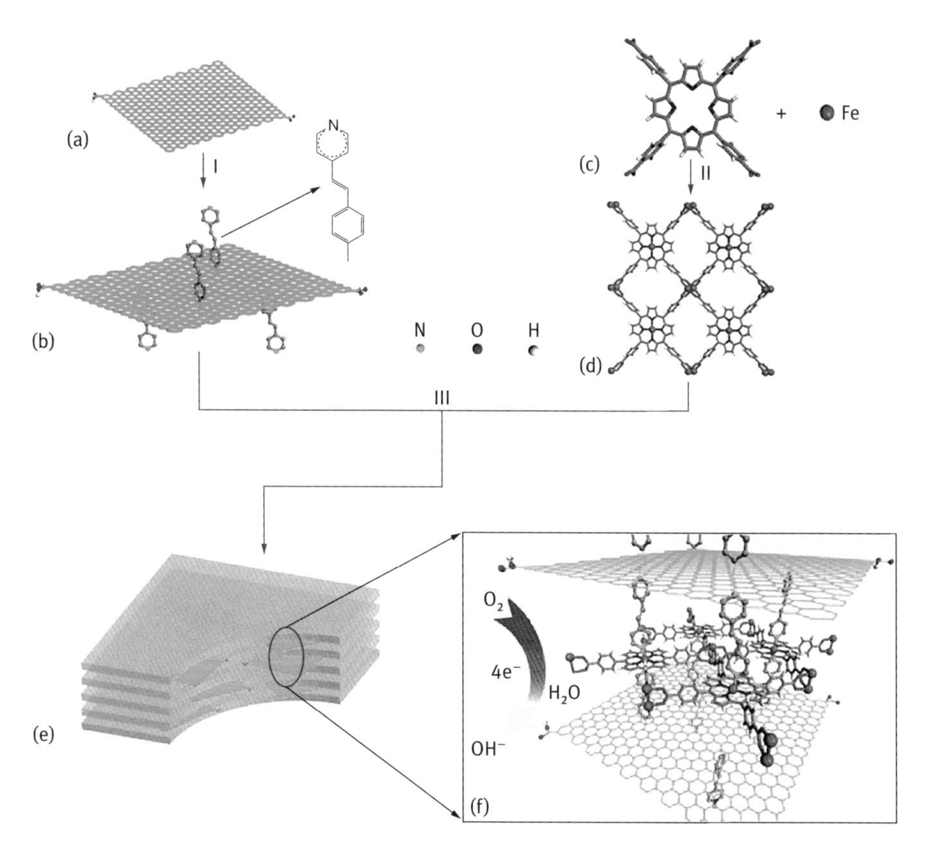

Fig. 6.15: (a) Synthetic routes to make graphene-porphyrin MOF: (a) G-dye synthesized from rGO sheets *via* diazotization with 4-(4-aminostyryl) pyridine, (b) (Fe–P)n MOF synthesized *via* reaction between TCPPs and Fe ions, (c) (G-dye-FeP)n MOF formed *via* reaction between (Fe–P)n MOF and G-dye (from [161]).

porphyrin in the MOF and increased its porosity. The increased porosity enhances the electrochemical charge transfer rate, showing enhanced oxygen reduction and this can useful as Pt-free electrodes for fuel cells.

The photocatalytic activity of zeolite titanosilicates (TS-1) is enhanced by more than 25 times on adding graphene [162]. *In situ* microwave-assisted solvothermal technique has been used to prepare the graphene-TS1 hybrid from a mixture of graphene and metal organic precursors. Addition of graphene has a profound and remarkable effect on the morphology of the TS-1 converting it from large (300 nm) hexagonal prisms to ellipsoidal particles with a small addition of graphene to 10 nm spherical particles at 20 wt% of graphene. The BET surface area of the hybrid was comparable to that of pristine TS1 while the adsorption of dye molecules decreased on addition of the graphene. However, the photocatalytic activity of the hybrids increases to 25 times that of pristine TS1 (Fig. 6.16(a)). Possible reasons for such synergistic activity of the hybrids are suggested to be: (1) prolonged lifetime of photoactive electrons and holes as a consequence of better charge separation at the graphene-TS1 interface, (2) higher reactivity of the edge atoms in graphene towards electron transfer than bulk graphene, (3) high degree of mesopores with dimensions suitable for diffusion of the large organic molecule, or (4) a possible different reaction pathway for the degradation of the organic molecule on graphene. Figure 6.16(b) shows a schematic representation of the possible mechanism of dye degradation in the graphene-TS1 hybrid. Combining the beneficial attributes of MOF-like porosity and the presence of transition metals can provide reactive adsorption sites for catalysis or contribute to magnetism.

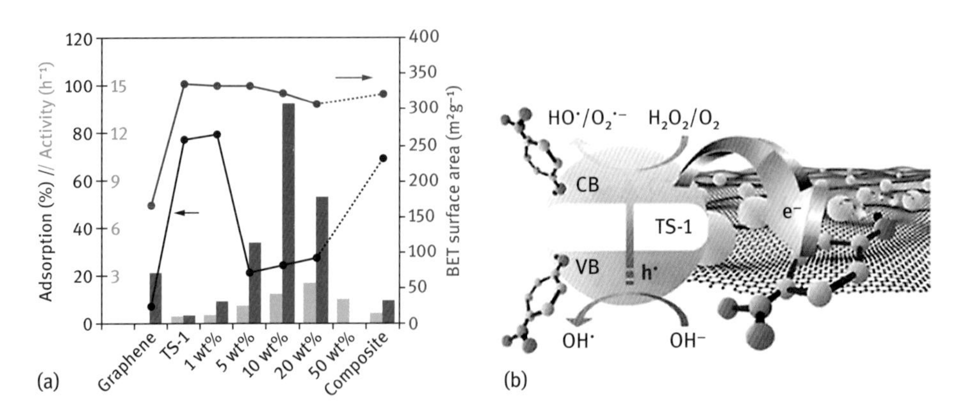

Fig. 6.16: (a) BET surface area (red line), percentage adsorption of dye molecules (black line) and photocatalytic activity (columns) for different hybrids with SWNT (green bars) and graphene (blue bars) and (b) possible mechanism for synergy in photocatalytic activity of the hybrid *via* electron transfer to graphene (from [162]).

6.9 Summary

The above sections should clearly demonstrate that the study of graphene and its hybrids, especially the latter, has only just begun. Hybrids of graphene with various inorganic nanostructures (nanoparticles, nanowires and nanosheets) are likely to possess many novel properties with potential applications. Graphene-MOF hybrids are yet to be explored extensively and they are found to reveal many useful properties.

Bibliography

[1] Novoselov, K. S.; Geim, A. K.; Morozov, S. V.; Jiang, D.; Zhang, Y.; Dubonos, S. V.; Grigorieva, I. V.; Firsov, A. A., *Science* **2004**, *306*, 666.

[2] Rao, C. N. R.; Matte, H. S. S. R.; Subrahmanyam, K. S., *Acc. Chem. Res.* **2013**, *46*, 149.

[3] Geim, A. K.; Novoselov, K. S., *Nature Mater.* **2007**, *6*, 183.

[4] Rao, C. N. R.; Sood, A. K.; Subrahmanyam, K. S.; Govindaraj, A., *Angew. Chem. Int. Ed.* **2009**, *48*, 7752.

[5] Schafhaeutl, C., *Phil. Mag.* **1840**, *16*, 570.

[6] Berger, C.; Song, Z.; Li, T.; Li, X.; Ogbazghi, A. Y.; Feng, R.; Dai, Z.; Marchenkov, A. N.; Conrad, E. H.; First, P. N.; de Heer, W. A. *J., Phys. Chem. B* **2004**, *108*, 9912.

[7] Juang, Z.-Y.; Wu, C.-Y.; Lo, C.-W.; Chen, W.-Y.; Huang, C.-F.; Hwang, J.-C.; Chen, F.-R.; Leou, K.-C.; Tsai, C.-H., *Carbon* **2009**, *47*, 2026.

[8] Reina, A.; Jia, X. T.; Ho, J.; Nezich, D.; Son, H. B.; Bulovic, V.; Dresselhaus, M. S.; Kong, J., *Nano Lett.* **2009**, *9*, 30.

[9] C. N. R. Rao; K. S. Subrahmanyam; H. S. S. Ramakrishna Matte; B. Abdulhakeem; A. Govindaraj; B. Das; P. Kumar; Ghosh, A.; Late, D. J., *Sci. Tech. Adv. Mater.* **2010**, *11*, 054502.

[10] Wang, J.; Zhu, M.; Outlaw, R. A.; Zhao, X.; Manos, D. M.; Holloway, B. C. *Carbon* **2004**, *42*, 2867.

[11] Subrahmanyam, K. S.; Vivekchand, S. R. C.; Govindaraj, A.; Rao, C. N. R., *J. Mater. Chem.* **2008**, *18*, 1517.

[12] Guoqing, X.; Wontae, H.; Namhun, K.; Sung, M. C.; Heeyeop, C., *Nanotechnology* **2010**, *21*, 405201.

[13] Andersson, O. E.; Prasad, B. L. V.; Sato, H.; Enoki, T.; Hishiyama, Y.; Kaburagi, Y.; Yoshikawa, M.; Bandow, S., *Phys. Rev. B* **1998**, *58*, 16387.

[14] Subrahmanyam, K. S.; Panchakarla, L. S.; Govindaraj, A.; Rao, C. N. R., *J. Phys. Chem. C* **2009**, *113*, 4257.

[15] Panchakarla, L. S.; Subrahmanyam, K. S.; Saha, S. K.; Govindaraj, A.; Krishnamurthy, H. R.; Waghmare, U. V.; Rao, C. N. R. *Adv. Mater.* **2009**, *21*, 4726.

[16] Zhiyong, W.; Nan, L.; Zujin, S.; Zhennan, G. *Nanotechnol.* **2010**, *21*, 175602.

[17] Dreyer, D. R.; Park, S.; Bielawski, C. W.; Ruoff, R. S. *Chem. Soc. Rev.* **2010**, *39*, 228.

[18] Ding, Y. H.; Zhang, P.; Zhuo, Q.; Ren, H. M.; Yang, Z. M.; Jiang, Y., *Nanotechnology*, *22*, 215601.

[19] Abdelsayed, V.; Moussa, S.; Hassan, H. M.; Aluri, H. S.; Collinson, M. M.; El-Shall, M. S.,*J. Phys. Chem. Lett.* **2010**, *1*, 2804.

[20] Kumar, P.; Subrahmanyam, K. S.; Rao, C. N. R., *International Journal of Nanoscience* **2011**, *10*, 559.

[21] Kumar,P.; Subrahmanyam, K. S.; Rao, C. N. R., *Mater. Express* **2011**, *1*, 252.

[22] Ramakrishna Matte, H. S. S.; Gomathi, A.; Manna, A. K.; Late, D. J.; Datta, R.; Pati, S. K.; Rao, C. N. R., *Angew. Chem. Int. Ed.* **2010**, *49*, 4059.

[23] Matte, H. S. S. R.; Plowman, B.; Datta, R.; Rao, C. N. R., *Dalton Trans.* **2011**, *40*, 10322.

[24] Cataldo, S.; Salice, P.; Menna, E.; Pignataro, B., *Ener. Environ. Sci.* **2012**, *5*, 5919.

[25] Tung, V. C.; Huang, J.-H.; Kim, J.; Smith, A. J.; Chu, C.-W.; Huang, J., *Ener. Environ. Sci.* **2012**, *5*, 7810.

[26] Yang, J.; Heo, M.; Lee, H. J.; Park, S.-M.; Kim, J. Y.; Shin, H. S., *ACS Nano* **2011**, *5*, 8365.

[27] Tung, V. C.; Chen, L.-M.; Allen, M. J.; Wassei, J. K.; Nelson, K.; Kaner, R. B.; Yang, Y., *Nano Lett.* **2009**, *9*, 1949.

[28] King, P. J.; Khan, U.; Lotya, M.; De, S.; Coleman, J. N. *ACS Nano* **2010**, *4*, 4238.

[29] Kim, Y.-K.; Min, D.-H., *Langmuir* **2009**, *25*, 11302.

[30] Rao, C. N. R.; Ramakrishna Matte, H. S. S.; Voggu, R.; Govindaraj, A., *Dalton Trans.* **2012**, *41*, 5089.

[31] Jeong, H. J.; Jeong, H. D.; Kim, H. Y.; Kim, J. S.; Jeong, S. Y.; Han, J. T.; Bang, D. S.; Lee, G.-W., *Adv. Func. Mater.* **2011**, *21*, 1526.

[32] Sukjae, J.; Houk, J.; Youngbin, L.; Daewoo, S.; Seunghyun, B.; Byung Hee, H.; Jong-Hyun, A., *Nanotechnol.* **2010**, *21*, 425201.

[33] Li, B.; Cao, X.; Ong, H. G.; Cheah, J. W.; Zhou, X.; Yin, Z.; Li, H.; Wang, J.; Boey, F.; Huang, W.; Zhang, H., *Adv. Mater.* **2010**, *22*, 3058.

[34] Rao, C. N. R.; Voggu, R., *Mater. Today* **2010**, *13*, 34.

[35] Moses, K.; Panchakarla, L. S.; Ramakrishna Matte, H. S. S.; Govindarao, B.; Rao, C. N. R., *Ind. J. Chem. A* **2011**, *50*, 1239.

[36] Du, J.; Cheng, H.-M., *Macromol. Chem. Phys.* **2012**, *213*, 1060.

[37] Zhu, Y.; Murali, S.; Cai, W.; Li, X.; Suk, J. W.; Potts, J. R.; Ruoff, R. S., *Adv. Mater.* **2010**, *22*, 3906.

[38] Huang, X.; Yin, Z.; Wu, S.; Qi, X.; He, Q.; Zhang, Q.; Yan, Q.; Boey, F.; Zhang, H., *Small* **2011**, *7*, 1876.

[39] Li, B.; Zhong, W.-H., *J. Mater. Sc.* **2011**, *46*, 5595.

[40] Kuilla, T.; Bhadra, S.; Yao, D.; Kim, N. H.; Bose, S.; Lee, J. H., *Progress. Polymer Sc.* **2010**, *35*, 1350.

[41] Xu, Y.; Hong, W.; Bai, H.; Li, C.; Shi, G., *Carbon* **2009**, *47*, 3538.

[42] Dongyu, C.; Mo, S., *Nanotechnol.* **2009**, *20*, 315708.

[43] Wu, Q.; Xu, Y.; Yao, Z.; Liu, A.; Shi, G., *ACS Nano* **2010**, *4*, 1963.

[44] Ramanathan, T.; Stankovich, S.; Dikin, D. A.; Herrera-Alonso, M.; Piner, R. D.; Adamson, D. H.; Schniepp, H. C.; Chen, X.; Ruoff, R. S.; Nguyen, S. T.; Aksay, I. A.; Prud'Homme, R. K.; Brinson, L. C.; Yongmin, K., *Nature Nanotechnol.* **2008**, *3*, 327.

[45] Prasad, K. E.; Das, B.; Maitra, U.; Ramamurty, U.; Rao, C. N. R., *Proc. Natl. Acad. Sci. USA* **2009**, *106*, 13186.

[46] Kim, I.-H.; Jeong, Y. G., *J. Polymer Sc. B* **2010**, *48*, 850.

[47] Kim, H.; Macosko, C. W., *Macromol.* **2008**, *41*, 3317.

[48] Xiao, X.; Xie, T.; Cheng, Y.-T., *J. Mater. Chem.* **2010**, *20*, 3508.

[49] Zhang, K.; Zhang, L. L.; Zhao, X. S.; Wu, J., *Chem. Mater.* **2010**, *22*, 1392.

[50] Lu, J.; Drzal, L. T.; Worden, R. M.; Lee, I., *Chem. Mater.* **2007**, *19*, 6240.

[51] Du, X. S.; Xiao, M.; Meng, Y. Z.; Hay, A. S., *Polymer. Adv. Tech.* **2004**, *15*, 320.

[52] Jacob George, J.; Bandyopadhyay, A.; Bhowmick, A. K., *J. Appl. Polymer Sc.* **2008**, *108*, 1603.

[53] Chen, L.; Lu, L.; Wu, D.; Chen, G., *Polymer Comp.* **2007**, *28*, 493.

[54] Barun, D.; Prasad, K. E.; Ramamurty, U.; Rao, C. N. R., *Nanotechnol.* **2009**, *20*, 125705.

[55] Wang, W.-p.; Liu, Y.; Li, X.-x.; You, Y.-z. *J. Appl. Polymer Sc.* **2006**, *100*, 1427.

[56] Pramoda, K. P.; Hussain, H.; Koh, H. M.; Tan, H. R.; He, C. B., *J. Polymer. Sc. A* **2010**, *48*, 4262.

[57] Zhou, X.; Wu, T.; Hu, B.; Yang, G.; Han, B. *Chem. Comm.* **2010**, *46*, 3663.
[58] Lee, C.; Wei, X.; Kysar, J. W.; Hone, J. *Science* **2008**, *321*, 385.
[59] Cheng, H. K. F.; Sahoo, N. G.; Tan, Y. P.; Pan, Y.; Bao, H.; Li, L.; Chan, S. H.; Zhao, J., *ACS Appl. Mater. Interf.* **2012**, *4*, 2387.
[60] Ramanathan, T.; Stankovich, S.; Dikin, D. A.; Liu, H.; Shen, H.; Nguyen, S. T.; Brinson, L. C., *J. Polymer. Sc. B* **2007**, *45*, 2097.
[61] Cadek, M.; Coleman, J. N.; Barron, V. K.; Hedicke, V.; Blau, W. J., *Appl. Phys. Lett.* **2002**, *81*, 5123.
[62] Behler, K. D.; Stravato, A.; Mochalin, V.; Korneva, G.; Yushin, G.; Gogotsi, Y., *ACS Nano* **2009**, *3*, 363.
[63] Maitra, U.; Prasad, K. E.; Ramamurty, U.; Rao, C. N. R., *Solid. State. Commun.* **2009**, *149*, 1693.
[64] Coleman, J. N.; Curran, S.; Dalton, A. B.; Davey, A. P.; McCarthy, B.; Blau, W.; Barklie, R. C., *Phys. Rev. B* **1998**, *58*, R7492.
[65] Du, X. S.; Xiao, M.; Meng, Y. Z., *Europ. Polym J.* **2004**, *40*, 1489.
[66] Mo, Z.; Shi, H.; Chen, H.; Niu, G.; Zhao, Z.; Wu, Y., *J. Appl. Polymer Sc.* **2009**, *112*, 573.
[67] Tang, Z.; Kang, H.; Shen, Z.; Guo, B.; Zhang, L.; Jia, D., *Macromol.* **2012**, *45*, 3444.
[68] Hu, L.; Tu, J.; Jiao, S.; Hou, J.; Zhu, H.; Fray, D. J., *Phys. Chem. Chem. Phys.* **2012**, *14*, 15652.
[69] Stankovich, S.; Dikin, D. A.; Dommett, G. H. B.; Kohlhaas, K. M.; Zimney, E. J.; Stach, E. A.; Piner, R. D.; Nguyen, S. T.; Ruoff, R. S., *Nature* **2006**, *442*, 282.
[70] Hu, H.; Chen, G., *Polymer Comp.* **2010**, *31*, 1770.
[71] Yu, A.; Ramesh, P.; Itkis, M. E.; Bekyarova, E.; Haddon, R. C., *J. Phys. Chem. C* **2007**, *111*, 7565.
[72] Liu, Y.; Deng, R.; Wang, Z.; Liu, H., *J. Mater. Chem.* **2012**, *22*, 13619.
[73] Niyogi, S.; Bekyarova, E.; Hong, J.; Khizroev, S.; Berger, C.; de Heer, W.; Haddon, R. C., *J. Phys. Chem. Lett.* **2011**, *2*, 2487.
[74] Sun, Z. Z.; James, D. K.; Tour, J. M., *J. Phys. Chem. Lett.* **2011**, *2*, 2425.
[75] Ghosh, A.; Rao, K. V.; George, S. J.; Rao, C. N. R., *Chem.- Eur. J.* **2010**, *16*, 2700.
[76] Ramakrishna Matte, H. S. S.; Subrahmanyam, K. S.; V. Rao, K.; George, S. J.; Rao, C. N. R., *Chem. Phys. Lett.* **2011**, *506*, 260.
[77] Ghosh, A.; Rao, K. V.; Voggu, R.; George, S. J., *Chem. Phys. Lett.* **2010**, *488*, 198.
[78] Kozhemyakina, N. V.; Englert, J. M.; Yang, G.; Spiecker, E.; Schmidt, C. D.; Hauke, F.; Hirsch, A., *Adv. Mater.* **2010**, *22*, 5483.
[79] Wojcik, A.; Kamat, P. V., *ACS Nano* **2010**, *4*, 6697.
[80] Subrahmanyam, K. S.; Ghosh, A.; Gomathi, A.; Govindaraj, A.; Rao, C. N. R., *Nanosci. Nanotechnol. Lett.* **2009**, *1*, 28.
[81] Samanta, S. K.; Subrahmanyam, K. S.; Bhattacharya, S.; Rao, C. N. R., *Chem.- Eur. J.* **2012**, *18*, 2890.
[82] Zhang, Z. X.; Huang, H. L.; Yang, X. M.; Zang, L., *J. Phys. Chem. Lett.* **2011**, *2*, 2897.
[83] Liu, L.-H.; Yan, M., *J. Mater. Chem.* **2011**, *21*, 3273.
[84] Yang, H.; Shan, C.; Li, F.; Han, D.; Zhang, Q.; Niu, L,. *Chem. Comm.* **2009**, 3880.
[85] Stankovich, S.; Piner, R. D.; Chen, X.; Wu, N.; Nguyen, S. T.; Ruoff, R. S., *J. Mater. Chem.* **2006**, *16*, 155.
[86] Xu, Y.; Bai, H.; Lu, G.; Li, C.; Shi, G., *J. Am. Chem. Soc.* **2008**, *130*, 5856.
[87] Wang, X.; Tabakman, S. M.; Dai, H., *J. Am. Chem. Soc.* **2008**, *130*, 8152.
[88] Liang, Y.; Wu, D.; Feng, X.; Müllen, K., *Adv. Mater.* **2009**, *21*, 1679.
[89] Xu, Y.; Wu, Q.; Sun, Y.; Bai, H.; Shi, G., *ACS Nano* **2010**, *4*, 7358.
[90] Wang, D.-W.; Li, F.; Zhao, J.; Ren, W.; Chen, Z.-G.; Tan, J.; Wu, Z.-S.; Gentle, I.; Lu, G. Q.; Cheng, H.-M., *ACS Nano* **2009**, *3*, 1745.
[91] Das, B.; Voggu, R.; Rout, C. S.; Rao, C. N. R., *Chem. Comm.* **2008**, 5155.

[92] R. Voggu; B. Das; C. S. Rout; C. N. R. Rao, *J. Phys. Condensed Mater*, **2008**, *20* 472204.
[93] K. S. Subrahmanyam; R. Voggu; A. Govindaraj; C. N. R. Rao, *Chem. Phys. Lett.*, **2009** *472*, 96.
[94] N. Varghese; A. Ghosh; R. Voggu; S. Ghosh; C. N. R. Rao, *J. Phys. Chem. C.*, **2009**, 16855.
[95] A. K. Manna; S. K. Pati *Chem. Asian. J.* **2009**, *4* 855.
[96] Niyogi, S.; Bekyarova, E.; Itkis, M. E.; McWilliams, J. L.; Hamon, M. A.; Haddon, R. C., *J. Am. Chem. Soc.* **2006**, *128*, 7720.
[97] Englert, J. M.; Dotzer, C.; Yang, G.; Schmid, M.; Papp, C.; Gottfried, J. M.; Steinraick, H.-P.; Spiecker, E.; Hauke, F.; Hirsch, A., *Nat. Chem.* **2011**, *3*, 279.
[98] Xu, Y. F.; Liu, Z. B.; Zhang, X. L.; Wang, Y.; Tian, J. G.; Huang, Y.; Ma, Y. F.; Zhang, X. Y.; Chen, Y. S., *Adv. Mater.* **2009**, *21*, 1275.
[99] Liu, Z. B.; Xu, Y. F.; Zhang, X. Y.; Zhang, X. L.; Chen, Y. S.; Tian, J. G., *J. Phys. Chem. B* **2009**, *113*, 9681.
[100] Liu, Y.; Zhou, J.; Zhang, X.; Liu, Z.; Wan, X.; Tian, J.; Wang, T.; Chen, Y., *Carbon* **2009**, *47*, 3113.
[101] Melucci, M.; Treossi, E.; Ortolani, L.; Giambastiani, G.; Morandi, V.; Klar, P.; Casiraghi, C.; Samori, P.; Palermo, V., *J. Mater. Chem.* **2010**, *20*, 9052.
[102] Liu, Z.; Robinson, J. T.; Sun, X.; Dai, H., *J. Am. Chem. Soc.* **2008**, *130*, 10876.
[103] Avinash, M. B.; Subrahmanyam, K. S.; Sundarayya, Y.; Govindaraju, T., *Nanoscale* **2010**, *2* 1762.
[104] Matte, H. S. S. R.; Jain, A.; George, S. J., *RSC Adv.* **2012**, *2*, 6290.
[105] Wildgoose, G. G.; Banks, C. E.; Compton, R. G., *Small* **2006**, *2*, 182.
[106] Voggu, R.; Pal, S.; Pati, S. K.; Rao, C. N. R., *J. Phys. Conden. Mat.* **2008**, *20*, 5.
[107] Voggu, R.; Suguna, P.; Chandrasekaran, S.; Rao, C. N. R., *Chem. Phys. Lett.* **2007**, *443*, 118.
[108] Ma, X.; Lun, N.; Wen, S., *Diamond. Related Mater.* **2005**, *14*, 68.
[109] Henley, S.; Watts, P.; Mureau, N.; Silva, S., *Appl. Phys. A Mater. Sci. Process.* **2008**, *93*, 875.
[110] Li, C.-H.; Yu, Z.-X.; Yao, K.-F.; Ji, S.-f.; Liang, J., *J. Mol. Catal. A.* **2005**, *226*, 101.
[111] Tessonnier, J.-P.; Pesant, L.; Ehret, G.; Ledoux, M. J.; Pham-Huu, C., *Appl. Catal. A.* **2005**, *288*, 203.
[112] Ye, X. R.; Lin, Y.; Wai, C. M., *Chemical Communications* **2003**, 642.
[113] Ye, X.-R.; Lin, Y.; Wang, C.; Engelhard, M. H.; Wang, Y.; Wai, C. M., *J. Mater. Chem* **2004**, *14*, 908.
[114] Kong, J.; Chapline, M. G.; Dai, H., *Adv. Mater.* **2001**, *13*, 1384.
[115] Muszynski, R.; Seger, B.; Kamat, P. V., *J. Phys. Chem. C* **2008**, *112*, 5263.
[116] Seger, B.; Kamat, P. V., *J. Phys. Chem. C* **2009**, *113*, 7990.
[117] Hong, W.; Bai, H.; Xu, Y.; Yao, Z.; Gu, Z.; Shi, G., *J. Phys. Chem. C* **2010**, *114*, 1822.
[118] Si, Y.; Samulski, E. T., *Chem. Mater.* **2008**, *20*, 6792.
[119] Wang, Y.; Ni, Z.; Hu, H.; Hao, Y.; Wong, C. P.; Yu, T.; Thong, J. T. L.; Shen, Z. X., *Appl. Phys. Lett.* **2012**, *97*, 163111.
[120] Ma, J.; Zhang, J.; Xiong, Z.; Yong, Y.; Zhao, X. S., *J. Mater. Chem* **2011**, *21*, 3350.
[121] Li, Y.; Fan, X.; Qi, J.; Ji, J.; Wang, S.; Zhang, G.; Zhang, F., *Nano Res.* **2010**, *3*, 429.
[122] Guo, S.; Dong, S.; Wang, E., *ACS Nano* **2009**, *4*, 547.
[123] Dong, L.; Gari, R. R. S.; Li, Z.; Craig, M. M.; Hou, S., *Carbon* **2010**, *48*, 781.
[124] Kamat, P. V., *J. Phys. Chem. Lett.* **2009**, *1*, 520.
[125] Subrahmanyam, K. S.; Manna, A. K.; Pati, S. K.; Rao, C. N. R., *Chem. Phys. Lett.* **2010**, *497*, 70.
[126] Das, B.; Choudhury, B.; Gomathi, A.; Manna, A. K.; Pati, S. K.; Rao, C. N. R., *Chem. Phys. Chem.* **2011**, *12*, 937.
[127] Zhang, H.; Lv, X.; Li, Y.; Wang, Y.; Li, J., *ACS Nano* **2009**, *4*, 380.
[128] Gopalakrishnan, K.; Joshi, H. M.; Kumar, P.; Panchakarla, L. S.; Rao, C. N. R., *Chem. Phys. Lett.* **2011**, *511*, 304.
[129] Zhang, X.-Y.; Li, H.-P.; Cui, X.-L.; Lin, Y., *J. Mater. Chem.* **2010**, *20*, 2801.

[130] Lv, X.-J.; Fu, W.-F.; Chang, H.-X.; Zhang, H.; Cheng, J.-S.; Zhang, G.-J.; Song, Y.; Hu, C.-Y.; Li, J.-H., *J. Mater. Chem.* **2012**, *22*, 1539.

[131] Wang, D.; Choi, D.; Li, J.; Yang, Z.; Nie, Z.; Kou, R.; Hu, D.; Wang, C.; Saraf, L. V.; Zhang, J.; Aksay, I. A.; Liu, J., *ACS Nano* **2009**, *3*, 907.

[132] Robel, I.; Bunker, B. A.; Kamat, P. V., *Adv. Mater.* **2005**, *17*, 2458.

[133] Sheeney-Haj-Ichia, L.; Wasserman, J.; Willner, I., *Adv. Mater.* **2002**, *14*, 1323.

[134] Rao, C. N. R.; Sood, A. K.; Subrahmanyam, K. S.; Govindaraj, A., *Angew. Chem. Int. Ed.* **2009**, *48*, 7752.

[135] Lin, Y.; Zhang, K.; Chen, W.; Liu, Y.; Geng, Z.; Zeng, J.; Pan, N.; Yan, L.; Wang, X.; Hou, J. G., *ACS Nano* **2010**, *4*, 3033.

[136] Kim, Y.-T.; Han, J. H.; Hong, B. H.; Kwon, Y.-U., *Adv. Mater.* **2010**, *22*, 515.

[137] Nethravathi, C.; Nisha, T.; Ravishankar, N.; Shivakumara, C.; Rajamathi, M., *Carbon* **2009**, *47*, 2054.

[138] Cao, A.; Liu, Z.; Chu, S.; Wu, M.; Ye, Z.; Cai, Z.; Chang, Y.; Wang, S.; Gong, Q.; Liu, Y., *Adv. Mater.* **2010**, *22*, 103.

[139] Wang, Y.; Yao, H.-B.; Wang, X.-H.; Yu, S.-H. *J., Mater. Chem.* **2011**, *21*, 562.

[140] Lightcap, I. V.; Kamat, P. V. *J., Am. Chem. Soc* **2012**, *134*, 7109.

[141] Winter, M.; Brodd, R. J., *Chem. Rev.* **2004**, *104*, 4245.

[142] Shiva, K.; Rajendra, H. B.; Subrahmanyam, K. S.; Bhattacharyya, A. J.; Rao, C. N. R., *Chem. Eur-J.* **2012**, *18*, 4489.

[143] Wang, H.; Casalongue, H. S.; Liang, Y.; Dai, H., *J. Am. Chem. Soc* **2010**, *132*, 7472.

[144] Paek, S.-M.; Yoo, E.; Honma, I., *Nano Lett.* **2008**, *9*, 72.

[145] Yao, J.; Shen, X.; Wang, B.; Liu, H.; Wang, G., *Electrochem. Comm.* **2009**, *11*, 1849.

[146] Wang, D.; Kou, R.; Choi, D.; Yang, Z.; Nie, Z.; Li, J.; Saraf, L. V.; Hu, D.; Zhang, J.; Graff, G. L.; Liu, J.; Pope, M. A.; Aksay, I. A., *ACS Nano* **2010**, *4*, 1587.

[147] Meduri, P.; Pendyala, C.; Kumar, V.; Sumanasekera, G. U.; Sunkara, M. K., *Nano Lett.* **2009**, *9*, 612.

[148] Shiva, K.; Rajendra, H. B.; Subrahmanyam, K. S.; Bhattacharyya, A. J.; Rao, C. N. R., *Chem. Euro. J* **2012**, *18*, 4489.

[149] Chang, K.; Chen, W., *Chem. Comm.* **2011**, *47*, 4252.

[150] Das, S. K.; Mallavajula, R.; Jayaprakash, N.; Archer, L. A., *J. Mater. Chem.* **2012**, *22*, 12988.

[151] Chang, K.; Chen, W., *J. Mater. Chem.* **2011**, *21*, 17175.

[152] Chang, K.; Chen, W.; Ma, L.; Li, H.; Li, H.; Huang, F.; Xu, Z.; Zhang, Q.; Lee, J.-Y., *J. Mater. Chem.* **2011**, *21*, 6251.

[153] Chang, K.; Chen, W., *ACS Nano* **2011**, *5*, 4720.

[154] Shiva, K.; Matte, H. S. S. R.; Rajendra, H. B.; Bhattacharyya, A. J.; Rao, C. N. R., *Nano Energy*, **2013**, *2*, 787–793.

[155] Yaghi, O. M.; Li, H., *J. Am. Chem. Soc.* **1995**, *117*, 10401.

[156] Petit, C.; Bandosz, T. J., *Adv. Mater.* **2009**, *21*, 4753.

[157] Petit, C.; Mendoza, B.; Bandosz, T. J., *Langmuir* **2010**, *26*, 15302.

[158] Jahan, M.; Bao, Q.; Yang, J.-X.; Loh, K. P., *J. Am. Chem. Soc.* **2010**, *132*, 14487.

[159] Lee, J. H.; Kang, S.; Jaworski, J.; Kwon, K.-Y.; Seo, M. L.; Lee, J. Y.; Jung, J. H., *Chem.Europ. J.* **2012**, *18*, 765.

[160] Huang, Z.-H.; Liu, G.; Kang, F., *ACS Appl. Mater.Inter.* **2012**, *4*, 4942.

[161] Jahan, M.; Bao, Q.; Loh, K. P., *J. Am. Chem. Soc.* **2012**, *134*, 6707.

[162] Ren, Z.; Kim, E.; Pattinson, S. W.; Subrahmanyam, K. S.; Rao, C. N. R.; Cheetham, A. K.; Eder, D., *Chem. Sci.* **2012**, *3*, 209.

Markus Antonietti, Li Zhao, and Maria-Magdalena Titirici

7 Sustainable carbon hybrid materials made by hydrothermal carbonization and their use in energy applications

The production of functional nanostructured carbons starting from cheap natural precursors using environmentally-friendly processes is a highly attractive subject in materials chemistry today. Plant biomass, used from ancient times for carbon production, has recently been rediscovered to produce functional carbonaceous materials, covering economic, environmental and societal issues. Besides the classical "fire" route to produce activated carbons from dendronic side products, the hydrothermal carbonization (HTC) process ("water") showed clear advantages to generate a variety of sustainable carbonaceous materials with attractive nanostructures and functionalization patterns for a wide range of applications. In this tutorial article we present the latest developments of the HTC technique. It will be shown that HTC does not only access carbonaceous materials under comparably mild hydrothermal conditions, but also replaces the more technical and structurally ill-defined charring by a controlled chemical process. It will be shown that this allows to tailor the final structure by the tools of colloid and polymer science, leading to very different morphologies with miscellaneous applications, here we consider carbon nanocomposites and energy hybrids.

7.1 Introduction

Until recently, synthesis of nanostructured carbon materials was usually based on very harsh conditions such as electric arc discharge techniques [1], chemical vapor deposition [2], or catalytic pyrolysis of organic compounds [3]. In addition (excluding activated carbons), only little research has been done to synthesize and recognize the structure of carbon materials based on natural resources. This is somewhat hard to understand, as carbon structure synthesis has been practiced from the beginning of civilization on the base of biomass, with the petrochemical age only being a late deviation. A refined approach towards advanced carbon synthesis based on renewable resources would be significant, as the final products provide an important perspective for modern material systems and devices.

It will be shown in this chapter that hydrothermal carbonization can turn into a model case for a process that turns low value and widely available biomass into interesting carbon nanostructures using environmentally benign steps. These low cost nanostructured carbon materials are then designed for applications in fields such as energy storage, energy conversion and catalysis. Besides controlling the chemistry of carbonization (*i.e.* C-C-linkage), two other important handles for the achievement of useful properties are the control over morphology – both at nano- and macroscale – and the control over functionality, which can be easily done by chemical means in hydrothermal carbonization.

The natural process of peat or coal formation is a low temperature, aqueous process. It is presumably not mainly biological, but chemical in nature and takes place on the timescale of some hundreds (peat) to millions (black coal) of years. As "coaling" is a rather natural experiment, there are countless trials in the literature to chemically imitate carbon formation from carbohydrates with faster chemical processes, and hydrothermal carbonization (HTC) was found to be especially promising. The first experiments of modern times were presumably carried out by Bergius, who described in 1913 the hydrothermal transformation of cellulose into coal-like materials [4]. More systematic investigations were performed by Berl and Schmidt in 1932, who already varied the source of biomass and treated the different samples in the presence of water at temperatures between 150 ° and 350 °C [5]. These two authors summarized with a series of papers in 1932 the knowledge about the emergence of coal at that time [6]. Later, Schuhmacher, van Krevelen *et al.* analyzed the influence of pH on the outcome of reaction and found serious differences in the decomposition schemes, with their deductions mainly based on mean composition data [7]. A rather current review of present knowledge on coal structure and its origin is found in [8].

The technique of hydrothermal carbonization was recently "rediscovered" by several working groups [9–11]. Since then, it became an important technique for the production of various carbonaceous materials and hybrids, usually applying mild temperatures (< 200 °C) in water inside closed recipients and under self-generated pressure.

In the first part of this chapter we will describe some general aspects of hydrothermal carbonization, using either carbohydrates or complex biomass to control structure formation in the presence of various catalysts and/or templates. In the second part, we then describe some of the most promising applications of these carbon/hybrid materials in energy applications.

7.2 Hydrothermal synthesis of carbonaceous materials

7.2.1 From pure carbohydrates

In order to gain some information about the fundamentals of the hydrothermal carbonization process, the hydrothermal carbonization of different carbohydrates and carbohydrate products was examined [12, 13]. For instance, hydrothermal carbons synthesized from diverse biomass (glucose, xylose, maltose, sucrose, amylopectin, starch) and biomass derivatives (HMF and furfural) were treated under hydrothermal conditions at 180 °C and were analyzed with respect to their chemical and morphological structures by SEM, ^{13}C solid-state NMR and elemental analysis. This was combined with GC-MS experiments on residual liquor solutions to analyze side products and unreacted species. It was demonstrated that under those conditions all hexose sugars, no matter their complexity, degrade into hydroxymethyl furfural, which finally condenses to a carbon-like material having morphological similarities and the same chemical and structural composition practically throughout all starting products. By contrast, xylose derivatives dehydrate into furfural, which in turn reacts to provide very similar carbon structures as obtained from pure furfural, which are however different from the hexose products (e.g., a higher content of all carbon aromatic structures).

^{13}C solid-state NMR also showed that starting from more complex, "real life" biomass instead of clean sugars does not change the outcome of the hydrothermal carbonization reaction too much, and remarkable similarities between the products of homologous series do occur, both with respect to morphology and local structure. Basically, the hydrothermal carbonization reaction takes place in three important steps: (1) dehydration of the carbohydrate to (hydroxymethyl)furfural; (2) polymerization towards polyfuranes (3) carbonization *via* further intermolecular dehydration, presumably Diels–Alder type rearrangements towards all-carbon aromatic systems.

The hydrothermal carbons obtained in the end from soluble, non-structural carbohydrates are micrometer sized, spherically shaped particle dispersions, containing a sp^2 hybridized backbone (also responsible for the brown to black color) decorated with a dense layer of polar oxygenated functionalities still remaining from the original carbohydrate. The presence of these surface groups offers the possibility of further functionalization and makes the materials more hydrophilic and well-dispersible in water. The size of the final particles depends mainly on the carbonization time and precursor concentration inside the autoclave, as well as additives and stabilizers potentially added to the primary reaction recipe. An SEM image of a model reaction illustrating this dispersion state is shown in Fig. 7.1.

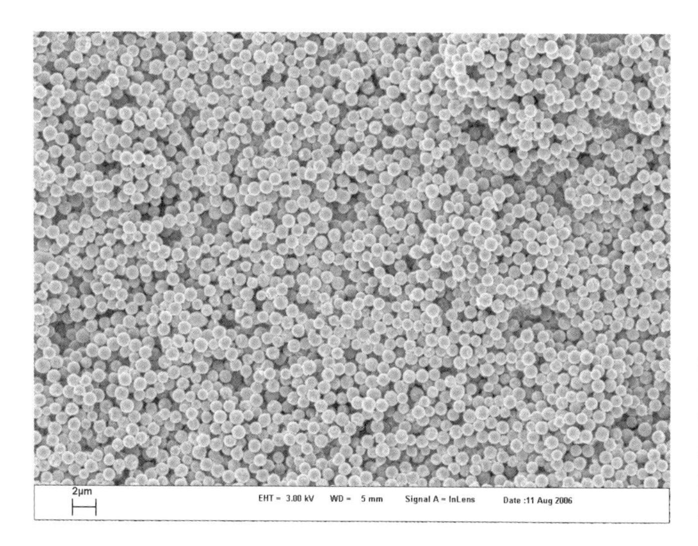

Fig. 7.1: Scanning electron micrographs of a typical carbon dispersion from glucose as a model system (scale bar 2 μm).

It was also found that the presence of some metal ions and borates can effectively accelerate the hydrothermal carbonization of starch, which shortens the reaction time to some hours. Thus, iron ions and iron oxide nanoparticles were shown to effectively catalyze the hydrothermal carbonization of starch (≤ 200 °C) and also had a significant influence on the morphology of the formed carbon nanomaterials [10]. In the presence of Fe^{2+} ions, both hollow and massive carbon microspheres could be obtained. In contrast, the presence of Fe_2O_3 nanoparticles leads to very fine, rope-like carbon nanostructures, reminding one of disordered carbon nanotubes.

7.2.1.1 Porous materials

The carbonaceous materials collected directly after hydrothermal carbonization have the characteristics of a polymer and possess practically no micropores. Therefore, a small surface area is obtained, as compared for instance to activated carbons. This is not necessarily good or bad, but for applications of these materials in catalysis and energy storage, the presence of controlled porosity at the nanoscale is needed. This is why a variety of techniques was applied to increase the surface area. For instance, if hydrothermal carbonization of carbohydrates takes place in the presence of various templates or additives, interesting pore systems can be imprinted. In the following paragraphs, we will first discuss various morphologies obtained using the HTC process in the presence of extraneous substances.

The first mesoporous hydrothermal carbons were produced by performing the reaction in the presence of nanostructures silica templates [9]. Thus, it was discovered that it is important to match the polarity of the template surface with the carbon precursor. Silica templates with a moderately hydrophobic surface lead to hollow carbon spheres, whereas demixing or macroporous carbon casts occur in the presence of templates of both more hydrophilic or hydrophobic surface. For mesoporous templates, mesoporous carbon shells are obtained after removal of the template, with the whole carbonaceous structure composed of *ca.* 8–16 nm sizes primary globular carbon nanoparticles. Furthermore, the moderately hydrophobic silica templates filled with 60 wt% carbon precursor resulted in mesoporous carbonaceous microspheres, whereas application of 30 wt% carbon precursor gave only small carbon spherules (6–10 nm in size), owing to the lack of interconnectivity between the primary particles in the coating. This nicely illustrates that the carbon coating process operates "patch-wise" *via* stable colloidal intermediates. For nonporous templates, hollow carbonaceous spheres with a robust carbon coating are observed. A replica of an ordered SBA 15 material has also been successfully produced [14]. In this material, the residing functional groups could also be successfully converted into amino groups. Some TEM pictures of these mesoporous ordered hydrophilic materials are given in Fig. 7.2.

The perfection of this nanocoating process following free shapes in three dimensions is nicely illustrated by the coating of nanowires. Uniform carbonaceous nanofibers of only some nanometers in diameter with high aspect ratio could be

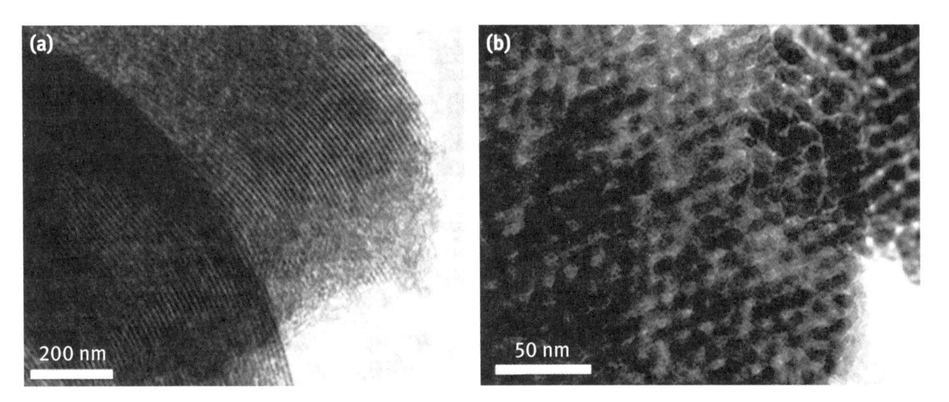

Fig. 7.2: (a) TEM micrograph of the carbon replica obtain by a 100 % pore filling of SBA-15 template; (b) TEM micrograph of the carbon replica obtained after a 25 % pore filling of SBA-15 template.

produced using the HTC of glucose in the presence of Te nanowires [15]. Well-defined ultralong carbon nanofibers are made by removing the Te nanowires from the Te@carbon-rich composite nanocables. The carbonaceous nanofibers have an average diameter of *ca.* 50 nm and up to tens or hundreds of micrometers in length (Fig. 7.3). Here too the particulate character of the coating process is nicely reflected in the nanoscale roughness of the carbon fibers.

A similar approach for the production of hydrothermal carbon nanotubes is the hydrothermal carbonization of glucose in the macrochannels of anodic alumina membranes [16]. Depending on the pore size of the membrane different hollow hydrother-

Fig. 7.3: (a) A general view of the nanofibers and (b) an enlarged SEM image. (c)–(d) TEM images of the nanofibers. Taken from [15] with kind permission of Wiley-VCH.

mal carbon nanotubes can be obtained. These nanotubes contain functional groups at their surface and therefore they can be further functionalized. Yu also reported on the production of hollow carbon spheres using the microwave-induced hydrothermal treatment of carbohydrates (starch) in the presence of H_2SeO_3 leading first to the production of core-shell Se/C colloids. After removing the Se core by thermal treatment, hollow C capsules were formed [17].

7.2.1.2 Hybrid materials

The coating of templates already indicated that carbonaceous hybrid materials on the nanoscale length can be produced when leaving the second material phase within the carbon frame. In the case of functional nanoparticles and nanostructures, this has tremendous importance for various applications in science and technology [18]. Consequently, hydrothermal carbonization has been also intensively used for the production of various hybrids.

Li *et al.* reported first on the decoration of hydrothermal carbon spheres obtained from glucose with noble metal nanoparticles [19]. They used the reactivity of as-prepared carbon microspheres to load silver and palladium nanoparticles onto their surfaces, both *via* surface binding and room-temperature surface reduction. Furthermore, it was also demonstrated that these carbon spheres can encapsulate nanoparticles in their cores with retention of the surface functional groups. Nanoparticles of gold and silver could be encapsulated deep in the carbon by *in situ* hydrothermal reduction of noble-metal ions with glucose (the Tollens reaction), or by using silver nanoparticles as nuclei for subsequent formation of carbon spheres. Some TEM images of such hybrid materials are shown in Fig. 7.4.

The production of Ag@Carbon nanocables was reported by Yu *et al.* [20]. In this case the authors used the hydrothermal carbonization of starch in the presence of Ag NO_3 leading to the one-step formation of carbon/Ag hybrid nanocables. Such silver–carbon nanocables can have a length as long as 10 mm and overall diameters of 1 micron with a 200–250 nm silver lining. When made at higher concentrations, they tend to fuse with each other (Fig. 7.5(a) and (b)). This method was extended to a polyvinyl alcohol(PVA)-assisted synthesis of flexible noble metal (Ag, Cu)@carbon composite microcables [21].

7.2.1.3 Nitrogen-doped carbons

Another gemstone in the portfolio of rational carbon synthesis is nitrogen-doped carbons. Recently, they became the subject of particular interest to researchers due to their remarkable performance in applications such as CO_2 sequestration [22], removals of contaminants from gas and liquid phases [23], environmental protection [24], catalysts and catalysts supports [25], or in electrochemistry as supercapacitors [26], cells and batteries to improve stability and the loading capacity of carbon.

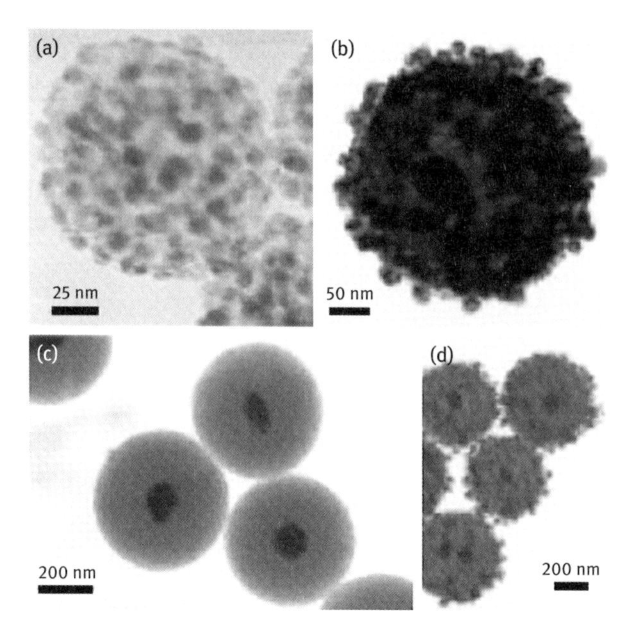

Fig. 7.4: (a) Carbon spheres loaded with silver nanoparticles at room temperature; (b) carbon spheres loaded with palladium nanoparticles by refluxing; (c) silver core of carbon spheres from encapsulation of silver nanoparticle seeds, (d) layered structure with an silver core, a platinum shell, and a carbon interlayer, formed by seeded encapsulation followed by the reflux method.

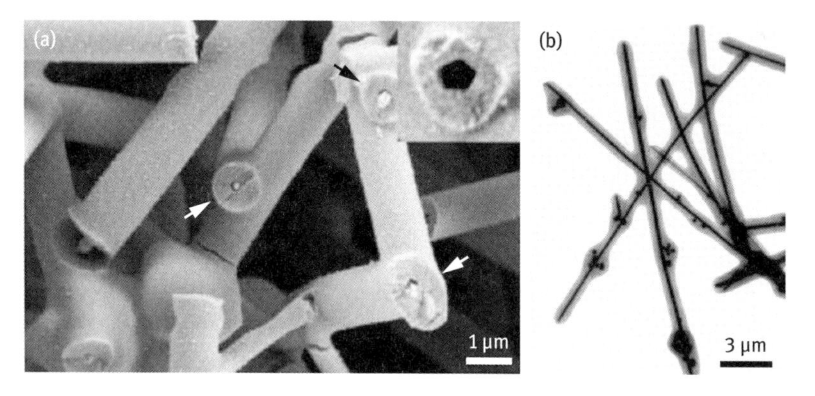

Fig. 7.5: (a) Typical SEM image of the nanoscale with encapsulated, silver nanowires. The insert shows an incompletely drilled tube with a pentagonal cross-section. (b) TEM image of typical silver–carbon nanocables formed after treating at 160 °C for 12 h: 5 g starch, 5 mmol AgNO3, pH 4.

It was therefore straightforward to apply HTC by including sustainable nitrogen sources with appropriate co-reactivity. The synthesis of well-defined carbonaceous nanostructures with significant N-doping level and a developed sponge-like mesopore system was recently demonstrated by using hydrothermal treatment at 180 °C of glucose in the presence of ovalbumin ("egg-white") [27]. This simple process at first provides a unique and sustainable pathway to control the primary particle stability in hydrothermal carbonization, resulting in the formation of nanometer-sized sponge scaffolds composed of spherical particles (20–50 nm) (Fig. 7.6). Another advantage is that the protein efficiently introduces structural nitrogen (as much as 8 wt%) into the carbonaceous scaffold.

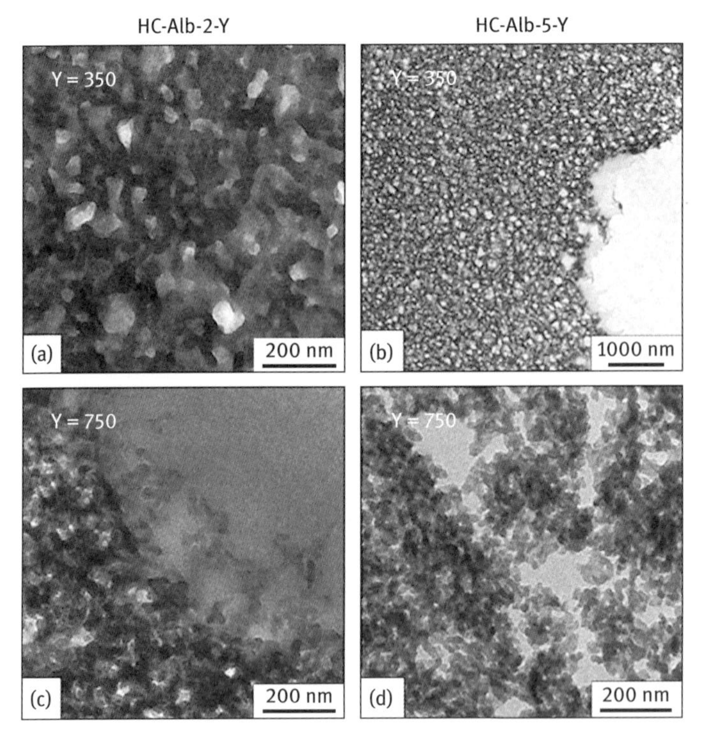

Fig. 7.6: TEM images for post-calcined samples HC-AlbX-Y, X = 2,5. Y (= 350, 750) indicates the calcination temperature in Celsius. Scale bars are 200 nm.

Classical characterization methods (gas sorption, TEM, SEM, FTIR, XPS and elemental analysis) were used to describe the resulting porous carbon structures. Temperature-dependent experiments have shown that all the various materials kept the nitrogen content almost unchanged up to 950 °C, while the thermal and oxidation stability was found to be significantly increased with N-doping as compared to all pure carbons. Last but not least, it should be emphasized that the whole material synthesis occurs in a remarkably energy and atom-efficient fashion from cheap and sustainable resources.

Another approach towards the synthesis of nitrogen rich carbons under green and sustainable conditions is the direct hydrothermal carbonization of nitrogen-containing carbohydrates, such as glucosamine or chitosane [28]. Again, carbonaceous materials with a high content of nitrogen were obtained.

7.2.2 From complex biomass

Simultaneous to the understanding of some basics of hydrothermal carbonization using pure carbohydrate models, the synthesis of hydrothermal carbon materials using raw biomass was continued. It has been analyzed whether complex biomass – hydrothermally carbonized – can also be directed to complex structural motifs with distinct surface polarities. Ideally, for this purpose one can use the structures and functionalization components already included in the biomass. We specifically selected waste biomass for material synthesis, starting products which are known to be hard to use otherwise, rich in ternary components, and applied different HTC conditions [29]. That way, one can avoid the food–raw materials competition, a prerequisite we regard as crucial for the development of a fully sustainable chemistry.

Diverse biomass compounds were heated in sealed autoclaves in the presence of acids as catalyst at 200 °C for 16 h, and essentially two kinds of reaction scenarios were found. For soft biomass, nontextured biomass, hydrophilic and water-dispersible carbonaceous nanoparticles in the size range of 20–200 nm were obtained. The occurrence as spherical particles indicates that the soft biomass was first liquefied, and then carbonized, which offers possibilities to chemically interfere with this complex process as discussed above.

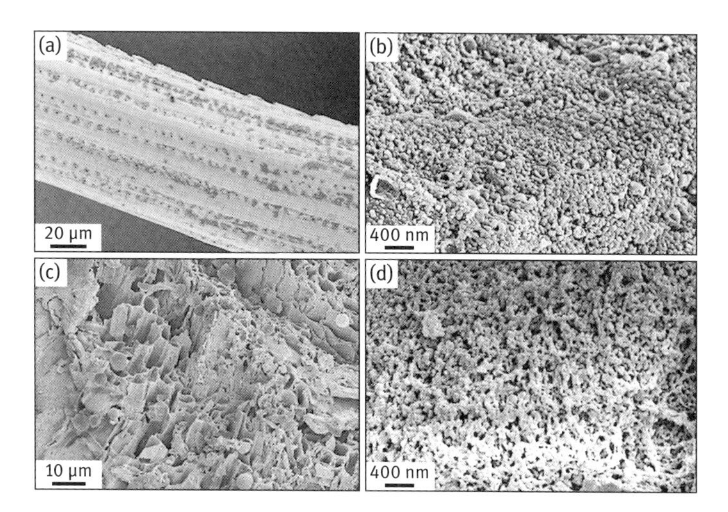

Fig. 7.7: HRSEM of pine needles (a) before and (b) after being hydrothermally carbonized at 200 °C for 12 h; (c) low-magnification SEM overview of a HTC-treated oak leaf; (d) high-magnification picture of the same HTC-treated oak leaf indicating its nanostructure.

For hard biomass made from crystalline cellulose, an “inversed” structure was found, with the carbon being the continuous phase, penetrated by a sponge-like continuous system of nanopores (representing the majority volume). These products are also hydrophilic due to the remainder of *ca.* 20 wt% functional oxygenated groups, and can be easily wetted with water. Such structures are ideal for water binding, capillarity, and ion exchange. By simple chemical means, the product is practically indistinguishable from the carbonaceous fraction of soil, but structurally much better defined. These morphologies are shown in Fig. 7.7.

7.2.3 Energy applications of hydrothermal carbons and their hybrids

The most appealing feature of hydrothermal carbonization is the fact that it represents an easy, green and scalable process allowing the production of various carbon and hybrid nanostructures with practical applications on a price base which is mostly well below any number of corresponding petrochemical processes. Even at this early stage, HTC materials have already found numerous applications in relevant fields of modern materials or chemical industry such as catalysis, water purification, energy storage and CO_2 sequestration. In the following paragraphs we will summarize some of the most remarkable examples in energy applications where HTC-based materials have proven to possess extraordinary properties which partly exceeds those of the "gold standards" of current materials.

7.2.3.1 Energy storage in Li-ion batteries

One of the most covered, for the non-expert unexpected, but promising applications of HTC materials is in the field of energy storage and new energy cycles. Reliable and affordable energy storage is a prerequisite for using renewable energy in remote locations, for integration into the energy grid and the development of a future decentralized energy supply system. Mobile electric energy storage is also among the most promising technology markets in the transport sector. However, the production of sustainable materials with promising properties for energy storage is a hardly touched upon, but most valuable topic in modern materials chemistry.

HTC materials have been used and structurally improved as electrodes in Li-ion batteries [30–32]. Rechargeable lithium-ion batteries are the technical leading solution and essential to portable electronic devices. Owing to the rapid development of such equipment there is an increasing demand for lithium-ion batteries with higher energy density and a longer lifetime.

For high energy density, the electrode materials in the lithium-ion batteries must possess high specific storage capacity and coulombic efficiency. Graphite and $LiCoO_2$ are normally used as anode and cathode, respectively, and have high coulombic efficiencies (typically > 95 %) but rather low capacities (372 and 145 mAhg-1, respectively) [53].

The first application of HTC as an anode in Li-ion batteries was first reported by Huang *et al.* [30]. After the hydrothermal carbonization of sugar, the resulting spherical particles were further carbonized under argon at 1000 °C for 5 h. The reversible lithium insertion / extraction capacity of this kind of material was about 400 $mAhg^{-1}$. This proves that HC materials post-treated at higher temperatures have specific capacities in the same range as commercial graphite, which was a promising starting point. Current values for carbon materials are however meanwhile around 800 mAh/g, combined with excellent stability and dynamic rate behavior [74]. However, other materials have been proposed in the last decade as anode materials, too. Among these, silicon has attracted great interest as a candidate to replace graphite owing to its numerous appealing features: it has the highest theoretical capacity ($Li_{4.4}Si$ ~ 4200 $mAhg^{-1}$) of all known materials, and Si is abundant, inexpensive, and safer than graphite [34].The practical use of Si powders as an anode in lithium-ion batteries is, however, massively hindered by two major drawbacks: the low intrinsic electric conductivity and the severe volume changes during Li insertion/extraction processes, leading to poor cycling performance [35].

In order to overcome these problems, hybridization of both materials (C and Si) in one electrode material by HTC seemed to be a promising option [75]. For this purpose, pre-formed silicon nanoparticles were dispersed into a dilute solution of glucose followed by hydrothermal treatment at 180 °C. The carbon-coated particles were then further treated at 750 °C in order to improve the conductivity and structural order of the carbon layer. It was shown that the hydrothermal treatment, following by high temperature carbonization, resulted in formation of a few nanometer thin layer of SiOx layer on the Si nanoparticles, effectively leading to a Si/SiOx/C nanocomposite. Some TEM micrographs of these materials are shown in Fig. 7.8.

It can be observed that the thickness of the layer forming a complete shell around the Si nanoparticles is around 10 nm (SiOx and C), whereas the diameter of core is around 30–40 nm, which is very similar to the size of the primary Si nanoparticles. In HRTEM images (Fig. 7.8(b)–(d)) it becomes evident that the Si nanoparticles are coated with a layer of silicon oxide and a layer of carbon of varying thickness. Strongly bent graphitic layers were observed on the surface of all of the particles (indicated by small arrows in Fig. 7.8(d)). HRTEM, Raman, and FTIR investigations confirm that the nanocomposite consists of a silicon core, a shell of thin amorphous SiOx, and a carbon layer. This structure turned out to be very interesting for lithium storage as it combines sufficient conductivity of graphene layers with polymer-like elasticity to withstand the deformation stresses throughout lithium insertion. The total amount of carbon in this sample was only 25 %, according to elemental analysis and TGA.

The lithium-storage properties of these Si@SiOx/C nanocomposite electrodes were investigated in different electrolyte systems and compared to pure Si nanoparticles. From all the analyzed systems, the Si@SiO*x*–C nanocomposite in conjunction with the solvent vinylene carbonate (VC) to form the solid-electrolyte interface showed the best lithium storage performance in terms of a highly reversible lithium-storage capacity (1100 mAh g^{-1}), excellent cycling performance, and high rate capability (Fig. 7.9).

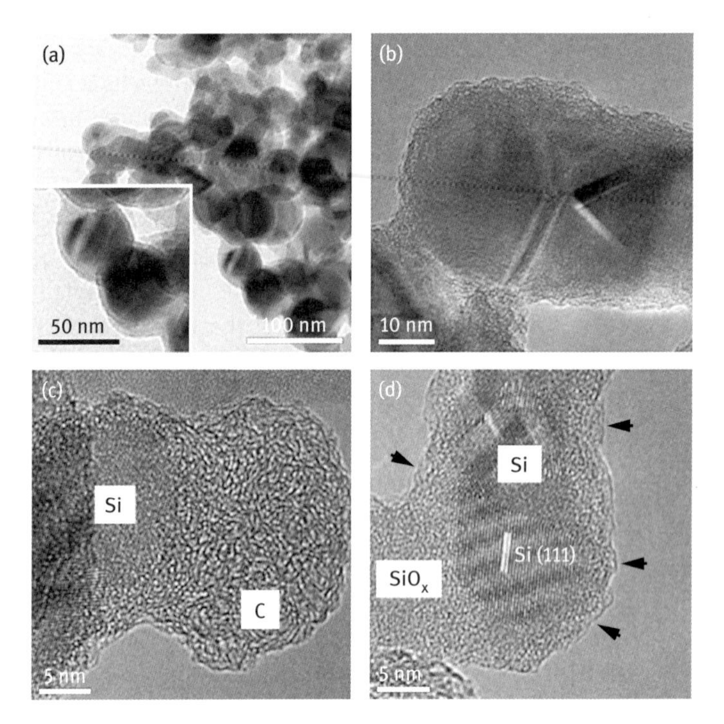

Fig. 7.8: TEM images of the Si@SiOx/C nanocomposite nanoparticles produced by hydrothermal carbonization of glucose and Si and further carbonization at 750 °C under N2. (a) Overview of the Si@SiOx/C nanocomposites and a TEM image at higher magnification (in the inset) showing uniform spherical particles; (b) HRTEM image clearly showing the core/shell structure; (c), (d) HRTEM image displaying details of the silicon nanoparticles coated with SiOx and carbon.

Besides the combination Si/C, there have been some other trials to improve the performance of Li-battery materials using the HTC process. For example, other authors produced NiO–C nanocomposite by dispersing the as-prepared net-structured NiO in glucose solution and performed subsequent carbonization under hydrothermal conditions at 180 °C [36]. The carbon in the composite was amorphous by X-ray diffraction analysis, and its content was 15 wt% calculated according to the energy dispersive X-ray spectroscopy (EDX). Transmission electron microscopy (TEM) image of the NiO–C nanocomposite showed that the NiO network was homogeneously filled by amorphous carbon. The reversible capacity of this NiO–C nanocomposite after 40 cycles is 429 mAh g^{-1}, which is again much higher than that of NiO (178 mAh g^{-1}). These improvements were again attributed to the carbon, which can enhance the conductivity of NiO, suppress the aggregation of active particles, and increase their structure stability during cycling.

Cao *et al.* produced HTC/Sn nanocomposites by hydrothermal treatment of sucrose and $SnCl_4$ solution [37]. After the carbon was removed, the electrochemical properties of the hollow SnO_2 materials were tested. This material has a high initial dis-

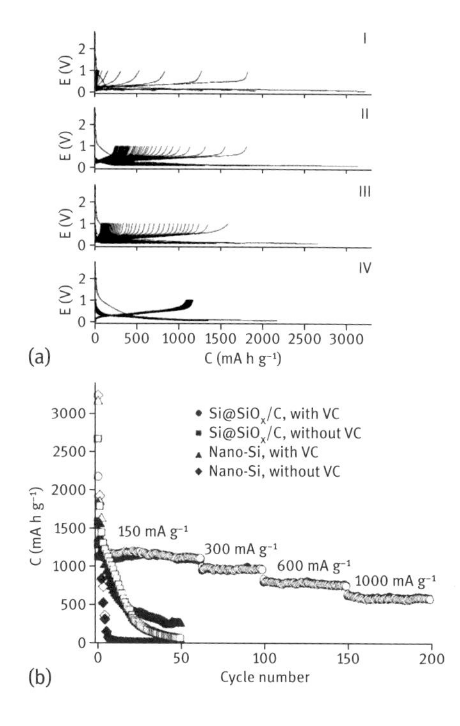

Fig. 7.9: (a) Galvanostatic discharge/charge curves (Li insertion, voltage decreases; Li extraction, voltage increases, respectively) of pure Si nanoparticles (I, II) and Si@SiOx/C nanocomposite (III, IV) electrodes cycled at a current density of 150 mAg^{-1} between voltage limits of 0.05–1 V in VC-free (I, III) and VC-containing (II, IV) 1 m $LiPF_6$ in EC/DMC solutions. (b) Cycling and rate performance of the same systems.

charge capacity of 764 $mAhg^{-1}$, close to its theoretical specific capacity, indicating a full utilization of all the SnO_2 crystallites.

Mesoporous SnO_2 microspheres were also prepared by hydrothermal treatment of pre-formed 3–7 nm sized SnO_2 nanoparticles in the presence of glucose [38]. Throughout the formation of carbon spheres, the nanoparticles are homogenously incorporated into the carbon matrix, as shown in Fig. 7.10(a). After the removal of the carbon matrix through calcination, these nanoparticles assemble together into mesoporous SnO_2 microspheres, where the porosity and high surface area are achieved by the interstitial porosity between the agglomerated nanoparticles (Fig. 7.10(b)–(d)). In terms of electrochemical applications, the micrometer-sized spheres enable easy handling in terms of separation or device formation in comparison with their nanosized constituents, while their resulting mesoporosity provides a high surface area network that could be easily penetrated by the electrolyte. This porosity in combination with the nanosized building blocks results in significant improvement of the electrochemical performance when compared with a nonporous, commercially available SnO_2 sample (Fig. 7.10(e)).

Despite quite some progress reported in improving the performance and lifetime of anode materials, a great deal of research needs to be dedicated to the improvement of the cathode in Li-ion batteries. This task was addressed by hydrothermal carbon coating techniques. Thus, Olivine $LiMPO_4$ (Me = Mn, Fe, and Co) cathodes with a thin carbon coating have been prepared by a rapid, one-pot, microwave-assisted hydrothermal process enabling a shortened reaction time (15 min) at 230 °C [39]. Again, glucose was chosen as a source of hydrothermal carbon. The resulting $LiMePO_4/C$ nanocomposites were characterized by X-ray diffraction, Raman spectroscopy, SEM, and TEM before and after heating at 700 °C for 1 h in an argon flow. The uniform nanocarbon coating enhanced the electronic conductivity and led to excellent electrochemical performance for $LiFePO_4/C$ in lithium cells. In contrast, both $LiMnPO_4/C$ and $LiCoPO_4/C$ exhibited in those experiments only inferior electrochemical performances even after carbon coating.

Related work on the structural optimization of hydrothermally coated $LiFePO_4$ is found in [40].

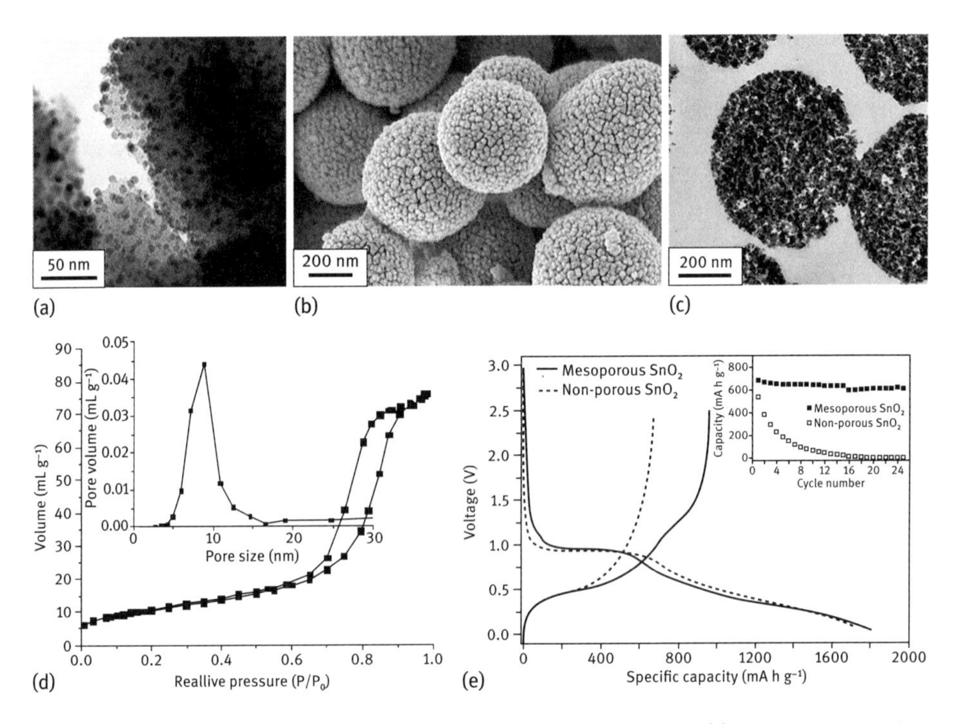

Fig. 7.10: (a) TEM image of SnO_2 nanoparticles dispersed in the HTC matrix; (b) SEM micrograph of the resulting mesoporous SnO_2 spheres after removal of the carbon matrix; (c) TEM micrograph of the same SnO_2 spheres, (d) adsorption isotherm and pore size distribution of the resulting SnO_2 spheres; (e) first discharge/charge profiles in lithium storage for the mesoporous SnO_2 and microsized SnO_2 samples cycled at a current density of 100 mA g^{-1}. The inset shows the cycling performance of both samples cycled between voltage limits of 0.05 and 1 V, revealing stable values above 600 mAh/g.

7.2.3.2 HTC-carbons for supercapacitors

Supercapacitors are currently being extensively studied as possible auxiliary energy storage devices to be used with rechargeable batteries. They combine the advantages of both dielectric capacitors and rechargeable batteries, which can deliver high power within a very small period and store high energy. Based on the charge-storage mechanism, supercapacitors can be divided into two categories: One is the electrical double-layer capacitor (EDLC), where the capacitance arises from the charge separation at an electrode/electrolyte interface; the other is the redox capacitor, where the capacitance comes from Faradaic reactions at the electrode/electrolyte surface [41–43].

Nanostructured carbon materials and hybrids are regarded as the first candidate as electrode materials for EDLC. Recently, several reports have been published on the improvement of the capacitive behavior of carbon materials as supercapacitors [43, 44]. Activated carbons with high surface areas are the most commonly used materials in supercapacitors; the large surface area of carbon can help to enhance capacity for charge accumulation at the electrode/electrolyte interface [45]. Generally, for the activated carbon with a specific surface area of 1000 $m^2\ g^{-1}$, the specific capacitance is 150 $F\ g^{-1}$ [46]. It has to be mentioned that the carbon electrodes are a relevant cost factor, since the activated carbons are typically synthesized from special petroleum coke, thus the development of lower cost carbons with high surface area and specific energy is one of the keys for the widespread application of supercapacitors. In addition, properties such as oxidation stability (lifetime) and high specific conductivities (to reduce Ohmic losses and heat) are key development parameters.

The presence of some heteroatoms containing chemical functionality, such as oxygen or nitrogen, gives carbon materials acid/base character and thus enhances the capacitance by the pseudocapacitive effect. Several reports have been published on the improved capacitive behavior of supercapacitor electrodes which were made from carbon materials enriched with nitrogen or oxygen surface functional groups [47–51]. These electrochemically active centers contribute to the overall capacitance, with pseudocapacitance that generally originates from the Faradaic interactions between the ions of electrolytes and the carbon electrode surface. The oxygen surface groups formed in most of the activated carbons during the activation process have an acidic character, thus introducing electron-acceptor properties into the carbon surface. On the other hand, nitrogen functionalities located at the periphery of graphene sheets, *i.e.*, pyridinic, pyridonic, and oxidized nitrogen, have generally basic characters, inducing electron-donor properties [52]. The origin of pseudocapacitance in nitrogen-enriched carbons is however yet to be clarified in detail. As an example for high oxygen-doped porous carbons, high specific capacitances in the order of 220 $F\ g^{-1}$ were reported [53]. The incorporation of nitrogen in carbons is classically achieved either by treating carbon materials with ammonia gas/air [51] or using nitrogen-containing carbon precursors, such as melamine [54], polyacrylonitriles [55], polyvinylpyridine [44], and quinoline-containing pitch [51].

The technique of hydrothermal carbonization for conversion of carbohydrates into carbon has great advantages for this application since it can maintain a high content of oxygen and nitrogen into the resulting materials. N-doped carbons generated from the biomass derivatives chitosan and glucosamine possess an increased electronic conductivity and an improved oxidation stability, but are typically nonporous with a low specific surface areas (< 20 $m^2\ g^{-1}$). For use in supercapacitors, porosity has to be introduced into such materials, especially microporosity, which is normally associated with high capacitance values [56].

A standard chemical activation protocol with KOH was applied onto hydrothermally synthesized nitrogen and oxygen containing carbons, using D-glucosamine as a natural model precursor [57]. Indeed, the so activated carbons had a much higher N_2 sorption capacity displaying type I sorption isotherms characteristic for microporous materials, peaking with a maximal surface area at 600 m^2/g . However, in this paper the nitrogen content also decreased with increasing amount of KOH activation agent applied. This goes well with older observations which found that obtaining carbon materials with large nitrogen contents and at the same time a high porosity was not an easy task since there was a substantial detriment in the nitrogen functionalities due to oxidation [51].

The observed small and interconnected pores are expected to perform as electrodes of supercapacitors. Cyclic voltammetry (CV) and galvanostatic charge/discharge curves were used to characterize the capacitive properties; the resulting data in simple acid (1 mol L^{-1} H_2SO_4) are shown in Fig. 7.11.

The cyclic voltammograms (Fig. 7.11(a)) show the typical capacitive behavior with a rectangular shape. The appearance of humps in the CV curves indicate that the capacitive response also contains elements of redox reactions, which relate to the heteroatom functionalities of the materials. The redox reactions also can be observed in the galvanostatic charge/discharge curve (Figure 7.11(b)). Unlike linear characteristics, a transition can be easily noticed between 0.4 V and 0.6 V in the acidic electrolyte. The small conversions in line slopes around 0.5 V correlate with the peaks in the CV curves. For galvanostatic charge/discharge at a high current load of 2 A g^{-1}, the Ohmic drop is rather limited, which proves good conductivity, and the ability of a quick charge propagation in carbon and the pore electrolyte system.

The specific capacitance was calculated from the galvanostatic charge/discharge curves as described in [57]. After chemical activation with KOH, although presenting still comparably low specific surface areas (below $m^2\ g^{-1}$), these carbons exhibit a surprisingly high capacitive performance as compared to previously reported activated carbons with the same surface area. Sample CA-GA-2 in an acidic electrolyte (Fig. 7.11) presents the best performance because it combines a medium surface area (571 $m^2\ g^{-1}$) with an acceptable N-doping level (4.4 %), and a specific capacitance of 300 F g^{-1} at a current density of 0.1 A g^{-1} could be determined. These data are comparable to activated carbons reported in the literature with a surface area of more than 3000 $m^2\ g^{-1}$ [58]. Even at a high current density of 4 A g^{-1}, the specific capacitance of CA-GA-2 is

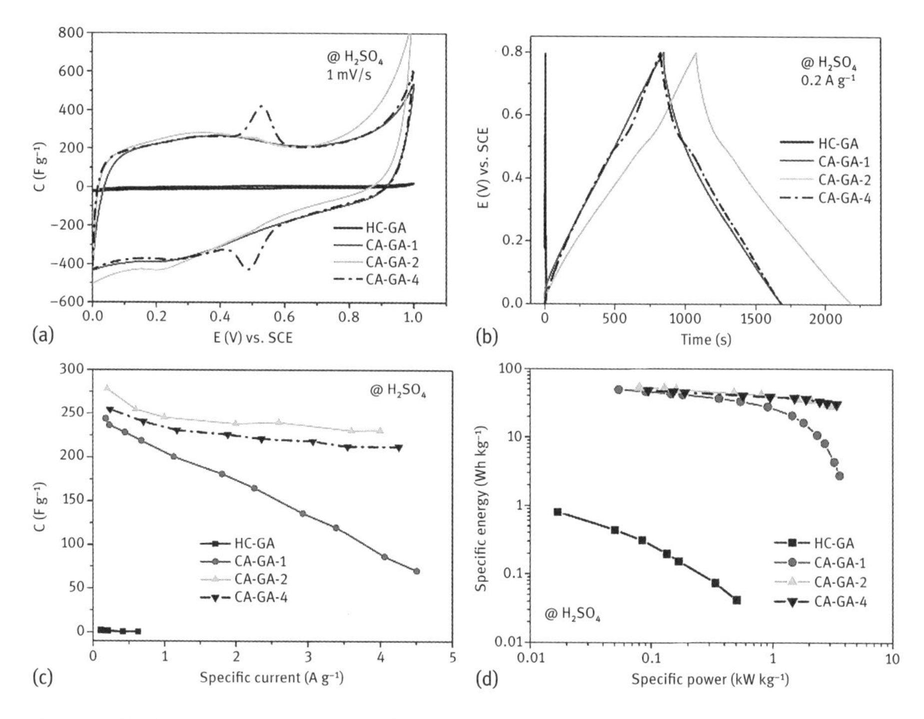

Fig. 7.11: Electrochemical performance of different carbons using a three-electrode cell in 1 mol L^{-1} H_2SO_4: (a) cyclic voltammograms at a scan rate of 1 mV s^{-1}, (b) galvanostatic charge/discharge curves at a current density of 0.2 A g^{-1}, (c) relationship of the specific capacitance with respect to the charge/discharge specific currents, and (d) Ragone plots.

still as high as 230 F g^{-1} (in 1 mol L^{-1} H_2SO_4), which makes such materials directly suited for applications.

Ragone plots are presented in Fig. 7.11(d). The highest specific energy of ~ 50 Wh kg^{-1} in 1 mol L^{-1} H_2SO_4 has been found for the sample CA-GA-2 at the specific power of 0.1 kW kg^{-1} with only weak fading at higher rates.

The cyclic stability (data are not shown) was also tested. After 2000 cycles, there is only a loss of 7 % in the acidic medium. This supports our view that N-doped carbons possess a significantly improved chemical stability.

7.2.3.3 Construction of material heterojunctions with carbon: "Dyads"

TiO_2 is one the most promising photocatalysts for the degradation of pollutants in air or water; however, it shows poor absorption of visible light. Using a one-step HTC hybridization technique, a nano-hybrid structure where hydrothermal carbon activates TiO_2 over the complete visible light region could be synthesized [59]. The resulting catalyst was carefully characterized, and it could be demonstrated that the titania/carbon nanocomposite is not only traditionally sensitized, but acts as a novel "dyad" structure, with an even improved hole reactivity of the titania while the electron is taken up by the carbon layer. This results in an improved photocatalytic activity over the complete spectral range.

Usually, when TiO_2 is irradiated by light with a wavelength (λ) < 350 nm, electrons are promoted across the band gap (> 3 eV) into the conduction band, leaving holes in the valence band [60–62]. These holes, due to the stability of TiO_2, have high oxidation power and can react with the adsorbed hydroxide ions to produce hydroxyl radicals, the main oxidizing species responsible for deep photo-oxidation of organic compounds [63]. However, the application of pure TiO_2 is limited, because it requires ultraviolet (UV) light, which makes up only a small fraction (< 5 %) of the total solar spectrum reaching the earth's surface.

To extend the optical absorption, both cation and anion doping have been performed, however with rather restricted success. Another very promising approach for introducing effective VIS-absorption is carbon doping, where the visible response strongly depends on the form of C in the TiO_2 lattice [64, 65]. For example, carbon-doped TiO_2 for water splitting has been reported by Khan *et al.* [66], as was accomplished *via* the controlled combustion of metallic Ti in a natural gas flame at 850 °C. Sakthivel and Kisch synthesized carbon-modified TiO_2 by hydrolysis of titanium tetrachloride with tetrabutylammonium hydroxide, followed by further heat treatment at 500 °C. This powder showed five times higher activity than nitrogen-doped TiO_2 in the degradation of 4-chlorophenol under artificial light (λ > 455 nm) [67]. Morawski *et al.* reported a new preparation method of carbon-TiO_2 by the carbonization of *n*-hexane deposited on TiO_2 at high temperatures, with the resulting materials providing five times higher catalytic efficiency than that of pure titanium dioxide [68], Recently, Ren *et al.* published a two-step low temperature procedure to produce C-doped TiO_2 with visible light photoactivity [69]. An active debate regarding the fundamental nature of the non-metal species (e.g. N, C, S) causing the visible-light absorption in such modified-TiO_2 materials has continued in the community, and two theses have coexisted for several years: (i) the non-metal substitutes a lattice atom (*i.e.* doping), and (ii) the non-metal forms chromophoric complexes at the surface (*i.e.* sensitization). Kisch and co-workers have suggested that the activity of urea-derived TiO_2-N in visible light was ascribed to the sensitization of TiO_2 by melon [70].

Sensitization of TiO_2 and AgCl by the plasmon state of the noble metals for visible light photocatalysis has also been documented. In these systems, the collective dipole oscillations of the surface plasmon are believed to create electron-hole pairs by interband transition [71]. Similarly, the surface of nanosized carbon materials can show collective polarization modes, and therefore, these optical absorption transitions are potentially feasible to sensitize TiO_2. Indeed, the subband structure of "black" carbon materials covers the whole solar visible spectrum from infrared to ultraviolet light. For carbon-modified TiO_2, various carbon species are supposed to exist on the surface because the occurrence of carbonization onto TiO_2 during the synthesis cannot be excluded.

To disclose this possibility of polarization activation of TiO_2 by a nearby carbon nanostructure and to construct thereby an active bulk heterojunction of two electronic materials, a C@TiO_2 hybrid was synthesized. To avoid carbon from doping into bulk TiO_2 lattice directly, the hybrid is synthesized at low temperature under solvothermal conditions by a one-step carbonization of furfural in the presence of a simultaneous sol-gel reaction of a Ti-precursor, allowing for the formation and co-assembly of carbon and TiO_2 into an interpenetrating C@TiO_2 nano-architecture. A structurally similar type of organic-inorganic composite nanostructure was reported by Spange, however by twin polymerization of hybrid monomers [72].

Figure 7.12 shows the morphology and inner structure of the solvothermally synthesized C@TiO_2 sample (a) and (c), as well as the pure TiO_2 sample obtained after calcination under air ((b) and (d)). In both cases, SEM micrographs (Fig. 7.12(a) and (b)) depict porous, spherically shaped colloidal particles with 2–4 μm diameter. The inner structure is composed of a sponge-like, crystalline titania scaffold, on which the carbon is either coated or deposited, leaving sufficient space for the solvent phase. The inner structure and porosity was confirmed by the TEM micrographs of microtomed samples, demonstrating in both cases a sponge-like, mesoporous architecture (Fig. 7.12(c) and (d)).

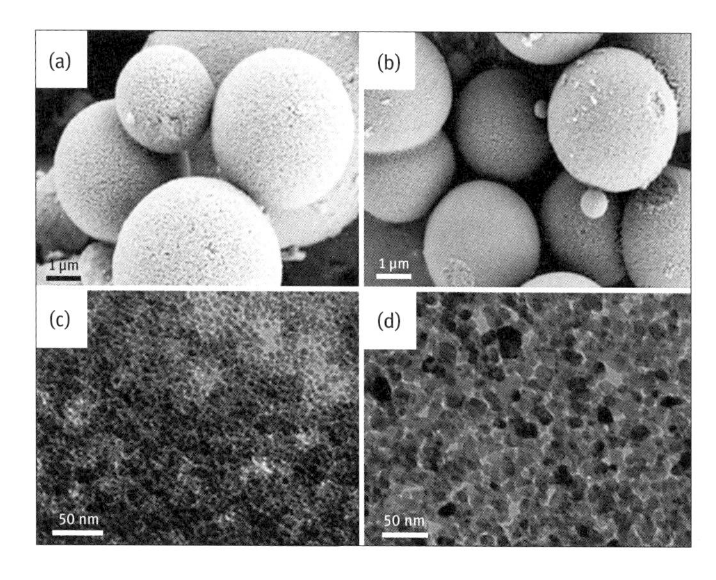

Fig. 7.12: SEM (upper row) and TEM (lower row) images of (a) and (c) C@TiO_2 and (b) and (d) calcined TiO_2 where the C-species have been removed.

Mesoporosity was also confirmed by N_2 gas sorption. The pore size was found to be smaller for C@TiO_2 material than for pure TiO_2, in good agreement with TEM image analysis. UV-VIS diffuse reflectance spectroscopy (DRUVS) demonstrates that C@TiO_2 material can adsorb significantly more light in the 420–800 nm regions as compared with pure TiO_2 (Fig. 7.13). The comparison with the simple mixture of hydrothermal carbon with TiO_2 (13 wt% C) and the pure carbon sample underlines that the anatase phase and the carbon phase are indeed tightly synergistically coupled. The titania band-gap is clearly smeared out to longer wavelengths, whilst the carbon spectrum has changed, with more absorption in the visible range observed concurrently with less extinction in the IR region. This different behavior defines the C@TiO_2 as a "dyad" structure, where the electronic coupling of two parts gives rise to synergistic properties arising from the beneficial interaction of the two. In other words: the two nanocomponents form a joint electronic system, with additional states bound to the interface.

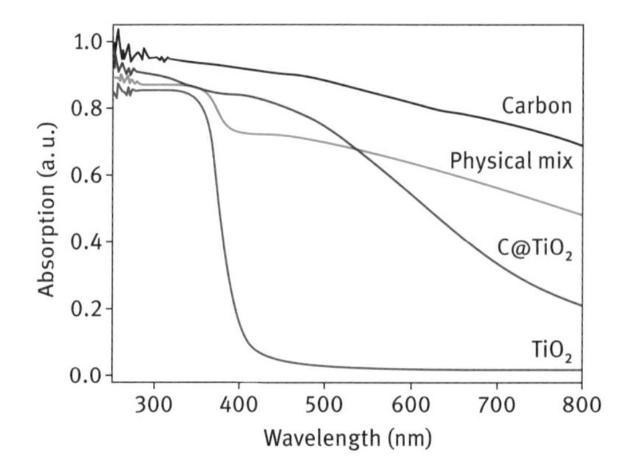

Fig. 7.13: UV-VIS diffuse reflectance spectra of modified and pure samples (a) TiO_2, (b) C@TiO_2, (c) physical mixture of TiO_2 and hydrothermal carbon and (d) pure carbon.

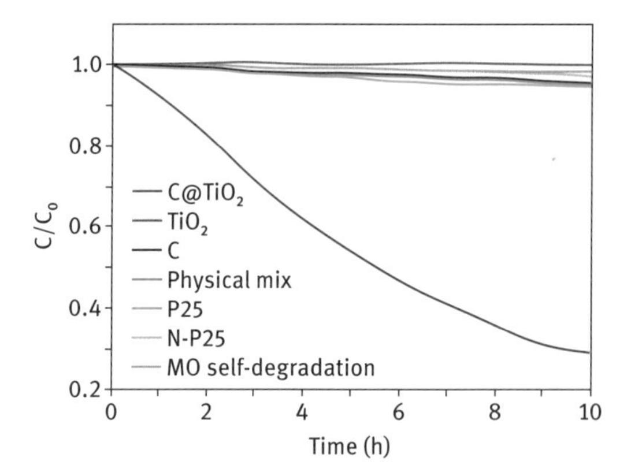

Fig. 7.14: Photocatalytic degradation of MO in the presence of C@TiO_2 and other samples under visible light irradiation ($\lambda > 420$ nm).

Since DRUVS spectra clearly show that the optical response of C@TiO_2 shifts into the visible light region, we hoped that the C@TiO_2 material presents photochemical activity in the visible light region. Therefore, the degradation of methyl orange (MO) in aqueous solution under visible light was investigated. Figure 7.14 shows the comparison between C@TiO_2 and the commercially available Degussa P25 (the gold standard for photodegradation), nitrogen-doped P25, as well as pure hydrothermal carbon, pure TiO_2 and the physical mixture of HTC and TiO_2. The C@TiO_2 shows by far the highest photocatalytic activity of all compared materials in visible light ($\lambda > 420$ nm), thus proving that indeed a special behavior is found due to the dyadic bonding between the carbon and TiO_2. Both the commercial P25 and N-doped P25 showed only an activity level comparable with the self-degradation of MO under the same irradiation conditions.

The specific mechanism of this reaction warrants some further discussion. It seems to be obvious that the C@TiO_2 can induce the formation of hydroxyl radicals under visible light irradiation ($\lambda > 420$ nm). However, this ability relies on the fact that the hole of the joint system is positioned near the titania valence band, as it needs ~ 2.3 V *vs.* normal hydrogen electrode (NHE) to drive the hydroxyl radical formation. This would be not possible for the usual "sensitizing" situation, which would fill the much less noble and therefore less reactive carbon subgaps. We proposed on the base of our data that the synergistic dyad structure of the C@TiO_2 material provides access to optically active charge transfer transitions where the electron is transferred to joint charge transfer states more located at the carbon, while the hole stays electronically and structurally near to titania, as only in this situation we can subsequently convert a surface bound water molecule into a hydroxyl radical. The generated hydroxyl radicals then subsequently photo-bleach the methyl orange (and all other organic compounds), while the carbon is protected by the previous reductive electron transfer. The presence of hydroxyl radicals was later in addition proven by the terephthalic acid photoluminescence probing assay.

Complementing the chemical reactivity experiments, the photocurrent generation of C@TiO_2 was also examined. Promisingly, the C@TiO_2 material was able to generate significant photocurrents under both UV and visible light irradiation. A typical n-type photocurrent was generated with UV ($\lambda > 320$ nm), while under visible light irradiation the material exhibits both n-type and p-type currents. We explained this by the fact that UV-absorption mainly activates pure titania bands, while the ambipolar, biphasic nature of C@TiO_2 is dominant under visible light. This allows both electrons and holes to contribute to charge transport. As expected, the photocurrent decreased when increasing the incident light wavelength, but there was still a noticeable photocurrent found at $\lambda = 780$ nm. Previously described C-doped TiO_2 materials only showed optical absorption bands lower than 650 nm [66, 69, 73]. Therefore, the photoelectrochemical activity at even longer wavelengths nicely supports the very usual "entanglement" of the two materials into a joint dyadic activation state.

7.3 Summary

Hydrothermal Carbonization (HTC), an alternative chemical pathway leading to a variety of carbonaceous materials, was presented in context with the generation of energy materials and energy hybrids. HTC is sustainable, as at least a majority of the precursors are biomass based, and the reaction takes place in pure water at mild temperatures. Furthermore, the resulting materials are solid particles or high surface area scaffolds, while their surfaces are decorated with polar functional groups, thus making them hydrophilic and functional.

After the description of chemical structure and control of meso-architecture and surface area, selected applications of such carbon materials as battery electrodes, supercapacitors, and in the design of controlled hybrid heterojunctions were presented. In the Li battery, coating or hybridization with hydrothermal carbon brought excellent capacities at simultaneous excellent stabilities and rate performances. This was exemplified by hybridization with Si, SnO_2 (both anode materials) as well as $LiFePO_4$ (a cathode material). In the design of supercapacitors, porous HTC carbons could easily reach the benchmark of optimized activated traditional carbons, with better stability and rate performance.

Describing the use of hydrothermal carbonization in the design of chemical heterojunctions or "dyads", we presented a C@TiO_2 illustrating that the entangled material interface can contribute physical properties which are synergistic, *i.e.* not included in each of the primary hybrid components. This was exemplified by an improved charge separation and the coupled photochemical effects, but the approach is much broader and will soon be massively found in the field of energy materials.

The high, partly superior performance of the biocarbon hybrids is worth underlining, as it is still the standard opinion that making processes more sustainable usually comes with a compromise in performance. This is clearly not the case for functional carbon materials and hybrids, which in our opinion is due to the fact that the more technological process of charring is replaced by a better controllable chemical process in a solvent, namely very hot water.

Research on HTC is just at its beginning, but has already proved to be extremely successful in the production of many different distinctive carbon hybrid structures rich in functionality. However, there are still many aspects be studied and considered in the future. New meaningful approaches with a high potential for practical impact are however near countless. Further studies will facilitate the design of novel carbonaceous structures, interestingly mostly already very close to beneficial applications for our daily lives.

Acknowledgments

The authors are grateful to the EnerChem project and all the research partners involved in this project house. MA thanks Shu-Hong Yu for a long cooperation on this subject. MMT thanks Rezan Demir Cakan and Niki Baccile for their collaboration.

Bibliography

[1] A. Thess, R. Lee, P. Nikolaev, H. J. Dai, P. Petit, J. Robert, C. H. Xu, Y. H. Lee, S. G. Kim, A. G. Rinzler, D. T. Colbert, G. E. Scuseria, D. Tomanek, J. E. Fischer, R. E. Smalley, *Science* **1996**, *273*, 483.

[2] A. Eftekhari, P. Jafarkhani, F. Moztarzadeh, *Carbon* **2006**, *44*, 1343.

[3] L. Gherghel, C. Kubel, G. Lieser, H. J. Rader, K. Müllen, *J. Am. Chem. Soc.* **2002**, *124*, 130.

[4] F. Bergius, H. Specht: "Die Anwendung hoher Drücke bei chemischen Vorgängen", Halle, 1913

[5] E. Berl, A. Schmidt, *Ann.Chem.* **1932**, *493*, 977

[6] J. P. Schuhmacher, H. A. van Vucht, M. P. Groanewege, L. Blom, D. W. van Krevelen, *Fuel* **1956**, *35*, 281

[7] J. P. Schuhmacher, F. J. Huntjens, D. W. vam Krevelen, *Fuel* **1960**, *39*, 223

[8] M. W. Haenel, *Fuel* **1992**, *71*, 1211

[9] M. Titirici, A. Thomas, M. Antonietti, *Advanced Functional Materials*, 17 (6), 1010, 2007

[10] Xianjin Cui, Markus Antonietti, and Shu-Hong Yu, *Small* 2006, 2, 756–759

[11] Qing Wang, Hong Li, Liquan Chen, Xuejie Huang, *Carbon* 39 (2001) 2211–2214

[12] M. M. Titirici, M. Antonietti, N. Baccile, *Green Chemistry*, 10, 2008,1204

[13] N. Baccile, G. Laurent, F Babonneau, F. Fayon, M. M Titirici, M. Antonietti, *J.Phys.Chem.* C, 113 (2009), 9644–9654

[14] M. M. Titirici, A. Thomas, M. Antonietti, *Journal of Materials Chemistry*, 17, 2007, 3412

[15] Hai-Sheng Qian, Shu-Hong Yu, Lin-Bao Luo, Jun-Yan Gong, Lin-Feng Fei, Xian-Ming Liu, *Chem. Mater.* **2006**, *18*, 2102–2108

[16] Shiori Kubo, Irene Tan, RJ White, M. Antonietti, M. M. Titirici, *Chem. Mater.* **2010**, 22, 6590–6597

[17] Jimmy C. Yu, Xianluo HuQuan Li, Zhi Zheng, and Yeming Xu, *Chem. Eur. J.* 2006, 12, 548–552

[18] Dominik Eder and Alan H. Windle, *Adv. Mater.* 2008, 20, 1787–1793

[19] Xiaoming Sun and Yadong Li, *Angew. Chem. Int. Ed.* 2004, 116, 607 –611

[20] S H Yu, X Cui, L. li, K. Li, B. Yu, M. Antonietti, H. Cöelfen, *Adv. Mater*, 2004, 16, 18

[21] Bin Deng, An-Wu Xu, Guang-Yi Chen, Rui-Qi Song, and Liuping Chen, *J. Phys. Chem. B* **2006**, *110*, 11711–11716

[22] A. Arenillas, T. C. Drage, K. Smith, C. E. Snape, *J. Anal. Appl. Pyrolysis* 74 (2005) 298–306

[23] A. S. Vinu, C. Anandan, P. Anand, K. Srinivasu, T. Ariga,T. Mori, *Microporous and Mesoporous Materials* 109 (2008) 398–404

[24] P. Zeng, Y. B. Yin, M. Bilek and D. McKenzie, *2nd International Conference on Advances of Thin Films and Coating Technology*, Singapore, 2004, p. 202

[25] H. Yoon, S. Ko and J. Jang, *Chem. Commun.*, 2007, 1468

[26] Conchi O. Ania, Volodymyr Khomenko, Encarnación Raymundo-Piñero, José B. Parra, and François Béguin, *Adv. Funct. Mater.* **2007**, 17, 1828–1836

[27] N. Baccile, M. Antonietti, M. M. Titirici, *CHEMSUSCHEM* **3** 2010, 246–253.

[28] L. Zhao, N. Baccile, S. Gross, Y. J. Zhang, W. Wei, Y. H. Sun, M. Antonietti, M.-M. Titirici, *Carbon* **2010**, *13*, 3778–3787.

[29] M. M. Titirici, A. Thomas, M. Antonietti, *Chemistry of Materials*, 19, 2007, 4205

[30] Qing Wang, Hong Li, Liquan Chen, Xuejie Huang, *Carbon* 39 (2001) 2211–2214

[31] Y-S Hu, R. Demir-Cakan, M.M. Titirici, J. O. Müller, R. Schlögl, M. Antonietti, J. Maier, *Angew. Chem.. Int. Ed.*, 2008, 47, (9), 1645

[32] R. Demir-Cakan, M. M Titirici, M. Antonietti, G. Cui, J. Maier, Y-S Hu *Chemical Communications*, 2008, 3759

[33] A. M. Cao, J. S. Hu, H. P. Liang, L. J. Wan, *Angew. Chem. Int. Ed.* 2005, 44, 4391–4395.

[34] Y. Wang, J. R. Dahn, *J. Electrochem. Soc.* 2006, 153, A2188–A2191.

[35] B. Gao, S. Sinha, L. Fleming, O. Zhou, *Adv. Mater.* 2001, 13, 816–819
[36] X.H. Huang, J.P. Tu, C.Q. Zhang, J.Y. Xiang, *Electrochemistry Communications* 9 (2007) 1180–1184
[37] Han X. Yang, Jiang F. Qian, Zhong X. Chen, Xin P. Ai, and Yu L. Cao, *J. Phys. Chem. C* **2007**, *111*, 14071
[38] M. M. Titirici, Y-S Hu, R. Demir-Cakan, J. Maier, M. Antonietti, *Chemistry of Materials*, 20(4), 1227, 2008
[39] A. Vadivel Murugan, T. Muraliganth, and A. Manthiram *Journal of The Electrochemical Society*, **156**, A79–A83, 2009
[40] Jelena Popovic, Rezan Demir-Cakan, Julian Tornow, Mathieu Morcrette, Dang Sheng Su, Robert Schlögl, Markus Antonietti, Maria-Magdalena Titirici, *Small* 2011, *7*, 1127–1135
[41] B. E. Conway, Ed., *Electromical Supercapacitors, Scientific Fundamental and Tchnological Applications*, Kluwer Academic/Plenum Publishers, New York **1997**.
[42] A. Burke, *J. Power Sources* **2000**, *91*, 37–50.
[43] A. G. Pandolfo, A. F. Hollenkamp, *J. Power Sources* **2006**, *157*, 11–27.
[44] E. Frackowiak, F. Beguin, *Carbon* **2001**, *39*, 937–950.
[45] R. Kotz, M. Carlen, *Electrochim. Acta* **2000**, *45*, 2483–2498.
[46] C. Vix-Guterl, E. Frackowiak, K. Jurewicz, M. Friebe, J. Parmentier, F. Béguin, *Carbon* **2005**, *43*, 1293–1302.
[47] C.-T. Hsieh, H. Teng, *Carbon* **2002**, *40*, 667–674.
[48] G. Lota, B. Grzyb, H. Machnikowska, J. Machnikowski, E. Frackowiak, *Chem. Phys. Lett.* **2005**, *404*, 53–58.
[49] K. Jurewicz, K. Babel, A. Ziolkowski, H. Wachowska, M. Kozlowski, *Fuel Process. Technol.* **2002**, *77*, 191–198.
[50] K. Jurewicz, K. Babel, A. Ziolkowski, H. Wachowska, *Electrochim. Acta* **2003**, *48*, 1491–1498.
[51] K. Jurewicz, K. Babel, A. Ziolkowski, H. Wachowska, *J. Phys. Chem. Solids* **2004**, *65*, 269–273.
[52] C. Vagner, G. Finqueneisel, T. Zimny, P. Burg, B. Grzyb, J. Machnikowski, J. V. Weber, *Carbon* **2003**, *41*, 2847–2853.
[53] E. Raymundo-Piñero, F. Leroux, F. Béguin, *Adv. Mater.* **2006**, *18*, 1877–1882.
[54] D. Hulicova, J. Yamashita, Y. Soneda, H. Hatori, M. Kodama, *Chem.f Mater.s* **2005**, *17*, 1241–1247.
[55] F. Beguin, K. Szostak, G. Lota, E. Frackowiak, *Adv. Mater.***2005**, *17*, 2380 ff..
[56] C. O. Ania, V. Khomenko, E. Raymundo-Piñero, J. B. Parra, F. Béguin, *Adv. Funct. Mater.* **2007**, *17*, 1828–1836.
[57] Zhao Li, Fan Li-Zhen Zhou Meng-Qi, H. Guan., S. Y. Qiao, M. Antonietti, M. M. Titirici, *Adv.Mater.* **2010**, *22*, 5202–5204
[58] E. Raymundo-Piñero, K. Kierzek, J. Machnikowski, F. Béguin, *Carbon* **2006**, *44*, 2498–2507.
[59] Li Zhao, Xiufang Chen, Xinchen Wang, Yuanjian Zhang, Wei Wei, Yuhan Sun, Markus Antonietti, Maria-Magdalena Titirici, *Adv. Mater.* **2010**, *22*, 3317–3321
[60] M. R. Hoffmann, S. T. Martin, W. Choi, D. W. Bahnemann, *Chemical Reviews* **1995**, *95*, 69–96.
[61] M. A. Fox, M. T. Dulay, *Chemical Reviews* **1993**, *93*, 341–357.
[62] J. C. Yu,J. Lin, D. Lo, S. K. Lam, *Langmuir* **2000**, *16*, 7304–7308.
[63] A. L. Linsebigler, G. Q. Lu, J. T. Yates, *Chemical Reviews* **1995**, *95*, 735–758.
[64] H. Kamisaka, T. Adachi, K. Yamashita, *Journal of Chemical Physics* **2005**, *123*.
[65] C. Di Valentin, G. Pacchioni, A. Selloni, *Chemistry of Materials* **2005**, *17*, 6656–6665.
[66] S. U. M. Khan, M. Al-Shahry, W. B. Ingler, *Science* **2002**, *297*, 2243–2245.
[67] S. Sakthvel, H. Kisch, *Angewandte Chemie International Edition* **2003**, *42*, 4908–4911.
[68] M. Janus, B. TrybaM. Inagaki, A. W. Morawski, *Applied Catalysis B: Environmental* **2004**, *52*, 61–67.

[69] W. J. Ren, Z. H. Ai, F. L. Jia, L. Z. Zhang, X. X. Fan, Z. G. Zou, *Applied Catalysis B-Environmental* **2007**, *69*, 138–144.
[70] D. Mitoraj, H. Kisch, *Angewandte Chemie International Edition* **2008**, *47*, 9975–9978.
[71] H. Raether, In *Springer Tracts in Modern Physics*; Springer: Berlin, 1980; Vol. 88.
[72] S. Spange, S. Grund, *Advanced Materials* **2009**, *21*, 2111–2116.
[73] H. Irie, Y. Watanabe, K. Hashimoto, *Chemistry Letters* **2003**, *32*, 772–773.
[74] Y. S. Hu, P. Adelhelm, B. M. Smarsly, S. Hore, M. Antonietti, J. Maier, *Adv.Funct.Mater*. **2007**, *17*, ***1873–1878***
[75] Y. S. Hu, R. Demir-Cakan, M. M. Titirici, J. O. Muller, R. Schlogl, M. Antonietti, J. Maier, *Ang. Chem. Int. Ed*. **2008**, *47*, 1645–1649

Juan J. Vilatela

8 Nanocarbon-based composites

This chapter reviews the development of nanocarbon-based composites over recent years and outlines future trends for these materials. Nanocarbons such as carbon nanotubes (CNTs) and graphene have an outstanding combination of mechanical, electrical and thermal properties in the tube or plane axis, respectively. The purpose of the integration of these nanocarbons in polymer matrices is to make composites where the nanocarbon properties are exploited on a macroscopic length scale. Nanocarbon-based composites can be roughly divided into three types. In the first, the nanocarbon is dispersed as a filler. At low volume fractions (< 1 %) this strategy results in large improvements in matrix properties, new possibilities for structural health monitoring and novel processing methods. The second type of composites consists of a hierarchical structure with both nanocarbons and macroscopic reinforcing fibers in the polymer matrix. The nanocarbons can be dispersed in the matrix, pre-placed on the fiber fabric or directly grown on the surface of individual fiber filaments in the case of CNTs. By placing the nanocarbons perpendicular to the fiber axis, critical interlaminar properties of structural composites can be significantly improved, such as interlaminar shear strength and electrical conductivity. The third route involves the pre-assembly of nanocarbons as a continuous film or fiber where the nanocarbons are preferentially aligned parallel to each other and the fiber/film axis so as to exploit their axial or in-plane properties. The macroscopic assemblies can be integrated in polymer matrices, for example as an array of fibers. These high volume fraction (> 20 %) composites have a complex structure that can be engineered across multiple length scales.

8.1 Introduction

The appeal of CNTs, graphene and other nanocarbons lies in their combination of outstanding physical properties, similar to those of in-plane graphite, but realized in a nanomaterial. The fact that one of the dimensions of the nanomaterial is on the atomic scale opens the possibility for molecular engineering by introducing dopants, functionalizing, *etc.*; whereas its other(s) dimension(s) on the mesoscale implies that the assembly of multiple nano building blocks becomes a powerful tool to engineer properties of the macroscopic ensemble. Table 8.1 compares the mechanical, electrical and thermal properties of graphite with those of common engineering materials: carbon fiber used in structural composites (high strength/stiffness and low density); copper, used in electrical applications and thermal management (high electrical and thermal conductivities); and steel, used in uncountable applications (combination of mechanical properties, ease of processing and cost). Other relevant properties of

Tab. 8.1: Properties of graphite and other materials.

	Graphite (in-plane)	Graphite (out of plane)	Carbon fiber[1)]	Copper	Steel
Volumetric density (g/cc)	2.1		1.8–2.2	8.9	7.8
Tensile strength (GPa)	130 [1]	–	1.4–7	0.2	1.9
Tensile stiffness (GPa)	1000 [1]	–	160–930	124	210
Electrical conductivity (S/m)	1x107 [2] – 13×10^7 [3]	10^3 [4]	7×10^3–8×10^5	59×10^6	6×10^4
Thermal conductivity (W/mK)	1000 [5] – 5000 [6]	13 [5]	7–1000	400	52
Specific surface (m2/g)	2600 (graphene)		0.2	–	–

[1)] From manufacturer's data sheets.

nanocarbons, such as optoelectronic, are discussed throughout this book but not in this chapter.

The in-plane properties of graphite can in principle be attained in individual layers of graphene or a carbon nanotube, provided that the material has a high degree of graphitization. Even then, their nanoscopic length scale implies that multiple "molecules" need to be integrated in order to produce a macroscopic material, typically in the form of a composite. Often, this assembly gives rise to a hierarchical structure, which can be engineered across different length scales. Naturally, the challenge is to obtain bulk properties as close to those of the nano building blocks. This chapter discusses different strategies of producing composites containing nanocarbons, their relative merits in exploiting the properties of the nanocarbon, and points to the more promising results and future trends.

8.2 Integration routes: From filler to other more complex structures

Nanocarbon composites can be broadly divided into three kinds, each with some possible subdivisions. Examples of these composites and their schematic representations are presented in Fig. 8.1. The first type corresponds to composites where the nanocarbon is used as a filler added to a polymer matrix analogous, for example, to rubber reinforced with carbon black (CB). The second consists of hierarchical composites with both macroscopic fibers and nanocarbon in a polymer, such as a carbon fiber laminate with CNTs dispersed in the epoxy matrix. The third type is macroscopic fibers based

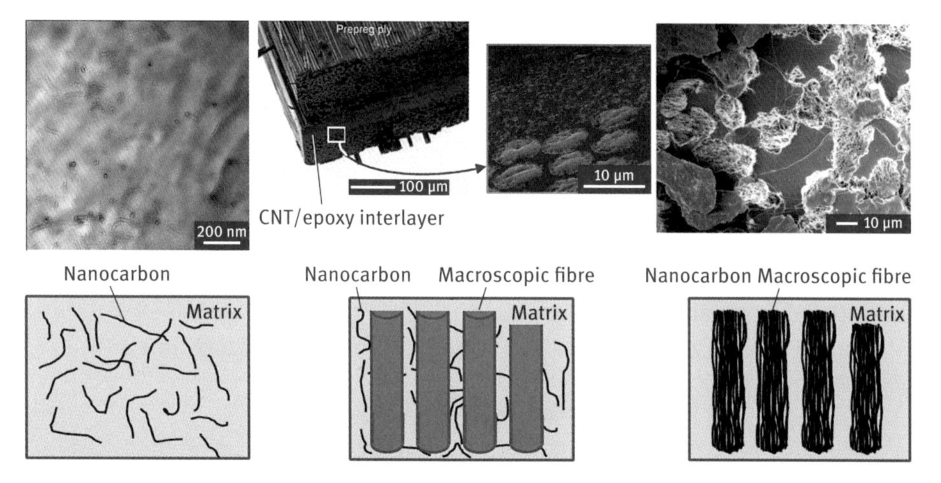

Fig. 8.1: Electron micrographs of different nanocarbon composite types (top) and their schematic representation (bottom). The nanocarbons can be dispersed as a filler (left), combined with macroscopic fibers in a hierarchical composite (middle), or assembled as a continuous nanostructured fiber (right). Micrographs from references [7, 8, 9], with kind permission from Elsevier (2010, 2008, 2009).

on nanocarbons, where the fiber can be a composite itself or can be integrated to form a fiber-reinforced polymer composites (FRPC) composite.

8.2.1 Filler route

A great deal of work has been done on incorporating nanocarbons as fillers in polymer matrices to form composites (see for example [10, 11]), this being the first composite manufacture route explored after carbon nanotubes became available in substantial quantities in the 1990s. The idea is to disperse the nanocarbon in the polymer matrix and obtain the maximum effect from the addition of the nanocarbon. For thermoplastics, dispersion and mixing is usually achieved during extrusion, whereas for thermosets it is typically done by shear mixing (*e.g.* calendering) and/or sonication of the nanocarbon in the resin. The lower volume fraction corresponds typically to nanocarbons directly dispersed in the polymer matrix, whereas at higher volume fraction the nanocarbons are often dispersed in solvent and then re-aggregated and mixed with the polymer matrix, which results in preferential orientation.

The filler route has proved to be very efficient to obtain isotropic composites with relatively large improvements in matrix properties at small mass (volume) fractions of nanocarbon. For example, electrical percolation in epoxy has been obtained with only 0.0025 wt% of multi-wall nanotubes (MWNTs) [12]. Similarly, a 2.7-fold increase in matrix modulus has been observed on addition of 0.6 vol% MWNTs to polyvinyl alcohol (PVA) [13]. Although more modest compared to the previous two examples, a

320 % increase in thermal conductivity relative to the plain matrix has been obtained in SWNT/epoxy composites [14].

8.2.2 Evaluation of reinforcement

Of great interest is the level of reinforcement obtained with nanocarbons at different volume fractions, and especially, how this reinforcement compares with other more conventional materials.

The efficiency of reinforcement can be derived from a rule of mixtures [15]:

$$E_c = \eta_0 \eta_1 E_f V_f + E_m \left(1 - V_f\right) \tag{8.1}$$

Where η_0 is the length efficiency factor related to the aspect ratio of the filler and thus the stress build-up along its length; it can take values from 0 to 1 and can be calculated using theoretical models. η_1 is an orientation factor that takes values of 1 for perfect alignment, 3/8 for alignment in the plane and 1/5 for random orientation. Equation (8.1) can be arranged as:

$$\frac{E_c - E_m}{V_f} = \eta_0 \eta_1 E_f + E_m$$

The left-hand side term represents the increment in matrix Young's modulus divided by the volume fraction (dE/dVf), all of which are experimentally determined. Additionally, since $E_f \gg E_m$ and when η_o and η_1 are reasonably large, the equation can be rewritten as:

$$\frac{dE}{dV_f} = \eta_0 \eta_1 E_f \tag{8.2}$$

In this simple form, this expression is a good first approximation to compare the experimental reinforcement achieved upon addition of filler to the matrix, to the theoretical prediction [11]. It provides a measure of how efficiently the properties of the nanofiller are exploited in the composite, but also enables the comparison with the level of reinforcement achieved using other fillers. Note, in addition, that equation (8.2) sets an upper limit between $E_f/5 = 200$ GPa and $E_f = 1000$ GPa, depending on whether the nanocarbon is randomly or perfectly oriented (without taking η_0 into account).

Figure 8.2 presents a plot of dE/dV_f at different volume fractions for a wide variety of CNT/polymer composites reported in the literature, including different polymer matrices, types of nanotubes, processing routes, and degrees of orientation. The graph also includes for comparison dE/dV_f for randomly dispersed short glass fiber and short CF in thermoplastics [16, 17], where the fibers have aspect ratios in the range of the CNTs (around 70000 and 40000, respectively). The upper limits of dE/dVf are also included for reference. There is a very large scattering of the data for CNT composites, which evidences differences in experimental preparation, which is probably due to different raw materials, processing methods and composite testing. In spite of the data scatter, the graph clearly shows the decrease in CNT reinforcement with increasing volume fraction. Above 20 vol% loading the increase in composite modulus per unit volume fraction of CNT is below 30 GPa and in the same range as the glass fiber and CF composites.

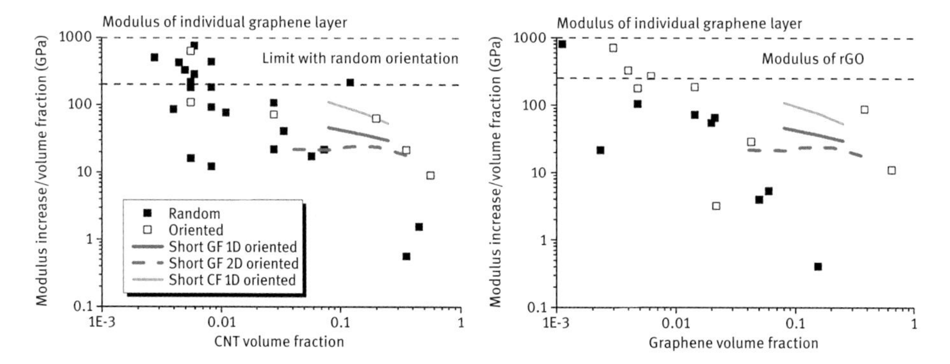

Fig. 8.2: Plot of increase in modulus against volume fraction for a wide variety of nanocarbon/polymer systems. Volume fractions calculated from reported mass fractions assuming specific gravities of 1.8, 2.1 and 1 for CNT, graphene and matrix, respectively. Short fiber data from [16] and [17]. Moduli of individual layer graphene (1TPa) and reduced-graphene oxide (250GPa) included for reference.

Graphene-polymer composites thus far show the same behavior as CNTs, as can be appreciated in Fig. 8.2(b), an equivalent plot with data polymer composites with the different graphene varieties: graphene oxide (GO), reduced graphene oxide (rGO) and graphite platelets. In spite of the significant scattering observed, attributed mainly to differences in sample nature and manufacturing route, there is a clear decrease in the level of reinforcement with increasing filler volume fraction.

The plots in Fig. 8.2 are an indication of the increasing difficulty in dispersing the filler at higher volume fraction. In addition to simple geometric constraints implying that high aspect ratio particles are difficult to individualize at high V_f due to their shape, the large surface to volume ratio of these nanocarbons means that the nanocarbon/polymer interface is also very large and that virtually all the polymer forms part of it. Thus with increasing V_f, "wetting" of the nanocarbon becomes increasingly difficult and so does their dispersion [18], which is a prerequisite for efficient reinforcement with nanofillers [19]. To highlight the importance of individualizing nanocarbons in the matrix for mechanical reinforcement, note that the interfacial shear strength of a polymer-nanocarbon interface (~ 1MPa [20]) is typically much higher than that of the nanocarbon-nanocarbon (~ 30kPa [21]) and thus, that efficient reinforcement requires avoiding nanocarbon aggregation.

Two additional consequences of the high aspect ratio of nanofillers are worth pointing out. One is the rise in viscosity of the polymer-nanocarbon system on addition of the nanocarbon, which often hinders processing of the mix and puts an upper limit on V_f. The other is that the large polymer interface formed in the composite can have superior properties to the bulk polymer, for example due to the nucleation of crystalline domains on the nanoparticles surface or through modified curing in the case of thermosets, which can be easily misinterpreted as nanocarbon reinforcement, particularly at low volume fractions (eq. (8.2) does not account for this). Some data points in Figs. 8.1 and 8.2 above the theoretical limit (taking into account η_1) are likely to be to some extent due to these effects.

8.2.3 Other properties

Due to their high aspect ratio, nanocarbons dispersed in a polymer matrix can form a percolating conductive network at very low volume fractions (< 0.1 %). The conductivity of a composite above the transition from an insulator can be described by the statistical percolation using an excluded volume model [22, 23] to yield the following expression:

$$\sigma = \sigma_0 \left(V_f - V_{fc}\right)^t \tag{8.3}$$

Where σ is the composite conductivity, σ_0 a proportionally coefficient, V_{fc} the percolation threshold and t an exponent that depends on the dimensionality of the system. For high aspect ratio nanofillers the percolation threshold is several orders of magnitude lower than for traditional fillers such as carbon black, and is in fact often lower than predictions using statistical percolation theory, this anomaly being usually attributed to flocculation [24] (Fig. 8.3).

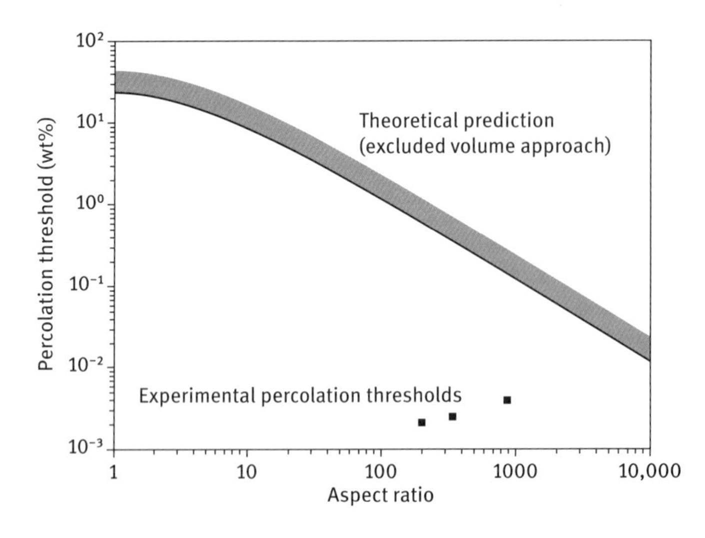

Fig. 8.3: Percolation threshold for different aspect ratio fillers and experimental values obtained with CNTs [25]. With kind permission from Elsevier (2004).

Unlike mechanical reinforcement, which benefits from individualization of the particles, achieving electrical conductivity in the polymer matrix requires some level of aggregation of the fillers. Thus, the electrical conductivity of the composite is determined by the establishment of a conducting network of nanofillers in the polymer matrix [25]. Typically the nanocarbons are first dispersed and then re-aggregated, the whole process controlled mainly by adjusting the viscosity of the system at the different stages. In spite of significant efforts to determine rational strategies to control structure properties in these systems [22], it remains difficult to take experimental results achieved with a polymer matrix and apply them to another one. This is particularly challenging for multi-component thermosets where the order in which nanocarbon, resin and hardener are mixed together and the viscosity at those stages introduce additional processing parameters that affect aggregation and therefore composite electrical conductivity.

At the high volume fraction end ($V_f > 10\,\%$), well above the percolation threshold, equation (8.3) can be approximated as

$$\sigma=\sigma_0 V_f^t \tag{8.4}$$

With t taking experimental values in the range 1–5 for both CNTs and graphene [26, 27, 28]. The upper limit of equation (8.4) is σ_0, which has been experimentally found to be as high as around 10^4 S/m for both CNTs [24] and graphene [29]. These values can be taken as indicators of the maximum conductivities attainable by dispersing nanocarbons in polymer matrices.

While conductivities of nanocarbons dispersed in polymers fall short of those of metals, a variety of applications can be unlocked by turning an insulating matrix into a conductor, which requires only small volume fractions that can therefore keep the system viscosity at a level compatible with composite processing techniques. Of particular interest are novel functionalities of these conductive matrices that exploit the presence of a conductive network in them, such as structural health monitoring (SHM) based on changes in electrical resistance of the nanocarbon network as it is mechanically deformed [30].

Figure 8.4(a) shows an example of piezoresistance measurements on a CNT/epoxy composite under tensile loading. In the elastic regime there is an exponential dependence of the resistance change against strain, due to the increase in interparticle distance and therefore the lower tunneling probability between contacts. Yet, often this dependence can be approximated with a linear relationship for practical purposes and with competitive gauge factor > 2. In the inelastic region this linear relationship no longer holds, with resistance in fact often decreasing with increasing deformation. After unloading a permanent change in electrical resistance can occur due to plastic deformation.

The example in Fig. 8.4(b) shows the electrical resistance changes for a sample in bending during repeated loading. Each individual loading cycle produce a clear, fairly linear signal, however, repeated cycling results in a non-recoverable change in resistance due to permanent deformations of the nanocarbon network [31], similar to the results obtained with CB composites (*e.g.* [32]). Overall, the results indicate a

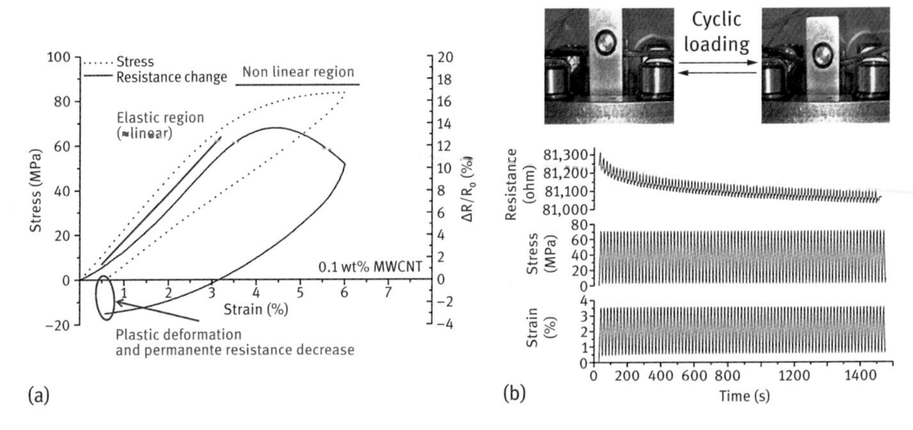

Fig. 8.4: Plots of relative change in electrical resistance against tensile deformation of a CNT/epoxy composite: (a) shows the various characteristics of the piezoresistivity of nanocarbon networks: linear resistance change in the elastic regime, nonlinear region after inelastic deformation and the permanent electrical resistance drop due to plastic deformation (image adapted from [30]); (b) shows the change in electrical resistance after several bending cycles. With kind permission from APS (2009).

strong dependence of composite piezoresistivity on factors such as matrix ductility, level of strain and type of deformation (tensile *vs.* compressive), and in general any property that affects the deformation of the network during composite loading. In this sense, the use of thermoset matrices which are typically more brittle and less prone to creep than thermoplastics is probably more suitable for the use of nanocarbons as sensors.

New, or improved processing of composites can also result from the presence of nanocarbons in a polymer matrix. The high electromagnetic absorption of graphitic nanocarbons has been widely studied as a route to download energy into the matrix and rapidly raise its temperature, for example, using microwave radiation to cure thermosets (with CB [33], and CNTs [34]) or to melt thermoplastics [35]. Alternatively, thermoset curing can be achieved by Joule heating as current is directly passed through the sample.

In all these methods heating occurs from within the material, which is intrinsically more energy efficient than oven-heating. The small distance between nanocarbons dispersed in the matrix, even at low volume fractions, implies that heat transfer to the matrix occurs very rapidly (rates of > 750 °C/min can be achieved) and uniformly throughout the curing part [36]. Figure 8.5 shows an example of a setup used to cure resistively a CNT/epoxy part with a predetermined heating ramp achieved by feeding the sample temperature to a controller that adjusts the current delivered to the sample. Figure 8.5(c) shows how this method can be used on an aerospace composite panel repaired by joining a CF laminate to it using CNT/epoxy as adhesive cured by Joule heating as current is passed between panels.

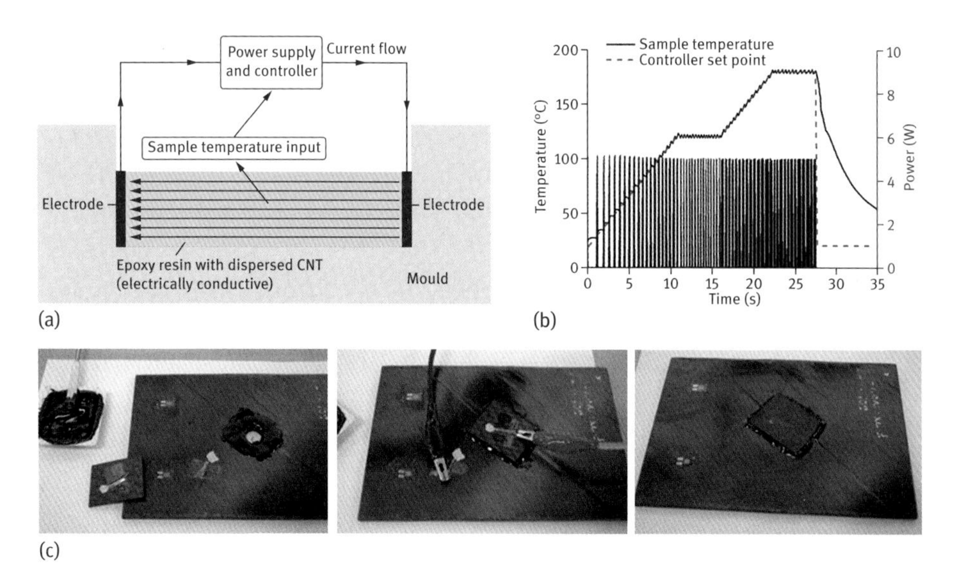

Fig. 8.5: Epoxy curing through resistive heating of nanocarbons dispersed in the matrix: (a) shows a schematic representation of the process; (b) experimentally obtained curing cycle and (c) repair of a structural composite panel using the conductive epoxy as resistively curable adhesive [36]. With kind permission from Elsevier (2013).

The research on nanocarbons dispersed in polymer matrices in recent years has shown that this route is very efficient at small volume fractions above electrical percolation, where it can be the basis for new composite functionalities in terms of processing and properties. It is also clear that there is an inherent difficulty in dispersing these nanoscopic objects at high volume fractions, which therefore limits composite absolute properties to a very small fraction of those of the filler. Independent of their absolute properties, composites based on dispersed nanocarbons have served as a test ground to understand better the basic interaction between nanocarbons and polymer matrices, often setting the foundation to study more complex composite structures, such as those discussed in the following sections.

8.3 Hierarchical route

Another approach to exploit the properties of nanocarbons consists in integrating them in standard fiber-reinforced polymer composites (FRPC). The rationale behind this route is to form a hierarchical composite, with the nanocarbon playing a role at the nanoscale and the macroscopic fiber providing mainly mechanical reinforcement. This strategy typically aims to give FRPCs added functionality, improve their interlaminar properties and increase the fiber surface area. The first two properties are critical for the transport industry, for example, where the replacement of structural metallic elements with much lighter carbon fiber composite (CFC) materials has raised new challenges in terms of the handling of anisotropic mechanical properties (*e.g.* interlaminar failure), and electrical conduction (lightning strike protection).

8.3.1 Structure and improvement in properties

Nanocarbons are generally integrated in FRPC by one of three methods: dispersing them in the polymer matrix and then processing the composite with the filler-containing resin; by pre-placing them at the interface between laminates; and by directly synthesizing the nanocarbon on the surface of the fiber before matrix infusion (not yet demonstrated with graphene). The different hierarchical structures produced by these methods are shown schematically and with examples in Fig. 8.6. The nanocarbon-containing composites have been typically characterized in terms of their interlaminar shear strength (ILSS) and fracture toughness (Mode I and II). Both properties have been significantly increased with addition of small quantities of CNTs (< 10 %); some significant results are presented in Table 8.2.

Since the first method consists essentially in modifying the resin used as matrix by adding nanocarbon filler, the limitations to the volume fraction discussed in the previous section apply to it. In fact, the tolerances on resin viscosity are even more stringent in this case as the resin has to be compatible with established infusion techniques used in composite production (resin transfer molding, pultrusion, *etc.*) which require viscosities below 500–100 cPs. Using small amounts of nanocarbon has resulted in significant increases in interlaminar properties (Table 8.2). Improvements in electrical conductivity following this route have also been obtained in CNT/glass fiber/epoxy composites, reaching values of 1 S/m (from 10–11 S/m) at a CNT mass fraction of only 0.01[40]. Graphene has so far seen little activity in this type of composite, presumably due to the limited availability of sufficient quantities; however, the results so far have shown fatigue (in bending) increases of three orders of magnitude in graphene/glass fiber/epoxy composites with only 0.02 wt% graphene [39].

Alternatively, the nanocarbon can be placed at the interface between laminates before infusion or composite consolidation, with the purpose of specifically improving interlaminar mechanical properties. This structure can be produced by various methods, including spraying a dispersion of nanocarbons in a volatile solvent onto composite pre-pregs, and by transfer of mats of aligned CNTs onto the pre-preg surface using epoxy as an adhesive. The latter strategy is particularly attractive since the CNTs end up at the interface between laminates and aligned in the direction where the reinforcement is sought [8], although it can result in an increase of interlaminar thickness. Examples of improvements in ILSS and fracture toughness by these methods are listed in Table 8.2.

In the third route of integration of nanocarbons in FRPCs, carbon nanotubes are radially attached to the surface of the fibers, typically grown *in situ* by chemical vapor

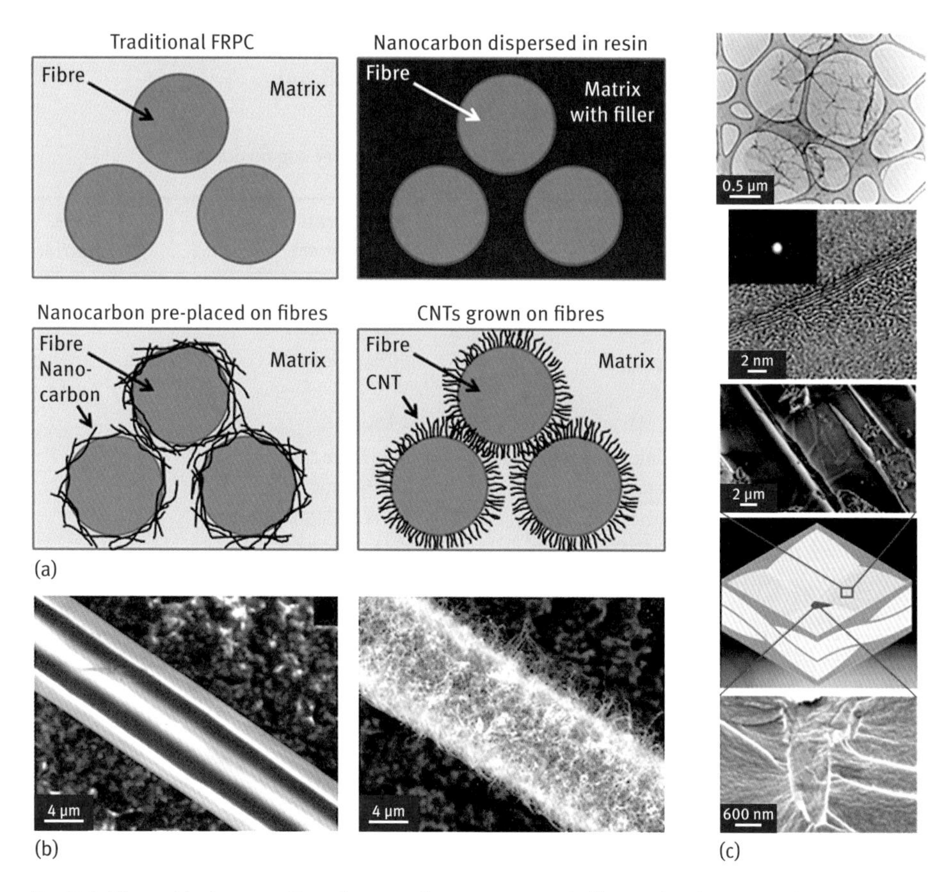

Fig. 8.6: Hierarchical composites of nanocarbon, macroscopic fiber and polymer matrix. (a) Schematic of different configurations of nanocarbons in FRPC (adapted from [37]). (b) Silica fibers with CNTs grown radially on their surface [38]. (c) Hierarchical composite of GF and GO dispersed in a thermoset matrix [39]. With kind permission from Elsevier (2010) and ACS (2010).

deposition (CVD) and demonstrated on a variety of ceramic and carbon fibers (see for example [41]). This process can be used continuously by feeding the fibers into a cold wall CVD reactor [42]. The reinforcement of fibers orthogonally is probably more suitable for CNTs, whose graphitic layers align perpendicular to the substrate on which they grow, whereas in graphene the layers are parallel to the substrate.

The macroscopic fibers with radially-protruding CNTs have a very high surface area, around 4 m^2/g compared to 0.1 m^2/g for the pristine material [38], which is readily wetted by the matrix through capillary forces, providing significant increases in interfacial shear strength (IFSS) and better interlaminar performance (Table 8.2). However, one of the remaining challenges is to achieve CNT growth on CFs without degrading their properties by the metal catalyst dissolving carbon from the fiber itself, commonly referred to as "pitting". This damage can be partially avoided by pre-depositing a protective thin layer onto which the CNTs are grown [43, 44].

Tab. 8.2: Property improvement in hierarchical nanocarbon/polymer composites [45].

System filler/fiber/matrix	Filler integration	Mechanical improvement	Mass fraction
CNT/glass/epoxy	Dispersed	20 % ILSS	0.001
CNT/glass/polyester	Dispersed	100 % Mode-I	0.01
CNT/CF/epoxy	Dispersed	60 % Mode I	
75 % Mode II	0.01		
CNT/CF/vinyl ester	Pre-placed on fiber	45 % ILSS	–
CNT/CF/epoxy	Pre-placed between laminates	150 % Mode I 200 % Mode II	–
CNT/CF/epoxy	Synthesized on fiber	27 % ILSS 50 % Mode I	–
CNT/alumina/epoxy	Synthesized on fiber	69 % ILSS 76 % Mode I	–
GO/glass/epoxy[39]	Dispersed and pre-placed on fiber	Greater fatigue life: 1200-fold in bending 4-fold in tension	0.02

The incorporation of nanocarbons in hierarchical composites can also result in large improvements in their electrical conductivity, and to a lesser extent in their thermal conductivity. For ceramic fibers both in-plane and out-of-plane electrical conductivities are increased by several orders of magnitude [41], whereas for CF the improvement is significant only perpendicular to the fiber direction due to the already high conductivity of the fiber itself [46]. The out-of-plane electrical conductivity of CNT/CF/epoxy composites is approaching the requirements for lightning strike protection in aerospace composites, thought to be around 1–10 S/m. Yet further improvements are required, as well as the evaluation of other composite properties relevant for this application, such as maximum current density and thermal conductivity.

8.3.2 Other properties

Another notable improvement in FRPCs by addition of nanocarbons is the use of the nanocarbon for non-destructive damage detection. The detection methods are based on the electrical conductivity of the network of CNTs in the polymer matrix. In one of them, the percolating CNTs are resistively-heated and thermally-imaged during mechanical deformation using a thermal camera. The damage propagation in the composite disrupts the CNT conductive network, therefore reducing local temperature and producing contrast in the thermal image [47]. An example corresponding to damage detection in a CNT/alumina fiber/epoxy composite during tensile loading is presented in Fig. 8.7(a).

Similarly, by directly measuring changes in electrical resistance during mechanical deformation it is possible to monitor crack propagation in hierarchical composites with non-conductive fibers and CNTs dispersed in the matrix [48]. Figure 8.7(b) shows

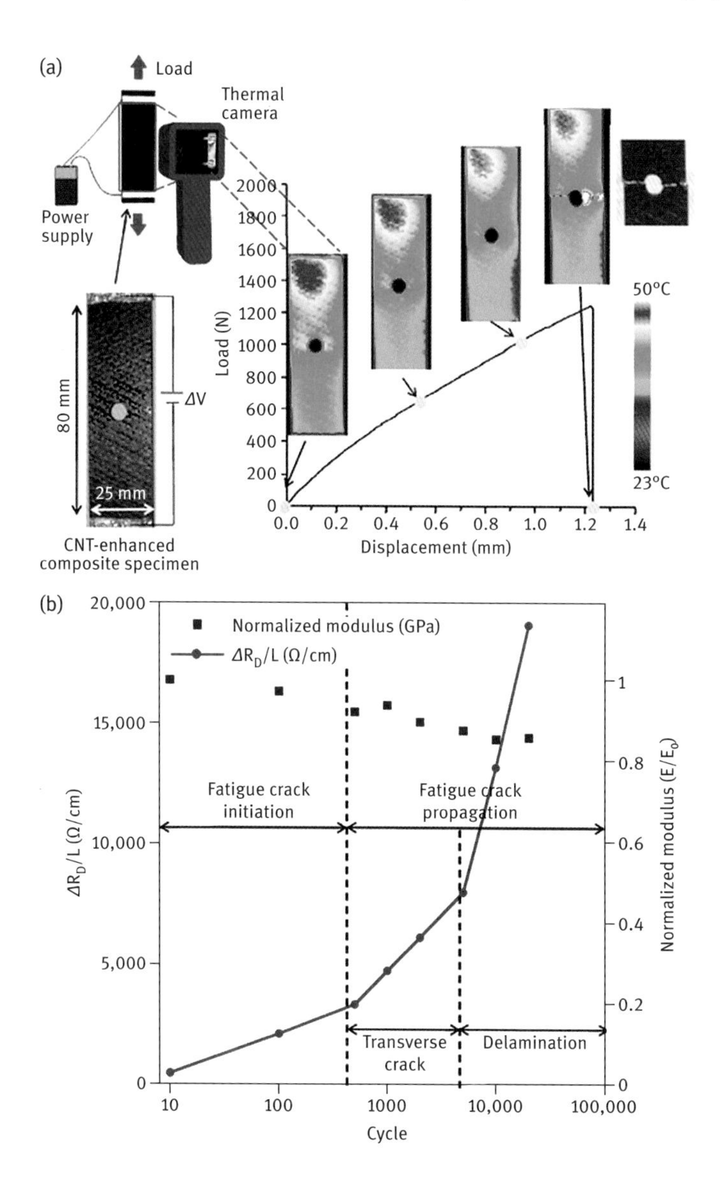

Fig. 8.7: Examples of hierarchical composites where the presence of CNTs is used for SHM. (a) Damage detection through thermal imaging of resistively-heated CNTs in an alumina composite [47] and (b) detection of crack propagation by monitoring electrical resistance (normalized by specimen length) in a CNT/glass fiber/epoxy composite [48]. With kind permission from IOP (2011) and Wiley (2006).

the increase in electrical resistance during fatigue loading of a CNT/GF/epoxy composite. The evolution of resistance can be accurately correlated to the different stages of failure of the composite structure.

These methods are based on the electrical conductivity of a low volume fraction of CNTs in the matrix and are therefore more suitable for non-conductive fibers (glass, polymeric). In hierarchical CF composites, electrical conduction is dominated by the properties of the macroscopic fibers present at much higher volume fraction, which cannot be decoupled from those of the nanocarbons.

Hierarchical composites produced by the addition of nanocarbons to standard FRPCs have tremendous potential. First, because the role of the nanocarbon is to produce only moderate improvements in the absolute properties of the material or to give it additional functionality, these effects being potentially attainable with low mass fraction of nanocarbons. Second, because the ethos itself of hierarchical composites means that rather than competing with well-established composites, nanocarbons are integrated into them to improve their performance and extend their application range.

A particularly promising area for these materials is structural supercapacitors that are both load-bearing and power storing. In these materials, CF fiber laminas act as electrodes and the polymer matrix as ion-conducting electrolyte [49]. The large surface area provided by CNTs grown on the surface of the CFs could potentially be of benefit for this application. It will also be interesting to see the likely emergence of hierarchical hybrids that combine macroscopic fibers with other nanostructure materials besides CNTs and graphene that thus provide added functionality.

8.4 Fiber route

An alternative route to nanocarbon-based composites is pre-assembling the nanocarbon to form a macroscopic structure and then integrating it into a polymer matrix to form composite using standard FRPC manufacturing methods. The most studied example consists of macroscopic fibers of CNTs assembled with preferential orientation of the graphitic layers parallel to each other and to the fiber axis, an arrangement which exploits the in-plane properties of the building blocks and which follows basic principles of polymer fiber manufacturing proposed around eighty years ago [50, 51] and used successfully to produce high-performance polymer fibers. Graphene-based fibers have also been produced recently after using previous knowledge from CNT fiber spinning [52]. The original schematic proposed by Staudinger and some examples of synthetic fibers, as well as an estimated date of their initial synthesis is presented in Fig. 8.8.

The last decade, since CNT fibers were first produced [53], has seen a large improvement in the properties of CNT fibers as a consequence of improved synthesis of the CNT building blocks and their assembly. For most fibers, axial properties improve when increasing the length of the aligned building blocks; longer CNTs result in higher

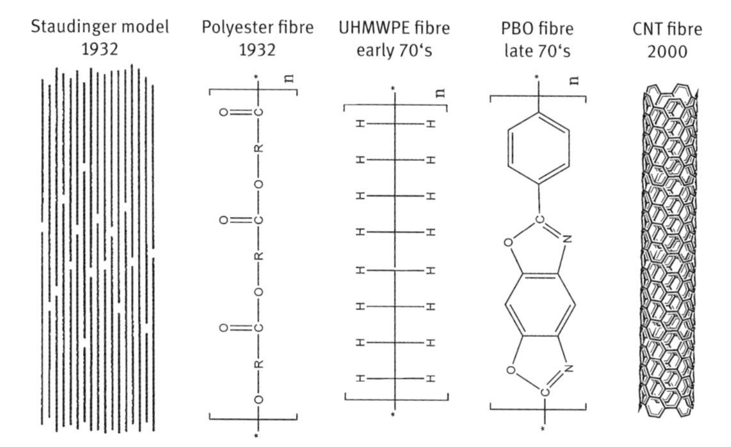

Fig. 8.8: The basic structure of high-performance polymer fibers (Staudinger's model [51]) and some examples of polymers and of a CNT used as building block for synthetic fibers. (Courtesy of H. Yue). With kind permission from Wiley (2006).

tensile strength and modulus [54, 55]. Similarly, there is a clear tendency towards CNTs with fewer layers since this type of CNTs maximize properties per unit mass and lead to better packing of the CNTs through polygonization [56] or radial collapse [57]. In the case of graphene, since the flake size approaches that of the fiber itself (around 15 microns), a similar improvement would require making highly elongated rectangular graphene sheets and preferentially aligning them in the fiber axis direction, similar in fact to already existing large diameter collapsed CNTs [58]. It is also worth pointing out that while the one-dimensional geometry of a fiber represents a natural embodiment to exploit the axial properties of CNT, it might not be equally efficient for a flat material such as graphene, for which assembly as a film could prove more efficient to produce high-performance composites. Graphene-based macroscopic fibers could have applications in other areas that benefit from the optoelectronic properties and monolayer structure of graphene, such as in electrochemistry and photocatalysis; these are discussed elsewhere [26].

8.4.1 Different assembly routes

There are three main methods to produce CNT-based macroscopic fibers: wet spinning, forest drawing and direct spinning from CVD. These are shown schematically in Fig. 8.9. So far, only the first has been applied to produce graphene fibers.

In wet spinning, the nanocarbons are first dispersed in a liquid and then injected into a coagulation bath where a large proportion of the dispersant is drawn out and a continuous fiber is formed. Of the several dispersants that can be used for this process, superacids are particularly promising. In superacids nanocarbons form ther-

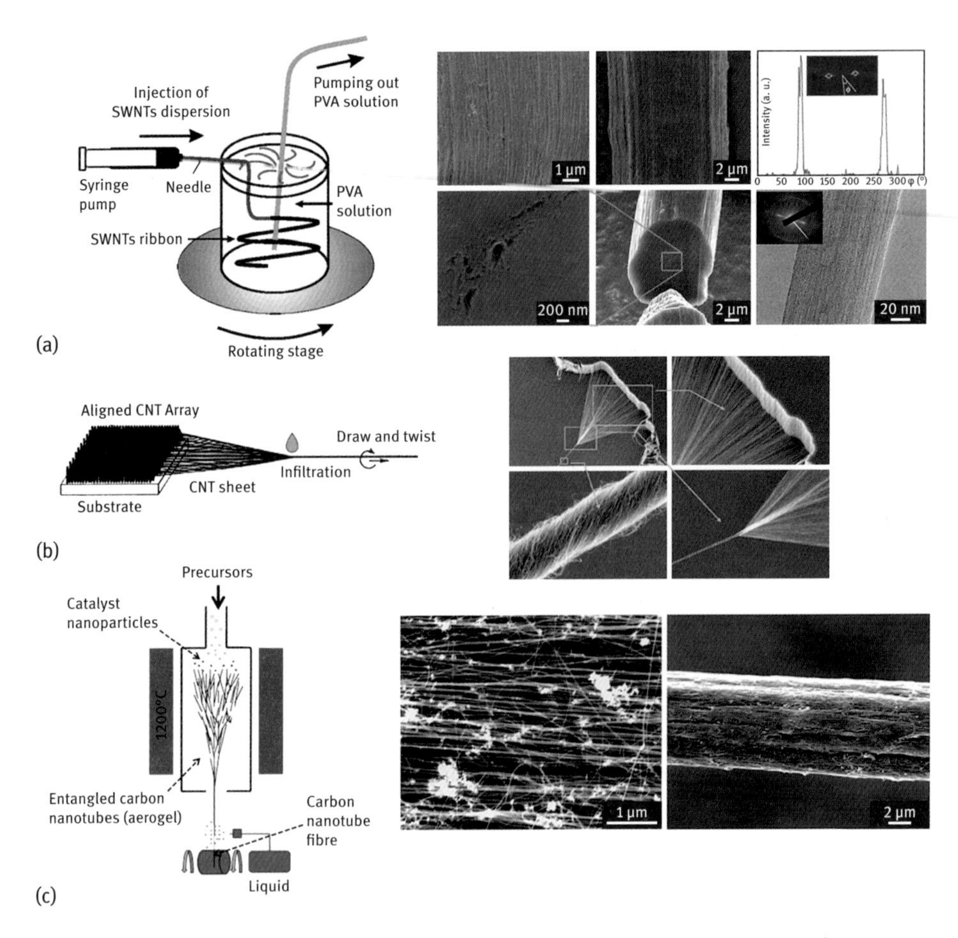

Fig. 8.9: Different methods for spinning CNT fibers and scanning electron micrographs of representative samples. (a) Wet spinning of nanocarbons dispersed in liquid, (b) drawing from a forest of aligned CNTs and (c) direct spinning from the gas phase during CNT synthesis by CVD. Images from references [53, 59, 60, 61, 62]. With kind permission from AAAS (2000, 2013), Elsevier (2007, 2011), Wiley (2010).

modynamically stable solutions, rather than dispersions, and can associate in liquid crystalline phases, that result in a high degree of alignment and packing in the final fibers [59]. Commercial high-performance synthetic fibers such as Zylon® and Kevlar®, made of rigid-rod polymers such as poly(*p*-phenylene-2,6-benzobisoxazole) (PBO) and poly(*p*-phenylene terephthalamide) PPTA are produced using a similar process. Figure 8.9 shows typical SEM images of CNT fibers produced using this method.

The second method, applicable only to CNTs, consists in drawing a fiber from an array of aligned CNTs by taking advantage of the van der Waals forces between neighboring CNTs in the array to withdraw them continuously from the substrate [63]. The drawing process forms a low density film that has to be twisted or densified with liquid to form a fiber. In spite of the efforts to identify the characteristics that make an array spinnable, the process still seems to rely on conditions specific to each laboratory. It is possible that by-products of the CVD CNT synthesis in the form of volatile heavy organic molecules could play a role in helping the spinning process, as some experimental observations would suggest [64]. Yet, this process is more widespread than the wet and direct spinning methods and consequently a vast part of the scientific development on CNT fibers has been done on this type of material.

The direct spinning method consists in drawing an aerogel of CNTs directly from the gas phase during CNT growth by CVD. The process is based on using a high reactor temperature (> 1000 °C), hydrogen as carrier gas, and often sulfur [65] to control carbon diffusion, which results in very fast growth rates (> 1 mm/s) of unusually long CNTs (around 1mm), which therefore entangle in the gas phase to form an aerogel that can be drawn out directly from the reactor as “elastic smoke”. The material spun directly from the reactor is a low density diaphanous web which can be densified on-line through capillary forces after contact with a liquid [66]. Unlike the other two processes, the direct spinning method combines all the production steps in one, from the synthesis of the building blocks to their assembly as a macroscopic fiber. On the one hand, the coupling of synthesis and spinning constrains the purity of the final material compared to powder CNTs, but on the other, it is likely to reduce productions costs compared to other fiber spinning methods. For example, work at Cambridge University has shown that it is possible to spin CNT fibers using natural gas as a carbon source;[1] thus, effectively turning one of the cheapest reactants available directly into a high-performance fiber.

8.4.2 Assembly properties and structure

Current axial properties of CNT and graphene (GO and rGO) fibers are presented in Table 8.3, corresponding to the highest value reported for each spinning method at gauge lengths > 5 mm and for non-treated samples; it also includes values for in-plane graphite (a graphene layer), aramid fiber, carbon fiber and copper. It is clear that with the exception of modulus, most properties are still a very small fraction of those of the ideal building block. Even if we were to assume that only the outer-most layer of the real building blocks contributes to the properties of the fibers, that would still leave significant room for improvement (theoretical limit ~ ideal building block/number of layers). Since most CNT fibers are already made up of long CNTs with few layers, forecoming improvements in mechanical properties are likely to come from engineering of the interface between nanocarbons (specific strength is proportional to length and shear strength[67]). Some recent experimental examples have shown increments in

1 Private communication from Dr F. Smail.

tensile properties after chemical functionalization [68] or irradiation [69]. Still, several aspects of the stress transfer between graphene layers remain unclear. Reported values for the shear strength of the graphene-graphene interface, for example, go from kPas to MPas (see [54] for a list of values), with some reports suggesting a negligible shear stress. Some of these discrepancies are probably due to the symmetry of graphite and the strong dependence of shear strength on crystallographic registry and sliding direction [70]. Ultimately, experimental and modeling work probably needs to be combined to determine the "effective" shear strength in a macroscopic assembly of closely packed nanocarbons. Similarly, improvement of electrical and thermal properties on the fiber scale will benefit from a better understanding of the interlayer charge and heat transfer processes, as well as from the growth of CNTs that specifically benefit these properties [71].

Tab. 8.3: Axial properties of CNT fibers at long gauge lengths, ideal graphene layer and other materials.

Fiber type (spinning method)	Specific strength (GPa/SG or N/tex[1])	Specific modulus (GPa/SG or N/tex)	Electrical conductivity (S/m)	Thermal conductivity (W/mK)
CNT fiber (wet spinning)	0.77 [59]	92 [59)	2.9×10^6 [59)	380 [59]
Graphene fiber (wet spinning)	0.18 [72]	8.7 [72]	2.5×10^4 [52]	–
CNT fiber (forest drawn)	1.9 [73][2]	195 [73]	9.7×10^4 [74]	45 [75]
CNT fiber (direct spinning)	1.5 [66]	87 [66]	8×10^5 [76]	40 [77]
Ideal graphene layer	62 [1]	476 [1]	1×10^7 [2] – 13×10^7 [3]	1000 [5] – 5000 [6]
Kevlar 49[3]	2.1	78	–	–
AS4 CF5[3]	2.5	128	5.9×10^4	6.8
Copper	0.02	14	5.9×10^7	400

[1] Tex is the fiber linear density expressed in g/km. N/tex, which is numerically equivalent to GPa/SG, is a unit widely used in the textile industry and very useful to characterize non-circular fibers.
[2] In GPa.
[3] From product data sheet.

In the comparison between CNT fibers and other materials in Table 8.3, the combination of properties of the former stands out. Copper, for example, has a higher electrical conductivity than the best CNT fiber by a factor of 20, but its density is 9 times higher and its tensile strength 5 times lower (in GPa); hence the interest in CNT fibers as lightweight electrical conductors.

Nanocarbon-based fibers also have a unique structure, different from that of traditional high-performance fibers (Fig. 8.9). Rather than being a solid block, they are comprised of multiple high aspect ratio elements that are often not perfectly packed. This fibrillar structure, together with the low shear strength between building blocks, gives these fibers exceptional flexibility in bending [62]. More importantly, the imperfect packing results in the internal volume of the fibers being accessible for infiltration of foreign molecules, such as gases, liquids and polymers, opening a whole range of possibilities for multiscale engineering and unlocking many potential applications (*e.g.* chemical sensing). This type of structure is not unique to nanocarbons, it derives from the assembly of nano building blocks into macroscopic materials; as larger volumes of other nano building blocks (inorganic nanowires, cellulose whiskers, *etc.*) become available in the future, porous macroscopic assemblies of them with this type of hierarchical structure will also become the subject of interest. The next section discusses the implications of this hierarchical structure for composites.

8.4.3 Assembly composites

When a porous assembly of nanocarbons, such as a CNT fiber, is embedded in a polymer matrix, the capillary forces arising from the porous structure draw polymer molecules into the structure. This polymer infiltration depends on several parameters, such as polymer molecular weight [9] and naturally, the pore structure. So far, there is evidence that epoxy infiltrates all CNT fibers studied so far, providing good adhesion to the matrix and therefore effective stress transfer to the load-bearing elements, often improving inter-tube stress transfer. The electron micrographs in Fig. 8.10 show examples of a CNT fiber cross-section in an epoxy composite completely coated with epoxy, and a fragmentation test to determine fiber/matrix IFSS. In the latter experiment, the values of IFSS obtained showed a strong dependence not only on the embedding matrix properties, but also on those of the choice of polymer infiltrated into the fiber [78].

The infiltration of polymer into the fiber implies that the fiber itself becomes a composite (Fig. 8.11) and that the identity of the reinforcing element is *a priori* unknown; it is a subunit of the fiber on a length-scale between the individual CNTs and the fiber itself. Thus, in fact, the properties of the fiber in the composite are not necessarily the same as those of the as-made fiber. Several reports suggest that the presence of the polymer improves the fiber mechanical properties [79], not just by filling up the empty space in it, but by directly increasing stress transfer to the CNTs, as observed by *in situ* Raman spectroscopy during fiber mechanical testing [80]. Yet, further understanding of the reinforcement mechanism in these systems is required, for instance, to clarify the compressive performance of CNT fibers in composites.

The structure of these macroassemblies of nano building blocks also raises interesting questions about optimum composite design. The fact that the porosity of these

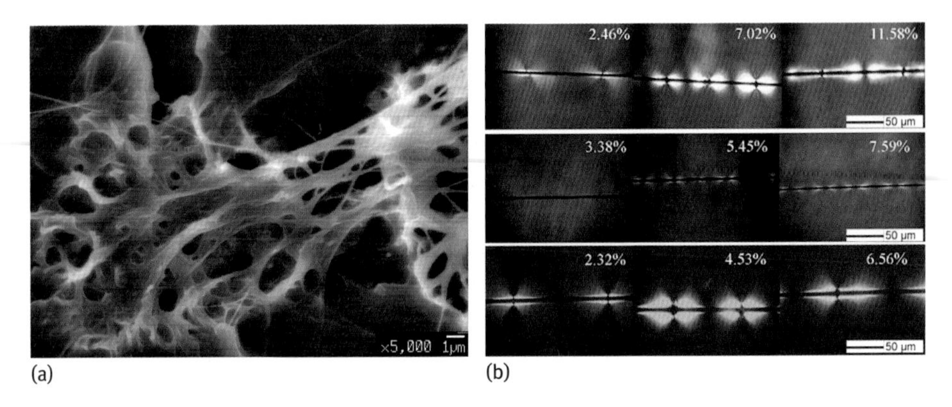

Fig. 8.10: The porous structure and high specific surface of CNT fibers enhances adhesion to polymer matrices: (a) shows the cross-section of fiber/epoxy fractured specimen, evidencing good wetting by the polymer [9]; (b) shows fragmentation tests on CNT fibers in epoxy, for fibers infiltrated with PVA (a) and PI (b) [78]. With kind permission from Elsevier (2009, 2011).

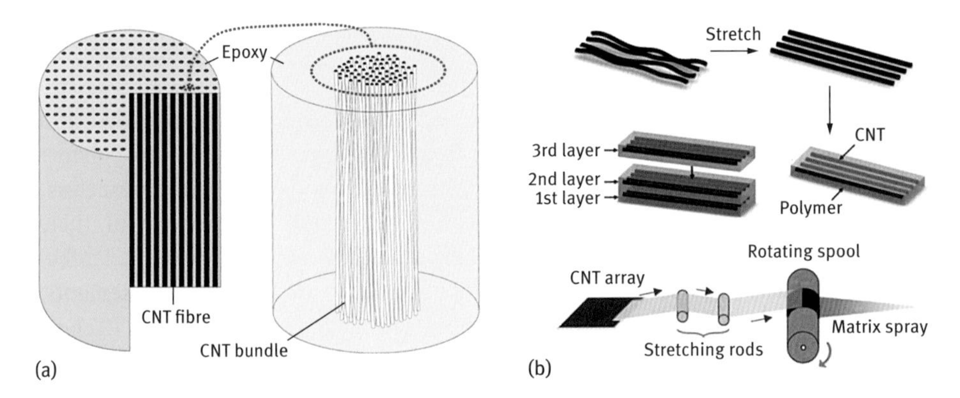

Fig. 8.11: (a) Schematic of the hierarchical structure of a composite with multiple CNT fibers, where each fiber is a composite itself [77]. (b) The intercalation of small amounts of polymer between CNTs during CNT fiber production is a promising method to fabricate structural composites based on nanocarbons [81]. With kind permission from Elsevier (2011) and Taylor and Francis (2012).

systems can be to some extent controlled, ranging from a low density network of entangled nanocarbons to a fully densified fiber, implies that it should also be possible to control the volume fraction of infiltrated polymer when one of these assemblies is integrated in a polymer matrix. Wang *et al.* have recently embedded films of CNTs spun from forest in bismaleimide (BMI), forming thin (5 μm) composites with over 45 vol% fraction of highly aligned MWNTs with an exceptional combination of longitudinal mechanical (3.9 GPa strength and 293 GPa modulus), electrical (1.2×10^5 S/m) and thermal (41 W/mK) properties [81].

It remains to be seen whether these properties can be achieved on thicker samples, but ultimately, these results show that a composite of 50 vol% CNT fiber in a hard polymer matrix can have better properties, at least mechanical ones, than a 100 vol% CNT fiber. The results suggest that the polymer in the fiber directly contributes to the stress transfer between fiber elements. Note also that the shear strength between nanocarbons is in the range of kPa, whereas the IFSS of a polymer/nanocarbon is of the order of MPas. Thus, it would appear that for mechanical properties, rather than a traditional composite geometry of multiple macroscopic fibers in a matrix, a more efficient architecture would correspond to highly aligned nanocarbons surrounded by a thin polymer layer. This architecture might not equally maximize electrical and thermal properties though.

These issues will be clarified as bigger quantities of CNT fiber become available in the near future as current scale-up developments (*e.g.* Nanocomp – Du Pont and Tortech) progress, making it possible to produce larger composites (at least in the mil-

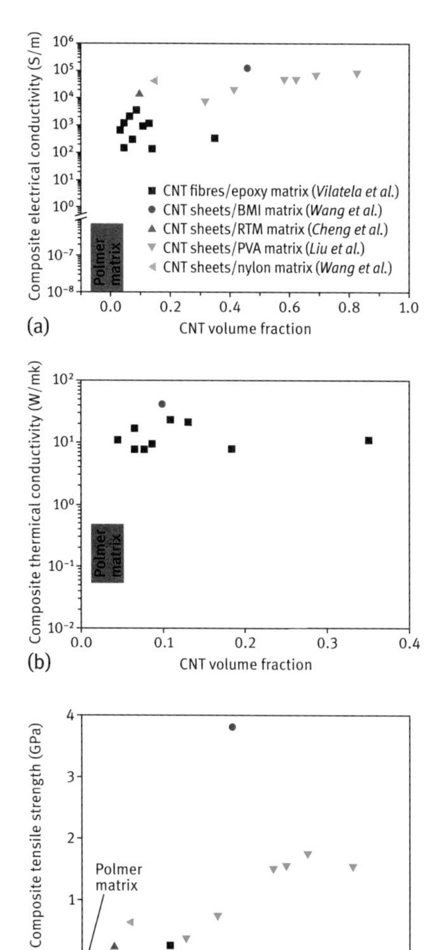

Fig. 8.12: Longitudinal properties of high volume fraction CNT fiber-based composites produced by various research groups and using different polymer matrices [81, 77, 79, 82]. For [79], volume fractions are calculated assuming specific gravities of 1.2 and 1 for the matrix and fiber, respectively.

limeter thickness range) with high volume fractions of CNT fibers. As a reference for future work, Fig. 8.12 presents longitudinal mechanical, electrical and thermal properties of composites with CNT fibers at different mass fractions.

8.4.4 Other properties of nanocarbon assemblies

In addition to their potential use as structural composites, these macroscopic assemblies of nanocarbons have shown promise as mechanical sensors [83], artificial muscles [84], capacitors [85], electrical wires [59], battery elements [85], dye-sensitized solar cells [86], transparent conductors [87], *etc.* What stands out is not only the wide range of properties of these type of materials but also the possibility of engineering them to produce such diverse structures, ranging from transparent films to woven fibers. This versatility derives from their hierarchical structure consisting of multiple nano building blocks that are assembled from bottom to top.

8.5 Summary

The extensive work of the last couple of decades using CNTs as fillers in polymer matrices shows the intrinsic difficulty in dispersing nanoscopic objects at high volume fractions. This clearly sets a limit on composite absolute properties, which in terms of mechanical properties is in the range of conventional composites with dispersed short glass or short CF fiber, but only a fraction of the properties of the CNT. The emerging data from graphene composites follows the same trend and confirms the fundamental difficulty in individualizing particles of nanometric size at volume fractions above ~ 1 vol% independent of their type. On the other hand, in the low volume fractions regime large increases in electrical conductivity are obtained as the electrical percolation is reached when a continuous network of nanocarbons is established in the matrix, typically at volume fractions below 0.1 vol% for thermosets, and which can serve as the basis for several new functions in the composite.

Turning a thermoset matrix into an electrical conductor implies that it can be cured through Joule heating, by simply passing current through it. This method is inherently more energy efficient than an oven-based process, but also, thanks to the small size (*i.e.* high specific surface) and even distribution of nanoheaters, much faster and more uniform. The method appears to have potential in the form of a resistively-curable adhesive for composite repair/joining. The nanocarbon network in a composite can also be used for structural health monitoring by direct measurement of resistance against mechanical deformation thanks to the network piezoresistivity. These applications clearly benefit from the shape and properties of nanocarbons and are thus more likely to compete with much cheaper traditional fillers (*e.g.* silica, carbon black, fiber glass).

More recently-developed hierarchical composites combining nanocarbons and macroscopic fibers aim at building from already established composite materials which have exceptional mechanical properties in the fiber direction, but much poorer properties orthogonal to the fibers. The presence of CNTs perpendicular to the fiber surface by direct CVD growth is an ideal arrangement to improve interlaminar properties. Recent results suggest that the bottleneck of achieving this growth without degradation of fiber properties has been overcome, therefore paving the way for first scale-up trials.

In addition to mechanical reinforcement, the presence of nanocarbons in these hierarchical composites can also be used for piezoresistive structural health monitoring or damage evaluation by thermal imaging. Other functions of the nanocarbon, for example in structural supercapacitors, are likely to emerge in the near future.

The assembly of nanocarbons into macroscopic fibers has resulted in materials with a combination of mechanical, electrical and thermal properties already superior to those commercially available, thus, beginning to make the promise of the potential of nanomaterials come true. This achievement comes mainly from the organized assembly of nano building blocks to benefit from their axial (in-plane for graphene) properties while also achieving high volume fractions through tight packing. An additional improvement of the properties of these materials and their composites is likely to come from their complex hierarchical structure and the possibilities it offers in terms of interfacial control and multiscale engineering. New exciting materials that combine these fibers with other nanomaterials such as metal oxides, also hold great promise in the future.

Acknowledgments

The author is grateful to R. Guzman-Villoria for useful discussions about hierarchical nanocarbon composites and to N. Krol for text editing and acknowledges financial support from MINECO (Spain) through its "Juan de la Cierva" Program and FP7-Marie Curie Action-CIG.

Bibliography

[1] Lee C, Wei X, Kysar JW, Hone J. Measurement of the elastic properties and intrinsic strength of monolayer graphene. *Science*. 2008 Jul 18;321(5887):385–8.

[2] Soule D. Magnetic field dependence of the hall effect and magnetoresistance in graphite single crystals. *Physical Review*. 1958;112(3):698–707.

[3] Pietronero L, Strassler S., Zeller H., Rice M., Electrical-conductivity of a graphite layer. *Physical Review B*. 1980;22(2):904–10.

[4] Dutta A. Electrical conductivity of single crystals of graphite. *Physical Review*. 1953;90(2):187–92.

[5] Hooker C, Ubbelohd AR, Young D. Anisotropy of thermal conductance in near-ideal graphite. *Proceedings of the Royal Society of London Series A. Mathematical and Physical Sciences*. 1965;284(1396):17 ff.

[6] Balandin AA, Ghosh S, Bao W, Calizo I, Teweldebrhan D, Miao F, et al. Superior thermal conductivity of single-layer graphene. *Nano Letters*. 2008 Mar;8(3):902–7.

[7] Logakis E, Pissis P, Pospiech D, Korwitz A, Krause B, Reuter U, et al. Low electrical percolation threshold in poly(ethylene terephthalate)/multi-walled carbon nanotube nanocomposites. *European Polymer Journal*. 2010 May;46(5):928–36.

[8] Garcia EJ, Wardle BL, Hart AJ. Joining prepreg composite interfaces with aligned carbon nanotubes. *Composites Part A: Applied Science and Manufacturing*. 2008;39(6):1065–70.

[9] Mora RJ, Vilatela JJ, Windle AH. Properties of composites of carbon nanotube fibers. *Composites Science and Technology*. 2009 Aug;69(10):1558–63.

[10] McNally T, Pötschke P. *Polymer-carbon nanotube composites: Preparation, properties and applications*. 1st ed. Cambridge: Woodhead Publishing Limited; 2011.

[11] Coleman JN, Khan U, Gun'ko YK. Mechanical reinforcement of polymers using carbon nanotubes. *Adv Mater*. 2006 Mar 17;18(6):689–706.

[12] Sandler JKW, Kirk JE, Kinloch IA, Shaffer MSP, Windle AH. Ultra-low electrical percolation threshold in carbon-nanotube-epoxy composites. *Polymer*. 2003 Sep;44(19):5893–9.

[13] Coleman JN, Cadek M, Blake R, Nicolosi V, Ryan KP, Belton C, et al. High-performance nanotube-reinforced plastics: Understanding the mechanism of strength increase. *Advanced Functional Materials*. 2004 Aug;14(8):791–8.

[14] Du FM, Guthy C, Kashiwagi T, Fischer JE, Winey KI. An infiltration method for preparing single-wall nanotube/epoxy composites with improved thermal conductivity. *Journal of Polymer Science Part B: Polymer Physics*. 2006 May 15;44(10):1513–9.

[15] Cox H. The elasticity and strength of paper and other fibrous materials. *British Journal of Applied Physics*. 1952;3(MAR):72–9.

[16] Fu SY, Lauke B, Mader E, Yue CY, Hu X. Tensile properties of short-glass-fiber- and short-carbon-fiber-reinforced polypropylene composites. *Composites Part A: Applied Science and Manufacturing*. 2000;31(10):1117–25.

[17] Thomason JL, Vlug MA. Influence of fiber length and concentration on the properties of glass fiber-reinforced polypropylene: 1. Tensile and flexural modulus. *Composites Part A: Applied Science and Manufacturing*. 1996;27(6):477–84.

[18] Shaffer M, Kinloch IA. Prospects for nanotube and nanofiber composites. *Composites Science and Technology*. 2004 Nov;64(15):2281–2.

[19] Schaefer DW, Justice RS. How nano are nanocomposites? *Macromolecules*. 2007 Nov 27;40(24):8501–17.

[20] Gong L, Kinloch IA, Young RJ, Riaz I, Jalil R, Novoselov KS. Interfacial stress transfer in a graphene monolayer nanocomposite. *Adv Mater*. 2010 Jun 25;22(24):2694–7.

[21] Soule DE, Nezbeda CW. Direct basal-plane shear in single-crystal graphite. *Journal of Applied Physics*. 1968;39(11):5122–39.

[22] Balberg I, Anderson C, Alexander S, Wagner N. Excluded volume and its relation to the onset of percolation. *Physical Review B*. 1984;30(7):3933–43.

[23] Celzard A, McRae E, Deleuze C, Dufort M, Furdin G, Mareche JF. Critical concentration in percolating systems containing a high-aspect-ratio filler. *Physical Review B*. 1996 Mar 1;53(10):6209–14.

[24] Bauhofer W, Kovacs JZ. A review and analysis of electrical percolation in carbon nanotube polymer composites. *Composites Science and Technology*. 2009 Aug;69(10):1486–98.

[25] Martin CA, Sandler JKW, Shaffer MSP, Schwarz M-K, Bauhofer W, Schulte K, et al. Formation of percolating networks in multi-wall carbon-nanotube–epoxy composites. *Composites Science and Technology*. 2004 Nov;64(15):2309–16.
[26] Vilatela JJ, Eder D. Nanocarbon composites and hybrids in sustainability: A review. *ChemSusChem*. 2012 Mar 12;5(3):456–78.
[27] Pang H, Chen T, Zhang G, Zeng B, Li Z-M. An electrically conducting polymer/graphene composite with a very low percolation threshold. *Materials Letters*. 2010 Oct 31;64(20):2226–9.
[28] Liang J, Wang Y, Huang Y, Ma Y, Liu Z, Cai F, et al. Electromagnetic interference shielding of graphene/epoxy composites. *Carbon*. 2009 Mar;47(3):922–5.
[29] Wang W, Liu Y, Li X, You Y. Synthesis and characteristics of poly(methyl methacrylate)/expanded graphite nanocomposites. *J Appl Polym Sci*. 2006 Apr 15;100(2):1427–31.
[30] Wichmann MHG, Buschhorn ST, Gehrmann J, Schulte K. Piezoresistive response of epoxy composites with carbon nanoparticles under tensile load. *Physical Review B*. 2009 Dec;80(24).
[31] Zhang R, Deng H, Valenca R, Jin J, Fu Q, Bilotti E, et al. Strain sensing behaviour of elastomeric composite films containing carbon nanotubes under cyclic loading. *Composites Science and Technology*. 2013 Jan 24;74(0):1–5.
[32] Flandin L, Hiltner A, Baer E. Interrelationships between electrical and mechanical properties of a carbon black-filled ethylene–octene elastomer. *Polymer*. 2001 Jan;42(2):827–38.
[33] Soesatyo B, Blicblau AS, Siores E. Effects of microwave curing carbon doped epoxy adhesive-polycarbonate joints. *International Journal of Adhesion and Adhesives*. 2000;20(6):489–95.
[34] Rangari VK, Bhuyan MS, Jeelani S. Microwave processing and characterization of EPON 862/CNT nanocomposites. *Materials Science and Engineering: B*. 2010 Apr 15;168(1–3):117–21.
[35] Wang C, Chen T, Chang S, Cheng S, Chin T. Strong carbon-nanotube-polymer bonding by microwave irradiation. *Advanced Functional Materials*. 2007 Aug 13;17(12):1979–83.
[36] Mas B, Fernández-Blázquez JP, Duval J, Bunyan H, Vilatela JJ. Thermoset curing through Joule heating of nanocarbons for composite manufacture, repair and soldering. *Carbon*. 2013 Nov;63(0):523–9.
[37] Qian H, Greenhalgh ES, Shaffer MSP, Bismarck A. Carbon nanotube-based hierarchical composites: a review. *J Mater Chem*. 2010;20(23):4751–62.
[38] Qian H, Bismarck A, Greenhalgh ES, Shaffer MSP. Carbon nanotube grafted silica fibers: Characterising the interface at the single fiber level. *Composites Science and Technology*. 2010 Feb;70(2):393–9.
[39] Yavari F, Rafiee MA, Rafiee J, Yu Z-Z, Koratkar N. Dramatic increase in fatigue life in hierarchical graphene composites. *Acs Applied Materials & Interfaces*. 2010 Oct;2(10):2738–43.
[40] Qiu J, Zhang C, Wang B, Liang R. Carbon nanotube integrated multifunctional multiscale composites. *Nanotechnology*. 2007 Jul 11;18(27).
[41] Veedu VP, Cao A, Li X, Ma K, Soldano C, Kar S, et al. Multifunctional composites using reinforced laminae with carbon-nanotube forests. *Nat Mater*. 2006 Jun;5(6):457–62.
[42] Guzmán de Villoria R, Hart AJ, Wardle BL. Continuous high-yield production of vertically aligned carbon nanotubes on 2D and 3D substrates. *ACS Nano*. 2011 May 17;5(6):4850–7.
[43] Tolbin AY, Nashchokin AV, Kepman AV, Dunaev AV, Malakho AP, Morozov VA, et al. Influence of conditions of catalytic growth of carbon nanostructures on mechanical properties of modified carbon fibers. *Fiber Chemistry*. 2012 Jul;44(2):95–100.
[44] Delmas M, Pinault M, Patel S, Porterat D, Reynaud C, Mayne-L'Hermite M. Growth of long and aligned multi-walled carbon nanotubes on carbon and metal substrates. *Nanotechnology*. 2012 Mar;23(10):105604–105604.

[45] Qian H, Greenhalgh ES, Shaffer MSP, Bismarck A. Carbon nanotube-based hierarchical composites: a review. *Journal of Materials Chemistry*. 2010;20(23):4751–62.

[46] Bekyarova E, Thostenson ET, Yu A, Kim H, Gao J, Tang J, et al. Multiscale carbon nanotube–carbon fiber reinforcement for advanced epoxy composites. *Langmuir*. 2007 Feb 28;23(7):3970–4.

[47] De Villoria RG, Yamamoto N, Miravete A, Wardle BL. Multi-physics damage sensing in nano-engineered structural composites. *Nanotechnology*. 2011 May 6;22(18).

[48] Gao L, Thostenson ET, Zhang Z, Byun J-H, Chou T-W. Damage monitoring in fiber-reinforced composites under fatigue loading using carbon nanotube networks. *Philosophical Magazine*. 2010 Apr 28;90(31–32):4085–99.

[49] Shirshova N, Qian H, Shaffer MSP, Steinke JHG, Greenhalgh ES, Curtis PT, et al. Structural composite supercapacitors. *Composites Part A: Applied Science and Manufacturing*. 2013 Mar;46(0):96–107.

[50] Carothers WH, Hill JW. Studies of polymerization and ring formation. xv. Artificial fibers from synthetic linear condensation superpolymers. *J Am Chem Soc*. 1932 Apr 1;54(4):1579–87.

[51] Die hochmolekularen organischen verbindungen: Kautschuk und Cellulose. By Prof. H.Staudinger. Pp. xv+540. Berlin: j.Springer, 1932. Paper, 49.60 rm.; bound, 52 rm. *J Chem Technol Biotechnol*. 1933 May 5;52(18):386–386.

[52] Xu Z, Gao C. Graphene chiral liquid crystals and macroscopic assembled fibers. *Nat Commun*. 2011 Dec 6;2:571.

[53] Vigolo B, Pénicaud A, Coulon C, Sauder C, Pailler R, Journet C, et al. Macroscopic fibers and ribbons of oriented carbon nanotubes. *Science*. 2000 Nov 17;290(5495):1331–4.

[54] Vilatela JJ, Elliott JA, Windle AH. A model for the strength of yarn-like carbon nanotube fibers. *Acs Nano*. 2011 Mar;5(3):1921–7.

[55] Behabtu N, Green MJ, Pasquali M. Carbon nanotube-based neat fibers. *Nano Today*. 2008 Oct;3(5–6):24–34.

[56] Ruoff RS, Tersoff J, Lorents DC, Subramoney S, Chan B. Radial deformation of carbon nanotubes by van der Waals forces. *Nature*. 1993 Aug 5;364(6437):514–6.

[57] Chopra NG, Benedict LX, Crespi VH, Cohen ML, Louie SG, Zettl A. Fully collapsed carbon nanotubes. *Nature*. 1995 Sep 14;377(6545):135–8.

[58] Motta M, Moisala A, Kinloch IA, Windle AH. High performance fibers from "dog bone" carbon nanotubes. *Adv Mater*. 2007 Nov 5;19(21):3721–6.

[59] Behabtu N, Young CC, Tsentalovich DE, Kleinerman O, Wang X, Ma AWK, et al. Strong, light, multifunctional fibers of carbon nanotubes with ultrahigh conductivity. *Science*. 2013 Jan 11;339(6116):182–6.

[60] Ghemes A, Minami Y, Muramatsu J, Okada M, Mimura H, Inoue Y. Fabrication and mechanical properties of carbon nanotube yarns spun from ultra-long multi-walled carbon nanotube arrays. *Carbon*. 2012 Oct;50(12):4579–87.

[61] Atkinson KR, Hawkins SC, Huynh C, Skourtis C, Dai J, Zhang M, et al. Multifunctional carbon nanotube yarns and transparent sheets: Fabrication, properties, and applications. *Physica B: Condensed Matter*. 2007 May 15;394(2):339–43.

[62] Vilatela JJ, Windle AH. Yarn-like carbon nanotube fibers. *Advanced Materials*. 2010 Nov 24;22(44):4959–4963.

[63] Jiang K, Li Q, Fan S. Nanotechnology: Spinning continuous carbon nanotube yarns. *Nature*. 2002 Oct 24;419(6909):801–801.

[64] Naraghi M, Filleter T, Moravsky A, Locascio M, Loutfy RO, Espinosa HD. A multiscale study of high performance double-walled nanotube–polymer fibers. *ACS Nano*. 2010 Oct 26;4(11):6463–76.

[65] Motta MS, Moisala A, Kinloch IA, Windle AH. The role of sulphur in the synthesis of carbon nanotubes by chemical vapour deposition at high temperatures. *Journal of Nanoscience and Nanotechnology*. 2008 May;8(5):2442–9.
[66] Koziol K, Vilatela J, Moisala A, Motta M, Cunniff P, Sennett M, et al. High-performance carbon nanotube fiber. *Science*. 2007 Dec 21;318(5858):1892–5.
[67] Vilatela JJ, Elliott JA, Windle AH. A model for the strength of yarn-like carbon nanotube fibers. *ACS Nano*. 2011 Feb 24;5(3):1921–7.
[68] Boncel S, Sundaram RM, Windle AH, Koziol KKK. Enhancement of the mechanical properties of directly spun CNT fibers by chemical treatment. *ACS Nano*. 2011 Nov 21;5(12):9339–44.
[69] Miao M, Hawkins SC, Cai JY, Gengenbach TR, Knott R, Huynh CP. Effect of gamma-irradiation on the mechanical properties of carbon nanotube yarns. *Carbon*. 2011 Nov;49(14):4940–7.
[70] Shibuta Y, Elliott JA. Interaction between two graphene sheets with a turbostratic orientational relationship. *Chemical Physics Letters*. 2011 Aug 25;512 (4–6):146–50.
[71] Sundaram RM, Koziol KKK, Windle AH. Continuous direct spinning of fibers of single-walled carbon nanotubes with metallic chirality. *Adv Mater*. 2011 Nov 16;23(43):5064–8.
[72] Cong H-P, Ren X-C, Wang P, Yu S-H. Wet-spinning assembly of continuous, neat, and macroscopic graphene fibers. Sci Rep [Internet]. 2012 Aug 30;2. Available from: http://dx.doi.org/10.1038/srep00613
[73] Zhang X, Li Q, Holesinger TG, Arendt PN, Huang J, Kirven PD, et al. Ultrastrong, Stiff, and Lightweight Carbon-Nanotube Fibers. Adv Mater. 2007 Dec 3;19(23):4198–201.
[74] Li QW, Li Y, Zhang XF, Chikkannanavar SB, Zhao YH, Dangelewicz AM, et al. Structure-dependent electrical properties of carbon nanotube fibers. *Adv Mater*. 2007 Oct 19;19(20):3358–63.
[75] Aliev AE, Guthy C, Zhang M, Fang S, Zakhidov AA, Fischer JE, et al. Thermal transport in MWCNT sheets and yarns. *Carbon*. 2007 Dec;45(15):2880–8.
[76] Stano KL, Koziol K, Pick M, Motta MS, Moisala A, Vilatela JJ, et al. Direct spinning of carbon nanotube fibers from liquid feedstock. *International Journal of Material Forming*. 2008 Jul;1(2):59–62.
[77] Vilatela JJ, Khare R, Windle AH. The hierarchical structure and properties of multifunctional carbon nanotube fiber composites. *Carbon*. 2012 Mar;50(3):1227–34.
[78] Liu Y-N, Li M, Gu Y, Zhang X, Zhao J, Li Q, et al. The interfacial strength and fracture characteristics of ethanol and polymer modified carbon nanotube fibers in their epoxy composites. *Carbon*. 2013 Feb;52(0):550–8.
[79] Liu W, Zhang X, Xu G, Bradford PD, Wang X, Zhao H, et al. Producing superior composites by winding carbon nanotubes onto a mandrel under a poly(vinyl alcohol) spray. *Carbon*. 2011 Nov;49(14):4786–91.
[80] Vilatela JJ, Deng L, Kinloch IA, Young RJ, Windle AH. Structure of and stress transfer in fibers spun from carbon nanotubes produced by chemical vapour deposition. *Carbon*. 2011 Nov;49(13):4149–58.
[81] Wang X, Yong ZZ, Li QW, Bradford PD, Liu W, Tucker DS, et al. Ultrastrong, Stiff and Multifunctional Carbon Nanotube Composites. *Materials Research Letters*. 2012 Oct 11;1(1):19–25.
[82] Cheng QF, Wang JP, Wen JJ, Liu CH, Jiang KL, Li QQ, et al. Carbon nanotube/epoxy composites fabricated by resin transfer molding. Carbon. 2010 Jan;48(1):260–6.
[83] Wu AS, Chou T-W, Gillespie JW, Lashmore D, Rioux J. Electromechanical response and failure behaviour of aerogel-spun carbon nanotube fibers under tensile loading. *J Mater Chem*. 2012;22(14):6792–8.
[84] Aliev AE, Oh J, Kozlov ME, Kuznetsov AA, Fang S, Fonseca AF, et al. Giant-stroke, superelastic carbon nanotube aerogel muscles. *Science*. 2009 Mar 20;323(5921):1575–8.

[85] Ren J, Li L, Chen C, Chen X, Cai Z, Qiu L, et al. Twisting carbon nanotube fibers for both wire-shaped micro-supercapacitor and micro-battery. *Adv Mater*. 2013 Feb 25;25(8):1155–9.
[86] Chen T, Qiu L, Cai Z, Gong F, Yang Z, Wang Z, et al. Intertwined aligned carbon nanotube fiber based dye-sensitized solar cells. *Nano Lett*. 2012 Apr 13;12(5):2568–72.
[87] Fraser IS, Motta MS, Schmidt RK, Windle AH. Continuous production of flexible carbon nanotube-based transparent conductive films. *Science and Technology of Advanced Materials*. 2010 Aug;11(4).

Robert Schlögl

9 Carbon-Carbon Composites

Carbon-Carbon Composites (C3) or in modern language "Carbon" form a family of compounded materials. A network formed from carbon fibers of various origins is bonded with pyrolytic carbon generated from a chemical reaction of a carbon-containing molecular precursor. The resulting C3 is a lightweight strong material also at high temperatures with chemical stability, high elasticity and conducting properties for heat and electrons. The material was developed for military and space applications and has recently found widespread civilian high-tech applications such as in brake systems and in electric vehicles. Its share in performance-structural materials is expected to grow substantially. Drawbacks are the very limited machinability, the difficulty in repairing C3 parts and the high specific energy input in the manufacturing process of C3, determining to a large extent its cost.

9.1 Introduction

Carbon in its sp^2 electronic hybridization [1] exhibits many desirable material properties [2] such as strength and stability combined with excellent transport properties. Its disadvantage is the large anisotropy: only within the bonding plane of the C=C bonds are these properties well developed. All the good material properties do not hold in the perpendicular direction where dispersive forces create only a weak binding in the third dimension. This leads to the well-known layer structure of graphitic carbon with its almost two-dimensional properties and, amongst other properties, to its lubricating function and to the abrasive properties used in pencils.

The introduction of defects into the network of hexagonal rings in graphene allows for a non-flat mesostructure occurring often as ribbons that can interlace and so form hard structures found in glassy carbon [3] or in pyrocarbon [4, 5] materials. This material is, however, not elastic, exhibits thus low tensile strength and can hardly be used for replacing steel structures. It is, however, of excellent thermal and also chemical stability. The interlaced ribbon structure is illustrated in medium-resolution TEM as "amorphous" with a cloudy contrast and exhibits many terminal carbon atoms that carry functional groups. If these groups are hydrogen or resonantly bonded oxygen atoms (quinones or heterocyclic) then a high chemical stability results, exceeding that [6, 7] of better ordered sp^2 carbon structures [8] such as nanotubes [9, 10].

The C3 family of materials [11–13] exhibits this chemical stability due to a highly functionalized fraction of sp^2 carbon [14–17], but in addition contains a carbon fiber backbone in its second bulk component. Carbon fibers [18–21] are the ordered variant of interlaced ribbons: in fibers the anisotropic sp^2 basic structural units are oriented in one direction [20] by various mechanisms during synthesis. The result is a high strength lightweight material [22] with mechanical properties exceeding those of steel fibers. Due to the anisotropy of the sp^2 bonding, these properties disappear in all sheer directions to the fiber axis. Carbon fibers or woven two-dimensional objects like felts are thus only mechanically stable along the fiber axis. Filling the voids between the carbon fibers with a strong material reinforces the fiber structure in all shear direction and gives isotropic mechanical stability to the resulting object.

For applications where only mechanical properties are relevant, it is often sufficient to use resins for the filling and we end up with carbon-reinforced polymer structures. Such materials [23] can be soft, like the family of poly-butadiene materials leading to rubber or tires. The transport properties of the carbon fibers lead to some limited improvement of the transport properties of the polymer. If carbon nanotubes with their extensive propensity of percolation are used [24], then a compromise between mechanical reinforcement and improvement of electrical and thermal stability is possible provided one solves the severe challenge of homogeneous mixing of binder and filler phases. For the macroscopic carbon fibers this is less of a problem, in particular when advanced techniques of vacuum infiltration of the fluid resin precursor and suitable chemical functionalization of the carbon fiber are applied.

Such polymer composites (that will not be treated in this chapter) can be used as precursors to the C3 materials where the polymer is converted into a carbon phase with a low content of heteroatoms. A well-developed sp^2 structure is desired, with its basic structural units being oriented perpendicular to the fiber axis. The required excellent mechanical and transport properties in the weak direction of the initial fiber can thus be delivered. This material is now called “carbon” and finds widespread application in energy-related structural material applications such as electric passenger cars, as construction material for airplanes and as the core structure of turbine blades for windmills and compression turbines.

9.2 Typology of C3 materials

C3 compounds discriminate themselves from other forms of high-performance carbon by the distinct presence of a fibrous filler phase and a binder phase. There are clear interfaces between the phases and their molecular properties [25–27] determine the macroscopic mechanical and transport properties. Whereas it is of great value to generate a structure that is homogeneous on the mesoscale in isotropic performance carbons, the desire and purpose of structuring in C3 compounds is to establish a clear interface structure, establishing the ability to tune their properties. The interfaces with perpendicular directions of the strong mechanical properties act as stoppers for microcracks due to their structural anisotropy. They further act as elastic spring elements due to their chemical binding structure: either stiff carbon-carbon bonds or more ductile carbon heteroatom-carbon bonds with finite bond angles can be introduced. A prominent example of such designed internal interfaces is the use of C3 materials in airplane parts. After some service time, cracks in large parts such as wing elements can be seen by the naked eye. They do not, however, deteriorate the mechanical overall properties due to the intermittent stopping of crack propagation by the anisotropy of the chemical binding in the fibers and in the binder phase.

The advantage that both filler and binder are of the same basic structural units, namely graphene flakes from sp^2 carbon, enables the matching of mechanical properties of binder and filler and removes interfacial instability, such as chemical corrosion or unmatched thermal expansion coefficients. The advantage of having the same chemical constitution in the basic structural units of two phases, but ordering them in different mesoscopic organization (filler with 1-dimensional fibers and binder with 2-dimensional flake structure), was recognized early [2]: at the time when the generation of homogeneous high performance carbon was being optimized.

In Fig. 9.1 a schematic diagram is shown of mesoscopic arrangements of binder and filler. The two sketches illustrate two modes of creating a mechanically effective composite. An isotropic carbon binder can be generated into which the fiber bundles are embedded. Alternatively, an attempt can be made [28] to covalently grow graphene basic structural units (BSU) onto the fibers in directions roughly perpendicular to the fiber axis and then fill the void space as effectively as possible with an isotropic carbon binder. This can be achieved by controlling the growth kinetics of the binder or by a sequential binder synthesis, possibly with different molecular precursors for the anisotropic BSU and the isotropic phase.

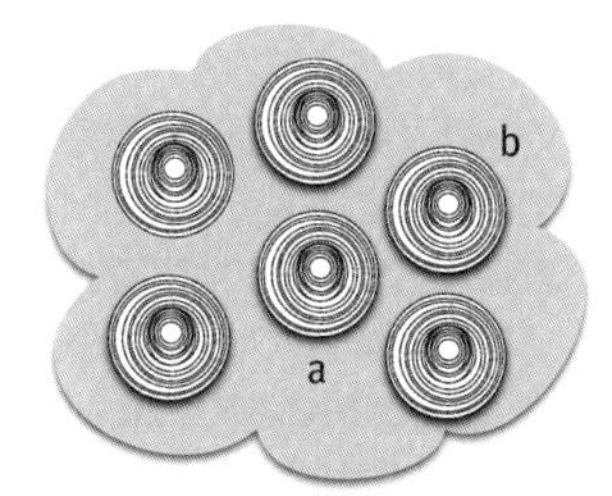

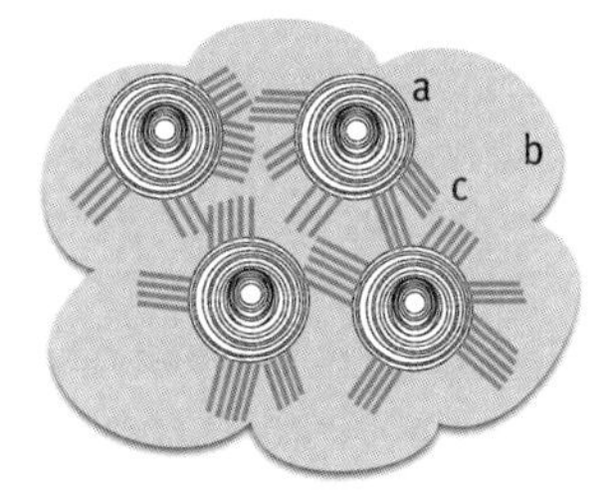

Fig. 9.1: Top view on two variants of C3 materials. The carbon fibers (a) themselves exhibit a complex inner microstructure that needs carful optimization for strength and stability. The isotropic filler phase (b) should be free of pores and other weak points caused by uneven distribution in the composite body. The ordered graphitic BSU (c) can provide a very strong but still flexible anchoring of the fibers in the isotropic matrix.

It is a chemical challenge to initiate the growth of BSU such that they all adhere to the carbon fiber filler. If unsuccessful, soot particles may result. These may form a granular non-interacting second phase which strongly counteracts the intended effect of a C3 material. The structural anisotropy of the carbon adhering to the fiber bundles has been evidenced convincingly by microscopy techniques [28] and serves as the design criterion for mechanistic considerations controlling the process [28–32] of carbon infiltration by gaseous reactants.

The resulting strong bonding between filler and binder requires careful adaptation of thermal expansion coefficients. If they are not very similar, the developing strain will easily crack the filler fibers [22] and thus reduce the mechanical properties during cycles of service temperature. The chemical background of adapting the thermal expansion coefficients lies in the molecular ordering of the basic structural units and also in the residual heteroatomic content from the carbonization of the polymer precursor. Careful thermal treatment and continuous mechanical strain during the carbonization are practical measures of control.

The result of such a combination of carbon phases in different relative arrangement can be seen in typical macroscopic properties reported in Table 9.1 in comparison to a ferritic steel structure of similar dimensions. The values are approximate and vary from product to product and of course depend strongly on the choices of starting materials and the perfection of synthesis.

Tab. 9.1: Characteristic selected material properties of C3. After ref. [12].

Ferritic steel	*Property*	*Carbon-carbon composite (C3)*
200	Young modulus [GPa]	300
300	Compressive strength [MPa]	100
700	Tensile strength [MPa]	900
2×10^{-5}	Thermal expansion [K^{-1}]	2×10^{-6}
20	Thermal conductivity [$Wm^{-1}K^{-1}$]	100

The reported data have been compiled from the literature [12]. The values were approximated to typical dimensions that can be significantly exceeded by high-end component materials and optimized treatments at the expense of enormously rising process cost. Such improved parts of C3 materials are employed in military and space applications. Only scant information can be found in the open literature about their synthesis and properties. The reader may be advised that the property profile can be strongly improved if cost is not a consideration.

The thermal properties of C3 materials at high temperatures are most remarkable if protected from oxidation. This issue is discussed below in more detail. If they are not oxidized, the C3 materials exhibit similar stability data as ceramics [22], in particular at temperatures above 1500 K where protective coatings applied behave like a plastic and close developing surface cracks against air attack. C3 materials expose the advantages of their hierarchical structure being present in both filler and binder phase and develop wood-like properties under ambient conditions. A descriptive parameter is the work of fracture. Ductile metals exhibit 10^5 J/m^2, wood and C3 materials 10^4 J/m^2 and ceramics about 10^2 J/m^2. Whereas most materials lose these properties rapidly with increasing temperature, the C3 materials withstand temperature loads to well above 1500 K. They are dimensionally stable like SiC ceramics without the disadvantages of brittleness and poor thermal conductivity. For these reasons, characteristic high-tech applications of C3 materials are tiles for re-entry space vehicles, high-performance brakes and critical heat exchangers in nuclear installations.

To achieve such properties and to maintain them under practical service conditions it is essential that all phases are synthesized to high qualitative homogeneity and that the structural arrangement shown in Fig. 9.1 prevails homogeneously throughout the macroscopic body of the part. Synthesis is thus a critical step since not only chemical precision but also macroscopic homogeneity are critical properties requiring carefully controlled and automated unit operations.

9.3 Synthesis

The synthesis of C3 components has to control multiple dimensions. At the molecular level, the sp^2 configuration, the amount of defects in the graphene BSU of fibers and binder and the amount, location and type of heteroatoms are relevant [10, 33, 34]. At the mesoscopic level [22], the homogeneous density of the composite and the intended relative orientation of the interfaces between binder and filler need carful control. Finally, at the macroscopic level [2, 35], the shape and dimensions of the final part need utmost attention as the C3 materials can hardly be machined, interconnected or repaired without loss of a good deal of their properties. In particular, cutting and drilling destroy the local microstructure and delaminate the binder from the filler phase. In this way rough surfaces will form with no mechanical stability forming excellent locations for chemical attack.

Joining of C3 compounds was successfully demonstrated [11] using a "cement" of Ti carbide basis and hot pressing as method. Such operation can be done with certain geometries and at enormous cost and is only possible in high-tech applications such as in nuclear reactors. For less-demanding applications, gluing with polymer materials is an option.

Modern C3 materials for automotive applications, such as components of the car body, are synthesized according to the flow scheme of Fig. 9.2. Here an integrated synthesis of both filler and binder components is taken as a cost-effective approach. In high-tech applications it is more customary to independently optimize the preparation of the fiber component [15, 19, 20, 36, 37] and then the C3 synthesis in separate processes with extensive quality control measures in-between.

It can be seen from Fig. 9.2 that several high-energy treatments are necessary during the synthesis. These treatments require high quality heating with electrical furnaces and so the energy bill largely determines the cost of the component. From a

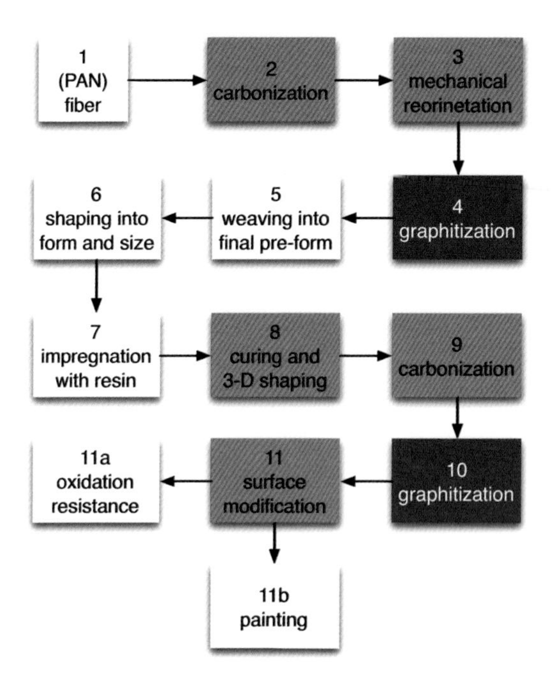

Fig. 9.2: Synthesis flowchart of modern “carbon” for essentially flat C3 compounds. Some of the steps (2,3: 8,9) may be connected into a continuous process. The energy intensity of the steps is indicated by the coloring scheme.

perspective of life cycle analysis, it is clear that it would be highly desirable to recycle such C3 components rather than to thermally remove them after service life. As it is difficult to reshape the material and its decomposition is impossible, one option is to cut them in smaller pieces and fabricate new C3 components with other shapes and lower mechanical requirements. Lower grade allocations could be reinforcement applications in polymer materials.

The production process indicated in Fig. 9.2 divides itself into 3 main processes. Steps 1–4 deal with the generation of the carbon fiber from a polymeric precursor. Here the main challenge is the generation of a high-strength material with the correct surface termination relating to the intended methods of binder attachment. The surface analysis of such systems [15, 23, 38, 39] is a challenge despite extensive studies and the application of a whole suite of techniques. For details of the surface modifications chemistry see other contributions to this book. A typical example [36] is the study of the effect of fiber activation by various treatment conditions and its extensive analytical differentiation using aramide precursor fibers. In addition to the conventional surface analysis of carbon fibers [18, 40] that can carry quite some pitfalls on the methodical level [41], the more reliable chemical derivatization techniques [42, 43] have been optimized by using fluorescence labeling techniques [44] originally developed in the life sciences.

Multiple forms of carbon fibers can be used. It has been found [38] that either fibers made from polymer precursors containing stoichiometric amounts of nitrogen such as polyacrlyonitrile (PAN) or fibers [19] from pure carbon with excellent ordering of the graphene BSU are useful. These starting materials mark the extremes of "mass market" and "high-tech" applications. The first steps in preparing the fibers from a PAN precursor have been studied for a long time [15, 20], in particular with respect to the retention of chemical functionality from its precursor and to the creation of a preferred orientation. The main result is that a careful slow and multi-step heat treatment is prerequisite for a stable and homogeneous result. The retention of mechanical strain during carbonization greatly supports the homogeneous orientation of the growing BSU within the fiber.

The second set of synthesis steps shown in Fig. 9.2 concerns the preparation of the green body for the C3 formation. Steps 5–8 are related to generating a flat or curved sheet from the carbon fiber and the precursor of the binder phase. This process can be complicated because the final part made from C3 cannot be changed in its form nor can it be interconnected to another C3 part by reasonably affordable techniques suitable for mass production. The quality of the final product is here decided by the homogeneity of the material distribution and the exact shaping.

A family of techniques has been developed for the formation and creation of the binder phase. In Fig. 9.2 only the most common method of impregnation and curing is shown with steps 7–10. There are alternative methods that partly deliver much higher quality products but they are far more complex and energy-intensive in their application. Figure 9.3 summarizes the main methodologies.

We recognize in Fig. 9.3 two families of methods. In the one designated CVI (chemical vapor infiltration) or CVD (chemical vapor deposition) the source of carbon is supplied continuously from the gas phase. At high temperatures above 1000K the supply molecules decompose catalytically at the fiber surface into carbon atoms and typically into hydrogen. It is now important to control the reaction such that the decomposition occurs only at the surface of the carbon fiber. The diffusion and desorption of the molecular precursor are fast and effective processes at reaction conditions so that the fraction of usefully reacting species is low. This gives rise to low nucleation and growth rates and requires extensive reaction times at high temperatures. Very careful kinetic modeling and designs of reactors [29, 30, 45, 46] are needed to optimize these processes with respect to product quality and energy consumption.

A critical factor here is the reactivity of the hydrogen by-product that is not only able to gasify the initial surface termination of the carbon fiber but also to etch away the newly formed pyrolytic carbon. This effect is desirable for optimization of the growing structure but additionally slows down the reaction.

Another factor for consideration [31, 32] is the porosity of the filler body. If the pores are too large, the gas diffusion is facilitated but the growth time is long and the danger of depositing granular detrimental soot is larger. If the pore structure is too dense, the diffusion into the deeper parts of the pores may be inhibited against the decomposition and deposition of the pyrocarbon: the result is a superficially dense product with deep voids and low mechanical stability being of low or insufficient mechanical quality.

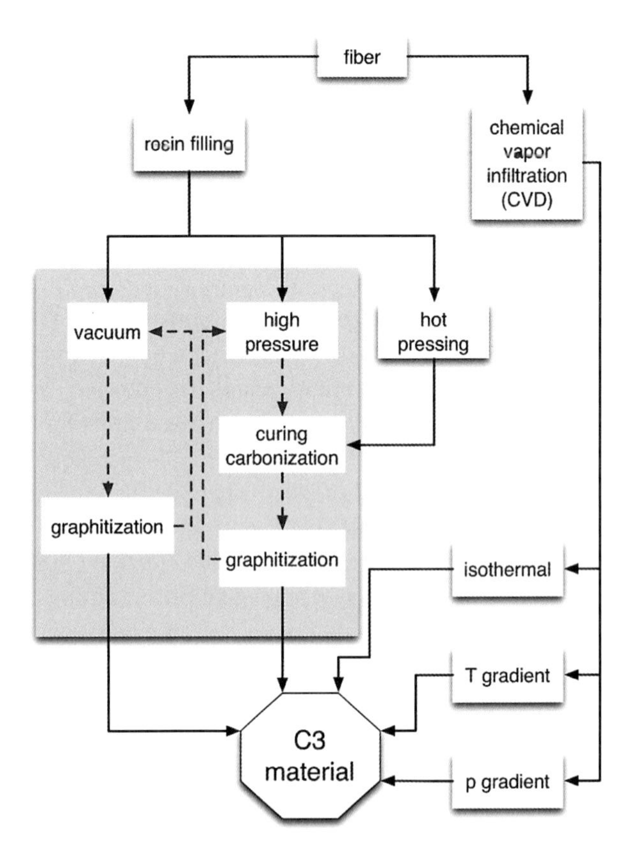

Fig. 9.3: Families of methods for creating C3 materials. The differences relate to the source of the binder carbon and to the method of application. The dashed lines indicate that in these processes repetitive graphitization and filling procedures can be applied to improve the density and homogeneity of the final material. Such repetitive procedures are not necessary when a continuous supply of the carbon source is applied as in the CVD procedures. The shaded processes are mainly applied in large-scale productions of C3 materials.

These aspects were carefully studied [29, 46–48] and analyzed in several macrokinetic models. They are essential for choosing the methods indicated in Fig. 9.3 and constructing the suitable combination of reactor and operation conditions. In Fig. 9.3 the CVD/CVI methods are designated according to the methodology for how a gradient in the chemical potential of the reaction is applied.

As an alternative to a continuous supply of the reactant forming the binder phase one may apply a green body as polymer-filler composite and carbonize/graphitize under inert conditions. This technically simpler method is conceptually demanding because the bonding partners leaving the polymer resin create void spaces and internal pressure that may disturb the ordering of the binder-filler relation. In addition, the missing atoms from the polymer binder will lead to voids and pores. This issue is tackled by repeating the impregnation and thermal treatments cycles as indicated in Fig. 9.3.

In summary, it may be stated that despite progress in identifying suitable molecular precursors [49–55] for the resin-type synthesis, the CVD or CVI methodologies deliver the best possible results [30, 45–47, 56, 57] for high-tech C3 materials. The molecular adhesion mechanism, the anisotropy of the binder carbon and the densification of the porous green body can be controlled to the best possible extent. The disadvantage is the complex process control of delivering high temperatures and carefully controlled gas supply at known chemical potentials with the excessive energy input during heating of the CVI reactor.

Summarizing the results of the kinetic studies on CVD, a reaction network can be constructed as exemplified in Fig. 9.4 for the case of methane as carbon source.

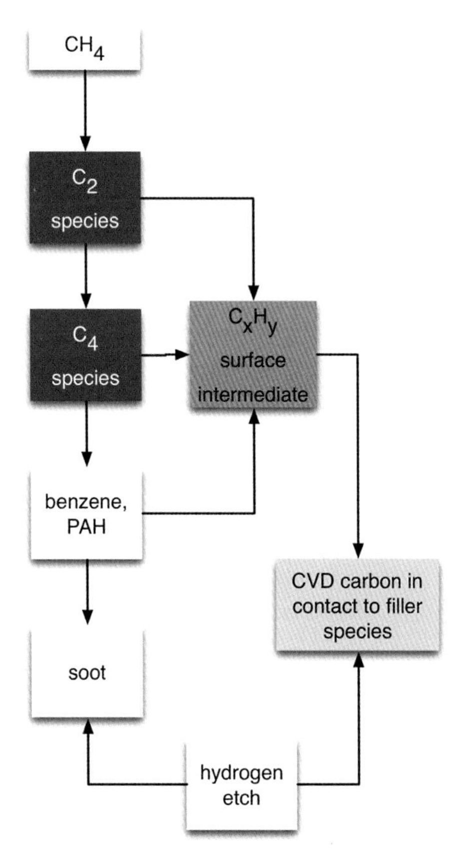

Fig. 9.4: Reaction network for the formation of CVD binder carbon during the synthesis of a C3 material from methane. The supporting data were collected from model experiments with carbon-coated ceramic channels and consecutive gas analysis as described [31, 32] in the literature. The colored boxes indicate the main reactive intermediates.

The network consists of a train of molecular condensation reactions occurring in the gas phase delivering reactive intermediates that form in homogeneous reaction molecular species with low reactivity for CVD (PAH) and soot depositing loosely on the filler material. At correct choice of the chemical potential the intermediates also react heterogeneously with the filler carbon surface and form an intermediate "carbon-hydrogen complex" that decomposes finally into the desired carbon binder. Although this network is useful as a descriptor for the overall process, much microscopic understanding is still missing. For example, the details of the interface reaction and also of the gas phase processes are not yet well described by experiments but rather approximated in order to arrive at the macrokinetic description used for choosing reaction parameters. In these models it was found that the effect of gasification of condensed carbon by the activated hydrogen, stemming either from the precursor molecule or even added deliberately, is beneficial for the quality of the product if present in the right abundance. This is because it tends to remove poorly-ordered and molecular species preferentially over well-developed carbon BSU. Excessive hydrogenation on the other hand can also passivate the reactive terminal sites of graphene layers or remove the anchoring heteroatoms of the filler (N-atoms from the PAN precursor) and thus deteriorate the mechanical, thermal and chemical properties of the otherwise well-ordered C3 product.

9.4 Identification of the structural features of C3 material

One application of C3 is the protection of structural wall elements in fusion devices against contact with the hydrogen plasma. The excellent high-temperature mechanical properties of C3 and its good thermal conductivity are the positive properties needed here in addition to the low contaminating effect of carbon atoms into the fusion plasma. The disadvantage however is the erosion upon the sputtering of hydrogen atoms. In addition to earlier models of purely physical erosion it was shown that chemical erosion, termed "chemical sputtering", is also a critical damaging mechanism. During these studies [58] several hydrogen etched samples of C3 materials were studied for their morphological properties. This etching occurs at defects of the material and by its morphology thus creates an excellent image of the spatial distribution of reactive defects in the sp^2 carbon structures. Figure 9.5 summarizes some results [59] taken from literature.

The large images in Fig. 9.5 show a representative cross-sectional view onto the material before and after etching. Before etching the perpendicular orientation of the anisotropic graphene layer packages in binder and filler can be clearly seen. The fiber exhibits a radial orientation, the binder a tangential orientation. In the deeper voids between the filler bundles one can see the packages of graphene layers forming interlaced structures with little isotropic carbon according to morphology.

This arrangement changes drastically after inspection of the etched sample. The deep large voids designate the areas of isotropic phase (see Fig. 9.1 phase (b)). It further highlights that the anisotropic well-crystallized graphene layers adhere strongly only in a thin rim around the fiber axis. The immediate interface is also quite reactive to

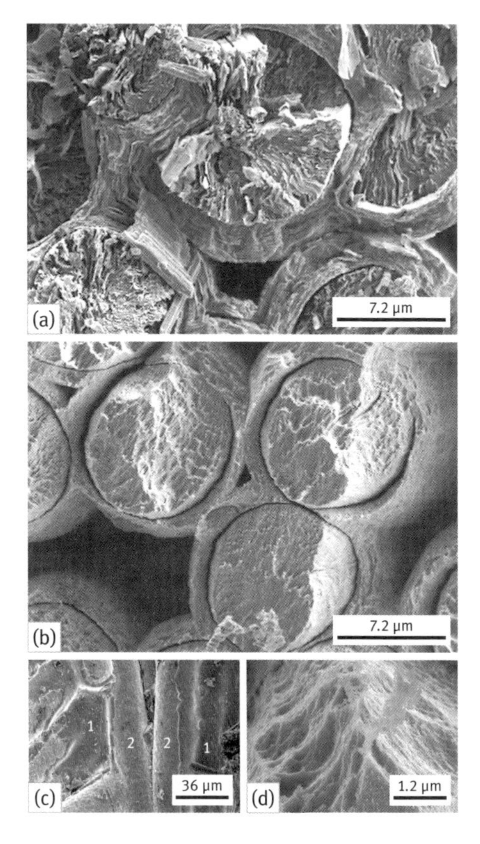

Fig. 9.5: SEM images of C3 material before and after hydrogen plasma treatment. For experimental details see [59]. (a) and (b) are top views before and after etching of the C3 material perpendicular to its outer surface, (c) is an in-plane view before etching and (d) represents a view along the long axis of a partly etched fiber.

hydrogen, as indicated by the perfect uniform thin gap between the two constitution phases.

The in-plane view onto the surface of the C3 sample in Fig. 9.4(c) shows a situation where the embedding of the filler (2) by the binder (1) can be recognized along the fiber axis. The wetting between the two phases is not perfect and varies from location to location highlighting the serious issue of preparing the molecular structure of the interface between binder and filler by functionalization [39, 60, 61] of the filler.

Figure 9.5(d) gives an impression about the topo-chemical nature of the hydrogen atom's attack on carbon. Even these highly reactive species attack carbon not in an isotropic form but react from the edges and thus decorate, after some extent of conversion, the planar shape of the BSU as stacks of graphene layers with uneven but identical outer shapes. The rounded protrusions into the edge structure arise from defect clusters that would manifest themselves in a perpendicular view as "etch pits".

9.5 Surface chemistry

The last operations (11a, 11b) in the synthesis workflow shown in Fig. 9.2 concern the surface finish. For low temperature applications this can be done by powder coatings where the sufficient electrical conductivity is useful for applying powder paints such as in automotive applications. In many high-tech applications the mechanical proper ties are needed at high and extreme temperatures and this often in conjunction with oxidizing atmospheres of either air (re-entry spacecraft) or in combustion aggregates such as rockets and turbines. In these applications it is essential that the carbon is protected from oxidation. This issue has been reviewed [62] for its great relevance in applications. The problem arises above 723 K in flowing air and needs to be suppressed up to about 1500 K marking the upper service temperature of C3 materials in air.

Carbon oxidation is a surface gas-solid process [63–66] that requires a bi-functional carbon. To activate oxygen it is essential that free electrons in the form of radicals or in weakly bound states are available. It should be noted that at high temperatures above 900 K not only oxygen but also H_2O and CO_2 serve as oxidants [67, 68] to carbon. For reacting the activated oxygen with carbon it is essential that prismatic edge sites of the graphene layer are available since the basal planes of carbon in defect-free form are fully non-reactive towards gasification. A perfectly flat defect-free sp^2 layer of graphene would thus be the optimal oxidation protection. As this cannot be realized in practice, several families of protective measures have been developed over the past. They group according to the scheme in Fig. 9.6.

Figure 9.6 shows that at the molecular scale either measure tends to seal the reactive positions by forming passivating carbon-heteroatomic structures as surface groups. This has been evaluated in great detail [69–71] with experimental and theoretical methods. The main drawback is that at temperatures above 1000K even these chemical groups become gradually unstable and begin forming carbides with the consequence that they lose their protective function at defect sites of the carbon matrix.

Another method at the molecular level is the inhibition of oxidation catalysis by alkali and transition metal impurities. In particular, alkali metal oxides in traces serve as effective catalyst with almost ubiquitous presence in technological environments. The mechanism of operation is well described in the literature [64, 72–77] despite its complex and multi-pathway behavior.

The protective function of boron and phosphorous [78, 79] can also be stabilized by the addition of storage phases releasing slowly molecular species [22] that migrate to the locations of oxidation driven by the local heat evolution of the beginning oxidative damage.

The other general strategy depicted in Fig. 9.6 is surface protection by a passivating coating. This coating must also adhere to the C3 material under thermal stress, it must not react itself with carbon and as a thin layer it must protect the carbon from oxygen diffusion. The combination of all these requirement is best fulfilled with glassy

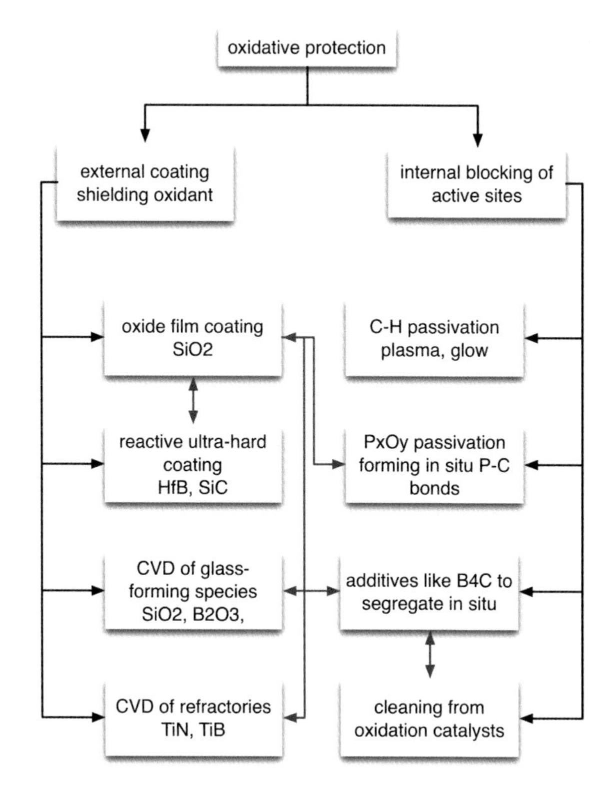

Fig. 9.6: Concepts of oxidation protection of C3 materials. In technical solutions several combinations of protective methods can be used (see colored arrows) at the expense of the price of the material.

SiO_2 or with other Si-based refractories such as Si_3N_4 and SiC. Other ultrahard coatings are also of use, but may require an intermediate buffer layer of Si oxide.

A typical figure of merit [22] is to suppress the oxidation rate of C3 materials of about 1×10^{-6} g/cm^2/s at 1000 K by an order of magnitude. This is achieved by all coatings mentioned in Fig. 9.6 at temperatures above 1073 K. At temperatures between 900 K and 1073 K, however, there lies a "valley of death" that is always passed through during heating and cooling of high-temperature service devices. In this range of temperatures the unmatched temperature expansion coefficients of coatings and C3 base create cracks that do not heal rapidly by plastic flow of the coating. For this reason buffer layers and the very careful coating by CVD methods [80] find application in order to maximize the adhesion of the coating layer. An independent design measure can be the use of woven two-dimensional filler structures instead of the mechanically preferable one-dimensional fiber bundle structure: the two-dimensional topology of the strain fields during thermal stress reduces the mechanical load onto the C3-coating interface and counteracts crack formation.

9.6 Summary

C3 materials exhibit a long development history since the fundamental insight that a carbon with a regular mesoscopic ordering of interfaces between objects made from strands and stacks of graphene form a material with a combination of extraordinary mechanical properties. A lightweight material with chemical stability like a ceramic and fracture properties like a metal or a biopolymer and transport properties like a metal is available through a process that can be controlled to a great extent. This arises from the insight into the kinetic events that are required in transforming molecules into a material.

At present, several classes of C3 material can be distinguished. There is the high-tech material made mostly from unidirectional fiber bundles and CVD/CVI binder formation. These materials exhibit complex shapes and are used under conditions of extreme environments. The second class relates to the application in automotive, wind turbine and airplane manufacturing where the lightweight properties allow replacement of steel and Al alloys. These usually more planar objects with complex shapes are made from filler fabrics and resin filled molecular precursors giving the parts as green body a good geometric stability and requiring more moderate energy input in their fabrication in comparison to gas phase infiltration.

The third family of research grade materials is less well defined and encompasses aerogels of carbon [81, 82] designed mesoscopic void structures in C3 with nanostructured fillers [51, 83], composites with nanocarbon fillers [24, 82, 84–88] and carbon-heterostructure [54, 89–94] compounds. The references stated here are only examples for a wide range of activities stemming from the efforts to synthesize novel nanostructured composites. These materials often exhibit unusual surface properties and are used in electrochemical and catalytic applications rather in the domain of traditional C3 compounds where mechanical properties dominate the application profile.

The reviewer states in conclusion that there is still substantial room for improvement in C3 materials both with respect to saving energy during synthesis and with respect to property profiles. A causal analysis of the interfacial qualities at the atomic level that goes beyond the current state of the art is required. At least in the open literature it is rare that structure-function relations are based upon an atomic scale knowledge of the material under study. Such improved knowledge may lead to sophisticated combined approaches in synthesis such as those that are beginning to emerge in the subfield of oxidation protection where the necessary atomic scale understanding is available.

Bibliography

[1] P. Scharff, *New carbon materials for research and technology*, 1998.
[2] E. Fitzer, *Carbon*, 1987, **25**, 163–190.

[3] S. R. Kelemen and H. Freund, *J.Catal.*, 1986, **102**, 80–91.
[4] B. Reznik and K. J. Huttinger, *Carbon*, 2002, **40**, 621–624.
[5] A. Oberlin, *Carbon*, 2002, **40**, 7–24.
[6] B. Henschke, H. Schubert, J. Blöcker, F. Atamny and R. Schlögl, *Thermochim. Acta*, 1994, **234**, 53–83.
[7] Z. Du, A. F. Sarofim and J. P. Longwell, *Energy & Fuels*, 1991, **5**, 214–221.
[8] F. Atamny, R. Schlögl, W. J. Wirth and J. Stephan, *Ultramicroscopy*, 1992, **42**, 660–667.
[9] B. Frank, A. Rinaldi, R. Blume, R. Schlögl and D. S. Su, *Chem. Mater.*, 2010, **22**, 4462–4470.
[10] B. Frank, J. Zhang, R. Blume, R. Schlögl and D. S. Su, *Angew. Chem. Int. Ed.*, 2009, **48**, 6913–6917.
[11] F. T. Lan, K. Z. Li, H. J. Li, Y. G. He, X. T. Shen and W. F. Cao, *Journal of Materials Science*, 2009, **44**, 3747–3750.
[12] T. Windhorst and G. Blount, *Materials & Design*, 1997, **18**, 11–15.
[13] R. Naslain, H. Hannache, L. Heraud, J. Y. Rossignol, F. Christin and C. Bernard, *Proceedings of the 4th European Conference on Chemical Vapour Deposition*, 1983, 293–304.
[14] C. Vix-Guterl, J. Dentzer and R. Gadiou, *Annales De Chimie-Science Des Materiaux*, 2005, **30**, 593–608.
[15] A. Dandekar, R. T. K. Baker and M. A. Vannice, *Carbon*, 1998, **36**, 1821–1831.
[16] J. Lahaye, *Fuel*, 1998, **77**, 543–547.
[17] P. Albers, K. Deller, B. M. Despeyroux, G. Prescher, A. Schafer and K. Seibold, *Journal of Catalysis*, 1994, **150**, 368–375.
[18] U. Zielke, K. J. Hüttinger and W. P. Hoffman, *Carbon*, 1996, **34**, 983–998.
[19] M. Endo and H. W. Kroto, *J.Phys.Chem.*, 1992, **96**, 6941–6944.
[20] W. Ruland, *Journal of Applied Physics*, 1967, **38**, 3585–3589.
[21] W. B. Downs and R. T. K. Baker, *Journal of Materials Research*, 1995, **10**, 625–633.
[22] J. E. Sheehan, K. W. Buesking and B. J. Sullivan, *Annual Review of Materials Science*, 1994, **24**, 19–44.
[23] N. Tsubokawa, *Bulletin of the Chemical Society of Japan*, 2002, **75**, 2115–2136.
[24] L. M. Manocha, H. Patel, S. Manocha, A. K. Roy and J. P. Singh, *Journal of Nanoscience and Nanotechnology*, 2009, **9**, 3119–3124.
[25] L. R. Radovic and F. RodriguezReinoso, in *Chemistry and Physics of Carbon, Vol 25*, 1997, vol. 25, pp. 243–358.
[26] N. M. Rodriguez, A. Chambers and R. T. K. Baker, *Langmuir*, 1995, **11**, 3862–3866.
[27] N. M. Rodriguez, M. S. Kim and R. T. K. Baker, *Journal of Physical Chemistry*, 1994, **98**, 13108–13111.
[28] B. Reznik, D. Gerthsen and K. J. Huttinger, *Carbon*, 2001, **39**, 215–229.
[29] W. Benzinger and K. J. Hüttinger, *Carbon*, 1996, **34**, 1465–1471.
[30] W. Benzinger and K. J. Hüttinger, *Carbon*, 1998, **36**, 1033–1042.
[31] W. G. Zhang, Z. J. Hu and K. J. Hüttinger, *Carbon*, 2002, **40**, 2529–2545.
[32] W. G. Zhang and K. J. Hüttinger, *Carbon*, 2003, **41**, 2325–2337.
[33] O. Khavryuchenko, B. Frank, A. Trunschke, K. Hermann and R. Schlögl, *J. Phys. Chem. C*, 2013, **117**, 6225–6234.
[34] G. Genti, M. Gangeri, M. Fiorello, S. Perathoner, J. Amadou, D. Béguin, M. J. Ledoux, C. Pham-Huu, M. E. Schuster, D. S. Su, J.-P. Tessonnier and R. Schlögl, *Catal. Today*, 2009, **147**, 287–299.
[35] S. N. Marinkovic, in *Advanced Materials and Processes: Yucomat Ii*, eds. D. P. Uskokovic, S. K. Milonjic and D. I. Rakovic, 1998, vol. 282–2, pp. 239–250.
[36] J. P. Boudou, P. Parent, F. Suarez-Garcia, S. Villar-Rodil, A. Martinez-Alonso and J. M. D. Tascon, *Carbon*, 2006, **44**, 2452–2462.

[37] S. Motojima, I. Hasegawa, S. Kagiya, K. Andoh and H. Iwanaga, *Carbon*, 1995, **33**, 1167–1173.
[38] E. Raymundo-Pinero, D. Carzorla-Amorós, A. Linares-Solano, J. Find, U. Wild and R. Schlögl, *Carbon*, 2002, **40**, 597–608.
[39] G. X. Zhang, S. H. Sun, D. Q. Yang, J. P. Dodelet and E. Sacher, *Carbon*, 2008, **46**, 196–205.
[40] U. Zielke, K. J. Hüttinger and W. P. Hoffman, *Carbon*, 1996, **34**, 999–1005.
[41] R. Schlögl, in *Handbook of Heterogeneous Catalysis*, eds. G. Ertl, H. Knözinger, F. Schüth and J. Weitkamp, Wiley VCH Verlag, Weinheim, 2008, vol. Vol. 1, pp. 357–427.
[42] H. P. Boehm, *Carbon*, 1994, **32**, 759–769.
[43] H. P. Boehm, *Carbon*, 2002, **40**, 145–149.
[44] X. Feng, N. Dementev, W. G. Feng, R. Vidic and E. Borguet, *Carbon*, 2006, **44**, 1203–1209.
[45] W. Benzinger and K. J. Hüttinger, *Carbon*, 1999, **37**, 1311–1322.
[46] W. Benzinger and K. J. Hüttinger, *Carbon*, 1999, **37**, 941–946.
[47] W. Benzinger, A. Becker and K. J. Huttinger, *Carbon*, 1996, **34**, 957–966.
[48] P. Delhaes, *Carbon*, 2002, **40**, 641–657.
[49] P. Buvat and F. Jousse, in *2001: A Materials and Processes Odyssey, Books 1 and 2*, eds. L. Repecka and F. F. Saremi, 2001, vol. 46, pp. 134–144.
[50] K.-S. Kim and S.-J. Park, *Journal of Electroanalytical Chemistry*, 2012, **673**, 58–64.
[51] K. Kraiwattanawong, N. Sano and H. Tamon, *Carbon*, 2011, **49**, 3404–3411.
[52] A. S. Mukasyan and J. D. E. White, *Ceramics International*, 2009, **35**, 3291–3299.
[53] R. Pratap, D. Parrish, P. Gunda, D. Venkataraman and M. K. Lakshman, *Journal of the American Chemical Society*, 2009, **131**, 12240–12249.
[54] Y. Wang, Y.-d. Xu, L.-l. Zhang and L.-f. Cheng, *Journal of Solid Rocket Technology*, 2007, **30**, 541–543, 551.
[55] N. Xiao, Y. Zhou, J. Qiu and Z. Wang, *Fuel*, 2010, **89**, 1169–1171.
[56] A. Becker and K. J. Hüttinger, *Carbon*, 1998, **36**, 177–199.
[57] A. Becker and K. J. Hüttinger, *Carbon*, 1998, **36**, 201–211.
[58] H. Grote, W. Bohmeyer, P. Kornejew, H. D. Reiner, G. Fussmann, R. Schlögl, G. Weinberg and C. H. Wu, *J. Nucl. Mater.*, 1999, **266–269**, 1059–1064.
[59] H. Grote, W. Bohmeyer, P. Kornejew, H. D. Reiner, G. Fussmann, R. Schlögl, G. Weinberg and C. H. Wu, *Journal of Nuclear Materials*, 1999, **266**, 1059–1064.
[60] J.-P. Tessonnier, A. Villa, O. Majoulet, D. S. Su and R. Schlögl, *Angewandte Chemie-International Edition*, 2009, **48**, 6543–6546.
[61] Q. Xiao, P. Wang and Z. Si, *Progress in Chemistry*, 2007, **19**, 101–106.
[62] D. W. McKee, *Carbon*, 1987, **25**, 551–557.
[63] H. Marsh and T. E. O'Hair, *Carbon*, 1969, **7**, 702–703.
[64] J. A. Moulijn and F. Kapteijn, *Carbon*, 1995, **33**, 1155–1165.
[65] A. A. Lizzio and L. R. Radovic, *Industrial & Engineering Chemistry Research*, 1991, **30**, 1735–1744.
[66] F. Atamny, J. Blöcker, B. Henschke, R. Schlögl, T. Schedel-Niedrig, M. Keil and A. M. Bradshaw, *J.Phys.Chem.*, 1992, **96**, 4522.
[67] K. J. Hüttinger and O. W. Fritz, *Carbon*, 1991, **29**, 1113–1118.
[68] K. J. Hüttinger and W. F. Merdes, *Carbon*, 1992, **30**, 883–894.
[69] Y. J. Lee and L. R. Radovic, *Carbon*, 2003, **41**, 1987–1997.
[70] A. A. Lizzio, H. Jiang and L. R. Radovic, *Carbon*, 1990, **28**, 7–19.
[71] L. R. Radovic and B. Bockrath, *Journal of the American Chemical Society*, 2005, **127**, 5917–5927.
[72] M. J. Illan-Gomez, C. S. M. de Lecea, A. Linares-Solano and L. R. Radovic, *Energy & Fuels*, 1998, **12**, 1256–1264.
[73] D. W. McKee, *Chemistry and Physics of Carbon*, 1981, **16**, 1–118.
[74] D. W. McKee, *Fuel*, 1983, **62**, 170–175.

[75] D. W. McKee, *Carbon*, 1987, **25**, 587–588.
[76] F. Kapteijn, R. Meier, S. C. van Eyck and J. A. Moulijn, ed. J. Lahaye, Ehrburger, P., Fundamental Issues in Control of Carbon Gasification,, Kluwer, Dordrecht, 1991, pp. 221–237.
[77] F. Kapteijn, O. Peer and J. A. Moulijn, *Fuel*, 1986, **65**, 1371–1376.
[78] D. W. McKee, C. L. Spiro and E. J. Lamby, *Carbon*, 1984, **22**, 507–511.
[79] D. W. McKee, C. L. Spiro and E. J. Lamby, *Carbon*, 1984, **22**, 285–290.
[80] K. L. Choy, *Progress in Materials Science*, 2003, **48**, 57–170.
[81] X.-g. Liu, *New Carbon Materials*, 2009, **24**, 282–288.
[82] C.-l. Qin, X. Lu, G.-p. Yin, X.-d. Bai and Z. Jin, *Transactions of Nonferrous Metals Society of China*, 2009, **19**, S738-S742.
[83] K. Kraiwattanawong, N. Sano and H. Tamon, *Microporous and Mesoporous Materials*, 2012, **153**, 47–54.
[84] I. V. Anikeeva, Y. G. Kryazhev, V. S. Solodovnichenko and V. A. Drozdov, *Solid Fuel Chemistry*, 2012, **46**, 271–274.
[85] K. Ariga, A. Vinu, Y. Yamauchi, Q. Ji and J. P. Hill, *Bulletin of the Chemical Society of Japan*, 2012, **85**, 1–32.
[86] Y. Kubota, Y. Sugi and T. Tatsumi, *Catalysis Surveys from Asia*, 2007, **11**, 158–170.
[87] J. Li, S. Ding, C. Zhang and Z. Yang, *Polymer*, 2009, **50**, 3943–3949.
[88] G. Qian-ming, L. Zhi, Z. Xiang-wen, W. Jian-jun, W. Ye and L. Ji, *Carbon*, 2005, **43**, 2426–2429.
[89] J. Cao, H. Q. Wang, J. L. Qi, X. C. Lin and J. C. Feng, *Scripta Materialia*, 2011, **65**, 261–264.
[90] Z. Chenxi and A. Manthiram, *Advanced Energy Materials*, 2013, **3**, 1008–1012.
[91] M. Sanchez-Sanchez, A. Manjon-Sanz, I. Diaz, A. Mayoral and E. Sastre, *Crystal Growth & Design*, 2013, **13**, 2476–2485.
[92] H. Watanabe, S. Kobayashi, M. Fukushima and S. Wakayama, *Jsme International Journal Series a-Solid Mechanics and Material Engineering*, 2006, **49**, 237–241.
[93] S. A. I. Steiner, B. T. F., B. C. Bayer, R. Blume, M. A. Worsley, W. A. MoberlyChan, E. L. Shaw, R. Schlögl, A. J. Hart, S. Hofmann and B. L. Wardle, *J. Am. Chem. Soc.*, 2009, **131**, 12144–12154.
[94] Y. Sungwoo, C. Yue, C. Yingwen, C. V. Varanasi and L. Jie, *Journal of Power Sources*, 2012, **218**, 140–147.

Teresa J. Bandosz

10 Graphite oxide-MOF hybrid materials

10.1 Introduction

Strict environmental regulations have resulted in the search for new efficient materials that can participate in environmental protection and remediation. So far, activated carbons have been considered as efficient adsorbents owing to their large surface area and the presence of a high volume of small pores. Carbonaceous materials are able to adsorb large quantities of various species, especially if the compounds to separate have a hydrophobic nature [1]. To increase the capacity of carbons to remove metal cations or polar species their surface has to be modified by impregnation with various metal salts or incorporation of heteroatoms such as nitrogen, oxygen sulfur, or phosphorus to their matrix [2].

Species on which research efforts in the separation sciences have recently focused are Toxic Industrial Compounds, TICs [3]. The list includes hydrogen sulfide, ammonia, nitrogen oxide, formaldehyde, sulfur dioxide and other small molecule species. These species are either gases or vapors at ambient conditions and when accidentally or deliberately released in open or confined spaces occupied by humans they may cause significant health problems or even death. Therefore, efforts to develop adsorbents capable of removing these species from ambient air are being intensified. The problem is not trivial since the physical adsorption forces are rather small at these conditions and even if some of these species can be physically adsorbed, there is a danger of their spontaneous release from the surface in the desorption process. Another factor to consider is the presence of water in ambient air since water can significantly interfere with the separation efficiency. Therefore, chemisorption and/or reactive adsorption are target variations of the adsorption process [4].

Reactive adsorption consists of at least two processes: physisorption of molecules on a surface and one or more reactions in which an adsorbate actively takes part in some of those reactions involving oxygen and water supplied to the system as ambient air. The goal of this process is either to convert toxic species into something nontoxic or toxic that is strongly retained on the surface, or to convert toxic species into nontoxic species removed from the surface of an adsorbent with the passing air stream [4]. In these processes both chemistry and porosity of an adsorbent surface are of paramount importance. The pores should be small enough to attract adsorbate molecules *via* either specific or nonspecific forces, but large enough to accommodate certain chemical species/functional groups promoting reactions in the pore system. In some cases a pore should be large enough to accommodate water molecules to promote dissociation and acid-base reactions in this specific nanoreactor it consists of [4, 5].

The above requirements directed our attention to the specific surface features of Metal Organic Frameworks (MOF) [6, 7] and graphite oxides (GO) [8, 9]. The former are known for their well-defined porosity, a high surface area and the presence of reactive sites either on metal or on the organic linkages. Their disadvantage in an application as adsorbents is that the pores are large in comparison with the TICs and the specific chemical constitution of the cages' walls results in rather weak physical adsorption forces. Graphite oxide, on the other hand, is rather a nonporous material from the view point of the nitrogen molecule [10]. Nevertheless, it is built of the dense array of graphene-like layers decorated with specific oxygen functional groups. These layers, when separated and assembled in the porous network, could provide strong forces for physical adsorption. Thus combining both MOF and GO can lead to unique materials exhibiting the best features of both components that could be important for removal of small molecule toxic species at ambient conditions.

10.2 Building blocks

10.2.1 Graphite oxide

Graphite oxides are well known materials and extensive reviews on their physical and chemical properties have been published by Dreyer and co-workers [8] and Zhu [9]. Unfortunately, the application of these materials as adsorbents has not been commonly explored or sufficiently addressed. They are obtained by oxidation of graphite. This process enables the incorporation of oxygen atoms to the basal planes and to the edges of graphene layers. These oxygen functional groups identified so far on the surface of GO are epoxide, keto and hydroxyl groups on the basal planes, and carboxylic groups on the edges [11, 12, 13]. A direct incorporation of oxygen atoms into the graphene layers was also observed in a recent study [14, 15]. Introduction of functional groups to the basal planes is accompanied by an increase in the distance between the graphene layers from about 0.34 nm to 0.6–1.2 nm. The wide range of interlayer distances encountered in GO is explained by the various degrees of oxidation and the hydration levels [16, 17]. Owing to its oxygen functional groups, GO has a hydrophilic character and molecules of water can easily be intercalated between the graphene layers. This hydrophilic character is also responsible for the easy dispersion of GO in water, alkaline solutions or alcoholic media [17, 18]. Not only does the oxidation of graphite enable the incorporation of oxygen groups, but it also leads to the formation of defects [18] (Fig. 10.1). These defects usually correspond to vacancies or adatoms in the graphene layers. Considering this, GO is commonly represented as distorted/corrugated graphene layers stacked in a more or less ordered fashion [19, 20, 21]. The most often used methods to oxidize graphite are those developed by Brodie [22] and Hummers [23], either in their original or modified versions. For detailed information on GO, their chemistry, morphology, and physical properties the reader is referred to the abovementioned reviews by Dreyer and co-workers [8] and Zhu and co-workers [9].

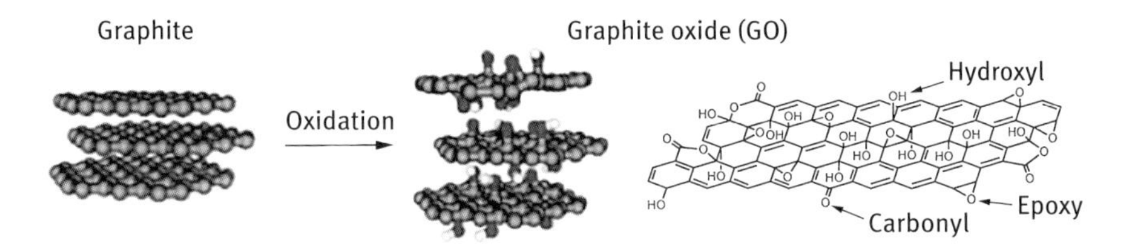

Fig. 10.1: Graphite oxide synthesis and its surface chemistry.

10.2.2 Metal Organic Frameworks: MOF-5, HKUST-1 and MIL-100(Fe)

Like graphite oxides, MOFs have been addressed extensively in the literature during the last ten years. Comprehensive reviews on their synthesis, properties and applications have been recently presented by Rowsell and Yaghi [6] and James [7]. In developing our idea of new materials, MOF-5 was the first choice of MOF as a composite component. It belongs to the class of MOF-n compounds [24, 25]. In these materials derivates of benzene carboxylate are organic linkers. MOF-5 contains zinc oxide clusters as the metallic centers and benzene dicarboxylic (BDC) as the organic bridges (Fig. 10.2). This material has a cubic structure and its surface area ranges from 700 to 4 000 m^2/g depending on the preparation method. Its structure has been fully described by Li and co-workers [25]. The lattice of this specific MOF has an ideal cubic shape with pores of size 0.152 nm [25]. The binding sites on zinc are saturated and the material, unfortunately, is not stable when stored in an atmosphere containing moisture [6, 7].

Knowing the instability and limited reactivity of MOF-5 another choice of this class of materials used in our study was HKUST-1 (Fig. 10.2). This is a copper-based MOF with copper ions as the metallic component and benzene tricarboxylate (BTC) as the organic bridges. Its structure has been investigated in detail by Chui and co-workers [26]. It comprises channels of about 0.9 nm in diameter surrounded by tetrahedral pockets of about 0.35 nm in diameter. Copper centers are unsaturated and therefore can be important sites for reactive adsorption. For the synthesis of this MOF a procedure adapted from Millward and co-workers was used [27]. The advantage of HKUDST-1 over MOF-5 is its much higher stability at ambient conditions. Even though HKUST-1 decomposes when in direct contact with water, it preserves its structure and thus porosity when exposed to air containing a moderate level of moisture [28].

The third MOF selected for our study was MIL-100 (Fe) (Fig. 10.2). Even though most MOFs have a microporous structure, this MOF belongs to MIL (Material of Institut Lavoisier) materials having pores in the range of nanometers [29–31]. In this compound, metallic iron (III) clusters are connected by benzenetricarboxylate organic linkers [31]. The nature of the iron offers several advantages compared to other metals owing to its low cost, redox properties, non-toxicity and environmental-friendly character. MIL-100(Fe) has a zeolite-like morphology [30–32] with three types

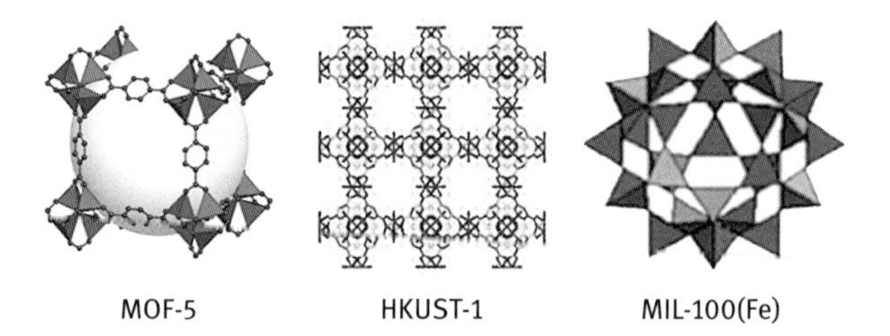

Fig. 10.2: The structure of MOFs used as composite components.

of pores/cages. Their internal diameters are of 0.87, 2.5 and 2.9 nm. The two latter pores are accessible through pentagonal windows with sizes about 0.47–0.55 nm and 0.86 nm respectively [30–32]. The synthesis of this MOF used in our work was adapted from that reported by Horcajada and co-workers [31].

10.3 Building the hybrid materials: Surface texture and chemistry

In the process of composite building, the GO component obtained using Hummers' method was dispersed in the solvents and chemicals used to obtain a particular MOF and the composite was synthesized with GO present in the synthesis mixture following the same steps as those used for the MOF synthesis [33–35]. To broaden the spectrum of materials' properties the content of GO varied between 5 and 50 %. These variations in composition were expected to have an effect on the porosity of the final composites. It was expected that oxygen from epoxy groups of exfoliated graphite oxide layers would be involved in formation of MOF units and thus the new pores would exist on the interface. They were expected to be in the micropore range with the sizes similar to our target TICs molecules. This effect was clearly seen for HKUST-1/ GO composites where an increase in the volume of these pores was seen based on the analysis of the amount of hydrogen adsorbed at 77 K and the porosity evaluated from the nitrogen adsorption isotherms (Fig. 10.3) [34]. For these materials the content of GO ranged between 5 and 46 %. When the measured structural parameters were compared to those that had been calculated assuming the physical mixtures of nonporous graphite oxide and porous MOF, the synergetic effect of an increase in the porosity was seen for the content of GO up to 20 %. Then the porosity decreased. That decrease was linked to staked GO layers and defects in the growing of MOF crystals. For these materials the surface area ranged between 650–1000 m^2/g with the volume of pores 0.35 to 0.57 cm^3/g. A synergetic effect of an increase in measured parameters of porous structure was also found in the case of MOF-5/GO composites [33]. For these materials, as in the case of the composite with HKUST-1, the high content of GO was not beneficial for porosity development. The measured surface areas were between 600 and 800 m^2/g and pore volume between 0.325 and 0.408 cm^3/g. The situation looked different for MIL-100/GO composites [35]. Here a small increase in porosity compared to the hypothetical mixture was found for very small content of GO (5 %) and then the apparent porosity of the composites with the higher contents of GO was even less than those calculated assuming physical mixtures of the components. A comparison of the extent of synergetic effect for the selected composites studied is presented in Fig. 10.3.

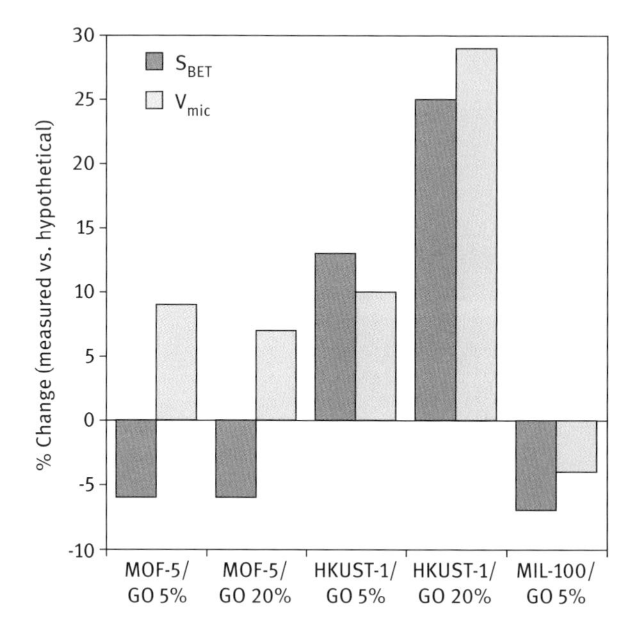

Fig. 10.3: Comparison of the extent of synergetic effect in porosity for chosen composites.

In order to explain the observed trend and thus the process of composite formation, other surface properties such as crystallographic structure and microscopic texture were examined. Even though the X-ray diffraction patterns for the composites of MOF-5 or HKUST-1 with GO showed the basic structure of MOF units was preserved, with an increase in the GO content a splitting of some peaks was found for MOF-5, suggesting the distortion of its cubic symmetry [33]. The diffraction peaks characteristics for d_{002} of GO were not seen, which is a clear indication of the exfoliation process, even with the high content of GO [34]. Interestingly, the situation looked different for the composites of GO and MIL-100 [35]. Here, due to the mesoporous character of MIL-100(Fe), most peaks for the MIL sample were observed in a small angle region [31]. For the composite with a small content of GO (4 %) most peaks from MIL were preserved, suggesting that the MOF structure was mainly preserved and the graphene layers from GO only induced slight distortion. No sign of GO was visible for this material. When the content of GO increased to 9 % most peaks from the MIL component disappeared and only small intensity peaks were seen at about 3.5, 4 and 4.2°. This was a clear indication that in that particular sample, GO prevented the proper crystallization of the MIL-100 units. This result was consistent with the porosity study and the reason for this behavior was investigated in detail taking into account the features of MIL-100 [36].

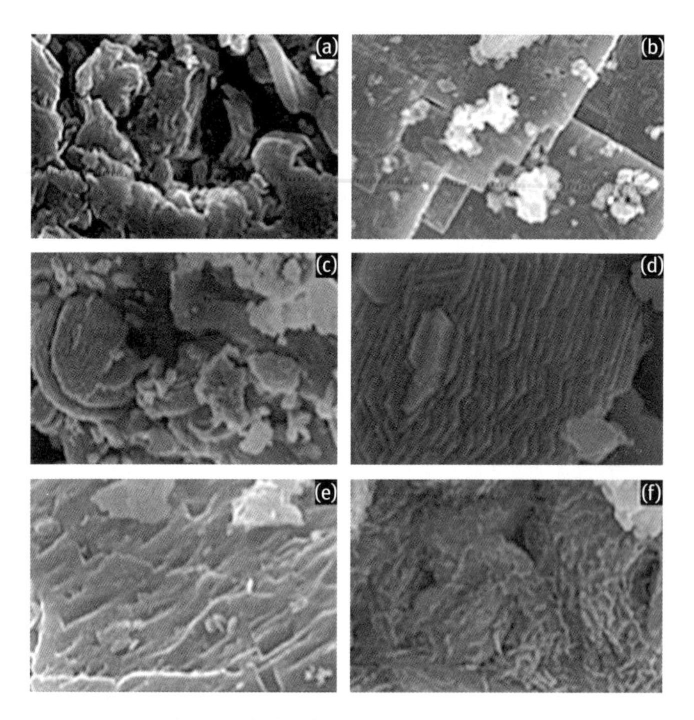

Fig. 10.4: SEM micrographs for the parent materials and the composites. (a): GO, (b): MOF-5, (c): MOF/ 5 % GO, (d): MOF/ 5 % GO, (e): MOF/ 10 % GO and (f): MOF/ 20 % GO (reprinted with permission from [33], © 2009, Wiley).

Investigation of the microscopic texture of the composites in comparison with those of the parent MOFs and GO sheds more light on the process of composite formation. The examples of SEM images for GO, MOF-5 and their composites with an increasing content of GO are presented in Fig. 10.4. The composite with 5 % GO shows very well defined platelets of sharp angles, all similar sizes, stacked together into one unit. A plausible interpretation of that micrograph was formation of small units of MOF-5 (platelets), which are separated by graphite oxide layers. To have such "organization" the chemical bonds, which will be discussed later in this chapter, have to be involved. An increase in the content of GO causes the distortion of the sharp edges of the stacked MOF crystals and at high content of GO a "worm-like" structure appears. The texture of the composite with HKUST-1 looks different [35]. As seen in the HRTEM images presented in Fig. 10.5, for the composite with low content of GO well-defined graphene-based layers with embedded HKUST-1 units are visible. When the content of GO increases, more graphene-based layers are present and a rather well-defined lattice image appears within the layers of GO. Even though electron beam illumination can cause the breakdown of HKUST-1 and can thus prevent any visualization of its lattice structure [37], the pattern observed represents the lattice image of HKUST-1. It is likely that distorted graphene-based layers present in the composite helped MOF to retain its crystalline structure by dissipating the electrostatic charges [38]. For the high contents of GO some HKUST-1 units are still visible within the GO distorted graphene-based layers (Fig. 10.5(c)) along with separate agglomerates of stacked distorted graphene-based layers (Fig. 10.5(c)). These analyses indicate that the two composite components are well mixed within our materials.

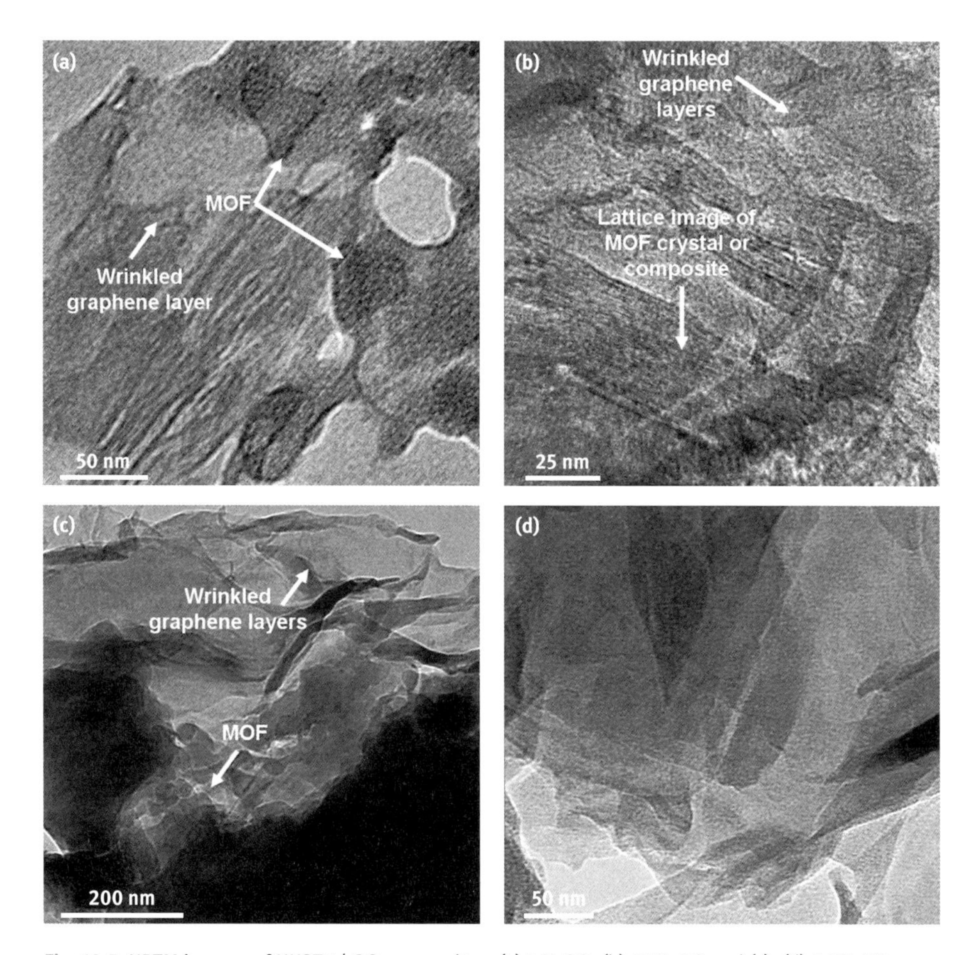

Fig. 10.5: HRTM images of HKST-1/ GO composites. (a) 5 % GO, (b) 18 % GO and (c), (d) 46 % GO (reprinted with permission from [34], © 2011, Elsevier).

An explanation for the organization of the composite components units discussed above was found in the results of a thermal analysis. The examples of differential thermogravimetric (DTG) curves for graphite oxide, HKUST-1 and their composites with various contents of GO are presented in Fig. 10.6. Since no significant weight loss was measured above 450 °C, only the temperature range between 30 °C and 450 °C is shown, even though the analyses were carried out up to 1000 °C. In the case of GO a peak at 100 °C corresponds to the removal of physically adsorbed water, while the second peak around 200 °C represents the decomposition of epoxy groups. The broad hump between 250 °C and 400 °C indicates the decomposition of carboxylic and sulfonic groups [10]. The DTG curve of HKUST-1 exhibits a small peak at about 100 °C, which corresponds to a dehydration step [39]. Additional molecules of water are released at ~ 300 °C and a collapse of the HKUST-1 structure is seen as an intense peak at ~ 350 °C [40]. These processes are accompanied by the release of CO_2 and result in the formation of copper oxide. The DTG curves for the composites look similar to the one of MOF with the exception of the peak at ~ 300 °C, which decreases in intensity for the composites with the lowest GO content and then vanishes for the composites with the higher graphite content. This suggests that some copper centers of the composites are present in a more hydrophobic environment than that in MOF. This environment has to originate from distorted graphene-based layers. A very interesting and distinguishing feature is the absence of the peak representing decomposition of GO epoxy groups on the DTG curves of the composites [41]. This is an important proof of the involvement of these groups in the formation of the composites and the coordination of the cupric ions to the oxygen atoms of the epoxy groups. This is consistent with the lack of a dehydration step at 300 °C discussed above. This hypothesis was verified by subjecting GO to the same synthesis process as for the composites but in the absence of copper nitrate and BTC. The DTG curves for this material exhibited a peak at 200 °C and thus the possibility of other reactions than those involving copper could be eliminated. A similar pattern was observed for the composites with MOF-5 and MIL-100 regardless of the content of GO component.

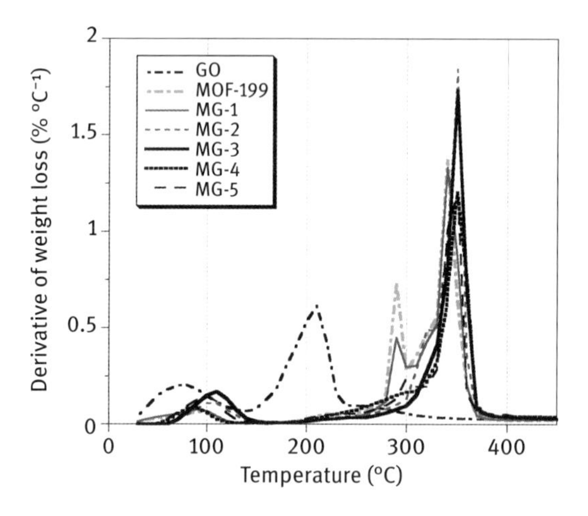

Fig. 10.6: Examples of the DTG curves for HKUST-1 (MOF-199), GO and the composites (MG) with 5 % (a), 10 % (b), 20 % (c), 40 % (d) and 50 % GO (reprinted with permission from [34], © 2011, Elsevier).

To further verify the importance of epoxy groups for the composite or composite formation and the effects of this process on the porosity of the final materials, an attempt was made to build the composite of HKUST-1 with various graphites [36]. In the resulting materials the synergetic effect of the increase in the porosity was not found and was linked to the absence of oxygen groups attached to the basal plane of graphite units. The size of graphite flakes was also found to affect the properties of the resulting hybrid materials.

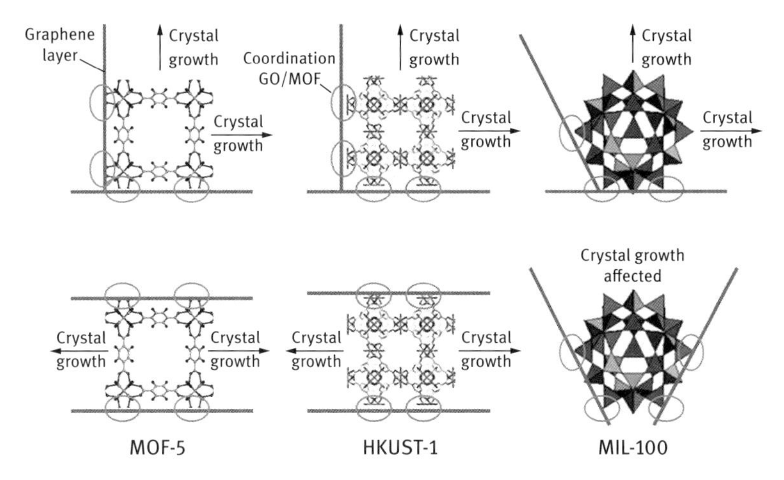

Fig. 10.7: Schematic comparison of the coordination between GO carbon layers and the MOF units for different types of MOF network: MOF-5, HKUST-1 and MIL-100 (reprinted with permission from [35], © 2011, Wiley).

The results presented above led us to develop models of our composites' formation. Apparently, the geometry of the MOF units and their chemistry affect the final properties of the composites (Fig. 10.7). In the case of MOF-5, the hexagonal nature of MOF resulted in the sandwich-type structure where small units of MOF detected using XRD are separated by the graphite oxide layers. The structure is more complex in the case of HKUST-HUST where binding sites for GO are located on both perpendicular sides of the MOF crystal. On the other hand, the near spherical shape of MIL-100 crystals enables bonding at various angles and thus prevents the proper growth of MOF units. As a result, this composite does not have the promising porous properties [35].

10.4 MOF-Graphite oxides composites as adsorbents of toxic gases

Since the main intent in synthesizing our materials was the development of new versatile adsorbents capable of effectively removing either acidic or basic TICs at ambient conditions, the dynamic breakthrough experiments were carried out at ambient conditions. The details of the homemade experimental setup are presented in [42–44]. Ammonia, hydrogen sulfide and nitrogen dioxide were our target TICs.

10.4.1 Ammonia

The most extensive study was carried out on ammonia removal. It is a small molecule with 0.31 nm of dynamic diameter [45] having pK_a of 23 [46]. The previous studies on modified carbons show its strong ability for reaction with surface acidic groups and also for complexation with metals [47]. The ammonia adsorption capacities on the composites of MOF-5 with GO of different composition are presented in Fig. 10.8. When the results were compared to the hypothetical ones calculated for the physical mixture of the MOF-5 and GO and their percentage of individual capacities [33, 48] the synergetic effect of composite formation was clearly seen in moist conditions. Here the measured capacity for the best performing samples was more than 100 % higher compared to the hypothetical one. Even though the adsorption capacity increased with an increase in the GO content (studied up 50 % of GO), with a GO content higher than 50 % a decrease in the performance is expected owing to the detrimental effect of such high GO content on the porosity of the materials [48]. Since zinc centers in MOF-5 and the composites are saturated, the main enhancement in the adsorption capacity in the case of these materials was attributed to adsorption in the additional pore space formed at the interface of MOF and GO [33, 48]. Interactions of ammonia with the MOF structure result in a collapse of its structure owing to the formation of hydrogen bonds between the hydrogen atoms of NH_3 and oxygen atoms of ZnO_4 sites. In the presence of water, the collapse of the structure is even more pronounced since water additionally leads to the destruction of this particular MOF *via* the "replacement" of oxygen atoms in the ZnO_4 tetrahedra by oxygen atoms from water [49]. The release of carboxylic acid creates additional acidic centers (besides those from GO [10]) for ammonia adsorption. Moreover, some ammonia can be dissolved in the water present in the pore system. This mechanism was confirmed by the extensive characterization of exhausted samples by FTIR, XRD and adsorption of nitrogen [33, 48]. The schematic view of the ammonia adsorption centers in the structure of the MOF-5/GO and HKUST/GO

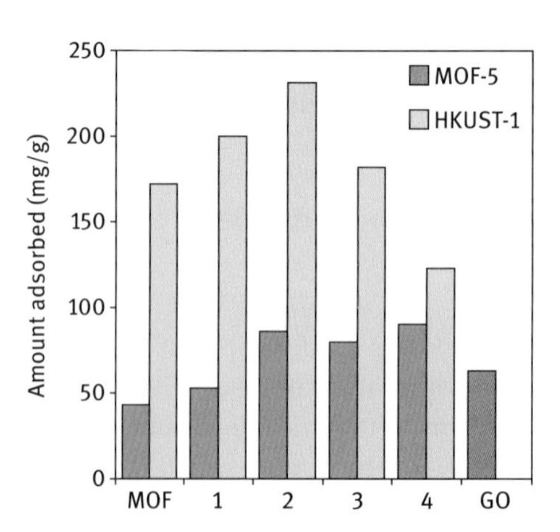

Fig. 10.8: Comparison of the ammonia adsorption capacities on MOF-5/GO composites and HKUST/GO composites (1–4 represent increasing GO content: 1 ~ 5 %, 2 ~ 10 %; 3 ~ 20 %; 4 ~ 40 %).

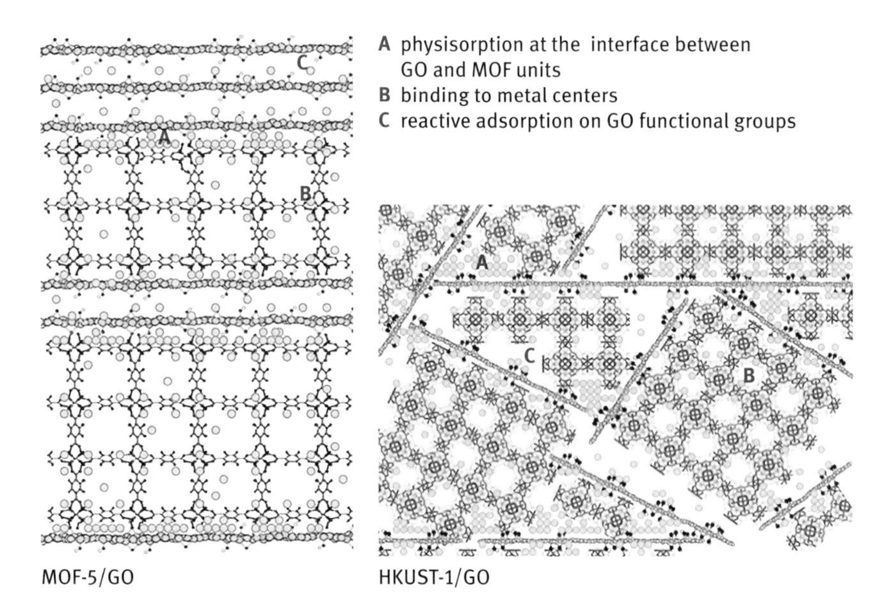

Fig. 10.9: Ammonia adsorption centers in MOG5-GO [35] composite and HKUST-1/GO composite [42].

composites is presented in Fig. 10.9 [48]. Ammonia was found to be strongly adsorbed on the surface of these materials.

HKUST-1 and its composites with GO presented much higher capacities for ammonia adsorption than those found on MOF-5/GO (Fig. 10.8) [42]. As mentioned above, these materials also show much higher stability to ambient water and therefore they have real potential to be applied as adsorbents of TICs. The maximum capacity for reactive adsorption of ammonia at ambient conditions was 23 wt%. And also here the synergetic effect of the composite formation was clearly evidenced. In the case of these materials, besides the development of the new porosity at the interface of both composite components and between the composite units, the MOF chemistry providing unsaturated copper sites for interactions with ammonia is also very important. This much more pronounced effect of an increase in porosity for these composites compared to the composites containing MOF-5 (Fig. 10.3) enhanced the physical adsorption of both NH_3 and water, and this could further contribute to dissolving ammonia in the water film. But the most marked effects on the performance of these materials are in their reactivity upon exposure to ammonia and water. During this process marked changes in the color of the bed were noticed [42, 50]. The first one was from dark blue to deep sky blue and then from deep sky blue to Maya blue. For the composite, the teal shades were also observed, not found for HKUST-1 alone [28]. The first change in color was attributed to the ammonia binding to the copper sites and the second to the formation of ammonia complex. The extensive FTIR, XRD and TA analyses suggested the formation of ammonia salts of benzene tricarboxylic acid $(NH_4)_3BTC$,

$Cu(NH_3)_2BTC_{2/3}$ and finally copper hydroxide in the presence of water. The formation of the BTC salts was supported by the collapse of the structure after interaction of ammonia with unsaturated copper centers. The release of BTC and copper oxide centers provides sites for reactive adsorption of ammonia during the course of the breakthrough experiments. Interestingly, even though the structure collapses, some evidence of the structural breathing of the resulting materials caused by reactions with ammonia was found, based on the ammonia adsorption at equilibrium and the analysis of the heat of interactions [51].

When MIL-100(Fe)/GO composites were studied as ammonia adsorbents, the ammonia breakthrough capacity in dry conditions (without prior drying of the samples) decreased with an increase in the GO content [35]. The improvement compared to MIL-100 (Fe) was found only for the composite with the smallest content of GO (~ 5 %). Interestingly, the capacities measured on these materials in dry conditions were higher than those in moist ones. This was attributed to the lack of microporosity where water could be adsorbed and thus could contribute to the dissolution of ammonia and acid-base reactions. Moreover, such behavior was explained by the specific coordination of metallic centers and the interactions of water with those centers. Following the analysis of MIL-100 (Cr) presented by Vimont and co-workers, we assumed that two main types of water also exist in MIL-100 (Fe) [52]. The first one corresponds to water molecules directly coordinated to the metallic sites, and the second to water molecules hydrogen-bonded to the first type of water molecules. Since the second type is the first to come into contact with ammonia, its lower Brønsted acidity is probably the cause for less ammonia being retained on the surface of non-dried samples than in the case of the dried samples. It has to be mentioned here that for the dried material, a (small) part of metallic centers might be also free of water and thus some ammonia molecules could be coordinated to the iron sites. The reason for the smaller capacity measured in moist conditions could be that a competition between ammonia and water molecules for adsorption on the pre-adsorbed water molecules (the first type). The existence of this competition was supported by the similarity in the energy associated with hydrogen bonding between two molecules of water and between a molecule of water and ammonia [53]. If the retention of ammonia on our materials occurred only *via* Brønsted interactions or coordination to the metallic sites as described above, then no significant change should have been observed in the exhausted adsorbents compared to the initial one. This was confirmed by analysis of the surface chemistry of the exhausted samples [35].

10.4.2 Nitrogen dioxide

Since of all the composites studied HKUST-1/GO composites were the most suitable materials as ammonia adsorbents, they were also tested as NO_2 and H_2S reactive adsorbents (Fig. 10.10). In the case of NO_2 adsorption in dry conditions, the measured

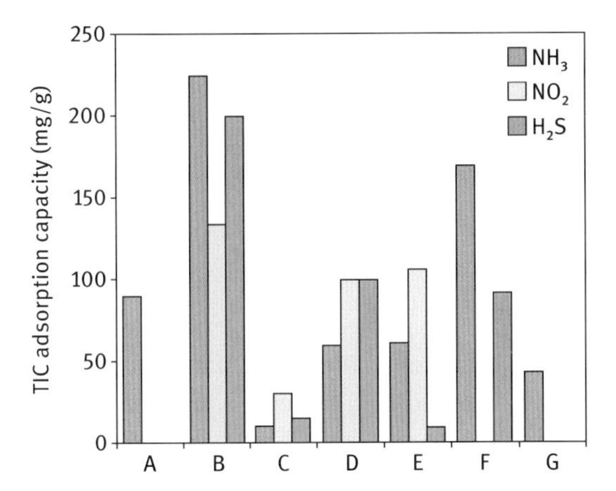

Fig. 10.10: Comparison of the TICs adsorption capacity on the best MOF/GO composites and activated carbons: (a) MOF5/GO; (b) HKUST-1/GO; (c) unmodified activated carbons; (d) modified activated carbons; (e) GO; (f) HKUST-1; (g) MOF-5.

NO_2 adsorption capacities were more than twice as high as those in moist conditions [43]. Generally, the addition of graphite oxide led to an enhancement in the NO_2 capacity in dry conditions, since all the composite materials displayed a better performance than that of HKUST-1. In moist conditions, only the composite with the highest contents of GO (20 %) performed better than HKUST-1. As in the case of ammonia adsorption, the measured performance was compared to the hypothetical one that had been calculated assuming the physical mixture of the components. A very strong positive effect was found in dry conditions. On the other hand, in moist conditions, the experimentally measured amounts of NO_2 adsorbed were smaller than the hypothetical ones for all composites except for the sample with 20 % of GO. This suggested that water strongly modifies the adsorption mechanism of NO_2 on HKUST-1 and on the composite materials in comparison with the mechanism, which takes place in dry conditions. Taking into account the polar character of the species present in the system, the competitive adsorption between water and NO_2 was proposed as a plausible explanation of the observed trend. This is supported by several reports indicating that both water [28, 54] and NO_2 [55, 56] can bind to copper. This process leads to a decrease in the number of adsorption centers. Therefore, in dry conditions where all these metallic sites are available for binding to NO_2, the optimum capacity can be reached. During adsorption of NO_2, a small quantity of NO was released ranging from 5 to 18 % of NO_2 supplied to the system [43]. Taking this into account, and based on the extensive characterization of surface chemistry indicating the formation of nitrates (mono- and bidentates) and the fact that (as in the case of ammonia) the structure collapsed after the adsorption process (especially when NO_2 was adsorbed in dry conditions), the adsorption mechanism was proposed [43]. In dry conditions bidentates are formed when Cu-O bonds break and the material loses its microporosity (see

reaction (10.1)),

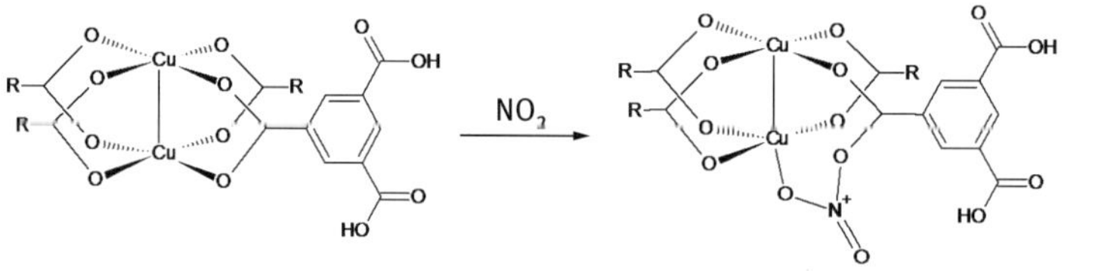

(10.1)

with R = benzene di-carboxylic.

Furthermore, an observed change in the coordination of the carboxylate ligands from benzene tri-carboxylic [57] likely involves a second NO_2 molecule and leads to the formation of a monodentate nitrate bound to the copper (see reaction (10.2)). This rearrangement may cause the appearance of carboxylic groups on the benzene tri-carboxylic linkage and the formation of NO, which is released in each case during NO_2 adsorption process in dry conditions (see reaction (10.3)).

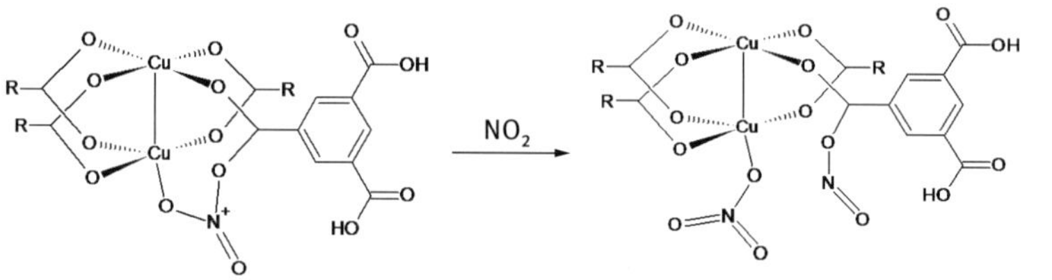

(10.2)

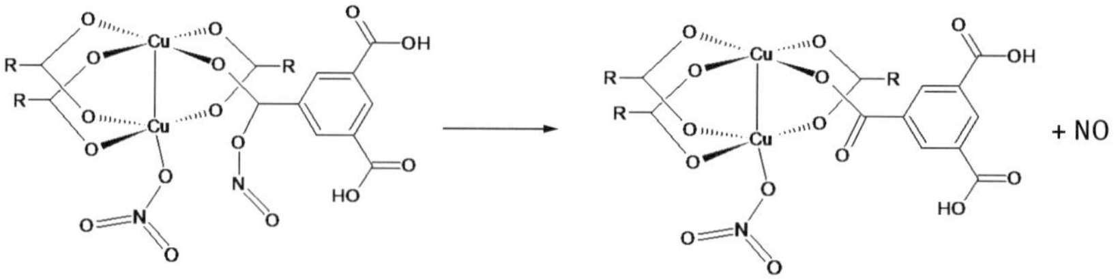

(10.3)

In moist conditions, the presence of water hinders the adsorption of NO_2 on the copper sites and physical adsorption in small micropores is the main mechanism of the removal process.

10.4.3 Hydrogen sulfide

The hydrogen sulfide adsorption on the HKUST-1/GO composites was only studied in moist conditions [44]. The choice was based on the presence of water in the ambient air and on the extensive studies of H_2S reactive adsorption on carbonaceous materials where water was important for dissociation of hydrogen sulfide before its further reaction with surface active sites [5]. Here up to 20 wt% of H_2S was adsorbed in the composite with the smallest content of GO. [44]. An increase in the GO content resulted in a decrease in the capacity but nevertheless it was higher than that on MOF alone. The synergetic affect was demonstrated by 30–100 % higher adsorption capacities in comparison with the hypothetical ones. The analysis of the products of surface reactions and an observed collapse in porosity suggested that in a first step, hydrogen sulfide binds to the copper sites (likely replacing a molecule of water) and then successive reactions leading to the formation of carboxylic acid and copper sulfide as well as a change in the coordination between copper and the oxygen of BTC take place [44]. This path is presented in reaction (10.4). The latter phenomenon induces a breaking of bonds that results in the decomposition of the HKUST-1 structure and a decrease in the porosity. As the number of available copper sites increases, the potential for H_2S to be adsorbed and form copper sulfide increases as well. The smaller adsorption on HKUST-1, where the concentration of copper sites is the highest, is only an apparent discrepancy. Since in the composites an additional pore space was created and the dispersive forces are stronger, physisorption in the micropores enhances the adsorption process there. Pre-adsorbed water also favors adsorption, causing dissolution of H_2S. The highest adsorption on the composite with the smallest contribution of GO was explained by the fact that small amount of GO does not significantly limit the sizes of HKUST-1 units with active copper centers and at the same time creates some volume of small pores where water can be strongly retained. Addition of more GO, even though it results in an increase in the volume of micropores, causes less water adsorption owing to the more hydrophobic character of GO compared to that of MOF.

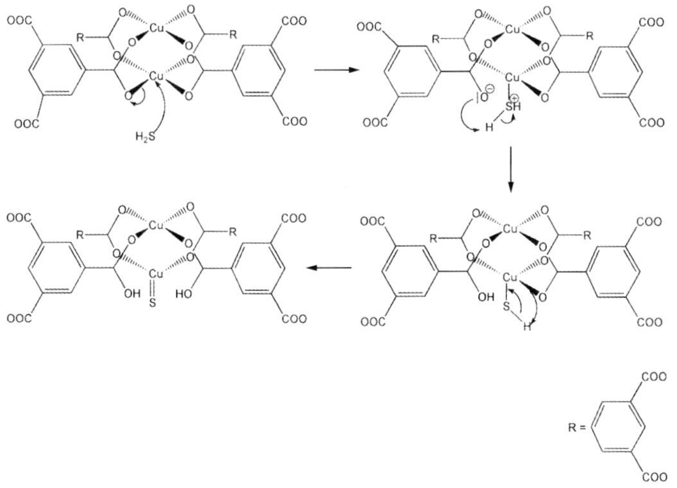

(10.4)

10.5 Beyond the MOF-Graphite oxides composites

The composites introduced above represent one particular group of materials. GO materials were demonstrated as excellent building units also for other composites. Some examples include Keggin cations [58], molybdenum and tungsten species [59], iron oxides [60, 61], silica [62], manganese oxides [63, 64], zirconium hydroxides [65], cobalt oxides [66], titanium oxides [67] or zinc (hydr)oxide [68–70]. All of them demonstrated superior performance in comparison with the parent materials with the range of applications starting from adsorbents through gas sensing [71] and energy storage [63, 66] to solar energy harvesting processes [67]. In all these applications the bonds of inorganic phase with the carbonaceous phase are important (Fig. 10.11). They provide sites of specific chemistry and also improve DC conductivity of the materials. The latter is important in energy-related applications [66–70].

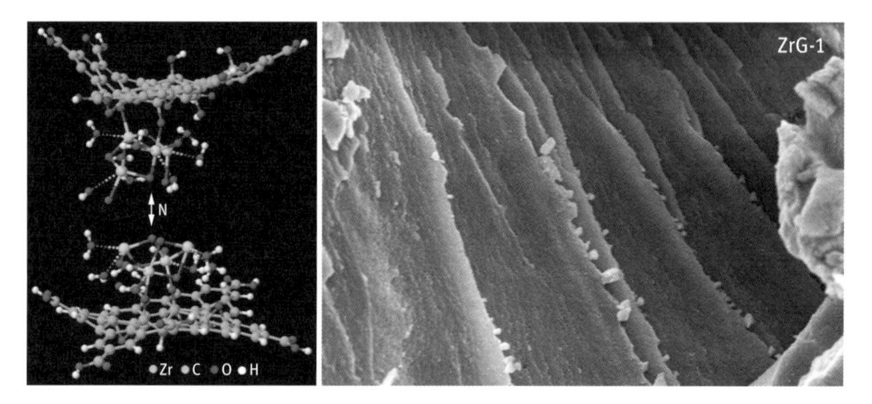

Fig. 10.11: $Zr(OH)_4$/GO composites (reprinted with permission from [65], © 2010, American Chemical Society).

Recently, the crucial role of graphite oxide and partially reduced graphite oxide was elucidated in their composites with semiconductors such a titanium oxide [67] or zinc oxide [68–70, 72]. These materials showed important photoactivity for water splitting [67, 72] and other surface reactions involving an oxidation process [68–70]. The main function of GO and reduced GO phase was indicated as separation of electrons and holes generated in a semiconductor by UV-Vis radiation and the prevention of their recombination. This significantly increased the quantum efficiency of the water splitting processes. Moreover, the graphite-based phase helped in electron transfer. This process was especially important for reactive adsorption of H_2S, [68] NO_2 [69], and SO_2 [70] on zinc (hydr)oxide /GO composites. In the case of the latter species, the water splitting process was indicated to take place at separated locations on the surface resulting in a very good performance of these materials as adsorbents [70]. Sulfides, sulfur and sulfates were the products of surface reactions. The formation of sulfides increased the photoactivity even more, changing the energy gap and creating heterojunctions on the surface. Another important function of GO in composites is its effect on activation of oxygen and thus in enhancing activation reactions [65, 68]. Activation and formation of superoxygen ions was caused by a specific oxygen functionality existing on its surface.

10.6 Summary

The composites described in this chapter present superior quality which is demonstrated by their surface properties and performance in comparison with the parent components, GO and MOF or other inorganic phases. The important aspect of these composite formations is taking advantage of the promising properties of both phases and the creation of the hybrid, which exhibits the surface features of both phases and, as a bonus, new unique properties created on the interface. Moreover, the specific behavior of the individual components when placed together can open the door for new applications, not foreseen in this concise chapter. One should see that the detailed characterization of these materials as adsorbents is only one example of their application, which we could explore in detail. Nevertheless, the zinc (hydr)oxide story, where the enhanced photoactivity and water splitting reactions were noticed while investigating the adsorption phenomena, is one more example of the "open book" of the usefulness of such new materials.

Bibliography

[1] Marsh H, Rodriguez-Resinoso F. *Activated Carbon*, Elsevier, Amsterdam, 2006; ISBN: 978-0-08-044463-5.
[2] Ania CO, Bandosz TJ. *Surface Chemistry of Activated Carbons and its Characterization In Activated Carbon Surfaces in Environmental Remediation,* Bandosz, T J, Ed. Elsevier, Oxford, 2006. 159–230.
[3] Hincal F, Erkegoglum P. Toxic Industrial Chemicals (TICs) – Chemical warfare without chemical weapons, *FABAD J. Pharm. Sci.* 2006, 31, 220–229.
[4] Bandosz TJ. Towards understanding reactive adsorption of small molecule toxic gases on carbonaceous materials, DOI 10.1016/j.cattod.2011. 08.017
[5] Bandosz TJ. On the adsorption/oxidation of hydrogen sulfide on unmodified activated carbons at ambient temperatures, *J. Coll. Interface*. 2002, 246, 1–20.
[6] Rowsell JLC, Yaghi OM. Metal–organic frameworks: A new class of porous materials, *Micro. Meso. Mater* 2004, 73, 3–14.
[7] James SL. Metal-organic frameworks, *Chem. Soc. Rev.*, 2003, 32, 276–288,
[8] Dreyer DR, Park S, Bielawski CW, Ruoff RS. The chemistry of graphene oxide, *Chem. Soc. Rev.* 2010, 39, 228–240.
[9] Zhu Y, Murali S, Cai W, Li X, Won Suk, J, Potts JR, Ruoff RS. Graphene and graphene oxide: Synthesis, properties, and applications, *Advance Mater.* 2010, 22, 3906–3924.

[10] Petit C, Seredych M, Bandosz TJ. Revisiting the chemistry of graphite oxides and its effect on ammonia adsorption, *J. Mater. Chem.* 2009, 19, 9077–9185.

[11] Szabó T, Berkesi O, Dékány I. DRIFT study of deuterium-exchanged graphite oxide, *Carbon* 2005, 43, 3186–3189.

[12] Hontoria-Lucas C, López-Peinado AJ, de D. López-González J, Rojas-Cervantes ML, Martín Aranda RM. Study of oxygen-containing groups in a series of graphite oxides: Physical and chemical characterization, *Carbon* 1995, 33, 1585–1592.

[13] Li JL, Kudin KN, McAllister MJ, Prud'homme RK, Aksay IA, Car R. Oxygen-driven unzipping of graphite materials, *Phys. Rev. Lett.* 2006, 96, 176101–176104.

[14] Buchsteiner A, Lerf A, Piepe Jr. Water dynamics in graphite oxide investigated with neutron scattering, *J. Phys. Chem. B* 2006, 110, 22328–22338.

[15] Jeong HK, Lee YP, Lahaye RJWE, Park MH, An KH, Kim IJ, Yang CW, Park CJ, Ruoff RS, Lee YH, Evidence of graphitic AB stacking order of graphite oxides, *J. Am. Chem. Soc.* 2008, 130, 1362–1366 .

[16] Szabo T, Tomabacz E, Illes E, Dekany I. Enhanced acidity and pH-dependent surface charge characterization of successively oxidized graphite oxides, *Carbon* 2006, 44, 537–545.

[17] Hirata M, Gotou T, Horiuchi S, Fujiwara M, Ohba M. Thin-film particles of graphite oxide 1: high-yield synthesis and flexibility of the particles, *Carbon* 2004, 42, 2929–2937.

[18] Hirata M, Gotou T, Ohba M. Thin-film particles of graphite oxide 1: Preliminary studies for internal micro fabrication of single particle and carbonaceous electronic circuits, *Carbon* 2005, 43, 503–510.

[19] Mkhoyan KA, Contryman AW, Silcox J, Stewart DA, Eda G, Mattevi C, Miller S, Chhowalla M, Atomic and electronic structure of graphene-oxide, *Nano Lett.*, 2009, 9, 1058 1063.

[20] Schniepp HC, Li JL,McAllister MJ, Sai H, Herrera-Alonso M, Adamson DH, Prud'homme RK, Car R, Saville DA, Aksay IA. Functionalized single graphene sheets derived from splitting graphite oxide, *J. Phys. Chem. B* 2006, 110, 8535–8539.

[21] Lerf A, He H, Forster M, Klinowski J. Structure of graphite oxide revisited, *J. Phys. Chem. B*, 1998, 102, 4477–4482.

[22] Brodie DC. Sur le poids atomique du graphite, *Ann. Chim. Phys.* 1860, 59, 466

[23] Hummers WS, Offeman RE. Preparation of graphite oxide, *J. Am. Chem. Soc.* 1958, 80, 1339–1339.

[24] Yaghi OM, O'Keeffe M, Ockwig NW, Chae HK, Eddaoudi M, Kim J. Reticular synthesis and the design of new materials, *Nature* 2003, 423, 705–714.

[25] Li H, Eddaoudi M, O'Keeffe M, Yaghi OM. Design and synthesis of an exceptionally stable and highly porous metal-organic framework, *Nature* 1999, 402, 276–279.

[26] Chui S, Lo SMF, Charmant JPH, Orpen AG, Williams ID. A chemically functionalizable nanoporous material $[Cu_3(TMA)_2(H_2O)_3]_n$, *Science* 1999, 283, 1148–1150.

[27] Millward AR, Yaghi OM Metal–Organic frameworks with exceptionally high capacity for storage of carbon dioxide at room temperature, *J. Am. Chem. Soc.* 2005, 127, 17998–17999.

[28] Peterson GW, Wagner GW, Balboa A, Mahle J, Sewell T, Karwacki CJ. Ammonia vapor removal by Cu(3)(BTC)(2) and Its characterization by MAS NMR, *J. Phys. Chem. C* 2009, 113, 13906–13917.

[29] Serre C, Millange F, Thouvenot C, Noguès M, Marsolier G, Louër D, Férey G. Very large breathing effect in the first nanoporous chromium(iii)-based solids: MIL-53 or $CrIII(OH)\cdot\{O_2C{-}C_6H_4{-}CO_2\}\cdot\{HO_2C{-}C_6H_4{-}CO_2H\}_x\cdot H_2O_y$. *J. Am. Chem. Soc.* 2002, 124, 13519–13526.

[30] Férey G, Serre C, Mellot-Draznieks C, Millange F, Surblé S, Dutour J, Margiolaki I. A hybrid solid with giant pores prepared by a combination of targeted chemistry, simulation, and powder diffraction, *Angew. Chem. Int. Ed.* 2004, 43, 6296–6301.

[31] Horcajada P, Surble S, Serre C, Hong DY, Seo YK, Chang JS, Greneche JM, Margiolaki I, Ferey G. Synthesis and catalytic properties of MIL-100(Fe), an iron(III) carboxylate with large pores, *Chem. Commun*. 2007, 2007, 2820–2822.
[32] Llewellyn PL, Bourrelly S, Serre C, Vimont A, Daturi M, Hamon L, De Weireld G, Chang JS, Hong DY, Kyu Hwang Y, Hwa Jhung S, Feìre, G. High uptakes of CO_2 and CH_4 in mesoporous metal-organic frameworks MIL-100 and MIL-101, *Langmuir* 2008, 24, 7245–7250.
[33] Petit C, Bandosz TJ. MOF-graphite oxide composites: combining the uniqueness of graphene layers and metal-organic frameworks, *Adv. Mater.* 2009, 21, 4753–4757.
[34] Petit C, Bandosz TJ. The synthesis and characterization of copper-based metal organic framework/graphite oxide composites, *Carbon* 2011, *49*, 563–572.
[35] Petit C, Bandosz TJ. Synthesis, characterization and adsorption properties of MIL(Fe) – graphite oxide composites: exploring the limits of materials' fabrication, *Adv. Function. Mater.* 2011, 21, 2108–2117.
[36] Petit C, Mendoza B, O'Donnell D, Bandosz TJ. Effect of graphite features on the properties of Metal–Organic Framework/Graphite hybrid materials prepared using an in situ process, *Langmuir* 2011, 27, 10234–10242.
[37] Houk RJT, Jacobs BW, El Gabaly F, Chang NN, Talec AA, Graham DD, House SD, Robertson IM, Allendorf MD. Silver cluster formation, dynamics, and chemistry in metal-organic frameworks, *Nano Lett.* 2009, 9(10), 3413–3418.
[38] Yang SJ, Choi JY, Chae HK, Cho JH, Nahm KS, Park CR. Preparation and enhanced hydrostability and hydrogen storage capacity of CNT@MOF-5 hybrid composite, *Chem. Mater*. 2009, 21, 1893–1897.
[39] Thomas KM. Adsorption and desorption of hydrogen on metal–organic framework materials for storage applications: comparison with other nanoporous materials, *Dalton Trans* 2009, 2009, 1487–1505.
[40] Seo YK, Hundal G, Jang IT, Hwang YK, Jun CH, Chang JS. Microwave synthesis of hybrid inorganic-organic materials including porous $Cu_3(BTC)_2$ from Cu(II)-trimesate mixture, *Micro. Meso. Mater*, 2009,119, 331–337.
[41] Vilatela JJ, Eder D. Nanocarbon composites and hybrids in sustainability: a review, *Chem. Sus. Chem*. 2012, 5(3), 456–478.
[42] Petit C, Mendoza B, Bandosz TJ. Reactive adsorption of ammonia on Cu-based MOF/graphene composites, *Langmuir* 26 (2010) 15302–15309.
[43] Levasseur B, Petit C, Bandosz TJ. Reactive adsorption of NO_2 on copper-based MOF and Graphite oxide-MOF composites, *ACS Appl. Mater. Interfaces* 2010, 3606–3613.
[44] Petit C, Mendoza B, Bandosz TJ. Hydrogen sulfide adsorption on metal-organic frameworks and metal-organic frameworks / graphite oxide composites, *ChemPhysChem* 2010, 11, 3678–3684.
[45] Thompson JC. Compressibility of metal-ammonia solutions, *Phys. Rev. A* 1971, *4*, 802–804.
[46] Slowiski EJ, Masterton WL. Qualitative analysis and the properties of the ions in aqueous solutions, *Brooks Cole*, 1990.
[47] Bandosz TJ, Petit C. On the reactive adsorption of ammonia on activated carbons modified by impregnation with inorganic compounds, *J. Coll. Interface Science* 2009, 338, 329–345.
[48] Petit C, Bandosz TJ Enhanced adsorption of ammonia on Metal-Organic Framework / graphite oxide composites: analysis of surface interactions, *Adv. Funct. Mater*. 2009, 19, 1–8.
[49] Greathouse JA, Allendorf MD. The interaction of water with MOF-5 simulated by molecular dynamics, *J. Am. Chem. Soc*. 2006, 128, 10678–10679.
[50] Petit C, Bandosz TJ. Exploring the coordination chemistry of MOF/graphite oxide composites and their applications as adsorbents, *Dalton Transactions*, in press, 2012.

[51] Petit C, Huang L, Jagiello J, Kenvin J, Gubbins KE, Bandosz TJ. Toward understanding reactive adsorption of ammonia on Cu-MOF/graphite oxide nanocomposites, *Langmuir* 2011, 27, 13043–13051

[52] Vimont A, Goupil J-M, Lavalley J-C, Daturi M, Surblé S, Serre C, Millange F, Férey, G, Audebrand N. Investigation of acid sites in a zeotypic giant pores chromium(III) carboxylate, *J. Am. Chem. Soc.* 2006, 128, 3218–3227.

[53] Marten B, Kim K, Cortis C, Friesner RA, Murphy RB, Ringnalda MN, Sitkoff D, Honig B. New model for calculation of solvation free energies: correction of self-consistent reaction field continuum dielectric theory for short-range hydrogen-bonding effects, *J. Phys. Chem.* 1996, 100, 11775–11788.

[54] Prestipino C, Regli L, Vitillo JG, Bonino F, Damin A, Lamberti C, Zecchina A, Solari PL, Kongshaug KO, Bordiga S. Local structure of framework Cu(II) in HKUST-1 Metallorganic framework: Spectroscopic characterization upon activation and interaction with adsorbates, *Chem. Mater.* 2006, 18, 1337–1346.

[55] Márquez-Alvare C, Rodríguez-Ramos I, Guerrero-Ruiz A. Removal of NO over carbon supported copper catalysts: II. Evaluation of catalytic properties under different reaction conditions, *Carbon* 1996, 34, 1509–1514.

[56] Ali IO. Preparation and characterization of copper nanoparticles encapsulated inside ZSM-5 zeolite and NO adsorption, *Mat. Sci. Eng. A* 2007, 459, 294–302.

[57] Nakamoto K. Complexes of alkoxides, alcohols, ethers, ketones, aldehydes, esters and carboxylic groups, in: *Infrared and Raman spectra of inorganic and coordination compounds*, Wiley, 2009, 62–67.

[58] Seredych M, Bandosz TJ, Adsorption of ammonia on graphite oxide / aluminum polycation and graphite oxide/zirconium polyoxycations composites, *J. Colloid Interface Sci.* 2008, 324 25–35.

[59] Petit C, Bandosz TJ. Grapite Oxide/Plyoxometalate nanocomposites as adsorbents of ammonia, *J. Phys Chem. C* 2009, 113, 3800–3809.

[60] Bashkova S, Bandosz T.J. Adsorption/reduction of NO_2 on graphite oxide/iron composites, *Ind. Chem. Eng. Res.* 2009, 48, 10884–10891.

[61] Morishige K, Hamada T. Iron oxide pillared graphite, *Langmuir* 2005, 21, 6277–6281.

[62] Matsuo Y, Tabata T. Preparation and characterization of silylated graphite oxide, *Carbon* 2005, 43, 2875–2882.

[63] Seredych M, Bandosz TJ. Manganese oxide and graphite oxide/MnO_2 composites as reactive adsorbents of ammonia at ambient conditions, *Microporous and Mesoporous Materials* 2012, 150, 55–63.

[64] Yu G, Hu L, Vosgueritchian M, Wang H, Xie X, McDonough JR, Cui X, Cui Y, Bao Z. Solution-processed graphene/MnO_2 nanostructured textiles for high-performance electrochemical capacitors, *NanoLett* 2011, *11*, 2905–2911.

[65] Seredych M, Bandosz TJ. Effects of surface features on adsorption of SO2 on graphite oxide/Zr(OH)4 composites, *J. Phys. Chem. C* 114 (2010) 14552–14560.

[66] Yang S, Cui G, Pang S, Cao Q, Kolb U, Feng X, Amier J, Mullen K. Fabrication of cobalt and cobalt oxide/graphene composites: Towards high-performance anode materials for lithium batteries, *ChemSusChem* 2010, 3, 236–239.

[67] Yang HB, Guo CX, Guai GH, Song QL, Jiang SP, Li CM. Reduction of charge recombination by an amorphous titanium oxide interlayer in layered graphene/quantum dots photochemical cells, *ACS Appl. Mater. Interfaces* 2011, 3, 1940–1945.

[68] Seredych M, Mabayoje O, Bandosz TJ. Visible-light-enhanced Interactions of hydrogen sulfide with composites of zinc(oxy)hydroxide with graphite oxide and graphene, *Langmuir* 2012, 28, 1337–1346.

[69] Seredych M, Mabayoje O, Bandosz TJ. Interactions of NO_2 with zinc (hydr)oxide/graphene phase composites: Visible light enhanced surface reactivity, *J. Phys. Chem. C* 2012, DOI 10.1021/jp211141j.
[70] Seredych M, Mabayoje O, and Bandosz TJ. Role of water splitting in visible light enhanced reactive adsorption of SO_2 on composites of Zinc(Oxy)Hydroxide with Graphite Oxide and Graphene, *Adv. Funct. Mater.*, submitted
[71] Zhan Z, Zheng L, Pan Y, Sun G, Li L. Self-powered, visible light photodetector based on thermally reduced graphene oxide-ZnO(rGO-ZnO) hybrid nanostructure, *J Mater. Chem.* DOI 10.1039/c1jm13920.
[72] Strelko VV, Kutz VS, Thrower PA. On the mechanism of possible influence of heteroatoms of nitrogen, boron and phosphorus in a carbon matrix on the catalytic activity of carbons in electron transfer reactions, *Carbon* 2000, 38, 1499–1502.

Part III: **Applications of nanocarbon hybrids**

Dang Sheng Su

11 Batteries/Supercapacitors: Hybrids with CNTs

11.1 Introduction

Energy must be storable for any kind of application at any time in a sustainable society using renewable energy resource. This requires high-capacity storage but also cost-effective technology. There is no single way that is universal for different applications, but electrochemical energy storage is currently the most versatile technology. It is flexible to cover energy demand at small and large scales and at variable time domains. Electrochemical energy storage systems are composed of electrodes and electrolyte as the main components assembled in a compact system. Depending on the storage mechanism, they may be divided as batteries or capacitors. Rechargeable batteries and supercapacitors are the representatives for such storage systems. The most important components in such system are electroactive materials for the construction of the electrode. Developing new electrode materials for high performance electrochemical storage system has been a core challenge in materials science and electrochemistry as reflected by the numerous publications in the literature [1, 2, 3].

Carbon nanotubes (CNTs) have been tested as electrode materials by many groups as early as the 1990s for electrochemical energy storage [4, 5, 6]. CNTs possess the following characteristics that could be of essential importance for electrochemical energy storage:

- CNTs differ from graphite in the bonding state of carbon atoms. The rolling up of the graphene sheet to form the tube causes a rehybridization of carbon bonding orbitals (nonplanar sp^2 configuration), thus leading to a splaying-out of the πdensity of the graphene sheet, whose magnitude is related to the diameter of the tube (see Chapter 1);
- CNTs differ from activated carbon or carbon black in the ordering degree. Activated carbons usually do not have a well-defined long-range ordering while each CNT is a rolled-up graphene layer with well-defined short-range and long-range ordering [7];
- CNTs have a different porous structure than activated carbon. The specific surface area of CNTs can range from 50 m^2/g (multi-walled CNTs with 50 graphene walls) to 1315 m^2/g (single-walled CNTs). Theoretically, the porous structure of CNTs is identical to the tubular structure of CNTs and the pore sizes of CNTs correspond to the inner diameters of opened CNTs and should have a narrow distribution. Activated carbons usually have a broad pore distribution covering micropore, mesopore and macropore;
- CNTs have higher thermal and electric conductivity than activated carbon or carbon black. The good electro-conductivity of CNTs is especially important when

they are used to support electro-active materials allowing a facile transfer of electron;
- The cavity of a CNT could play a special role for charge storage due to the special tubular structure of CNTs.

These characteristics of CNTs have stimulated various works to use them as electrode materials for batteries and supercapacitors, as reviewed by numerous publications [8, 9, 10]. Due to the special properties, CNTs have a multiple storage mechanism for lithium ions [8]. Carefully engineering functionalized CNTs (Chapter 3) by some special techniques could provide the electrode for a high power lithium battery [11], but currently the most promising application of CNTs in electrochemical storage systems is their use as additives rather than as active materials in electrodes [12, 13]. The recent tendency is to design carbon hierarchical materials including CNT-based hybrids for electrochemical applications. The hybridization of CNTs with different types of electroactive materials (nanocarbons such as graphene, carbon nanofibers, or nanomaterials of metal, metal oxide, metal sulfides or other compounds, see Chapter 5) combine the advantage of both components and is crucial for enhancing the advanced functions in various applications [14]. Hybrid materials can enable versatile and tailor-made properties with performances far beyond those of the individual materials. A new class of CNT-based functional hybrids is highly desirable for optimizing the optical, electrical and catalytic properties of CNTs and their counterpart and enhancing its performance in electrochemistry [15, 16].

This chapter deals with the potential and perspective of CNT-based carbon and inorganic hybrid materials for battery and capacitor applications.

11.2 Application of hybrids with CNTs for batteries

11.2.1 Lithium ion battery

The Li-ion battery (LIB) is the example for the great successes of modern electrochemistry [17, 18]. LIBs have been commercialized and used in portable electronic systems. An LIB consists of an anode and a cathode that are capable of reversibly hosting Li in ionic form. The electrolyte contains lithium salts dissolved in an organic carbonate solution. The storage capacity of a battery is determined by the amount of Li that can be stored reversibly in the two electrodes. The current LIBs have an energy storage capacity of no more than 250 Wh kg^{-1}. The main challenges for high-performance batteries for transport applications and stationary storage of energy on a large scale are:
(1) Increasing the gravimetric and/or volumetric energy density;
(2) Improving the high rate performance;
(3) Enhancing the safety for applications in various environment and boundary conditions.

The current material for the anode of an LIB is graphite that can accommodate lithium ions by intercalation in a charging state. Every six carbon atoms accommodate one lithium ion (LiC_6). The capacity of Li uptake is 372 $mAhg^{-1}$. Differing from graphite, nanostructured carbons have more possibility to store lithium [19]. The storage possibilities of lithium in CNTs are multiple through the following processes, as it is partly illustrated in Fig. 11.1(a) [8]:

(1) Intercalation (LiC_6 stoichiometry);
(2) Adsorption and accumulation on the outer surface;
(3) Adsorption and accumulation in the inner channel when CNTs are open;
(4) Storage in the void space between bundles of tubes.

The Li storage in CNT materials can be additionally increased through:

(5) Adsorption on the surface of CNTs *via* reaction with heteroatoms or functional groups;
(6) Storage in forms of 1, 2, 4 and 5 in/on the graphitic or amorphous co-components in CNT materials (intended or unintentional from impurities).

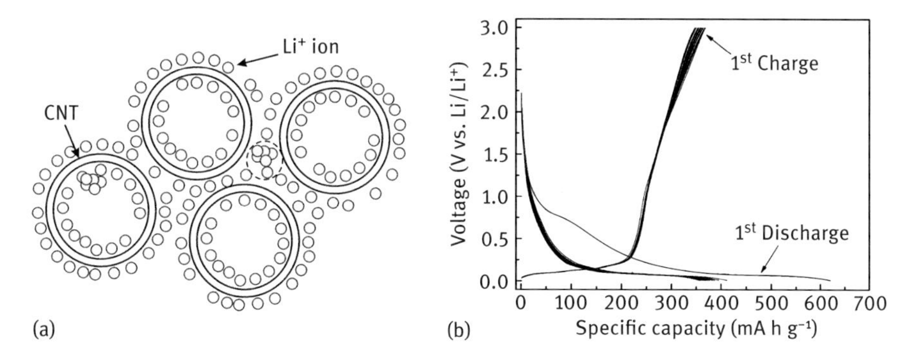

Fig. 11.1: (a) Schematic cross-sectional view of various lithium storage positions in an open CNT (drawing after M. Endo). Lithium ions that are intercalated into the graphitic layers or adsorbed/reacted with defects or functional groups are not shown. (b) Galvanostatic discharge (Li insertion, voltage decreases)/charge (Li extraction, voltage increases) curves of CNT sample cycled at a rate of C/5 in 1M $LiPF_6$ in EC/DMC solution (courtesy of Dr. Yongsheng Hu, PR China, reprinted from Su *et al.* [8] with kind permission from Wiley VCH (2010)).

CNTs have been tested as anode materials for LIBs [4, 5, 6, 20]. Fig. 11.1(b) shows the discharge (Li insertion)/charge (Li extraction) curves of an electrode comprising CNTs with highly-ordered graphitic structure, cycled in 1M $LiPF_6$ ethylene carbonate (EC)/dimethyl carbonate (DMC) (1 : 1 by volume) at a rate of C/5. Flat plateaus can be observed at low voltages in the discharge and charge curves, and are ascribed to Li intercalation–deintercalation between graphitic layers of CNTs. Sloped regions in the discharge–charge curves are ascribed to Li *via* other mechanisms mentioned above.

A large irreversible capacity is observed in the first discharge and charge process, which is due to the formation of a solid-electrolyte interface (SEI). The first charge capacity is about 360 $mAhg^{-1}$ and the reversible capacity stabilizes at 350 $mAhg^{-1}$ after 20 cycles. It is obvious from Fig. 11.1(b) that a significant part of capacity is not obtained at a constant potential. At constant voltage, the capacity is only 200 $mAhg^{-1}$. In fact, the performance of CNTs in Li-ion batteries as shown in Fig. 11.1(b) is only an exception for CNTs with well-graphitized wall structure. The diversity of CNTs' microstructure and texture [21] as well as surface functionality combined with the multiple storage mechanisms renders the reported Li storage capacities highly variable, but usually with a huge irreversible capacity and hysteresis [22]. The Li intake can be as high as 1400 $mAh\ g^{-1}$ or only a few hundred $mAhg^{-1}$, with rather different irreversible and reversible capacities [8]. This is expected since the abovementioned storage mechanisms of Li in CNT-composed electrodes are strongly dependent on textural parameters and the degree of graphitization of the CNTs used [22].

Although CNTs provide a new possibility for lithium ion storage, process optimization is, however, still required to explore the multiple lithium storage mechanism to get high-power lithium batteries. For instance, redox of surface oxygen-containing functional groups on CNTs by lithium ions in organic electrolytes can occur reversibly at high power, and thus provide a possibility for high energy and high power electrode materials. This has been maximally explored for the functionalized CNTs assembled by a layer-by-layer technique to an additive-free and densely packed electrode [11]. This is a good example that there must be a concerted cooperation between materials scientists and electrochemists to develop materials for next generation batteries. CNT-based hybrid materials could be an interesting candidate for such efforts and have shown some promising perspectives.

11.2.1.1 CNT- based hybrids for the anode

CNT-based carbon hybrids

The good electrical conductivity and high graphitization degree of CNTs are the main reasons for the construction of CNT-based carbon hybrid materials (Fig. 11.2). CNTs can be container or supporter for CNFs or CNTs of smaller size (Fig. 11.2(a) and (b)), can "intercalate" graphene vertically or horizontally (Fig. 11.2(c) and (d)). These are only four representative CNT-based carbon hybrids. This kind of hybrid structure is one example of the hierarchical ordering of nanocarbons allowing the combination of otherwise conflicting properties, such as high storage capacity, high mechanical stability, and good electrical contact [8]. CNFs and graphenes can store Li *via* the above mentioned multiple mechanism and exhibit a high initial storage capacity [23, 24], but they suffer from volumetric changes during charging–discharging and have a large irreversible loss of Li ions that is not only limited to the first few cycles. CNTs of highly graphitized structure are mechanically stable and robust. A combination of these carbons forming hybrid materials would ideally increase the storage capacity and improve the cycle life. This is evidenced by Zhang *et al.* using CNT-CNF hybrid: CNTs (50 nm in diameter) are used as nanocontainers to confine CNFs (5–6 nm in diameter) [25]. The CNT-CNF hybrid (denoted as CNFs@CNTs in the following) concept is shown schematically in Fig. 11.2(a). It can be synthesized by (1) selective deposition of the active metal (Co) in the channel of the CNTs, and (2) growth of CNFs *via* catalytic chemical vapor deposition.

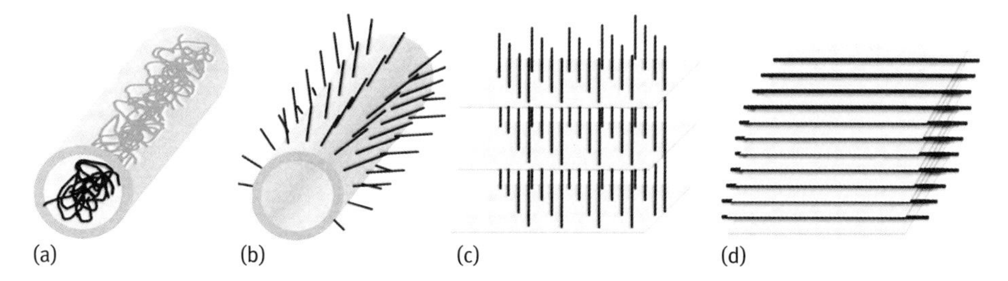

Fig. 11.2: Schematic illustration of several CNT-based carbon hybrids.

The electrochemical performance of electrolyte at a rate of C/5 is given in Fig. 11.3(a). The discharge and charge curves of CNFs@CNTs in 1M $LiPF_6$ EC/DMC (1 : 1 by volume) exhibited extended flat plateaus. The sloped region in the discharge/charge curves is attributed to Li insertion/extraction into the disordered structure of the CNFs@CNTs electrode. During 120 cycles, the reversible capacity of CNFs@CNTs remained approximately constant at 410 $mAhg^{-1}$ over a potential region between 0 and 3 V, while it gradually decreased to 258 $mAhg^{-1}$ when the electrode was formed from commercial CNTs (Fig. 11.3(b)). The obtained stability of CNFs@CNTs over less-graphitized carbon materials mainly arose from steric hindrance effects which suppress the diffusion of big electrolyte molecules through wall defects. Their outstanding cycling performance in combination with their high storage capacity shows the advantage of having CNT-CNF hybrids rather than units of CNTs or CNFs for LIBs.

CNT-graphene hybrid materials [26, 27, 28, 29] could deliver synergistic effects between two different graphitic components with respective unique properties. The electrical conductivity along the vertical direction of graphene nanosheets (GNS) can be improved by bridging GNS vertically with CNTs (Fig. 11.2(c)) and graphene layers can also stabilized by aligned CNTs vertically or horizontally in-between the graphene planes in the CNT-GNS hybrid composite (Fig. 11.2(c),(d)). This hybrid structure can additionally prevent the restacking of graphene layers and facilitate the diffusion of lithium ions into CNTs with large aspect ratio. Such CNT-GNS hybrid nanostructure was of importance to the battery performance enhancements including capacity, cyclability and rate capability. Chen *et al.* demonstrated that the electrode composed of CNT-GNS hybrid can display highly reversible capacities of 573 mAh g^{-1} at a small current of 0.2C and 520 mAh g^{-1} at a large current of 2C [30].

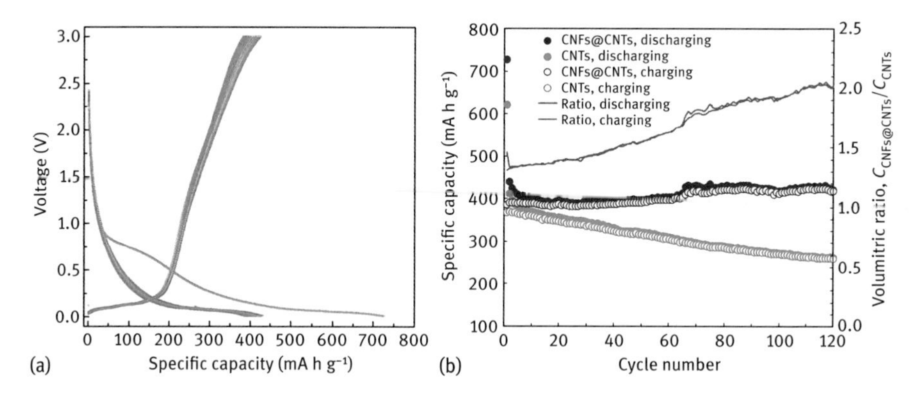

Fig. 11.3: Electrochemical performance of CNFs@CNTs. (a) Galvanostatic discharge/charge (Li insertion/extraction, voltage decrease/increase) curves of CNFs@CNTs at a cycling rate of C/5 in 1M $LiPF_6$ in 1 : 1 (v/v) ethylene carbonate (EC)/dimethyl carbonate (DMC); (b) comparison of the electrochemical performance of pristine CNTs and CNFs@CNTs in 1M $LiPF_6$ in EC/DMC solution (reprinted with permission from [25]).

Fan and co-workers also reported that CNT-graphene hybrid material, prepared *via* a CVD approach in a fluidized bed reactor, shows high reversible capacity (667 mAh g^{-1}), high-rate performance (189 mAh g^{-1} at 6 mA cm^{-2}), and good cycling stability [31]. Graphene platelets arranged almost perpendicularly to the fiber axis form nanochannels that are favorable for lithium ion diffusion from different orientations. The obtained hybrid materials contain many cavities, open tips, and graphene platelets with exposed edges, providing more extra space for Li^+ storage, in agreement with the abovementioned multiple storage mechanism in nanocarbons. In addition, 3D interconnected architectures facilitate the collection and transport of electrons during the cycling process. The electrochemical performance of the CNT-graphene hybrid is superior to that when CNTs, CNFs, graphene are used individually [31].

CNT-based hybrids, when carefully engineered, have additional advantages that could be used for high-performance batteries. Vertically aligned CNTs (VACNTs) possess the advantages of a high degree of order, good controllability, and easy manipulation. Graphene paper (GP) has good conductivity and can be used to collect current released during charging/discharging of the electrodes in a LIB. Li *et al.* grew VACNTs directly onto a freestanding and ~ 3 µm thick GP by using CVD [32]. A VACNT-GP hybrid material is obtained (Fig. 11.4(a)). GP serves as current collector and VACNT for Li storage. Figure 11.4(b) shows the discharge capacity versus the cycle number for a cell with the VACNT/GP film as the anode. A stable discharge capacity of 290 mAh g^{-1} at 30 mA g^{-1} was achieved after 40 cycles, with a coulombic efficiency of ~ 95 %. This is higher than the values previously reported for films composed of randomly entangled CNTs [33] showing the advantages of hybrid materials with a high degree of ordering. The bare GP exhibited a much lower reversible capacity of 155 mAh g^{-1} after 20 cycles. When the current rate was increased step by step from 60 mA g^{-1} to 3.6 A g^{-1}, capacities of 265 and 55 mAh g^{-1} were achieved respectively. As is shown in Fig. 11.4(b), when the current rate was reset to 30 mA g^{-1}, even a slight increase (6.9 %) in lithium storage capacity to 310 mAh g^{-1} was observed, indicating a good rate performance and excellent electrochemical stability of the VACNT-GP electrode. This high-rate capability is attributed to the unique structure of the VACNT-GP hybrid electrode with a large number of transport paths provided by the VACNTs rooted firmly on the conducting GP. Although the capacity of the VACNT-GP electrode is somehow lower than some types of novel carbons, the most highlighted advantage of such a hybrid carbon electrode for LIB is the integration of conductive active materials with flexible and stable current collector.

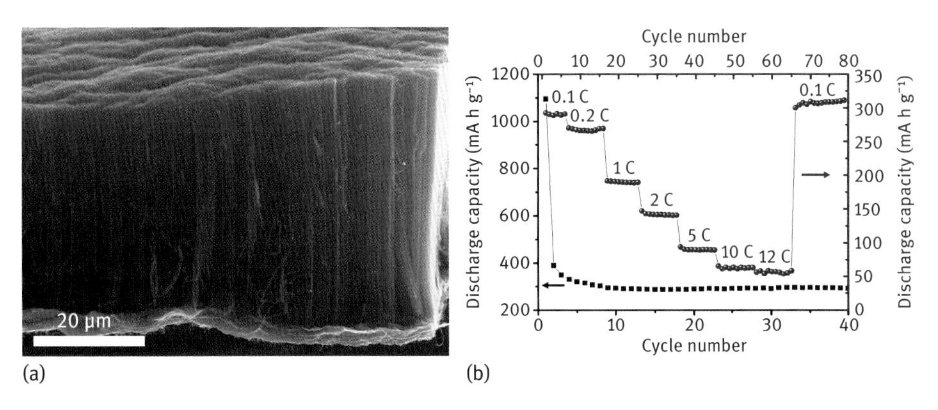

Fig. 11.4: (a) Cross-section SEM observation of the VACNT/GP film; (b) cyclic performance and high-rate capability of the VACNT/GP film anode of an LIB (reprinted with permission from Shisheng Li, *et al.*, Adv. Energy Mater. 2011, 1, 486–490).

All these CNT-based carbon-carbon hybrids give examples of carbon materials with a hybrid nanoarchitecture, combining two or more building blocks in a desired fashion, in which the advantages of each unit are explored while the disadvantages are suppressed. We observe here a storage behavior comparable to or better than that of good graphite. Graphite is however less sensitive to cycling kinetics. This sensitivity is reduced in CNT-based hybrids due to the nanostructuring and the highly accessible nature of the system. There is still room for substantial improvement of the capacity and the storage kinetics through dimensional optimization and through improvement of the structural ordering of the storage medium by optimizing the deposition kinetics and by changing the catalyst system in the growth of CNT-based hybrids.

CNT-based inorganic hybrids

CNT-based inorganic hybrid is another kind of emerging material with promising performance as anode materials of LIBs [15]. The appealing reason is the fact that some metals or semiconductors can react with Li forming alloys, Li_xM_y, through electrochemical processes [34]. The reaction involves a large number of atoms per formula unit, is partially reversible and thus can be used for lithium storage. For instance, $Li_{4.4}Sn$ has a gravimetric and volumetric capacity of 993 mAh g^{-1} and 1000 mAh cm^{-3}, *vs.* 372 mAh g^{-1} and 855 mAh cm^{-3}, respectively, for graphite. The corresponding values for $Li_{4.4}Si$ are 4200 mA h g^{-1} and 1750 mA h cm^{-1}. However, the diffusion rate of Li ions in solid state reaction is low, and the accommodation of a large amount of Li through electrochemical alloy formation (lithiation) also results in a large volume expansion-contraction of the host materials [35]. The volumetric change with large mechanical strain leads to deterioration of the electrode (cracking, crumbling, or eventually pulverization) and causes a substantial loss of capacity in addition to a shortened lifetime of only a few charge-discharge cycles [35].

Carbon-based nanocomposite concepts have been successfully developed to limit or reduce these adverse effects and at the same time enhance the electron or ion transport [8]. CNT is an ideal building block in the carbon-inorganic composite/hybrid due to its mechanical, physical, chemical properties as mentioned above. CNTs are apparently superior to other carbonaceous materials such as graphite or amorphous carbon and are more adaptable to the homogeneous dispersion of nanoparticles than other carbonaceous materials [36].

CNT-based hybrids of Si [37], Sn [38], SnO_2 [2, 39], SnO_2-Au [40] have been tested as anode materials for LIBs. It is reported that an electrode composed of CNT-Si hybrids can have a reversible specific capacity of 940 mAh g^{-1}, a coulombic efficiency of 80 % for the first cycle and good cyclic performance [37]. The improvement of the cyclic performance for silicon particles is related to maintaining the electronic contact during cycling and forming a thick SEI film on it. A much higher reversible stable capacity of about 2050 mAh g^{-1} with very good rate capability and an acceptable first cycle irreversible loss of about 20 % was reported for CNT-Si hybrid materials when CNTs are vertically aligned with nanoscale amorphous/nanocrystalline Si droplets deposited on the surface [41]. An Sn/C composite structure, namely Sn@carbon encapsulated in bamboo-like hollow carbon nanofibers, as a potential anode material for LIBs displays a high reversible capacity of 737 mAh g^{-1} after 200 cycles at 0.5 C [42]. It also exhibits a reversible discharge capacity as high as 480 mAh g^{-1} when cycled at 5 C.

The fact that CNT-based inorganic hybrids can enhance the cycle life of the electrode is impressively demonstrated by the work of Part and co-workers [43]. They fabricated a CNT-SnSb hybrid by means of reductive precipitation of metal chloride salts. The anodic performances of the CNT-SnSb hybrid, CNTs and SnSb alloys in the potential range from 0.05 to 1.5 V (vs Li/Li^+) reveal the advantage of using the CNT-based hybrid (Fig. 11.5(a)): the CNT-SnSb hybrid has good cyclic performance, a high reversible specific capacity (> 480 mAh g^{-1}), and high coulombic efficiency (> 95 %) up to the 50th cycle (Fig. 11.5(b)). The nanosized SnSb alloy contributed to the increases in Li^+ storage, whereas randomly distributed CNTs contributed to maintaining the electronic conduction around the SnSb alloy particles, as well as accommo-

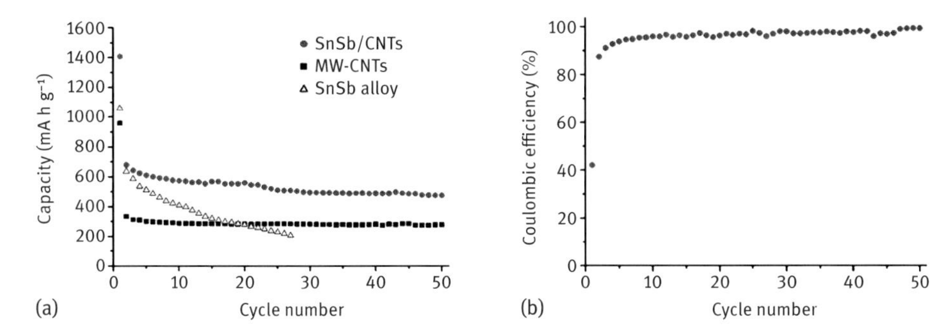

Fig. 11.5: Anode performance of the SnSb-CNT nanocomposite: (a) cyclic performance of SnSb-CNT nanocomposite and normal CNTs up to the 50th cycle at the same current density, 100 mA g^{-1}; and (b) the coulombic efficiency of SnSb-CNT nanocomposite electrode up to the 50th cycle. Reprinted with permission from [43].

dating the volume variation that was mainly induced by the reaction between Li^+ and SnSb.

Recently, A. Goyal and co-workers showed that CNT-Cu_2O hybrid material is a good candidate for flexible LIBs [44]. They prepared binder-free hybrid electrodes by conformally coating CNTs with Cu_2O *via* electrodeposition and then embedded the resulting architecture into a porous poly(vinylidene fluoride)-hexafluoropropylene (PVDF-HFP)–SiO_2 polymer electrolyte membrane. Aligned CNT arrays are preferable to be used for flexible LIBs since they offer both mechanical flexibility and robustness. CNTs are energy storage materials, but have limited capacity. The introduction of Cu_2O improves their capacity. The presence of CNTs in the core of the hybrid system acts as not only an electrical cable but also as a buffer to absorb these volume expansions during the conversion reaction of Cu_2O to Cu and Li_2O. The synergistic presence of high-capacity transition metal oxides and conductive CNTs results in twice the reversible areal capacity of 2.3 mAh cm^{-2} as compared to 1.2 mAh cm^{-2} for pure CNTs. CNT-MoO_2 coaxial hybrid arrays have also been synthesized and applied in LIBs showing promising performance [45, 46].

CNT-based inorganic hybrid materials are part of carbon-based inorganic hybrid materials as anodic electrodes in LIBs. The concept has been proven to be successful at least at laboratory scale, and is promising as a potential alternative to replace graphite-based anodes. However, little is known about the interface structure between CNT and the supported active materials, and thus the electron transfer between the two components. More detailed fundamental research on the interface and interaction between CNTs and active materials at atomic level is needed for a better understanding of the abovementioned improvement.

11.2.1.2 CNT-based hybrids for the cathode

The positive electrode (cathode) of an LIB consists of a host framework into which the mobile (working) cation is inserted reversibly [47, 48]. $LiCoO_2$ has been the dominant cathode material for lithium-ion batteries. Other lithium transition-metal phosphates ($LiMPO_4$, M = Fe, Co, Ni, Mn) and lithium transition-metal oxide spinels (LiM_2O_4, M = Mn, Ni) have been the focus of many research works to developing high performance LIBs [49, 50]. One of the main drawbacks of such lithium-containing compounds is its low electronic conductivity and low Li-ion diffusivity during charge/discharge. The most promising solution to overcome the inherently bad conductivity is to combine the active materials with nanocarbons [51, 52].

CNTs provide a unique opportunity for realizing such an improvement, since they display good electrical conductivity along their tube axis and a high surface area suitable for tethering of electroactive compounds, especially those with nanoparticulate morphology. An electrode containing CNTs as conducting agent is superior to one containing carbon black in terms of both high-rate (1 C) performance and cycle life [53]. This improvement is due largely to the resilience of the CNT aggregates that form conductive bridges between particles of the active material. These resilient bridges maintain intimate contacts between the particles even when the composite expands on cycling. By contrast, similar but rigid bridges of carbon black in the cathode are broken on cycling. Kang's group prepared a composite cathode using CNT-$LiFePO_4$ hybrid and obtained a 155 mA h g^{-1} capacity at C/10 rate in the first cycle [54]. The formation of a three-dimensional network wiring by CNTs improves the electronic conductivity of the composite cathode enabling $LiFePO_4$ to discharge high capacity at fast rate with high efficiency.

Schneider and co-workers studied especially the influence of the arrangement of CNTs on the electrochemical performance [55]. They prepared hybrid materials composed of $LiMPO_4$ (M = Fe or Co) with CNTs by tethering lithium phosphate olivine on isolated stochastically disordered CNTs as well as on ordered 3D CNT arrays *via* solution-based impregnation routes. Hybrids with stochastically arranged CNTs exhibit better electrochemical performance when compared to the ordered 3D MWCNT/$LiCoPO_4$ composite architecture, explained by the better connection, and thus electronic conductivity, between the $LiCoPO_4$ nanoparticles and the stochastically ordered CNTs in the former composite. This finding differs from other studies [56] using aligned CNTs array composite.

The concept to use CNT-$LiMPO_4$/LiM_2O_4 composite/hybrid to increase the rate performance (not the energy density) of LIBs gives some interesting and promising aspects. But the work in this research area is by far not as systematic as for the anodic side. As discussed above, the performance depends on many factors, and there is still no study on how the microstructure and aspect ratio of CNTs influence the performance of the obtained cathode. One other technical problem to use such composite/hybrid cathode materials is the impurity in CNTs remaining after the CVD process.

Just a trace of such a metallic impurity may make the CNTs not useful as additive for campsite/hybrid cathodes of LIBs.

11.2.2 Lithium sulfur battery

Lithium-sulfur (Li-S) batteries use lithium metal as anode, sulfur composite as cathode and could have very high theoretical capacity (1675 $mAh \cdot g^{-1}$) and specific energy (2567 $Wh \cdot kg^{-1}$) [57]. The energy density of a Li-S battery is at least five times higher than LIBs and sulfur is cheaper, abundant and environmental compatible. But the Li-S battery as a new technology faces numerous challenges and practical exploitation is still a long way off. The challenges to be solved mainly include the insulating nature of sulfur that retards its reduction, leading to low power, and a short cyclic life due to the dissolution of polysulfide intermediates in the electrolyte [57].

Carbon-sulfur hybrid materials, i.e., porous carbons [58, 59] or CNTs having nanosized S in the pores or channels, are the most promising solution for the Li-S battery to increase the electronic and ionic conductivity of sulfur or sulfide, and prevent, to a great extent, the solubility of the polysulfide ions formed on reduction of S or upon oxidation of insoluble sulfides [60]. An intimate contact between carbon and sulfur is essential [61].

Single-walled and multi-walled CNTs have been used to construct CNT-S hybrid material as a cathode in a Li-S battery. A SWCNT-S hybrid containing coaxial nanocables exhibited reversible capacities of 676 and 441 mAh g^{-1} for the first discharging cycle at a current rate of 0.5 C, 100th at 1.0 C, respectively [62]. These capacities are much higher than the corresponding capacities of the MWCNT-S cathode, due to the high dispersion of S on the SWCNTs with high specific surface area. By introducing polyethylene glycol (PEG) as a physical barrier to trap the highly polar polysulfide species, the cycling stability of the reversible capacities is expected to be further improved [62].

Significant improvement is achieved by using the SWCNT-graphene hybrid (Fig. 11.6(a)) to form a hybrid with sulfur. Zhao *et al.* prepared SWCNT-graphene hybrid in a high-temperature CVD process using FeMgAl LDH flakes to achieve the *in situ* deposition of both SWCNTs and graphene [63]. SWCNT-G hybrids, with SWCNTs grown on the graphene plane being available after the removal of the as-calcined FeMgAl layered double oxide flakes. Covalent C-C bonding between the graphene and SWCNTs is suspected. SWCNT-G hybrid has a 3D electrical conductive network (Fig. 11.6(a)) to accommodate S and exhibited excellent rate performance for Li-S batteries (Fig. 11.6(b)). A reversible capacity of 928 mAh g^{-1} can be achieved at 1 C current rate with a S loading amount of 60 wt%. At a very high current rate of 5 C, a capacity as high as *ca.* 650 mAh g^{-1} can be preserved even after 100 cycles with a coulombic efficiency of *ca.* 92%, which is very promising as potential electrode material for Li-S batteries.

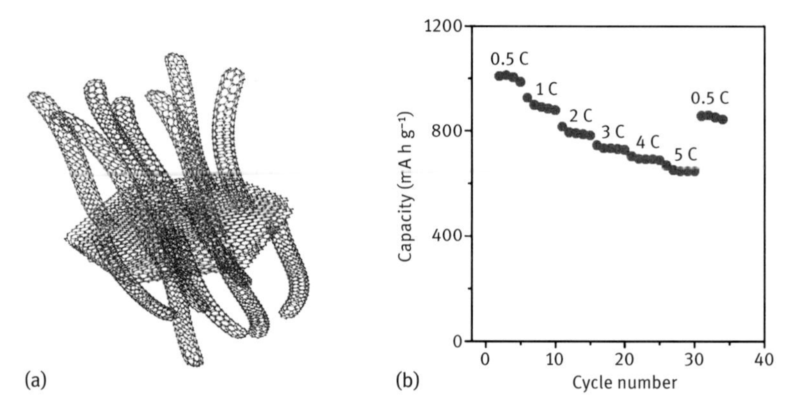

Fig. 11.6: (a) Schematic illustration of SWCNT-G hybrids used for (SWCNT-G)-S hybrids; (b) rate performance of the (SWCNT-G)-S hybrid electrode for Li-S cells. Reprinted with permission from [63].

These examples and several additional ones published recently in the literature on these topics [64] show that carbon materials, especially the carbon-carbon hybrids, play a crucial role in developing high performance Li-S batteries. The conductive nanocarbon framework precisely constraining sulfur within its pores/channels generates essential electrical contact to the insulating sulfur and thus provides access to Li^+ ingress/egress for reactivity with the sulfur. The kinetic inhibition to diffusion within the framework and the sorption properties of the carbon aid in trapping the polysulfides formed during redox [65]. It remains to optimize the amount of sulfur to get a maximal performance of the hybrid electrode.

11.2.3 Lithium air battery

The Li-air battery is another option to go beyond the horizon of Li-ion batteries [64]. In a Li-air cell, either based on aqueous or non-aqueous electrolytes, the negative electrode is lithium metal and the positive electrode consists of porous carbon with catalyst [66]. In both cases, on discharge, the Li-metal anode is oxidized, releasing Li^+ into the electrolyte, and the process is reversed on charge. At the positive electrode, O_2 from the air enters the porous cathode, dissolves in the electrolyte within the pores and is reduced at the electrode surface on discharge. Based on the reversible reaction $2Li + O_2 \leftrightarrow Li_2O_2$, an Li-air battery can have a specific energy of 3505 Wh kg^{-1}, which is much larger than those of the $C/LiMn_2O_4$ (or $C/LiCoO_2$, $C/LiFePO_4$) series of rechargeable LIBs [64].

The positive electrode should have a porous structure to support the transport of O_2 to the electrolyte/electrode interface as much as possible through the gas phase rather than by the slower diffusion in the electrolyte. CNT materials with mesopores and space between the entangled CNTs have been tested as electrode materials for Li-air batteries [67]. A nitrogen-doped CNT electrode has a specific discharge capacity of 866 mAh g^{-1}, which is about 1.5 times that of CNTs [67]. The heteroatom nitrogen doping may play a role in oxygen reduction, but the detailed fundamental mechanism is under study.

CNT-based inorganic hybrids have been tested to catalyze the anion formation in the air electrode of Li-air batteries [68, 69]. CNTs serve to support the catalyst and provide a surface for the redox reaction to occur. Air cathode composed of CNT-MnO_2 hybrids can facilitate the oxygen reduction and evolution reactions, effectively improving the energy efficiency and cyclic ability [68]: a low charge potential of 3.8 V and a considerable capacity of 1768 mAh/$g_{(carbon)}$ (796 mAh/$g_{(electrode)}$) are obtained. CNT-Co_3O_4 hybrids were also fabricated for use in the air electrodes of Li-air batteries exhibiting a high discharge capacity and low overvoltage during the charge-discharge process, which indicates that the hybrid is also potentially a good catalyst for the air electrode [69]. These show again that the CNT-based hybrids materials could potentially be a good catalyst for the air electrode by exhibiting better electrochemical properties.

The main task for the oxygen cathode side is to design a good gas-diffusion electrode resisting oxidation/corrosion when high voltages are accessed on charging. The presence of both electrochemical and chemical side reactions, in which gases such as water and carbon dioxide in the ambient air were involved, may need some conceptual change to fully explore the high potential of Li-air batteries, for instance, to avoid organic liquid electrolyte and to overcome some intrinsic problem of Li-air systems [70].

Kitaura and co-workers designed an all-solid state Li-air battery using CNT-inorganic hybrid (Fig. 11.7) [71, 72]. A solid state electrolyte is used to obtain a good safety of the devices, but a CNT-LAGP ($Li_{1+x}Al_yGe_{2-y}(PO_4)_3$) hybrid is necessary since only CNTs can make the continuous electron conduction paths based on their high aspect ratio [67, 68]. LAGP particles contacting CNTs form another Li-ion conduction path (Fig. 11.7) [71]. The electron conduction path, lithium ion conduction path and air gas diffusion path through a nanopore inside the CNT and among LAGP sintered particles construct three continuous path structures in the air electrode. The all-solid state lithium-air cells were successful in discharging and charging. The first discharge and charge capacities of an all-solid state Li-air battery were about 1700 mA h g^{-1} and 900 mA h g^{-1}, respectively, at a current density of 500 mA g^{-1} in the voltage range of 2.0–4.2 V (vs. Li/Li^+). The decomposition of organic liquid electrolyte during discharging and the oxidation of the decomposition product during charging that result in large polarization are avoided in such an all-solid state Li-air battery using CNT-LAGP hybrid as a cathode [73]. This is a good example showing a CNT-based hybrid for the innovative design of energy storage systems.

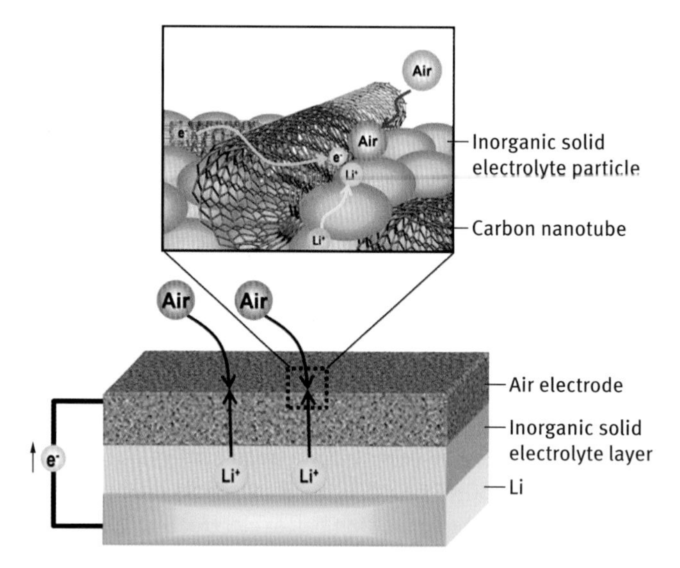

Fig. 11.7: Schematic diagram of an all-solid state lithium-air battery using lithium anode, an inorganic solid electrolyte, and an air electrode composed of carbon nanotubes and solid electrolyte particles. Reprinted with permission from Hirokazu Kitaura *et al.*, Energy Environ. Sci., 2012, 5, 9077 [71].

11.3 Application of hybrids with CNTs in supercapacitor

Supercapacitors, or electrical double layer capacitors (EDLC), are another technology for chemical energy storage, but based on a different principle with respect to the battery [74]. It stores charge electrostatically through the reversible adsorption of ions of the electrolyte onto active materials [75]. The electrode must be electrochemically stable and expose a high surface area. A pseudocapacitor stores the charge not electrostatically, but through a charge transfer that produces a redox reaction (Faradaic reaction) on functional groups or heteroatoms in carbon materials. Both storage mechanism (EDLCs and pseudocapacitance) are simultaneously present in many supercapacitor devices/materials, although one of the two storage mechanisms dominates. Supercapacitors can have a high power density up to 10 $kW \cdot kg^{-1}$ that can be achieved in very short times (few seconds), but the specific energy (about 5 $Wh \cdot kg^{-1}$) is lower than Li-based batteries. The performance of the current supercapacitors must be substantially improved with the aim of increasing the energy density, while keeping the long cycling life to meet the tough requirements of applications in hybrid electric vehicles and large industrial storage.

The following three concepts may provide possible solutions for the new generation supercapacitors [36].

(1) Developing nanocarbons with high specific surface area (> 2000 m^2/g) and with tailored pore size which allow a facile ion transfer [76, 77, 78]. The key to obtain

high capacitance by charging the double layer is to use the high accessible specific surface of electron-conducting electrodes to adsorb ions;

(2) Modifying the surface characteristics to enhance the charge storage. There is a broad protocol to modify the surface characteristics of carbon materials, especially the nanostructured carbon materials, to have excellent adsorption/desorption behavior;

(3) Designing a specific material architecture. 3D hierarchical carbon [79, 80], 3D aperiodic [79, 81, 82] or highly-ordered hierarchical carbons are representative samples with multimodal pore structure to optimize the performance of the capacitors. The micropore, mesopore and macropore structure of such three-dimensional hierarchical carbons are generally perfectly interconnected.

CNTs as electrode show voltammograms with a regular box-like shape, characteristic of a pure electrostatic attraction and have a capacitance up to 120 F g^{-1} (the value depends on the scanning rate) [83]. Supercapacitors with SWCNT electrodes exhibit a gravimetric capacitance between 20 and 300 F g^{-1} [84, 85, 86, 87] depending on the specific surface area they have. The major disadvantage of these nanocarbons for supercapacitor applications is their low specific surface area. The agglomeration of CNTs and restacking of graphene sheets due to van der Waals interaction reduce not only the theoretically very high specific surface area, but also hinder the facile diffusion of ions. All these shortcomings can be overcome when CNTs are used to build a hybrid with other carbon or inorganic units, as discussed below.

11.3.1 CNT-based carbon hybrid for supercapacitors

The abovementioned CNT-CNF hybrid has been tested for supercapacitive performance [25]. A specific capacitance of approximately 70 Fg^{-1} at a current density of 148 mAg^{-1} in 1M H_2SO_4 is obtained. At higher current densities of 370 and 740 mAg^{-1}, capacitance values of about 48 and 40 F g^{-1} were obtained, respectively. These values are not higher than those of activated carbon in H_2SO_4 solution, mainly due to the relatively low specific surface area of 347 m^2 g^{-1}. This different performance of the CNT-CNF hybrid in LIB and in EDLC shows different requirements of these two electrochemical storage systems. The type of hybrid must be designed to match the demand of the applications.

CNT-graphene hybrid is another kind of CNT-based composite that has been tested for supercapacitor applications. Graphene should have, at least theoretically, high specific surface area and should be a good candidate for supercapacitors, but they tend to stack to several layers decreasing the specific surface area. Spanning CNTs between the graphenes (Fig. 11.2(c), (d)) can prevent such restacking and preserve the space between the neighboring graphene sheets, maintaining its effective surface area. The addition of CNTs between graphene layers can provide an additional ion transport pathway. Chen and co-workers fabricated CNT-graphene 3D hybrids (schematically shown in Fig. 11.2(d)) by a hydrothermal route [88]. Galvanostatic cycling of supercapacitor electrodes at different constant current densities of 100, 250, 500 and 1000 mA g^{-1} revealed that the discharge curves are linear in the total range of potential with constant slopes, showing nearly perfect capacitive behavior (Fig. 11.8 for the hybrid with a G:CNT ratio of 10 : 1). The specific capacitances are 187, 181, 170, and 167 F g^{-1} at the corresponding current density. The 3D hybrid CNT-graphene hybrid holds ~ 90 % of their specific capacitance at the high current density (Fig. 11.8(b)), which is evidence of good rate capability. The space between the graphene sheets was spanned by the connected CNTs, which facilitated the electrolyte ions to move quickly at a high current density and exhibited good rate performance. The 3D hybrid gave the best result with a specific capacitance of 318 F g^{-1} [88].

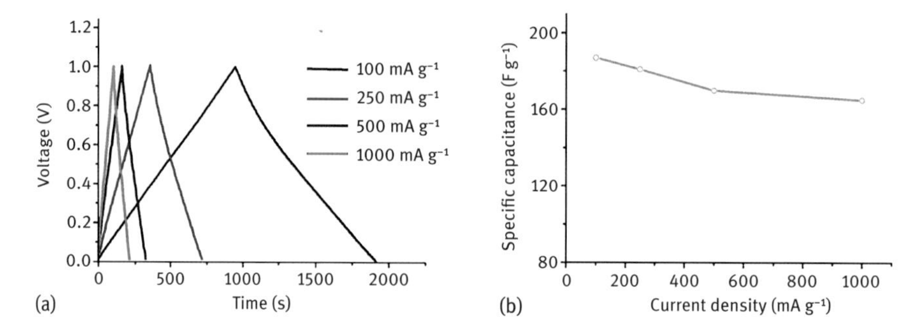

Fig. 11.8: (a) Galvanostatic charge-discharge curve of the supercapacitor using G/CNTs (10 : 1) as the active material at constant current densities of 100, 250, 500, and 1000 mA g^{-1} using 30 wt% KOH electrolyte. (b) Specific capacitance of 3D hybrid G/CNTs (10 : 1) measured at different current densities. Reprinted with permission from [88].

The electrochemical performance of the CNT-based hybrid depends not only on its architecture/morphology, but also on the microstructure and type of the building blocks. The same group prepared SWCNT-graphene hybrid by arc discharge method to get a hybrid with less structural defects in the graphene and SWCNTs, also to have graphene instead of graphene oxides [89]. They show that the supercapacitor prepared using this hybrid material has a low capacitance of 70 F g^{-1} if it is not activated with KOH. But a high performance of 350 F g^{-1} at 2.7 V (in the $TEABF_4$/AN electrolyte) is achieved after 8000 cycles. This “electro-activation” during charging/discharge may indicate that the interface and the channels in the hybrids are not ideal for ion transport. Further study is still needed. A CNT-graphene hybrid with vertically aligned CNTs in-between the graphene planes has a maximum specific capacitance of 385 F g^{-1} at a scan rate of 10 mV s^{-1} in 6 M KOH aqueous solution [27].

The strategies mentioned to improve supercapacitor performance also require other elements to obtain high performance devices. For instance, the right selection of the organic or inorganic electrolyte is essential when such CNT hybrids are used because of the need to fine match between the pore size and that of the solvated ions [90, 91]. The corrosion of nanocarbon at high voltage is another critical element to obtain high energy density supercapacitors with long life. Until supercapacitors can be applied for large-scale energy applications is still a long way off.

11.3.2 CNT-based inorganic hybrid for supercapacitors

Metal oxides (for instance RuO_2, IrO_2) are typical electrode materials for supercapacitors based on pseudocapacitance [92]. Metal oxides usually exhibit poor electrical conductivity that can cause a low power density. Many efforts have been made to develop composite materials consisting of metal oxide nanoparticles and carbon as electrode materials [93, 94, 95]. It is not unexpected that CNT-based inorganic hybrids are also tested for pseudocapacitic supercapacitors [74]. The redox reactions can be irreversible as compound formation between electroactive species and the electrode material may occur, thus pseudocapacitors often suffer from a lack of stability during long-time cycling [8].

Many cheaper transition metal oxides have been also considered as promising materials for supercapacitors [96, 97, 98]. When used as electrode material NiO has high resistivity, which is a serious drawback for practical applications to supercapacitors. Forming CNT-NiO hybrids could enhance the electrode conductivity through a CNT network and increase the specific surface area of NiO when finely dispersed on CNTs. The specific capacitance of the supercapacitor is directly related to the specific surface area of electrodes. CNTs have high conductivity and are good support for metal oxide particles. Lee and co-workers prepared CNT-NiO composite electrodes with 10 wt% CNTs [99]. Figure 11.9(a) and (b) shows the cyclic voltammetry behavior of the bare NiO and CNT-NiO (10 % CNTs) nanocomposite, respectively. Both electrodes have the

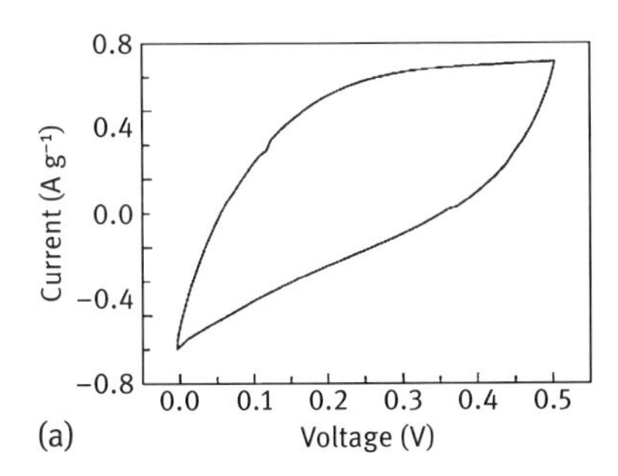

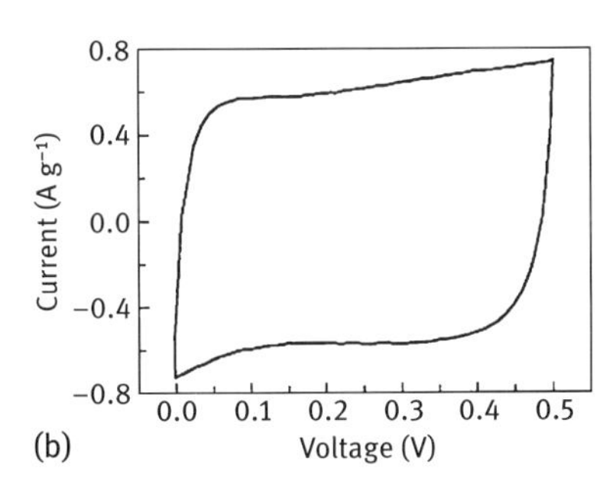

Fig. 11.9: Electrochemical properties of supercapacitors using the bare NiO and NiO/CNT (10 %) composite electrodes. The cyclic voltammetry(CV) behavior of (a) the bare NiO electrodes and (b) the NiO/CNT (10 %) composite electrodes in 2M KOH aqueous solution (sweep rate, 10 mV/s) (reprinted with permission from Y. Lee *et al.*, Synthetic Metals, 150, 2005, 153–157).

characteristic of a capacitor with constant charging and discharging rates over a complete cycle. The CNT-NiO nanocomposite electrode shows more rectangular shape in the CV curve and is closer to an ideal capacitor than the bare NiO. The presence of a CNT network in the NiO significantly improved (i) the electrical conductivity of the host NiO by the formation of a conducting network of CNTs and (ii) the active sites for the redox reaction of the metal oxide by increasing its specific surface area. Enhanced performance as electrode materials of a supercapacitor was also reported for CNT-RuO_2 hybrid materials [100].

11.4 Summary

CNTs are an emerging family of advanced nanomaterials with profound impact on electrochemical energy storage. The uniqueness of their nanoscopic and macroscopic properties has inspired mimicking research in other kinds of nanomaterials. The knowledge and experience acquired from CNT-related studies have also been useful in the current research on graphene-based materials. Concerning the requisites of electrochemical energy storage, the main characteristics that are usually considered regarding CNTs are the geometry (open-end tube, close-end tube, wall thickness, length), chemistry (surface functionalization, lattice substitution), physics (metallic, semiconducting) and interface with building blocks (covalent, van der Waals, electrostatic, π-π interaction, *etc.*). Synthetic chemistry of CNT-based hybrids has been the most investigated topic; and this has led to the production of a vast variety of nanohybrid materials with interesting properties in electrochemical energy storage. Unfortunately, regarding the practical implementation of CNT-based hybrids, this innovative lab-oriented synthetic chemistry can hardly meet the actual demands from industry. The lack of systematic understanding of the properties of inorganic building blocks and their interaction with CNTs has been a main block on the way towards the real use of these hybrid nanomaterials. Regardless of the attractiveness of synthetic chemistry, the mass production of CNTs has already achieved significant success in the battery and supercapacitor industry, though most CNTs have been used as conducting additives to replace traditional carbon black. Revisiting the more than a decade long history of CNTs and CNT-based hybrids for applications in batteries and supercapacitors, it is easy to notice that new efforts have to be devoted to the areas beyond synthetic chemistry, such as the intrinsic properties of the building blocks and the hybrid, as well as to the mass production strategy of those advanced high performance hybrid materials. It is too early to conclude whether CNT-based hybrids can or cannot be of value for batteries or supercapacitors until breakthrough innovations and progress can be achieved for technology, but it brings a new aspect in the research of electrochemical energy storage.

Acknowledgment

The author thanks Dr. Da-Wei Wang from Australia for many constructive discussions and reading of the manuscript, and Mr. Daming Wang and Dr. Bingsen Zhang, China, for technical help.

Bibliography

[1] N. S. Choi, Z. H. Chen, S. A. Freunberger, X. L. Ji, Y. K. Sun, K. Amine, G. Yushin, L. F. Nazar, J. Cho, P. G. Bruce, *Angew. Chem. Int. Edit.* **2012**. 51, 9994–10024
[2] Z. H. Wen, Q. Wang, Q. Zhang, J. H. Li, *Adv. Funct. Mater.* **2007**. 17, 2772–2778
[3] J. B. Goodenough, K.-S. Park, *J. Am. Chem. Soc.* **2013**. 135, 1167–1176
[4] J. C. Withers, R. O. Loutfy, T. P. Lowe, *Fullerene Sci. Technol.* **1997**. 5, 1–31
[5] V. A. Nalimova, D. E. Sklovsky, G. N. Bondarenko, H. Alvergnat Gaucher, S. Bonnamy, F. Beguin, *Synth. Met.* **1997**. 88, 89–93
[6] F. Leroux, K. Metenier, S. Gautier, E. Frackowiak, S. Bonnamy, F. Beguin, *J. Power Sources* **1999**. 81, 317–322
[7] Editor ed. *Carbon Nanotube Science*, Cambridge University Press, Cambridge **2009**
[8] D. S. Su, R. Schlogl, *ChemSusChem* **2010**. 3, 136–168
[9] C. Peng, S. W. Zhang, D. Jewell, G. Z. Chen, *Prog. Nat. Sci.* **2008**. 18, 777–788
[10] Editor ed. *Recent Advances in Supercapacitors, Transworld Research Network*, **2006**
[11] S. W. Lee, N. Yabuuchi, B. M. Gallant, S. Chen, B. S. Kim, P. T. Hammond, Y. Shao-Horn, *Nat. Nanotechnol.* **2010**. 5, 531–537
[12] A. G. Pandolfo, A. F. Hollenkamp, *J. Power Sources* **2006**. 157, 11–27
[13] C. Sotowa, G. Origi, M. Takeuchi, Y. Nishimura, K. Takeuchi, I. Y. Jang, Y. J. Kim, T. Hayashi, Y. A. Kim, M. Endo, M. S. Dresselhaus, *ChemSusChem* **2008**. 1, 911–915
[14] G. Centi, S. Perathoner, *Eur. J. Inorg. Chem.* **2009**. 3851–3878
[15] D. Eder, *Chem. Rev.* **2010**. 110, 1348–1385
[16] J. J. Vilatela, D. Eder, **2012**. 5, 456–478
[17] S. K. Yoshino A., Nakajima T., J. Patent, Editor. **1985**.
[18] K. Zaghib, M. Dontigny, A. Guerfi, P. Charest, I. Rodrigues, A. Mauger, C. M. Julien, *J. Power Sources* **2011**. 196, 3949–3954
[19] N. A. Kaskhedikar, J. Maier, *Adv. Mater.* **2009**. 21, 2664–2680
[20] J. R. Dahn, T. Zheng, Y. H. Liu, J. S. Xue, *Science* **1995**. 270, 590–593
[21] J. P. Tessonnier, D. Rosenthal, T. W. Hansen, C. Hess, M. E. Schuster, R. Blume, F. Girgsdies, N. Pfander, O. Timpe, D. S. Su, R. Schlogl, *Carbon* **2009**. 47, 1779–1798
[22] E. Frackowiak, S. Gautier, H. Gaucher, S. Bonnamy, F. Beguin, *Carbon* **1999**. 37, 61–69
[23] H. Habazaki, M. Kiriu, H. Konno, *Electrochem. Commun.* **2006**. 8, 1275–1279
[24] E. Yoo, J. Kim, E. Hosono, H. Zhou, T. Kudo, I. Honma, *Nano Lett.* **2008**. 8, 2277–2282
[25] J. Zhang, Y. S. Hu, J. P. Tessonnier, G. Weinberg, J. Maier, R. Schlogl, D. S. Su, *Adv. Mater.* **2008**. 20, 1450–1457
[26] D. Yu, L. Dai, **2009**. 1, 467–470
[27] Z. J. Fan, J. Yan, L. J. Zhi, Q. Zhang, T. Wei, J. Feng, M. L. Zhang, W. Z. Qian, F. Wei, *Adv. Mater.* **2010**. 22, 3723–3756
[28] V. C. Tung, L. M. Chen, M. J. Allen, J. K. Wassei, K. Nelson, R. B. Kaner, Y. Yang, *Nano Lett.* **2009**. 9, 1949–1955
[29] S. Q. Chen, W. K. Yeoh, Q. Liu, G. X. Wang, *Carbon* **2012**. 50, 4557–4565

[30] S. Q. Chen, P. Chen, Y. Wang, *Nanoscale* **2011**. 3, 4323–4329
[31] Z. J. Fan, J. Yan, T. Wei, G. Q. Ning, L. J. Zhi, J. C. Liu, D. X. Cao, G. L. Wang, F. Wei, *ACS Nano* **2011**. 5, 2787–2794
[32] S. S. Li, Y. H. Luo, W. Lv, W. J. Yu, S. D. Wu, P. X. Hou, Q. H. Yang, Q. B. Meng, C. Liu, H. M. Cheng, *Adv. Energy Mater.* **2011**. 1, 486–490
[33] G. Maurin, F. Henn, B. Simon, J. F. Colomer, J. B. Nagy, *Nano Lett.* **2001**. 1, 75–79
[34] M. Wakihara, O. Yamamoto (eds). *Lithium Ion Battery: Fundamentals and Performance*, Wiley-VCH, Weinheim **1998**
[35] C. M. Park, J. H. Kim, H. Kim, H. J. Sohn, *Chem. Soc. Rev.* **2010**. 39, 3115–3141
[36] D. S. Su, Centi, G., *Journal of energy chemistry* **2013**. in press
[37] J. Shu, H. Li, R. Z. Yang, Y. Shi, X. J. Huang, *Electrochem. Commun.* **2006**. 8, 51–54
[38] Y. Wang, M. Wu, Z. Jiao, J. Y. Lee, *Chem. Mat.* **2009**. 21, 3210–3215
[39] H. X. Zhang, C. Feng, Y. C. Zhai, K. L. Jiang, Q. Q. Li, S. S. Fan, *Adv. Mater.* **2009**. 21, 2299–2321
[40] G. Chen, Z. Y. Wang, D. G. Xia, *Chem. Mat.* **2008**. 20, 6951–6956
[41] W. Wang, P. N. Kumta, *ACS Nano* **2010**. 4, 2233–2241
[42] Y. Yu, L. Gu, C. L. Wang, A. Dhanabalan, P. A. van Aken, J. Maier, *Angew. Chem.-Int. Edit.* **2009**. 48, 6485–6489
[43] M. S. Park, S. A. Needham, G. X. Wang, Y. M. Kang, J. S. Park, S. X. Dou, H. K. Liu, *Chem. Mat.* **2007**. 19, 2406–2410
[44] A. Goyal, A. L. M. Reddy, P. M. Ajayan, *Small* **2011**. 7, 1709–1713
[45] H. Pan, B. Liu, J. Yi, C. Poh, S. Lim, J. Ding, Y. Feng, C. H. A. Huan, J. Lin, **2005**. 109, 3094–3098
[46] A. L. M. Reddy, M. M. Shaijumon, S. R. Gowda, P. M. Ajayan, *Nano Lett.* **2009**. 9, 1002–1006
[47] B. Xu, D. N. Qian, Z. Y. Wang, Y. S. L. Meng, *Materials Science & Engineering R-Reports* **2012**. 73, 51–65
[48] X. Z. C. M. Hayner, H. H. Kung, *Annual Reviews* **2012**. 3, 445–471
[49] R. Marom, S. F. Amalraj, N. Leifer, D. Jacob, D. Aurbach, *J. Mater. Chem.* **2011**. 21, 9938–9954
[50] R. Pitchai, V. Thavasi, S. G. Mhaisalkar, S. Ramakrishna, *J. Mater. Chem.* **2011**. 21, 11040–11051
[51] S. Xin, Y. G. Guo, L. J. Wan, *Accounts of Chemical Research* **2012**. 45, 1759–1769
[52] L. Dimesso, C. Forster, W. Jaegermann, J. P. Khanderi, H. Tempel, A. Popp, J. Engstler, J. J. Schneider, A. Sarapulova, D. Mikhailova, L. A. Schmitt, S. Oswald, H. Ehrenberg, *Chem. Soc. Rev.* **2012**. 41, 5068–5080
[53] K. Sheem, Y. H. Lee, H. S. Lim, *J. Power Sources* **2006**. 158, 1425–1430
[54] X. L. Li, F. Y. Kang, X. D. Bai, W. Shen, *Electrochem. Commun.* **2007**. 9, 663–666
[55] J. J. Schneider, J. Khanderi, A. Popp, J. Engstler, H. Tempel, A. Sarapulova, N. N. Bramnik, D. Mikhailova, H. Ehrenberg, L. A. Schmitt, L. Dimesso, C. Forster, W. Jaegermann, *Eur. J. Inorg. Chem.* **2011**. 4349–4359
[56] H. Zhang, G. P. Cao, Z. Y. Wang, Y. S. Yang, Z. J. Shi, Z. A. Gu, *Electrochim. Acta* **2010**. 55, 2873–2877
[57] X. L. Ji, L. F. Nazar, *J. Mater. Chem.* **2010**. 20, 9821–9826
[58] J. Schuster, G. He, B. Mandlmeier, T. Yim, K. T. Lee, T. Bein, L. F. Nazar, *Angew. Chem.-Int. Edit.* **2012**. 51, 3591–3595
[59] N. Jayaprakash, J. Shen, S. S. Moganty, A. Corona, L. A. Archer, *Angew. Chem.-Int. Edit.* **2011**. 50, 5904–5908
[60] Y. J. Choi, Y. D. Chung, C. Y. Baek, K. W. Kim, H. J. Ahn, J. H. Ahn, *J. Power Sources* **2008**. 184, 548–552
[61] D. W. Wang, G. M. Zhou, F. Li, K. H. Wu, G. Q. Lu, H. M. Cheng, I. R. Gentle, *Phys. Chem. Chem. Phys.* **2012**. 14, 8703–8710

[62] H. L. Wang, Y. Yang, Y. Y. Liang, J. T. Robinson, Y. G. Li, A. Jackson, Y. Cui, H. J. Dai, *Nano Lett.* **2011**. 11, 2644–2647

[63] X. Z. C. Hayner, H. H. Kung, *Annual Reviews* **2012**. 3, 445–471

[64] P. G. Bruce, S. A. Freunberger, L. J. Hardwick, J. M. Tarascon, *Nat. Mater.* **2012**. 11, 19–29

[65] X. Liu, *Nat. Mater.* **2009**. 8

[66] K. M. Abraham, Z. Jiang, *J. Electrochem. Soc.* **1996**. 143, 1–5

[67] Y. L. Li, J. J. Wang, X. F. Li, J. Liu, D. S. Geng, J. L. Yang, R. Y. Li, X. L. Sun, *Electrochem. Commun.* **2011**. 13, 668–672

[68] 68. J. X. Li, N. Wang, Y. Zhao, Y. H. Ding, L. H. Guan, *Electrochem. Commun.* **2011**. 13, 698–700

[69] T. H. Yoon, Y. J. Park, *Nanoscale Res. Lett.* **2012**. 7, 1–4

[70] T. Zhang, H. S. Zhou, *Angew. Chem.-Int. Edit.* **2012**. 51, 11062–11067

[71] H. Kitaura, H. S. Zhou, *Energy Environ. Sci.* **2012**. 5, 9077–9084

[72] H. Kitaura, H. S. Zhou, *Adv. Energy Mater.* **2012**. 2, 889–894

[73] S. A. Freunberger, Y. H. Chen, Z. Q. Peng, J. M. Griffin, L. J. Hardwick, F. Barde, P. Novak, P. G. Bruce, *J. Am. Chem. Soc.* **2011**. 133, 8040–8047

[74] L. L. Zhang, X. S. Zhao, *Chem. Soc. Rev.* **2009**. 38, 2520–2531

[75] Editor ed. *Encyclopedia of Nanoscience and Nanotechnology*, American Scientific Publishers, **2004**

[76] Y. Gogotsi, A. Nikitin, H. H. Ye, W. Zhou, J. E. Fischer, B. Yi, H. C. Foley, M. W. Barsoum, *Nat. Mater.* **2003**. 2, 591–594

[77] R. K. Dash, G. Yushin, Y. Gogotsi, *Microporous Mesoporous Mat.* **2005**. 86, 50–57

[78] H. Nishihara, H. Itoi, T. Kogure, P. X. Hou, H. Touhara, F. Okino, T. Kyotani, *Chem.-Eur. J.* **2009**. 15, 5355–5363

[79] D. W. Wang, F. Li, M. Liu, G. Q. Lu, H. M. Cheng, *Angew. Chem.-Int. Edit.* **2008**. 47, 373–376

[80] C. M. Chen, Q. Zhang, X. C. Zhao, B. S. Zhang, Q. Q. Kong, M. G. Yang, Q. H. Yang, M. Z. Wang, Y. G. Yang, R. Schlogl, D. S. Su, *J. Mater. Chem.* **2012**. 22, 14076–14084

[81] D. C. Wu, F. Xu, B. Sun, R. W. Fu, H. K. He, K. Matyjaszewski, *Chem. Rev.* **2012**. 112, 3959–4015

[82] C. H. Huang, Q. Zhang, T. C. Chou, C. M. Chen, D. S. Su, R. A. Doong, *ChemSusChem* **2012**. 5, 563–571

[83] E. Frackowiak, S. Delpeux, K. Jurewicz, K. Szostak, D. Cazorla-Amoros, F. Beguin, *Chem. Phys. Lett.* **2002**. 361, 35–41

[84] C. Y. Liu, A. J. Bard, F. Wudl, I. Weitz, J. R. Heath, *Electrochem. Solid State Lett.* **1999**. 2, 577–578

[85] E. Frackowiak, K. Jurewicz, S. Delpeux, F. Beguin, *J. Power Sources* **2001**. 97–8, 822–825

[86] J. N. Barisci, G. G. Wallace, R. H. Baughman, *Electrochim. Acta* **2000**. 46, 509–517

[87] K. H. An, W. S. Kim, Y. S. Park, Y. C. Choi, S. M. Lee, D. C. Chung, D. J. Bae, S. C. Lim, Y. H. Lee, *Adv. Mater.* **2001**. 13, 497–503

[88] Y. Wang, Y. Wu, Y. Huang, F. Zhang, X. Yang, Y. Ma, Y. Chen, **2011**. 115, 23197

[89] Y. Wu, S. W. Finefrock, H. Yang, **2012**. 1, 651–653

[90] J. Chmiola, G. Yushin, Y. Gogotsi, C. Portet, P. Simon, P. L. Taberna, *Science* **2006**. 313, 1760–1763

[91] C. Largeot, C. Portet, J. Chmiola, P. L. Taberna, Y. Gogotsi, P. Simon, *J. Am. Chem. Soc.* **2008**. 130, 2730–2739

[92] A. Rudge, J. Davey, I. Raistrick, S. Gottesfeld, J. P. Ferraris, *J. Power Sources* **1994**. 47, 89–107

[93] H. T. Liu, P. He, Z. Y. Li, Y. Liu, J. H. Li, *Electrochim. Acta* **2006**. 51, 1925–1931

[94] Y. Lei, C. Fournier, J. L. Pascal, F. Favier, *Microporous Mesoporous Mat.* **2008**. 110, 167–176

[95] J. Li, X. Y. Wang, Q. H. Huang, S. Gamboa, P. J. Sebastian, *J. Power Sources* **2006**. 160, 1501–1505

[96] J. P. Zheng, T. R. Jow, *J. Electrochem. Soc.* **1995**. 142, L6–L8

[97] K. C. Liu, M. A. Anderson, *J. Electrochem. Soc.* **1996**. 143, 124–130
[98] V. Srinivasan, J. W. Weidner, *J. Electrochem. Soc.* **1997**. 144, L210–L213
[99] J. Y. Lee, K. Liang, K. H. An, Y. H. Lee, *Synth. Met.* **2005**. 150, 153–157
[100] X. qin, *Carbon* **2004**. 42, 451–453

Zhong-Shuai Wu, Xinliang Feng, and Klaus Müllen

12 Graphene-metal oxide hybrids for lithium ion batteries and electrochemical capacitors

The development of novel hybrid materials based on nanocarbons and metal oxides for electrochemical energy storage devices has recently attracted great attention due to the increasing global demand for energy. In this chapter, we review recent efforts dedicated to the fabrication of graphene-metal oxide hybrids as high-performance electrodes for lithium ion batteries and electrochemical capacitors. Several key structural models of graphene-metal oxide hybrids are described, including anchored, encapsulated, sandwich, layered, and mixed models. These structural models benefit from the kinetics of charge transfer, accommodate structural changes during the change/discharge processes, and give rise to enhanced electrochemical properties, such as high capacity, rate capability, cycling stability, and energy/power densities. By studying the structure-property relationships of different hybrids, one can conclude that there are synergetic effects between graphene and metal oxides. This chapter offers new insights into the design and fabrication of novel high-performance graphene-metal oxide electrode materials for energy storage devices.

12.1 Introduction

Graphene is one-atom-thick sheet of sp^2 bonded carbon atoms in a honeycomb crystal lattice structure [1–4]. In 2004, Geim and Novoselov first reported the existence of two-dimensional (2D) ultrathin carbon exfoliated by cleaving graphite with adhesive tape. Since then, a wide range of research activities has focused on the synthesis, physical attributes, and applications of graphene. It is the thinnest known material in the world and the mother of all graphitic carbon materials, which can be rolled up into one-dimensional carbon nanotubes (CNTs), stacked into three-dimensional (3D) graphite, and wrapped into a zero-dimensional fullerene [3].

Graphene is a zero-gap semiconductor and its electrical properties are more aptly described by the unconventional Dirac-like equation than the Schrödinger equation [2]. Graphene possesses unique electronic properties such as ballistic transport, room-temperature quantum Hall effect, and ultrafast charge mobility (200,000 $cm^2\ V^{-1}\ S^{-1}$). Moreover, single-layer graphene has excellent mechanical properties with a Young's modulus of 1.0 TPa and a stiffness of 130 GPa, high optical transmittance of ~ 97.7 %, superior thermal conductivity of 5000 $W\ m^{-1}\ K^{-1}$, and high specific surface area of 2620 $m^2\ g^{-1}$. These unique features make graphene superior to CNTs, graphite, metals, and semiconductors for use singly or as a component in materials for certain applications.

To date, several intriguing strategies, including micromechanical cleavage [1], epitaxial growth [5, 6], chemical exfoliation [7–11], chemical vapor deposition [12–14], bottom-up organic synthesis [15, 16], and electrochemical exfoliation [17] have been reported for the production of graphene materials. Each synthetic method has its advantages and disadvantages in terms of scalability and quality of the resulting graphene materials [18, 19]. Among them, chemical exfoliation of graphite is considered as a promising route towards the low-cost, large-scale production of graphene [7]. For example, oxidation of graphite by a strong oxidant (such as $KMnO_4$ or CrO_3) in H_2SO_4 solution generally produces graphene oxide (GO) [20]. Chemical [21–24] or thermal [10, 25–27] reductions are required to generate reduced GO (rGO), which can partially recover the electronic properties of graphene. Other than the easy bulk synthesis, special emphasis has been placed on oxygen-containing groups located at the edges and surfaces of GO and rGO, which provide the opportunity for tailor-made chemical modification. Due to the easy dispersion of GO and rGO in solution, functionalization of graphene with different metals, metal oxides, and polymers can be realized to afford targeted graphene-based hybrid materials [28]. The major breakthrough applications of graphene expected in the coming years include graphene-based electronics and optoelectronics, graphene-filled polymer hybrids, transparent and flexible conducting electrodes, and electrochemical energy storage and conversion devices. Additionally, graphene also holds promise for use in sensors, dye-sensitized solar cells, field emission [29], catalysts, and hydrogen storage.

12.2 Graphene for LIBs and ECs

Lithium ion batteries (LIBs) and electrochemical capacitors (ECs) are two important energy storage devices that can complement each other. LIBs work slowly but provide high energy density whereas ECs offer high power density, but suffer from lower energy density [30].

LIBs are highly rechargeable batteries in which lithium ions move from the negative electrode to the positive electrode during discharge, and return when charging [31]. Typically, LIBs use an intercalated lithium compound as the electrode material, compared to the metallic lithium used in non-rechargeable lithium batteries. Currently, LIBs are one of the most popular types of rechargeable batteries for portable electronics, with one of the highest energy densities, no memory effect, and only a slow loss of charge when not in use. Research activity has mainly been directed at improving the traditional LIB technology, focusing on energy density, durability, cost, and intrinsic safety.

ECs are another promising electrical energy storage device with a higher energy density than electrical capacitors, and a better rate capability and cycling stability than LIBs [32]. Carbon-based electric double layer capacitors and metal oxide- or polymer-based pseudocapacitors are two main types of ECs. The charge-storage model of carbon-based electric double layer capacitors is an electrostatic process with fast charge adsorption and separation at the interface between the electrode and electrolyte. The pseudocapacitor involves a redox reaction between the electrode materials and electrolyte ions [33]. Overall, the development of high-performance electrode materials is crucial for increasing the energy density and power density.

Graphene has attracted wide attention for use in LIBs and ECs because of its advantageous features such as large surface area, good flexibility, good chemical and thermal stability, wide potential windows, and rich surface chemistry. Yoo *et al.* reported that graphene nanosheets (GNS) from rGO exhibit a specific capacity of ~ 540 mAh g^{-1}, much higher than that of graphite. Further, they found that lithium storage capacities increase after embedding CNTs (730 mAh g^{-1}) or fullerene (784 mAh g^{-1}) into the graphene layers [34]. Lian *et al.* reported that the capacity obtained from high-quality graphene is ~ 1264 mA h g^{-1} at a current density of 100 mA g^{-1} [35]. The results of various experiments strongly suggest that the lithium storage of graphene is highly dependent on its structural parameters, including its specific surface area, interlayer spacing, surface functional groups, electrical conductivity, disorder degree, and electrode/electrolyte wettability [36–38]. Graphene is also considered to be a promising electrode candidate for ECs. For example, previous studies showed that graphene-based ECs exhibit excellent performance with a specific capacitance of 75 F g^{-1}, an energy density of 31.9 Wh kg^{-1} in ionic liquid electrolyte [39], and a specific capacitance of 135 and 99 F g^{-1} in aqueous and organic electrolyte, respectively [40]. These characteristics highlight the great potential of graphene as a new electrode material in LIBs and ECs. For practical application in LIBs and ECs, however, several challenges remain to be overcome. For example, one intractable issue regarding the use of graphene is that chemically-derived graphene commonly suffers from serious agglomeration and restacking after drying because of van der Waals interactions between GS, thereby lowering the electrochemical performance of graphene. Additionally, some issues concerning Li storage mechanisms, the absence of a voltage plateau and the large irreversible capacity of LIBs, remain unresolved.

12.3 Graphene-metal oxide hybrids in LIBs and ECs

12.3.1 Typical structural models of graphene-metal oxide hybrids

Graphene is a promising 2D building block in hybrids functionalized with a variety of inorganic materials, such as metal oxides [41–44]. The fabrication of graphene-metal oxide hybrids is considered to be a practical pathway to fully utilize the advantages of graphene in LIBs and ECs [41]. In the hybrids, metal oxide particles attached to graphene can not only suppress the agglomeration and restacking of graphene, but also enhance the accessible activated surface area for energy storage. On the other hand, graphene as a supporting material can sufficiently induce nucleation, growth, and formation of fine metal oxide microstructures with uniform dispersion and controlled morphology on the graphene surface [41–44]. The resulting hybrids can thus gain a perfect integrated structure with mixed electron and ion conductive networks for energy storage. In addition, the size effect of the metal oxide and interfacial interactions between two components often causes a synergistic effect in such graphene-metal oxide hybrids [41].

Several important structural models of graphene-metal oxide hybrids are proposed in terms of the integration format of graphene with metal oxide, as shown in Fig. 12.1. Specifically, these models include the anchored, encapsulated, sandwich-like, layered, and mixed models [41]. The main functions of graphene and metal oxides in these hybrids are: (i) graphene acts as a 2D support for homogeneously anchoring or dispersing metal oxides with controllable size, morphology, and crystallinity; (ii) the nano-/microparticles of metal oxides efficiently suppress the restacking of graphene; (iii) graphene serves as a 2D or 3D conductive network to enhance the electrical properties and charge transfer of pure oxides; (iv) graphene suppresses the volume change and agglomeration of metal oxides during the electrochemical reaction process; and (v) oxygenated groups on graphene provide new chemical functionality and compatibility to allow for easy processing of metal oxides in the hybrid, and thus enhance interfacial contact between graphene and metal oxides.

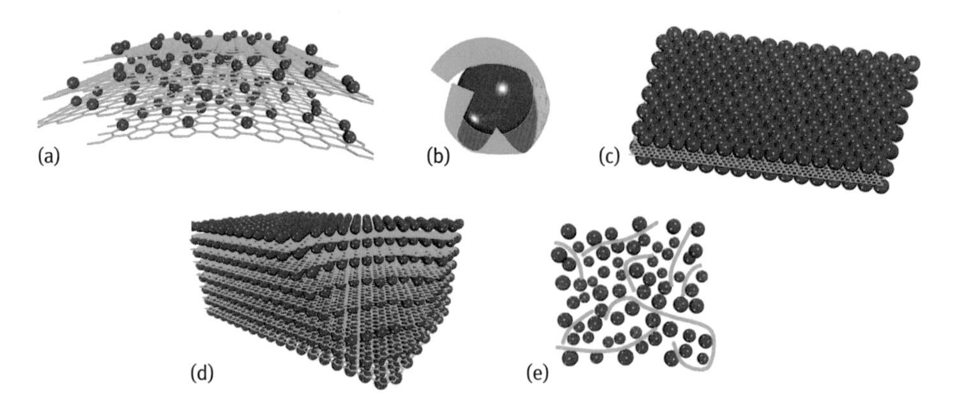

Fig. 12.1: Main structural models of graphene-metal oxide hybrids. (a) Anchored model: oxide particles are anchored to the graphene surface. (b) Encapsulated model: oxide particles are encapsulated by graphene. (c) Sandwich-like model: graphene is sandwiched between the metal oxide layers. (d) Layered model: a structure composed of alternating layers of oxide nanoparticles and graphene. (e) Mixed model: graphene and oxide particles are mechanically mixed and graphene sheets form a conductive network among the oxide particles. Red: metal oxide; Blue: graphene. Reprinted with permission from [41]. Copyright 2012, Elsevier B.V.

12.3.2 Anchored model

Recent progress revealed that graphene is a good 2D support for the growth of metal oxide nanoparticles on the graphene edges and surfaces due to the presence of oxygen-containing groups [42–44]. Thus, different metal oxide nanoparticles with controllable size, morphology, and crystallinity can be homogeneously anchored on graphene. This structural feature can be defined as an "anchoring model", and indeed, it is one of the most important models in the design and synthesis of graphene-metal oxide hybrids for LIBs and ECs due to its simplicity and easy operation.

The oxygen-containing groups on graphene play a critical role in controlling the size, shape, and dispersion of metal oxide nanoparticles. For example, Wu *et al.* reported that small RuO_2 nanoparticles with a size of 5 ~ 20 nm are homogeneously deposited on rGO with a C/O ratio of ~ 10. In sharp contrast, without the presence of graphene, the as-prepared hydrous RuO_2 powder tends to agglomerate and form big particles that are hundreds of nanometers, even to tens of micrometers, in size under the same experimental conditions [45]. Further comparison between oxygen-rich GO (with a C/O ratio of < 3) and oxygen-free CVD-grown graphene for anchoring RuO_2 revealed that highly uniform RuO_2 nanoparticles with a size of less than 5 nm can be formed on GO, while only large RuO_2 particles with a size ranging from tens to hundreds of nanometers are sparsely distributed on the surface of CVD-grown graphene [45]. A similar effect has been demonstrated by Ren *et al.* for graphene hybrids with photoactive zeolite particles, such as TS-1 [46a]. The presence of graphene decreased the size of the TS-1 particles by more than 2 orders of magnitudes to about 10 nm. One explanation for this small size may be the heat sink effect, which suppresses grain growth by dissipating local excess heat formed during crystallization and phase transformation treatments [45b]. Graphene also had a peculiar effect on the shape of the TS-1 particles. Dai *et al.* further confirmed that the size, morphology, and crystallinity of nanocrystals can be tailored by the oxygen content of the graphene (Fig. 12.2) [47]. For example, graphene with a low oxygen content of ~ 5 % results in single-crystal nanoplates with a well-defined shape. In contrast, the highly oxidized GO (~ 20 % of oxygen) results in the formation of irregularly shaped nanocrystals. In principle, two types of interfacial interactions between metal oxide and GO are involved in the synthesis. One is the reactive chemisorption on functional groups (such as HO-C=O and -OH) that bridge metal centers with carboxyl or hydroxyl groups at oxygen-defective sites; the other is the van der Waals interactions between the pristine region of graphene and metal oxides. Therefore, graphene-metal oxide hybrids with different sizes, nanostructures (such as nanoneedles [48], nanowires [47], and nanosheets [47]) and crystallinity influence the electrochemical behavior of LIBs and ECs.

Based on the anchored model, graphene hybrids with different metal oxides, such as SnO_2 [49–56], Co_3O_4 [44, 57–60], Mn_3O_4 [61], MnO [62], Fe_2O_3 [63, 64], Fe_3O_4 [65–67], MoO_3 [68], TiO_2 [69, 70], CuO [71], and Cu_2O [72] have been reported for LIBs, while SnO_2 [50], Co_3O_4 [73, 74], RuO_2 [45], TiO_2 [75], MnO_2 [47, 48, 76, 77], Mn_3O_4 [78], and

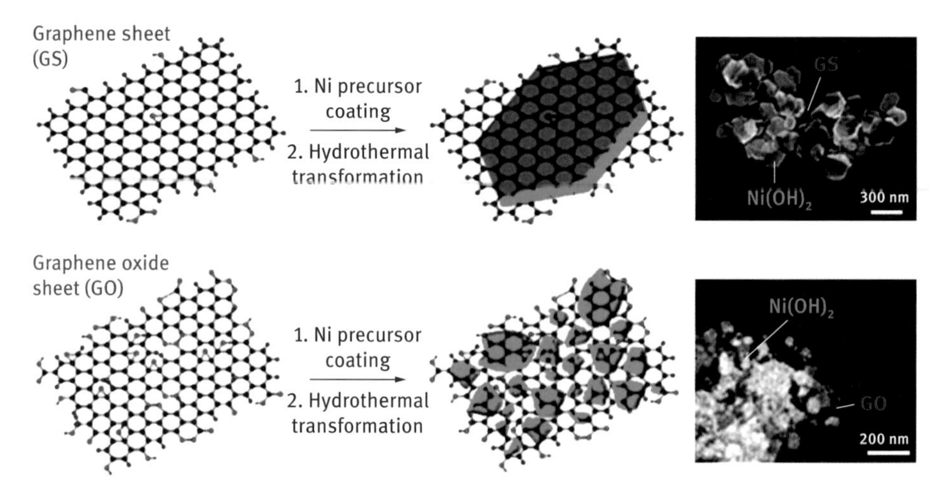

Fig. 12.2: $Ni(OH)_2$ nanocrystal growth on a graphene sheet (GS, upper) and graphite oxide sheet (GO, lower) with 5 % and 20 % oxygen content, respectively. Reprinted with permission from [47]. Copyright 2010, American Chemical Society.

ZnO [79] have been used in ECs. These hybrids all show enhanced electrochemical performance in terms of the high reversible capacity of LIBs or specific capacitance of ECs, rate capability, and cycling performance.

In the case of LIBs, a typical example is the graphene-Co_3O_4 hybrid [44]. The first discharge and charge capacities are 2179 and 955 mAh g^{-1} for graphene, 1105 mAh g^{-1} and 817 mAh g^{-1} for Co_3O_4, and 1097 and 753 mAh g^{-1} for graphene-Co_3O_4 hybrid electrodes. The graphene-Co_3O_4 hybrid, however, exhibits much better cycling performance than the respective graphene and Co_3O_4 (Fig. 12.3). The reversible capacity of graphene and Co_3O_4 particles decreased from 955 to 638 mAh g^{-1} and from 817 to 184 mAh g^{-1}, respectively, for up to 30 cycles. In contrast, the reversible capacity of the graphene-Co_3O_4 hybrid slightly increased upon cycling, and reached ~ 935 mAh g^{-1} after 30 cycles (Fig. 12.3). Similarly, Kang *et al.* reported the *in situ* fabrication and Li storage of a graphene-Co_3O_4 hybrid comprising 5 nm sized Co_3O_4 particles uniformly decorated on graphene. They obtained a capacity of more than 800 mAh g^{-1} at a current rate of 200 mA g^{-1} and more than 550 mAh g^{-1} at 1000 mA g^{-1} [58]. Chen *et al.* reported the microwave-assisted synthesis of a graphene-Co_3O_4 hybrid that had a large capacity of 1235 mAh g^{-1} at 89 mA g^{-1} and a capacity of 931 mAh g^{-1} at a large rate of 4450 mA g^{-1} [57]. High-capacity graphene-SnO_2 hybrids have been also synthesized for LIBs. Paek *et al.* reported a 3D graphene-SnO_2 hybrid with a delaminated structure by assembling SnO_2 nanoparticles on GS in an ethylene glycol solution (Fig. 12.4), which displays a higher capacity (810 mAh g^{-1}) and better cycling performance than that of bare SnO_2 (550 mAh g^{-1}) [49]. Kim *et al.* reported echinoid-like SnO_2 nanoparticles uniformly deposited on GS through electrostatic attractions. This hybrid also has high capacity and good cycling behavior compared to commercial SnO_2. A reversible

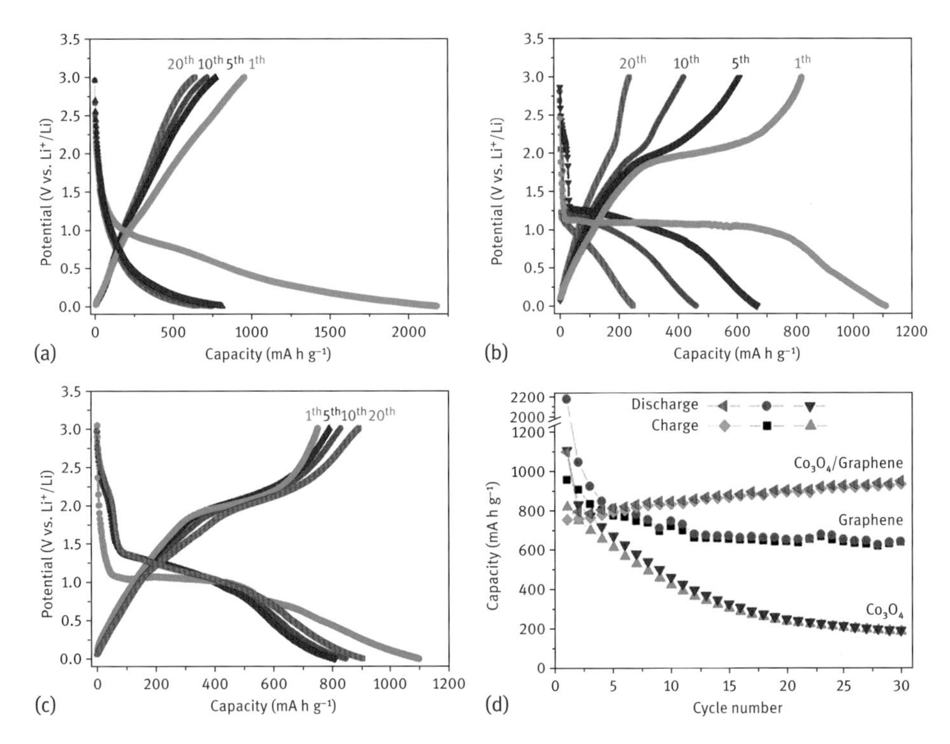

Fig. 12.3: Galvanostatic charge-discharge curves of (a) graphene, (b) Co_3O_4, and (c) Co_3O_4/graphene hybrid cycled at the 1st, 5th, 10th, and 20th between 3 and 0.01 V (vs. Li^+/Li) at a current density of 50 mA g^{-1}. (d) Comparison of the cycling performance of graphene, Co_3O_4, and Co_3O_4/graphene hybrid. Reprinted with permission from [44]. Copyright 2010, American Chemical Society.

capacity of 634 mAh g^{-1} was retained with a coulombic efficiency of 98 % after 50 cycles [80].

The superior Li-storage performance of metal oxide-graphene hybrid electrodes can be explained as follows [44, 49]. First, GS act as flexible 2D carbon supports for homogeneously anchoring metal oxide nanoparticles. The GS not only provide an elastic buffer space to accommodate the volume expansion/contraction of nanoparticles during Li insertion/extraction process, but also efficiently prevent the agglomeration of nanoparticles and the cracking or crumbling of electrode materials upon continuous cycling. Second, good electrical conductivity of graphene in hybrids serves as conductive channels to decrease the inner resistance of LIBs. Further, the presence of nanoparticles between GS effectively prevents the aggregation of GS, and consequently maintains the high active surface area of hybrids. Therefore, a synergetic effect between GS and metal oxide is responsible for the excellent electrochemical performance of the overall electrode.

The concept with the anchored model is also suitable for synthesizing hybrid electrodes for high-performance ECs. For example, Wu *et al.* reported the synthesis

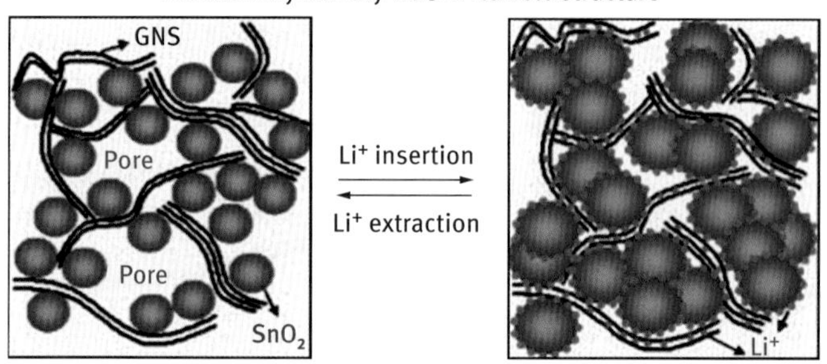

Fig. 12.4: Schematic illustration of the synthesis and structure of GS/SnO_2 (GNS/SnO_2). Reprinted with permission from [49]. Copyright 2010, American Chemical Society.

of hydrous RuO_2 nanoparticles (5–20 nm) anchored on GS as a hybrid electrode for ECs (Fig. 12.5) [45]. In contrast to the use of graphene or RuO_2, the fabricated ECs based on graphene-RuO_2 hybrid have high specific capacitance (~ 570 F g^{-1}), enhanced rate capability, excellent electrochemical stability (~ 97.9 % retention after

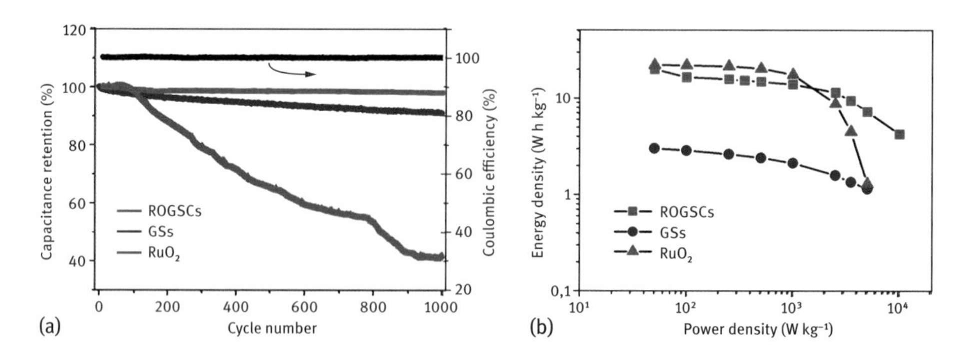

Fig. 12.5: (a) Cycling stability of the as-prepared GSs, RuO_2, and graphene-RuO_2 hybrid (ROGSCs) measured with a two-electrode cell at a current density of 1 A g^{-1} (right: coulombic efficiency of ROGSCs), and (b) Ragone plot for the as-prepared GSs, RuO_2, and ROGSCs-based ECs. Reprinted with permission from [45]. Copyright 2010, American Chemical Society.

1000 cycles), and high energy density (20.1 Wh kg^{-1}) at a low operation rate (100 mA g^{-1}) or high power density (10000 W kg^{-1}). As another example, Fan *et al.* reported the microwave-assisted synthesis of GS/Co_3O_4 hybrid, with well-dispersed Co_3O_4 nanoparticles (3–5 nm) on GS, which greatly improves the electrochemical utilization of Co_3O_4 and double-layer capacitance from graphene [14]. Specifically, the GS/Co_3O_4 hybrid exhibits high specific capacitance (243 F g^{-1} at 10 mVs^{-1}) and excellent cycle life (~ 95.6 % after 2000 cycles).

12.3.3 Encapsulated model

Metal oxides such as SnO_2, Co_3O_4, and Fe_3O_4 with a high theoretical specific capacity are considered promising high-energy anode candidates for LIBs. These materials generally undergo severe structural and volume changes caused by the conversion reaction (such as Co_3O_4 and Fe_3O_4) or Li-alloy reaction (such as SnO_2) during charge and discharge processes, however, leading to the pulverization of electrodes and severe capacity loss [81]. To suppress the volume change and improve the cycling performance of these metal oxides, a promising strategy is to modify their surfaces by thin carbon coating, resulting in carbon-wrapped metal oxides. Due to the structural flexibility, GS serve as a promising matrix for the encapsulation or wrapping of different metal oxide particles. Commonly, it can be referred to as an “encapsulated model” or a “wrapped model”.

Zhou *et al.* reported a hybrid of graphene-wrapped Fe_3O_4 particles (GS/Fe_3O_4) as anode materials for high performance LIBs (Fig. 12.6) [82]. For pristine Fe_3O_4 particles, morphology studies suggest that they become smaller and agglomerated after 30 cycles, indicative of the pulverization of particles during the cycling process. In sharp contrast, the Fe_3O_4 particles in the hybrid are closely wrapped between the graphene layers and the particle size is closely maintained before (196 nm) and after (213 nm) the cycling process. Thereby, graphene-wrapped Fe_3O_4 nanoparticles show a highly reversible specific capacity of 1026 mAh g^{-1} after 30 cycles, which is much higher than that of commercial Fe_3O_4 particles (475 mAh g^{-1}), which have a rapidly fading capacity [82]. Moreover, the GS/Fe_3O_4 hybrid electrode exhibits excellent cycle performance with no capacity loss for 80 cycles (Fig. 12.6(c)–(f)). Similar enhanced electrochemical performance was also observed for other hybrids of graphene-wrapped metal oxides, such as TiO_2 [83], NiO [84], MoO_2 [85], and V_2O_5 [86].

Another remarkable example is to confine individual metal oxide nanoparticles wrapped by very thin graphene layers (1 to 5 layers). Recently, our group reported the fabrication of graphene-encapsulated metal oxide (GE–MO) by the co-assembly of negatively-charged GO and positively-charged oxide nanoparticles (Fig. 12.7) [87]. The assembly process is driven by the mutual electrostatic interactions of the two species. The resulting GE–MO possesses flexible and ultrathin graphene shells that effectively enwrap the oxide nanoparticles. This unique hybrid architecture can (i) suppress the

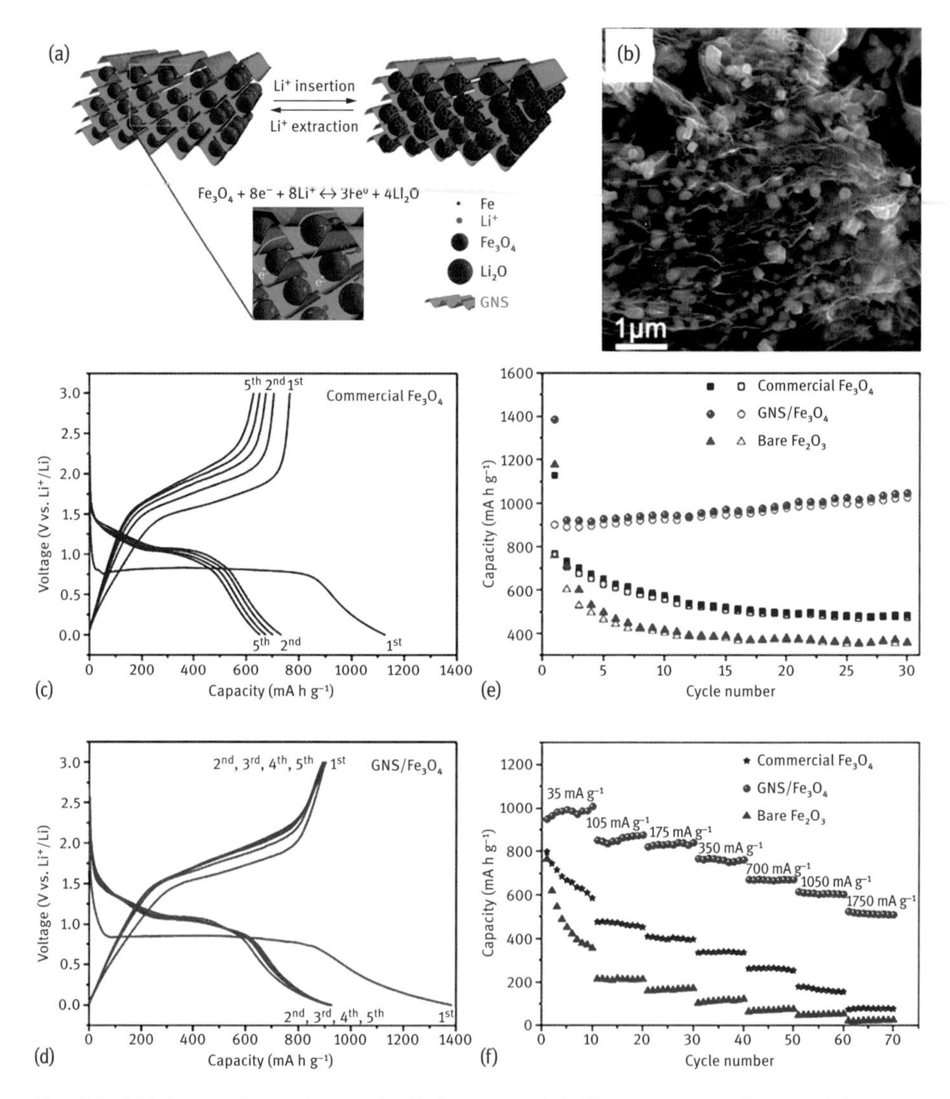

Fig. 12.6: (a) Schematic illustration of a flexible interwoven hybrid structure consisting of flexible GS (GNS) and Fe_3O_4 particles. (b) SEM image of the cross-section of GS/Fe_3O_4 (GNS/Fe_3O_4) hybrid. (c) and (d) The discharge/charge profiles of (c) commercial Fe_3O_4 particles and (d) GS/Fe_3O_4 hybrid. (e) Cycling performance of the commercial Fe_3O_4 particles, GS/Fe_3O_4 hybrid, and bare Fe_2O_3 particles at 35mA g^{-1}. Solid symbols: discharge, open symbols: charge and (f) rate performance of the commercial Fe_3O_4 particles, GS/Fe_3O_4 hybrid, and bare Fe_2O_3 particles at different current densities. Reprinted with permission from [82]. Copyright 2010, American Chemical Society.

aggregation of oxide nanoparticles, (ii) accommodate volume changes during the cycle processes, (iii) give rise to a high oxide content in the composite (up to 91.5 % by weight), and (iv) maintain a high electrical conductivity of the overall electrode. Thus, graphene-encapsulated electrochemically active Co_3O_4 nanoparticles (GE–Co_3O_4) ex-

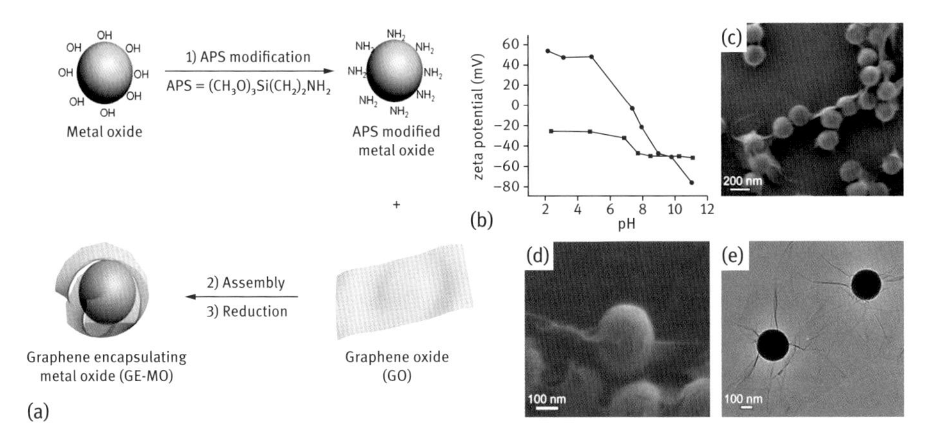

Fig. 12.7: (a) Fabrication schematic of a graphene-encapsulated metal oxide. (b) Zeta potentials of APS-modified silica (●) and graphene oxide (■) in aqueous solutions with various pH values. (c) and (d) typical SEM, and (e) transmission electron microscopy (TEM) images of graphene-encapsulated silica spheres. Reprinted with permission from [87]. Copyright 2010, John Wiley & Sons, Inc.

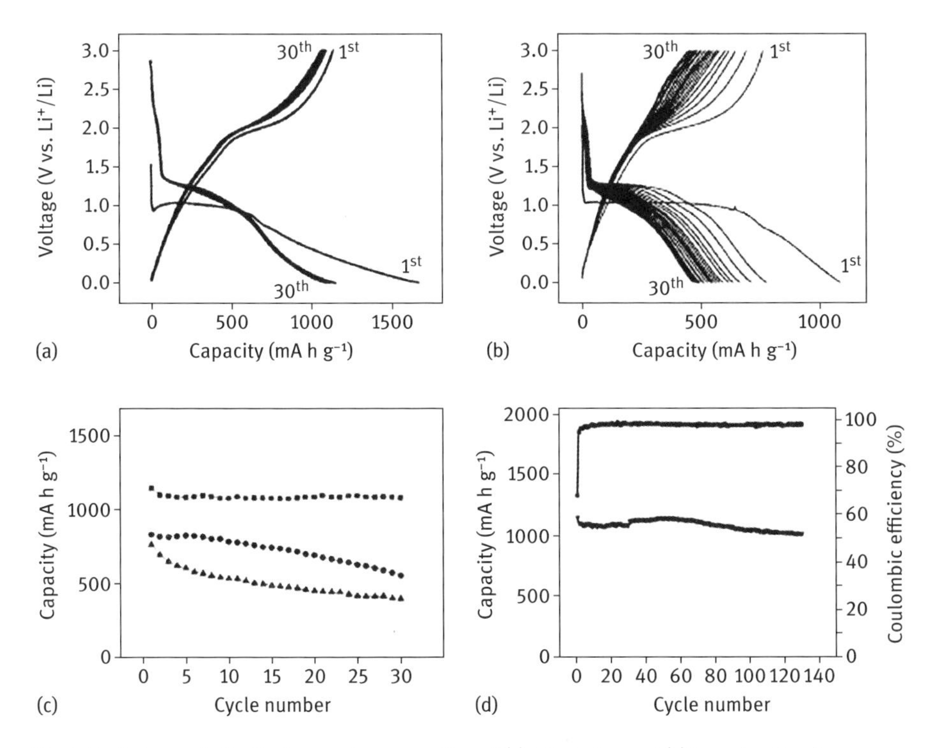

Fig. 12.8: Galvanostatic discharge-charge profiles of (a) GE-Co_3O_4 and (b) bare Co_3O_4 electrodes (current density=74 mAg^{-1}). (c) Comparison of cycle performance of GE-Co_3O_4 (■), mixed Co_3O_4/graphene hybrid (●), and bare Co_3O_4 electrodes (▲) over 30 cycles, and (d) cycle performance and coulombic efficiency of the GE-Co_3O_4 electrode during 130 cycles (current density = 74 mAg^{-1}). Reprinted with permission from [87]. Copyright 2010, John Wiley & Sons, Inc.

hibit a very high reversible capacity of 1100 $mAhg^{-1}$ in the first 10 cycles, and over 1000 $mAhg^{-1}$ after 130 cycles, with excellent cycle performance (Fig. 12.8). A similar enhanced Li-storage performance was also revealed by other graphene-encapsulated metal oxides such as Fe_3O_4 [88, 89] and Fe_2O_3 [64] for LIBs.

12.3.4 Sandwich-like model

The deposition of mesoporous metal oxides on the graphene surface in a 2D manner can result in graphene-based porous hybrid nanosheets. Such sandwich-like materials can inherit some features of graphene, such as a large aspect ratio, very thin thickness, high surface area, and high monodispersity. Therefore, the hybrid nanosheets used in electrochemical energy storage generally exhibit a short diffusion length of ions and increased electrochemically active sites. Further, the GS confined in metal ox-

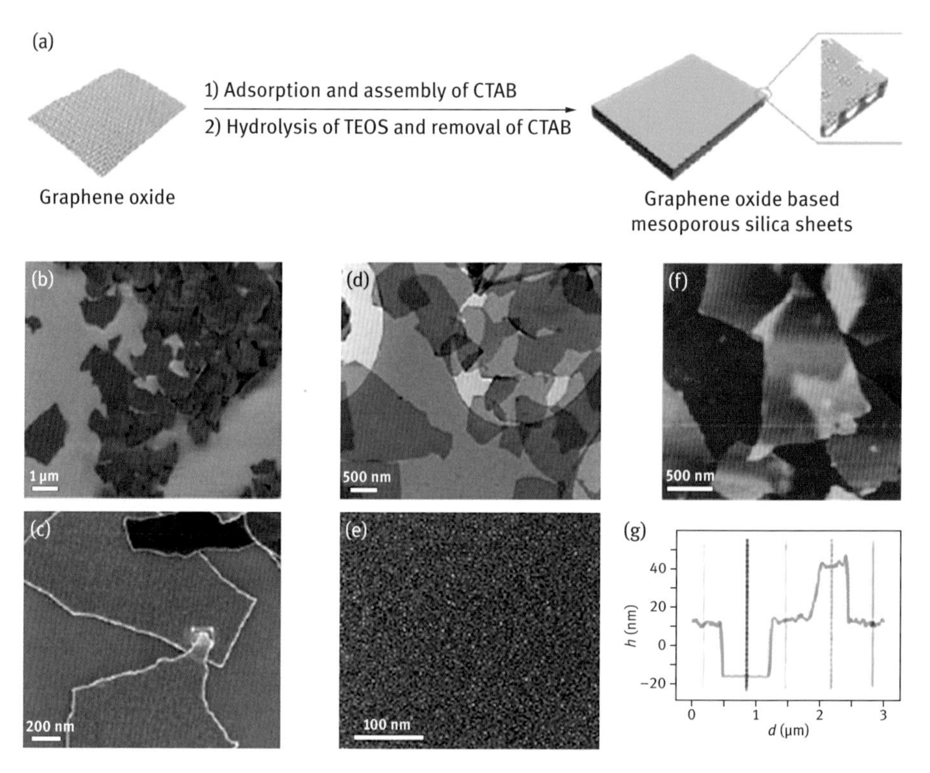

Fig. 12.9: Graphene-oxide-based mesoporous silica (GM-silica) sheets. (a) Fabrication process for GM-silica sheets. (b), (c) Typical SEM and (d), (e) TEM images reveal the flat GM-silica sheets with sizes from 200 nm to several micrometers having a mesoporous structure. (f) Representative atomic force microscopy image and (g) corresponding thickness analysis taken around the white line in (f) reveal a uniform thickness of 28 nm for GM-silica sheets. Reprinted with permission from [90]. Copyright 2010, John Wiley & Sons, Inc.

ide layers can provide highly conductive pathways to facilitate electron transport and simultaneously enhance the capacity contribution in the whole electrode. Our group recently developed a novel strategy for fabricating graphene-based 2D sandwich-like hybrids in which graphene is used as the template [91]. Specifically, graphene-based mesoporous silica (GM-silica) sheets are produced in gram scale by traditional sol-gel chemistry. The GS is fully separated by a mesoporous silica shell (Fig. 12.9) [90, 92, 93]. The unique 2D porous features of GM-silica sheets render them a promising universal template for the creation of various functional sheets, such as graphene-based mesoporous carbon (GM-C) [90], carbon nitride [92], and metal oxide (Co_3O_4 [90], TiO_2 [93]) nanosheets, in which graphene is individually dispersed. Furthermore, this strategy can be readily applied to the fabrication of a variety of novel 2D nanosheet materials to promote their broad applications in catalysts, sensors, supercapacitors, dye-sensitized solar cells, and batteries. For example, graphene-based mesoporous carbons prepared from GM-silica sheets by a nanocasting method have a high first reversible capacity (915 mAh g^{-1}) at a rate of C/5. After 30 cycles, both discharge and charge capacities of GM-C sheets are stable at ~ 770 mAh g^{-1}, with 84 % capacity retention. Moreover, GM-C sheets exhibit excellent rate performance. The reversible capacities are as high as 540 and 370 mAhg^{-1}, when the discharge-charge rate is increased to 1C and 5C, respectively. Another example is the fabrication of sandwich-like graphene-based mesoporous titania (G-TiO_2) nanosheets (Fig. 12.10). The resulting hy-

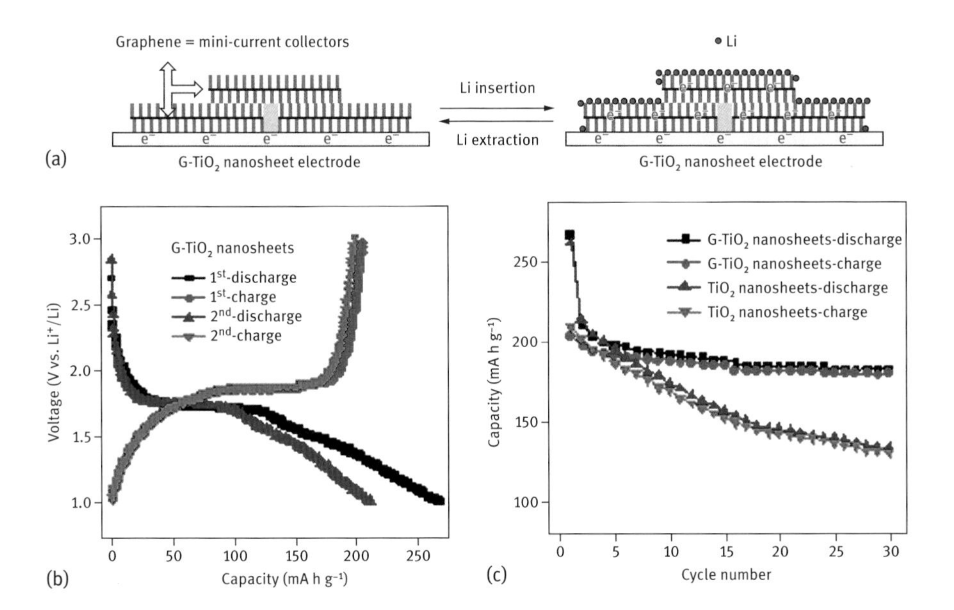

Fig. 12.10: (a) Lithium insertion and extraction in G-TiO_2 nanosheets, where graphene acts as a mini-current collector during discharge-charge processes. (b) First two charge-discharge curves of G-TiO_2 nanosheets at a current density of C/5. (c) Cycle performance of G-TiO_2 and TiO_2 nanosheets at a current density of C/5. Reprinted with permission from [93]. Copyright 2011, John Wiley & Sons, Inc.

brid nanosheets exhibit a large specific capacity (269 mAh/g), and a high rate capability with the reversible capacity retained at 162 and 123 mAh g^{-1} at rates of 1 C and 10 C, respectively [93].

12.3.5 Layered model

A well-defined layer structure of integrated metal oxide and conductive GS is known as a "layered model". In this structure, thin-film electrodes on arbitrary substrates and free-standing thin film electrodes are fabricated as binder-free, additive-free, or even flexible electrode energy storage devices [94–96]. On the one hand, these thin film electrodes can not only increase the volumetric/gravimetric energy density of the related devices, but they also reduce potential side reactions between liquid electrolyte and active materials to enhance the cycle stability. Further, thin film electrodes exhibit excellent mechanical feature, which is important for suppressing the volume change of the metal oxide nanoparticles confined within the graphene layers.

Much effort has been devoted to applying layer-by-layer (LbL) deposition and other assembly techniques to prepare layered graphene-based hybrid films with metal oxides, CNTs, or polymers in a controllable manner. The LbL technique offers an easy and inexpensive method of forming multiple layers and allows for the incorporation of a variety of materials within the film structures. Therefore, the LbL assembly method provides a versatile bottom-up nanofabrication technique for fabricating metal oxide/graphene hybrids with a layered structure for electrochemical energy storage. For example, Li *et al.* reported the construction of multilayer films by electrostatic LBL self-assembly, applying poly(sodium 4-styrenesulfonate) mediated GS (PSS-GS), manganese dioxide (MnO_2) sheets, and poly(diallyldimethylammonium; PDDA) as building blocks (Fig. 12.11(a), (b)) [94]. The prepared multilayer film electrodes for ECs were examined using cyclic voltammetry and galvanostatic charge/discharge in 0.1 M Na_2SO_4 electrolyte (Fig. 12.11(c)–(f)). The specific capacitance of the ITO/(PDDA/PSS-GS/PDDA/MnO_2)$_{10}$ electrode reached 263 F g^{-1} at 0.283 A g^{-1}. Moreover, these film electrodes also show good cycling stability and high coulombic efficiency [94]. Yu *et al.* reported the fabrication of free-standing LBL-assembled hybrid graphene-MnO_2 nanotube thin films by an ultrafiltration technique [95]. Such graphene-MnO_2 films have excellent cycling and rate capabilities for LIBs. Specifically, they obtained 495 mAh/g at 100 mA/g after 40 cycles. In contrast, pristine MnO_2 electrodes show only 140 mAh/g at 80 mA/g after 10 cycles. The enhanced electrochemical performance is attributed to the thin layer of graphene that provides not only conductive pathways accelerating the conversion reaction of MnO_2, but also buffer layers to maintain the electrical contact with MnO_2 nanotubes during lithium insertion/extraction [95]. Nevertheless, it should be pointed out that the major drawbacks of the LBL method lie in the lack of control for nanoscale spatial precision, the time requirement, and the difficulties inherent in the bulk synthesis of the materials.

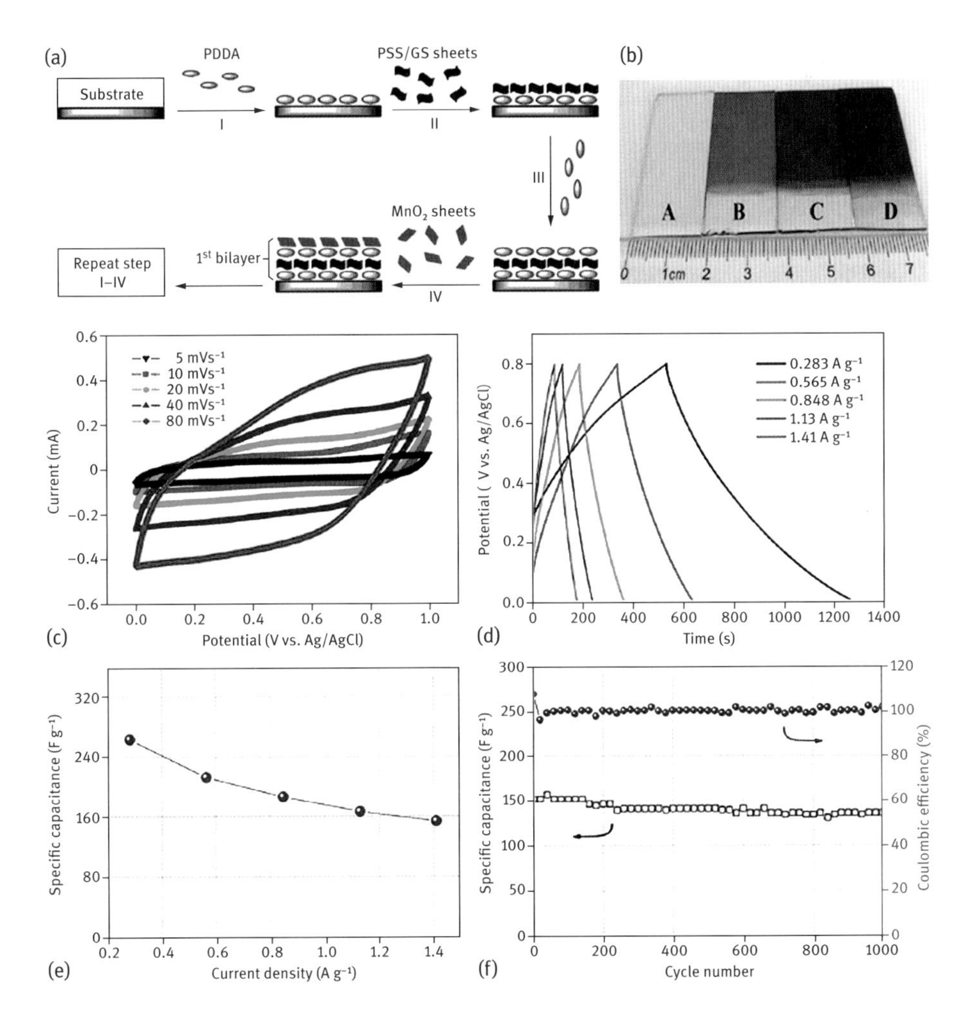

Fig. 12.11: A schematic view for constructing multilayer films on substrate. (b) Photographs of multilayer films of ITO/(PDDA/PSS-GS/PDDA/MnO_2)$_n$, n = 0, 5, 10, and 15 for A, B, C, and D, respectively. (c) Cyclic voltammetry curves of ITO/(PDDA/PSS-GS/PDDA/MnO_2)$_{10}$ electrode at different scan rates. (d) Charge-discharge behavior of an ITO/(PDDA/PSS-GS/PDDA/MnO_2)$_{10}$ electrode at different current densities. (e) Variation of specific capacitances with current density for ITO/(PDDA/PSS-GS/PDDA/MnO_2)$_{10}$ electrode. (f) Specific capacitance and coulombic efficiency change of ITO/(PDDA/PSS-GS/PDDA/MnO_2)$_{10}$ electrode versus the number of charge/discharge cycles at a current density of 1.41 A g^{-1}. Reprinted with permission from [94]. Copyright 2011, The Royal Society of Chemistry.

Flexible free-standing graphene hybrid films with layered structures were recently designed as additive/binder-free electrodes for energy storage devices. For example, Wang *et al.* developed a new strategy based on the ternary self-assembly of metal oxides, surfactants, and graphene to produce well-controlled, ordered layer-structured hybrids for LIBs (Fig. 12.12) [96]. The anionic surfactants were first used to adsorb

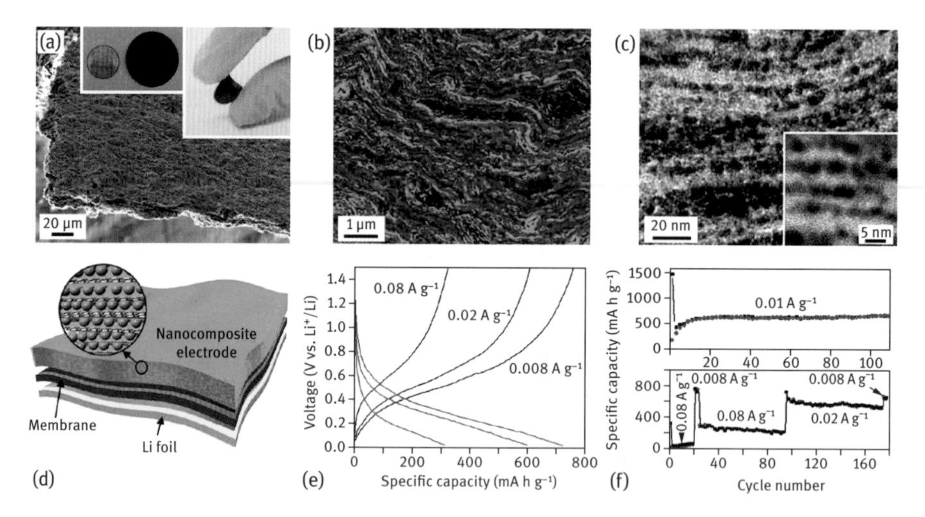

Fig. 12.12: (a) SEM image of a self-assembled free-standing graphene-SnO_2 hybrid electrode. Photographs in the inset show a disk-like 3 cm diameter graphene-SnO_2 hybrid electrode on the left and a folded electrode on the right. (b) Cross-sectional SEM image of the free-standing graphene-SnO_2 hybrid electrode. (c) Cross-sectional TEM images of a graphene-SnO_2 hybrid film. Inset shows a high-resolution TEM image in the hybrid film with alternating layers of nanocrystalline SnO_2 and graphene materials. (d) A Li-ion battery configuration directly using a freestanding metal oxide/graphene hybrid film as an electrode. (e) Charge/discharge profiles of a graphene-SnO_2 hybrid electrode between 0.02 and 1.5 V at 0.008, 0.02, and 0.08 A/g, respectively. (f) (Top) Specific capacity of SnO_2 as a function of charge/discharge cycles in the graphene-SnO_2 hybrid electrode at 0.01 A/g. (Bottom) Specific capacity of SnO_2 as a function of charge/discharge cycles in the graphene-SnO_2 hybrid at various charge/discharge current densities of 0.008, 0.08, and 0.02 A/g, respectively. Reprinted with permission from [96]. Copyright 2010, American Chemical Society.

and assemble onto the surface of GO. Subsequently, surfactant micelles functionalized graphene were used as building blocks for the self-assembly. The surfactants on the graphene bound to the metal cations, forming an ordered nanocomposite. Metal oxides were then crystallized between GS, producing ordered hybrids with alternating layers of graphene-graphene stacks and metal oxide nanocrystals. Alternatively, metal oxides and surfactants were assembled into a hexagonal mesoporous metal oxide/graphene hybrids using a nonionic block copolymer surfactant. Finally, layered metal oxide/graphene hybrids were fabricated as free-standing flexible films with thicknesses ranging from 5 to 20 µm by vacuum filtration. A steady specific capacity of 760 mAh/g for the graphene-SnO_2 electrode was obtained at a current density of 0.008 A/g, which is close to the theoretical capacity (780 mAh/g). The graphene-SnO_2 film electrode showed a good rate capability, e.g., specific capacity of 225 and 550 mAh/g at current densities of 0.08 and 0.02 A/g, respectively. This ternary self-assembly approach can also be applied to fabricate different layered materials of graphene with metal oxides (such as NiO, MnO_2) for applications beyond LIBs and ECs [96].

12.3.6 Mixed models

Apart from the previous models in which the morphology and structure of metal oxide particles composited with graphene can be controlled, the simplest model is based on a physical blending of metal oxide and graphene (normally ≤10 wt%) either in powder or in solution. Although this model lacks the structural and morphologic control of the hybrid materials, the so-called "mixed model" is particularly useful for cathode materials in LIBs, in which graphene mainly acts as a conductive component for the composites [97, 98]. For example, Ding *et al.* reported the synthesis of nanostructured graphene-$LiFePO_4$ hybrids by co-precipitation of $(NH_4)_2Fe(SO_4)_2 6H_2O$, $NH_4H_2PO_4$, and LiOH in aqueous solution of graphene, in which graphene is used as an additive to improve the electrical conductivity. The graphene-$LiFePO_4$ hybrids with only 1.5 wt% graphene delivered a specific capacity of 160 mA h g^{-1}, which is much higher than that of $LiFePO_4$ (113 mA h g^{-1}), when used as cathode material in LIBs [97]. Similar results were found for other hybrids of metal oxides such as $Li_4Ti_5O_{12}$ [98, 99], TiO_2 [100], and $Li_3V_2(PO_4)_3$ blended with graphene [101].

The mixed model of graphene and metal oxides (MnO_2, Co_3O_4, *etc.*) has also been applied to ECs. Wu *et al.* reported the preparation of a MnO_2 nanowire/graphene composite (MGC) by solution-phase mixing of GS and MnO_2 nanowires. They studied asymmetric ECs based on MGC as a positive electrode and graphene as a negative electrode in an aqueous Na_2SO_4 solution (Fig. 12.13) [47]. Notably, the superior electrical conductivity of graphene makes the MnO_2/graphene hybrid promising for use as a Faradic electrode in asymmetric ECs. Specifically, these devices exhibit a superior energy density of 30.4 Wh kg^{-1}, much higher than those of symmetrical ECs based on

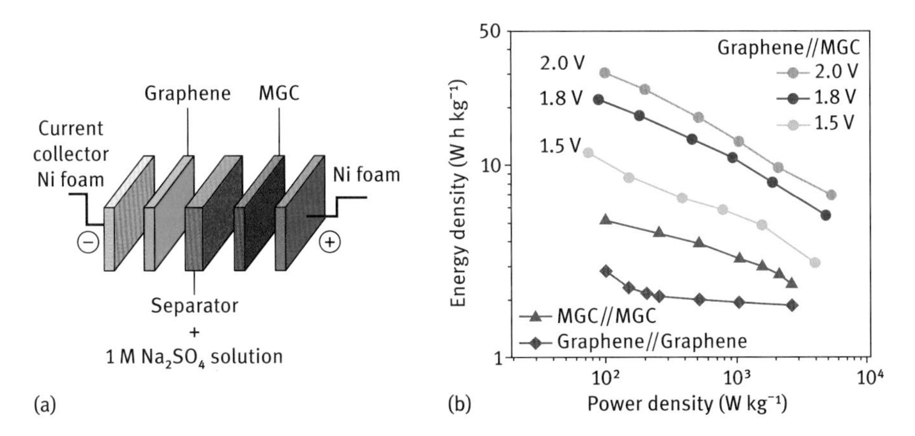

Fig. 12.13: (a) Schematic illustration of the assembled structure of asymmetric ECs based on MGC as a positive electrode and graphene as a negative electrode. (b) Ragone plot related to energy and power densities of graphene//MGC asymmetric ECs with various voltage windows, graphene//graphene and MGC//MGC symmetric ECs. Reprinted with permission from [47]. Copyright 2010, American Chemical Society.

graphene//graphene (2.8 Wh kg^{-1}) and MGC//MGC (5.2 Wh kg^{-1}), and significantly higher than those of other aqueous MnO_2-based asymmetric ECs (Fig. 12.13) [47].

12.4 Summary

Recent progress in graphene-metal oxide hybrid materials for LIBs and ECs are summarized in this chapter. Anchored, encapsulated, sandwich-like, layered, and mixed structures are the main proposed models for the design and development of graphene-metal oxide hybrids. Through the precise control of the morphologies and microstructures of these hybrids, excellent electrochemical performance in LIBs and ECs can be achieved. Despite the fact that research in this field is still in a relatively initial stage, graphene-metal oxide hybrids show great potential for applications in LIBs and ECs, such as enhanced capacity/capacitance, improved rate capability, excellent cycling stability, and high energy and power densities. We believe that through further improvement and continuous exploitation of various materials, different novel graphene-metal oxide hybrids will be fabricated with feasible applications, such as in portable electronics, electric vehicles, and hybrid electric vehicles.

Acknowledgments:

This work was financially supported by the ERC grants on NANOGRAPH, 2DMATER, ENERCHEM, DFG Priority Program SPP 1459, BMBF LiBZ, ESF Project GOSPEL (Ref Nr: 9–EuroGRAPHENE–FP–001), EU Project GENIUS, and MOLESOL.

Bibliography

[1] K. S. Novoselov, A. K. Geim, S. V. Morozov, D. Jiang, Y. Zhang, S. V. Dubonos, I. V. Grigorieva, A. A. Firsov, *Science* **2004**, *306*, 666.
[2] K. S. Novoselov, A. K. Geim, S. V. Morozov, D. Jiang, M. I. Katsnelson, I. V. Grigorieva, S. V. Dubonos, A. A. Firsov, *Nature* **2005**, *438*, 197.
[3] A. K. Geim, K. S. Novoselov, *Nat. Mater.* **2007**, *6*, 183.
[4] A. K. Geim, *Science* **2009**, *324*, 1530.
[5] C. Berger, Z. M. Song, T. B. Li, X. B. Li, A. Y. Ogbazghi, R. Feng, Z. T. Dai, A. N. Marchenkov, E. H. Conrad, P. N. First, W. A. de Heer, *J. Phys. Chem. B* **2004**, *108*, 19912.
[6] C. Berger, Z. M. Song, X. B. Li, X. S. Wu, N. Brown, C. Naud, D. Mayo, T. B. Li, J. Hass, A. N. Marchenkov, E. H. Conrad, P. N. First, W. A. de Heer, *Science* **2006**, *312*, 1191.
[7] S. Park, R. S. Ruoff, *Nat. Nanotechnol.* **2009**, *4*, 217.
[8] K. P. Loh, Q. L. Bao, P. K. Ang, J. X. Yang, *J. Mater. Chem.* **2010**, *20*, 2277.
[9] J. N. Coleman, *Adv. Funct. Mater.* **2009**, *19*, 3680.
[10] Z. S. Wu, W. Ren, L. Gao, B. Liu, C. Jiang, H. M. Cheng, *Carbon* **2009**, *47*, 493.
[11] Z. S. Wu, W. C. Ren, L. B. Gao, B. L. Liu, J. P. Zhao, H. M. Cheng, *Nano Res.* **2010**, *3*, 16.

[12] X. S. Li, W. W. Cai, J. H. An, S. Kim, J. Nah, D. X. Yang, R. Piner, A. Velamakanni, I. Jung, E. Tutuc, S. K. Banerjee, L. Colombo, R. S. Ruoff, *Science* **2009**, *324*, 1312.

[13] A. Reina, X. T. Jia, J. Ho, D. Nezich, H. B. Son, V. Bulovic, M. S. Dresselhaus, J. Kong, *Nano Lett.* **2009**, *9*, 30.

[14] Z. P. Chen, W. C. Ren, L. B. Gao, B. L. Liu, S. F. Pei, H. M. Cheng, *Nat. Mater.* **2011**, *10*, 424.

[15] L. J. Zhi, K. Müllen, *J. Mater. Chem.* **2008**, *18*, 1472.

[16] X. Y. Yang, X. Dou, A. Rouhanipour, L. J. Zhi, H. J. Rader, K. Müllen, *J. Am. Chem. Soc.* **2008**, *130*, 4216.

[17] N. Liu, F. Luo, H. X. Wu, Y. H. Liu, C. Zhang, J. Chen, *Adv. Funct. Mater.* **2008**, *18*, 1518.

[18] C. N. R. Rao, A. K. Sood, K. S. Subrahmanyam, A. Govindaraj, *Angew. Chem. Int. Ed.* **2009**, *48*, 7752.

[19] Y. W. Zhu, S. Murali, W. W. Cai, X. S. Li, J. W. Suk, J. R. Potts, R. S. Ruoff, *Adv. Mater.* **2010**, *22*, 3906.

[20] W. Hummers, R. Offman, *J. Am. Chem. Soc.* **1958**, *80*, 1339.

[21] S. Stankovich, D. A. Dikin, R. D. Piner, K. A. Kohlhaas, A. Kleinhammes, Y. Jia, Y. Wu, S. T. Nguyen, R. S. Ruoff, *Carbon* **2007**, *45*, 1558.

[22] X. B. Fan, W. C. Peng, Y. Li, X. Y. Li, S. L. Wang, G. L. Zhang, F. B. Zhang, *Adv. Mater.* **2008**, *20*, 4490.

[23] H. J. Shin, K. K. Kim, A. Benayad, S. M. Yoon, H. K. Park, I. S. Jung, M. H. Jin, H. K. Jeong, J. M. Kim, J. Y. Choi, Y. H. Lee, *Adv. Funct. Mater.* **2009**, *19*, 1987.

[24] S. F. Pei, J. H. Du, J. P. Zhao, W. C. Ren, H. M. Cheng, *Carbon* **2010**, *48*, 4466.

[25] H. C. Schniepp, J. L. Li, M. J. McAllister, H. Sai, M. Herrera-Alonso, D. H. Adamson, R. K. Prud'homme, R. Car, D. A. Saville, I. A. Aksay, *J. Phys. Chem. B* **2006**, *110*, 8535.

[26] M. J. McAllister, J. L. LiO, D. H. Adamson, H. C. Schniepp, A. A. Abdala, J. Liu, M. Herrera-Alonso, D. L. Milius, R. CarO, R. K. Prud'homme, I. A. Aksay, *Chem. Mater.* **2007**, *19*, 4396.

[27] Z. S. Wu, W. Ren, L. Gao, J. Zhao, Z. Chen, B. Liu, D. Tang, B. Yu, C. Jiang, H. M. Cheng, *ACS Nano* **2009**, *3*, 411.

[28] T. Ramanathan, A. A. Abdala, S. Stankovich, D. A. Dikin, M. Herrera-Alonso, R. D. Piner, D. H. Adamson, H. C. Schniepp, X. Chen, R. S. Ruoff, S. T. Nguyen, I. A. Aksay, R. K. Prud'homme, L. C. Brinson, *Nat. Nanotechnol.* **2008**, *3*, 327.

[29] Z. S. Wu, S. Pei, W. Ren, D. Tang, L. Gao, B. Liu, F. Li, C. Liu, H. M. Cheng, *Adv. Mater.* **2009**, *21*, 1756.

[30] A. S. Arico, P. Bruce, B. Scrosati, J. M. Tarascon, W. Van Schalkwijk, *Nat. Mater.* **2005**, *4*, 366.

[31] M. Armand, J. M. Tarascon, *Nature* **2008**, *451*, 652.

[32] P. Simon, Y. Gogotsi, *Nat. Mater.* **2008**, *7*, 845.

[33] C. Liu, F. Li, L. P. Ma, H. M. Cheng, *Adv. Mater.* **2010**, *22*, 28.

[34] E. Yoo, J. Kim, E. Hosono, H. Zhou, T. Kudo, I. Honma, *Nano Lett.* **2008**, *8*, 2277.

[35] P. C. Lian, X. F. Zhu, S. Z. Liang, Z. Li, W. S. Yang, H. H. Wang, *Electrochim. Acta* **2010**, *55*, 3909.

[36] D. Y. Pan, S. Wang, B. Zhao, M. H. Wu, H. J. Zhang, Y. Wang, Z. Jiao, *Chem. Mater.* **2009**, *21*, 3136.

[37] Z. S. Wu, W. C. Ren, L. Xu, F. Li, H. M. Cheng, *ACS Nano* **2011**, *5*, 5463.

[38] Z. S. Wu, L. L. Xue, W. C. Ren, F. Li, L. Wen, H. M. Cheng, *Adv. Funct. Mater.* **2012**, *22*, 3290.

[39] S. R. C. Vivekchand, C. S. Rout, K. S. Subrahmanyam, A. Govindaraj, C. N. R. Rao, *J. Chem. Sci.* **2008**, *120*, 9.

[40] M. D. Stoller, S. J. Park, Y. W. Zhu, J. H. An, R. S. Ruoff, *Nano Lett.* **2008**, *8*, 3498.

[41] Z. S. Wu, G. M. Zhou, L. C. Yin, W. C. Ren, F. Li, H. M. Cheng, *Nano Energy* **2012**, *1*, 107.

[42] P. V. Kamat, *J. Phys. Chem. Lett.* **2010**, *1*, 520.

[43] I. V. Lightcap, T. H. Kosel, P. V. Kamat, *Nano Lett.* **2010**, *10*, 577.

[44] Z. S. Wu, W. C. Ren, L. Wen, L. B. Gao, J. P. Zhao, Z. P. Chen, G. M. Zhou, F. Li, H. M. Cheng, *ACS Nano* **2010**, *4*, 3187.

[45] Z. S. Wu, D. W. Wang, W. Ren, J. Zhao, G. Zhou, F. Li, H. M. Cheng, *Adv. Funct. Mater.* **2010**, *20*, 3595.

[46] a) Z. Ren, E. Kim, S. W. Pattinson, K. S. Subrahmanyam, C. N. R. Rao, A. K. Cheetham and D. Eder, *Chem. Sci.*, **2012**, *3*, 209. b) b) D. Eder, I. A. Kinloch, A. H. Windle, *J. Mater. Chem.* **2008**, *18*, 2036.

[47] Z. S. Wu, W. C. Ren, D. W. Wang, F. Li, B. L. Liu, H. M. Cheng, *ACS Nano* **2010**, *4*, 5835.

[48] S. Chen, J. W. Zhu, X. D. Wu, Q. F. Han, X. Wang, *ACS Nano* **2010**, *4*, 2822.

[49] S. M. Paek, E. Yoo, I. Honma, *Nano Lett.* **2009**, *9*, 72.

[50] J. Yao, X. P. Shen, B. Wang, H. K. Liu, G. X. Wang, *Electrochem. Commun.* **2009**, *11*, 1849.

[51] L. S. Zhang, L. Y. Jiang, H. J. Yan, W. D. Wang, W. Wang, W. G. Song, Y. G. Guo, L. J. Wan, *J. Mater. Chem.* **2010**, *20*, 5462.

[52] M. Zhang, D. Lei, Z. F. Du, X. M. Yin, L. B. Chen, Q. H. Li, Y. G. Wang, T. H. Wang, *J. Mater. Chem.* **2011**, *21*, 1673.

[53] P. Lian, X. Zhu, S. Liang, Z. Li, W. Yang, H. Wang, *Electrochim. Acta* **2011**, *56*, 4532.

[54] X. Wang, X. Zhou, K. Yao, J. Zhang, Z. Liu, *Carbon* **2011**, *49*, 133.

[55] S. Ding, D. Luan, F. Y. C. Boey, J. S. Chen, X. W. Lou, *Chem. Commun.* **2011**, *47*, 7155.

[56] B. Zhao, G. Zhang, J. Song, Y. Jiang, H. Zhuang, P. Liu, T. Fang, *Electrochim. Acta* **2011**, *56*, 7340.

[57] S. Q. Chen, Y. Wang, *J. Mater. Chem.* **2010**, *20*, 9735.

[58] H. Kim, D. H. Seo, S. W. Kim, J. Kim, K. Kang, *Carbon* **2011**, *49*, 326.

[59] B. J. Li, H. Q. Cao, J. Shao, G. Q. Li, M. Z. Qu, G. Yin, *Inorg. Chem.* **2011**, *50*, 1628.

[60] J. Zhu, Y. K. Sharma, Z. Zeng, X. Zhang, M. Srinivasan, S. Mhaisalkar, H. Zhang, H. H. Hng, Q. Yan, *J. Phys. Chem. C* **2011**, *115*, 8400.

[61] H. L. Wang, L. F. Cui, Y. A. Yang, H. S. Casalongue, J. T. Robinson, Y. Y. Liang, Y. Cui, H. J. Dai, *J. Am. Chem. Soc.* **2010**, *132*, 13978.

[62] C.-T. Hsieh, C.-Y. Lin, J.-Y. Lin, *Electrochim. Acta* **2011**, *56*, 8861.

[63] X. Zhu, Y. Zhu, S. Murali, M. D. Stoller, R. S. Ruoff, *ACS Nano* **2011**, *5*, 3333.

[64] W. Zhou, J. Zhu, C. Cheng, J. Liu, H. Yang, C. Cong, C. Guan, X. Jia, H. J. Fan, Q. Yan, C. M. Li, T. Yu, *Energy Environ. Sci.* **2011**, *4*, 4954.

[65] M. Zhang, D. N. Lei, X. M. Yin, L. B. Chen, Q. H. Li, Y. G. Wang, T. H. Wang, *J. Mater. Chem.* **2010**, *20*, 5538.

[66] J. Su, M. Cao, L. Ren, C. Hu, *J. Phys. Chem. C* **2011**, *115*, 14469.

[67] J. Zhou, H. Song, L. Ma, X. Chen, *RSC Adv.* **2011**, *1*, 782.

[68] J. Hu, A. Ramadan, F. Luo, B. Qi, X. Deng, J. Chen, *J. Mater. Chem.* **2011**, *21*, 15009.

[69] D. W. Choi, D. H. Wang, V. V. Viswanathan, I. T. Bae, W. Wang, Z. M. Nie, J. G. Zhang, G. L. Graff, J. Liu, Z. G. Yang, T. Duong, *Electrochem. Commun.* **2010**, *12*, 378.

[70] S. Ding, J. S. Chen, D. Luan, F. Y. C. Boey, S. Madhavi, X. W. Lou, *Chem. Commun.* **2011**, *47*, 5780.

[71] B. Wang, X.-L. Wu, C.-Y. Shu, Y.-G. Guo, C.-R. Wang, *J. Mater. Chem.* **2010**, *20*, 10661.

[72] B. Li, H. Cao, G. Yin, Y. Lu, J. Yin, *J. Mater. Chem.* **2011**, *21*, 10645.

[73] W. Zhou, J. Liu, T. Chen, K. S. Tan, X. Jia, Z. Luo, C. Cong, H. Yang, C. M. Li, T. Yu, *Phys. Chem. Chem. Phys.* **2011**, *13*, 14462.

[74] J. Yan, T. Wei, W. M. Qiao, B. Shao, Q. K. Zhao, L. J. Zhang, Z. J. Fan, *Electrochim. Acta* **2010**, *55*, 6973.

[75] A. K. Mishra, S. Ramaprabhu, *J. Phys. Chem. C* **2011**, *115*, 14006.

[76] J. Zhang, J. Jiang, X. S. Zhao, *J. Phys. Chem. C* **2011**, *115*, 6448.

[77] H. Huang, X. Wang, *Nanoscale* **2011**, *3*, 3185.

[78] B. Wang, J. Park, C. Y. Wang, H. Ahn, G. X. Wang, *Electrochim. Acta* **2010**, *55*, 6812.
[79] J. Wang, Z. Gao, Z. Li, B. Wang, Y. Yan, Q. Liu, T. Mann, M. Zhang, Z. Jiang, *J. Solid State Chem.* **2011**, *184*, 1421.
[80] H. Kim, S. W. Kim, Y. U. Park, H. Gwon, D. H. Seo, Y. Kim, K. Kang, *Nano Res.* **2010**, *3*, 813.
[81] M. S. Whittingham, *Chem. Rev.* **2004**, *104*, 4271.
[82] G. M. Zhou, D. W. Wang, F. Li, L. L. Zhang, N. Li, Z. S. Wu, L. Wen, G. Q. Lu, H. M. Cheng, *Chem. Mater.* **2010**, *22*, 5306.
[83] J. S. Chen, Z. Wang, X. C. Dong, P. Chen, X. W. Lou, *Nanoscale* **2011**, *3*, 2158.
[84] W. Lv, F. Sun, D.-M. Tang, H.-T. Fang, C. Liu, Q.-H. Yang, H.-M. Cheng, *J. Mater. Chem.* **2011**, *21*, 9014.
[85] Y. Sun, X. Hu, W. Luo, Y. Huang, *ACS Nano* **2011**, *5*, 7100.
[86] H. Liu, W. Yang, *Energy Environ. Sci.* **2011**, *4*, 4000.
[87] S. B. Yang, X. L. Feng, S. Ivanovici, K. Müllen, *Angew. Chem. Int. Ed.* **2010**, *49*, 8408.
[88] J. Z. Wang, C. Zhong, D. Wexler, N. H. Idris, Z. X. Wang, L. Q. Chen, H. K. Liu, *Chem-Eur J* **2011**, *17*, 661.
[89] D. Chen, G. Ji, Y. Ma, J. Y. Lee, J. Lu, *ACS Appl. Mater. Inter.* **2011**, *3*, 3078.
[90] S. B. Yang, X. L. Feng, L. Wang, K. Tang, J. Maier, K. Müllen, *Angew. Chem. Int. Ed.* **2010**, *49*, 4795.
[91] D. Q. Wu, F. Zhang, P. Liu, X. L. Feng, *Chem. Euro. J.* **2011**, *17*, 10804.
[92] S. B. Yang, X. L. Feng, X. C. Wang, K. Müllen, *Angew. Chem. Int. Ed.* **2011**, *50*, 5339.
[93] S. B. Yang, X. L. Feng, K. Müllen, *Adv. Mater.* **2011**, *23*, 3575.
[94] Z. P. Li, J. Q. Wang, X. H. Liu, S. Liu, J. F. Ou, S. R. Yang, *J. Mater. Chem.* **2011**, *21*, 3397.
[95] A. P. Yu, H. W. Park, A. Davies, D. C. Higgins, Z. W. Chen, X. C. Xiao, *J. Phys. Chem. Lett.* **2011**, *2*, 1855.
[96] D. H. Wang, R. Kou, D. Choi, Z. G. Yang, Z. M. Nie, J. Li, L. V. Saraf, D. H. Hu, J. G. Zhang, G. L. Graff, J. Liu, M. A. Pope, I. A. Aksay, *ACS Nano* **2010**, *4*, 1587.
[97] Y. Ding, Y. Jiang, F. Xu, J. Yin, H. Ren, Q. Zhuo, Z. Long, P. Zhang, *Electrochem. Commun.* **2010**, *12*, 10.
[98] N. Zhu, W. Liu, M. Q. Xue, Z. A. Xie, D. Zhao, M. N. Zhang, J. T. Chen, T. B. Cao, *Electrochim. Acta* **2010**, *55*, 5813.
[99] Y. Shi, L. Wen, F. Li, H. M. Cheng, *J. Power Sources* **2011**, *196*, 8610.
[100] N. Li, G. Liu, C. Zhen, F. Li, L. L. Zhang, H. M. Cheng, *Adv. Funct. Mater.* **2011**, *21*, 1717.
[101] H. Liu, P. Gao, J. Fang, G. Yang, *Chem. Comm.* **2011**, *47*, 9110.

John Robertson

13 Nanocarbons for field emission devices

13.1 Introduction

Field emission is the emission of electrons from a solid under an intense electric field, usually at ambient temperatures. It occurs by the quantum mechanical tunneling of electrons through a potential barrier (Fig. 13.1). This leads to an exponential dependence of emission current density J on the local electric field, as given by the Fowler–Nordheim equation,

$$J = a\frac{(F)^2}{\phi}.\exp\left(-\frac{b\phi^{3/2}}{F}\right)$$

Here, ϕ (eV) is the barrier height, F is the local field, and a and b are constants. The typical field required for emission from solids is of order 1000 V/μm. The easiest way to create such a high field is by field enhancement at a sharp tip, so that the local field F is many times larger than the applied field $F', F = \beta F'$. β is a dimensionless geometrical field enhancement factor given by h/r for a tip, where h is the height and r is its radius.

$$\beta = h/r$$

Thus, carbon nanotubes are naturally of interest due to their very large aspect ratio, h/r.

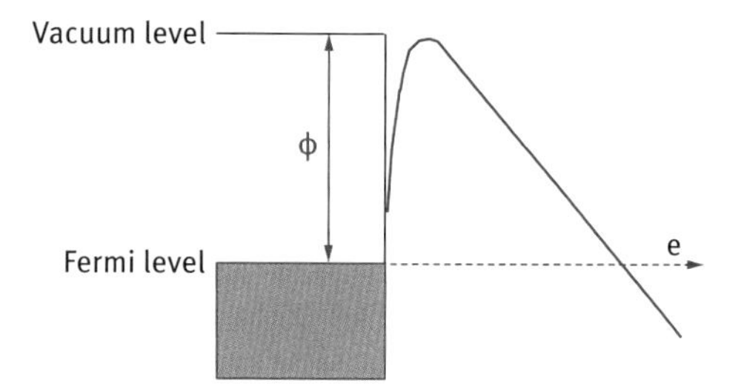

Fig. 13.1: Band diagram of field emission from a solid surface, showing the triangle tunnel barrier.

The early field emission systems were 'Spindt tips' fabricated from Si or Mo [1, 2]. There was extensive development of field emission systems based on these materials. Carbon has a number of natural advantages over other materials for field emission [3, 4]; it is not easily poisoned by typical residual gases in the vacuum system. Second, it has a high cohesive energy, so that it is not eroded by ion bombardment from residual gas ions. Third, carbon can carry the highest current density of any solid, about 10^9 A/cm^2, due ultimately to the high strength of the C-C bonds. Fourth, many carbons have a negative temperature coefficient of resistance, so that they become less resistive if heated up by the large local currents, whereas this leads to thermal runaway in Mo or doped Si tips. Finally, carbon can be cheap.

Many types of carbon have been studied from the perspective of field emission, perhaps in the wrong order of usefulness. Initially, there was great interest in diamond, because its hydrogen terminated surfaces have a negative electron affinity (NEA) [5–7], meaning that its conduction band edge lies above the vacuum level and electrons can in principle be emitted without barrier from its conduction band into the vacuum. However, this just passes the problem backwards; there is now a large potential barrier for excitation of electrons from the back electrode into the diamond's conduction band. This is because there is no shallow donor for diamond, and it cannot be doped n-type easily. The Fermi level never lies close in energy to the vacuum level, so the practical value of a low *work function* is never achieved.

The next point to realize is that the best emitter is a metal. Many forms of carbon initially studied are semiconductors or even insulators, including nanodiamond [8–11] and diamond-like carbon (DLC) [12–13,4]. Combine this with local field enhancement means that there is never uniform emission from a flat carbon surface, it emits from local regions of field enhancement, such as grain boundaries [8–11] or conductive tracks burnt across the film in a 'forming' process akin to electrical breakdown [13]. Any conductive track is near-metallic and is able to form an internal tip, which provides the field enhancement within the solid state [4]. Figure 13.2 shows the equipotentials around an internal tip due to grain boundaries or tracks inside a less conductive region.

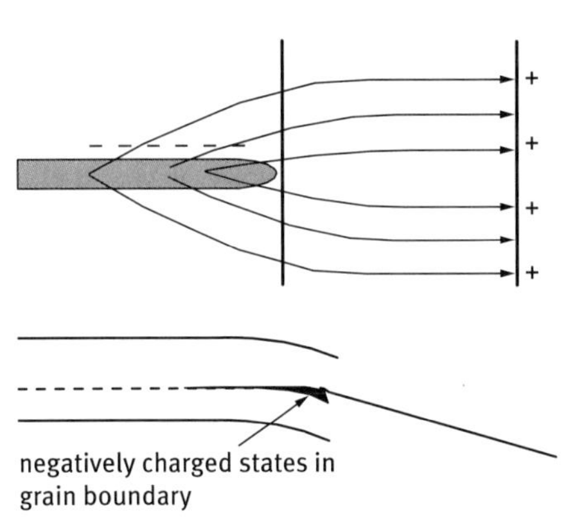

Fig. 13.2: Upper panel showing field lines from a conductive grain boundary inside a solid, under field emission. Lower panel shows the band bending and charging of gap states at the grain boundary.

Thus, emission can be increased by using nanodiamond grain boundaries or DLC with conductive tracks. However, these systems suffer from irreproducibility and instability. Eventually, sp^2 carbon systems such as carbon nanotubes (CNTs), nanowalls and nanocarbons were studied, which are the best systems, because they are metallic and have the desired shape for field enhancement.

13.2 Carbon nanotubes – general considerations

Carbon nanotubes can be single walled (SWNT) or multi-walled (MWNT). SWNTs can be metallic or semiconducting, depending on their chirality. MWNTs contain several walls, so in combination they will tend to be metallic. In addition, as the band gap of semiconducting tubes varies inversely with their diameter, the larger diameter of MWNTs means that effectively most walls are metallic.

The early work of Rinzler *et al.* [14], DeHeer [15] and Chernozatonskii *et al.* [16] showed the promise of CNTs for field emission. However making something technically useful is more complicated. Due to their ability to carry large current densities, each CNT emitter can carry up to about 10 μA [17–19]. This is a large current. There are a few applications where emission from a single good emitter is enough (*e.g.* Transmission electron microscope cathodes). But many applications need higher total currents, so that current must be shared between many emitters [20–21]. This is less easy than it seems.

Figure 13.3 shows schematically an array of CNT field emitters with different spacings, together with the equipotentials. In principle, the CNTs should be vertically aligned, but sometimes they will be aligned imperfectly. We see that for widely-spaced aligned CNTs, the equipotential distribution follows that of a series of isolated CNTs. The field enhancement at the top of each CNT is roughly that for an isolated tube. This regime holds for spacings d greater than twice the CNT height, $d > 2h$ (Fig. 13.3(a)). On the other hand, as the spacing reduces, the equipotentials become flatter, and the field enhancement declines (Fig. 13.3(b)). This is 'tip screening' [22]. Thus, field enhancement becomes less, and becomes a function of both the tip's aspect ratio and the spacing. This means that in practice, it becomes much less easy to control. Finally, orientational disorder also leads to field screening (Fig. 13.3(c)).

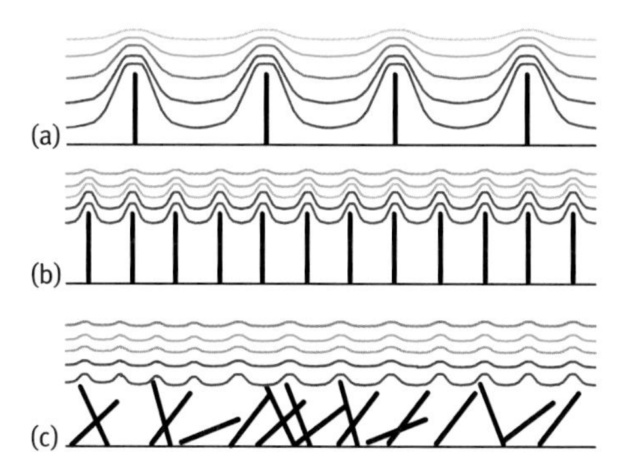

Fig. 13.3: Equipotentials around various arrays of carbon nanotubes (a) Field enhancement at an array of widely spaced nanotube emitters, (b) closely spaced emitters, showing the field screening, (c) field screening at a randomly oriented nanotube forest.

There are two approaches from here, either grow well-controlled arrays of vertically aligned CNTs and attempt good design and performance [23–25], or grow random CNT forests and hope for the best [20, 21].

Nilsson *et al.* [26, 27] showed how the total emission current from a less than perfect array of CNT emitters will depend on the distribution of the field enhancement factors, β. The β probability distribution follows a roughly bell-shaped distribution. Clearly, the high β values matter most, so that these follow an effectively exponential decreasing probability distribution (Fig. 13.4),

$$N(\beta) = N_0 \exp(-\beta/k)$$

Although it may be intended that current should be shared between 10^6 emitters, in practice there is huge current crowding so that emission occurs from only a few of them, perhaps 1 or 10 in 10^5. This has two severe consequences. First, the emission site density is much lower than might be expected from the nanotube density. Second, the operating emitters can easily exceed their maximum currents and burn out, passing the emission onto those with the next highest β value, and so on. This leads to unstable emission behavior [27].

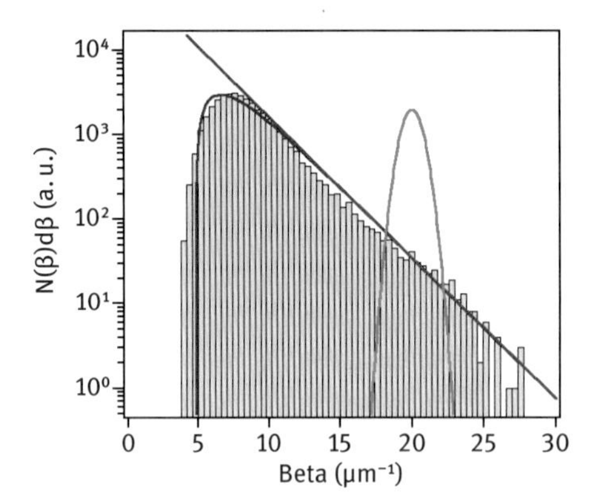

Fig. 13.4: Distribution of field enhancement factors in a CNT emitting, showing exponential decay at high enhancement factors.

One solution is to have a huge reserve number of emitters, as in the randomly grown spaghetti-like or forests of CNTs. The other is to attempt to grow a highly regular array of vertically oriented emitters, each of the same spacing and height (Fig. 13.5). Teo *et al.* [25] have examined this. It is possible but it requires high precision. Even in highly regular arrays of CNTs, only about 1 in 10^4 CNTs emit. The CNTs are grown by plasma enhanced chemical vapor deposition (PECVD) as the bias field causes the CNTs to be aligned vertical to the substrate.

A second way to promote current sharing between emitters is to use a ballast resistor [28]. This resistor is in series with the CNTs and the voltage drop across it acts to share current between the emitters [29]. The idea is that some fraction of the total applied voltage between cathode and anode should be dropped across the ballast resistor, so that there is less across the emitter itself. Ballast resistors increase the emission site density [29]. It acts like negative feedback.

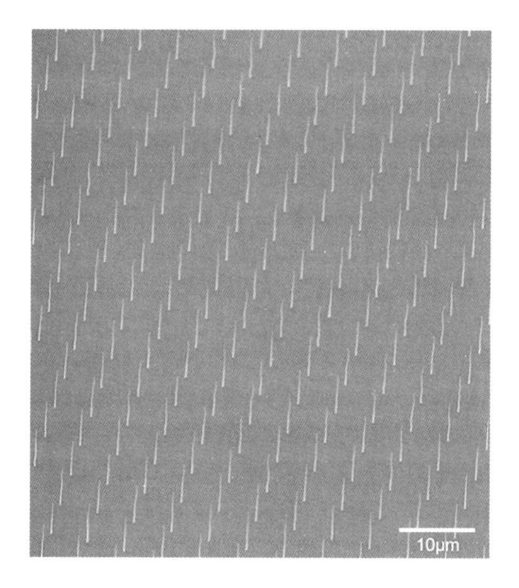

Fig. 13.5: SEM image showing an accurately patterned array of CNTs grown by PECVD (courtesy K. Teo).

There are two geometries for a ballast resistor, a thin film under the CNT (vertical resistor), or a thin film at the side of the CNT (horizontal resistor). Now, the local field for field emission from carbon is extremely high, 10^9 V/m, which may be 10 times larger than the breakdown field of many insulators. Given the dimensions of an emission system, it is likely that the thickness of any ballast resistor thin film will be smaller than that of the net cathode-anode distance. Thus, the applied field is likely to cause breakdown in any vertical ballast resistor design. Thus, horizontal ballast resistors are a safer option. However, they are more difficult to implement because they must be placed between each emitter and the current feed.

A final point about CNT field emitters is that MWNTS are preferred to SWNTs [25]. Nanotubes do have a very high Young's modulus in tension. But their small diameter of SWNTs means that they are quite flexible and can bend easily. Their stiffness varies as the moment of inertia, or diameter to the fourth power. MWNTs are much stiffer, and this is more useful for emitter structures. It is not useful, if CNTs bend over under various influences such as applied fields or vibrations, and touch adjacent emitters. They should remain rigidly in their designed position.

CNTs are also valuable as field emitters because they have a small virtual source size [30], a high brightness, and a small positive temperature coefficient of resistance [31]. The latter means that they can run hot under high emission currents, but not go into thermal runaway. Emission from nanotubes can be visualized by electron holography in a TEM [32].

Nanotubes must be conditioned to optimize the emission. Absorbates should be removed by running a high current through them to desorb them [17]. The stability can then be verified [18]. Eventually the current per nanotube saturates [19].

13.2.1 Field emission from nanocarbons

Nanocarbon emitters behave like variants of carbon nanotube emitters. The nanocarbons can be made by a range of techniques. Often this is a form of plasma deposition which is forming nanocrystalline diamond with very small grain sizes. Or it can be deposition on pyrolytic carbon or DLC run on the borderline of forming diamond grains. A third way is to run a vacuum arc system with ballast gas so that it deposits a porous sp^2 rich material. In each case, the material has a moderate to high fraction of sp^2 carbon, but is structurally very inhomogeneous [29]. The material is moderately conductive. The result is that the field emission is determined by the field enhancement distribution, and not by the sp^2/sp^3 ratio. The enhancement distribution is broad due to the disorder, so that it follows the Nilsson model [26] of emission site distributions. The disorder on nanocarbons makes the distribution broader. Effectively, this means that emission site density tends to be lower than for a CNT array, and is less controllable. Thus, while it is lower cost to produce nanocarbon films, they tend to have lower performance.

13.2.2 Emission from nanowalls and CNTs walls

Carbon nanowalls are an unusual form of nanocarbon that can be produced by plasma deposition with a certain set of deposition parameters. They are graphitic but with their graphitic planes perpendicular to the substrate. The walls are a finite thickness, not a few layers thick like graphene, and the walls tend to be quite wiggly [33, 34]. They have not been extensively studied.

On the other hand they do have a high aspect ratio, and are quite suitable for field emission applications. Their structure means that the field emitting sites have more redundancy than simple CNTs, so the emitting site could pass along a wall, giving a higher stability.

A second structure to consider is a synthetic nanowall structure consisting of a closely packed forest of CNTs patterned into walls, on a square, hexagonal or triangular lattice (Fig. 13.6) [35]. These also allow for emission site redundancy. They have the added advantage that they can accommodate the lateral ballast resistor between a current source and the emitting walls.

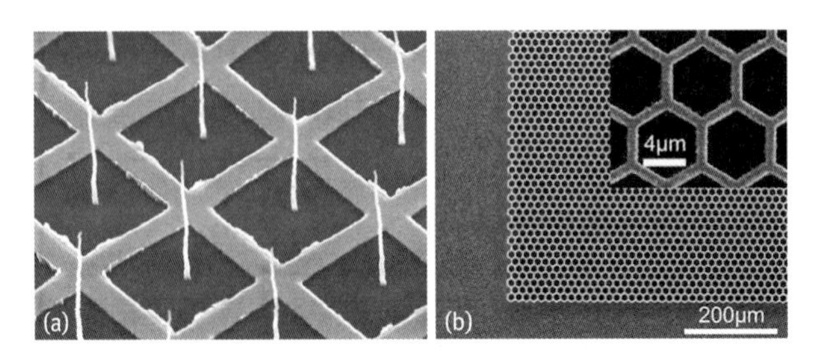

Fig. 13.6: (a) CNT array with lateral ballast resistors to the square grid, (b) honeycomb patterned mesh structure of CNT emitters (courtesy W. I. Milne).

13.3 Applications

We now consider some specific applications [36].

13.3.1 Field emission electron guns for electron microscopes

Electron microscopes, whether scanning electron microscopes (SEM), transmission electron microscopes (TEM), or scanning transmission electron microscopes (STEM) require high brightness electron sources. Perhaps the most demanding case is the STEM. This is a specialized but useful application for CNTs [37–42]. The parameter of interest is the reduced brightness, the emission current density per steradian per eV ($A/m^2/Sr/eV$). In the electron trajectories, the electrons originate from a virtual source of diameter d, behind the actual source [39, 40]. The electron energy spread ΔE is also a critical parameter, as this should be minimized to keep the chromatic aberration small.

The performances of various different emitter technologies are compared in Table 13.1 [39, 40]. The different technologies are thermionic emitters of tungsten (W) or lanthanum hexaboride (LaB_6, a low work function conducting ceramic), Schottky emitters based on Si, and cold cathode emitters which are W field emitters. A single MWNT is used for this application (Fig. 13.5). It is seen that the CNT field emitter out performs other electron sources on the basis of reduced brightness (B) and virtual source size (r). The energy spread ΔE is low, similar to that of metal field emitters. Thermionic emitters have large energy spreads due to their high operating temperatures. They have smaller brightness, because thermionic emission is fundamentally not such a high brightness emission. The stability of thermionic sources is good. Initially the stability and lifetime of field emission sources was poor. This is because atoms and molecules would absorb on the emitter surfaces, and vary the work function, thus varying the current. A good vacuum system with local pumping was necessary to fix this. The emitters are effectively triode type, with an emitter field controlled by a local grid/anode, and the total acceleration voltage controlled in the microscope.

Tab. 13.1: Performances of different types of electron guns; source size (r), reduced brightness (B), energy spread (ΔE), stability, and operating temperature.

	Thermionic, W	Thermionic, LaB_6	Schottky	Cold FE	CNT FE
r (nm)	10,000	10,000	20	3	10
B ($A/Sr/m^2/V$)	10^6	10^7	10^8	10^8	$3x10^9$
ΔE (eV)	1.5	1.0	0.7	0.2	0.3
Stability (%)	< 1	< 1	< 1	5	0.5
Operating T (C)	2200	1500	1700	25	25–400

The first CNT field emitter tips were hand-made [37], formed by fishing a CNT from the spaghetti CNT mat, and then attaching it to a tungsten base with carbon glue. A short CNT could be made by burning it off, or twisting and cutting. A high temperature anneal by emitting current is useful to optimize the walls and bonding in the emission tip [41]. This can be studied by *in situ* microscopy [42].

The CNTs are now grown onto suitable bases by CVD (Fig. 13.7) [43], a metal catalyst is deposited by FIB or other patterning method, and then the CNT is grown. One CNT generally grows from each catalyst spot. The growth is often in the plasma-enhanced regime, to provide the correct alignment.

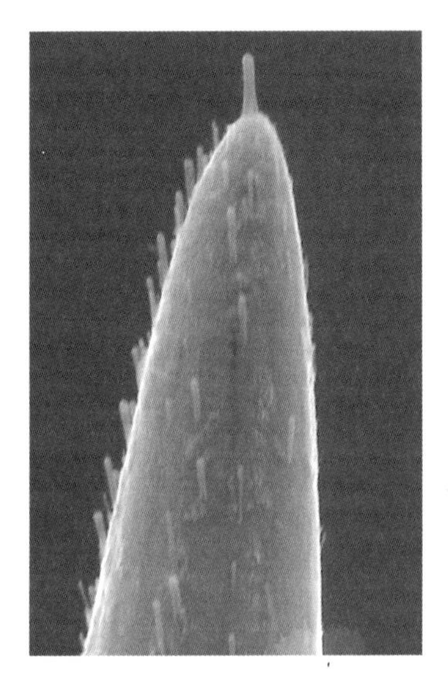

Fig. 13.7: Carbon nanotube field emitter grown on metal tip (courtesy M. Mann).

13.3.2 Displays

The field emission display held a particular fascination for this field, because of its potentially large market [36, 44–49]. Field emission displays (FEDs) are flat panel displays, which are a flat panel equivalent of the cathode ray tube (CRT), but in which each pixel is addressed by its own electron beam from a field emitter, rather than having a beam scanned across it as in the CRT (Fig. 13.8) [44]. The emitters can be diode or triode type. The triode type is the most elegant, the diode type is lower cost.

There are a number of competing flat panel technologies, such as active matrix liquid crystal displays (AMLCD), plasma displays (PD), FEDs, organic light emitting diodes (OLEDs) and electro-luminescent displays (compared in Table 13.2). The advantages of the FED are high brightness, high power efficiency (black = off), the ability to work over a wide temperature range, wide viewing angle, and high video rate, compared to the AMLCD. The AMLCD is a light switch, so that its light source is always on and it is not particularly energy efficient. LCs have a limited operating temperature

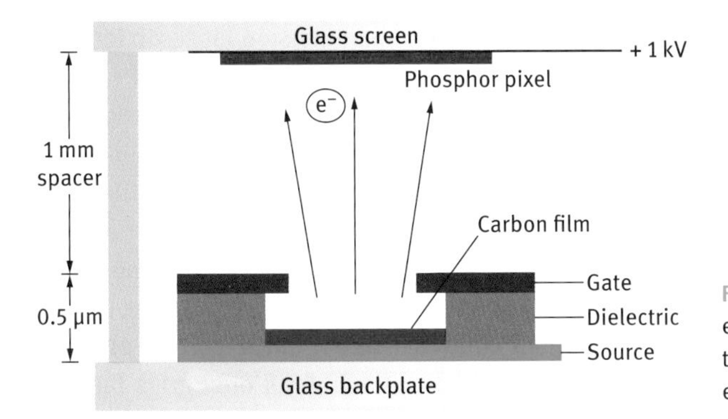

Fig. 13.8: Schematic of a field emission display showing triode system with carbon emitter film.

Tab. 13.2: Performance metrics of display technologies, more blobs the better.

	AMLCD	OLED	Plasma	FED
Low cost	XXXX		X	
Weight	XX	XXX		XX
Size	XX			
Viewing angle	XX	XXX	XX	XXX
Video rate	X	X		XX
Brightness			XX	XX

range, and initially they had a rather narrow viewing angle, which was corrected by 'in plane switching'. Their response was also rather slow. Nevertheless, with a large active investment, one by one, the LCD industry overcame each of these disadvantages, so that now the AMLCD is by far the dominant display. The plasma display had numerous advantages such as wide viewing angle being an emissive display. But it is heavy, and has a large power consumption. It has been beaten off. The problem of the FED was that its manufacturing process required phosphor technology, spacers, gettering, vacuum sealing, *etc.* The manufacturing yield was never high enough. Many types of CNT arrangement were tried, for example with CNT-glass matrix [45], back gates [46, 47], control of CNT length to ensure no shorting between tip and gate [48], and testing the degradation process [49].

13.3.3 Microtriodes and E-beam lithography

The continued downscaling of transistor sizes is essential to the continuation of Moore's law. Lithography has always been a limiting factor. Optical lithography is limited by the optical resolution

$$R = \lambda/NA$$

where NA is the numerical aperture and λ is the wavelength. It is not possible to increase λ further, because excimer lasers are at their maximum energy for such molecules. Nevertheless, R could be decreased by using immersion lenses to increase the NA. This has allowed a continued lowering of R.

An alternative is to use electron beam lithography, whose basic resolution is of order 4 Å. However, e-beam lithography is a serial addressing system, rather than a parallel system, so that we must write a 2D image as a series of lines, rather than a 2D pattern, and this takes a much longer time.

To increase printing rates an idea is to develop parallel e-beam lithography [36, 50–56]. This would use electrically addressable two-dimensional arrays of electron sources. Each source would be a field emitter inside a CMOS control element. The electron sources in this case are quite complex, having not only grids but also focusing electrodes [36].

The sources are constructed as triodes with control gates, either one or many CNTs grown in the cavity, and focusing electrodes (Fig. 13.9). Images of CNTs grown in triode structures are shown in Fig. 13.10 [36, 56]. The processing schedule is given in Fig. 13.11. It is important to control the CNT length and growth time, to ensure that the CNTs do not short circuit to the grid electrode.

However, again this application did not succeed. Optical lithography continues. First by steppers with half or third spacing. Now, a new form of lithography is close to implementation, known as extreme-UV lithography. This uses coherent UV from a plasma of Sn^{4+} ions, which emits at around 100 eV.

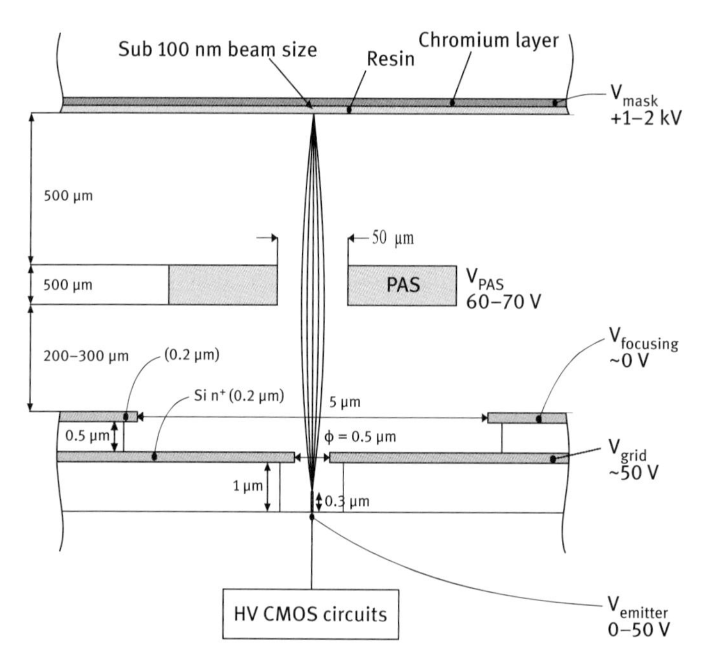

Fig. 13.9: Schematic diagram of a triode emitter, including electron focusing electrode, for a parallel lithography system.

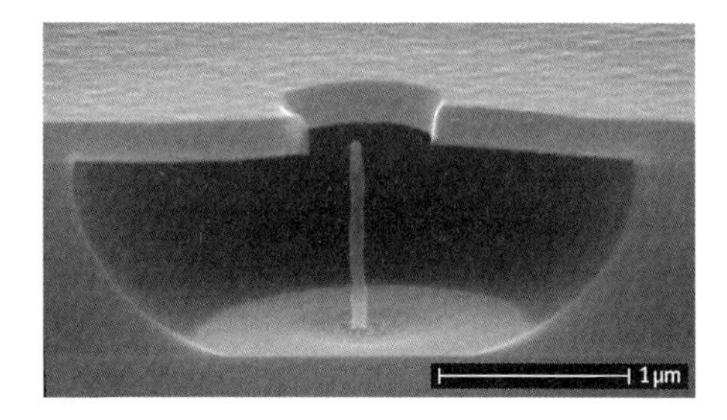

Fig. 13.10: A triode emitter structure, showing single CNT emitter.

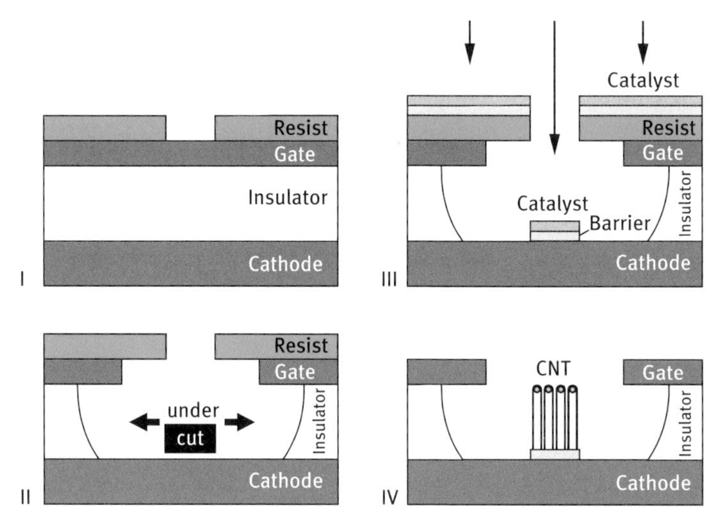

Fig. 13.11: Process path to fabricate the triode emitter structure, showing layer growth, etching, catalyst deposition, and CNT growth steps.

13.3.4 Microwave power amplifiers

High frequency power amplifiers are still a regime where the superiority of semiconductor devices is not paramount. This is because the electron velocity is a semiconductor is orders of magnitude less than in the vacuum. A fast device needs very short source-drain distance and this can be incompatible with high power dissipation. Larger devices are possible for vacuum tubes. The standard tube, a traveling wave tube, would have a thermionic cathode, and the input signal would modulate the resulting electron beam *via* an inductive coupler. The most elegant solution of a grid modulation is not possible because the grid would get too hot due to thermal radiation from the cathode [36, 53, 54]. On the other hand, a field emission cathode can be modulated by a grid.

This type of traveling wave tube with cold cathode has been developed and met reasonable performance. However, a good system pumping around the cathode was difficult to achieve in the small enclosed spaces.

13.3.5 Ionization gauges

A further device is ionization gauges using field emission [57, 58].

13.3.6 Pulsed X-ray sources and tomography

X-rays are generated by the impingement of high voltage electron beams on a metal. At present, the electron beams are created by thermionic sources. However, a number of applications require rapidly pulsed X-ray beams. These applications include X-ray tomography, where a 3-dimensional or time-resolved image can be built up by combining the X-ray images from many pulsed sources (Fig. 13.12). This can be used in medical imaging or security scanning.

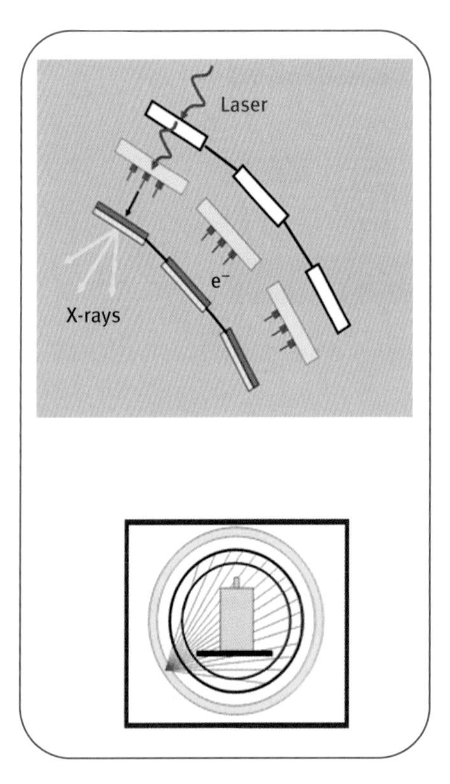

Fig. 13.12: Schematic of laser controlled X-ray imager.

The point is that thermionic electron sources are not good for pulsed electron sources because the filament cannot be switched on and off quickly. Thus to make a tomographic image of the body, the equipment would rotate a series of thermionic X-ray sources around a gantry. However, the sources are extremely heavy (due to the high current density and weight of the anode). X-ray images of the body need high X-ray intensities at moderate acceleration voltages because of the low contrast of different components of the body (they all contain water). Thus, a time resolved X-ray tomography of the body to observe heart beats or blood flow ideally needs field emission electron guns for X-ray sources which can now be fixed in space. The image is then assembled by computer from the separate time-resolved images.

Another application is for X-ray security machines. Here, the contrast between different materials is large, so the X-ray intensity is less high, but a high acceleration voltage is needed, as the penetration depth should be large. Thus a number of pulsed X-ray devices have been made, following the early work of Zhou *et al.* [59–62].

For medical images, the current density is high, as mentioned above. The anode therefore requires cooling. The standard arrangement would have a cathode and grid at near ground potential and the anode at a high potential, leading to engineering problems for cooling. Thales has developed an optically controlled electron gun. The CNTs are growth on n-i-p photodiodes, either Si or GaAs, which are switched on by a laser. The laser isolates the control circuitry and allows the cathode and grid to be floated at high potential, and the anode to be grounded, and be cooled (Fig. 13.13) [63].

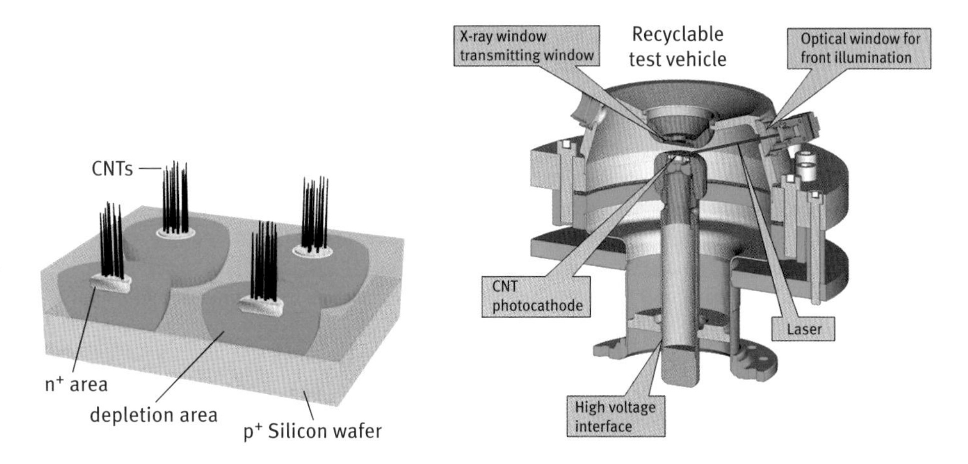

Fig. 13.13: Schematic of CNTs grown on n-i-p photodiodes, and test vehicle (courtesy P. Legagneux).

13.4 Summary

The field emission properties of carbon nanotube forests and single nanotubes are described. Controlled emission is possible for aligned CNT arrays where the spacing is twice the CNT height, as grown by plasma enhanced chemical vapor deposition. This leads to the maximum field enhancement factor. For random forests, the field enhancement obeys an exponential distribution, leading to a lower emission site density and imperfect current sharing. Ballast resistors can help alleviate this problem. Random nanocarbons perform less well than CNTs. Some applications are covered. Electron guns for scanning electron microscopes and transmission electron microscopes are a successful application, but small market. There is potential for CNT emitters in X-ray imaging systems. Many of the other proposed application areas have not fulfilled their promise.

Acknowledgments

The author is grateful to W. I. Milne, M. T. Cole, K. Teo, P. Legageneux, and M. Mann for the various images.

Bibliography

[1] C A Spindt, I Brodie,L Humphrey, E R Westerberg, J. Appl. Phys. **47** 5248 (1973)
[2] P R Schwoebel, I Brodie, J. Vac. Sci. Technol. B **13** 1391 (1995)
[3] J E Jaskie, MRS Bulletin **21** (March 1996) p59
[4] J Robertson, J. Vac. Sci. Technol. B **17**, 659 (1999)
[5] F J Himpsel, J S Knapp, J A VanVechten, D E Eastman, Phys. Rev. B **20** 624 (1979)
[6] B B Pate, P M Stefan, C Binns, P J Jupiter, M L Shek, I Lindau, W E Spicer, J Vac Sci Technol **19** 249 (1981)
[7] J.B. Cui, J Ristein, L Ley, Phys. Rev. Lett. **81**, 429 (1998).
[8] O Groning, O M Kuttel, E Schaller, P Groning, L Schlapbach, App. Phys. Lett. **69** 476 (1996)
[9] O Gröning, O M Küttel, P Gröning, L Schlapbach, J. Vac. Sci. Technol. B **17** 1064 (1999)
[10] A V Karabutov, V.D. Frolov, S.M. Pimenov, and V.I. Konov, Diamond Relat. Mater. **8**, 763 (1999)
[11] L Zhang, T Sakai, N Sakuma, T Ono, Diamond Related Mats 10 829 (2001)
[12] B S Satyanarayana, A Hart, WI Milne, and J Robertson, App. Phys. Lett. **71**, 1430 (1997)
[13] T W Mercer, N J DiNardo, J B Rothman, M P Siegal, T A Friedmann, L J MartinezMiranda, App. Phys. Lett. **72** 2244 (1998)
[14] Rinzler, A. G., Hafner, J. H., R E Smalley, Science, **269**, 1550 (1995)
[15] W A deHeer, A Chatelain, D Ugarte, Science **270** 1179 (1995)
[16] L A Chernozatonskii, et al, Chem. Phys. Lett. **233** 63 (1995)
[17] K A Dean, B R Chalamala, App. Phys. Lett. **75** 3017 (1999)
[18] K A Dean, B R Chalamala, J. App. Phys. **85** 3832 (1999)
[19] K A Dean, B R Chalamala, App. Phys. Lett. **76** 375 (1999)
[20] W Zhu, C Bower, O Zhou, G Kochanski, S Jin, App. Phys. Lett. **75** 873 (1999)
[21] O Groning, O M Kuttel, C Emmenegger, P Groning, L Schlapbach, *J Vac. Sci. Technol. B*, **18**, 665–678 (2000).
[22] L Nilsson, O Groning, C Emmenegger, O Kuttel, E Schaller, L Schlapbach, H Kind, J M Bonard, K Kern, App. Phys. Lett. **76** 2071 (2000)
[23] V Semet, V T Binh, P Vincent, D Guillot, K B K Teo, M Chhowalla, G A J Amaratunga, W I Milne, P Legagneux, D Pribat, App. Phys. Lett. **81** 343 (2002)
[24] K B K Teo, M Chhowalla, Applied Physics Letters **80**, pp. 2011–2013 (2002)
[25] K B K Teo, S B Lee, M Chhowalla, V Semet, V T Binh, O Groning, M Catignolles, G Pirio, P Legagneux, D Pribat, D G Hasko, H Ahmed, G Amaratunga, W I Milne, Nanotechnology **14** 204 (2003)
[26] L Nilsson, O Groning, P Groning, O Kuttel, L Schlapbach, J. App. Phys. **90** 768 (2001)

[27] L Nilsson, O Groning, O Kuttel, P Groning, L Schlapbach, J Vac. Sci. Technol. B **20** 326 (2002)
[28] J D Levine, J. Vac. Sci. Technol. **B 14**, 2008 (1996)
[29] J B Cui, J Robertson, W I Milne, J App Phys **89** 3490 (2001)
[30] J M Bonard, K A Dean, Physical Review Letters, **89**, 197602 (2002)
[31] S T Purcell, P Vincent, Phys. Rev. Lett., **88**, 105502 (2002)
[32] J Cumings, A Zettl, M R McCartney, J C H Spence, Phys. Rev. Lett. **88** 054804 (2002)
[33] A T H Chuang, B O Boskovic, J Robertson, Diamond Related Mats **15** 1103 (2006)
[34] Y Zhang, J Du, S Tang, S Deng, J Chen, N Xu, Nanotechnology **23** 015202 (2012)
[35] C Li, M Mann, D Hasko, W Lei, B Wang, D Chu, D Pribat, G Amaratunga, W I Milne, App Phys Lett **97** 113107 (2010)
[36] W I Milne, K B K Teo, G Amaratunga, P Legagneux, L Gangloff, J P Schnell, V Semet, V T Binh, O Groning, J. Mat. Chem. 14 933 (2004)
[37] de Jonge, N., Lamy, Y. (2003). Controlled mounting of individual multiwalled carbon nanotubes on support tips, *Nano Letters*, **3**, pp. 1621–1624
[38] N De Jonge, M Allioux, *Appl. Phys. Lett.*, **85**, pp. 1607–1609 (2004)
[39] N de Jonge, M Allioux, J T Oostveen, K B K Teo, W I Milne, Phys Rev Lett **94** 186807 (2005)
[40] N de Jonge, J M Bonard, Philos Trans Roy. Soc. A **362** 2239 (2004)
[41] N de Jonge, M Doytcheva, M Allioux, M Kaiser, S A M Mentink, K B K Teo, R G Lacerda, W I Milne, Adv. Mats. **17** 451 (2005)
[42] M Doytcheva, M Kaiser, N de Jonge, Nanotechnology 17 3226 (2006)
[43] M Mann, Y Zhang, K B K TEo, T Wells, M M ElGomati, W I Milne, Microelec. Eng. **87** 1491 (2010)
[44] W B Choi, D S Chung, J H Kang, H Y Kim, Y W Jin, I T Han, Y H Lee, J E Jung, N S Lee, G S Park, J M Kim, App. Phys. Lett. **75** 3129 (1999) FED
[45] J E Jung, J M Kim, Physica B **323** 71 (2002)
[46] Y S Choi, J M Kim, Diamond Related Mats **10** 1705 (2001)
[47] D S Chung, J M Kim, App. Phys. Lett. **80** 4045 (2002)
[48] Y C Kim, J M Kim, App. Phys. Lett. **92** 283112 (2008)
[49] J H Lee, J M Kim, D H Cloe, App. Phys. Lett. **89** 253115 (2006)
[50] K B K Teo, M Chhowalla, J. Vac. Sci Technol.B, **21**, 693 (2003)
[51] L Gangloff, E Minoux, Nano Letters, **4**, 1575 (2004). triodes
[52] E Minoux, O Groning, K B K Teo, S Dalal, L Gangloff, J P Schnell, L Hudanski, I Y Y Bu, P Vincent, P Legagneux, G Amaratunga, W I Milne, Nano Lett. **5** 2135 (2005)
[53] K B K Teo, E Minoux, *Nature*, **437**,968 (2005)
[54] L Hudanski, E Minoux, L Gangloff, K B K Teo, J P Schnell, S Xavier, J Robertson, W I Milne, D Pribat, P Legagneux, Nanotechnology, **19** 105201 (2009)
[55] M A Guillorn, M D Hale, V I Merkulov, M L Simpson, G Y Eres, H Cui, A A Puretzky, D B Geohegan, App. Phys. Lett. **81** 2860 (2002)
[56] A V Melechko, V I Merkulov, T E Knight, M A Guillorn, K L Klein, D H Lowndes, M Simpson, J. App. Phys. **97** 041301 (2005)
[57] I M Choi, and S Y Woo, Applied Physics Letters, **87** 173104 (2005)
[58] I M Choi, S Y Woo, Applied Physics Letters, **90** 023107 (2007)
[59] G Z Yue, Q Qui, B Gao, Y Cheng, L Zhang, H Shimoda, S Chang, J P Lu, O Zhou, App. Phys. Lett. **81** 355 (2002)
[60] J Zhang, G Yang, Y Z Lee, S Chang, J P Lu, O Zhou, App. Phys. Lett. **89** 064106 (2006)
[61] G Cao, L M Burk, Y Z Lee, X C Coplon, S Sultana, J Lu, O Zhou, Mecidal Phys. **37** 5306 (2010)
[62] X Qian, A Tucker, E Gidcomb, O Zhou, Medical Phys. **39** 2090 (2012)
[63] F Andrianiazy, J P Mazellier, L Gangloff, P Legagneux, P Ponnard, N Martinez, X L Han, J F Lampin, Proc. International Vacuum Nanoelectronics Conf. (2012), p 46

Panagiotis Trogadas and Peter Strasser

14 Carbon, carbon hybrids and composites for polymer electrolyte fuel cells

Carbon has unique characteristics that make it an ideal material for use in a wide variety of applications ranging from metal refining to electrocatalysis and fuel cells. In polymer electrolyte fuel cells (PEFCs), carbon is used as a gas diffusion layer, electrocatalyst support and oxygen reduction reaction (ORR) electrocatalyst. When used as electrocatalyst support, amorphous carbonaceous support materials suffer from enhanced oxidation rates at high potentials over time. This drawback has prompted an extensive effort to improve the properties of amorphous carbon and to identify alternate carbon-based materials to replace carbon blacks. Alternate support materials are classified in carbon nanotubes and fibers, mesoporous carbon and graphene. A comparative review of all these supports is provided. Work on catalytically active carbon hybrids is focused on the development of precious group metal (PGM)-free electrocatalysts that will significantly reduce the cost without sacrificing catalytic activity. Of the newer electrocatalysts, nitrogen/metal-functionalized carbons and composites are emerging as possible contenders for commercial PEFCs. Nitrogen-doped carbon hybrids with transition metals and their polymer composites exhibit high ORR activity and selectivity and these catalytic properties are presented in detail in this chapter.

14.1 Introduction

Carbon is unique among chemical elements since it exists in different forms and microtextures transforming it into a very attractive material that is widely used in a broad range of electrochemical applications. Carbon exists in various allotropic forms due to its valency, with the most well-known being carbon black, diamond, fullerenes, graphene and carbon nanotubes. This review is divided into four sections. In the first two sections the structure, electronic and electrochemical properties of carbon are presented along with their applications. The last two sections deal with the use of carbon in polymer electrolyte fuel cells (PEFCs) as catalyst support and oxygen reduction reaction (ORR) electrocatalyst.

14.2 Carbon as electrode and electrocatalyst

14.2.1 Structure and properties

14.2.1.1 Carbon black

Carbon black consists of spherical particles (diameter less than 50 nm) that may aggregate and form agglomerates (~250 nm diameter) [1]. The carbonaceous particles have para-crystallite structures [1] consisting of parallel graphitic layers with 0.35–0.38 nm interplanar spacing. The crystalline graphitic portion of carbon black has sp^2 hybridization involving a triangle in-plane formation of sp^2 orbitals while the fourth p_z orbital lies normal to this plane forming weaker delocalized π bonds with other neighboring carbon atoms [2–4]. Chemical or gas treatment of carbon black at high temperatures (800–1100 °C) and high pressure leads to leads to the formation of activated carbon black; this form of carbon is characterized by larger and more crystalline graphitized carbon particles (~ 20–30 μm) with distinct micro-porosity and varying BET surface area (200–1200 m^2g^{-1}) [1].

14.2.1.2 Diamond

Diamond is another natural allotrope of carbon that has cubic structure in which each atom uses sp^3 orbitals to form four strong, covalent bonds. These bonds contribute to the hardness and high melting point of diamond [1]. On diamond {1,0,0} the surface atoms reconstruct and form π-bonded, symmetric dimers arranged in rows (~2.5 A inner distance). This causes a further splitting of the dimer orbitals and the formation of a band gap (~1.3 eV) between occupied and unoccupied surface states [5]. In contrast, on diamond {1,1,0} and {1,1,1} surfaces, the bonds are arranged in symmetric π-bonded chains without dimerization. Even though the inner distance between the π-orbitals (~1.5 and 1.4 A for {1,1,0} and {1,1,1} surfaces respectively) is smaller than the distance between π-dimers on {1,0,0} surface, the distance between the π-bonded chains is large (3.6 and 4.4 A respectively). As a result, a π-electron system is formed along the chains and these surfaces obtain metallic character [5].

14.2.1.3 Fullerenes

The most stable fullerene, C_{60}, is built of carbon atoms with sp^2 hybridization assembled in a truncated icosahedron. It consists of 60 carbon atoms made up of 20 hexagons and 12 pentagons, which result in the quasi-spherical structure of the molecule (Fig. 14.1) [6]. The molecular structure of fullerene results in a closed-shell electronic structure with highly degenerate π- and σ-derived molecular electronic states including the π-derived fivefold degenerate highest occupied molecular orbital (HOMO) and the triply degenerate lowest unoccupied molecular orbital (LUMO) [6, 7]. According to the isolated pentagon rule, separate 60-atom cages are held together by weak forces in order to avoid coming into contact with their surface pentagons and thus increase the stability of fullerene.

Fig. 14.1: Structure of the C_{60} fullerene molecule consisting of 20 hexagons and 12 pentagons with carbon atoms at each corner (Reprinted from [6] with permission from Elsevier).

14.2.1.4 Carbon nanotubes (CNTs, see Chapter 1)

CNTs consist of cylinders made of graphite layers that are closed at both ends. Both single-walled (SWCNTs) and multi-walled nanotubes (MWCNTs) exist, with diameters of a few nanometers and lengths of the order of 1 mm [1]. Nanotubes are classified by the wrapping vector (n,m) corresponding to the direction and distance in which the graphite layer is wrapped up [6] (Fig. 14.2). The armchair (n,n) and the zigzag (n,0) tubes that have reflection planes are achiral while all other tubes with independent *n* and *m* are chiral. Metallic nanotubes are characterized by wrapping vectors where $n - m = 3 \cdot l\,(l = 0, 1, 2, 3, etc.)$ while all the other tubes are semiconductors [8, 9]. An interesting property of both fullerenes and carbon nanotubes is their ability to entrap atoms of other elements within their molecular structure. In this respect hydrogen storage in carbon nanotubes is of particular interest to fuel cell developers.

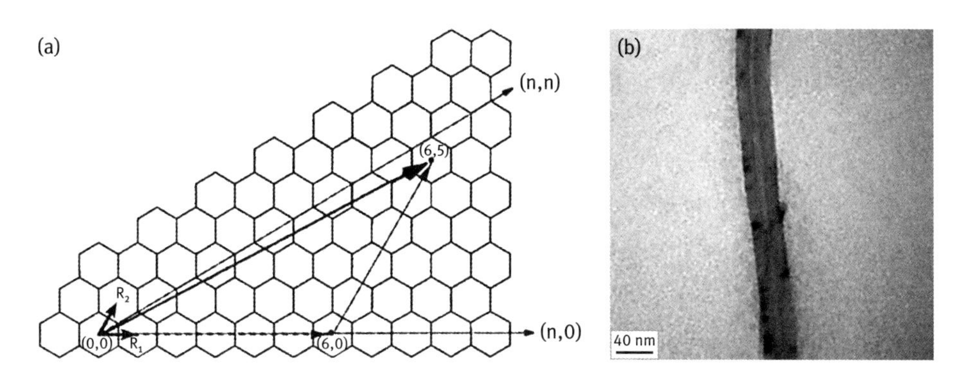

Fig. 14.2: a) Structure of the 2-D graphite layer (nanotube precursor). The primitive lattice vectors R_1 and R_2 are depicted. CNTs can be envisaged as wrapped up graphite layers whereby the wrapping direction and distance are given by a single vector. The wrapping vectpr (6,5) is shown for illustration (Reproduced from [10]); b) Transmission electron microscopy (TEM) image of CNT (Reprinted from [11] with permission from Elsevier).

14.2.1.5 Carbon nanofibers (CNFs)

CNF is an industrially produced derivative of carbon formed by the decomposition and graphitization of rich organic carbon polymers (Fig. 14.3). The most common precursor is polyacrylonitrile (PAN), as it yields high tensile and compressive strength fibers that have high resistance to corrosion, creep and fatigue. For these reasons, the fibers are widely used in the automotive and aerospace industries [1]. Carbon fiber is an important ingredient of carbon composite materials, which are used in fuel cell construction, particularly in gas-diffusion layers where the fibers are woven to form a type of carbon cloth.

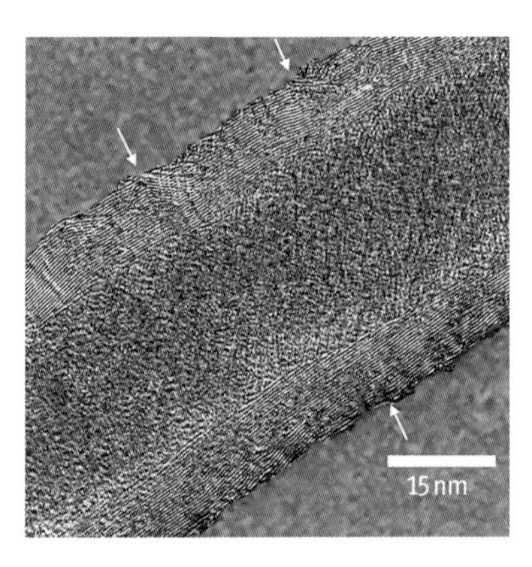

Fig. 14.3: High resolution TEM micrograph of CNF (Reprinted with permission from [12] – © 2004 American Chemical Society).

14.2.2 Electrochemical properties

During recent years, extensive research has been focused on the application of carbons as electrode materials because of their accessibility, physico-chemical properties, processability and relatively low cost. Carbon electrodes are thermally and mechanically stable, chemically resistant in different solutions (from strongly acidic to basic) as well as chemically inert [13–15].

Carbon blacks are promising electrode materials due to their relatively high activities and long lifetimes in contrast to the lower activity or rapid deactivation of the other carbonaceous materials [16–20]. These catalytic characteristics of carbon blacks are attributed to their microstructure that has many active sites consisting of edges and defects in nanosized graphitic layers [19–21].

Carbon black possesses time-varying catalytic characteristics [16, 17, 20, 22]. Catalytic deactivation starts at the beginning of the reaction and it continues gradually without reaching a steady state making the determination of the reaction kinetic parameters indefinite. Thus, it is important to establish an evaluation method of activation energies for carbon blacks which exhibit time-varying catalytic characteristics.

14.2.2.1 Surface reactions

The electrochemistry of carbon black can be described by two types of reaction, namely intercalation and surface reactions.

In the *intercalation reactions*, ions (anions X^- or cations M^+) penetrate into the van der Waals gaps between the ordered carbon layers resulting in the enlargement of their inter-layer distance [23, 24]. The corresponding charges are conducted by carbon and accepted into the carbon host lattice.

$$\begin{aligned} C_n + M^{\oplus} + e^- &\Leftrightarrow M^{\oplus}C_n^- \quad \text{Cathodic reduction} \\ C_n + X^- &\Leftrightarrow C_n^{\oplus}X^- + e^- \quad \text{Anodic oxidation} \end{aligned} \tag{14.1}$$

Intercalation reactions (14.1) represent the ideal case; there is an increase in the inter-layer distance while the carbon atom arrangement within the layers remains unchanged. However, during intercalation of cations from polymer [25] and solid [26] electrolytes, ternary phases ($M^{\oplus}(\text{solv})_y C_n^- / C_n^{\oplus}(\text{solv})_y X^-$) are produced because the solvent from the electrolyte is also accepted into the carbon lattice.

$$\begin{aligned} C_n + M^{\oplus} + e^- + \text{ysolv} &\Leftrightarrow M^{\oplus}(\text{solv})_y C_n^- \quad \text{Cathodic reduction} \\ C_n + X^- + \text{ysolv} &\Leftrightarrow C_n^{\oplus}(\text{solv})_y X^- + e^- \quad \text{Anodic oxidation} \end{aligned} \tag{14.2}$$

From the degree of solvation it is possible to distinguish between two-dimensional and three-dimensional solvated phases (Fig. 14.4) [23].

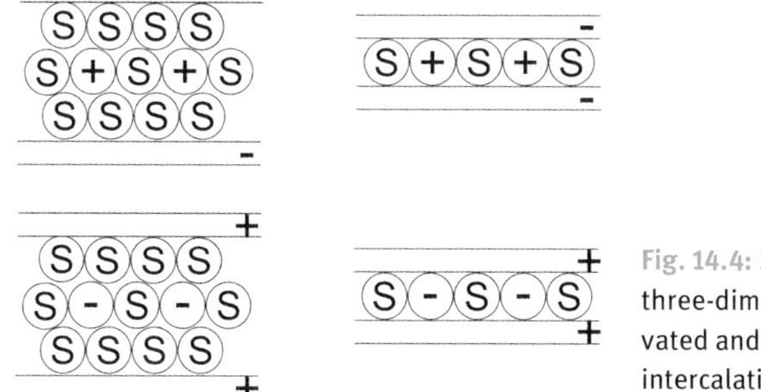

Fig. 14.4: Schematic diagram showing the three-dimensional and two-dimensional solvated and unsolvated cationic and anionic intercalation compounds; S = Solvent [23].

In the two-dimensional solvates, the distance between the carbon layers is determined either by the size of the solvent molecule [27] or by the size of the ions [28, 29].

The *surface reactions* of carbon black are generally reactions of C atoms at the edges of the layers or at other lattice defects, or reactions of functional groups which are bound to such C atoms. Surface reactions are of particular interest with respect to the performance of carbon as double-layer capacitor [30–33]. The presence of oxygen groups on the surface of carbon affects its electrochemical response. The faradaic current is significantly increased with an increase in the number of surface oxygen groups present resulting in an increase in capacitance [30]. The formation of CO surface complexes (hydroxyl, carbonyl, quinone) are responsible for the enhancement in capacitance and can undergo the quinone/hydroquinone redox pair mechanism (14.3) [31].

$$C_xO + H^+ + e^- \Leftrightarrow C_xOH \tag{14.3}$$

14.2.3 Applications

The above mentioned unique properties of carbon have transformed it into an attractive material that is widely used in various electrolytic processes in the chemical industry. Several applications of carbon are reviewed below.

14.2.3.1 Reduction of nitrobenzene and nitraniline

Porous carbon electrodes were used in the field of electrolyte-soluble (p-nitraniline) [34] and electrolyte -insoluble (nitrobenzene) [35] organic cathodic depolarizers.

The use of porous carbon electrodes allows accurate control of the ratio of oxidized and reduced forms of p-nitriline by adjusting the electrolyte flow through the electrode. As a result, the reduction of p-nitraniline to p-phenylenediamine was carried out continuously at current densities three-fold greater than the current densities obtained from batch processes [34] and ~45 % of the total yield was obtained, depending on the p-nitriline used [34].

Similar results were obtained when an insoluble depolarizer was used. Nitrobenzene was in contact with the carbon electrode resulting in higher system performance (voltage reached ~0.58 V [35] when the carbon-nitrobenzene system was used) [35].

14.2.3.2 Production of aluminum and chlorine

Chlorine production

At the beginning of the last century, electrochemical production of chlorine became important as compared with the chemical methods used in industry [36]. The electrolytic evolution and dissolution of chlorine on graphite electrodes has been studied to obtain fundamental information about its mechanism.

The electrochemical reaction occurs at the surface of graphite anode [37–39]. At potentials lower than 1.25 V, chlorine is formed by a Volmer/Heyrovsky mechanism with the latter being the rate determining step. Chloride ions are initially discharged on surface sites "*" that are not covered by chlorine atoms (Volmer reaction (14.4a)), followed by the discharge of chloride ions on adsorbed chlorine ions (Heyrovsky reaction (14.4b)) [39]:

$$Cl^- \rightarrow Cl^*_{ad} + e^- \tag{14.4a}$$

$$Cl^*_{ad} + Cl^- \rightarrow Cl_2 + e^- \tag{14.4b}$$

At higher potentials, the anode surface is covered by carbon oxide leading to the extinction of active sites for the chlorine electrode process and to potential overshoot [37, 40].

Triaca and co-workers [41] studied the electrolytic evolution and dissolution of chlorine at graphite electrodes in molten lithium chloride by the galvanostatic transient method. A low inductance cell designed to minimize the oscillations produced in the voltage/time transient by residual inductances in the circuit [41] allowed the use of high current densities up to 200 Acm-2. The anodic evolution results agree with the Volmer/Heyrovsky mechanism mentioned above. However, the cathodic dissolution reaction is limited by diffusion of chlorine in the melt to the electrode surface (at high current densities) and by dissociation of the chlorine on the surface followed by charge transfer at low current densities [41].

$$\begin{aligned} Cl_2 + C &\rightarrow 2 \cdot Cl \cdots C \\ Cl \cdots C + e^- &\leftrightarrow Cl^- + C \end{aligned} \tag{14.5}$$

Janssen [38] has also studied the mechanism of chlorine evolution on different types of graphite anodes (Acheson and pyrolytic graphite electrodes). Two different surface compounds, the l_p and h_p, were formed on all the graphite electrodes investigated at low and high potentials respectively. Chlorine was formed according to the Volmer–Heyrovsky mechanism on all electrode types. The current density, transfer coefficient and activation energy for the commercial graphite and Acheson electrode were similar, while these values deviated strongly for the pyrolytic graphite electrodes [38].

Aluminum production

Aluminum is produced according to the Hall–Heroult process [42–44]. At the cathode, Al_xF_y species are reduced and lead to liquid aluminum. As the electrolysis proceeds, the metal from the aluminum oxide precipitates at the bottom of the cell. At the anode, oxygen evolution takes place producing carbon dioxide/monoxide and hence resulting in current and performance losses [42–44].

Sum and Skyllas-Kazacos [44] studied the deposition and dissolution of aluminum in an acidic cryolite melt. The graphite electrode was preconditioned (immersed in cryolite melt) to saturate the surface of the electrode in sodium before aluminum deposition could be observed. Current reversal chronoamperometry was used to measure the rate of aluminum dissolution in the acidic melt. Dissolution rate was mass transport controlled [45] and in the order of $0.8 \cdot 10^{-7}$ and $1.8 \cdot 10^{-7}$ $molcm^{-2}s^{-1}$ at 1030 °C and 980 °C respectively [44].

14.2.3.3 Metal refining

Gold and silver recovery

The ability of activated carbon to absorb gold and silver from solution has been commercially exploited since the beginning of the 1970s [46, 47] and the techniques used can be classified in: (i) carbon in pulp (CIP); (ii) carbon in leach (CIL); and (iii) carbon in column (CIC) [46, 47].

The major advantage of CIL over CIP is the lower capital cost of the plant and the higher surface area of screening, even though CIL has lower metallurgical efficiency. The activated carbons used in these two processes are required to have a high rate of gold extraction and loading capacity as well as resistance to abrasion. In the CIC process, less resistance to abrasion carbon is required since the abrasive effect in this system is not pronounced.

The abovementioned extraction techniques are more efficient than the Merill–Crowe process that was used until then [47], since they offer high recovery rates of precious metals in low grade solutions and low operational and capital costs [46].

Grain refining of Mg-Al and Al-Ti alloys

Grain refinement is an important technology for improving the mechanical properties of magnesium alloys which are used as light materials in the automotive industry to reduce fuel consumption [48]. Grain refinement produces alloys with high extrudability and rollability as well as excellent resistance to hot tearing, resulting in cost reduction during the alloy production process [48–51].

The main grain refining methods for magnesium alloys can be classified into:

(i) ferric chlorine inoculation ($FeCl_3$) [52];
(ii) carbon inoculation [53–57];
(iii) superheating [58, 59]; and
(iv) addition of solute elements [60–63].

Of these four methods, carbon inoculation is the most effective way to refine the grains of Mg-Al alloys since it has low cost and operating temperature as well as less fading [48, 49]. The most commonly accepted theory is the Al_4C_3 nuclei hypothesis [49, 64, 65]; Al_4C_3 particles formed in the Mg-Al melt act as nuclei for the Mg grains during solidification. However, the mechanism of grain refinement in Mg-Al base alloys by carbon addition has not yet been fully understood [49].

Du and co-workers [48, 66] studied the effect of carbon on the grain refinement of Mg-3Al alloy. High grain refining efficiency was obtained when these alloys were refined by carbon. A further increase in efficiency was obtained by the combination of 0.2 wt% C and less than 0.2 wt% of a solute element (Ca, Sr) [48, 66]. Addition of a higher Ca amount would increase the brittleness of the alloy [60, 66]. Similar results were demonstrated when 0.2 wt% Sr was added in the alloys instead of Ca [48].

Moreover, grain refinement is also used in aluminum foundries to produce a grain structure with improved mechanical and cosmetic properties [67]. The common material used is Al-Ti-B alloys consisting of soluble Al_3Ti and insoluble TiB_2 particles in the aluminum matrix [68]. However, TiB_2 particles tend to agglomerate leading to internal cracking of the alloys and formation of surface defects [67]. Hence, carbon was introduced into the melt instead of boron to form Al-Ti-C alloys. A series of grain-refined samples with constant Ti concentration (3 wt%) varying carbon content was synthesized by Birol [67] to study the effect of carbon on grain efficiency and it was discovered that the Al-3Ti-0.15C alloy had the best performance.

14.2.3.4 Photoelectrochemical water splitting

Photoelectrochemical water splitting (capturing and storing solar energy in the chemical bond of hydrogen) is a promising way to produce clean energy and is expected to play a significant role in the future with the depletion of fossil fuel supplies [69–72]. Many materials have been investigated to discover a suitable photoanode [73–79]; among them titanium oxide (TiO_2), which is promising due to its low cost, chemical inertness, UV photoactivity and chemical stability [69, 70, 80–82]. However, due to its large band gap (~ 3–3.2 eV) and short carrier diffusion length, the conversion efficiency of titanium oxide is reduced [69]. In an attempt to shift the optical response of titanium oxide from ultraviolet to visible spectra range, titanium oxide was doped with nitrogen [83], carbon [69, 70, 84–86] and sulfur [87]. The incorporation of carbon into titanium oxide resulted in the formation of an intragap band (~1.3–1.6 eV) [84, 88] into the band gap of TiO_2 that can be used to absorb photons at low band gap energy [85, 86, 89]. This effect resulted in high photoconversion efficiency under visible light [86, 88]. Moreover, carbon modified TiO_2 synthesized by wet process using glucose solution as the carbon source demonstrated a 13-fold increase in photocatalytic activity compared to titanium oxide while it increased only eight-fold when tetrabutylammonium hydroxide was used as the carbon source [86].

Apart from titanium oxide, two other carbon-modified semiconductors were studied in water photoelectrolysis due to their low band gap energy, namely iron (Fe_2O_3) and tungsten oxide (WO_3) [70, 90]. Carbon-modified iron oxide demonstrated promising photoconversion efficiency, ~ 4 % and 7 % for modified oxides synthesized in oven and by thermal oxidation respectively [90]. Also, carbon-modified tungsten oxide (C-WO_3) photocatalysts exhibited a ~ 2 % photoconversion efficiency [70].

14.2.3.5 Electrolytic synthesis of hydrogen peroxide (H_2O_2)

Another application of carbon and carbon hybrids is their use as electrode material in proton exchange membrane (PEM) electrochemical flow reactor for the production of hydrogen peroxide (H_2O_2).

Tatapudi and Fenton [91] investigated three different catalysts, namely gold, graphite and activated carbon, on the production of H_2O_2 by oxygen reduction at the cathode of a PEM flow reactor. Graphite (10 $mgcm^{-2}$ catalyst loading with 20 % Teflon binder) produced the highest amount of peroxide (25 mgL^{-1} at 2.5 V *vs.* standard hydrogen electrode (SHE)) while gold and activated carbon produced lower peroxide at higher applied voltages (~3.5 V *vs.* SHE) than the graphite. It was also reported that an increase in the percentage of Teflon binder results in an enhanced electrode resistance leading to lower cell current and hence lower cell performance [91, 92].

Mesoporous nitrogen-doped carbon was also used as catalyst for the production of hydrogen peroxide [93]. H_2O_2 was produced in a three electrode setup at 1600 rpm rotation speed and ~325 $\mu g cm^{-2}_{geo}$ catalyst loading. H_2O_2 concentration (Fig. 14.5) was approximately 20 mgL^{-1} at constant potential of 0.1 V (*vs.* reversible hydrogen electrode (RHE)) showing that nitrogen-doped carbon is a promising catalyst for the electrochemical production of H_2O_2 [93].

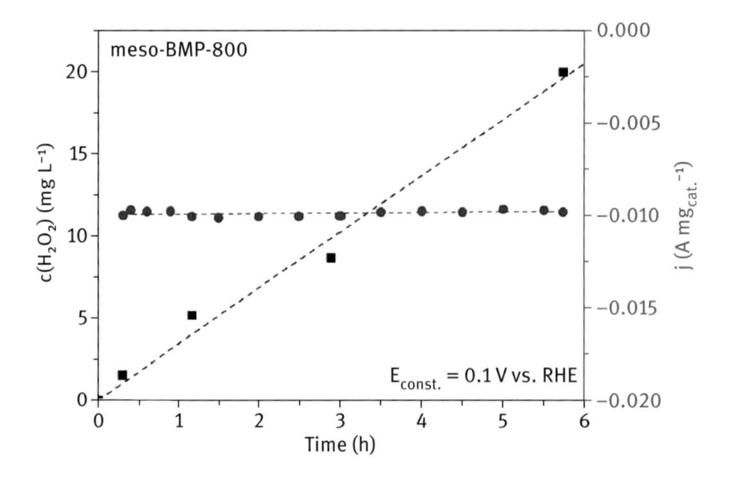

Fig. 14.5: Photometric determination of H_2O_2 as a function of time. Current (circle) and concentration (square) behavior with time for the electrochemical H_2O_2 production is shown (Reprinted with permission from [93] – © 2012 American Chemical Society).

Moreover, the effect of operating conditions (applied current, electrolyte concentration, air flow rate and pH) on the amount of electrogenerated H_2O_2 was investigated [94]:

(i) *Effect of applied current.* H_2O_2 production rate is increased as the applied current is increased from 60 mA to 100 mA. However, above 100 mA the peroxide production rate decreases. At 60 mA the cell potential is 3 V (*vs.* saturated calomel electrode (SCE)) while at 100, 200, 300 and 500 mA it is 3.4, 4.3, 5.4 and 6.6 V (*vs.* SCE) respectively [94]. At potentials higher than 4.3 V (*vs.* SCE) the reduction of oxygen leads to the formation of water [95] and not H_2O_2 and thus the amount of electrogenerated peroxide is reduced [94].

(ii) *Effect of electrolyte concentration.* Sodium sulfate (Na_2SO_4) is used as the electrolyte solution; low electrolyte concentration results in low conductivity of the solution and thus higher cell potential is needed to reach the required current density [94]. Cell potential was 6.7 V and 3.4 V (*vs.* SCE) for sodium sulfate concentrations of 0.025 and 0.05 M. Thus, increasing the electrolyte solution enhanced the electrogeneration rate of H_2O_2. However, increasing the sodium sulfate concentration from 0.05 M to 0.1 M had no significant effect on the hydrogen peroxide electrochemical generation [94].

(iii) *Effect of pH.* At pH values below 3, reduction of H_2O_2 to water and hydrogen gas evolution takes place (14.6) reducing the amount of H_2O_2 generated in the medium, while at pH values above 3, proton concentration decreases and hence the H_2O_2 production rate is reduced as well [94].

$$H_2O_2 + 2H^+ + 2e^- \rightarrow 2H_2O$$
$$2H^+ + 2e^- \rightarrow H_2 \quad (14.6)$$

(iv) *Effect of air flow rate.* Air was used instead of pure oxygen due to its low cost. An increase in the air flow rate results in an increase of the amount of H_2O_2 produced. High flow rate leads to high mass transfer rate of dissolved oxygen and hence enhanced peroxide production [94].

Based on these results, it was concluded that the optimum conditions for H_2O_2 electrogeneration are 0.05 M sodium sulfate, 100 mA applied current, pH = 3 and 2.5 $Lmin^{-1}$ air flow rate [94]. These conditions were used to investigate the electrochemical generation of H_2O_2 of three different carbon-based cathode materials: graphite, activated carbon on graphite (AC/graphite) and carbon nanotubes on graphite (see Chapter 9). The amount of electrogenerated H_2O_2 using CNTs-graphite was three times higher than that of AC/ graphite and seven times higher than that of bare graphite (Fig. 14.6). CNTs/graphite has large specific area and high amount of mesoporous pores leading to facile oxygen reduction on the cathode and hence increased H_2O_2 production [94].

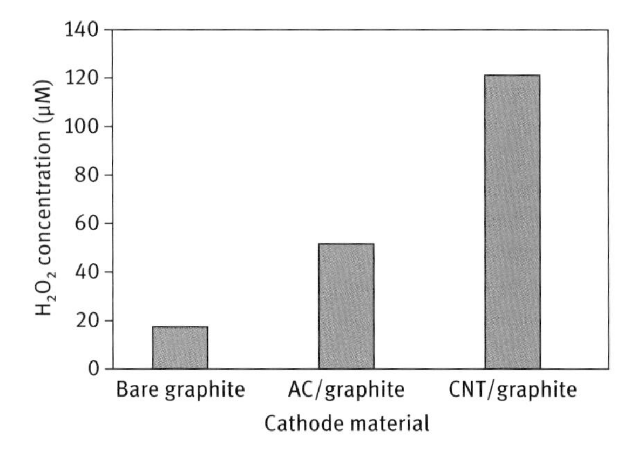

Fig. 14.6: Electrogenerated H_2O_2 on bare graphite, AC/graphite and CNTs/graphite. Operating conditions: 180 min electrolysis, room temperature, 0.05 M Na_2SO_4, 100 mA, pH = 3, air flow rate = 2.5 $Lmin^{-1}$ (Reprinted from [94] with permission from Elsevier).

14.3 Carbon, carbon hybrids and carbon composites in PEFCs

14.3.1 Carbon as structural component in PEFCs

A PEFC consists of two electrodes in contact with an electrolyte membrane (Fig. 14.7). The membrane is designed as an electronic insulator material separating the reactants (H_2 and O_2/air) and allowing only the transport of protons towards the electrodes. The electrodes are constituted of a porous gas diffusion layer (GDL) and a catalyst (usually platinum supported on high surface area carbon) containing active layer. This assembly is sandwiched between two electrically conducting bipolar plates within which gas distribution channels are integrated [96].

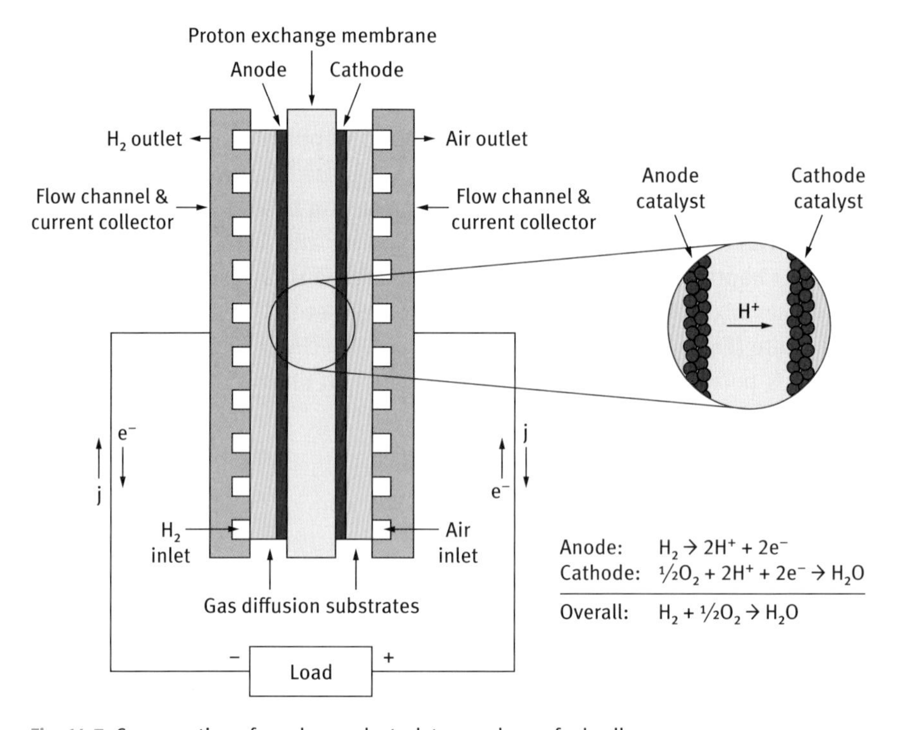

Fig. 14.7: Cross section of a polymer electrolyte membrane fuel cell.

The function of the electrolyte membrane is to facilitate transport of protons from anode to cathode and to serve as an effective barrier to reactant crossover. The electrodes host the electrochemical reactions within the catalyst layer and provide electronic conductivity, and pathways for reactant supply to the catalyst and removal of products from the catalyst [96]. The GDL is a carbon paper of 0.2–0.5 mm thickness that provides rigidity and support to the membrane electrode assembly (MEA). It incorporates hydrophobic material that facilitates the product water drainage and prevents the gas paths from flooding [96]. The most widely used carbon in GDL is Vulcan (XC-72R) [97] and acetylene carbon black. The latter with low pore volume and optimized thickness demonstrated better cell performance than Vulcan [98]. The active layer consists of catalyst particles, ionomer and pore spaces, which form a three-phase boundary where the electrochemical reaction takes place. A good electrode has to effectively facilitate the trade-off between enabling high catalytic activity, retaining enough water to guaranty good proton conductivity in the ionomer phase, and having an optimal pore size distribution to facilitate rapid gas transport [96]. All the components of the MEA need to be stable (under both chemical and mechanical stress) for several thousands of hours in the fuel cell under the prevailing operating and transient conditions.

14.3.2 Carbon as PEFC catalyst support

In a fuel cell, the electrocatalysts generate electrical power by reducing the oxygen at the cathode and oxidizing the fuel at the anode [1]. Pt and Pt alloys are the most commonly used electrocatalysts in PEFCs due to their high catalytic activity and chemical stability [99–103].

Carbon black (Vulcan XC-72R) is the most commonly used support for Pt and Pt-alloy catalysts for fuel cells due to its high surface area (~ 250 $m^2 g^{-1}$) and low cost. It consists of spherical graphite particles (less than 50 nm diameter) [100] with ~ 0.35 nm interplanar spacing [100]. However, carbon black is susceptible to oxidation under high potentials (greater than 1.2 V *vs.* SHE) resulting in active surface area loss [104] and alteration of pore surface characteristics (generation of surface oxides CO_{ad}) [105]. In automotive applications, high potentials occur during (i) *complete fuel starvation* of a cell in the stack; and (ii) during *partial fuel starvation* on the anode side of a cell in the stack due to limited supply of hydrogen [106, 107] or the presence of hydrogen-air over the active area during start-up and shutdown of the fuel cell [108]. During complete fuel starvation, the anode potential of the starved cell is greater than the cathode, resulting in water electrolysis and carbon oxidation on the anode to form the required electrons and protons for the oxygen reduction reaction at the cathode [108]. During partial fuel starvation, oxygen is present in the hydrogen deficient areas on the anode causing a decrease of the in-plane membrane potential and hence potential shifts [108, 109]. The abovementioned high potential conditions result in facile oxidation of the carbon support and thus significant performance losses [110].

$$\begin{aligned} &C + H_2O \rightarrow CO_{ad} + 2H^+ + 2e^- \\ &CO_{ad} + H_2O \rightarrow CO_2 + 2H^+ + 2e^- \end{aligned} \tag{14.7}$$

Alternative support materials are being investigated to replace carbon black as support in order to provide higher corrosion resistance and surface area. These supports can be classified into (i) carbon nanotubes and fibers; (ii) mesoporous carbon; and (iii) multi-layer graphene and they are presented in detail in the following section.

14.3.2.1 Carbon nanotubes (CNTs) and fibers (CNFs)

CNTs (single-walled or multi-walled) are the most common material used as catalyst support in PEFCs. SWCNTs have large surface areas while MWCNTs are more conductive than SWCNTs [111, 112].

Pristine CNTs are chemically inert and metal nanoparticles cannot be attached [111]. Hence, research is focused on the functionalization of CNTs in order to incorporate oxygen groups on their surface that will increase their hydrophilicity and improve the catalyst support interaction (see Chapter 3) [111]. These experimental methods include impregnation [113, 114], ultrasound [115], acid treatment (such as H_2SO_4) [116–119], polyol processing [120, 121], ion-exchange [122, 123] and electrochemical deposition [120, 124, 125]. Acid-functionalized CNTs provide better dispersion and distribution of the catalysts nanoparticles [117–120].

Conjugated polymers such as polypyrrole (Ppy) and polyaniline (PANI) form covalent bonds between Pt atoms and N atoms in PANI enabling strong adhesion of Pt nanoparticles onto the polymer [126–128]. Electron-conducting PANI was used to bridge the Pt nanoparticles and CNT walls with the presence of platinum-nitride (Pt-N) bonding and π-π bonding. The synthesized PANI binds across the CNT as a result of π-π bonding while Pt nanoparticles (~2-4 nm diameter) are loaded onto the CNT due to polymer stabilization and existence of Pt-N bonding [127] (Fig. 14.8). Accelerated degradation tests revealed high electroactivity and electrochemical stability of Pt-PANI/CNT catalysts compared to non-functionalized MWCNTs and commercial carbon black supports [127].

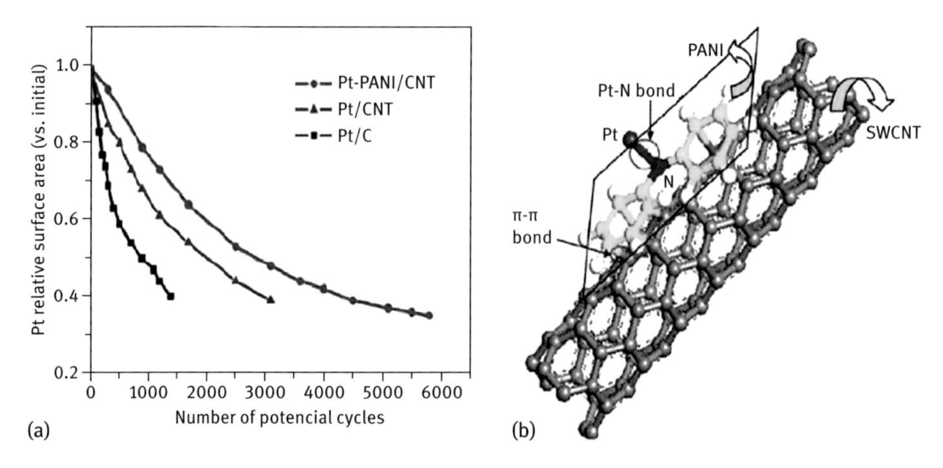

Fig. 14.8: (a) ECA of the catalysts as a function of the number of potential cycles. (b) Schematic showing molecular interactions in the synthesized Pt-PANI/CNT catalyst (Reprinted from [127] with permission from Elsevier).

Another technique to modify the surface of CNTs is the use of ultrasound as it enables the formation of smaller and more uniform nanoparticles [116]. Yang and co-workers

[116] compared the effect of ultrasonic treatment and reflux on the surface functionalization of MWCNTs. Ultrasonically fabricated MWCNTs demonstrated higher active surface area and improved CO tolerance due to better dispersion and utilization of the electrocatalyst [116].

Tang and co-workers [129] synthesized Pt-CNT layer acting both as gas diffusion and catalyst layer. CNT was grown *in situ* on carbon paper followed by sputter deposition of the Pt catalyst. A maximum power density of 595 $mWcm^{-2}$ (Pt loading ~ 0.04 $mgcm^{-2}$) was achieved (Fig. 14.9), which was higher than the power density of both Pt-Vulcan and Pt-CNT-Vulcan electrodes tested (~435 and 530 $mWcm^{-2}$ respectively) [130]. Hydroxyl radicals cannot easily penetrate the rigid structure of CNTs, while carbon black has an excess of dangling bonds and defects that oxygen atoms can easily penetrate into [130, 131]. Dangling bonds form surface oxides leading to higher corrosion rate [131, 132].

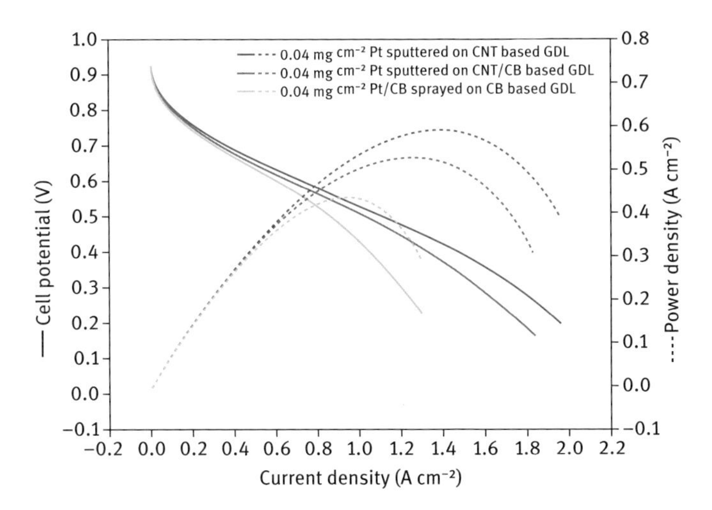

Fig. 14.9: Polarization curves of MEA Pt-CNT, Pt-Vulcan and Pt-CNT-Vulcan based MEAs (Reprinted from [130] with permission from Elsevier).

The polarization losses of Pt supported on MWCNTs was also examined [132, 133]. Pt-MWCNTs exhibited higher corrosion resistance and retention of electrochemical active surface area compared to Pt-Vulcan electrocatalysts. Only the outer graphene layers of MWCNTs were damaged, forming defects on the surface of nanotubes [134]. Moreover, MWCNTs prevented the water flooding of cathode catalyst layer by maintaining the electrode structure and hydrophobicity for a long period under continuous anodic potential stress. Ohmic losses were found to be the major cause contributing to the performance loss [133].

The degree of graphitization also plays a significant role on the corrosion rate of MWCNTs; the number of defects on the surface of nanotubes decreases as graphitization % increases resulting in higher electrochemical stability [134]. The higher graphitic content has also been linked to a stronger interaction between metal and carbon support. An increase in the degree of graphitization results in stronger π-sites (sp^2 hybridized carbon) on the support (which are the anchoring sites for the electrocatalyst), strengthening the metal-support interaction [135].

The stability during potential cycling and ORR activity of Pt (20 wt%) supported on MWCNTs and carbon black was also investigated [136]. Two different potential cycling conditions were used, namely lifetime (0.5 to 1.0 V *vs.* RHE) and start-up (0.5 to 1.5 V *vs.* RHE). Pt supported on MWCNTs catalyst exhibited a significantly lower drop in normalized electrochemically active surface area (ECA) values compared to Pt supported on Vulcan (Fig. 14.10), showing that MWCNTs possess superior stability to commercial carbon black under normal and severe potential cycling conditions [137].

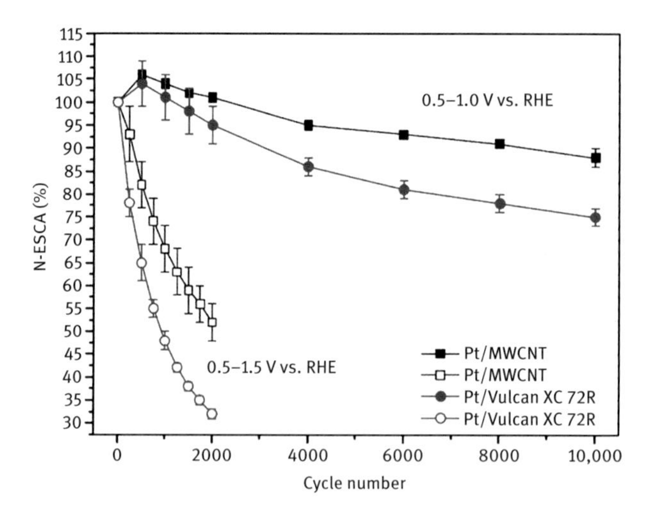

Fig. 14.10: Normalized ECA loss due to voltage cycling (0.5–1.0 V *vs.* RHE (solid) and 0.5–1.5 V *vs.* RHE (hollow); 50 mVs^{-1} scan rate) for Pt/MWCNT (square) and commercial Pt/Vulcan XC-72R (circle) (Reprinted with permission from [137] – © 2010 Royal Society of Chemistry).

Carbon nanofibers. CNFs are also used as catalyst support in PEFCs. Unlike CNTs, CNFs have a very thin or no hollow cavity. Their diameters are larger than CNTs and can be classified into three types: (i) ribbon-like CNF; (ii) platelet CNF; and (iii) herringbone depending upon the orientation of the nanofibers with respect to the growth axis [100]. Herringbone CNFs are known to have intermediate characteristics between parallel and platelet types, thereby exhibiting higher catalytic activity than the parallel and better durability than the platelet forms [137]. The main difference between CNTs and CNFs lies on the exposure of active edge planes. A predominant basal plane is exposed in CNTs while only the edge planes with anchoring sites for the electrocatalyst are exposed in CNFs [100].

Pt (5 wt%) supported on platelet and ribbon graphite nanofibers exhibited similar activities to those observed by Pt (25 wt%) on carbon black [138]. This phenomenon was attributed to the crystallographic orientations adopted by the catalyst particles dispersed on graphitic nanofiber structures [139]. Also, the electrocatalysts supported on CNFs were less susceptible to CO poisoning than Pt supported on carbon black.

Li and co-workers [139] synthesized Pt (5–30 wt%) nanoparticles supported on CNFs using a modified ethylene glycol method. Pt-CNF based MEAs with 50 wt% Nafion® exhibited higher cell performance than the carbon black based MEAs with an optimized 30 wt% Nafion® content (Fig. 14.11). This was attributed to the larger length to diameter ratio of CNFs that allows the formation of conductive networks in the Nafion® matrix [140].

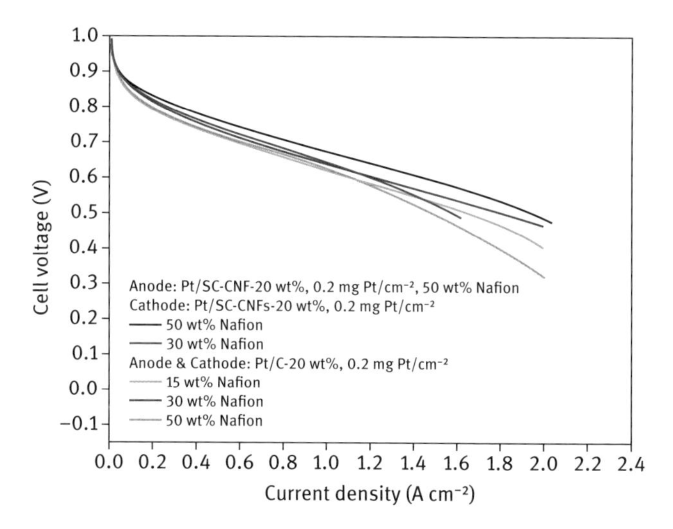

Fig. 14.11: Polarization curves of Pt/C (BASF-fuel cell, 20 wt.%) or Pt/SC-CNFs (homemade, 20 wt.%) based MEAs with varying Nafion® amount in the cathode. Operating conditions: 70 °C cell temperature, 100 % relative humidity (RH); anode/cathode: H_2/O_2, 200 $mLmin^{-1}$ flow rate (Reprinted from [140] with permission from Elsevier).

14.3.2.2 Mesoporous carbon (MC)

Despite the advantages offered by CNTs and CNFs, there are still many obstacles (cost, synthesis methods) to overcome to allow large-scale production. Another type of catalyst support material is mesoporous carbon that provides high surface area and conductivity [100, 141]. It can be classified into ordered (OMC) and disordered (DOMC) mesoporous carbon [100]. OMCs have been extensively used as catalyst support materials for fuel cells [140, 142–146]. The large surface area and 3D connected monodispersed mesospheres facilitate diffusion of the reactants, making them very attractive materials as catalyst supports [100].

The effect of MCs pore morphology on the electrocatalytic activity of Pt has been also studied [147]. Pt (20 wt%) was supported on OMC (CMK-3) and disordered wormhole mesoporous carbon (WMC) using a microwave polyol process. Both support materials had similar pore characteristics (~4 nm pore size) apart from their pore morphology. It was discovered that CMK-3 support provided more electrochemically active Pt support sites and higher active surface area than WMC leading to superior ORR activity and fuel cell performance. This enhanced catalytic activity was attributed to the highly-ordered structure and good 3D interconnection of the nanospacings of carbon nanorods, resulting in higher catalyst utilization efficiency compared to WMCs (Fig. 14.12) [148]. Hence, pore morphology of catalyst support plays an important role in the activity of the supported electrocatalyst.

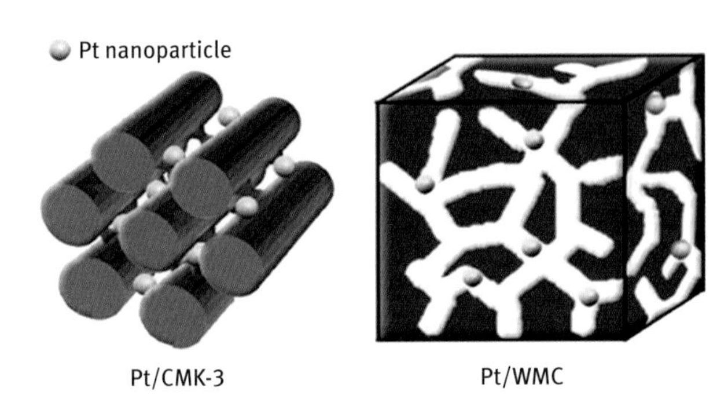

Fig. 14.12: Effect of pore morphology of the carbon support (CMK-3 and WMC) on the activity of Pt electrocatalyst (Reprinted from [148] with permission from Elsevier).

The effect of surface chemistry of OMCs (CMK-3) has also been investigated [148]. OMCs were functionalized by diluted nitric acid (HNO_3) to modify the morphological and textural properties of the support *via* the introduction of surface oxygen groups. The average size of Pt nanoparticles on the CMK-3 supports was approximately 7–8 nm, while larger catalyst particles (~22–23 nm) were observed (Fig 14.13) when CMK-3 supports were treated with concentrated HNO_3. Nitrogen adsorption-desorption isotherms revealed lower specific surface area and total pore volume for CMK-3 support (treated with concentrated HNO_3) which led to a significant increase of the catalyst particle size. Polarization and power density curves (Fig. 14.14) exhibited better electrocatalytic behavior for CMK-3 based electrodes (~13-27 $mWcm^{-2}$) compared to commercial E-TEK electrodes (9.5 $mWcm^{-2}$), despite their lower electrical conductivity and larger Pt nanoparticles [149]. Ohmic and mass transfer losses were greater for supports functionalized with concentrated HNO_3 due to their decreased electrical conductivity, higher agglomeration and lower specific surface area [149].

Hayashi and co-workers [145] built an ideal triple phase boundary inside the mesopores of carbon support in order to examine the electrochemical reactions occurring in nanoscale. Depending on the solvent used (2-propanol) to dilute Nafion®, the reactivity toward oxygen reduction was different. Nafion® dissolved in 2-propanol was able to penetrate deeper into the mesopores and contact with more Pt particles

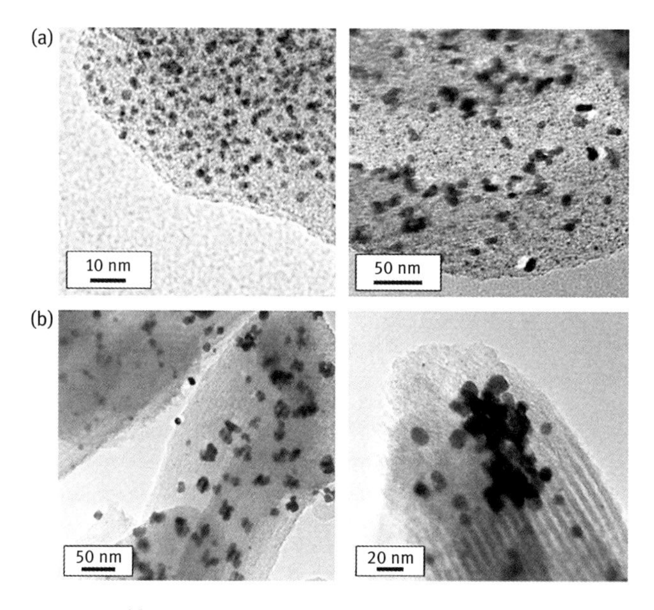

Fig. 14.13: (a) TEM images for Pt supported on CMK-3 catalysts treated with diluted HNO3 for 2 h (Pt/CMK-3 Nd2); (b) TEM images for Pt supported on CMK-3 catalysts treated with concentrated HNO3 for 2 h (Pt/CMK-3 Nc2) (Reprinted from [149] with permission from Elsevier).

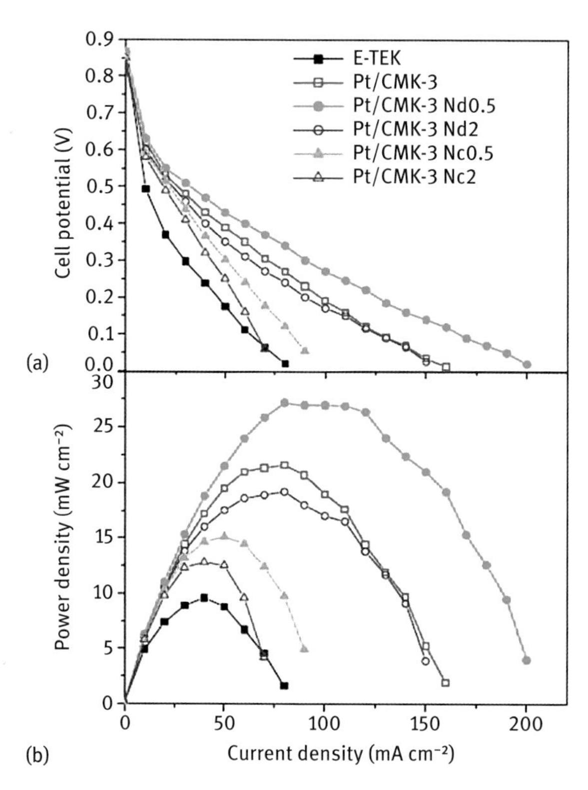

Fig. 14.14: Polarization (a) and power density (b) curves of Pt supported on CMK-3 catalysts with different surface chemistry at the anode side of a PEFC working at room temperature and atmospheric pressure (Reprinted from [149] with permission from Elsevier).

resulting in higher oxygen reduction reactivity. By changing the Pt precursor, ORR current started to increase at more positive potential, indicating enhanced ORR activity (Fig. 14.15). Platinum(II)acetylacetonate had been chosen as the new precursor since it is soluble in various organic solvents, increasing the number of Pt particles deposited on the mesopores [145–147].

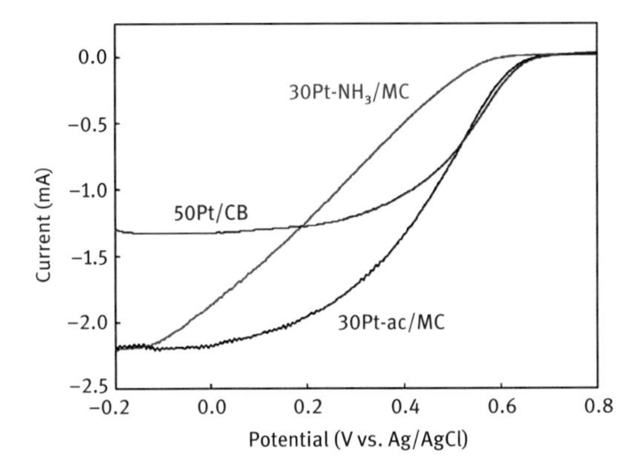

Fig. 14.15: Linear sweep voltammograms of (50 wt%) Pt/Carbon black, (30 wt%) Pt-NH3/MC, and (30 wt%) Ptac/MC (catalyst loading 14 μgcm^{-2}). Operating conditions: 1600 rpm rotating speed and 20 mVs^{-1} sweep rate (Reprinted from [145] with permission from Elsevier).

MCs have also been doped with nitrogen and their stability was investigated [150]. These catalysts were subjected to long-term potential cycling (10,000 cycles between 0.5 and 1 V *vs.* RHE, 50 mVs^{-1} scan rate) and their ECA results were compared to Pt sup-

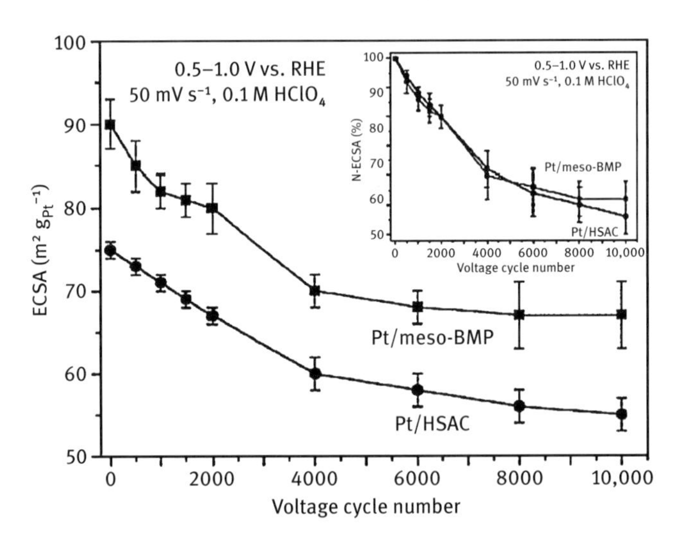

Fig. 14.16: ECA evolution (inset: normalized ECA) for Pt supported on nitrogen-doped MC (square) and Pt supported on high surface area carbon black (circle). Operating conditions: voltage cycling between 0.5 to 1.0 V *vs.* RHE, 50 mVs^{-1} scan rate, room temperature, 0.1 M $HClO_4$ (Reprinted with permission from [151] – © 2012 Wiley).

ported on high surface area carbon black (Fig. 14.16). The ECA values of Pt supported on nitrogen-doped MCs were higher that Pt-carbon black before and after the testing (90 m^2g^{-1} to 67 m^2g^{-1} and 75 m^2g^{-1} to 55 m^2g^{-1} respectively). The mean particle size growth was the main factor of the ECA loss as well as carbon corrosion [151].

14.3.2.3 Multi-layer graphene

Graphene is also used as catalyst support in PEFCs as it offers high conductivity, facile electron transfer and large surface area [151, 152]. The planar structure of graphene allows its edge and basal planes to interact with the nanoparticles of the electrocatalyst [100].

Jafri and co-workers [153] synthesized nitrogen-doped multi-layer graphene by thermal exfoliation of graphitic oxide followed by nitrogen plasma treatment. Pt-multi-layer graphene (PtG) and Pt-nitrogen-doped multi-layer graphene (PtNG) based MEAs demonstrated a maximum power density of 390 and 440 $mWcm^{-2}$ respectively. The improved performance of PtNG was attributed to the formation of pyrrolic nitrogen defects that increased the anchoring sites for the deposition of Pt on the surface leading to increased electrical conductivity and carbon-catalyst binding [154].

14.3.2.4 Interaction between the carbon support and the catalyst

Carbon support plays a vital role in the preparation and performance of catalysts since it influences the shape, size and dispersion of catalyst particles as well as the electronic interactions between catalyst and support [154, 155].

Carbon support is a heterogeneous surface consisting of a mixture of basal and edge planes exposed to the surface [156]. Metal particles on the heterogeneous surface are in dispersed state increasing their stability; particles are arranged on the carbon surface along the edges of graphene network at edge steps or intercrystalline boundaries [155]. The charge transfer from metal (catalyst) to carbon support is equal to the number of surface states in the carbon support [157].

(i) *Effect of the nature of the support.* The size and morphology of platinum particles depend on the nature of the support. Analysis of X-ray diffractograms (XRD) of different supports prepared by the same method and with the same metal loading demonstrated that Pt crystallites supported on CNFs were ~3 nm while Pt crystallites supported on Vulcan and OMCs were approximately 5.6 and 7.6 nm [158]. Thus, higher crystalline grades of the support are conductive to smaller platinum particle size and higher crystalline structures, which are associated with a strong metal-carbon interaction [159].

    The crystalline structure of the metal is also affected by the metal-support interaction. Metal particles supported on CNFs have a highly crystalline structure due to strong metal-support interaction [155], whereas Pt particles supported on Vulcan and OMCs have a more dense globular morphology due to weak metal-support interaction [159]. Hence, the more amorphous the carbon support, the higher the platinum size and the more dense the globular morphology [159].

(ii) *Effect of the surface chemistry of the support*. Figure 14.17 shows the relationship between the total number of surface oxygen groups of the support and the Pt crystallite size for each carbon material tested. An increase in the platinum crystallite size is observed as the number of surface oxygen groups increases [159].

In summary, carbon black is extensively used as catalyst support in PEFCs due to its low cost and high availability. Under long-term durability testing, carbon blacks undergo a morphological transformation and catalyst agglomerates are formed. Chemical activation of carbon is required in order to increase the number of anchoring sites for catalyst particles on its surface resulting in reduced catalyst utilization and low

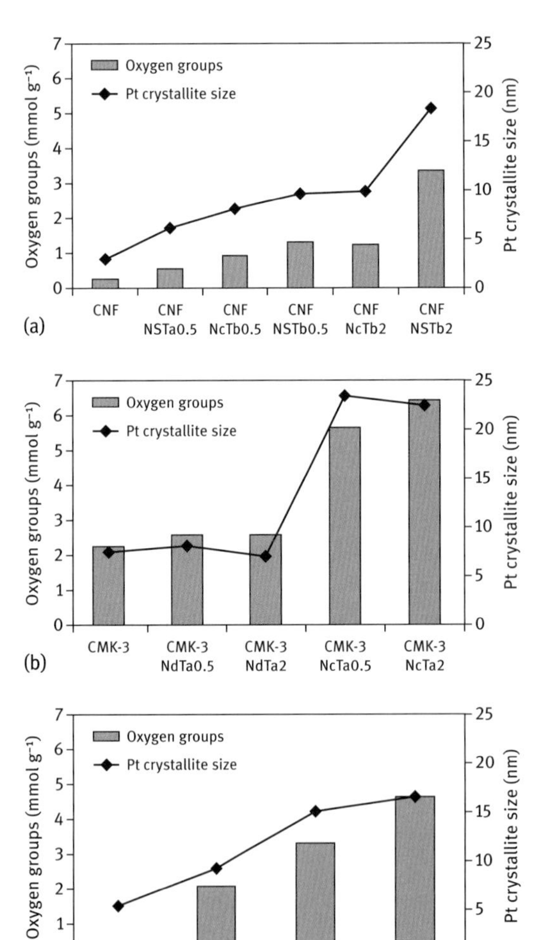

Fig. 14.17: Effect of the surface chemistry of the support on catalyst properties: relationship between the total number of surface oxygen groups of the support and the average Pt crystallite size for oxidized and non-oxidized (a) CNF; (b) CMK-3; and (c) Vulcan (Reprinted from [159] with permission from Elsevier).

thermal and electrochemical stability. Thus, new carbon support materials are being investigated and are categorized into carbon nanotubes, multi-layer graphene, mesoporous carbon and carbon nanofibers. The issue of low catalyst utilization was solved by the introduction of CNTs. Their high crystallinity, surface area and long-term stability made them an attractive substitute for carbon black. Multi-layer graphene as support provides facile electron transfer and high conductivity while the high surface area and tailored mesoporosity of ordered MCs allow high metal dispersion in the support. CNFs do not require any chemical pretreatment due to the presence of highly active edge planes on which the catalyst nanoparticles can be attached. Even though the properties on the new support materials are extremely promising, further PEFC testing is required with respect to ORR activity and long-term stability.

14.3.3 Carbon hybrids and composites as ORR electrocatalysts

14.3.3.1 Nitrogen-doped carbon

PEFCs are an attractive alternative power source for mobile and stationary applications characterized by low emissions, good energy conversion efficiency and high power density. One of the main obstacles towards the commercialization of this technology is the high cost of component materials (catalyst, membrane, etc.) [160–162].

To circumvent this issue, extensive research has been carried out to either reduce the Pt catalyst usage (or improve catalyst utilization) [163, 164] or identify alternative non-noble metal catalysts with similar catalytic activity [165]. A promising alternative electrocatalyst is nitrogen-doped carbon nanostructures [166, 167]. Nitrogen doping of carbon nanostructure materials can be done directly during the synthesis of porous carbon materials [168–175] or with treatment of already synthesized carbon nanostructures with nitrogen containing precursors (NH_3) [176–178].

Nitrogen-doped CNFs [179] prepared *via* chemical vapor deposition (CVD) demonstrated improved ORR activity. It was attributed to the presence of edge plane defects and nitrogen functionalities within the CNF structure [180]. Figure 14.18 shows the measured hydroperoxide (HO_2^-) decomposition rates at N-doped and non-doped CNFs according to the chemical decomposition of HO_2^- reaction

$$HO_2^- \leftrightarrow \frac{1}{2}O_2 + OH^- \tag{14.8}$$

In potassium nitrate solution (1 M KNO_3, pH ~7) the forward reaction rate for N-doped CNFs was approximately $5 \cdot 10^{-6}$ cms^{-1} and $3 \cdot 10^{-8}$ cms^{-1} for non-doped CNFs. In potassium hydroxide solution (1 M KOH, pH ~ 14), the forward reaction rate was ~ $1.8 \cdot 10^{-5}$ cms^{-1} and $1 \cdot 10^{-7}$ cms^{-1} for N-doped and non-doped CNFs respectively. Thus, in both neutral and alkaline solutions, N-doped CNFs exhibited a 100-fold increase in H_2O_2 decomposition [180], with values similar to Pt black [181].

Nitrogen doping has repeatedly been reported to increase the basic nature [182, 183] and catalytic activity [184, 185] of the graphitic carbon which is attributed to the

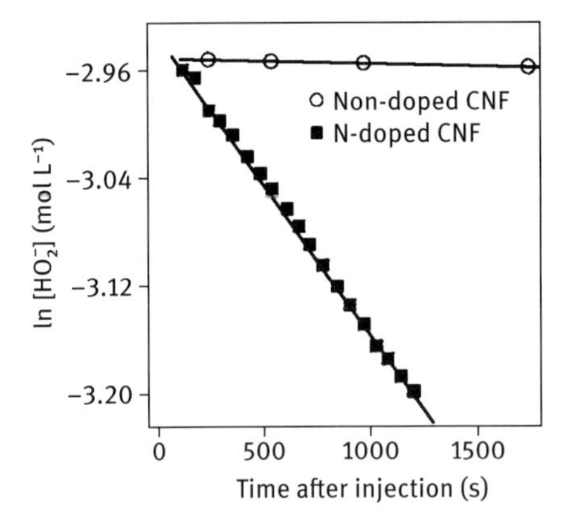

Fig. 14.18: Gasometric analysis of the heterogeneous decomposition of hydrogen peroxide at non-doped CNFs (circle) and N-doped CNFs (square) in 1 M KNO3 (Reprinted with permission from [180] – © 2005 American Chemical Society).

presence of nitrogen [186]. N-doping in a basal plane of carbon may be more favorable than that in edge sites because the number of available doping sites is greater in the basal plane than in the edge [187]. Geng and co-workers [188] studied the relationship between structure and activity in nitrogen-doped CNTs. ORR activity in acid and alkaline solutions increased as the nitrogen content on the surface was increasing (Fig. 14.19). Similar results were obtained by density functional calculations [11] demonstrating that oxygen adsorption becomes more energetically favorable as the number of nitrogen around C=C bonds increases.

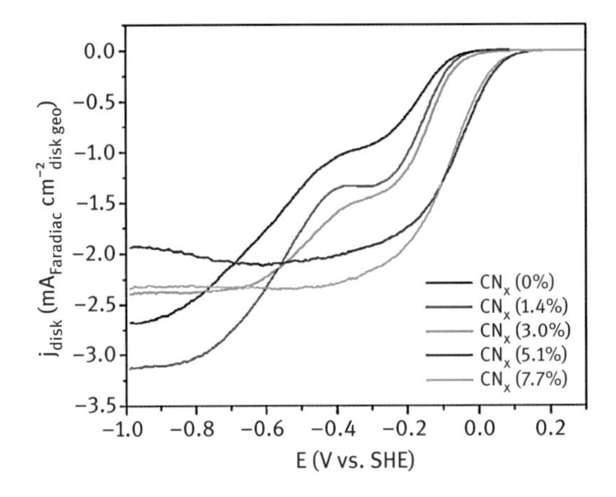

Fig. 14.19: Polarization curves of oxygen reduction on nitrogen-doped CNTs with different nitrogen content. Operating conditions: 0.1 M KOH, 5 mVs^{-1} scan rate, 1600 rpm rotation speed, 160 μgcm^{-2} catalyst loading (Reprinted from [189] with permission from Elsevier).

The basicity of the carbon surface derives from the π electron decolization in graphite carbon layers and the antioxidant character of carbon forming oxygen-containing groups [180]. Delocalized π electrons are capable of nucleophilic attack giving Lewis basicity to the carbon [180, 189].

14.3.3.2 Nitrogen-doped carbon hybrids with non-precious metals

Non-noble metals supported on nitrogen-doped carbon supports are synthesized by pyrolysis of nitrogen [190–196] and transition metal (such as Fe, Co) [194, 197–210] precursors at high temperatures (~ 600–900°C). Nitrogen precursors are ammonia (NH_3) [191, 192] or acetonitrile (CH_3CN) [193, 194, 197], while transition metal precursors include metal salts [194, 197], metal [197, 198], Prussian metal [199, 201] and N_4 macrocyclic complexes [200, 202–208, 211]. Dodelet and co-workers [211] determined the structural parameters of carbon black that are important to maximize its catalytic activity during pyrolysis in NH_3. These are the average particle diameter of carbon black, the amount of disordered phase and the mean size of graphene layers that constitute the graphitic crystallites in carbon black. The highest catalytic activity is obtained for the smallest particle diameter ($d_{particle}$), the largest amount of disordered phase and the largest mean size of graphene layers [212].

For Fe-based nitrogen-doped catalysts, the catalytic activity is directly related to iron content; the number of catalytic sites is increasing as Fe content is increasing until all phenanthroline nitrogens are coordinated with iron [212].

Two different catalytic sites exist at all pyrolysis temperatures in Fe-based catalysts made with Fe salt or prophyrin, namely FeN_4/C and FeN_2/C [213, 214]. FeN_2/C is an iron ion coordinated to two pyridinic nitrogen atoms while FeN_4/C is coordinated to four pyrrole nitrogen atoms. Rotating disk electrode (RDE) measurements revealed that FeN_2/C has higher activity and selectivity than FeN_4/C due to the better electrical contact of FeN_2/C catalytic site with the graphene plane [214]. However, further investigation is needed to fully comprehend the reasons for the higher activity of FeN_2/C catalytic sites.

Zelenay and co-workers [215] used cyanamide as the nitrogen precursor to synthesize transition metal-nitrogen-doped carbon ORR catalysts. Cyanamide forms graphitic C_3N_4 under pyrolysis with high nitrogen content resulting in high catalytic activity. The best performing catalyst was obtained after heat treatment at 1050 °C yielding an open circuit voltage of 1 V (*vs.* SHE) and ~ 105 $mAcm^{-2}$ current density at 0.8 V *vs.* SHE (Fig. 14.20). Cyanamide reduces sulfur evolution from the iron source ($FeSO_4$) during heat treatment indicating that an interaction between cyanamide and sulfate-derived species exists and stabilizes the sulfur through the formation of C-S bonds. However, this mechanism is still being investigated [216].

Iron and cobalt nitrogen-doped catalysts ($TM_xC_{1-x-y}N_y$; TM = Fe, Co; $0 < x < 0.09$; $0 < y < 0.5$) were also prepared by sputter deposition [216, 217]. Grazing-incidence X-ray diffraction and scanning electron microscopy (SEM) were used to observe the structural changes as a function of temperature [218]. At temperatures above 700 °C, the catalyst transformed to a heterogeneous mixture of partially graphitized nitrogen-containing carbon and Fe_3C or β-Co accompanied by a rapid decrease in nitrogen content [218]. Fe-N-C catalyst exhibited the highest activity at 800 °C since the thermal annealing at this temperature causes graphitization and retains sufficient amount of nitrogen to form active sites [218].

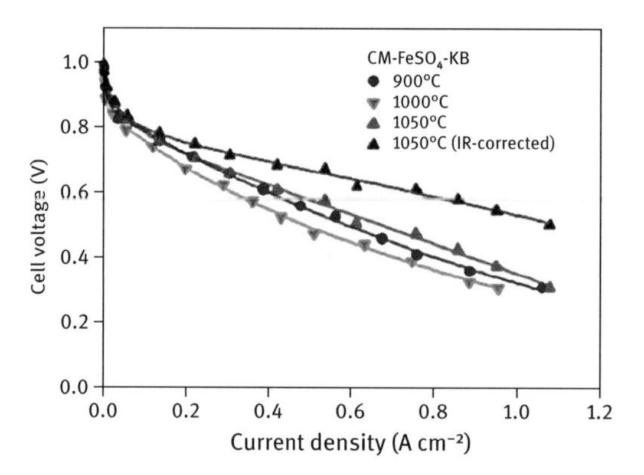

Fig. 14.20: Fuel cell polarization curves for CM-FeSO4-KB (cyanamide-$FeSO_4$-Ketjeblack) ORR catalysts obtained at different heat-treatment temperatures. Operating conditions: H_2/O_2, 100 % RH, 4 mgcm^{-2} transition metal catalyst loading (Reprinted from [216] with permission from Elsevier).

However, the instability of transition metal nitrogen-doped carbon catalysts in the highly acidic environment of PEFCs still remains a challenge. The origin of instability in acidic medium is due to the formation of H_2O_2 (product of the incomplete reduction of oxygen) [218]. Lefevre and Dodelet [219] submitted iron-nitrogen-doped carbon catalysts to peroxide treatment to quantify the effect of peroxide on their catalytic activity. Even 5 vol% of peroxide is detrimental to their activity due to the loss of iron content from their catalytic sites (Fig. 14.21) [219]. A partial solution to this instability issue is the heat treatment of catalysts at high temperatures. Even though stability is

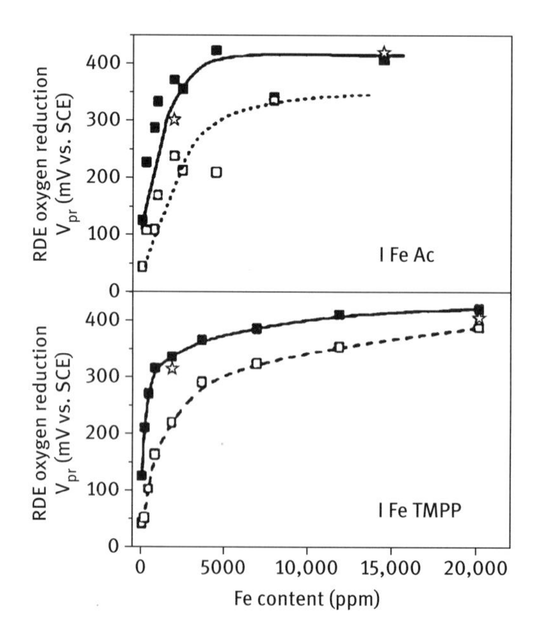

Fig. 14.21: Effect of H_2SO_4 (stars) and of H_2SO_4 + 5 vol% H_2O_2 (open squares) on the initial catalytic activity (dark squares) for type FeAc (iron acetate) and IFeTMPP (Fe tetramethoxyphenylporphyrin) catalysts (Reprinted from [219] with permission from Elsevier).

improved, the catalytic activity is decreased due to the loss of nitrogen content [219]. Thus, further research is needed to develop stable transition metal-nitrogen-doped carbon catalysts in the acidic environment of PEFC.

14.3.3.3 Polymer Composites with Nitrogen-doped Carbon Hybrids

Research is also focused on non-precious metal catalysts synthesized with conjugated heterocyclic conducting polymers such as Ppy, Pani and poly (3-methylthiophene (P3MT)) [166, 219–223].

The ORR activity and stability of carbon-supported materials with or without cobalt based on different heterocyclic polymers were investigated [221]. Energy dispersive atomic X-ray spectrometry (EDAX), Fourier transform infrared spectroscopy (FTIR) and Thermogravimetric analysis (TGA) studies demonstrated that the addition of cobalt modifies the chemical structure of all carbon-supported materials studied by the introduction of Co-N (Ppy-C-Co and Pani-C-Co) and Co-S (P3MT-C-Co) bonds [221]. The addition of cobalt results in improved electrocatalytic activity of C-P3MT and C-Pani catalysts even though their thermal stability is slightly reduced at lower temperatures due to the degradation of cobalt oxides (Fig. 14.22(a)) [221]. The potential at which ORR occurs was determined to be the highest for Ppy-C-Co (-0.2 to 1 V *vs.* normal hydrogen electrode (NHE)), followed by Ppy-C and P3MT-C-Co. Potentiostatic measurements were conducted for 48 hrs to determine the stability of these materials at the maximum potential in the ORR peak (obtained from cyclic voltamogramm). Ppy-C-Co and P3MT-C-Co were stable (after a small current decrease during the first 2 hrs) while Ppy-C exhibited a continuous current decrease throughout the experiment (Fig. 14.22(b)) [221]. Based on these results, it was reported that Ppy-C-Co catalysts was the most suitable material for use in PEFCs, even though the potential range in which ORR occurs needs to be further improved.

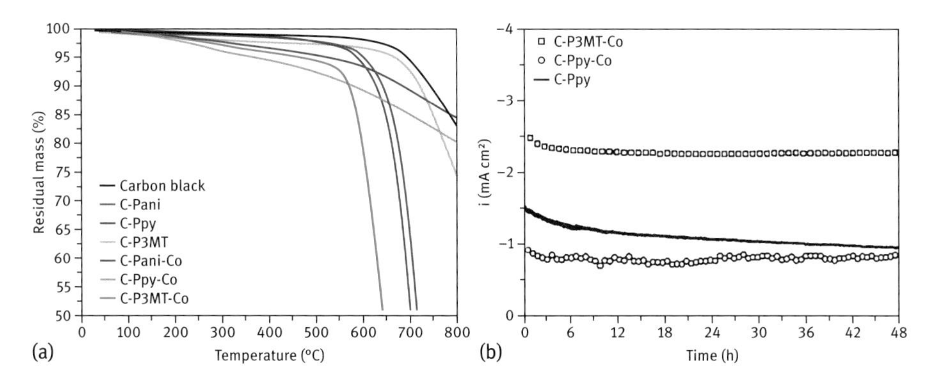

Fig. 14.22: (a) TGA curves of carbon-supported materials with or without cobalt obtained in nitrogen atmosphere; (b) Chronoamperometry results for Ppy-C-Co, P3MT-C-Co and Ppy-C in 0.5 M H_2SO_4 under oxygen (Reprinted from [221] with permission from Elsevier).

The activity of Ppy-C-Co catalyst was thoroughly examined by the synthesis of pyrolyzed and unpyrolyzed Ppy-C-Co with oxidative polymerization (80 °C). Pyrolyzed Ppy-C-Co catalyst demonstrated significantly improved ORR activity over the unpyrolyzed catalyst indicating that the heat treatment enhanced the catalytic activity. X-ray photoelectron spectroscopy (XPS) revealed that new nitrogen peaks corresponding to pyrrolic and graphitic nitrogen were observed. Both these groups are ORR active, increasing the catalytic activity and changing the ORR mechanism from a two-electron (H_2O_2 production) to a four-electron reduction (H_2O production) process compared to unpyrolyzed catalyst [224]. The Koutecky–Levich slope of unpyrolyzed Ppy-C-Co was close to the slope of the theoretical two-electron transfer reaction wereas pyrolyzed Ppy-C-Co catalysts had slope similar to the theoretical four-electron transfer reaction. Based on these plots, the calculated ORR electron number (n) was 2.4 for the unpyrolyzed Ppy-C-Co and 3.2 for the pyrolyzed sample (Fig. 14.23).

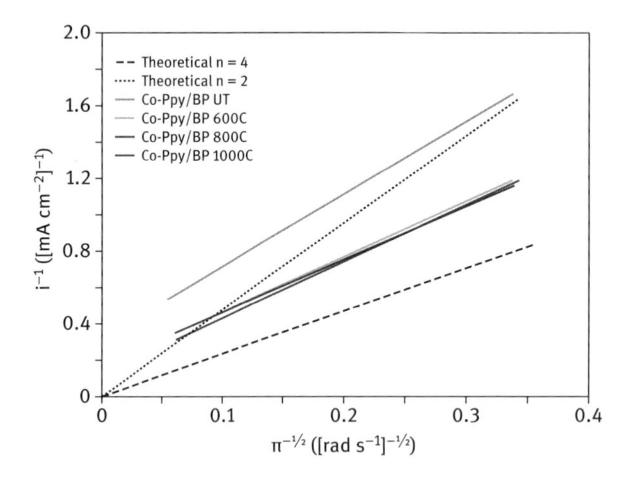

Fig. 14.23: Koutecky–Levich plots for the ORR of unpyrolyzed and pyrolyzed Ppy-C-Co catalysts at 0.3 V (*vs.* RHE) in 0.5 M H_2SO_4 under saturated O_2 and 0.122 $mgcm^{-2}$ Ppy-C-Co loading. The current densities were normalized to the geometric area (Reprinted from [224] with permission from Elsevier).

The effect of the weight and molar ratio of Ppy:Co on the ORR activity of these catalysts was recently investigated [224]. Ppy:Co molar ratio of 4 increased the ORR activity as it favors the formation of complexes in which Co is coordinated to 3 or 4 nitrogen atoms. Calcination of these complexes led to the generation of CoN_{x-2} sites characterized by high ORR activity since the coordination on Co in them favors the adsorption of oxygen on them. The ORR activity of Ppy-Co-C (molar ratio 4 and weight ratio 2) was adequate, being 200 mV lower than pristine Pt-C, and produced less H_2O_2 than Pt [225].

The electrocatalytic activity of composite carbon-supported materials based on heterocyclic polymers and nickel (Pani-C-Ni, Ppy-C-Ni, P3MT-C-Ni) was also studied

[222]. FTIR revealed that the modification with nickel altered the polymer bonds creating Ni-AN (for Pani and Ppy-based catalysts) and Ni-AS bonds (for P3MT -based catalysts), while TGA showed that the incorporation of nickel improved the thermal stability of the catalysts at high temperatures (above 400 °C). The Tafel slopes were very similar for all catalysts tested (~0.11-0.2); however, Ppy-C-Ni had the highest exchange current density (~ $4 \cdot 10^{-5}$ $mAcm^{-2}$) followed by P3MT-C-Ni and Ppy-C (~ $3.5 \cdot 10^{-5}$ and $2.5 \cdot 10^{-5}$ $mAcm^{-2}$ respectively). Based on these results, it was concluded that Ppy-C-Ni was the most suitable catalyst for ORR in acidic medium [222].

Popov and co-workers [225] synthesized N-doped ordered porous carbon (CN_x) *via* a nanocasting process using polyacrylnitrile (Pan) as the carbon and nitrogen precursor and mesoporous silica as hard template. The non-precious metal ORR catalysts were prepared by pyrolyzing CN_x at 1000 °C in argon after iron acetate was adsorbed onto it, followed by post-treatments. RDE measurements demonstrated a ~ 0.88 V (*vs.* NHE) onset potential for ORR in 0.5 M H_2SO_4, while a current density of 0.6 Acm^{-2} at 0.5 V was obtained in a H_2/O_2 PEFC using 2 $mgcm^{-2}$ catalyst loading [225].

To sum up, nitrogen doping of nanostructured carbon is a promising strategy to obtain low cost PEFC catalysts with good cell performance. Nitrogen-doped carbon with or without transition metal catalysts exhibits enhanced catalytic activity and durability toward ORR. Nitrogen is the indispensable element in many non-Pt metal catalysts (Fe, Co, etc.) for potential application in PEFCs. Research is also focused on the development of non-precious metal catalysts synthesized with conjugated heterocyclic conducting polymers (such as polypyrrole). Initial results show adequate ORR activity and fuel cell performance, but further research is needed to increase the stability in acidic medium and the potential range in which ORR occurs.

14.4 Summary

Carbon is widely used in the catalytic processes of the chemical industry due to its unique characteristics, such as chemical inertness, high surface area and porosity, good mechanical properties and low cost. It is used for the production of chlorine and aluminum, in metal refining (gold, silver, and grain refinement of Mg-Al alloys) as well as for the electrolytic production of hydrogen peroxide and photoelectrochemical water splitting.

Carbon is a vital component of PEFCs as well and it can serve as structural component (gas diffusion layer), catalyst support and ORR electrocatalyst. However, carbon is oxidized at high potentials leading to fuel cell performance losses and hence new carbon support materials such as carbon nanotubes and nanofibers as well as mesoporous carbon and multi-layer graphene are being investigated. They offer high crystallinity, surface area, increased fuel cell performance and stability. However, further PEFC testing is required with respect to their ORR activity, long-term stability (continuous cycling and accelerated degradation tests) and conductivity.

Pt is currently used as the electrocatalyst for oxygen reduction in PEFCs, but its high cost and limited resources have led to an extensive search for alternate, non-noble metal electrocatalysts. Among them, nitrogen doping of nanostructured carbon (with or without transition metals) demonstrated enhanced catalytic activity and durability. Nitrogen doped non-precious metal electrocatalysts with conducting polymers (polypyrrole, polyaniline) also exhibited adequate ORR activity and fuel cell performance. However, further research is needed toward their stability in highly acidic medium and broadening of potential range in which oxygen reduction occurs.

Nomenclature

Abbreviation	Definition
AC	Activated carbon
BET	Brunauer, Emmett and Teller
CIC	Carbon in column
CIL	Carbon in leach
CIP	Carbon in pulp
CNF	Carbon fiber
CNT	Carbon nanotube
CVD	Chemical vapor deposition
DOMC	Disordered mesoporous carbon
ECA	Electrochemically active surface area
EDAX	Energy dispersive atomic X-ray spectrometry
FTIR	Fourier transform infrared spectroscopy
GDL	Gas diffusion layer
HOMO	Highest occupied molecular orbital
LUMO	Lowest unoccupied molecular orbital
MC	Mesoporous carbon
MEA	Membrane electrode assembly
MWCNT	Multi-walled carbon nanotube
OMC	Ordered mesoporous carbon
ORR	Oxygen reduction reaction
NHE	Normal hydrogen electrode ($E = 0$ V)
P3MT	Poly (3-methylthiophene)
Pan	Polyacrylonitrile
Pani	Polyaniline
PEFC	Polymer electrolyte fuel cell
PEM	Proton exchange membrane
Ppy	Polypyrrole
Pt	Platinum
PtG	Pt-multi-layer graphene

PtNG	Pt-nitrogen-doped multi-layer graphene
RDE	Rotating disk electrode
RH	Relative humidity
RHE	Reversible hydrogen electrode (E = 0 V)
SCE	Saturated calomel electrode (E = +0.24 V *vs.* NHE)
SEM	Scanning electron microscopy
SHE	Standard hydrogen electrode (E = 0 V)
SWCNT	Single-walled carbon nanotube
TEM	Transmission electron microscopy
TGA	Thermogravimetric analysis
Vol	Volume percent (%)
wt	Weight percent (%)
WMC	Wormhole mesoporous carbon
XPS	X-ray photoelectron spectroscopy
XRD	X-ray diffraction

Bibliography

[1] A. L. Dicks, *J. Power Sources* **2006**, *156*, 128.
[2] N. M. R. Peres, F. Guinea, A. H. Castro Neto, *Annals of Physics* **2006**, *321*, 1559.
[3] J. P. Xanthakis, *Diamond and Related Materials* **2000**, *9*, 1369.
[4] S. Logothetidis, *Diamond and Related Materials* **2003**, *12*, 141.
[5] J. Ristein, *Diamond and Related Materials* **2000**, *9*, 1129.
[6] M. Knupfer, *Surface Science Reports* **2001**, *42*, 1.
[7] R. C. Haddon, L. E. Brus, K. Raghavachari, *Chem. Phys. Lett.* **1996**, *125*, 459.
[8] J. W. Mintmire, B. I. Dunlap, C. T. White, *Phys. Rev. Lett.* **1992**, *68*, 631.
[9] N. Hamada, S. I. Sawada, A. Oshiyama, *Phys. Rev. Lett.* **1992**, *68*, 1579.
[10] J. W. Mintmire, C. T. White, *Phys. Rev. Lett.* **1998**, *81*, 2506.
[11] D. Geng, H. Liu, Y. Chen, R. Li, X. Sun, S. Ye, S. Knights, *J. Power Sources* **2011**, *196*, 1795.
[12] S. Maldonado, K. J. Stevenson, *J. Phys. Chem. B* **2004**, *108*, 11375.
[13] E. Frackowiak, F. Beguin, *Carbon* **2001**, *39*, 937.
[14] S. Bok, A. A. Lubguban, Y. Gao, S. Bhattacharya, V. Korampally, M. Hossain, R. Hiruvengadathan, K. D. Gillis, S. Gangopadhyay, *J. Electrochem. Soc.* **2008**, *155*, K91.
[15] R. Ryoo, S. H. Joo, M. Kruk, M. Jaroniec, *Adv. Mater.* **2001**, *13*, 677.
[16] Y. Kameya, K. Hanamura, *Chem. Eng. J.* **2011**, *173*, 627.
[17] N. Muradov, *Catal. Commun.* **2001**, *2*, 89.
[18] N. Muradov, F. Smith, A. T. Raissi, *Catal. Today* **2005**, *102-103*, 225.
[19] E. K. Lee, S. Y. Lee, G. Y. Han, B. K. Lee, T. J. Lee, J. Y. Jun, K. J. Yoon, *Carbon* **2004**, *42*, 2641.
[20] S. Y. Lee, B. H. Ryu, G. Y. Han, T. J. Lee, K. J. Yoon, *Carbon* **2008**, *46*, 1978.
[21] S. Y. Lee, J. H. Kwak, G. Y. Han, T. J. Lee, K. J. Yoon, *Carbon* **2008**, *46*, 342.
[22] J. L. Pinilla, L. Suelves, M. J. Lazaro, R. Moliner, *Chem. Eng. J.* **2008**, *138*, 301.
[23] J. O. Besenhard, H. P. Fritz, *Angew. Chem.* **1983**, *22*, 950.
[24] M. S. Whittingham, *J. Electroanal. Chem.* **1981**, *118*, 229.
[25] R. Yazami, P. Touzain, *J. Power Sources* **1983**, *9*, 365.
[26] P. Pfluger, V. Geiser, S. Stolz, H. J. Guntherodt, *Synth. Met.* **1981**, *3*, 27.
[27] J. Amiell, P. Delhaes, F. Beguin, R. Setton, *Mater. Sci. Eng.* **1977**, *31*, 243.

[28] D. Billaud, A. Pron, F. L. Vogel, *Synth. Met.* **1980**, *2*, 177.
[29] A. Chenite, D. Billaud, *Carbon* **1982**, *20*, 120.
[30] C. Hsieh, H. Teng, *Carbon* **2002**, *40*, 667.
[31] M. J. Bleda-Martınez, D. Lozano-Castello, E. Morallon, D. Cazorla-Amoros, A. Linares-Solano, *Carbon* **2006**, *44*, 2642.
[32] D. Lozano-Castello, D. Cazorla-Amoros, A. Linares-Solano, S. Shiraishi, H. Kurihara, A. Oya, *Carbon* **2003**, *41*, 1765.
[33] K. Kinoshita, J. A. S. Bett, *Carbon* **1973**, *11*, 403.
[34] N. M. Winslow, *Trans. Electrochem. Soc.* **1941**, *80*, 121.
[35] N. M. Winslow, *Trans. Electrochem. Soc.* **1945**, *88*, 81.
[36] M. Janes, *Trans. Electrochem. Soc.* **1947**, *92*, 23.
[37] F. Hine, M. Yasuda, *J. Electrochem. Soc.* **1974**, *121*, 1289.
[38] L. J. J. Janssen, *Electrochim. Acta* **1974**, *19*, 257.
[39] L. J. J. Janssen, J. G. Hoogland, *Electrochim. Acta* **1970**, *15*, 941.
[40] P. Drossbach, *J. Electrochem. Soc.* **1956**, *103*, 700.
[41] W. E. Triaca, C. Solomons, J. OM. Bockris, *Electrochim. Acta* **1968**, *13*, 1949.
[42] F. C. Frary, *J. Electrochem. Soc.* **1948**, *94*, 31.
[43] P. Mandin, R. Wuthrich, H. Roustan, *ECS Trans.* **2009**, *19*, 1.
[44] E. Sum, M. Skyllas-Kazacos, *Electrochim. Acta* **1991**, *36*, 811.
[45] E. Sum, M. Skyllas-Kazacos, *J. Appl. Electrochem.* **1989**, *19*, 485.
[46] C. A. Fleming, *Hydrometallurgy* **1992**, *30*, 127.
[47] A. C. Q. Ladeira, M. E. M. Figueira, V. S. T. Ciminelli, *Minerals Engineering* **1993**, *6*, 585.
[48] J. Du, J. Yang, M. Kuwabara, W. Li, J. Peng, *J. Alloys and Compounds* **2009**, *470*, 134.
[49] Y. M. Kim, C. D. Yim, B. S. You, *Scr. Mater.* **2007**, *57*, 691.
[50] Y. C. Lee, A. K. Dahle, D. H. StJohn, *Metal. Mater. Trans. A* **2000**, *31*, 2895.
[51] D. H. StJohn, M. Qian, M. A. Easton, P. Cao, Z. Hidebrand, *Metal. Mater. Trans. A* **2005**, *36*, 1669.
[52] P. Cao, Q. Ma, D. H. StJohn, *Scr. Mater.* **2004**, *51*, 125.
[53] Q. Ma, P. Cao, *Scr. Mater.* **2005**, *52*, 415.
[54] P. Cao, Q. Ma, D. H. StJohn, *Scr. Mater.* **2005**, *53*, 841.
[55] Y. H. Liu, X. F. Liu, X. F. Bian, *Mater. Lett.* **2004**, *58*, 1282.
[56] L. Lu, A. Kdahle, D. H. StJohn, *Scr. Mater.* **2005**, *53*, 517.
[57] Q. L. Jin, J. P. Eom, S. G. Lim, W. W. Park, B. S. You, *Scr. Mater.* **2003**, *49*, 1129.
[58] T. Motegi, *Mater. Sci. Eng. A* **2005**, *413-414*, 408.
[59] P. Cao, Q. Ma, D. H. StJohn, *Scr. Mater.* **2007**, *56*, 633.
[60] P. Li, B. Tang, E. G. Kandalova, *Mater. Lett.* **2005**, *59*, 671.
[61] B. Jing, S. Yangshan, X. Shan, X. Feng, Z. Tianbai, *Mater. Sci. Eng. A* **2006**, *419*, 181.
[62] K. J. Hirai, H. T. Somekawa, Y. N. Takigawa, K. J. Higashi, *Mater. Sci. Eng. A* **2005**, *403*, 276.
[63] X. Q. Zeng, Y. X. Wang, W. J. Ding, A. A. Luo, A. K. Sachdev, *Metall. Mater. Trans. A* **2006**, *37*, 1333.
[64] L. Lu, A. K. Dahle, D. H. StJohn, *Scr. Mater.* **2006**, *54*, 2197.
[65] M. Qian, P. Cao, *Scr. Mater.* **2005**, *52*, 415.
[66] J. Du, J. Yang, M. Kuwabara, W. Li, J. Peng, *J. Alloys and Compounds* **2009**, *470*, 228.
[67] Y. Birol, *J. Alloys and Compounds* **2006**, *422*, 128.
[68] B. S. Murty, S. A. Kori, M. Chakraborty, *Int. Mater. Rev.* **2002**, *47*, 3.
[69] C. Cheng, Y. Sun, *App. Surf. Sc.* **2012**, *263*, 273.
[70] Y. A. Shaban, S. U. M. Khan, *Chem. Phys.* **2007**, *339*, 73.
[71] X. B. Chen, S. S. Mao, *Chem. Rev.* **2007**, *107*, 2891.
[72] A. Fujishima, K. Honda, *Nature* **1972**, *238*, 37.

[73] S. U. M.Khan, M. Al-Shahry, W. B. Ingler Jr., *Science* **2002**, *297*, 2243.
[74] S. Licht, B. Wang, S. Mukerji, T. Soga, M. Umeno, H. Tributsch, *J. Phys. Chem. B* **2000**, *104*, 8920.
[75] O. Khaselev, J. R. Turner, *Science* **1998**, *280*, 425.
[76] S. U. M. Khan, J. Akikusa, *J. Phys. Chem. B* **1999**, *103*, 7184.
[77] O. N. Srivastava, R. K. Karn, M. Misra, *Int. J. Hydrogen Energy* **2000**, *25*, 495.
[78] J. Akikusa, S. U. M. Khan, *Int. J. Hydrogen Energy* **2002**, *27*, 863.
[79] J. G. Yu, H. G. Yu, B. Cheng, X. J. Zhao, J. C. Yu, W. K. Ho, *J. Phys. Chem. B* **2003**, *107*, 13871.
[80] P. Hartmann, D. K. Lee, B. M. Smarsly, J. Janek, *ACS Nano* **2010**, *4*, 3147.
[81] C. Cheng, Y. Y. Tay, H. H. Hng, H. J. Fan, *J. Mater. Res.* **2011**, *26*, 2254.
[82] C. Cheng, S. K. Karuturi, L. J. Liu, J. P. Liu, H. X. Li, L. T. Su, A. I. Y. Tok, H. J. Fan, *Small* **2012**, *8*, 37.
[83] R. Asahi, T. Morikawa, T. Ohwaki, K. Aoki, Y. Taga, *Science* **2001**, *293*, 269.
[84] C. Xu, Y. A. Shaban, W. B. Ingler Jr., S. U. M. Khan, *Solar Energy Materials & Solar Cells* **2007**, *91*, 938.
[85] Y. Nakano, T. Morikawa, T. Ohwaki, Y. Taga, *Appl. Phys. Lett.* **2005**, *87*, 052111.
[86] C. Xu, R. Killmeyer, M. L. Gray, S. U. M. Khan, *Appl. Catal. B* **2006**, *64*, 312.
[87] T. Umebayashi, T. Yamaki, H. Itoh, K. Asai, *Appl. Phys. Lett.* **2002**, *81*, 454.
[88] Y. A. Shaban, S. U. M. Khan, *Int. J. Hydrogen En.* **2008**, *33*, 1118.
[89] C. Xu, R. Killmeyer, M. L. Gray, S. U. M. Khan, *Electrochem. Commun.* **2006**, *8*, 1650.
[90] S. U. Khan, Y. A. Shaban, *ECS Trans.* **2006**, *3*, 87.
[91] P. Tatapudi, J. M. Fenton, *J. Electrochem. Soc.* **1993**, *140*, L55.
[92] S. J. Ridge, R. E. White, Y. Tsou, R. N. Beaver, G. A. Eisman, *J. Electrochem. Soc.* **1989**, *136*, 1902.
[93] T. P. Fellinger, F. Hasche, P. Strasser, M. Antonietti, *J. Am. Chem. Soc.* **2012**, *134*, 4072.
[94] A. R. Khataee, M. Safarpour, M. Zarei, S. Aber, *J. Electroanal. Chem.* **2011**, *659*, 63.
[95] A. Ozcan, Y. Sahin, M. A. Oturan, *Chemosphere* **2008**, *73*, 737.
[96] N. Yousfi-Steiner, P. Mocoteguy, D. Candusso, D. Hissel, A. Hernandez, A. Aslanides, *J. Power Sources* **2008**, *183*, 260.
[97] E. Antolini, R. R. Passos, E. A. Ticianelli, *J. Power Sources* **2002**, *109*, 477.
[98] L. R. Jordan, A. K. Shukla, T. Behrsing, N. R. Avery, B. C. Muddle, M. Forsyth, *J. Power Sources* **2000**, *86*, 250.
[99] S. Sharma, B. G. Pollet, *J. Power Sources* **2012**, *208*, 96.
[100] A. Halder, S. Sharma, M. S. Hegde, N. Ravishankar, *J. Phys. Chem. C* **2009**, *113*, 1466.
[101] J. Kua, W. A. Goddard III, *J. Am. Chem. Soc.* **1999**, *121*, 10928.
[102] J. Chen, M. Wang, B. Liu, Z. Fan, K. Cui, Y. Kuang, *J. Phys. Chem. B* **2006**, *110*, 11775.
[103] J.C. Meier et al., *Beilstein J. Nanotechnol.*, **2014**, *5*, 44.
[104] E. Antolini, *J. Mater. Sci.* **2003**, *38*, 2995.
[105] K. H. Kangasniemi, D. A. Condit, T. D. Jarvi, *J. Electrochem. Soc.* **2004**, *151*, E125.
[106] K. Mitsuda, T. Murahashi, *J. Electrochem. Soc.* **1990**, *137*, 3079.
[107] J. P. Meyers, R. M. Darling, *J. Electrochem. Soc.* **2006**, *153*, A1432.
[108] A. Taniguchi, T. Akita, K. Yasuda, Y. Miyazaki, *J. Power Sources* **2004**, *130*, 42.
[109] C. A. Reiser, L. Bregoli, T. W. Patterson, J. S. Yi, J. D. Yang, M. L. Perry, T. D. Jarvi, *Electrochem. Solid State Lett.* **2005**, *8*, A273.
[110] S. Maass, F. Finsterwalder, G. Frank, R. Hartmann, C. Merten, *J. Power Sources* **2008**, *176*, 444.
[111] G. Che, B. B. Lakshmi, E. R. Fisher, C. R. Martin, *Nature* **1998**, *393*, 346.
[112] D. B. Mawhinney, V. Naumenko, A. Kuznetsova, J. T. Yates Jr., J. Liu, R. Smalley, *J. Am. Chem. Soc.* **2000**, *122*, 2383.

[113] T. Matsumoto, T. Komatsu, K. Arai, T. Yamazaki, M. Kijima, H. Shimizu, Y. Takasawa, J. Nakamura, *Chem. Comm.* **2004**, 840.
[114] X. Li, I. Hsing, *Electrochim. Acta* **2006**, *51*, 5250.
[115] C. Yang, X. Hu, D. Wang, C. Dai, L. Zhang, H. Jin, S. Agathopoulos, *J. Power Sources* **2006**, *160*, 187.
[116] T. S. Ebbesen, H. Hiura, M. E. Bisher, M. M. J. Treacy, J. L. Shreeve-Keyer, R. C. Haushalter, *Adv. Mater.* **1996**, *8*, 155.
[117] Z. Liu, X. Lin, J. Y. Lee, W. Zhang, M. Han, L. M. Gan, *Langmuir* **2002**, *18*, 4054.
[118] C. H. Wang, H. Y. Du, Y. T. Tsai, C. P. Chen, C. J. Huang, L. C. Chen, K. H. Chen, H. C. Shih, *J. Power Sources* **2007**, *171*, 55.
[119] P. V. Dudin, P. R. Unwin, J. V. Macpherson, *J. Phys. Chem. C* **2010**, *114*, 13241.
[120] S. L. Knupp, W. Li, O. Paschos, T. M. Murray, J. Snyder, P. Haldar, *Carbon* **2008**, *46*, 1276.
[121] Z. Liu, J. Y. Lee, W. Chen, M. Han, L. M. Gan, *Langmuir* **2004**, *20*, 181.
[122] K. Yasuda, Y. Nishimura, *Mater. Chem. Phys.* **2003**, *82*, 921.
[123] Y. Shao, G. Yin, J. Wang, Y. Gao, P. Shi, *J. Power Sources* **2007**, *161*, 47.
[124] T. M. Day, P. R. Unwin, J. V. Macpherson, *Nano Lett.* **2007**, *7*, 51.
[125] T. M. Day, P. R. Unwin, N. R. Wilson, J. V. Macpherson, *J. Am. Chem. Soc.* **2005**, *127*, 10639.
[126] D. He, C. Zeng, C. Xu, N. Cheng, H. Li, S. Mu, M. Pan, *Langmuir* **2011**, *27*, 5582.
[127] H. S. Oh, K. Kim, H. Kim, *Int. J. Hydrogen Energy* **2011**, *36*, 11564.
[128] Y. Zhao, X. Yang, J. Tian, F. Wang, L. Zhan, *J. Power Sources* **2010**, *195*, 4634.
[129] Z. Tang, C. K. Poh, K. K. Lee, Z. Tian, D. H. C. Chua, J. Lin, *J. Power Sources* **2010**, *195*, 155.
[130] Y. Shao, G. Yin, J. Zhang, Y. Gao, *Electrochim. Acta* **2006**, *51*, 5853.
[131] H. T. Fang, C. G. Liu, C. Liu, F. Li, M. Liu, H. M. Cheng, *Chem. Mater.* **2004**, *16*, 5744.
[132] S. Park, Y. Shao, R. Kou, V. V. Viswanathan, S. A. Towne, P. C. Rieke, J. Liu, Y. Lin, Y. Wang, *J. Electrochem. Soc.* **2011**, *158*, B297.
[133] L. Li, Y. Xing, *J. Electrochem. Soc.* **2006**, *153*, A1823.
[134] J. Wang, G. Yin, Y. Shao, Z. Wang, Y. Gao, *J. Phys. Chem. C* **2008**, *112*, 5784.
[135] F. Coloma, A. Sepulveda-Escribano, J. Fierro, F. Rodriguez-Reinoso, *Langmuir* **1994**, *10*, 750.
[136] F. Hasche, M. Oezaslan, P. Strasser, *Phys. Chem. Chem. Phys.* **2010**, *12*, 15251.
[137] J. S. Zheng, X. S. Zhang, P. Li, J. Zhu, X. G. Zhou, W. K. Yuan, *Electrochem. Commun.* **2007**, *9*, 895.
[138] C. A. Bessel, K. Laubernds, N. M. Rodriguez, R. T. K. Baker, *J. Phys. Chem. B* **2001**, *105*, 1115.
[139] W. Li, M. Waje, Z. Chen, P. Larsen, Y. Yan, *Carbon* **2010**, *48*, 995.
[140] J. Ding, K. Y. Chan, J. Ren, F. Xiao, *Electrochim. Acta* **2005**, *50*, 3131.
[141] C. Galeano et al, *J. Am. Chem. Soc.* **2012**, *134*, 20457.
[142] F. Su, J. Zeng, X. Bao, Y. Yu, J. Y. Lee, X. S. Zhao, *Chem. Mater.* **2005**, *17*, 3960.
[143] H. Chang, S. H. Joo, C. Pak, J. *Mater. Chem.* **2007**, *17*, 3078.
[144] A. Hayashi, H. Notsu, K. Kimijima, J. Miyamoto, I. Yagi, *Electrochim. Acta* **2008**, *53*, 6117.
[145] A. Hayashi, K. Kimijima, J. Miyamoto, I. Yagi, *J. Phys. Chem. C* **2009**, *113*, 12149.
[146] K. Kimijima, A. Hayashi, S. Umemura, J. Miyamoto, K. Sekizawa, T. Yoshida, I. Yagi, *J. Phys. Chem. C* **2010**, *114*, 14675.
[147] S. Song, Y. Liang, Z. Li, Y. Wang, R. Fu, D. Wu, P. Tsiakaras, *Appl. Catal. B: Environ.* **2012**, *98*, 132.
[148] L. Calvillo, M. Gangeri, S. Perathoner, G. Centi, R. Moliner, M. J. Lazaro, *Int. J. Hydrogen Energy* **2011**, *36*, 9805.
[149] B. Liu, S. Creager, *Electrochim. Acta* **2010**, *55*, 2721.
[150] F. Hasche, T. P. Fellinger, M. Oezaslan, J. P. Paraknowitsch, M. Antonietti, P. Strasser, *ChemCatChem* **2012**, *4*, 479.
[151] S. Liu, J. Wang, J. Zeng, J. Ou, Z. Li, X. Liu, S. Yang, *J. Power Sources* **2010**, *195*, 4628.

[152] S. Guo, S. Dong, E. Wang, *ACS Nano* **2009**, *4*, 547.
[153] R. I. Jafri, N. Rajalakshmi, S. Ramaprabhu, *J. Mater. Chem.* **2010**, *20*, 7114.
[154] X. Yu, S. Ye, *J. Power Sources* **2007**, *172*, 133.
[155] M. Kim, J. N. Park, H. Kim, S. Song, W. H. Lee, *J. Power Sources* **2006**, *163*, 93.
[156] B. Viswanathan, *Catal. Today* **2009**, *141*, 52.
[157] E. Hegenberger, N. Wu, J. Phillips, *J. Phys. Chem.* **1987**, *91*, 5067.
[158] L. Calvillo, V. Celorrio, R. Moliner, M. Lazaro, *Mater. Chem. Phys.* **2011**, *127*, 335.
[159] L. Calvillo, M. Gangeri, S. Perathoner, G. Centi, R. Moliner, M. J. Lazaro, *J. Power Sources* **2009**, *192*, 144.
[160] Y. Shao, J. Sui, G. Yin, Y. Gao, *Applied Catalysis B: Environmental* **2008**, *79*, 89.
[161] K. Prehn, A. Warburg, T. Schilling, M. Bron, K. Schulte, *Composites Science and Technology* **2009**, *69*, 1570.
[162] H. A. Gasteiger, S. S. Kocha, B. Sompalli, F. T. Wagner, *Appl. Catal. B Environ.* **2005**, *56*, 9.
[163] S. Litster, G. McLean, *J. Power Sources* **2004**, *130*, 61.
[164] C. Wang, M. Waje, X. Wang, J. M. Tang, R. C. Haddon, Y. Yan, *Nano Lett.* **2004**, *4*, 345.
[165] R. Bashyam, P. Zelenay, *Nature* **2006**, *443*, 63.
[166] R. A. Sidik, A. B. Anderson, N. P. Subramanian, S. P. Kumaraguru, B. N. Popov, *J. Phys. Chem. B* **2006**, *110*, 1787.
[167] P. H. Matter, L. Zhang, U. S. Ozkan, *J. Catal.* **2006**, *239*, 83.
[168] M. Glerup, J. Steinmetz, D. Samaille, O. Stephan, S. Enouz, A. Loiseau, S. Roth, P. Bernier, *Chem. Phys. Lett.* **2004**, *387*, 193.
[169] R. Droppa, P. Hammer, A. C. M. Carvalho, M. C. dos Sanots, F. Alvarez, *J. Non-Cryst. Solids* **2002**, *299*, 874.
[170] J. C. Hummelen, B. Knight, J. Pavlovich, R. Gonzalez, F. Wudl, *Science* **1995**, *269*, 1554.
[171] M. Yudasaka, R. Kikuchi, Y. Ohki, S. Yoshimura, *Carbon* **1997**, *35*, 195.
[172] M. Terrones, R. Kamalakaran, T. Seeger, M. Ruhle, *Chem. Comm.* **2000**, 2335.
[173] E. G. Wang, *Adv. Mater.* **1999**, *11*, 1129.
[174] C. C. Tang, Y. Bando, D. Golberg, F. F. Xu, *Carbon* **2004**, *42*, 2625.
[175] C. B. Cao, F. L. Huang, C. T. Cao, J. Li, H. Zhu, *Chem. Mater.* **2004**, *16*, 5213.
[176] L. Q. Jiang, L. Gao, *Carbon* **2003**, *41*, 2923.
[177] S. C. Roy, A. W. Harding, A. E. Russell, K. M. Thomas, *J. Electrochem. Soc.* **1997**, *144*, 2323.
[178] F. Jaouen, M. Lefevre, J. P. Dodelet, M. Cai, *J. Phys. Chem. B* **2006**, *110*, 5553.
[179] S. Maldonado, K. J. Stevenson, *J. Phys. Chem. B* **2005**, *109*, 4707.
[180] H. Sjostrom, S. Stafstrom, M. Boman, J. E. Sundgren, *Phys. Rev. Lett.* **1995**, *75*, 1336.
[181] R. Venkatachalapathy, G. P. Davila, J. Prakash, *Electrochem. Commun.* **1999**, *1*, 614.
[182] S. Biniak, G. Szymanski, J. Siedlewski, A. Swiatkowski, *Carbon* **1997**, *35*, 1799.
[183] G. S. Szymanski, T. Grzybek, H. Papp, *Catal. Today* **2004**, *90*, 51.
[184] S. Matzner, H. P. Boehm, *Carbon* **1998**, *36*, 1697.
[185] M. C. Huang, H. S. Teng, *Carbon* **2003**, *41*, 951.
[186] J. Ozaki, T. Anahara, N. Kimura, A. Oya, *Carbon* **2006**, *44*, 3358.
[187] Y. Okamoto, *Appl. Surf. Sci.* **2009**, *256*, 335.
[188] D. Geng, H. Liu, Y. Chen, R. Li, X. Sun, S. Ye, S. Knights, *J. Power Sources* **2011**, *196*, 1795.
[189] C. Leon, J. M. Solar, V. Calemma, L. R. Radovic, *Carbon* **1992**, *30*, 797.
[190] C. Medard, M. Lefevre, J. P. Dodelet, F. Jaouen, G. Lindbergh, *Electrochim. Acta* **2006**, *51*, 3202.
[191] R. Cote, G. Lalande, D. Guay, J. P. Dodelet, G. Denes, *J. Electrochem. Soc.* **1998**, *145*, 2411.
[192] F. Jaouen, F. Charreteur, J. P. Dodelet, *J. Electrochem. Soc.* **2006**, *153*, A689.
[193] G. Faubert, R. Cote, J. P. Dodelet, M. Lefevre, P. Bertrand, *Electrochim. Acta* **1999**, *44*, 2589.
[194] M. Bron, S. Fiechter, M. Hilgendorff, P. Bogdanoff, *J. Appl. Electrochem.* **2002**, *32*, 211.

[195] U.I. Kramm et al, *Phys. Chem. Chem. Phys.* **2012**, *14(33)*, 11673.
[196] J. Herranz et al, *J. Phys. Chem. C* **2011**, *115(32)*, 16087
[197] K. Sawai, N. Suzuki, *J. Electrochem. Soc.* **2004**, *151*, A682.
[198] G. Lalande, G. Faubert, R. Cote, D. Guay, J. P. Dodelet, L. T. Weng, P. Bertrand, *J. Power Sources* **1996**, *61*, 227.
[199] G. Lalande, D. Guay, J. P. Dodelet, S. A. Majetich, M. E. McHenry, *Chem. Mater.* **1997**, *9*, 784.
[200] A. Biloul, O. Contamin, G. Scarbeck, M. Savy, D. Vandenham, J. Riga, J. J. Verbist, *J. Electroanal. Chem.* **1992**, *335*, 163.
[201] A. L. Bouwkamp-Wijnoltz, W. Visscher, J. A. R. van Veen, E. Boellaard, A. M. van der Kraan, S. C. Tang, *J. Phys. Chem. B* **2002**, *106*, 12993.
[202] S. L. Gojkovic, S. Gupta, R. F. Savinell, *J. Electroanal. Chem.* **1999**, *462*, 63.
[203] S. L. Gojkovic, S. Gupta, R. F. Savinell, *J. Electrochem. Soc.* **1998**, *145*, 3493.
[204] P. Gouerec, M. Savy, J. Riga, *Electrochim. Acta* **1998**, *43*, 743.
[205] P. Gouerec, A. Bilou, O. Contamin, G. Scarbeck, M. Savy, J. M. Barbe, R. Guilard, *J. Electroanal. Chem.* **1995**, *398*, 67.
[206] P. Gouerec, A. Biloul, O. Contamin, G. Scarbeck, M. Savy, J. Riga, L. T. Weng, P. Bertrand, *J. Electroanal. Chem.* **1997**, *422*, 61.
[207] G. Faubert, R. Cote, D. Guay, J. P. Dodelet, G. Denes, P. Bertrand, *Electrochim. Acta* **1998**, *43*, 341.
[208] G. Faubert, R. Cote, D. Guay, J. P. Dodelet, G. Denes, C. Poleunis, P. Bertrand, *Electrochim. Acta* **1998**, *43*, 1969.
[209] A. Serov et al, *Electrochim. Acta* **2013**, *109*, 433.
[210] V. Di Noto et al, *Appl. Catal. B: Env.* **2012**, *111–112*, 185.
[211] F. Charreteur, F. Jaouen, S. Ruggeri, J. P. Dodelet, *Electrochim. Acta* **2008**, *53*, 2925.
[212] M. Lefevre, J. P. Dodelet, P. Bertrand, *J. Phys. Chem. B* **2000**, *104*, 11238.
[213] M. Lefevre, J. P. Dodelet, P. Bertrand, *J. Phys. Chem. B* **2002**, *106*, 8705.
[214] F. Jaouen, S. Marcotte, J. P. Dodelet, G. Lindbergh, *J. Phys. Chem. B* **2003**, *107*, 1376.
[215] H. T. Chung, C. M. Johnston, K. Artyushkova, M. Ferrandon, D. J. Myers, P. Zelenay, *Electrochem. Comm.* **2010**, *12*, 1792.
[216] E. B. Easton, A. Bonakdarpour, J. R. Dahn, *Electrochem. Solid State Lett.* **2006**, *9*, A463.
[217] E. B. Easton, R. Z. Yang, A. Bonakdarpour, J. R. Dahn, *Electrochem. Solid State Lett.* **2007**, *10*, B6.
[218] M. Lefevre, J. P. Dodelet, *Electrochim. Acta* **2003**, *48*, 2749.
[219] R. Othman, A. L. Dicks, Z. Zhu, *Int. J. Hydrogen En.* **2012**, *37*, 357.
[220] W. Martinez Millan, T. Toledano Thompson, L. G. Arriaga, M. A. Smit, *Int. J. Hydrogen En.* **2009**, *34*, 694.
[221] W. Martınez Millan, M. A. Smit, *J. App. Pol. Sc.* **2009**, *112*, 2959.
[222] K. Lee, L. Zhang, H. Lui, R. Hui, Z. Shi, J. Zhang, *Electrochim. Acta* **2009**, *54*, 4704.
[223] A. Serov et al, *Appl. Catal. B: Env.* **2012**, *127*, 300.
[224] D. Nguyen-Thanh, A. I. Frenkel, J. Wang, S. O'Brien, D. L. Akins, *Applied Catalysis B: Environmental* **2011**, *105*, 50.
[225] G. Liu, X. Li, P. Ganesan, B. N. Popov, *Applied Catalysis B: Environmental* **2009**, *93*, 156.

Benjamin Frank

15 Nanocarbon materials for heterogeneous catalysis

15.1 Introduction

Heterogeneous catalysis is of paramount importance in many areas of the chemical and energy industries as more than 90 % [1] of the chemical manufacturing processes in use throughout the world utilize catalysis and primarily heterogeneous catalysis. Catalytically active metal (oxide) nanoparticles are often dispersed on a mechanically stable high surface area support to maximize and stabilize their specific surface area as well as to fine-tune their properties and minimize the materials usage, *e.g.*, of expensive noble metals. Among the different types of supports used in heterogeneous catalysis the carbon materials attract a permanent interest due to their specific characteristics, which are mainly: (i) resistance to acidic/basic media and thermal sintering, (ii) control over macroscopic shape, porosity, and surface chemistry, and (iii) disposal or recovery of precious metals by support burning, which results in a low environmental impact. The use of, *e.g.*, graphite, carbon black, activated carbon, activated carbon fibers, glassy carbon, pyrolytic carbon, or polymer-derived carbon has a long history in this field of research [2–4]. Carbon materials providing a distinct structure on the nanometer scale, hereinafter referred to as nanocarbons, are equipped with further exceptional physical and chemical properties. Thus, the discovery of low-dimensional nanostructured carbon allotropes, such as carbon nanotubes (CNTs), nanofibers (CNFs), graphene, or fullerenes, ushered in a new era of cutting-edge applications for elemental carbon with a wide-ranging impact also on heterogeneous catalysis. The development of large-scale synthesis processes for multi-walled CNTs [5, 6] stimulated a vital research activity in academia and industry. It is recognized that carbon as a catalyst support as well as a catalyst on its own offers unparalleled flexibility in tailoring catalyst properties to specific needs. Carbon-supported noble metals for fine chemicals production are well established, but only a few large-volume processes currently use these systems.

Besides the practical application, the diversity of nanostructured carbon allotropes makes nanocarbon also an ideal model system for the investigation of structure-function correlations in heterogeneous catalysis. Nanocarbons can be tailored in terms of their hybridization state, curvature, and aspect ratio, *i.e.*, dimensions of stacks of basic structural units (BSU), Chapters 1 and 2. The preferred exposition of two types of surfaces, which strongly differ in their physico-chemical behavior, *i.e.*, the basal plane and prismatic edges, can be controlled. Such controlled diversity is seldom found for other materials giving carbon a unique role in this field of basic research. The focus of this chapter is set on the most prominent representatives of the nanocarbon family, CNTs and CNFs, which are equipped with the highest potential for industrial application.

15.2 Relevant properties of nanocarbons

15.2.1 Textural properties and macroscopic shaping

The textural properties of nanocarbons are strongly related to their structure. Several schemes for the classification of carbonaceous materials have been proposed [7, 8], however, for nanocarbons the multiscale diversity renders a phase diagram-like simplification rather impossible. In general, the BSU of intrinsically anisotropic sp^2 carbons is a section of graphene with edge defects and a curvature induced by non-six-membered carbon rings. Thus, for a systematic classification the following factors have to be considered: (i) hybridization state of the carbon atoms, (ii) stacking, (iii) curvature, and (iv) exposed surface [9]. A systematic classification based on these characteristics is given in Fig. 15.1. Clearly, the lower-symmetry "exotic" nanocarbons, such as bamboo-like CNTs or nanohorns, are difficult to classify in such a simplified scheme as is the case for heterogeneous systems like vapor-grown CNFs or diamond dots encapsulated in nanoonions.

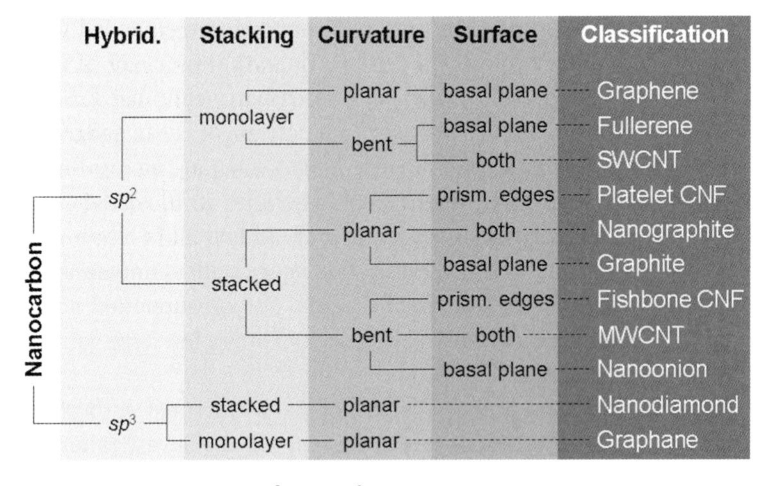

Fig. 15.1: Classification of sp^2- and sp^3-hybridized nanostructured carbon materials. CNTs are considered as open-end tubes, thus also exposing prismatic edge surfaces.

Graphene is a single carbon layer of the graphite structure, describing its nature by analogy to a polycyclic aromatic hydrocarbon of quasi-infinite size [10]. Isolated "graphene sheets" provide an exceptionally high specific surface area of ~ 2675 $m^2 g^{-1}$, however, are very sensitive to re-graphitization *via* agglomeration, scrolling, or wrapping, likely induced under "non-ideal" conditions [11]. Thus, it must be mentioned that most reports in catalytic science nominally dealing with graphene in fact handle a "multilayer-graphene" like material comprising several graphene layers. This not only affects the textural properties such as stiffness and structural stability but surely also its catalytic properties. The fluffy appearance of graphene-based materials such as graphene oxide hampers their application in fixed-bed catalytic processes due to the difficulty to immobilize or formulate the material without losing a substantial part of its valuable properties. Methods for the production of supported atomically free-standing graphene sheets [12] still need to be developed for application in heterogeneous catalysis. Instead, unsupported graphene and graphene oxide materials can be used in dispersed form in liquid phase batch processes, where they are recovered by filtration or centrifugation after the catalytic reaction.

For application in heterogeneous catalysis the fibrous quasi-one-dimensional nanocarbons such as CNTs and CNFs are of particular interest owing to their open pore structure between entangled fibers allowing the facile diffusion of reactant molecules to the catalytically active sites (Fig. 15.2). Especially in the case of SWCNTs and thin MWCNTs, the formation of a secondary stacked structure can result in the presence of an appreciable fraction of mesopores in the material. The absence of micropores in closed CNTs and certain types of CNFs helps to avoid diffusional intraparticle mass transfer limitations having a negative effect on both activity and product selectivity. Oxidative treatment of this material can open the tube caps to make the inner cavity of CNTs, which potentially is in the micropore range of < 2 nm, accessible to the reactants. Here, micropores are also generated by etching and defect generation in the outer tubular surface. However, microporosity is less relevant for the nanocarbon properties as is the case for traditional high surface area carbons such as activated carbon. If not densified by compression, the large content of macropores and pore volumes up to 3 cm^3 g^{-1} allow for good access of the reactants to the active sites of the catalyst. The typical range of specific surface areas of CNTs and CNFs is 20–

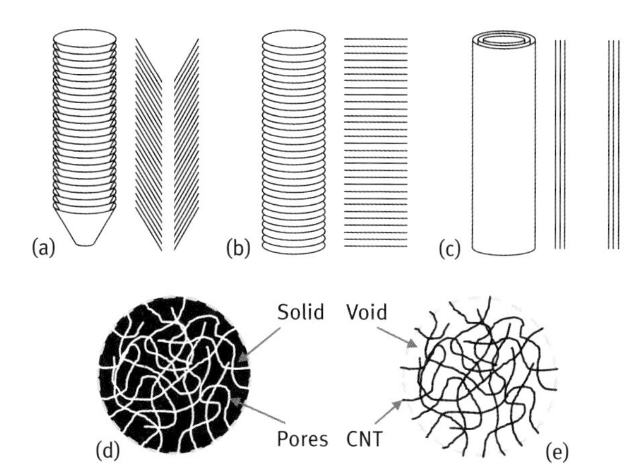

Fig. 15.2: Simplified structures of common quasi-1D carbon nanostructures in perspective and cross sectional view. (a) fishbone CNF, (b) platelet CNF, (c) multi-walled CNT. Schematic representation of pore structures in (d) activated carbon particle and (e) CNT agglomerate.

300 m^2 g^{-1}, whereas the outer diameter of the commercially available material ranges from 10 to 110 nm (SWCNTs: 0.7–2 nm) [5]. These parameters are correlated with each other, thus cannot be varied independently. Owing to the synthesis process, CNTs and CNFs typically form 10–500 μm sized agglomerates, which can be defibrated in the post-treatment processes, *e.g.*, by milling or ultrasonic treatments. Another feature of CNTs is their high thermal conductivity of up to 3000 W m^{-1} K^{-1} (in plane direction), which enables them to damp local hot spots during calcination and/or activation treatments as well as during exothermic reactions, thereby stabilizing small metal (oxide) clusters decorating the sidewalls of the tubules.

Ill-defined carbon materials that provide a distinct nanostructure, such as spherical particles in the case of soot and carbon black, or hexagonally ordered cylindrical pores in the case of ordered mesoporous carbons, are not discussed here. Surface chemical, thus catalytic properties of these material are closer to carbon black or activated carbon [13], which is frequently reviewed [2–4]. Here, the higher degree of sp^3 hybridization often results in a higher reactivity, however, at lower selectivity, as compared to nanocarbons exposing large basal plane fractions of the overall surface.

For application in flow reactors the nanocarbons need to be immobilized to ensure ideal flow conditions and to prevent material discharge. Similar to activated carbon, the material can be pelletized or extruded into millimeter-sized mechanically stable and abrasion-resistant particles. Such a material based on CNTs or CNFs is already commercially available [17]. Adversely, besides a substantial loss of macroporosity, the use of an (organic) binder is often required. This material inevitably leaves an amorphous carbon overlayer on the outer nanocarbon surface after calcination, which can block the intended nanocarbon surface properties from being fully exploited. Here, the more elegant strategy is the growth of nanocarbon structures on a mechanically stable porous support such as carbon felt [15] or directly within the channels of a microreactor [14, 18] (Fig. 15.3(a),(b)), which could find application in the continuous production of fine chemicals. Pre-shaped bodies and surfaces can be

Fig. 15.3: (a) Microreactor platelet coating with epitaxially grown CNFs. Reprinted from Ref. [14], Copyright (2010), with permission from Elsevier. (b,c) SEM images of immobilized CNTs (insets: optical photographs of macroscopic bodies). (b) Nanocarbon growth on carbon felt. Reprinted from Ref. [15], Copyright (2006), with permission from Elsevier; (c) pervasive transformation of copolymer resin beads [16] (Courtesy of J. Zhang, Chinese Academy of Sciences, Ningbo).

impregnated with a solution that contains the catalyst precursor. The composite is then subjected to typical chemical vapor deposition (CVD) conditions for CNT growth. Also macroporous natural lava without post-treatment has been successfully applied for this purpose [19], opening a sustainable way to nanocarbon-based catalyst production. Another innovative method is the pervasive carbonization and graphitization of iron-impregnated copolymer-resin beads to form millimeter-sized spherical bodies of tightly entangled CNTs [16] (Fig. 15.3(c)). The nanocarbon growth in macroscopic confined space can further lead to self-supporting monoliths with very low bulk densities as low as 190 kg m^{-3} [20], however, with poor mechanical stability. Thin and/or membrane-like arrays of vertically aligned CNTs are some of the latest developments with potential for heterogeneous catalysis [18, 21]. Large domains of aligned CNTs can also be obtained by self-assembly, leading to mechanically highly stable cm-sized pellets with a bulk density of more than 620 kg m^{-3} [22]. Entangled CNF bodies are reported to display a bulk crushing strength of 1 MPa [23], which makes them suitable for use as catalyst supports in fixed-bed reactors applications.

15.2.2 Surface chemistry and functionalization

The surface chemistry of sp^2 nanocarbons is dominated by their unique intrinsic anisotropy. The basal plane and the prismatic edges combine metallic and oxidic properties within one crystallite, thus defining the overall properties of the respective nanocarbon. It is impressively illustrated in the angle-dependent X-ray absorption spectroscopy of single-crystal graphite (Fig. 15.4 (c)) showing the $1s \rightarrow \sigma^*$ excitation at 291.1 eV in highest intensity at incidence along the *c*-direction, whereas the $1s \rightarrow \pi^*$ excitation at 285.5 eV arises with increasing incidence angles. The (0001) basal plane is chemically inert; however, electron-rich and thus serves as potential adsorption site in heterogeneous catalysis applications. Its electron density can be tuned and/or localized by heteroatom doping, *e.g.*, B or N, or by curvature inducing a certain pyramidalization angle > 90° between σ- and π-orbitals to generate electron-rich sites for reactant activation or for anchoring points for metal (oxide) clusters or organic functional groups. Thus it is, from a general point of view, possible to finely adjust the nucleo- or electrophilicity of edge heteroatoms to enter regimes of catalytic activity.

This approach, however, requires the absence of ill-defined carbon deposits originating from defect-induced soot formation on the surface of nanocarbons during their synthesis. Pyrolytic structures often counteract the control over activity and selectivity in catalytic applications of well-defined nanocarbons by offering an abundance of highly reactive sites, however, in maximum structural diversity. Although some nanocarbons are equipped with a superior oxidation stability over disordered carbons [25], such amorphous structures can further induce the combustion of the well-ordered sp^2 domains by creating local hotspots. Thermal or mild oxidative treatment,

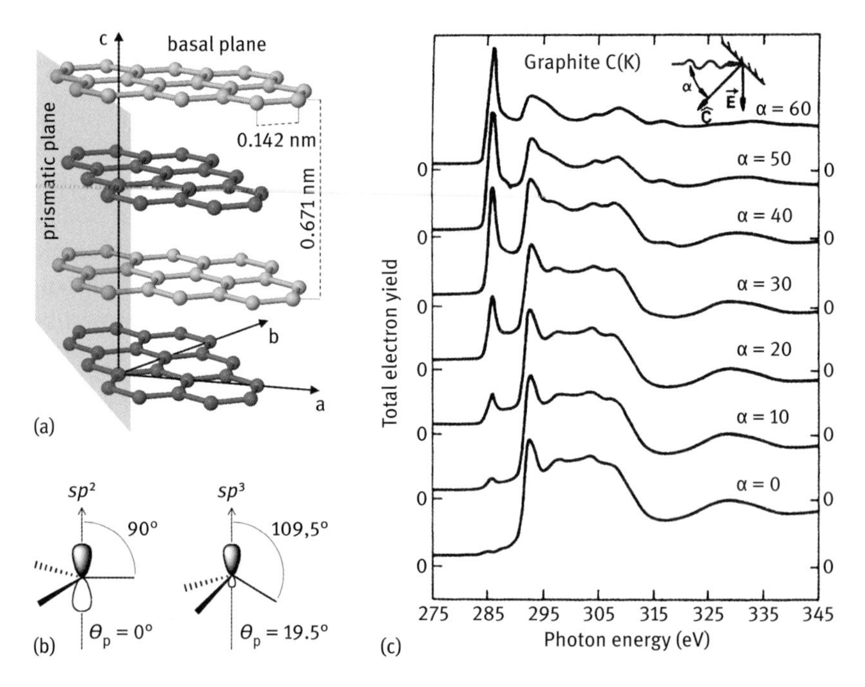

Fig. 15.4: (a) Structure of graphite with crystal translation vectors and (b) Illustration of the pyramidalization angle (c) C(K)-edge photoabsorption spectra of single-crystal graphite at various angles of incidence α, between the surface normal (c-axis) and the Poynting vector of the light [24]. Copyright (1986) by the American Physical Society.

as described below, can selectively remove or graphitize such debris. Also ultrasonic treatment was proven to be a facile and non-destructive method to remove metals and pyrolytic carbon from the outer surface of commercial CNT material [26].

Contrarily to the basal plane, the prismatic edges terminating the graphene layers as well as defects in the basal plane are highly reactive and usually saturated, with H, O, or N atoms being the key players in most of the catalytic applications of carbon materials (see Chapter 19). The abundance of groups illustrated in Fig. 15.5 is well known from organic chemistry. These functionalities define the acidity/basicity and also the hydrophilic character of the nanocarbons.

A broad variety of oxygen groups spontaneously forms in contact with air or can be introduced to the initially C-H saturated edge sites, *e.g.*, by aggressive treatment in concentrated nitric acid. This procedure is commonly applied to remove impurities such as amorphous carbon deposits or metal (oxide) residues from the well-ordered nanocarbon surface [5]. Also O_2 (at elevated temperatures), O_3, H_2O_2, HNO_3 vapor, or HNO_3/H_2SO_4 mixtures are successfully applied for this purpose. These oxidizing agents have in common that they leave a chemically poorly-defined surface in terms of a type-specific functionalization. In general, oxidative treatments result in a predominantly acidic surface with carboxylic acids, anhydrides, phenols and lactones as

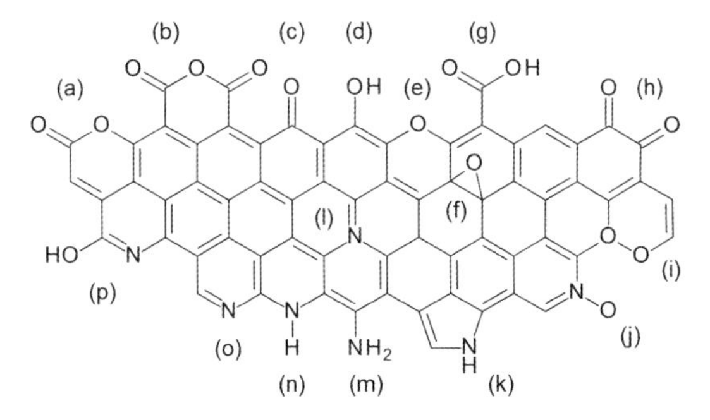

Fig. 15.5: Oxygen and nitrogen surface groups on carbon. (a) lactone, (b) carboxylic anhydride, (c) ketone, (d) phenol, (e) ether, (f) epoxide, (g) carboxylic acid, (h) quinone, (i) peroxide, (j) pyridine N-oxide, (k) pyrrole, (l) quaternary N, (m) primary amine, (n) secondary amine, (o) pyridine, (p) pyridone.

the predominant species. These can be easily removed by thermal treatment or reduction to disclose the basic properties of thermally stable ketone and quinone groups (Fig. 15.6). Nitric acid generates a high fraction of carboxylic groups [27], whereas hydrogen peroxide produces predominantly phenols [28]. In heterogeneous catalysis these two species are important as anchoring point for metal ions during ion-exchange preparation and as redox-active center for oxidation catalysis, respectively. In general, the higher the treatment temperature and the stronger the oxidizing agent, the more oxygen groups are attached to the carbon surface. This, however, is accompanied by nanocarbon decomposition under such harsh reaction conditions. A highly-specific functionalization can only be achieved by multi-step organochemical syntheses.

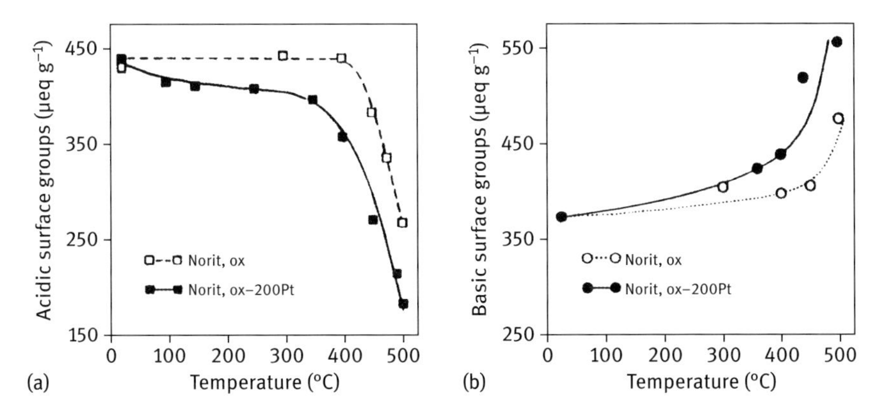

Fig. 15.6: Reduction of acidic surface sites on treatment with H_2 (left) and concurrent increase in basic surface sites (right). Open symbols: Activated carbon Norit, oxidized with O_2,: Filled symbols: Norit loaded with 200 μmol g^{-1} Pt. Reprinted from Ref. [29], Copyright (1994), with permission from Elsevier.

Nitrogen atoms also preferentially occupy edge sites of the BSU, however, they can also be incorporated in quaternary position into the graphene sheet (Fig. 15.5) and thus into the bulk of the nanocarbons. Bulk and surface N-doping succeeds by use of N-containing precursor molecules for nanocarbon growth [30]. The resulting materials are characterized by the pronounced basicity originating from amine, pyridine, or pyrrole groups. A post-synthetic near-surface N-doping can be achieved by thermal treatment of surface-oxidized nanocarbons, *e.g.*, in ammonia, leading to O/N substitution at defective carbon sites. Also plasma etching in nitrogen atmosphere has been successfully applied [31]. However, similar to the oxygen groups, a distinct type of nitrogen-containing groups can hardly be generated *via* these methods. Very often, a heterogeneous composition of O- and N-containing groups is formed during the nanocarbon synthesis/modification, which finally governs its overall acidic/basic nature.

Suitable characterization techniques for surface functional groups are temperature-programmed desorption (TPD), acid/base titration [29], infrared spectroscopy, or X-ray photoemission spectroscopy, whereas structural properties are typically monitored by nitrogen physisorption, electron microscopy, or Raman spectroscopy. The application of these methods in the field of nanocarbon research is reviewed elsewhere [5, 32].

15.2.3 Confinement effect

The inner cavity of carbon nanotubes stimulated some research on utilization of the so-called *confinement effect* [33]. It was observed that catalyst particles selectively deposited inside or outside of the CNT host (Fig. 15.7) in some cases provide different catalytic properties. Explanations range from an electronic origin due to the partial sp^3 character of basal plane carbon atoms, which results in a higher π-electron density on the outer than on the inner CNT surface (Fig. 15.4(b)) [34], to an increased pressure of the reactants in nanosized pores [35]. Exemplarily for inside CNT deposited catalyst particles, Bao *et al.* observed a superior performance of Rh/Mn/Li/Fe nanoparticles in the ethanol production from syngas [36], whereas the opposite trend was found for an Ru catalyst in ammonia decomposition [37]. Considering the substantial volume shrinkage and expansion, respectively, in these two reactions, such results may indeed indicate an increased pressure as the key factor for catalytic performance. However, the activity of a Ru catalyst deposited on the outside wall of CNTs is also more active in the synthesis of ammonia, which in this case is explained by electronic properties [34].

It is suggested that the confined space may increase the frequency of adsorption of reactant molecules with the active sites and thus may increase the catalytic turnover of particles located inside the CNTs as compared to those outside. However, the CNT wall will inevitably block a certain amount of active surface of the metal nanoparti-

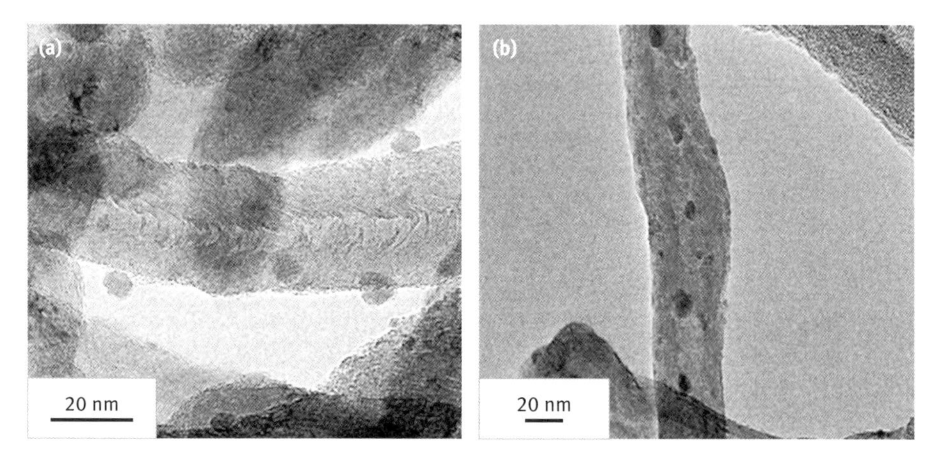

Fig. 15.7: MnO_x nanoparticles located (a) outside and (b) inside of the CNT support. Reprinted from Ref. [38], Copyright (2012), with permission from Elsevier.

cles inside a CNT. Owing to the huge aspect ratio of CNTs, the in-and-out transfer of reactant and product molecules can be significantly lowered by a small opening in the CNT tip. On the other hand this confined space hampers agglomeration of the metal clusters, which easily occurs on the outer surface of CNTs at high temperatures [39]. Thus, the concept and origin of confinement effect are still controversially discussed. The preparation of catalyst nanoparticles being selectively deposited on the inner or outer wall of the CNTs is described in Section 15.4.1.1.

15.3 Nanocarbon-based catalysts

The activity of elemental carbon as a metal-free catalyst is well established for a couple of reactions, however, most literature still deals with the support properties of this material. The discovery of nanostructured carbons in most cases led to an increased performance for the abovementioned reasons, thus these systems attracted remarkable research interest within the last years. The most prominent reaction is the oxidative dehydrogenation (ODH) of ethylbenzene and other hydrocarbons in the gas phase, which will be introduced in a separate chapter. The conversion of alcohols as well as the catalytic properties of graphene oxide for liquid phase selective oxidations will also be discussed in more detail. The third section reviews individually reported catalytic effects of nanocarbons in organic reactions, as well as selected inorganic reactions.

15.3.1 Dehydrogenation of Hydrocarbons

The ODH of ethylbenzene to styrene is a highly promising alternative to the industrial process of non-oxidative dehydrogenation (DH). The main advantages are lower reaction temperatures of only 300–500 °C and the absence of a thermodynamic equilibrium. Coke formation is effectively reduced by working in an oxidative atmosphere, thus the presence of excess steam, which is the most expensive factor in industrial styrene synthesis, can be avoided. However, this process is still not commercialized so far due to insufficient styrene yields on the cost of unwanted hydrocarbon combustion to CO and CO_2, as well as the formation of styrene oxide, which is difficult to remove from the raw product.

It was shown as early as 1969 that organic polymers containing quinone functionalities successfully catalyze gas phase ODH-type reactions in alternating feed mode [40]. Other redox-active polymers such as (carbonized) polynaphthoquinone [41] and trimerized phenathrenequinone [42] were shown to catalyze the ODH of ethylbenzene with high turnover rates pointing at the diketonic carbonyl group to be the central part of the active site in these reactions (Fig. 15.8(a)–(c)). Furthermore, electron conductivity and charge stabilization provided by a sufficiently large aromatic moiety in the polymer backbone are required to facilitate the electron transfer process at the active quinone site. The reduced counterpart of quinone active sites is the hydroquinone formed by hydrogen atom abstraction from the organic substrate molecule. Quinone regeneration with gas phase oxygen has been followed by infrared spectroscopy showing the recovery of the carbonyl band located at 1660 cm^{-1} [40]. Furthermore, the study of the ODH performance using 4-methyl-4-ethyl-1-cyclohexene as the substrate, which forms different reaction products depending on the acid/base properties of the catalyst, indicates that quinone-catalyzed ODH involves a homolytic C-H bond break rather than an ionic intermediate [40]. For ethylbenzene as the substrate, a transition state with the phenyl group in co-planar orientation with respect to the π-conjugated

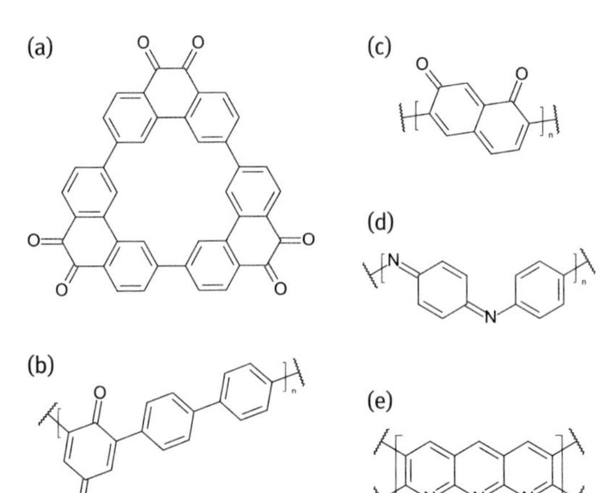

Fig. 15.8: Redox-active polymers for heterogeneous dehydrogenation catalysis. (a) trimerized phenanthrene quinone [42], (b) benzoquinone biphenyl copolymer [41], (c) polynaphthoquinone [40], (d) polyaniline [43], (e) pyrolyzed polyacrylonitrile [44].

system of the catalyst is suggested, which is then stabilized by weak π-π/van der Waals interaction [41]. The full reaction mechanism is still unclear; however, this geometry could also induce a concerted abstraction of α- and β-hydrogen atoms in ethylbenzene (see also Fig. 15.12(b)).

The use of pyrolyzed polyacrylonitrile (PPAN) and polyaniline (PAni) (Fig. 15.8(d), (e)) as catalysts for the ODH of ethylbenzene should only be mentioned here for the sake of completeness. Although first results were quite promising [45], this concept has so far not been followed in terms of N-doped nanocarbon catalyst development. This is most likely due to the poor self-oxidation resistance as a result of polar C-N bonds.

In any case the operation temperature of such model-type polymer catalysts is quite low due to uncontrollable decomposition/carbonization of the fragile polymer structure. Its large scale application is permitted by high costs of these materials. Alkhazov *et al.* were the first to report the application of cheap activated carbons in the ODH of ethylbenzene [46] and indicated reaction temperatures to be drastically lower than those typically required over metal oxide catalysts. Although their catalysts provided a significant ash content of ~ 10 %, later studies with pure carbon materials [47–49] supported the thesis of elemental carbon being the active phase for this reaction. It is interesting to note that ongoing studies on traditionally used acidic catalysts such as zeolites, alumina, or metal phosphates in their active state typically pointed at a thin coke layer on the metal surface [50], indicating that even on these systems the active phase comprises elemental carbon rather than redox-active or acidic transition metal sites.

The susceptibility of styrene to polymerization requires (i) a low ethylbenzene concentration in the feed gas and (ii) an abundance of meso- and macropores in the carbon catalyst. Besides poor oxidation stability the latter is the reason for substantial deactivation of traditional microporous activated carbon catalysts under reaction conditions (Fig. 15.9). Small pores are rapidly blocked by coke deposition making large parts of the original catalyst surface inaccessible for the reactant molecules, whereas activated carbons with wider pores perform more stably [51]. The considerable impact of textural properties is most evidently shown by the nonlinear relationship between the specific surface area of the catalyst and its activity in the ODH of ethylbenzene. When approaching a certain level of surface area, which is associated with a mean (micro)pore diameter of 1.2 nm, the ethylbenzene turnover remains constant highlighting the importance of the transport properties provided by an appropriate pore structure [52]. Only at larger pores is the overall catalyst performance controlled by the surface chemistry. Similar to the model polymers (Fig. 15.8(a)–(c)) the activity depends on the surface concentration of quinone groups. Pereira *et al.* reported a linear correlation with a positive offset, which they explained by the fact that a certain fraction of active sites is generated *in situ* on the carbon surface by oxygen present in the reacting mixture [47].

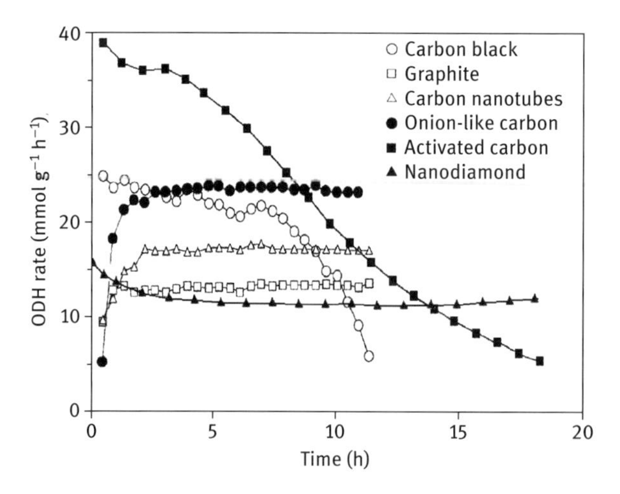

Fig. 15.9: Stability of various (nano)carbons in the ODH of ethylbenzene (EB) [53, 54]. ○, □, △: 20 mg, 10 ml min^{-1}, 2.16 % EB, EB/O_2 = 1 : 1, 823 K; ●: as before, 790 K; ■, ▲: 5 mg, 12.5 ml min^{-1}, 2.8 % EB, EB/O_2 = 2 : 1, 723 K. Reprinted from Ref. [55], Copyright (2009), with permission from DGMK.

Some long-term catalytic studies of differently structured (nano)carbon catalysts are compared in Fig. 15.9. Apparently, the superior stability is provided by a certain long-range ordering of carbon atoms in the nanocarbon structure, whereas amorphous carbons such as activated carbon or carbon black rapidly deactivate within hours by combustion and/or micropore blocking. Instead, well-ordered graphite, CNTs, onion-like carbon, and nanodiamond show a stable catalytic performance after a short initial equilibration period. It is interesting to note that nanocarbons, if referring to their specific surface area, share similar kinetic parameters for the ODH of ethylbenzene, which points at a similar active surface comprising active oxygen atoms embedded in surface defects. For differently sized CNTs and nanodiamond, Zhang *et al.* determined apparent activation energies in the range of 68–75 kJ mol^{-1}, areal conversion rates of 46–49 μmol m^{-2} h^{-1}, and reaction rate orders of ethylbenzene and oxygen of 0.51–0.61 and 0.30–0.34, respectively [53]. Consistently, other studies comparing the performances of (nano)carbon materials typically end up with differences in styrene productivity of factor of two maximum [56, 57], which is likely explained by carbon anisotropy, different degrees of surface functionalization or micropore plugging. The quantification of *ex situ* X-ray photoelectron spectra reveals roughly 4–7 % surface oxygen after reaction [53]. TEM and electron-energy loss spectroscopy (EELS) analyses show an active few layer sp^2 coating even on the bulk sp^3 hybridized nanodiamond catalyst. Thus, depending on the pristine state of the catalyst surface, a pronounced initial activation suggests a surface reconstruction process to form stable active sites and a deactivation in the initial stage of catalysis might arise from removal of unstable active groups [53]. Such an active carbon layer should not be confused with disordered coke, that is frequently observed from the very early studies [46], depending on the reaction conditions. A specific analysis of catalytic properties of surface defects and coke deposits shows that the latter are merely inactive and/or non-selective catalysts [58]. Contrarily, the coke layer, which forms *in situ* on metal-based catalysts, behaves different due to the oxidic support providing a growth directing matrix. Here, the formation of catalytically active defective sp^2 carbon with a higher degree of structural order furthermore results from the higher reaction temperatures applied for such systems.

Nanocarbons as selective and more oxidation-resistant alternatives to amorphous carbons give access to the ODH of light alkanes. Here the C-H bond to be activated is typically as stable as 410–420 kJ mol^{-1}, which requires reaction temperatures beyond 673 K (Fig. 15.10). A kinetic investigation revealed that even CNTs cannot provide sufficient long-term stability under such conditions [25], thus the carbon catalysts need protection by B_2O_3 or P_2O_5 [59]. This modification not only reduces catalyst combustion (Fig. 15.11(a),(b)) but also increases the alkene selectivity, as evidenced for the ODH of ethane [60], propane [61, 62] (Fig. 15.11(c)), *n*-butane [63], or isobutane [64]. From the economic point of view the activation of these molecules is even more interesting as lower alkenes are at present produced by expensive steam cracking of naphtha. Instead, short-chain alkanes are a cheap alternative feedstock and ODH reaction conditions are much milder than current processes. The production of propylene and ethylene could be decoupled. However, in contrast to ethylbenzene ODH the current state of the art of alkene productivities *via* ODH assigns to nanocarbon the task of unrevealing basic structure-reactivity relationships for knowledge-based improvement of high-performing metal-based catalysts, rather than being an alternative to such systems itself [65].

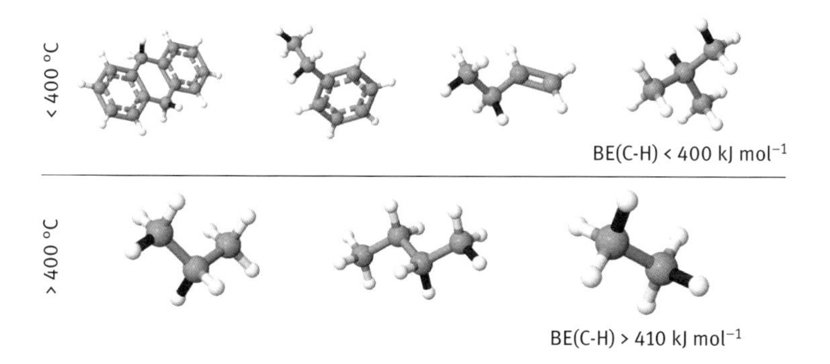

Fig. 15.10: Hydrocarbon substrates reported for carbon-catalyzed ODH.

In general, the mechanistic model established for the ODH of ethylbenzene can be adapted to alkanes (Fig. 15.11(d)). *In situ* XPS analyses of CNT catalysts during ODH of *n*-butane show sensitive response of the band located at 531.2 eV, which is assigned to carbonyl oxygen [63]. If oxygen is removed from the atmosphere, the intensity of this band decreases, however, it rapidly regenerates as oxygen is re-introduced into the XPS chamber. The redox mechanism, which can be separated into alkane conversion on the oxidized carbon surface and regeneration of the latter by gas phase oxygen, was also confirmed in the alternating feed mode for the ODH of propane [61]. By means of kinetic analyses it is shown that the catalyst modification with B_2O_3 and P_2O_5 slows down the activation of oxygen, which apparently reduces the formation of unselectively acting electrophilic oxygen species [60, 61]. However, the increase in alkene selectivity is accompanied by a drastic loss of overall activity (Fig. 15.11(c)), and an optimum loading has to be identified for each application [64].

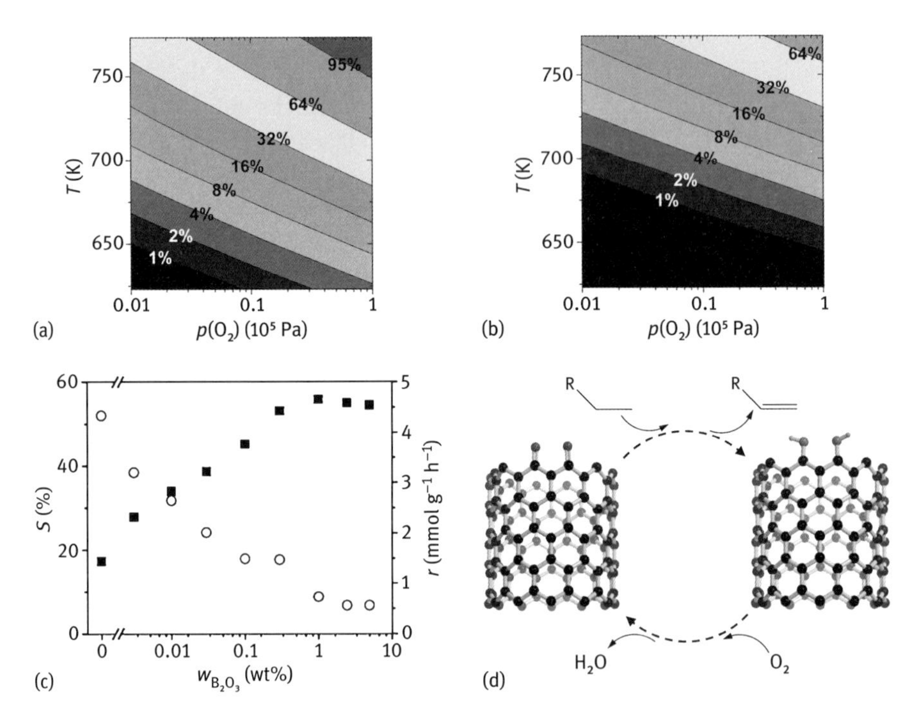

Fig. 15.11: (a),(b) Fraction of CNT catalysts combusted after 24 h time on stream in a O_2/He gas mixture: (a) CNTs, (b) 5 wt% P_2O_5/CNTs. (a),(b) Reprinted with permission from [25]. Copyright (2011) American Chemical Society. (c) Catalytic performance of B_2O_3-modified CNT catalysts in the ODH of propane. Propene selectivity at 5 % propane conversion (■) and reaction rate (o) as a function of B_2O_3 loading. (d) Reaction scheme of CNT-catalyzed ODH. (c),(d) Reprinted with permission from [61]. Copyright (2009) Wiley VCH.

Besides ODH processes, a few reports about non-oxidative dehydrogenation (DH) over carbon catalysts also exist. At the reaction temperature of 823–873 K, propane is reported to react to propylene and hydrogen in high yield (30–40 %) over ordered mesoporous carbon, which was shown to be much more active than graphitic and/or nanostructured carbon (CNTs) [66]. On the other hand, a hybrid catalyst system for dehydrogenation of ethylbenzene comprising nanodiamond cores and highly curved, defective graphene shells was recently introduced for the DH of ethylbenzene at 823 K, giving a styrene yield of 20–30 % [67]. These processes are characterized by high alkene selectivities and a sufficient process stability indicating the absence of severe coking. Both authors suggest the nucleophilic C=O bonds on the catalyst surface to be the active sites. Phenol groups formed during DH spontaneously dehydrogenate at such high reaction temperature to release H_2, which is considered as the rate-limiting step of the overall transformation.

15.3.2 Dehydrogenations of alcohols

The dehydrogenation of alcohols to the corresponding ketones/aldehydes has also been investigated over nanocarbon catalysts. In general, this reaction requires both Lewis acidic and basic sites and the product spectrum of the reaction of 2-propanol is considered as a sensitive probe for the acid/base properties of a catalyst surface under reaction conditions [68]. Here, acidic sites favor the dehydration of the alcohol leading to the formation of propylene, whereas acetone is an indicator for (additional) surface basicity. The control over the product spectrum by impregnation of (activated) carbon with mineral acids or alkali hydroxides [69] shall not be part of this chapter as the nanostructure and functionalization is less relevant. As indicated in Section 15.2.2 it is difficult to prepare a homogeneous surface functionalization on carbon without demanding multi-step synthesis protocols even when using heteroatoms such as N, P, or S (see Section 15.4.2). Thus, considering only oxygen functionalities, a broad product spectrum covering alkenes, ketones, ethers, esters and acetals is typically found. For instance, Szymanski *et al.* reported the reactions of ethanol, 2-propanol, and 2-butanol over nitric acid treated polymeric carbons [70–72]. Corresponding to the abundance of carboxylic acid groups, they mainly detected as dehydration products alkenes and ethers. However, ketones and aldehydes as dehydrogenation products were also detected to a considerable fraction.

The selectivity to ketones/aldehydes can be substantially increased when adding oxygen to the reacting mixture, thus turning the nature of the process oxidative. The investigation of gas-phase ODH of ethanol over carbon indicated a promising performance with conversions up to 70 % with a selectivity to acetaldehyde and ethyl acetate of 91 % [73]. However, using cyclohexanol as the substrate the vital surface chemistry on activated carbon further leads to consecutively formed ODH products benzene and phenol [74]. Weinstein *et al.* oxidatively dehydrogenated ethanol and 2-propanol with high yields of acetaldehyde and acetone, respectively, over graphitic CNFs [75, 76]. They investigated three types of CNFs, with varying orientations of the graphene sheets and found that fishbone fibers produced higher conversions of ethanol compared to platelet and ribbon fibers, which yielded similar results to one another. The product spectrum sensitively depends on the presence and concentration of oxygen in the reacting gas so it was concluded, in analogy with the ODH of ethylbenzene, that the oxygen groups terminating the prismatic edge sites of the graphene planes are responsible for the catalytic activity [75]. The mechanism depicted in Fig. 15.12(a) is suggested for the acid-base catalyzed dehydrogenation of secondary alcohols over carbon catalysts [70], whereas Fig. 15.12(b) represents a suggestion for the redox model of cyclohexanol ODH over quinone groups [74], similar to the C-H activation in hydrocarbons (Section 15.3.1). The catalyst regeneration can occur spontaneously under release of hydrogen (DH), or in oxidative manner under the release of water (ODH), which is thermodynamically more favorable.

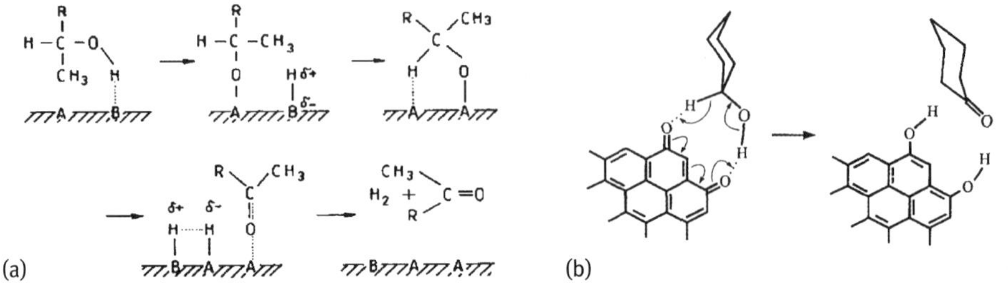

(a) (b)

Fig. 15.12: (a) Acid-base mechanism of alcohol dehydrogenation. Reprinted with permission from [70]. Copyright (1993) Pergamon (Elsevier). (b) Redox mechanism of cyclohexanol ODH. Reprinted with permission from [74]. Copyright (1998) Elsevier.

A recent research report by Dreyer *et al.* [77] called attention to the metal-free graphene-like materials. They investigated graphite oxide (GO) prepared according to the Hummers method [78, 79]. GO is a hydrophilic few-layer graphene-based material with a wide interlamellar *d*-spacing of up to 8 Å due to a pervasive honeycomb lattice decoration with alcohol and epoxide functionalities, as well as bulky carboxylic acid terminations of the GO sheets (see Fig. 15.5). A high activity and selectivity of this material in a number of fine chemical oxidations, hydrations, and coupling reactions is reported. For instance, the selective oxidation of benzyl alcohol to benzaldehyde succeeds with quantitative conversion over GO and only $< 7\,\%$ selectivity to the overoxidized product benzoic acid, depending on the reaction temperature. Such reactions occur under mild reaction conditions enabling the transformation of numerous organic substrates. Typically, liquid phase, ambient pressure and temperatures below 150 °C as well as the presence of gas phase oxygen (air) serve for excellent results. The reaction stops in inert nitrogen atmosphere pointing at atmospheric O_2 as the terminal oxidant and the catalytic interaction of GO. The authors suggest that oxygen functionalities are the key players in the catalytic oxidation, being regenerated by dissolved oxygen.

Table 15.1 summarizes the types of selective oxidations and hydrations, as well as corresponding catalytic performances of GO therein, as published by Bielawski's

Tab. 15.1: Conversion of organic substrates to the respective carbonyl compounds when reacted with graphite oxide [77].

	Substrate	Product	Conv.		Substrate	Product	Conv.
1	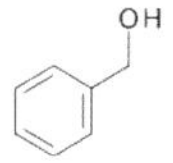 OH	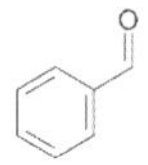 O	>98 %	7	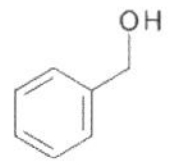 OH	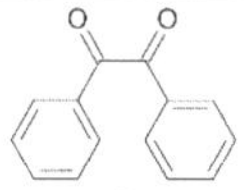 O O	56 %
2	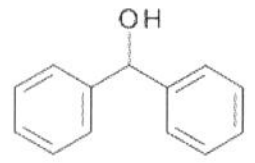 OH	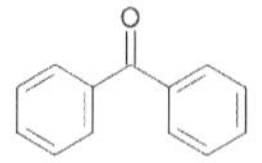 O	>98 %	8	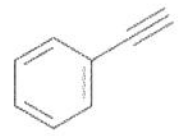	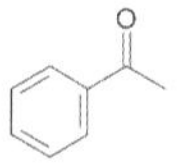 O	>98 %
3	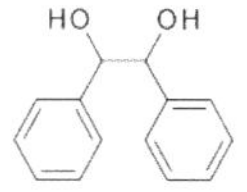 HO OH	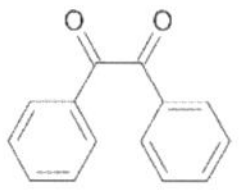 O O	96 %	9	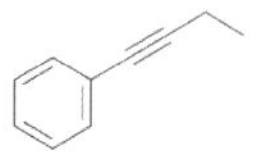	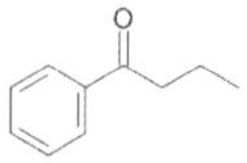 O	52 %
4	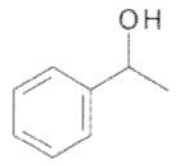 OH	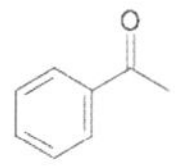 O	26 %	10	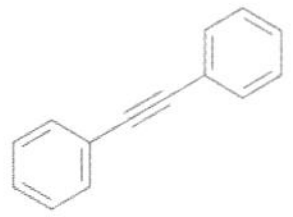	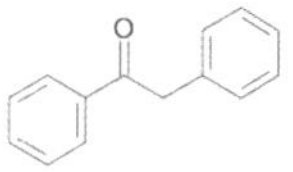 O	41 %
5	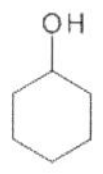 OH	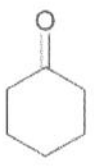 O	>98 %	11	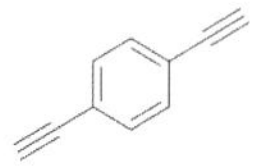	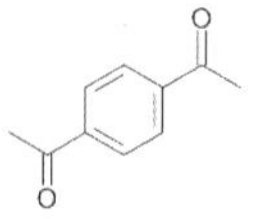 O O	26 %
6	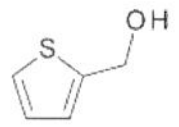 S OH	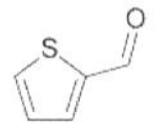 S O	18 %	12	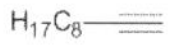 $H_{17}C_8$	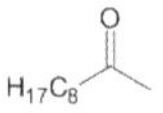 $H_{17}C_8$ O	27 %

group. The standard reaction conditions are 200 wt% catalyst (GO), 100 °C and a reaction time of 24 h [77]. A broad variety of primary and secondary alcohols as well as aryl and alkyl alkynes can be converted into the corresponding aldehydes and ketones. The selective oxidation of *cis*-stilbene to benzil (Table 15.1, entry 7) may find application in the Wacker-like transformations, which currently requires the use of Se-, Ce-, or Cr-based catalysts [80]. The heterogeneous catalyst can be easily removed from the reaction mixture by filtration and, depending on reaction conditions applied, re-used several times [77].

The positive impact of N-doping on graphene-like sp^2 carbon catalysts was recently reported by two groups for the aerobic oxidation of differently substituted benzyl alcohols [81] and of aryl alkanes [82]. Both catalyst systems show excellent yields of the corresponding benzylic aldehydes. The reports, however, differ in the suggestion of oxygen activation in the form of a peroxo complex, which can form directly at the quaternary N atom [81] or at the carbon atom located in neighbored *ortho*-position according to the hexagonal atom arrangement [82]. The latter, being supported by STM observations and theoretical calculations, appears more likely.

15.3.3 Other reactions

The selective insertion or removal of oxygen atoms is one of the ultimate challenges in heterogeneous catalysis. Nitrogen-doped CNTs were reported to successfully catalyze the aerobic oxidation of cyclohexane to adipic acid and its precursors cyclohexanol/cyclohexanone in the liquid phase (Tab. 15.2) [83]. The carbon catalysts showed a reaction rate 2–10 times higher than those of Au/ZSM-5 and $FeAlPO_x$ catalysts. The reaction is also catalyzed by nitrogen-free CNTs, however, when nitrogen atoms are used to dope graphitic domains, the whole reaction can be altered to one-step production of adipic acid to give an adipic acid selectivity as high as 60 % and a combined alcohol/ketone selectivity of 27 % at a conversion higher than 45 %. For their tests, Yu *et al.* grew bamboo-like N-CNTs (N/C = 4.5 %, XPS) by CVD in NH_3 atmosphere using aniline as the carbon and nitrogen source. An abundance of pyridinic nitrogen was identified to boost the catalytic performance. A radical mechanism, as is generally accepted for this reaction, was proven by addition of the radical scavenger *p*-benzoquinone, which totally suppressed the reaction to give a conversion as low as 0.8 %. It was suggested from the results of DFT calculations that the positive effect of nitrogen doping is due to the enhanced stabilization of the $C_6H_{11}OO\cdot$ peroxyl radical, which is an important intermediate in cyclohexane oxidation, in a π-π stacking complex with the CNTs [83]. High performance was also observed by the same research group for CNTs which were filled with iron [84]. A charge transfer from the metallic core to the catalytically active carbon surface can be expected from first-principles calculations [85] and is manifested in a correlation of the catalyst UPS work function with catalytic activity [84]. The oxidation of cyclohexane to cyclohexanone using H_2O_2 as the oxidant was previously shown to be catalyzed by B- and F-doped mesoporous carbon nitride polymers [86]. Here, a selectivity to cyclohexanone of 91 % is obtained, however, at a conversion of only 8 %.

Tab. 15.2: Organic reactions catalyzed by nanocarbon materials.

Catalyst	Substrate	Conv.	Products	Sel.	Ref.
N-CNTs Fe@CNTs		45 % 37 %	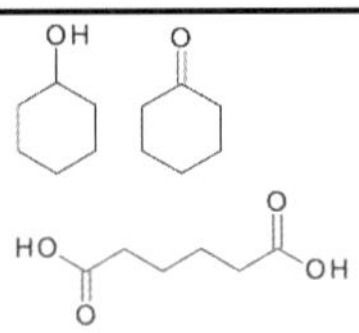 OH, O HO, O, O, OH	86 % 86 %	[83] [84]
CNTs	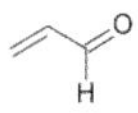 O, H	24 %	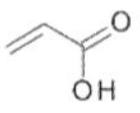 O, OH	90 %	[88]
N-CNTs C_{60}	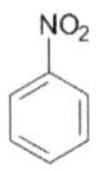 NO_2	100 % 100 %	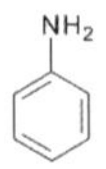 NH_2	100 % 92 %	[89] [90]

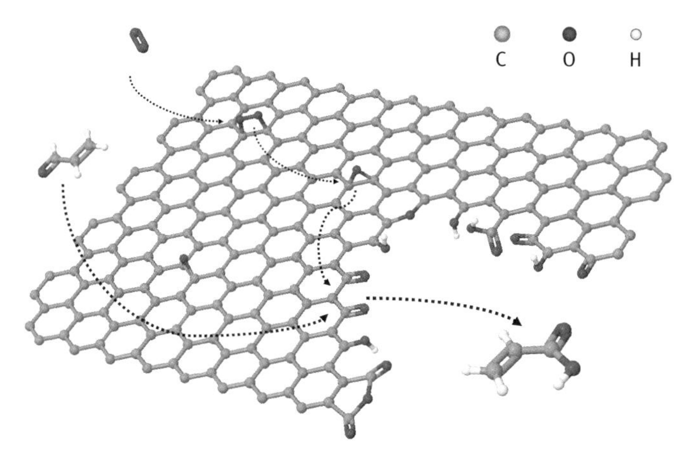

Fig. 15.13: Illustration of the selective oxidation of acrolein to acrylic acid on carbon. Adapted with permission from [87]. Copyright (2011) Wiley VCH.

The discrimination of reaction mechanism and active site requirements was intended in a report by Frank *et al.* on the application of sp^2 nanocarbons for the selective oxidation of acrolein to acrylic acid (Fig. 15.13) [87]. By means of isotope and kinetic studies as well as testing differently structured nanocarbons with various surface structures and functional groups exposed, they could highlight the importance of both the prismatic edges and the basal planes for selective turnover. The (curved) basal plane was suggested to dissociatively activate oxygen, whereas quinone and carboxyl groups terminating the graphite sheets serve as active sites. Consequently, multi-walled CNTs appeared as the most active and selective catalysts, whereas other nanocarbons such as fishbone CNFs, nanographite, nanodiamond, or nanoonions showed a poor performance. At 300 °C and in the presence of up to 40 vol% steam in the reacting mixture, acrolein conversion and acrylic acid selectivity could be increased up to 24 and 90 %, respectively [88].

While most of the reactions catalyzed by (nano)carbon are oxidations, a few examples of reductions are also reported. The almost quantitative reduction of differently substituted nitroarenes to the corresponding anilines over graphite was reported as early as 1985, however, using hydrazine as a very strong reducing agent [91]. Using gaseous hydrogen, the reaction also proceeds on C_{60} and C_{70} fullerenes [90]. The reaction is promoted by UV or visible light, allowing for mild reaction conditions (1 atm H_2, RT). However, in the dark the reaction requires 4–5 MPa H_2 and 140–160 °C to give high conversions. Milano-Brusco *et al.* reported a high performance in nitroarene hydrogenation over industrial NCNTs [89]. The interaction of nitrogen-rich carbonaceous solids with hydrogen has been studied, as computational screening proved that aromatic N=N double bonds featured molecular orbitals suited to activate dihydrogen [92]. Nitrobenzonitrile, nitrophenol, chloronitrobenzene, nitrobenzamide and other aromatic nitro-compounds comprising reducible groups in position 2 or 4 of the aromatic compound were also hydrogenated using metal-free nitrogen-doped CNTs as catalyst, typically operating at 150 °C and 40 bar with a reaction time of 120 min. Regarding the nitrogen functional groups, a linear correlation between the ratio of pyridinic to quaternary nitrogen atoms and the achievable conversion into aniline is found [93]. Unfortunately less is known about the reaction mechanism of these reactions as is the case for selective oxidations and dehydrogenations.

15.4 Nanocarbon as catalyst support

The utilization of large surface areas and, to a certain extent, controllable surface properties make carbon materials an ideal support for finely dispersed catalyst nanoparticles, as discussed in Section 15.2. The special features of nanocarbons for this purpose will be highlighted in the following section. Starting with the controlled synthesis of a variety of nanocarbon-inorganic hybrids, some examples will be discussed, where the superior catalytic performance arises from the unique properties of the nanostructured support.

15.4.1 Catalyst preparation strategies

In general, there are two possibilities to prepare nanocarbon-supported metal(oxide) catalysts. The *in situ* approach grows the catalyst nanoparticles directly on the carbon surface. The *ex situ* strategy utilizes pre-formed catalyst particles, which are deposited on the latter by adsorption [94]. Besides such solution-based methods, there is also the possibility of gas phase metal (oxide) loading, *e.g.*, by sputtering [95], which is used for preparation of highly loaded systems required for electrochemical applications not considered here.

15.4.1.1 *In situ* synthesis of nanocarbon-supported catalysts

The deposition of the metal catalyst precursor on the nanocarbon surface followed by its transformation into the catalytically active inorganic material is the most frequently reported method of catalyst preparation. The precursor, typically in the form of metal cations or complex anions, is typically loaded on the nanocarbon host by incipient wetness impregnation or by complexation with surface functional groups. The latter act as nucleation centers for the metal precursor. Consequently, the zeta potential of the carbon surface and the pH of the impregnating solution are of crucial importance to ensure a sufficient metal-support interaction. Nanosized clusters of noble metals such as Rh, Pd, or Pt are typically generated by chemical reduction of the deposited precursor with H_2, $NaBH_4$, N_2H_4, or by microwave irradiation, whereas oxidic precursors of V, Mn, or Mo based catalysts are typically formed by activation under mildly oxidizing conditions. Supported transition metal carbides such as Mo_2C can be synthesized by high-temperature treatment in a CH_4/H_2 atmosphere, where the nanocarbon support potentially acts as the carbon source for the carburization. The small particle diameter and narrow size distribution, which are typically the most important factors for the catalytic performance, can be controlled by the type of solvent and metal precursor, loading, temperature, and reaction time. Some examples of nanocarbon-supported catalysts synthesized this way are listed in Table 15.3. The preparation recipes depend on the type of metal to be deposited rather than on the nanocarbon. A rich surface functionalization of the carbon support prior to its impregnation, usually by HNO_3 treatment, is reported for most systems. Consequently, in the case of graphene-like materials the highly functionalized graphite oxide (GO) is preferred for metal impregnation followed by the collective reduction of both the support and metal precursor [11, 96]. Regarding graphene and GO systems, the deposition of metal (oxide) clusters can substantially alter the chemical, optical, and electronic properties of the support by deformation and disruption of the sp^2 carbon atoms, which can be reduced by appropriate synthesis strategies involving surfactants or polymers [97].

Tab. 15.3: Selected examples for nanocarbon-inorganic hybrid catalysts.

Catalyst	Precursor	Treatment	Ref.
Pd/rGO[a)]	$Pd(OAc)_2$	H_2, RT	[11]
Pd/GO	$Pd(NH_3)_4(NO_3)_2$	H_2, 60 °C	[98]
Pd/C_{60}	$Pd(OAc)_2(PPh_3)_2$	H_2, 250 °C	[99]
Pd/CNO[b)]	H_2PdCl_4	H_2, RT	[100]
Pd/SWCNTs	$Pd(OAc)_2$	DMF, 95 °C	[101]
Pd/N-CNTs	Na_2PdCl_4	$NaBH_4$, RT	[102]
Pd/MWCNTs	H_2PdCl_4	KBH_4, RT	[103]
Rh/MWCNTs	$Rh_2Cl_2(CO)_4$	H_2, 300 °C	[104]
Ru/MWCNTs	$Ru(NO)(NO_3)_3$	NH_3, 450 °C	[37]
V_2O_5/MWCNTs	$VO(O\textit{i}Pr)_3$	air, 350 °C	[105]
MoO_3/MWCNTs	$(NH_4)_6Mo_7O_{24}$	air, 450 °C	[106]
MnO_2/MWCNTs	$Mn(NO_3)_2$, $KMnO_4$	air, 110 °C	[107]
Mo_2C/MWCNT	$(NH_4)_6Mo_7O_{24}$	CH_4/H_2, 800 °C	[108]

a) reduced GO; b) carbon nanoonions

The selective deposition of catalyst particles on the inner or on the outer walls of CNTs is the prerequisite for the investigation or utilization of the confinement effect, as discussed in Section 15.2.3. Wet chemistry methods making use of the capillary effect are most effective; however, they depend on surface functionalization and tube diameter. In any case, CNT caps as well as radial carbon sheets and walls blocking parts of the inner CNT cavity have to be removed prior to impregnation, *e.g.*, by mild oxidative treatment. The impregnation of this material with a limited amount of liquid can lead to the intended result [109], but fails if the nanocarbon material is heterogeneously structured, *e.g.*, in case of a broad CNT diameter distribution [110]. A straightforward synthesis strategy for selective deposition is the two-step biphasic impregnation illustrated in Fig. 15.14 [111].

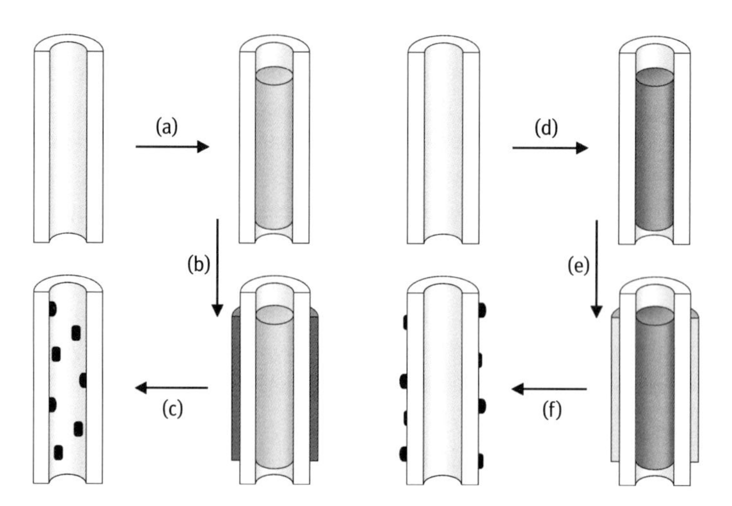

Fig. 15.14: Illustration of selective deposition strategies for catalyst nanoparticles (left) on the inner and (right) on the outer surface of CNTs according to Ref. [111]. For inside deposition the CNTs are (a) impregnated with an ethanolic solution of the metal precursor, followed by washing with distilled water to protect the outer surface, and (c) subsequent drying and final treatment to form the catalyst nanoparticles. For outside deposition the CNTs are (d) impregnated with an organic solvent to block the inner tubule, followed by (e) impregnation with an aqueous solution of the metal precursor and (f) subsequent drying and final treatment.

The first step of selectively depositing catalyst nanoparticles inside CNTs is their impregnation with a solution of the catalyst precursor. In the ideal case the solvent has a low surface tension, *e.g.*, ethanol, to enable facile penetration into the CNT. Distilled water has also been reported for this purpose, however, with the aid of ultrasonic treatment [112]. The precursor solution can be added in excess or in defined volume. In any case, some weakly adsorbed metal species will remain on the outer CNT surface, which must be removed by washing with an appropriate solvent [111, 112]. A high selectivity to inside deposition is reported for this strategy, which is independent of the CNT diameter.

A facile strategy for the selective deposition of catalyst particles on the outer wall of CNTs is the *ex situ* synthesis described in Section 15.4.1.2. For selective *in situ growth* of the catalyst on the outer wall of CNTs, an apparently easy method is using closed (untreated) CNTs. However, their poor functionalization mostly results in a poor dispersion of the metal. The hollow core of the CNTs, once opened by oxidative treatment/functionalization, thus has to be blocked with a low-tension organic solvent, *e.g.*, xylene [38], to prevent its filling with the metal precursor solution. Carbon surfaces have a very good affinity to organic solvents, thus the aqueous solutions cannot penetrate into the CNT due to the higher liquid/solid interface energy [111]. After calcination of such doubly impregnated CNTs, the catalyst particles are located mainly on the outer side of the CNTs.

15.4.1.2 *Ex situ* synthesis of nanocarbon-supported catalysts

Pre-formed metal (oxide) nanoparticles can be attached to the nanocarbon support by covalent bonding or by electrostatic attraction. The synthesis of colloidal nanoparticles, *e.g.*, in citric acid, microemulsion, or ionic liquid media, is numerously reviewed [113–115]. To achieve good dispersion and high stability of the supported system *via* metal-support interaction, the nanocarbon surface must be rich in appropriate functional groups to anchor the catalyst particles. The most facile way is utilizing the abundance of carboxylic acid groups generated by nitric acid treatment, which enables the adsorption of MgO, TiO_2, or $Zr(SO_4)_2$ nanoparticles [94]. Nitrogen atoms incorporated into the nanocarbon structure can also serve as an ideal anchor for catalyst nanoparticles, as shown for Pd nanoparticles supported on N-CNTs for H_2O_2 synthesis [116]. Pre-formed nanoparticles advantageously provide controllable and well-defined properties in terms of diameter, size distribution, and shape. When adsorbing on hollow nanocarbons such as CNTs they predominantly adsorb on the outer surface. Their major disadvantage, however, is a typically strongly adsorbing organic overlayer comprising capping and/or reducing additives such as surfactants or polymers, which have to be thoroughly removed prior to catalytic application.

A specific functionalization of the carbon surface can further improve the adsorption and catalytic behavior of catalyst nanoparticles. For instance, oxygen surface functionalities or surface defect structures can be utilized as an anchor for covalent attachment of aliphatic linkers, which carry, *e.g.*, amine or thiol functionalities at the end, by organochemical strategies. Such specific decoration not only ensures the fixation of the inorganic nanoparticles to minimize leaching in liquid phase applications but also helps to control the hydrophilic/lipophilic character of the heterogeneous catalyst to optimize its interaction with substrate molecules. Tessonnier *et al.* compared two methods to attach amine groups on multi-walled CNTs [117]. One route starts from carboxylic groups generated by initial nitric acid treatment. By reaction with thionyl chloride these are converted to the corresponding acid chlorides. The reaction with ethylenediamine leads to amine groups, which are covalently attached to the CNT structure *via* an amide group and a short aliphatic linker. The other route starts from H-saturated sp^3 defects, which are deprotonated with *n*-butyl lithium. Bromotriethylamine easily attaches to such carbanions by electrophilic attack under formation of a C-C bond. These systems were synthesized as heterogeneous base catalysts themselves for biodiesel synthesis *via* transesterification; however, such selectively decorated nanocarbons would also represent a suitable support for catalyst nanoparticles. For instance, above mentioned amidation route has been used for attaching thiol groups on oxidized single-walled CNTs to tether colloidal Au nanoparticles [118]. The generation of such functional groups does not inevitably require advanced organic chemistry but these can also be attached *via* π-π stacking. For instance, a pyrene moiety with a long aliphatic chain terminated by a thiol group can act as a linker between graphitic CNTs and Au nanoparticles. In this case the pyrene unit is the anchor, which attaches to the aromatic basal plane of the nanocarbon *via* van der Waals interaction [119].

15.4.2 Applications in heterogeneous catalysis

Considering the advantageous support properties of nanocarbons discussed in Section 15.2, numerous studies have been carried out on different catalytic reactions. To highlight the positive impact of the nanostructure, this section focuses on reports that give a reference catalyst based on a conventional carbonaceous support such as activated carbon, carbon black, or graphite, rather than metal oxides.

15.4.2.1 Fine chemical synthesis

Fine chemical transformations are typically performed in liquid phase and batch mode with nanosized noble metals such as Ru, Rh, Pd, or Pt supported on carbon or metal oxides being the mostly used catalysts. The range of reactions covers hydrogenations of carbon-carbon as well as carbon-heteroatom multiple bonds, dehydrogenations, (de)hydrations, reductions and oxidations of functional groups, or C-C coupling reactions such as Suzuki or Heck reactions. The liquid reaction medium challenges the catalyst stability in terms of leaching to prevent both the loss of expensive catalyst and the contamination of the reaction product with catalyst nanoparticles. Specifically, the well-functionalized nanocarbons provide a sufficient metal-support interaction to minimize leaching. The nature and degree of functionalization furthermore allows a certain degree of selectivity control due to the facile electron transfer from the nanocarbon to the metal (oxide) nanoparticle. The open pore structure of CNTs and CNFs substantially reduces mass transfer limitations of large organic molecules even in liquid phase to reduce non-selective total reductions or oxidations, where a partial transformation is intended.

A typical probe reaction for estimating catalytic properties in selective hydrogenations is the hydrogenation of cinnamaldehyde. This molecule contains both a C=C and a C=O double bond, thus the formation of hydrocinnamaldehyde and/or cinnamyl alcohol by reduction of the one or the other, or the formation of phenyl propanol in the case of complete reduction may indicate the potential of the catalyst for other fine chemical transformations. Indeed, this reaction was one of the first to be tested by CNT-supported catalysts [120]. Noble metals show a high activity in this reaction and the main product over Pd and Rh is hydrocinnamaldehyde. Instead, the α,β unsaturated alcohol, which is of particular interest because of the importance of such compounds in the fine chemical industry, can be obtained over Ir, Pt, Ru or over bimetallic systems such as Pt-Ru. Using 1 wt% Rh as the active phase, the activity on the CNT-supported catalyst is three times higher than on activated carbon, although the latter provides an almost 4-fold specific surface area [104]. For Ru-based catalysts, the impact of the support on product selectivity is remarkable [121]. Using similar degrees of Ru dispersion, cinnamyl alcohol selectivities of > 90 %, 30–40 %, and 20–30 % are obtained when using CNTs, activated carbon, and alumina, respectively, as the support materials [120, 122, 123]. For Pd nanoparticles located inside CNTs, the catalytic performance is explained with different degrees of functionalization on the inner and outer walls of CNTs, as well as with the confinement effect [124]. Indeed, heat treatment of the catalysts leading to partial defunctionalization can change the product spectrum, as found for Pt-Ru supported on single- and multi-walled CNTs, where mainly cinnamyl alcohol is formed [125]. Instead, such tunable properties are not observed for the corresponding activated carbon-supported system, which gives only low selectivities to the desired unsaturated alcohols. In general, a high selectivity to the unsaturated alcohol is explained in terms of a transfer of the π-electrons from the graphitic planes to the metal particles. In this way the charge density on the metal increases, thus decreasing the probability for the C=C bond activation. Also the preparation method, particle size, and solvent effects, as well the general reaction conditions must be considered when comparing the data assembled in Fig. 15.15. Table 15.4 illustrates that the extraction of pure *nanocarbon* support effects from present literature is a still a challenging task.

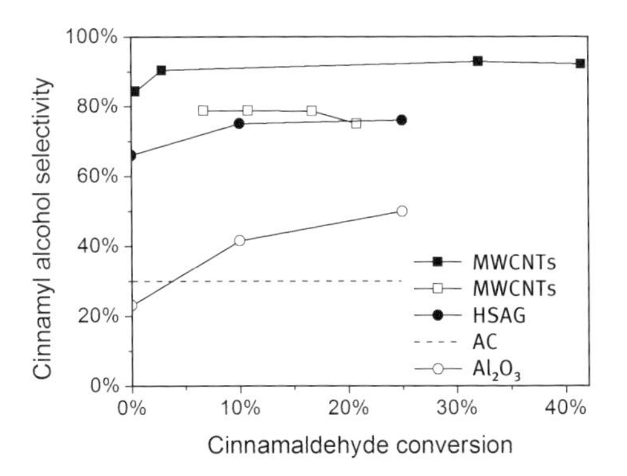

Fig. 15.15: Cinnamyl alcohol selectivity as a function of conversion in the selective hydrogenation of cinnamaldehyde over differently supported Ru catalysts. Reactions conditions as listed in Tab. 15.4. Data compiled from literature [120,122,123,126]. HSAG: high-surface area graphite; AC: activated carbon.

Tab. 15.4: Catalyst properties and reaction conditions for the Ru-catalyzed reduction of cinnamaldehyde.

Loading/wt%	Support	$d_p^{a)}$/nm	S/m^{2}g^{-1}	T/K	$p(H_2)$/MPa	Ref.
0.2	MWCNT	3–7	27	383	4.5	[120]
2	MWCNT	–	22.5	393	5	[126]
0.5	AC	< 3	~ 1000	333	0.1	[123]
0.9	HSAG	–^{e}	200	383	4.5	[122]
0.68	Al_2O_3	2.7	300	383	4.5	[122]

a) Catalyst nanoparticle size from TEM analysis; b) initial substrate concentration; e) H_2 chemisorption indicates very large Ru particles; AC: activated carbon; HSAG: high surface area graphite.

Graphene has also been successfully applied as a support for noble metal nanoparticles [127, 128]. For instance, Au/graphene was proven to be efficient in Suzuki coupling even if the reaction was performed in water and under aerobic conditions [129]. Here, highly dispersed ~ 3 nm sized Au nanoparticles were prepared by facile reducing chloroauric acid in the presence of sodium dodecyl sulfate, which is used as both a surfactant and a reducing agent. In general, the performance of such graphene-supported systems is explained by the exceptionally high surface area, conductivity, and by the intimate interaction of metal nanoparticles with surface functional groups. Similar examples utilizing, *e.g.*, Pd, Pt, Ag, as well as alloys and mixtures thereof, for C-C coupling reactions, selective oxidations, or reductions are frequently reported [11, 130–133]. The more prominent field of application for these materials, however, is in electrochemistry.

The catalytic properties of metal nanoparticles deposited on carbon surfaces can also be tuned by heteroatom doping. For instance, Xia *et al.* showed that N-doped CNTs used as a support for Pd nanoparticles result in smaller particle size, better dispersion, and higher Pd surface area as compared to the undoped CNTs [134]. NCNT-supported Pd nanoparticles are much harder to reduce, thus a defined amount of oxygen- and nitrogen-containing groups on the nanocarbon surface helps to adjust and stabilize a suitable oxidation state (Pd^{n+}/Pd^0 ratio), which gives control over the catalytic performance of the composites. For instance, in the selective hydrogenation of 1,5-cyclooctadiene the presence of pyridinic nitrogen atoms in the CNT support results in higher conversions and also in higher selectivities to partially hydrogenated cyclooctene for both Pd and Pt based catalysts [134, 135]. The influence of nitrogen doping on CNT-supported Pd catalysts was also investigated in the selective oxidation of benzyl alcohol. Arrigo *et al.* found that the metal nanoparticles deposited on such heteroatom-modified support change their shape and turn from spherical to flattened geometry as the nitrogen and oxygen content of the carbon support increases [102] (Fig. 15.16). Such a strong metal-support interaction as visualized by TEM substantially reduces leaching, which is the major problem in this reaction. The interaction documented improves adsorption properties and results in higher conversion in

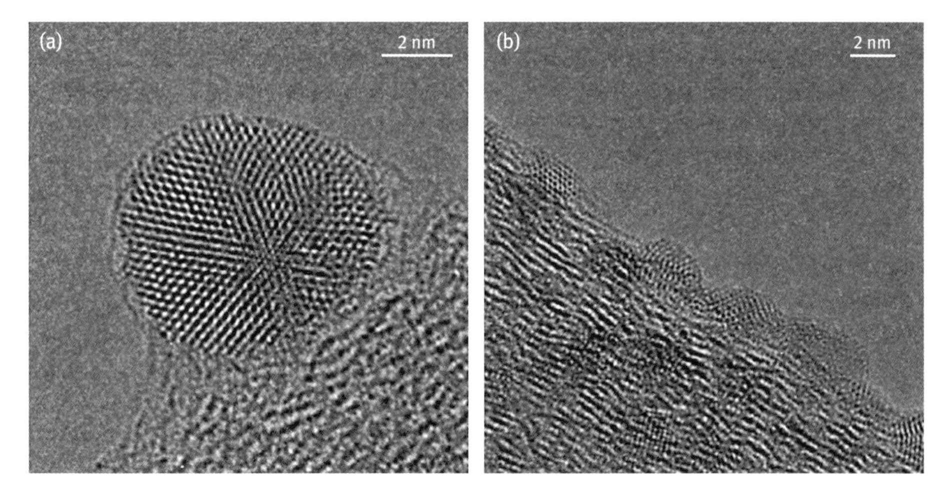

Fig. 15.16: HRTEM images of Pd nanoparticles deposited (a) on the CNF surface and (b) on the N-doped CNF surface (right) showing spherical and flattened geometries, respectively. Courtesy of R. Arrigo, FHI Berlin.

the order N-doped CNTs > oxidized CNTs > pristine CNTs, whereas the selectivity to benzaldehyde is less affected (65–75 % at 90 % conversion). The type of solvent appears to be critical here. Under solventless conditions, Pd supported on pristine CNTs and on activated carbon, respectively, provide quite similar catalytic performances [136]. However, in 1 : 1 dilution with cyclohexane as the solvent, the selectivity to benzaldehyde over CNT-supported Pd is reported to achieve 92 %, whereas the activated carbon-supported system provides only 74 % [137].

15.4.2.2 Fischer Tropsch synthesis

Considerable attention has been paid to the application of CNTs as the catalyst support for Fischer Tropsch synthesis (FTS), mainly driven by utilization of the confinement effect (Section 15.2.3). In general, this process is a potential alternative to synthesize fuel (alkanes) or basic chemicals like alkenes or alcohols from syngas, which can be derived from coal or biomass. The broad product spectrum, which can be controlled only to a limited extent by the catalyst, prohibited its industrial realization so far, however, it is considered an important building block for future energy and chemical resource management based on renewables.

A systematic study of differently supported Ru catalysts showed that carbon catalysts provide very high selectivities to higher hydrocarbons (C_{10}-C_{20}) and the CNT-supported catalyst is among the most active systems of all [138]. In parts this is related to the inertness of carbon preventing the formation of hardly reducible mixed metal oxides with the support, such as $CoAl_2O_4$ [139, 140], which is, besides coking, the main reason for catalyst deactivation. The carbon surface functionalized with oxygen groups leads to a better and more stable dispersion of easily reducible metal nanoparticles [138, 141]. Some problems with the carbonaceous support with long-term application could arise from metal-catalyzed gasification (methanation) under high pressure H_2 at elevated temperatures [142]. Among the different carbon supports tested, the CNTs provide the most promising performances. For instance, Xiong *et al.* showed for 3 wt% Ru/C catalyst that the activity in terms of CO conversion follows the order CNT > ordered mesoporous carbon >activated carbon, which is explained by macrokinetic effects controlled by the pore structures of the catalysts [143]. Owing to the fact that Ru nanoparticles are located mainly outside CNTs, the selectivity to short chain C_1-C_3 alkanes is highest here, whereas the formation of higher alkanes is favored in the microporous (activated carbons) and mesoporous systems.

The confinement effect of catalyst particles located inside CNTs has been explained by different approaches:
- a better reducibility of metal oxide particles located inside the CNTs [39]
- increased reactants' partial pressures in nm-sized pores [35]
- different electronic properties of inner and outer surfaces [34, 36]
- forced multiple contact of reactants with catalysts [39]
- limitation of particle growth (sintering) by CNT walls [39]

The product spectrum of FTS in terms of functional groups can be tuned by appropriate doping. For instance, an extensive study of FTS on CNT-supported bimetallic Co/Fe catalysts showed that the monometallic Co catalyst exhibited high selectivity (85.1 %) towards C_{5+} liquid hydrocarbons. The addition of small amounts of Fe did not significantly change the product selectivity. However, the bimetallic Co-Fe/CNT catalysts proved to be attractive in terms of alcohol formation (Tabs. 15.1 and 15.5) [144]. Also Pan *et al.* achieved a high selectivity to ethanol over an Mn-promoted Rh catalyst [36]. The up to five-fold higher activity of catalyst particles located inside CNTs compared to outside deposition was explained by electronic properties. In particular, it was suggested that due to the electron deficiency of the concavely-shaped inner CNT wall, the MnO promoter is more oxophilic to favor CO adsorption at the O atom, thus enhancing its dissociation rate. Also CNTs have been shown to promote the transfor-

Tab. 15.5: Catalytic performances of Co-Fe/CNT catalysts in FTS [144].

Catalyst	*X*(CO)/%	*S*(alc.)/%	alkene/alkane
10Co/CNT	47	2.3	0.56
10Co0.5Fe/CNT	54	4.1	0.96
10Co1Fe/CNT	48	5.4	1.08
10Co2Fe/CNT	35	22.0	1.32
10Co4Fe/CNT	29	26.3	1.48
10Fe/CNT	11	10.3	1.95

a) Reaction conditions: 220 °C, 2 MPa, $H_2/CO = 2$, 60 ml min^{-1} g^{-1}.

mation of metals into metal carbides. This could be related to higher activities and C_{5+} selectivity in FT synthesis [145].

It was also shown that tubular carbon supports can prolong the lifetime of FTS catalysts, which is the result of deposition of catalytic sites on the interior surfaces of the CNTs resulting in decreased sintering of the metal particles and therefore a more stable catalyst [145].

15.4.2.3 Ammonia decomposition

Due to the high hydrogen storage capacity of the ammonia molecule (17.7 wt% equal to an energy density of 4,318 Wh kg^{-1}), its decomposition is intensely investigated for CO_x-free hydrogen production for mobile fuel cell applications [146]. However, compared with the well-established Haber–Bosch process for ammonia synthesis, its decomposition is underdeveloped and requires substantial improvements before it can be considered as a practical contribution to the energy supply toolbox.

The most active catalysts for NH_3 decomposition are based on Ru, however, cheaper Fe, Co, Ni and alloy systems are also intensely investigated [148]. The impact of the support material is remarkable. In a study by Au *et al.*, Ru/CNTs performed better than all oxide-supported systems, whereas activated carbon resulted in one of the lowest NH_3 conversions (Tab. 15.6) [147]. The dispersion of the active component as well as basicity [147] and conductivity [149] of the support are discussed as the relevant factors for high catalytic efficiency. However, the difference between CNT and activated carbon support is still remarkable. Thus it is not surprising that even the residual catalyst material on commercial MWCNTs, which is basically based on Fe and Co, results is a high catalytic performance in NH_3 decomposition [150].

Tab. 15.6: Catalytic performances of differently supported Ru catalysts in NH3 splitting. 773 K; 150,000 ml h^{-1} g_{cat}^{-1} [147].

Catalyst	S_{BET}/ m^2 g^{-1}		Ru loading/ µmol g_{cat}^{-1}	$d_{p,TEM}$/ nm		Dispersion/%	H_2 formation/ mmol min^{-1} g^{-1}
	catalyst	support		range	average		
Ru/CNTs	224	168	464	2–5	3.9	25.6	27.7
Ru/AC	1220	1130	469	2–5	4.1	24.4	17.6
Ru/MgO	24	13	477	2–15	9.1	11.0	22.4
Ru/Al_2O_3	159	108	480	3–16	8.7	11.5	18.1
Ru/TiO_2	6	3	469	–	–	–	18.9
Ru/ZrO_2	30	19	475	–	–	–	16.2

Zheng *et al.* investigated the NH_3 splitting over Ru catalysts supported on differently treated CNTs. They observed a negative particle size effect, *i.e.*, the H_2 formation rate decreases at higher dispersion of the active component [37]. It was also shown that the activity decreases with an increasing degree of disorder on the carbon surface as quantified by Raman spectroscopy, which would explain the poor support properties of activated carbon in this reaction. The confinement effect on this reaction was considered by depositing Ru nanoparticles selectively on the inner and outer surfaces of the CNTs. Ru-in particles are inherently smaller than Ru-out particles and therefore no positive confinement effect was observed. The authors noted, however, that Ru-in particles are inherently less well ordered and exhibit specifically more defects, which is in line with the trend that Ru-in particles of the same size are more active than Ru-out particles [37]. It remains unclear whether the particle size, which tends to increase with low metal-support interaction, or electronic properties are responsible for the good performance of highly graphitized carbon supports. The elimination of acidic groups from the surfaces, prior to catalyst preparation, and/or the surface graphitization of the materials produced a higher catalytic activity during the reaction. The catalytic activity of Ru particles can also be significantly improved when supported on carbon nanotubes doped with nitrogen [151].

15.5 Summary

In general, there is a still growing interest in nanocarbons and nanocarbon-inorganic hybrids for catalytic applications owing to the encouraging pioneering studies and intense research within the past 20 years. Future studies must aim at the better control of nanostructure (geometry) and uniform surface functionalization to better understand these key aspects of metal-support interaction and how they are linked to catalytic reactivity [152]. Surprisingly, the harsh and rather unselective oxidation, *e.g.*, with HNO_3 reflux, is still the most common method to prepare the carbon surface for grafting of transition metals. More target-oriented studies in this field could provide a rational basis for the development of advanced catalysts. It has been numerously shown that the influence of carbon supports on the immobilized catalysts cannot be seen as a consequence of a single physical or chemical property.

The upscaling of nanocarbon production, while considering purity and homogeneity aspects in the kilogram scale, is a very important factor. Except for multi-walled CNTs no such process has been industrialized so far. If commercially available, high-purity nanocarbon materials provide exorbitant costs of up to several hundred Euro per milligram, which in most cases is inacceptable for applied research. As a consequence, nanocarbons are often synthesized in-house, which substantially lowers the comparability among different studies. Synthesis, characterization, and application must go hand-in-hand to exploit the full potential of nanocarbons.

Regarding the catalytic reactions, a great potential emerges in the conversion of syngas to higher hydrocarbons due to the confinement effect observed in CNTs. Also the reports in fine-chemical transformations are very promising and currently day-to-day complemented by new classes of reactions. The metal-free CNT-catalyzed ODH of ethylbenzene is currently investigated in pilot-scale [152]. The acid-base resistivity and open pore structure of 1D nanocarbons such as CNTs and CNFs give them a great potential in reactions related to biomass conversion. We can expect a vital development in the field of nanocarbon applications in heterogeneous catalysis.

Bibliography

[1] J. M. Thomas, W. J. Thomas, *Principles and Practice of Heterogeneous Catalysis*, VCH, **1997**.
[2] H. Jüntgen, *Fuel* **1986**, *65*, 1436–1446.
[3] L. R. Radovic, F. Rodríguez-Reinoso, in *Chemistry and Physics of Carbon, Vol. 25* (Ed.: P.A. Thrower), Marcel Dekker, New York, **1997**, pp. 243–358.
[4] P. Serp, J. L. Figueiredo, *Carbon Materials for Catalysis*, John Wiley & Sons, **2009**.
[5] J.-P. Tessonnier, D. Rosenthal, T. W. Hansen, C. Hess, M. E. Schuster, R. Blume, F. Girgsdies, N. Pfänder, O. Timpe, D. S. Su, et al., *Carbon* **2009**, *47*, 1779–1798.
[6] Q. Zhang, J.-Q. Huang, M.-Q. Zhao, W.-Z. Qian, F. Wei, *ChemSusChem* **2011**, *4*, 864–889.
[7] R. B. Heimann, S. E. Evsvukov, Y. Koga, *Carbon* **1997**, *35*, 1654–1658.
[8] O. Shenderova, V. Zhirnov, D. Brenner, *Crit. Rev. Solid State Mater. Sci.* **2002**, *27*, 227–356.
[9] R. Schlögl, in *Handbook of Heterogeneous Catalysis* (Eds.: G. Ertl, H. Knözinger, F. Schüth, J. Weitkamp), Wiley VCH, **2008**, pp. 357–427.
[10] E. Fitzer, K.-H. Kochling, H. P. Boehm, H. Marsh, *Pure Appl. Chem.* **1995**, *67*, 473–506.
[11] G. M. Scheuermann, L. Rumi, P. Steurer, W. Bannwarth, R. Mu¨lhaupt, *J. Am. Chem. Soc.* **2009**, *131*, 8262–8270.
[12] M. H. Gass, U. Bangert, A. L. Bleloch, P. Wang, R. R. Nair, A. K. Geim, *Nat. Nanotechnol.* **2008**, *3*, 676–681.
[13] L. Li, Z. H. Zhu, Z. F. Yan, G. Q. Lu, L. Rintoul, *Appl. Catal. A* **2007**, *320*, 166–172.
[14] L. Martínez-Latorre, S. Armenise, E. Garcia-Bordejé, *Carbon* **2010**, *48*, 2047–2056.
[15] J. J. Delgado, R. Vieira, G. Rebmann, D. S. Su, N. Keller, M. J. Ledoux, R. Schlögl, *Carbon* **2006**, *44*, 809–812.
[16] J. Zhang, R. Wang, E. Liu, X. Gao, Z. Sun, F.-S. Xiao, F. Girgsdies, D. S. Su, *Angew. Chem. Int. Ed.* **2012**, *51*, 7581–7585.
[17] J. Ma, D. Moy, A. Chishti, J. Yang, *Method For Preparing Supported Catalysts From Metal Loaded Carbon Nanotubes*, **2005**, U.S. Patent US20060142149.
[18] N. Ishigami, H. Ago, Y. Motoyama, M. Takasaki, M. Shinagawa, K. Takahashi, T. Ikuta, M. Tsuji, *Chem. Commun.* **2007**, 1626–1628.
[19] D. S. Su, X.-W. Chen, *Angew. Chem. Int. Ed.* **2007**, *46*, 1823–1824.
[20] J. Amadou, D. Begin, P. Nguyen, J. P. Tessonnier, T. Dintzer, E. Vanhaecke, M. J. Ledoux, C. Pham-Huu, *Carbon* **2006**, *44*, 2587–2589.
[21] B. J. Hinds, N. Chopra, T. Rantell, R. Andrews, V. Gavalas, L. G. Bachas, *Science* **2004**, *303*, 62–65.
[22] H. Xie, P. V. Pikhitsa, Y. J. Kim, W. Youn, I. S. Altman, J. G. Nam, S. J. Lee, M. Choi, *J. Appl. Phys.* **2006**, *99*, 104313–104313-6.
[23] K. P. De Jong, J. W. Geus, *Catal. Rev. Sci. Eng.* **2000**, *42*, 481.
[24] R. A. Rosenberg, P. J. Love, V. Rehn, *Phys. Rev. B* **1986**, *33*, 4034–4037.
[25] B. Frank, A. Rinaldi, R. Blume, R. Schlögl, D. S. Su, *Chem. Mater.* **2010**, *22*, 4462–4470.
[26] A. Rinaldi, B. Frank, D. S. Su, S. B. A. Hamid, R. Schlögl, *Chem. Mater.* **2011**, *23*, 926–928.
[27] W. Xia, C. Jin, S. Kundu, M. Muhler, *Carbon* **2009**, *47*, 919–922.

[28] N. V. Qui, P. Scholz, T. Krech, T. F. Keller, K. Pollok, B. Ondruschka, *Catal. Commun.* **2011**, *12*, 464–469.

[29] H. P. Boehm, *Carbon* **1994**, *32*, 759–769.

[30] A. Wolf, V. Michele, L. Mleczko, J. Assmann, S. Buchholz, *Method for Producing Nitrogen-doped Carbon Nanotubes*, **2010**, U.S. Patent 20100276644.

[31] K. Esumi, M. Sugiura, T. Mori, K. Meguro, H. Honda, *Colloids Surf.* **1986**, *19*, 331–336.

[32] D. S. Su, in *Carbon Nanomaterials* (Ed.: C.S.S.R. Kumar), Wiley VCH, Weinheim, Germany, **2011**, pp. 35–67.

[33] X. Pan, X. Bao, *Acc. Chem. Res.* **2011**, *44*, 553–562.

[34] S. Guo, X. Pan, H. Gao, Z. Yang, J. Zhao, X. Bao, *Chem. Eur. J.* **2010**, *16*, 5379–5384.

[35] Y. Long, J. C. Palmer, B. Coasne, M. Śliwinska-Bartkowiak, K. E. Gubbins, *Phys. Chem. Chem. Phys.* **2011**, *13*, 17163–17170.

[36] X. Pan, Z. Fan, W. Chen, Y. Ding, H. Luo, X. Bao, *Nat. Mater.* **2007**, *6*, 507–511.

[37] W. Zheng, J. Zhang, B. Zhu, R. Blume, Y. Zhang, K. Schlichte, R. Schlögl, F. Schüth, D. S. Su, *ChemSusChem* **2010**, *3*, 226–230.

[38] Y. Su, B. Fan, L. Wang, Y. Liu, B. Huang, M. Fu, L. Chen, D. Ye, *Catal. Today* **n.d.**, DOI 10.1016/j.cattod.2012.04.063.

[39] R. M. M. Abbaslou, A. Tavassoli, J. Soltan, A. K. Dalai, *Appl. Catal. A* **2009**, *367*, 47–52.

[40] J. Manassen, S. Khalif, *J. Catal.* **1969**, *13*, 290–298.

[41] Y. Iwasawa, H. Nobe, S. Ogasawara, *J. Catal.* **1973**, *31*, 444–449.

[42] J. Zhang, X. Wang, Q. Su, L. Zhi, A. Thomas, X. Feng, D. S. Su, R. Schlögl, K. Müllen, *J. Am. Chem. Soc.* **2009**, *131*, 11296–11297.

[43] T. Hirao, M. Higuchi, I. Ikeda, Y. Ohshiro, *J. Chem. Soc., Chem. Commun.* **1993**, 194–195.

[44] J. Manassen, J. Wallach, *J. Am. Chem. Soc.* **1965**, *87*, 2671–2677.

[45] P. N. Degannes, D. M. Ruthven, *Can. J. Chem. Eng.* **1979**, *57*, 627–630.

[46] T. G. Alkhazov, A. E. Lisovskii, Y. A. Ismailov, A. I. Kozharov, *Kinet. Catal.* **1978**, *19*, 482–485.

[47] M. F. R. Pereira, J. J. M. Órfão, J. L. Figueiredo, *Appl. Catal. A* **1999**, *184*, 153–160.

[48] M. F. R. Pereira, J. J. M. Orfão, J. L. Figueiredo, *Appl. Catal. A* **2000**, *196*, 43–54.

[49] M. F. R. Pereira, J. J. M. Órfão, J. L. Figueiredo, *Appl. Catal. A* **2001**, *218*, 307–318.

[50] A. E. Lisovskii, C. Aharoni, *Catal. Rev. Sci. Eng.* **1994**, *36*, 25.

[51] A. Guerrero-Ruiz, I. Rodríguez-Ramos, *Carbon* **1994**, *32*, 23–29.

[52] M. F. R. Pereira, J. J. M. Órfão, J. L. Figueiredo, *Coll Surf A* **2004**, *241*, 165–171.

[53] J. Zhang, D. S. Su, A. Zhang, D. Wang, R. Schlögl, C. Hébert, *Angew. Chem. Int. Ed.* **2007**, *46*, 7319–7323.

[54] D. Su, N. I. Maksimova, G. Mestl, V. L. Kuznetsov, V. Keller, R. Schlögl, N. Keller, *Carbon* **2007**, *45*, 2145–2151.

[55] B. Frank, J. Zhang, X. Liu, M. Morassutto, R. Schomäcker, R. Schlögl, D. S. Su, in *Proc. DGMK Conf. 2009–2: Production and Use of Light Olefins*, Dresden, **2009**, pp. 163–170.

[56] M. F. R. Pereira, J. L. Figueiredo, J. J. M. Órfão, P. Serp, P. Kalck, Y. Kihn, *Carbon* **2004**, *42*, 2807–2813.

[57] B. Frank, M. E. Schuster, R. Schlögl, D. S. Su, *Angew. Chem. Int. Ed.* **2012**, DOI 10.1002/anie.201206093.

[58] J. J. Delgado, X.-W. Chen, B. Frank, D. S. Su, R. Schlögl, *Catal. Today* **2012**, *186*, 93–98.

[59] D. W. McKee, *Chem. Phys. Carbon* **1991**, *23*, 174–232.

[60] B. Frank, M. Morassutto, R. Schomäcker, R. Schlögl, D. S. Su, *ChemCatChem* **2010**, *2*, 644–648.

[61] B. Frank, J. Zhang, R. Blume, R. Schlögl, D. S. Su, *Angew. Chem. Int. Ed.* **2009**, *48*, 6913–6917.

[62] Z.-J. Sui, J.-H. Zhou, Y.-C. Dai, W.-K. Yuan, *Catal. Today* **2005**, *106*, 90–94.

[63] J. Zhang, X. Liu, R. Blume, A. Zhang, R. Schlögl, D. S. Su, *Science* **2008**, *322*, 73–77.

[64] V. Schwartz, H. Xie, H. M. Meyer III, S. H. Overbury, C. Liang, *Carbon* **2011**, *49*, 659–668.
[65] F. Cavani, N. Ballarini, A. Cericola, *Catal. Today* **2007**, *127*, 113–131.
[66] L. Liu, Q.-F. Deng, B. Agula, X. Zhao, T.-Z. Ren, Z.-Y. Yuan, *Chem. Commun.* **2011**, *47*, 8334–8336.
[67] J. Zhang, D. S. Su, R. Blume, R. Schlögl, R. Wang, X. Yang, A. Gajović, *Angew. Chem. Int. Ed.* **2010**, *49*, 8640–8644.
[68] K. Tanabe, M. Misono, Y. Ono, H. Hattori, *New Solid Acids and Bases: Their Catalytic Properties*, Elsevier, **1989**.
[69] J. Bedia, J. M. Rosas, D. Vera, J. Rodríguez-Mirasol, T. Cordero, *Catal. Today* **2010**, *158*, 89–96.
[70] G. S. Szymanski, G. Rychlicki, *Carbon* **1993**, *31*, 247–257.
[71] G. S. Szymański, G. Rychlicki, A. P. Terzyk, *Carbon* **1994**, *32*, 265–271.
[72] G. S. Szymański, G. Rychlicki, *Carbon* **1991**, *29*, 489–498.
[73] G. C. Grunewald, R. S. Drago, *J. Am. Chem. Soc.* **1991**, *113*, 1636–1639.
[74] I. F. Silva, J. Vital, A. M. Ramos, H. Valente, A. M. B. do Rego, M. J. Reis, *Carbon* **1998**, *36*, 1159–1165.
[75] R. D. Weinstein, A. R. Ferens, R. J. Orange, P. Lemaire, *Carbon* **2011**, *49*, 701–707.
[76] A. R. Ferens, R. D. Weinstein, R. Giuliano, J. A. Hull, *Carbon* **2012**, *50*, 192–200.
[77] D. R. Dreyer, H.-P. Jia, C. W. Bielawski, *Angew. Chem. Int. Ed.* **2010**, *49*, 6813–6816.
[78] W. S. Hummers, R. E. Offeman, *J. Am. Chem. Soc.* **1958**, *80*, 1339–1339.
[79] H. P. Boehm, A. Clauss, G. O. Fischer, U. Hofmann, *Z. Naturforsch. B* **1962**, *17*, 150–153.
[80] H.-P. Jia, D. R. Dreyer, C. W. Bielawski, *Tetrahedron* **2011**, *67*, 4431–4434.
[81] J. Long, X. Xie, J. Xu, Q. Gu, L. Chen, X. Wang, *ACS Catal.* **2012**, *2*, 622–631.
[82] Y. Gao, G. Hu, J. Zhong, Z. Shi, Y. Zhu, D. S. Su, J. Wang, X. Bao, D. Ma, *Angew. Chem. Int. Ed.*, DOI 10.1002/anie.201207918 **n.d.**
[83] H. Yu, F. Peng, J. Tan, X. Hu, H. Wang, J. Yang, W. Zheng, *Angew. Chem. Int. Ed.* **2011**, *50*, 3978–3982.
[84] X. Yang, H. Yu, F. Peng, H. Wang, *ChemSusChem* **2012**, *5*, 1213–1217.
[85] H. Gao, J. Zhao, *J. Chem. Phys.* **2010**, *132*, 234704–234704-7.
[86] Y. Wang, J. Zhang, X. Wang, M. Antonietti, H. Li, *Angew. Chem. Int. Ed.* **2010**, *49*, 3356–3359.
[87] B. Frank, R. Blume, A. Rinaldi, A. Trunschke, R. Schlögl, *Angew. Chem. Int. Ed.* **2011**, *50*, 10226–10230.
[88] B. Frank, R. Blume, A. Rinaldi, A. Trunschke, R. Schlögl, in *Proc. DGMK Conf. 2011–2: Catalysis – Innovative Applications in Petrochemistry and Refining*, Dresden, **2011**, pp. 211–216.
[89] J. S. Milano-Brusco, H. Zang, A. Wolf, W. Leitner, *Chem. Ing. Tech.* **2010**, *82*, 1334–1334.
[90] B. Li, Z. Xu, *J. Am. Chem. Soc.* **2009**, *131*, 16380–16382.
[91] H. H. Byung, H. S. Dae, Y. C. Sung, *Tetrahedron Lett.* **1985**, *26*, 6233–6234.
[92] P. Makowski, F. Goettmann, A. Thomas, M. Antonietti, in *14th International Congress on Catalysis*, Seoul, Korea, **2008**.
[93] A. Wolf, V. Michele, J. Assmann, L. Mleczko, *Catalyst and Process for Hydrogenating Organic Compounds*, **2011**, U.S. Patent 20110130592.
[94] D. Eder, *Chem. Rev.* **2010**, *110*, 1348–1385.
[95] P. Albers, K. Seibold, A. J. McEvoy, J. Kiwi, *J. Phys. Chem.* **1989**, *93*, 1510–1515.
[96] G. Goncalves, P. A. A. P. Marques, C. M. Granadeiro, H. I. S. Nogueira, M. K. Singh, J. Graìcio, *Chem. Mater.* **2009**, *21*, 4796–4802.
[97] B. F. Machado, P. Serp, *Catal. Sci. Technol.* **2011**, *2*, 54–75.
[98] Á. Mastalir, Z. Király, Á. Patzkó, I. Dékány, P. L'Argentiere, *Carbon* **2008**, *46*, 1631–1637.
[99] R. Yu, Q. Liu, K.-L. Tan, G.-Q. Xu, S. C. Ng, H. S. O. Chan, T. S. A. Hor, *J. Chem. Soc., Faraday Trans.* **1997**, *93*, 2207–2210.

[100] F. M. Yasin, R. A. Boulos, B. Y. Hong, A. Cornejo, K. S. Iyer, L. Gao, H. T. Chua, C. L. Raston, *Chem. Commun.* **2012**, *48*, 10102–10104.

[101] S. Santra, P. Ranjan, P. Bera, P. Ghosh, S. K. Mandal, *RSC Adv.* **2012**, *2*, 7523–7533.

[102] R. Arrigo, S. Wrabetz, M. E. Schuster, D. Wang, A. Villa, D. Rosenthal, F. Girsgdies, G. Weinberg, L. Prati, R. Schlögl, et al., *Phys. Chem. Chem. Phys.* **2012**, *14*, 10523–10532.

[103] Z. Bai, H. Yan, F. Wang, L. Yang, K. Jiang, *Ionics* **2012**, DOI 10.1007/s11581-012-0779-8.

[104] R. Giordano, P. Serp, P. Kalck, Y. Kihn, J. Schreiber, C. Marhic, J.-L. Duvail, *Eur. J. Inorg. Chem.* **2003**, *2003*, 610–617.

[105] D. Wang, J.-P. Tessonnier, M. Willinger, C. Hess, D. S. Su, R. Schlögl, in *EMC 2008 14th European Microscopy Congress 1–5 September 2008, Aachen, Germany* (Eds.: S. Richter, A. Schwedt), Springer Berlin Heidelberg, Berlin, Heidelberg, **n.d.**, pp. 317–318.

[106] Z. Bai, P. Li, L. Liu, G. Xiong, *ChemCatChem* **2012**, *4*, 260–264.

[107] K. Mette, A. Bergmann, J.-P. Tessonnier, M. Hävecker, L. Yao, T. Ressler, R. Schlögl, P. Strasser, M. Behrens, *ChemCatChem* **2012**, *4*, 851–862.

[108] X. Li, D. Ma, L. Chen, X. Bao, *Catal. Lett.* **2007**, *116*, 63–69.

[109] J. Zhang, Y.-S. Hu, J.-P. Tessonnier, G. Weinberg, J. Maier, R. Schlögl, D. S. Su, *Adv. Mater.* **2008**, *20*, 1450–1455.

[110] H. Ma, L. Wang, L. Chen, C. Dong, W. Yu, T. Huang, Y. Qian, *Catal. Commun.* **2007**, *8*, 452–456.

[111] J.-P. Tessonnier, O. Ersen, G. Weinberg, C. Pham-Huu, D. S. Su, R. Schlo¨gl, *ACS Nano* **2009**, *3*, 2081–2089.

[112] Q. Fu, G. Weinberg, D. S. Su, *New Carbon Mater.* **2008**, *23*, 17–20.

[113] G. Schmid, *Nanoparticles: From Theory to Application*, John Wiley & Sons, **2011**.

[114] D. F. L. Fedlheim, C. A. J. Foss, *Metal Nanoparticles: Synthesis, Characterization, and Applications*, CRC Press, **2002**.

[115] M. Grzelczak, J. Pérez-Juste, P. Mulvaney, L. M. Liz-Marzán, *Chem. Soc. Rev.* **2008**, *37*, 1783–1791.

[116] S. Abate, R. Arrigo, M. E. Schuster, S. Perathoner, G. Centi, A. Villa, D. Su, R. Schlögl, *Catal. Today* **2010**, *157*, 280–285.

[117] J. P. Tessonnier, A. Villa, O. Majoulet, D. S. Su, R. Schlögl, *Angew. Chem. Int. Ed.* **2009**, *48*, 6543–6546.

[118] B. R. Azamian, K. S. Coleman, J. J. Davis, N. Hanson, M. L. H. Green, *Chem. Commun.* **2002**, 366–367.

[119] L. Liu, T. Wang, J. Li, Z.-X. Guo, L. Dai, D. Zhang, D. Zhu, *Chem. Phys. Lett.* **2003**, *367*, 747–752.

[120] J. M. Planeix, N. Coustel, B. Coq, V. Brotons, P. S. Kumbhar, R. Dutartre, P. Geneste, P. Bernier, P. M. Ajayan, *J. Am. Chem. Soc.* **1994**, *116*, 7935–7936.

[121] P. Kluson, L. Cerveny, *Appl. Catal. A* **1995**, *128*, 13–31.

[122] B. Coq, P. S. Kumbhar, C. Moreau, P. Moreau, M. G. Warawdekar, *J. Mol. Catal.* **1993**, *85*, 215–228.

[123] S. Galvagno, G. Capannelli, *J. Mol. Catal.* **1991**, *64*, 237–246.

[124] J.-P. Tessonnier, L. Pesant, G. Ehret, M. J. Ledoux, C. Pham-Huu, *Appl. Catal. A* **2005**, *288*, 203–210.

[125] H. Vu, F. Gonçalves, R. Philippe, E. Lamouroux, M. Corrias, Y. Kihn, D. Plee, P. Kalck, P. Serp, *J. Catal.* **2006**, *240*, 18–22.

[126] J. Qiu, H. Zhang, X. Wang, H. Han, C. Liang, C. Li, *React. Kinet. Catal. Lett.* **2006**, *88*, 269–276.

[127] P. V. Kamat, *J. Phys. Chem. Lett.* **2010**, *1*, 520–527.

[128] X. Huang, X. Qi, F. Boey, H. Zhang, *Chem. Soc. Rev.* **2012**, *41*, 666–686.

[129] Y. Li, X. Fan, J. Qi, J. Ji, S. Wang, G. Zhang, F. Zhang, *Mater. Res. Bull.* **2010**, *45*, 1413–1418.

[130] E. Yoo, T. Okata, T. Akita, M. Kohyama, J. Nakamura, I. Honma, *Nano Lett.* **2009**, *9*, 2255–2259.

[131] C.-H. Liu, X.-Q. Chen, Y.-F. Hu, T.-K. Sham, Q.-J. Sun, J.-B. Chang, X. Gao, X.-H. Sun, S.-D. Wang, *ACS Appl. Mater. Interfaces* **2013**, *5*, 5072–5079.
[132] S. Guo, S. Dong, E. Wang, *ACS Nano* **2010**, *4*, 547–555.
[133] J. Chai, F. Li, Y. Hu, Q. Zhang, D. Han, L. Niu, *J. Mater. Chem.* **2011**, *21*, 17922–17929.
[134] P. Chen, W. Xia, M. C. Ly, M. Muhler, in *15th International Congress on Catalysis*, Munich, Germany, **2012**.
[135] C. Li, X. Zhang, C. Liang, P. Chen, W. Xia, M. Muhler, in *15th International Congress on Catalysis*, Munich, Germany, **2012**.
[136] A. Villa, D. Wang, P. Spontoni, R. Arrigo, D. Su, L. Prati, *Catal. Today* **2010**, *157*, 89–93.
[137] A. Villa, D. Wang, N. Dimitratos, D. Su, V. Trevisan, L. Prati, *Catal. Today* **2010**, *150*, 8–15.
[138] J. Kang, S. Zhang, Q. Zhang, Y. Wang, *Angew. Chem. Int. Ed.* **2009**, *48*, 2565–2568.
[139] H. Xiong, M. A. M. Motchelaho, M. Moyo, L. L. Jewell, N. J. Coville, *J. Catal.* **2011**, *278*, 26–40.
[140] A. Tavasoli, R. M. Malek Abbaslou, A. K. Dalai, *Appl. Catal. A* **2008**, *346*, 58–64.
[141] M. Trépanier, A. Tavasoli, A. K. Dalai, N. Abatzoglou, *Fuel Proc. Technol.* **2009**, *90*, 367–374.
[142] J. Liu, J. Shen, X. Gao, L. Lin, *J. Therm. Anal. Calorim.* **1993**, *40*, 1245–1252.
[143] K. Xiong, J. Li, K. Liew, X. Zhan, *Appl. Catal. A* **2010**, *389*, 173–178.
[144] A. Tavasoli, M. Trépanier, R. M. Malek Abbaslou, A. K. Dalai, N. Abatzoglou, *Fuel Proc. Technol.* **2009**, *90*, 1486–1494.
[145] W. Chen, Z. Fan, X. Pan, X. Bao, *J. Am. Chem. Soc.* **2008**, *130*, 9414–9419.
[146] T. V. Choudhary, C. Sivadinarayana, D. W. Goodman, *Catal. Lett.* **2001**, *72*, 197–201.
[147] S.-F. Yin, Q.-H. Zhang, B.-Q. Xu, W.-X. Zhu, C.-F. Ng, C.-T. Au, *J. Catal.* **2004**, *224*, 384–396.
[148] A. Boisen, S. Dahl, J. K. Nørskov, C. H. Christensen, *J. Catal.* **2005**, *230*, 309–312.
[149] S.-F. Yin, B.-Q. Xu, C.-F. Ng, C.-T. Au, *Appl. Catal. B* **2004**, *48*, 237–241.
[150] J. Zhang, M. Comotti, F. Schüth, R. Schlögl, D. S. Su, *Chem. Commun.* **2007**, 1916–1918.
[151] F. R. García-García, J. Álvarez-Rodríguez, I. Rodríguez-Ramos, A. Guerrero-Ruiz, *Carbon* **2010**, *48*, 267–276.
[152] C. J. Shearer, A. Cherevan, D. Eder, *Adv. Mater.* **2014**, *26*(15), 2295–2318.
[153] R. Schlogl, G. Mestl, *Catalyst Comprising Nanocarbon Strutures For The Production Of Unsaturated Hydrocarbons*, **2005**, U.S. Patent US 20080071124.

Gabriele Centi and Siglinda Perathoner

16 Advanced photocatalytic materials by nanocarbon hybrid materials

Nanocarbon-semiconductor hybrid materials are finding increasing interest for applications in photocatalysis and as photoanodes, because nanocarbons enable improvements of the properties of the semiconductor particles. They can (i) provide a more efficient nanoarchitecture and stabilization of semiconductor nanoparticles, often with unconventional morphologies, (ii) facilitate more efficient electron collection and transport, as well as charge separation, and (iii) introduce additional functionalities relevant for photocatalytic behavior. This concise review introduces general aspects of this field, and selected recent relevant examples to provide indications of critical aspects for further consideration since the current lack of a more systematic approach is hindering the full exploitation of the potential of this new class of nanomaterials. The largest part of the activities in this field is related to photocatalytic depuration of waste emissions, but the applications related to sustainable energy applications, such as DSSC cells and devices to produce solar fuels (H_2 from water photoelectrolysis/splitting and hydrocarbons/alcohols from CO_2), is an area of emerging relevance.

16.1 Introduction

Scientific interest in nanocarbon hybrid materials to enhance the properties of photocatalysts and photoactive electrodes has been growing rapidly [1–8]. The worldwide effort to find new efficient and sustainable solutions to use renewable energy sources has pushed the need to develop new and/or improved materials able to capture and convert solar energy, for example in advanced dye-sensitized solar cells – DSSC (where the need to improve the photovoltaic performance has caused interest in using nanocarbons for a better cell design [9, 10]) or in advanced cells for producing solar fuels [11–13].

An efficient harvesting and conversion/use of solar light requires developing materials with specific nanoarchitectures, which are also necessary for an efficient charge transport and separation [2, 14]. Nanocarbons exist in a wide range of morphologies, predominantly with carbon atoms in sp^2 configurations although sp and sp^3 C atoms also are present due to defects [6, 7], and often govern the material properties. Well-known types of nanostructures are carbon nanotubes (CNT), graphene and fullerene, but this class of materials comprises many more types of carbon materials such as nanofibers, nanocoils, nanodiamond, nanohorns, nanoonion, *etc*. Nanocarbons can be considered the more extensively studied class of nanomaterials for which a tailored control in the nanoarchitecture and properties is possible [15]. In addition, these materials often show high electrical conductivity, even if the properties can go from insulating to conductive materials passing though semiconductor as well, depending on the nanostructure, type of defects and modalities of preparation [16].

On the other hand, metal oxide semiconductors, such as TiO_2 used typically as photocatalysts and in photoanodes [17, 18], often show a poor efficiency related to charge-recombination at grain boundary and limited electron transport. Even if significant advances have been made in controlling and optimizing their nanoarchitecture, producing for example well-ordered bidimensional nanostructured films based on an aligned array of TiO_2 nanotubes [19, 20] showing enhanced properties in terms of light harvesting and electron transfer [21], charge transport and recombination still remain a critical factor.

Realizing hybrid materials based on nanocarbons and TiO_2 (or other semiconductors) offers great potential to overcome these limitations and prepare optimized nanomaterials, even though the lack of understanding of the critical factors sometimes causes contrasting results. For example, Gray and co-workers [22] studying the photocatalytic properties of TiO_2-CNTs composites in the photooxidation of phenol observed a promotion for SWCNT (single-wall CNT) and large TiO_2 particles (100 nm), but an inhibition for smaller (5 nm) particles. MWCNT (multi-wall CNT) instead deactivate (with respect to Degussa P25 TiO_2) both small and large TiO_2 particles, but the effect is enhanced in the smaller ones. The behavior derives from the combination of the promotion of charge-separation in TiO_2-CNTs composites, and the negative effect induced on the creation of trapping sites which is enhanced in TiO_2 particles (due to the larger interfacial area) and in MWCNT with respect to SWCNT (due to the higher number of defects in the former).

The use of nanocarbon-semiconductor hybrid materials thus offers a great potential to the design and development of novel/improved photocatalysts and photoanodes, but it is necessary to have a detailed understanding of the many factors which determine the overall properties. This chapter will analyze these aspects presenting a concise analysis of the topic with selected relevant developments in the field, mainly in the last few years.

16.1.1 Hybrid *vs.* composite nanomaterials

Although the terms nanocomposite and hybrid are often used to define similar materials, we will use the classification indicated by Vilatela and Eder [1]. Nanocomposites are multiphase materials, in which one phase is dispersed in a second phase, resulting in a combination of the individual properties of the component materials. The volume fraction of the nanocarbon is typically less than a few percent. Nanocarbon hybrids are instead formed by both components with similar volume fractions. The inorganic compound (such as semiconductor nanoparticles) is deposited onto the surface of the nanocarbon. The interface between the two phases often determines the properties of the nanomaterial. Photocatalytic or photoanode materials based on a semiconductor such as titania and nanocarbons should be defined as hybrid nanomaterials [23] (according to the above definition) rather than nanocomposite as is often the case.

A key aspect related to the concept of hybrid materials is that the resulting material possesses properties which do not derive from the linear addition of the properties of the pristine materials. Many studies have shown how this is the common case in semiconductor-nanocarbons hybrid materials, for example the case of CNT-TiO_2 hybrids [23]. Figure 16.1 shows schematically the mechanism of electron transfer from a TiO_2 nanoparticle to an MWCNT. Due to the formation of a heterojunction and an electron transfer from the titania to the CNT, a modification of the characteristics of the titania nanoparticles occurs, for example reducing the processes of recombination of holes and electrons upon photoexcitation. The effect is effectively more complex (as discussed later in a more detail) and the presence of defects at the interface acting as exciton-like trap states is rather important. In nanostructured photoelectrodes, electron traps in TiO_2 arise mainly from "incorrectly" coordinated titanium ions found at intrinsic defect sites and surface sites, and near intercalated species. The nature of the TiO_2/CNTs interface (which in turn strongly depends also on the specific nature of CNT, particularly the presence of defect sites) determines this effect and in fact the UV-Visible spectrum of TiO_2 nanoparticles supported over CNTs indicates the formation of these trapping sites, depending on the "nature" of CNTs. Recent studies have shown that low mobility in polycrystalline TiO_2 nanotubes may also result from a single sharp resonance arising from exciton-like trap states [24]. A similar effect could be expected in TiO_2-CNTs nanostructured films. However, also the inverse processes of electron injection into the conduction band of TiO_2 can occur [23].

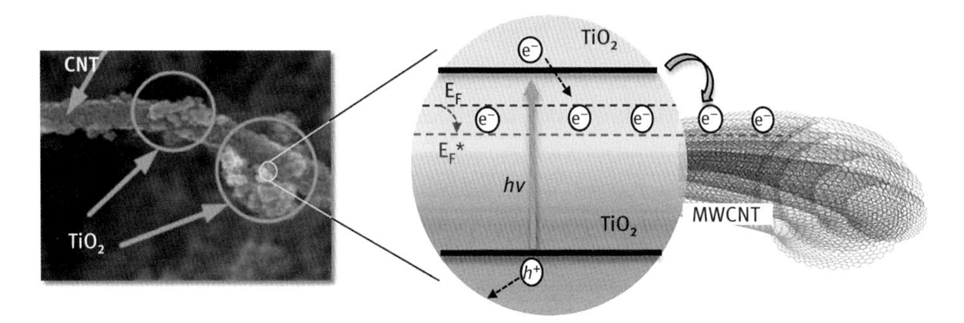

Fig. 16.1: Schematic model of the electron transfer from TiO_2 nanoparticles to CNTs. The inset shows a SEM image of TiO_2 nanoparticles supported over a MWCNT. Adapted with permission from Centi and Perathoner [7].

16.1.2 Use of nanocarbon hybrid materials in photoreactions

Recently, much activity has been dedicated to preparing nanocarbon-semiconductor materials and investigating their uses, especially in the photocatalytic removal of pollutants both in liquid and gas phase. Eder and co-workers [1, 23] in their comprehensive reviews on nanocarbon hybrids have discussed various examples for enhanced performances of hybrids in photocatalysis and other applications. Leary and Westwood [5] have specifically discussed the enhancement of TiO_2 photocatalytic properties by realizing hybrids with nanocarbons (CNTs, fullerene, graphene and other nanocarbons). Inagaki [3] has reviewed the promotion of the photocatalytic performance of TiO_2 by carbon coating, while Ng *et al.* [25] discussed the application of semiconductor-carbon nanomaterials especially for the photooxidation of pollutants. Faria and co-workers [26] have analyzed the application of graphene-TiO_2 hybrid nanomaterials as photocatalysts for the photodegradation of pollutants and micro-organisms. Various reviews have thus described the advances in this field with a focus on photocatalysis for environmental applications.

In general, the addition of nanocarbons has promoted the photocatalytic properties of semiconductors (TiO_2 was the most studied) for various reasons, ranging from a reduced rate of charge recombination to providing a better (hydrophobic) local microenvironment that enhances the local concentration of organic species. Carbon atoms may also diffuse into TiO_2 lattice inducing a slight shift in the band edge towards the visible region and in this way promoting activity. There are thus different ways by which photocatalytic activity of semiconductor particles can be enhanced by making hybrids with nanocarbons:

(i) band-gap tuning and photosensitization (as mentioned above, some nanocarbon materials may also behaves as semiconductors or quantum dots [16]);
(ii) minimization of charge carrier recombination by electron transfer and Fermi-level equilibration;
(iii) creation of a better local microenvironment;
(iv) creation of an overstructure which facilitates mass and charge fast diffusion;
(v) promotion of reactants adsorption.

It is thus evident that the characteristics of nanocarbons (conductivity, local structure, presence of defects and functional groups, morphology, *etc.*) are critical to determining the properties of the hybrid nanomaterial with the semiconductor. However, most of the literature studies put emphasis on the analysis of semiconductor characteristics, while often nanocarbons are only described in generic terms (CNT, for example). Yet, it is well known how the properties of nanocarbons can be considerably different from case to case (depending on details in preparation), even if the structure is formally the same (MWCNT, for example).

There are also various examples in the literature showing that photocatalytic performances are determined by the characteristics of the nanocarbon. Ouzzine *et al.*

[27] have shown how the crystallinity of carbon nanofibers (CNF) used to prepare TiO_2/CNF hybrid materials strongly influences photocatalytic performances in the conversion of low concentrations of propene. Shi *et al.* [28] also showed how the characteristics of carbon fibers used as support for TiO_2 determine the features of titania (morphology, *etc.*) and the photocatalytic behavior. However, the results are mostly phenomenological and there is in general a lack of detail about the correlation between features of the nanocarbon and photocatalytic performances of the hybrid material. In addition, as briefly outlined above, there are many parameters which can influence the photocatalytic behavior. A systematic approach to exploit these aspects to prepare optimized photocatalysts is lacking.

An optimal design of the hybrid photoactive material is much more critical when these nanomaterials are used in applications of sustainable energy (DSSC cells and devices to produce solar fuels, *e.g.* H_2 from water *via* photoelectrolysis or photocatalytic water splitting, or hydrocarbons/alcohols from CO_2). In fact, these applications are more demanding with respect to environmental applications (related to the removal of pollutants or micro-organisms from gaseous or liquid streams), because solar fuels, for example, have a higher energy than the reactants, while the contrary occurs in the environmental area (production of CO_2 from organics, for example). The latter processes are the most investigated, but from the perspective of applications the former are more relevant, due to a bigger potential market.

16.2 Nanocarbon characteristics

It is not in the scope of this chapter to discuss in detail the characteristics of nanocarbons, but to only remark on some of their features which can be relevant and have to be considered for a correct design and understanding of nanocarbon-semiconductor hybrid materials.

Nanocarbons indicate a broad range of carbon materials having a tailored nanoscale dimension and functional properties, which significantly depend on their nanoscale features. CNT and graphene belong to this class of materials comprising many more type of carbon materials, such as nanofibers, nanocoils, nanodiamonds, nanohorns, nanoonion, fullerene, *etc.* Nanocarbon definitions also include ordered mesoporous carbon materials obtained, for example, by the replica method from mesoporous silica or other ordered oxides, which are then removed after the template synthesis of the mesoporous carbon [29, 30]. The ordered mesoporous structure of these materials, equivalent to that of silica-based materials from which they are derived, offers rather interesting properties to develop valuable photocatalytic materials. However, there are few studies in this direction. Park and co-workers [31] showed that anatase TiO_2 photocatalysts supported on the ordered mesoporous carbon CMK-3 exhibit higher efficiency in removing the Rhodamine 6G dye from waste solutions with respect to a photocatalyst based on commercial P25 TiO_2 on activated carbon.

Liu *et al.* [32] reported the characteristics and reactivity of highly ordered mesoporous carbon-titania hybrid materials synthesized *via* organic-inorganic-amphiphilic co-assembly followed by *in situ* crystallization. In the degradation of Rhodamine B these materials also show enhanced properties due to the dispersion/stabilization of small titania nanocrystals and the adsorptive capacity of the nanocarbon.

A general issue is that these nanocarbons are often only discussed in terms of a class of materials based on their shape (CNT, *etc.*). However, the growing understanding of these materials [16, 33], of their controlled synthesis [34], and of the interfacial phenomena during interaction between nanocarbons and semiconductor particles [1, 6, 8, 23, 35] has clearly indicated that in addition to the relevant role given from the possibility to tune nanoarchitecture (and related influence on mass and charge transport, as well as on microenvironment [36]) the specific nanocarbon characteristics, surface chemistry and presence of defect sites determine the properties.

In addition to the different carbon nanostructures available [37] (carbon nanofibers, nanotubes, nanocoils, nanohorns, nanodiamond, nanoonions, graphene, *etc.*), various doped nanocarbon materials are available (particularly with N and B) [38–41]. Growing knowledge is also available on the assembling of low-dimensional carbon nano-objects into three-dimensional architectures (films, hollow spherical capsules, or hollow nanotubes) [36, 42], a research area further extending the possibilities to design specific nanoarchitectures for the hybrid materials. The direct growth of ordered arrays of carbon nanotubes or nanofibers (carpet-type) from a nanopatterned catalyst array or using inorganic templates [43] has opened new possibilities to develop advanced photocatalyst/-anode architectures [14].

A great variety of crystalline and disordered structures is known for carbon materials since three different hybridizations are possible: sp^3 (diamond), sp^2 (graphene) and sp^1 (carbyne). Between these limiting cases, a rich variety of carbon materials exists. Some are known and have been commercialized for a long time (graphite, carbon blacks and activated carbons) and some for about half a century (carbon fibers, glass-like carbons and pyrolytic carbons). Carbon fibers are analogous to graphite, but the arrangement of graphitic elements follows the fibrous habit. Glassy carbon contains ribbon-like graphitic domains of sp^2 hybridized carbon atoms stacked in layers, but these are irregularly twisted and entangled. Graphite is formed by stacked graphene layers, but often the layers are irregularly turned around the z-axis and shifted against each other in the xy-direction. These structures are called turbostratic, and are often present in CNTs. The order may be increased by high temperature thermolysis and postgraphitization, leading to pyrolytic graphite. However, the area of carbon materials received a large push from the discovery of CNTs about 20 years ago and fullerene more recently.

Fullerenes are cage-like carbon structures which derive from a graphene sheet where a few six-membered rings are replaced by five-membered ones which forces the layer into a bent shape. Placing the pentagons at suitable positions, a spherical structure of 60 carbon atoms is obtained – the buckminsterfullerene. Fullerenes with different numbers of carbon atoms are also possible. Multi-walled fullerenes, where the carbon cages are concentrically arranged one inside another, are indicated as carbon onions. A graphene sheet may be rolled up with its edges connected in a butt joint to generate a single-walled carbon nanotube (SWCNT). Several of these tubes may fit one into another to make a multi-walled carbon tube (MWCNT).

In CNTs, the rolled graphene sheets are oriented along the main z-axis, but different tubular carbon materials also exist. Bamboo-like carbon nanotubes are relatively straight-sided tubes divided into sections by "knots" made from single or multiple graphene sheets. Cup-stacked carbon nanotubes consist of columnar stacks of hollow, truncated cones, *i.e.* a superstructure of nanocones. The structure is similar to that of carbon nanofibers (fishbone) which also have the external surface with the graphene layers oriented in the 40–85° range with respect to the main z-axis. Carbon nanohorns are short single-walled closed carbon nanotubes (2–6 nm in diameter and about 50 nm in length) with a conical tip and an internal angle of ca. 20° situated on one end. Several of them may cluster to form structures that have been compared to dahlia blossoms in the literature. Helical carbon nanotubes are carbon nanotubes with coiled or helical shapes (resembling a microscopic spiral spring) due to the presence of defects that give rise to the twisting.

16.2.1 The role of defects

Three are the classes of carbon nanotubes: (i) zigzag carbon nanotubes, (ii) armchair carbon nanotubes and (iii) chiral carbon nanotubes. They differ in the way of which the basic graphene sheet is rolled up. In the ideal CNTs structure, the carbon nanotubes do not differ from the arrangement in graphite that also features equivalent atoms. Consequently, one might assume the π-electrons to be delocalized across the whole cylindrical structure. At the end of the nanotube, however, the equivalence of bonds is broken. The cylinder is deformed toward the ends and the degree of π-electron delocalization decreases in the entire system. Several resonance structures are possible, including localized double bonds or six-membered rings without aromatic character. Alternating bonds are most frequent near the ends of the tube while their number decreases toward the center.

Considering the ends of an open SWCNT, the outermost carbon atoms must be saturated by attaching either functional groups or a cap of carbon atoms, obtaining in the latter a closed SWCNT. The shape of caps in closed SWCNTs is highly variable. Both approximately hemispheric caps and pointed structures are found, and even concave domains have been observed. Pentagonal defects are possible with a convex curvature and even an inverse curvature (seven-membered rings). These defects lead to beak-shaped caps that exhibit considerably smaller diameters than the tubes themselves. In MWCNT, five-membered rings are also crucial to the formation of the ending cap or open structure of the nanotube, such as the formation of toroid ends. Defects of individual tubes may influence the MWCNT structure in other ways too. The simultaneous existence of five- and seven-membered rings, for example, makes a bent nanotube. In multi-walled systems, these elbow structures are found in-between encapsulated regions. The flexure is thus confined to a few graphene layers.

Defects in MWCNTs are always present. We can briefly differentiate between topological defects which lead to rehybridization (C5 and C7 rings instead of C6 lead to rehybridization between sp^2 and sp^3) and incomplete bonding defects (vacancies, dislocation) (Fig. 16.2). Functionalization or doping with heteroelements may add further modifications with respect to the ideal ordered structure, but are also the sites which allow for anchoring supported metals or metal oxides, or to functionalize the CNTs with organic groups.

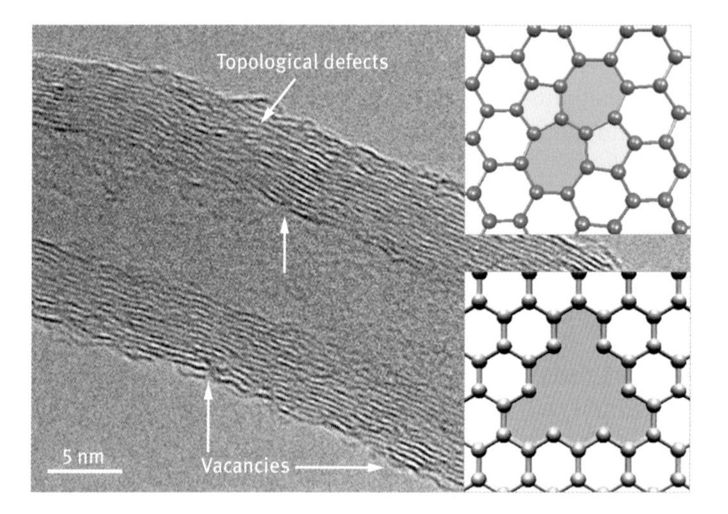

Fig. 16.2: TEM image of a CNT with indications of topological defects and vacancies. Reproduced with permission from Wiley VCH (2011) [7].

This short excursus on the different types of nanocarbon materials, although not exhaustive, illustrates the rich variety of the possible situations and nanostructures that derive from often small changes in the synthesis conditions. When a nanocarbon is indicated generically, without proving detailed data on its features, it is often a sample which may be intrinsically different from another apparently similar one. Often the sample itself is not homogeneous, presenting a variety of different local situations. These different characteristics reflect on the different interaction with supported semiconductor particles, and in turn in local inhomogeneities, difficulties in reproducibility and possible contradictory results. The characteristics of nanocarbons are thus a very relevant parameter to consider, although often underestimated.

16.2.2 Modification of nanocarbons

The properties of nanocarbon materials can be further modified by substituting part of the carbon atoms by inserting heteroatoms such as N and B. Finally, they can be further modified by post-treatment [23] (most common is by oxidation to create surface oxygen functional groups, such as carboxylic, carbonyl and phenolic groups) [44], and further functionalized by organic groups (at the reactive tips, or in the outer sidewall and the cylindrical inner surface of the tube, particularly at defect sites), or by metal or metal oxide particles [23, 45]. The rich chemistry of carbon-based (nano)materials offers thus a large range of new possibilities, but this is the intrinsic difficulty and the current limit in the investigation of these materials. Materials with significant structural/surface differences and sometimes differences between the individual elements in the same batch are often assumed similar. This is one of the main current limits in the analysis of the literature data, due to a lack of (i) detailed characteristics of the nanotubes, especially in terms of fine nanostructure and defect state, and (ii) understanding of the relation between these characteristics and the functional properties determining the performances of these materials. On the other hand, this richness in the properties and characteristics of nanocarbons shows that the potential of exploiting the properties of nanocarbon-semiconductor hybrids is still far from being achieved.

16.2.3 New aspects

There are also a number of new features recently discovered in nanocarbon materials which open new perspectives. Due to the large recent interest in graphene, a number of recent studies focused on unzipping CNT into graphene nanoribbons. By using encapsulated Co particles it is possible to achieve a precise cutting, repairing, and interconnecting of the nanoribbons [46]. Depending on the cutting angle, graphene nanoribbons show different chirality, which determines a change in the energy gaps and the presence of one-dimensional edge states with unusual magnetic structure [47]. This type of site is probably present in other types of nanocarbon species. This is an area not yet studied, but it may be expected that these sites would influence quantum-confined states in quantum dots, the band edges in the semiconductor particles, polarons and light adsorption, and charge recombination. It is thus expected that the properties of supported (small) semiconductor particles and quantum dots could be significantly influenced.

Under electron irradiation (or by other mechanisms) it is possible to generate carbon vacancies leading to the formation of extended defect domains (with the presence of pentagonal and heptagonal, and even four-membered carbon rings) showing semiconductor character. This is the mechanism of formation of semiconductor properties in quantum-dot carbon nanoparticles or graphene nanoribbon. The mechanism of charge transfer and nanocarbon-semiconductor interaction would be largely influenced by the presence of an insulating, conductive or semiconductor behavior of the nanocarbon. If the adsorption for semiconductor nanocarbon dots is in the visible region, and in some way tunable, it is also evident that this is a very important area to design and develop advanced photocatalytic materials active in the visible region.

16.2.4 Nanocarbon quantum dots

Li *et al.* [48] have recently shown that CQD/Cu_2O (CQD = C quantum dots) hybrid nanomaterials with protruding nanostructures on the surface show photocatalytic behavior with (near) IR light. Zhang *et al.* [49] prepared N-doped CQD by a solvothermal route and these materials displayed tunable luminescence due to different N contents. Yu and Kwak [50] developed photocatalysts based on CQDs and mesoporous hematite (α-Fe_2O_3), where the CQDs play a pivotal role in improving the photocatalytic activity under visible light irradiation. Yu *et al.* [51] developed ZnO/CQDs hybrid materials (by a one-step hydrothermal reaction) showing superior photocatalysts for the degradation of benzene and methanol under visible light. Mirtchev *et al.* [52] developed colloidally stable CQDs for their utilization as sensitizers in nanocrystalline TiO_2-based solar cells.

These examples show the great potential of using CQD and nanocarbon materials with tailored semiconducting properties to develop new types of photoactive materials, especially active in the visible region. Cao *et al.* [53] have reviewed the topic of creating energy band gaps within graphene and other carbon materials to impart fluorescence emissions. Many experimental techniques to introduce band gaps have been used, such as cutting graphene sheets into small pieces or manipulating the π electronic network to form quantum-confined sp^2 "islands" in a graphene sheet, which apparently involves the formation or exploitation of structural defects. In fact, defects in graphene materials not only play a critical role in the creation of band gaps for emissive electronic transitions, but also contribute directly to the bright photoluminescence emissions observed in these materials. Researchers have found similar defect-derived photoluminescence in carbon nanotubes and small carbon nanoparticles (CQDs), although Cao *et al.* [53] noted that the emissions properties of these different related carbon nanomaterials had not yet systematically examined, nor had their mechanistic origins been clearly proven. There are similarities between electrons confined in the conjugated π-domains in graphene and in nanoscale semiconductor particles, with similar size/dimension-dependent electronic energy band gaps and corresponding variations in fluorescence colors (Fig. 16.3) [53, 54].

CQDs can be efficient chromophores for photon harvesting and photoconversion [55] and it was shown that photogenerated electrons could be transferred to gold or platinum nanoparticles for the photocatalytic conversion of carbon dioxide and splitting of water for hydrogen generation [55, 56]. In both cases, CQDs similar to titania

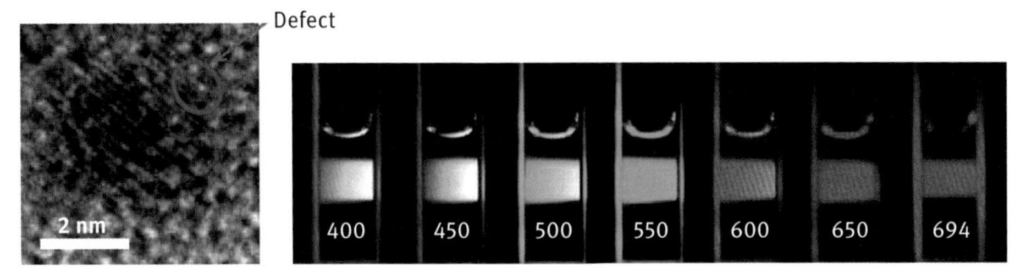

Fig. 16.3: A carbon nanoparticle with surface defects (left), and emission color variations in carbon dots depending on size and type of defects. Adapted from Cao *et al.* [53] and Sun *et al.* [54] with kind permission from American Chemical Society (2006 and 2013).

nanoparticles act as a photovoltaic nanodevice and the surface metal particles as a short circuit electrode for water (photo)electrolysis and CO_2 reduction by the photogenerated protons/electrons. CQDs thus resemble nanoscale semiconductors in terms of photoinduced redox processes, in which the radiative recombinations in the absence of quenchers result in bright photoluminescence. For the defect-derived photoluminescence emissions in graphene materials, similar redox-driven quenching processes with both electron donors and acceptors have been reported [57–59].

CQDs are more robust that most of the sensitizers traditionally used to promote visible-light activity of TiO_2, such as the Ru complexes typically used in DSSC cells [60, 61], or traditional visible-light active quantum dots such as CdS [62]. On the other hand, the sensitizer mechanism is more effective than other strategies used for promoting TiO_2 visible-light activity such as band gap engineering, nanostructuring, and doping [63, 64]. The field of using CQDs as sensitizers for nanocrystalline TiO_2 solar cells has only emerged recently [52], but it is very promising. This area can be integrated with the use of nanocarbon materials to enhance charge transport properties [65, 66] and to develop improved and transparent counter-electrods [9, 67, 68].

There is thus a bright future for carbon nanomaterials for DSSC applications [69–71]. CNT, graphene, and their hybrids are highly prospective materials to replace transparent conducting oxide (TCO) layers and counter electrodes in DSSCs. Moreover, carbon nanomaterials enable improvement of the performance of absorbing layers in working photoanodes by enhancing the light absorption and electron transport across the semiconducting nanostructured film, besides their possible role as sensitizers as discussed above. The possibility of using nanocarbons as components for printable solar cells [69] is also relevant since these are expected to play an important role in the future solar-cell market (flexible third-generation solar cells).

16.3 Mechanisms of nanocarbon promotion in photoactivated processes

Figure 16.1 briefly introduced some aspects of the mechanism of nanocarbon promotion in photoactivated processes, but it is useful to briefly analyze in more detail these aspects, because a better understanding of these features is of fundamental relevance for a more rational design of solar cells and photocatalysts.

It should also be briefly recalled that semiconductors can be added to nanocarbons in different ways, such as using sol-gel, hydrothermal, solvothermal and other methods (see Chapter 5). These procedures lead to different sizes and shapes in semiconductor particles resulting in different types of nanocarbon-semiconductor interactions which may significantly influence the electron-transfer charge carrier mobility, and interface states. The latter play a relevant role in introducing radiative paths (carrier-trapped-centers and electron-hole recombination centers), but also in strain-induced band gap modification [72]. These are aspects scarcely studied, particularly in relation to nanocarbon-semiconductor (TiO_2) hybrids, but which are a critical element for their rational design.

As indicated from first-principles band structure calculations [72], the band gap of TiO_2 in the anatase phase can be effectively reduced by applying stress along a soft direction. Figure 16.4 shows the crystal cells of anatase TiO_2, and with a red thick line the axis in representative octahedra of anatase TiO_2 which, under stress, induces a change in the band gap of anatase TiO_2. Figure 16.4(c) shows the effect of application of different types of stress (hydrostatic, epitaxial, and uniaxial stress) on the variation of band gap (ΔE). These results evidence that band gap change is much more significant when stress is applied along a weak direction of anatase, *i.e.* it depends on the specific crystallographic features of TiO_2 nanoparticles which, in turn, are depen-

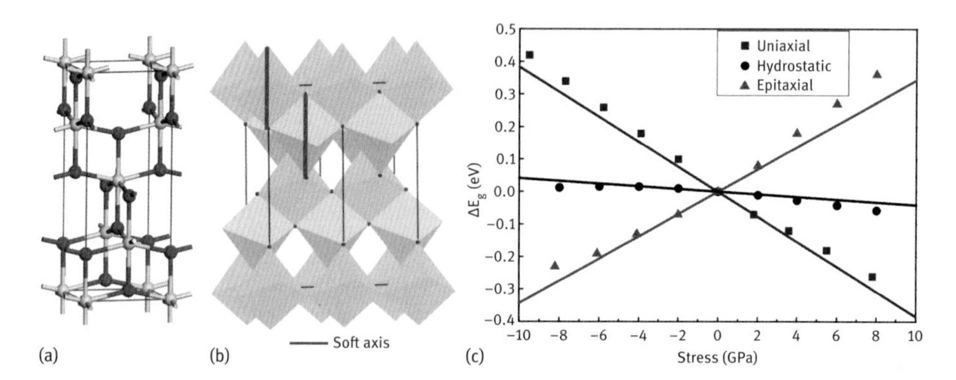

Fig. 16.4: (a) The crystal cells of anatase TiO_2. (b) View of the octahedra-packing of anatase TiO_2; the thick red lines indicate the axis in representative octahedra which, under stress, results in a change of band gap. (c) Band gap variation in anatase TiO_2 under hydrostatic, epitaxial, and uniaxial stress. Adapted from Yin *et al.* [72] with kind permission from American Institute of Physics (2010).

dent on the interaction with nanocarbon (as discussed later). The effect was noted in the literature, although not explained. As an example, Faria and co-workers [73] observed that the introduction of CNT into the TiO_2 matrix leads to an increase in the absorption edge (decrease in the band gap energy) and that the increase in light absorption is proportional to the CNT loading, but depends on the preparation method of the samples. They attributed the effect to the creation of an electronic interphase interaction between CNT and TiO_2, which is a generic statement underlying the phenomena discussed above. A parallel increase in the rate of photocatalytic degradation reactions is observed, underlining the importance of a better understanding of the above effects. This should include a full photophysical characterization of the photoinduced processes, because interface states will also introduce carrier-trapped-centers and electron-hole recombination centers which instead could negatively influence performances.

A further relevant mechanism of modification of photobehavior is related to C-doping of TiO_2 nanocrystals. This is an area to which relevant attention has recently been given. Liu *et al.* [74] showed that carbon doped TiO_2 nanoparticles, synthesized by a modified sol-gel route based on the self-assembly technique exploiting oleic acid as a pore directing agent and carbon source, show an apparent red shift in the optical absorption edge by 0.5 eV, 2.69 eV compared to the 3.18 eV of reference anatase TiO_2. XPS showed the substitution of carbon for oxygen atoms and ESR spectroscopy revealed the formation of two carbon related paramagnetic centers in C-TiO_2, whose intensity was markedly enhanced under visible light illumination, pointing out to the formation of localized states within the anatase band gap, following carbon doping. The doped materials show enhanced photocatalytic activity under visible light ($\lambda > 420$ nm) irradiation. Chen *et al.* [75] also reported the enhanced visible photocatalytic activity of titania-silica photocatalysts by carbon doping, indicating that the carbon narrowed the energy band gap, which induced visible light absorption.

These examples show that the study of C-doped TiO_2 materials is still largely phenomenological, although significant advances have been made in understanding the mechanism of promotion of TiO_2 by C-doping. A significant contribution in this direction has been made by Pacchioni and co-workers [76]. Using density functional theory (DFT) calculations within the generalized gradient corrected approximation, they investigated various structural models of carbon impurities in both the anatase and rutile polymorphs of TiO_2 and analyzed the associated modifications of the electronic band structure. At low carbon concentrations, under oxygen-poor conditions, substitutional (to oxygen) carbon and oxygen vacancies are favored, whereas under oxygen-rich conditions, interstitial and substitutional (to Ti) C atoms are preferred. Higher carbon concentrations undergo an unexpected stabilization caused by multi-doping effects, interpreted as inter-species redox processes. Carbon impurities result in modest variations of the band gap but induce several localized occupied states in the gap, which may account for the experimentally observed red shift of the absorption edge toward the visible. They also indicate that carbon doping may favor the forma-

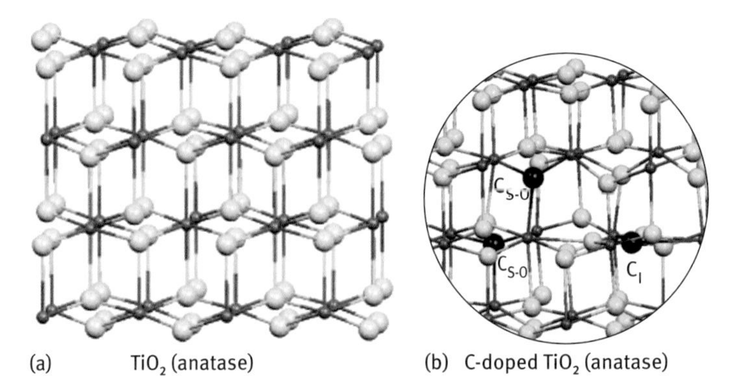

Fig. 16.5: (a) Supercell models for bulk anatase (96 atoms) and (b) partial geometry model for three carbon impurities in the anatase supercell. The yellow spheres represent O atoms, the small brown spheres represent Ti atoms, and the black spheres represent the carbon impurities. Adapted from Pacchioni *et al.* [76] with kind permission from the American Chemical Society (2005).

tion of oxygen vacancies in bulk TiO_2. Figure 16.5 shows the supercell models for bulk anatase (96 atoms) and the model for three carbon impurities in the anatase supercell.

Carbon doping in TiO_2 anatase (the most active from a photocatalytic perspective) results in a series of localized occupied states in the band gap whose density and nature are dependent on the interaction with the oxide matrix, the dopant concentration, and the presence of oxygen vacancies. These states explain the observed absorption edge shift toward the visible (vis region, up to almost 1.7 eV) of carbon-doped anatase TiO_2, with respect to the undoped material, but also indicate that no linear correlation is expected between band gap shift and visible-light photoactivity, because these localized states change the process photophysics (lifetime of charge separation, nonradiative paths, *etc.*). The observation that the type of C-species depends on oxygen atmosphere during preparation introduces an interesting aspect that has not yet been considered.

An aspect important to recall is that the mechanism of photoactivity of TiO_2 (and other semiconductors as well) in producing solar fuels (H_2 from water) may be properly described as photoelectrolysis rather that water splitting, production of photocurrent and short circuit electrolysis at surface sites, particularly metal nanoparticles deposited typically on TiO_2 to improve largely the performances. The charge carrier density, which may be improved by C-doping, is thus an important factor, because it results in an enhancement of photogenerated current. In agreement, Chen and Sun [77] observed on vertical-aligned carbon-doped TiO_2 nanowire arrays that the carbon doped TiO_2 photoanode yields a ~70 % enhancement in the photocurrent density in comparison to that of the pristine TiO_2. Incident-photon-to-electron conversion efficiency and other characterization techniques confirmed that the photocurrent improvement is mainly in the UV light region, and thus related to the increased charge carrier densities.

Doping may influence not only the bulk processes (as outlined above), but also the surface photoprocesses. Taziwa *et al.* [78] showed that surface phonons of TiO_2 are influenced by C-doping. Chen *et al.* [79] reported that in addition to interstitial carbon in the crystal lattice, graphite-like carbon species form on the surface of TiO_2 acting as sensitizer and contributing largely to the excellent photocatalytic performance of carbon modified TiO_2. Li *et al.* [80] also showed that doped carbon exists in the form of deposited carbonaceous species on the surface of TiO_2, acting as sensitizers to promote photodegradation efficiency of organic pollutants under visible-light irradiation.

The effects discussed above are also relevant in the presence of nanocarbon-TiO_2 hybrids, even if not typically considered, because the nanocarbons often show the presence of patches of amorphous carbon species which may react with TiO_2 nanocrystals, producing bulk- and surface C-doped TiO_2 nanoparticles during thermal treatments. Conductive nanocarbon species also play a role in enhancing charge transfer (to a conductive substrate in photoanodes, for example), or act as electron storage to enhance charge separation. In the presence of two different sites, their effective electronic coupling through the nanocarbon enables Z-scheme photocatalysis applications [81, 82]. Kudo and co-workers [82] showed the effectiveness of reduced graphene oxide as a solid electron mediator for water splitting in a Z-scheme photocatalysis system composed from an O_2-evolving photocatalyst ($BiVO_4$) and an H_2-evolving photocatalyst ($Ru/SrTiO_3$:Rh) coupled by a photoreduced graphene oxide shuttling the photogenerated electrons. Fu *et al.* [83] showed that in a Z-scheme mechanism (two separate steps of light absorption to better use full visible light spectrum) a directional electron transfer is necessary. Nanocarbons could be thus an excellent system to provide this directional (vectorial) charge transport, and be solid-state electron mediators alternative to classical redox solutions used in Z-schemes.

16.4 Advantages of nanocarbon-semiconductor hybrid materials

There are thus multiple effects by which the properties of the nanocarbon-semiconductor hybrid material can be different from the simple physical mixture of the two components [1]: The nanocarbon offers an effective way for an efficient dispersion of the semiconductor, thus preventing agglomeration, but also providing a hierarchical structure [15] for efficient light harvesting and eventually easy access from gas/liquid phase components (in photocatalytic reactions) or electrolyte (in DSSC).

The interfacial interaction between the nanocarbon and semiconductor particle stabilizes different morphologies for the supported semiconductor nanoparticles. For example, CNT induces the nucleation of 1D vanadium oxide nanostructures, with the nuclei growing into long free-standing nanorods [84]. Eder and Windle [85] showed how CNTs can affect the morphology of the deposited TiO_2 and the phase transformation from anatase to rutile. Wang *et al.* [86] showed that flower-like TiO_2, made of

TiO_2 rods with a size of ca. 1 µm could be nucleated and grow over the CNT surface under hydrothermal conditions (Fig. 16.6(a)). Battiston *et al.* [87] showed that TiO_2 nanopetals could be grown on single-wall carbon nanohorns (SWCNH) through two sequential vapor techniques, metal-organic chemical vapor deposition (MOCVD) and magnetron sputtering (Fig. 16.6(b)).

Fig. 16.6: Three examples of flower-like nanostructures for TiO_2 nanoparticles grown over CNTs: (a) Wang *et al.* [86]; (b) Battiston *et al.* [87]; (c) Liu and Zeng [88]. Adapted from the cited references.

Liu and Zeng [88] showed that [001]-oriented petal-like TiO_2 mesocrystals could be grown onto MWCNTs in aqueous solution (Fig. 16.6(c)). These examples show how in the presence of CNTs unusual morphologies for TiO_2 nanocrystals could be prepared, with clear relevance for their photocatalytic and/or photoanode activity due to both different crystallographic faces exposed and different type of interface with nanocarbons on which they are supported.

The interaction of semiconductor with nanocarbon induces a modification of the intrinsic properties of semiconductor particles (band gap, charge carrier density, lifetime of charge separation, non-radiative paths, *etc.*) [1] as well as of the surface properties which were discussed in detail in the previous section.

Carbon nanodots and other carbon surface species may act as efficient solid-state sensitizers to promote visible light absorption. In addition, various nanocarbons may act as semiconductors enabling to realize interesting Z-schemes for extended visible light activity.

Nanocarbons may show excellent electronic conductivity, up to 1000 times higher than copper wires, although defect characteristics are rather important and can strongly determine conductivity (see Chapter 1). Nanocarbons in hybrids with semiconductors can thus provide an efficient conductive network, even though the interface with the semiconductor is a critical factor not always considered. They can act also as a sink for electrons (Fermi levels of the nanocarbons are generally below the conduction band minimum of most semiconductors), enhancing lifetime of charge separation. Due to their thermal conductivity, they can also help in maintaining a more uniform temperature of the semiconductor-nanocarbon hybrid upon irradiation.

The nanocarbon-TiO_2 Schottky junction plays a crucial role in the photovoltaic characteristics of a semiconductor electrode [89]. According to the thermionic emission theory, the variation of the photocurrent over the voltage of the cells strongly depends on the height of the Schottky barrier. When the output voltage is low, the intrinsic 1-dimensional C nanostructures in CNTs can facilitate electron transport. With the voltage of the cell increasing, the energy dissipation on the Schottky junction increases dramatically and CNTs gradually lose the role of electron transport channels. At the high voltage range, however, leakage of electrons *via* the CNTs becomes predominant. By virtue of the charge transport channels of CNTs, increments of 44 % in photocurrent at short circuit condition and 18.7 % in the overall energy conversion were achieved [89].

Nanocarbons such as CNTs are well known for their enhanced adsorption properties for some gases and thus may act as additional adsorbents for gas species which then diffuse to the TiO_2-nanocarbon phase boundary to undergo conversion. This is an important aspect not only in environmental applications, but also in the use of these systems for solar fuels production. It is worth noting that these systems provide unique properties to tune properties by applying voltage. Abdi *et al.* [90] showed that in TiO_2/CNT hybrid materials a p-n junction was formed at the interface and consequently, a super-hydrophilic surface is achieved by applying an electrical bias voltage. This effect increases photocatalytic behavior.

The framework provided by the nanocarbon opens up the possibility of producing micro/nano-environments (similar to those present in enzymes) which may be used to enhance the photocatalytic behavior.

Nanoconfinement of semiconductor particles inside CNTs is another opportunity scarcely explored. It is well established in catalysts such as zeolite and mesoporous ordered-silica materials that the presence of size constraints in the growth of hosted particles largely influence the catalytic reactivity [91, 92]. CNTs, but also ordered mesoporous carbon materials and other nanocarbon materials having ordered nanocavities, show analogous characteristics, but very different chemical properties of the inner surface with respect to silica-type materials. The interaction with nanoparticles located inside these nanocavities is thus different from that present inside silica-type materials. Chen *et al.* [93] showed that titania confined inside CNTs stabilizes TiO_2 nanoparticles with a remarkable higher concentration of oxygen vacancies in comparison with nanoparticles dispersed on the outer surface of CNTs. This effect is responsible for extending the photoresponse of TiO_2 from the UV to the visible-light region. The CNT-confined TiO_2 exhibited improved visible-light activity in the degradation of methylene blue (MB) relative to the outside titania and compared with commercial P25 TiO_2.

Another effect involves charge transport resistivity at the semiconductor-semiconductor interface. The charge transport of the TiO_2 photoelectrode, limited by its poor conductivity (about 0.1 $cm^2\ V^{-1}\ s^{-1}$) [94], is the rate-determining step for the power-conversion efficiency in DSSCs [95]. As mention above, an usual strategy to improve charge transport is to add CNTs to the DSSC photoelectrode. It could be expected that the effect is proportional to the conductivity of the CNT, but Guai *et al.*

[95] observed different results by preparing photoanodes of TiO_2 hybrids with pristine (p), metallic (m) and semiconducting (s)-SWCNTs. Devices based on TiO_2/s-SWCNT hybrids outperform those with plain TiO_2 and TiO_2 hybrids with p- or m-SWCNTs. Optical and morphological characterization of the photoanodes and SWCNTs revealed no significant structural differences that could affect the performance enhancement. Through systematic electrochemical impedance and current–voltage analysis, they showed that only the incorporation of s-SWCNTs reduced the charge transport resistivity, while effectively suppressing charge recombination showing the highest power-conversion efficiency due to its semiconducting nature.

The performance of a DSSC is dependent on a number of parameters, including charge transport, charge transfer, and recombination rates for eventual charge separation. While the addition of SWCNTs in general improves power-conversion efficiency (PCE) with respect to a plain TiO_2 photoelectrode, due to reduced charge transport resistivity, only semiconducting material (s-SWCNT) shows the additional positive effect of suppressing charge recombination leading thus to a synergistic effect.

Energy-band diagrams for the two cases of metallic (m) and semiconducting (s) SWCNT/TiO_2 hybrid photoanodes (Fig. 16.7) [95] may explain the reason for this effect. When the photogenerated electrons are transferred to TiO_2 from the excited dye molecules, they are readily shuttled along the s-SWCNTs to the FTO conducting glass. Since electrons are less likely to be transported back along the semiconducting carbon nanotubes to cause I^{3-} reduction for charge recombination, this results in a more effective charge-separation process (device SC), and thus more photoelectrons are effectively transported and collected at the FTO terminal. This in turn leads to a much larger photocurrent compared to m- and p-SWCNT-based DSSCs. Furthermore, the effective accumulation of photoelectrons within the photoanode tends to result in a positive shift of the E_{nf} (quasi-Fermi level). This shift causes an increased V_{oc} (open-circuit voltage) for the s-SWCNT-based DSSCs, since it is dependent on the Fermi levels between the photoanode and the redox couples [96, 97]. On the other hand, although

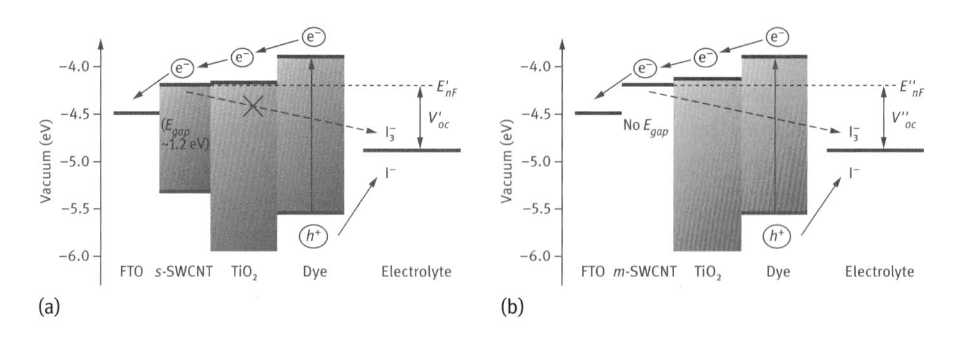

Fig. 16.7: Energy-band diagrams of DSSCs with incorporated (a) semiconducting (s-) SWCNTs and (b) metallic (m-) SWCNTs. The solid and dashed arrows represent desired charge transport and undesired recombination processes. Adapted from Guai *et al.* [95].

m-SWCNTs are able to transport the photoelectrons more rapidly due to better mobility than s-SWCNTs, the charge transport tends to be disrupted with an increased back reaction with the electrolyte too (Fig. 16.7(b)). Consequently, fewer photoelectrons are effectively collected at the FTO terminal for DSSCs with TiO_2/m-SWCNT composite photoanodes, resulting in smaller photocurrents and poorer performances.

A further role of nanocarbon is to provide an optimal nanoarchitecture in the photoanode for light absorption. A flat semiconductor surface is not ideal for maximizing light adsorption, while a porous nanostructured film would be preferable, even if specific studies are limited. In general, for a typical titania film thickness of 10–15 microns, porosity should be around 50 % [69]. The hybrid with nanocarbon could provide this optimal architecture, providing also to minimize TiO_2-TiO_2 grain boundary which enhances the rate of charge recombination.

16.5 Nanocarbon-semiconductor hybrid materials for sustainable energy

Previous sections have demonstrated how nanocarbon-semiconductor hybrid materials provide a number of potential advantages for the development of advanced devices for a sustainable use of renewable energy. Two of the more relevant areas are: (i) to improve the performances of DSSCs and (ii) to develop novel cells for producing solar fuels.

Various examples of the use of nanocarbons in DSSCs have been discussed in the previous sections. Nanocarbons find more extensive applications in DSSCs than just for the photoelectrode [69], for example (i) in developing novel transparent counter electrodes, as cited before, (ii) in developing transparent and flexible conductive substrates, and (iii) in developing novel electrolytes (all solid-state electrolytes). Thus, in addition to developing cells with improved efficiency (above the current limit of about 12.5 % efficiency), a main relevant target is to produce at competitive costs flexible DSSCs, which have a potential larger field of use than current rigid cells. Printable DSSCs are expected to dominate the future solar-cell market and the use of nanocarbons will be a key element to realize this technology [98]. However, cell efficiencies of nanocarbon-based DSSCs are still unsatisfactory (less than about 6 %) [99, 100], due to the difficulties in effectively controlling the many aspects discussed here.

A final comment should be dedicated to the use of nanocarbon-semiconductor hybrids in advanced devices for producing solar fuels (H_2 from water and products of CO_2 hydrogenation such as methanol) [11, 12], a relevant area for the future sustainability of energy. This is a growing area of interest, because the demanding requirements for these reactions imply a further need to design optimized systems. A number of studies have been reported in this area recently. Bai *et al.* [101] reported the preparation of C-doped TiO_2 nanorod spheres with a large percentage (95 %) of rutile (110) facets and showing an interesting hierarchical structure. In a MeOH-H_2O sacrificial reagent system in the presence of Cu^{2+}, the TiO_2 nanorod spheres show good behavior in generating H_2 under UV light irradiation, although data on the activity without sacrificial agents and under visible light irradiation would be necessary to understand the applicability of these systems. Ahmmad *et al.* [102] also reported that nanocarbons enhance the activity of TiO_2 in H_2 photogeneration, but in the presence of sacrificial agents (alcohols). Li *et al.* [103] reported that TiO_2 nanotubes/MWCNTs hybrids promote photocatalytic hydrogen evolution, but using Na_2S and Na_2SO_3 as sacrificial reagents and UV-visible light irradiation. Ge and Han [104] reported hybrids of MWNTs and graphitic carbon nitride (g-C_3N_4), showing a visible light hydrogen evolution rate of 7.6 μmol h^{-1}, but using methanol as sacrificial agent. These examples show the need to make an effort towards design systems active with visible light and in the absence of sacrificial agents, as well as to be more productive.

Yu *et al.* [105] reported instead visible light photocatalytic H_2 production from water on CNT modified by $Cd_{0.1}Zn_{0.9}S$ quantum dots. Under visible light, an H_2-production rate of 78.2 μmol h^{-1} with apparent quantum efficiency (QE) of 7.9 % at 420 nm was shown. These are interesting performances, but it is known that these quantum dots do not show enough long term stability.

In addition, the possibility to exploit the production of H_2/O_2 from water requires having physically separate evolution of hydrogen and oxygen for safety reasons, minimizing back reaction and avoiding the costs of H_2/O_2 separation. It is thus already necessary from the start to study the reaction in photoelectrocatalytic (PEC) devices [11] differently from a large part of actual literature studies. Therefore, notwithstanding the promising potential in using nanocarbons to develop advanced electrodes for water photoelectrolysis (a more correct indication than water splitting, as remarked before), a more systematic effort is still lacking.

The protons/electrons produced in water oxidation at a photoanode side of a PEC device could be used (on the cathode side) to reduce CO_2 to alcohols/hydrocarbons (CH_4, CH_3OH, HCOOH, *etc.*). In this way, an artificial leaf (photosynthesis) device could be developed [11]. While nanocarbon materials containing iron or other metal particles show interesting properties in this CO_2 reduction [106], it is beyond the scope of this chapter to discuss this reaction here. It is worthwhile, however, to mention how nanocarbon materials can be critical elements to design both anode and cathode in advanced PEC solar cells. Nanocarbons have also been successfully used for developing photocatalysts active in the reduction of CO_2 with water [107].

16.6 Summary

There are many different mechanisms through which the nanocarbon-semiconductor hybrid materials may show enhanced properties and which have to be exploited for the design of optimized materials used as photocatalysts or photoanodes. The presence, absence or efficacy of these enhancement mechanisms depends greatly on the specific nature of nanocarbon, the preparation method of the hybrid material and the nanocarbon-semiconductor interface. Doping and/or addition of surface modifiers can give rise to further mechanisms, but their efficient use requires to start from a better understanding of all the complex phenomena discussed above.

There is thus a great potential for using nanocarbon-semiconductor hybrids to develop improved devices, particularly for the more challenging area of using renewable energy, while most of the current studies are centered on environmental applications, for which there is limited need to develop advanced photocatalysts.

Acknowledgments

This contribution has been realized in the frame of projects PON01_01725 "New Photovoltaic Technologies" and PON/ENERGETIC which are gratefully acknowledged by the authors.

Bibliography

[1] Vilatela, J. J.; Eder, D., Nanocarbon composites and hybrids in sustainability: A Review. *ChemSusChem* **2012**, *5*: 456–478.
[2] Centi, G.; Perathoner, S., The Role of Nanostructure in Improving the Performance of Electrodes for Energy Storage and Conversion. *Eur. J. Inorg. Chem.* **2009**, *26*: 3851–3878.
[3] Inagaki, M., Carbon coating for enhancing the functionalities of materials. *Carbon* **2012**, *50*: 3247–3266.
[4] Toyoda, M.; Tsumura, T.; Tryba, B.; Mozia, S.; Janus, M.; Morawski, A.W.; Inagaki, M., Carbon materials in photocatalysis, *Chem. and Phy. of Carbon* **2013**, *31*: 171–267.
[5] Leary, R.; Westwood, A., Carbonaceous nanomaterials for the enhancement of TiO_2 photocatalysis. *Carbon* **2011**, *49*: 741–772.
[6] Centi, G.; Perathoner, S., Problems and perspectives in nanostructured carbon-based electrodes for clean and sustainable energy. *Catal. Today* **2010**, *150*: 151–162.
[7] Centi, G.; Perathoner, S., Carbon Nanotubes for Sustainable Energy Applications. *ChemSusChem* **2011**, *4*: 913–925.
[8] D'Souza, F.; Ito, O., Photosensitized electron transfer processes of nanocarbons applicable to solar cells. *Chem. Soc. Rev.* **2012**, *41*: 86–96.
[9] Wu, M.; Ma, T., Platinum-Free Catalysts as Counter Electrodes in Dye-Sensitized Solar Cells. *ChemSusChem* **2012**, *5*: 1343–1357.
[10] Chen, H.-Y.; Kuang, D.-B.; Su, C.-Y., Hierarchically micro/nanostructured photoanode materials for dye-sensitized solar cells. *J. Mater. Chem.* **2012**, *22*: 15475–15489.
[11] Bensaid, S.; Centi, G.; Garrone, E.; Perathoner, S.; Saracco, G., Towards Artificial Leaves for Solar Hydrogen and Fuels from Carbon Dioxide. *ChemSusChem* **2012**, *5*: 500–521.
[12] Centi, G.; Perathoner, S., Towards Solar Fuels from Water and CO_2. *ChemSusChem* **2010**, *3*: 195–208.
[13] Liang, Y. T.; Vijayan, B. K.; Gray, K. A.; Hersam, M. C., Minimizing Graphene Defects Enhances Titania Nanocomposite-Based Photocatalytic Reduction of CO_2 for Improved Solar Fuel Production. *Nano Lett.* **2011**, *11*: 2865–2870.

[14] Su, D.; Centi, G., Carbon nanotubes for energy applications, In *Nanoporous Materials for Energy and the Environment*, Rios, G.; Centi, G.; Kanellopoulos, N. (eds.), Pan Stanford Publishing Pte. Ltd.: **2012**, 173–202.
[15] Su, D. S.; Perathoner, S.; Centi, G., Catalysis on nano-carbon materials: Going where to? *Catal. Today* **2012**, *186*: 1–6.
[16] Su, D.S.; Perathoner, S.; Centi, G., Nanocarbons for the Development of Advanced Catalysts. *Chem Rev.* **2013**, *113*, 5782–5816.
[17] Poudel, P.; Qiao, Q., One dimensional nanostructure/nanoparticle composites as photoanodes for dye-sensitized solar cells. *Nanoscale* **2012**, *4*: 2826–2838.
[18] Mohamed, A. E. R.; Rohani, S., Modified TiO_2 nanotube arrays (TNTAs): progressive strategies towards visible light responsive photoanode, a review. *Energy Environ. Sci.* **2011**, *4*: 1065–1086.
[19] Roy, P.; Berger, S.; Schmuki, P., TiO_2 Nanotubes: Synthesis and Applications. *Angew. Chem., Int. Ed.* **2011**, *50*: 2904–2939.
[20] Centi, G.; Perathoner, S., Nano-architecture and reactivity of titania catalytic materials. Part 2. Bidimensional nanostructured films. *Catalysis (RSC)* **2009**, *21*: 82–130.
[21] Centi, G.; Perathoner, S., Nanostructured titania thin films for solar use in energy applications, In *Nanoporous Materials for Energy and the Environment*, Rios, G.; Centi, G.; Kanellopoulos, N. (eds.), Pan Stanford Publishing Pte. Ltd.: **2012**, 257–282.
[22] Yao, Y.; Li, G.; Ciston, S.; Lueptow, R. M.; Gray, K. A., Photoreactive TiO_2/carbon nanotube composites: synthesis and reactivity. *Environ. Sci. Technol.* **2008**, *42*: 4952–4957.
[23] Eder, D., Carbon Nanotube-Inorganic Hybrids. *Chem. Rev.* **2010**, *110*: 1348–1385.
[24] Richter, C.; Schmuttenmaer, C. A., Exciton-like trap states limit electron mobility in TiO_2 nanotubes. *Nat. Nanotechnol.* **2010**, *5*: 769–772.
[25] Ng, Y. H.; Ikeda, S.; Matsumura, M.; Amal, R., A perspective on fabricating carbon-based nanomaterials by photocatalysis and their applications. *Energy Environ. Sci.* **2012**, *5*: 9307–9318.
[26] Morales-Torres, S.; Pastrana-Martinez, L.M.; Figueiredo, J.L.; Faria, J.L.; Silva, A.M.T., Design of graphene-based TiO_2 photocatalysts – a review. *Environ. Sci. Pollut. Res.* **2012**, *19*: 3676–3687.
[27] Ouzzine, M.; Lillo-Rodenas, M. A.; Linares-Solano, A., Carbon nanofibers as substrates for the preparation of TiO_2 nanostructured photocatalysts. *Appl. Catal., B* **2012**, *127*: 291–299.
[28] Shi, J.-W.; Cui, H.-J.; Chen, J.-W.; Fu, M.-L.; Xu, B.; Luo, H.-Y.; Ye, Z.-L., TiO_2/activated carbon fibers photocatalyst: Effects of coating procedures on the microstructure, adhesion property, and photocatalytic ability. *J. Colloid Interface Sci.* **2012**, *388*: 201–208.
[29] Liang, C.; Li, Z.; Dai, S., Mesoporous carbon materials: synthesis and modification. *Angew. Chemie, Int. Ed.* **2008**, *47*: 3696–3717.
[30] Wan, Y.; Shi, Y.; Zhao, D., Supramolecular Aggregates as Templates: Ordered Mesoporous Polymers and Carbons. *Chem. Mat.* **2008**, *20*: 932–945.
[31] Park, I.-S.; Choi, S. Y.; Ha, J., High-performance titanium dioxide photocatalyst on ordered mesoporous carbon support. *Chem.Phys. Lett.* **2008**, *456*: 198–201.
[32] Liu, R.; Ren, Y.; Shi, Y.; Zhang, F.; Zhang, L.; Tu, B.; Zhao, D., Controlled Synthesis of Ordered Mesoporous C-TiO_2 Nanocomposites with Crystalline Titania Frameworks from Organic-Inorganic-Amphiphilic Coassembly. *Chem. Mat.* **2008**, *20*: 1140–1146.
[33] Su, D.S.; Schlögl, R., Nanostructured Carbon and Carbon Nanocomposites for Electrochemical Energy Storage. *ChemSusChem* **2010**, *3*: 136–168.
[34] Itami, K., Toward controlled synthesis of carbon nanotubes and graphenes. *Pure and Appl. Chem.* **2012**, *84*: 907–916.

[35] Katsukis, G.; Romero-Nieto, C.; Malig, J.; Ehli, C.; Guldi, D.M., Interfacing Nanocarbons with Organic and Inorganic Semiconductors: From Nanocrystals/Quantum Dots to Extended Tetrathiafulvalenes. *Langmuir* **2012**, *28*: 11662–11675.

[36] Centi, G.; Perathoner, S., Creating and mastering nano-objects to design advanced catalytic materials. *Coord. Chem. Rev.* **2011**, *255*: 1480–1498.

[37] Resasco, D.E., Carbon Nanotubes and Related Structures. In *Nanoscale Materials in Chemistry*, 2nd Ed., Klabunde K. J.; Richards R. M. (eds.), John Wiley & Sons, Inc., Hoboken, NJ, USA **2009**, 443–491.

[38] Yu, S.-S.; Zheng, W.-T., Effect of N/B doping on the electronic and field emission properties for carbon nanotubes, carbon nanocones, and graphene nanoribbons. *Nanoscale* **2010**, *2*: 1069–1082.

[39] Panchakarla, L.S.; Govindaraj, A.; Rao, C.N.R., Boron- and nitrogen-doped carbon nanotubes and grapheme. *Inorg. Chimica Acta* **2010**, *363*: 4163–4174.

[40] Stoyanov, S.R.; Titov, A.V.; Kral, P., Design and Modeling of Transition Metal-doped Carbon Nanostructures. *Coord. Chem. Rev.* **2009**, *253*: 2852–2871.

[41] Ayala, P.; Arenal, R.; Ruemmeli, M.; Rubio, A.; Pichler, T., The doping of carbon nanotubes with nitrogen and their potential applications. *Carbon* **2010**, *48*: 575–586.

[42] Lee, S.-H.; Lee, D.-H.; Lee, W.-J.; Kim, S.-O., Tailored Assembly of Carbon Nanotubes and Graphene. *Adv. Funct. Mat.* **2011**, *21*: 1338–1354.

[43] Janowska, I.; Hajiesmaili, S.; Begin, D.; Keller, V.; Keller, N.; Ledoux, M.-J.; Pham-Huu, C., Macronized aligned carbon nanotubes for use as catalyst support and ceramic nanoporous membrane template. *Catal. Today* **2009**, *145*: 76–84.

[44] Xia, W.; Wang, Y.; Bergsträßer, R.; Kundu, S.; Muhler, M., Surface characterization of oxygen-functionalized multi-walled carbon nanotubes by high-resolution X-ray photoelectron spectroscopy and temperature-programmed desorption . *Appl. Surface Science* **2007**, *254*: 247–250.

[45] Karousis, N.; Tagmatarchis, N.; Tasis, D., Current progress on the chemical modification of carbon nanotubes. *Chem. Rev.* **2010**, *110*: 5366–97.

[46] Wang, M.-S.; Bando, Y.; Rodriguez-Manzo, J.A.; Banhart, F.; Golberg, D., Cobalt nanoparticle-assisted engineering of multi-wall carbon nanotubes. *ACS Nano* **2009**, *3*: 2632–2638.

[47] Tao, C.; Jiao, L.; Yazyev, O.V.; Chen, Y-C.; Feng, J.; Zhang, X.; Capaz, R.B.; Tour, J.M.; Zettl, A.; Louie, S.G., Spatially resolving edge states of chiral graphene nanoribbons. *Nature Phys.* **2011**, *7*: 616–620.

[48] Li, H.; Liu, R.; Liu, Y.; Huang, H.; Yu, H.; Ming, H.; Lian, S.; Lee, S.-T.; Kang, Z., Carbon quantum dots/Cu_2O composites with protruding nanostructures and their highly efficient (near) infrared photocatalytic behavior. *J. Mat. Chem.* **2012**, *22*: 17470–17475.

[49] Zhang, Y.-Q.; Ma, D.-K.; Zhuang, Y.; Zhang, X.; Chen, W.; Hong, L.-L.; Yan, Q.-X.; Yu, K.; Huang, S.-M., One-pot synthesis of N-doped carbon dots with tunable luminescence properties. *J. Mat. Chem.* **2012**, *22*: 16714–16718.

[50] Yu, B. Y.; Kwak, S.-Y., Carbon quantum dots embedded with mesoporous hematite nanospheres as efficient visible light-active photocatalysts. *J. Mat. Chem.* **2012**, *22*: 8345–8353.

[51] Yu, H.; Zhang, H.; Huang, H.; Liu, Y.; Li, H.; Ming, H.; Kang, Z., ZnO/carbon quantum dots nanocomposites: one-step fabrication and superior photocatalytic ability for toxic gas degradation under visible light at room temperature. *New J. Chem.* **2012**, *36*: 1031–1035.

[52] Mirtchev, P.; Henderson, E.J.; Soheilnia, N.; Yip, C.M.; Ozin, G.A., Solution phase synthesis of carbon quantum dots as sensitizers for nanocrystalline TiO_2 solar cells. *J. Mat. Chem.* **2012**, *22*: 1265–1269.

[53] Cao L.; Meziani M. J; Sahu S.; Sun Y.-P., Photoluminescence Properties of Graphene versus Other Carbon Nanomaterials. *Acc. Chem. Res.* **2013**, *46*, 171–180.

[54] Sun, Y.-P.; Zhou, B.; Lin, Y.; Wang, W.; Fernando, K. A. S.; Pathak, P.; Harruff, B. A.; Wang, X.; Wang, H.; Luo, P. G.; Yang, H.; Chen, B.; Veca, L. M.; Xie, S.-Y., Quantum-Sized Carbon Particles for Bright and Colorful Photoluminescence. *J. Am. Chem. Soc.* **2006**, *128*: 7756–7757.

[55] Xu, J.; Sahu, S.; Cao, L.; Anilkumar, P.; Tackett, K.N., II; Qian, H.J.; Bunker, C.E.; Guliants, E.A.; Parenzan, A.; Sun, Y.-P., Carbon Nanoparticles as Chromophores for Photon Harvesting and Photoconversion. *ChemPhysChem* **2011**, *12*: 3604–3608.

[56] Cao, L.; Sahu, S.; Anilkumar, P.; Bunker, C. E.; Xu, J. A.; Fernando, K. A. S.; Wang, P.; Guliants, E. A.; Tackett, K. N., II; Sun, Y.-P., Carbon Nanoparticles as Visible-Light Photocatalysts for Efficient CO_2 Conversion and Beyond. *J. Am. Chem. Soc.* **2011**, *133*: 4754–4757.

[57] Gupta, V.; Chaudhary, N.; Srivastava, R.; Sharma, G. D.; Bhardwaj, R.; Chand, S., Luminescent Graphene Quantum Dots for Organic Photovoltaic Devices. *J. Am. Chem. Soc.* **2011**, *133*: 9960–996.

[58] Shen, J.; Zhu, Y.; Chen., C.; Yang, X.; Li, C., Facile Preparation and Upconversion Luminescence of Graphene Quantum Dots. *Chem. Commun.* **2011**, *47*: 2580–2582.

[59] Shen, J.; Zhu, Y.; Yang, X.; Zong, J.; Zhang, J.; Li, C., One-pot Hydrothermal Synthesis of Graphene Quantum Dots Surface-passivated by Polyethylene Glycol and Their Photoelectric Conversion under Near-infrared Light. *New J. Chem.* **2012**, *36*: 97–101.

[60] Xie, P.; Guo, F., Molecular design of ruthenium complexes for dye-sensitized solar cells based on nanocrystalline TiO_2. *Current Org. Chem.* **2011**, *15*: 3849–3869.

[61] Lee, J.-K.; Yang, M., Progress in light harvesting and charge injection of dye-sensitized solar cells. *Materials Science & Eng., B: Adv. Funct. Solid-State Mat.* **2011**, *176*: 1142–1160.

[62] Mora-Sero, I.; Gimenez, S.; Fabregat-Santiago, F.; Gomez, R.; Shen, Q.; Toyoda, T.; Bisquert, J., Recombination in Quantum Dot Sensitized Solar Cells. *Acc. Chem. Res.* **2009**, *42*: 1848–1857.

[63] Rehman, S.; Ullah, R.; Butt, A. M.; Gohar, N. D., Strategies of making TiO_2 and ZnO visible light active. *J. Hazar. Mat.* **2009**, *170*: 560–569.

[64] Pelaez, M.; Nolan, N. T.; Pillai, S. C.; Seery, M. K.; Falaras, P.; Kontos, A. G.; Dunlop, P. S. M.; Hamilton, J.W. J.; Byrne, J. A.; O'Shea, K., A review on the visible light active titanium dioxide photocatalysts for environmental applications. *Appl. Catal., B: Env.* **2012**, *125*: 331–349.

[65] Chen, J.; Lei, W.; Zhang, X. B., Enhanced electron transfer rate for quantum dot sensitized solar cell based on CNT-TiO_2 film. *J. Nanoscience and Nanotechn.* **2012**, *12*: 6476–6479.

[66] Jang, Y. H.; Xin, X.; Byun, M.; Jang, Yu J.; Lin, Z.; Kim, D. H., An Unconventional Route to High-Efficiency Dye-Sensitized Solar Cells via Embedding Graphitic Thin Films into TiO_2 Nanoparticle Photoanode. *Nano Letters* **2012**, *12*: 479–485.

[67] Paul, G. Sa.; Kim, J. H.; Kim, M.-S.; Do, K.; Ko, J.; Yu, J.-S., Different Hierarchical Nanostructured Carbons as Counter Electrodes for CdS Quantum Dot Solar Cells. *ACS Applied Materials & Interfaces* **2012**, *4*: 375–381.

[68] Ito, S.; Mikami, Y., Porous carbon layers for counter electrodes in dye-sensitized solar cells: recent advances and a new screen-printing method. *Pure and Applied Chem.* **2011**, *83*: 2089–2106.

[69] Brennan, L. J.; Byrne, M. T.; Bari, M.; Gun'ko, Y.K., Carbon nanomaterials for dye-sensitized solar cell applications: a bright future. *Adv. Energy Materials* **2011**, *1*: 472–485.

[70] Guo, C. X.; Guai, G. H.; Li, C. M., Graphene based materials: enhancing solar energy harvesting. *Adv. Energy Materials* **2011**, *1*: 448–452.

[71] Mora-Sero, I.; Bisquert, J., Breakthroughs in the Development of Semiconductor-Sensitized Solar Cells Full. *J. Phys. Chem. Lett.* **2010**, *1*: 3046–3052.

[72] Yin, W.-J.; Chen, S.; Yang, J.; Gong, X.ŋG.; Yan Y.; Wei, S.ŋH., Effective band gap narrowing of anatase TiO_2 by strain along a soft crystal direction, *Appl. Phys. Lett.* **2010**, *96*: 221901–3.
[73] Sampaio, M.J.; Silva, C.G.; Marques, R.R.N.; Silva, A.M.T. ; Faria, J.L., Carbon nanotube–TiO_2 thin films for photocatalytic applications. *Catal. Today* **2011**, *161*: 91–96.
[74] Liu, G.; Han, C.; Pelaez, M.; Zhu, D.; Liao, S.; Likodimos, V.; Ioannidis, N.; Kontos, A.G.; Falaras, P.; Dunlop, P.S. M., Synthesis, characterization and photocatalytic evaluation of visible light activated C-doped TiO_2 nanoparticles. *Nanotechn.* **2012**, *23*: 294003/1–10.
[75] Chen, Q.; Shi, H.; Shi, W.; Xu, Y.; Wu, D., Enhanced visible photocatalytic activity of titania-silica photocatalysts: effect of carbon and silver doping. *Catal.Science & Techn.* **2012**, *2*: 1213–1220.
[76] Di Valentin, C.; Pacchioni, G.; Selloni, A., Theory of Carbon Doping of Titanium Dioxide. *Chem. Mater.* **2005**, *17*: 6656–6665.
[77] Cheng, C.; Sun, Y., Carbon doped TiO_2 nanowire arrays with improved photoelectrochemical water splitting performance. *Appl. Surface Science* **2012**, *263*: 273–276.
[78] Taziwa, R.; Meyer, E. L.; Sideras-Haddad, E.; Erasmus, R. M.; Manikandan, E.; Mwakikunga, B. W., Effect of carbon modification on the electrical, structural, and optical properties of TiO_2 electrodes and their performance in labscale dye-sensitized solar cells. *Int. J. Photoenergy* **2012**, 904323/1–9.
[79] Chen, C.; Long, M.; Zeng, H.; Cai, W.; Zhou, B.; Zhang, J.; Wu, Y.; Ding, D.; Wu, D., Preparation, characterization and visible-light activity of carbon modified TiO2 with two kinds of carbonaceous species. *J. Mol. Catal. A: Chem.* **2009**, *314*: 35–41.
[80] Li, Y.-F.; Xu, D.; Oh, J.I.; Shen, W.; Li, X.; Yu, Y., Mechanistic Study of Codoped Titania with Nonmetal and Metal Ions: A Case of C + Mo Codoped TiO_2. *ACS Catal.* **2012**, *2*: 391–398.
[81] Qian, Z.; Pathak, B.; Nisar, J.; Ahuja, R., Oxygen- and nitrogen-chemisorbed carbon nanostructures for Z-scheme photocatalysis applications. *J. Nanoparticle Res.* **2012**, *14*: 895/1–895/7.
[82] Iwase, A.; Ng, Y. H.; Ishiguro, Y.; Kudo, A.; Amal, R., Reduced Graphene Oxide as a Solid-State Electron Mediator in Z-Scheme Photocatalytic Water Splitting under Visible Light. *J. Am. Chem, Soc.* **2011**, *133*: 11054–11057.
[83] Fu, N.; Jin, Z.; Wu, Y.; Lu, G.; Li, D., Z-Scheme Photocatalytic System Utilizing Separate Reaction Centers by Directional Movement of Electrons. *J. Phys. Chem. C* **2011**, *115*: 8586–8593.
[84] Chen, X.-W.; Zhu, Z.; Haevecker, M.; Su, D.S.; Schlögl, R., Carbon nanotube-induced preparation of vanadium oxide nanorods: Application as a catalyst for the partial oxidation of n-butane. *Materials Res. Bull.* **2007**, *42*: 354–361.
[85] Eder, D.; Windle, A.H., Morphology control of CNT-TiO_2 hybrid materials and rutile nanotubes. *J. Mater. Chem.* **2008**, *18*: 2036–2043.
[86] Wang, W.; Lu, C.; Ni, Y.; Su, M.; Xu, Z., Hydrothermal synthesis and enhanced photocatalytic activity of flower-like TiO2 on carbon nanotubes. *Mater. Lett.* **2012,** *79*: 11–13.
[87] Battiston, S.; Minella, M.; Gerbasi, R.; Visentin, F.; Guerriero, P.; Leto, A.; Pezzotti, G.; Miorin, E.; Fabrizio, M.; Pagura, C., Growth of titanium dioxide nanopetals induced by single wall carbon nanohorns. *Carbon* **2010**, *48*: 2470–2477.
[88] Liu, B.; Zeng, H. C., Carbon Nanotubes Supported Mesoporous Mesocrystals of Anatase TiO_2. *Chem. Mater.* **2008**, *20*: 2711–2718.
[89] Chen, J.; Li, B.; Zheng, J.; Zhao, J.; Zhu, Z., Role of Carbon Nanotubes in Dye-Sensitized TiO_2-Based Solar Cells. *J. Phys. Chem. C* **2012**, *116*: 14848–14856.
[90] Abdi, Y.; Khalilian, M.; Arzi, E., Enhancement in photoinduced hydrophilicity of TiO_2/CNT nanostructures by applying voltage. *J. Phys. D: Appl. Phys.* **2011**, *44*: 255405/1–6.

[91] Centi, G.; Perathoner, S.; Vazzana, F., Catalysis using guest single and mixed oxides in host zeolite matrices. in *Catalysis by Unique Metal Ion Structures in Solid Matrices*, Centi, G.; Bell, A.T.; Wichterlová, B. (eds.), Springer-Verlag. Dordrecht (The Netherlands), **2001**, 165–186.

[92] Cejka, J.; Centi, G.; Perez-Pariente, J.; Roth, W. J., Zeolite-based materials for novel catalytic applications: Opportunities, perspectives and open problems. *Catal. Today* **2012**, *179*: 2–15.

[93] Chen, W.; Fan, Z.; Zhang, B.; Ma, G.; Takanabe, K.; Zhang, X.; Lai, Z., Enhanced Visible-Light Activity of Titania via Confinement inside Carbon Nanotubes. *J. Am. Chem.Soc.* **2011,** 133: 14896–14899.

[94] Tiwana, P.; Docampo, P.; Johnston, M.B.; Snaith, H.J.; Herz, L.M., Electron Mobility and Injection Dynamics in Mesoporous ZnO, SnO_2, and TiO2 Films Used in Dye-Sensitized Solar Cells. *ACS Nano* **2011,** *5*: 5158–5166.

[95] Guai, G. H.; Li, Y.; Ng, C. M.; Li, C. M.; Chan-Park, M. B., TiO_2 Composing with Pristine, Metallic or Semiconducting Single-Walled Carbon Nanotubes: Which Gives the Best Performance for a Dye-Sensitized Solar Cell. *ChemPhysChem* **2012**, *13*: 2566–2572.

[96] Boschloo, G.; Hagfeldt, A., Characteristics of the Iodide/Triiodide Redox Mediator in Dye-Sensitized Solar Cells. *Acc. Chem. Res.* **2009**, *42*: 1819–1826.

[97] Hagfeldt, A.; Boschloo, G.; Sun, L.; Kloo, L.; Pettersson, H., Dye-Sensitized Solar Cells. *Chem. Rev.* **2010,** *110*: 6595–6663.

[98] Lei, B.-X.; Fang, W.-J.; Hou, Y.-F.; Liao, J.-Y.; Kuang, D.-B.; Su, C.-Y., All-solid-state electrolytes consisting of ionic liquid and carbon black for efficient dye-sensitized solar cells. *J. Photochem. and Photobiol., A: Chem.* **2010,** *216*: 8–14.

[99] Khamwannah, J.; Noh, S. Y.; Frandsen, C.; Zhang, Y.; Kim, H.; Kong, S. D.; Jin, S., Nanocomposites of TiO_2 and double-walled carbon nanotubes for improved dye-sensitized solar cells. *J. Renewable and Sustainable Energy* **2012,** *4*: 023116/1–9.

[100] Kim, D. Y.; Kim, J.; Kim, J.; Kim, A.-Y.; Lee, G.; Kang, M., The photovoltaic efficiencies on dye sensitized solar cells assembled with nanoporous carbon/TiO_2 composites. *J. Ind. and Eng. Chem.* **2012,** *18*: 1–5.

[101] Bai, H.; Liu, Z.; Sun, D. D., Facile preparation of monodisperse, carbon doped single crystal rutile TiO_2 nanorod spheres with a large percentage of reactive (110) facet exposure for highly efficient H_2 generation. *J. Mat. Chem.* **2012,** *22*: 18801–18807.

[102] Ahmmad, B.; Kusumoto, Y.; Somekawa, S.; Ikeda, M., Carbon nanotubes synergistically enhance photocatalytic activity of TiO_2. *Catal. Comm.* **2008,** *9*: 1410–1413.

[103] Li, H.; Zhang, X.; Cui, X.; Lin, Y., TiO_2 nanotubes/MWCNTs nanocomposite photocatalysts: synthesis, characterization and photocatalytic hydrogen evolution under UV-vis light illumination. *J. Nanoscience and Nanotechn.* **2012,** *12*: 1806–1811.

[104] Ge, L.; Han, C., 160. Synthesis of MWNTs/g-C_3N_4 composite photocatalysts with efficient visible light photocatalytic hydrogen evolution activity. *Appl. Catal., B: Env.* **2012,** *117–118*: 268–274.

[105] Yu, J.; Yang, B.; Cheng, B., Noble-metal-free carbon nanotube-$Cd_{0.1}Zn_{0.9}S$ composites for high visible-light photocatalytic H_2-production performance. *Nanoscale* 2012, *4*: 2670–2677.

[106] Ampelli, C.; Centi, G.; Passalacqua, R.; Perathoner, S. Synthesis of solar fuels by a novel photoelectrocatalytic approach. *Energy & Env. Science* **2010,** *3*: 292–301.

[107] Wang, Y.; Zhang, C.; Kang, S.; Li, B.; Wang, Y.; Wang, L.; Li, X., Simple synthesis of graphitic ordered mesoporous carbon supports using natural seed fat. *J. Mat. Chem.* **2011,** *21*: 14420–14423.

Jiangtao Di, Zhigang Zhao, and Qingwen Li

17 Electrochromic and photovoltaic applications of nanocarbon hybrids

Nanocarbons, in particular carbon nanotubes and graphene, are gaining increasing interest in the development of novel and high-performance optoelectronic materials and devices. The rich surface functionality of nanocarbons has made them ideal scaffolds for anchoring foreign semiconductor nanoparticles or conjugated polymers to form hybrid structures. The recent progress in achieving some scalable macro-assemblies of nanocarbons like transparent network, aligned film and fibers *etc.*, also helps pave up more routes for the fabrication of devices toward practical applications. This chapter provides a brief review on the synthesis of optoelectronic nanocarbon hybrids and focuses on their applications in electrochromic and photovoltaic devices.

17.1 Introduction

Electrochromism is the phenomenon that an electroactive material shows reversible changes in its optical properties (transmittance, absorbance, and reflectance), when a small voltage is applied, while photovoltaics is a method of converting sunlight into electrical current using semiconductor-based optoelectronic materials. Although these two systems follow different working mechanisms, they share some similarities in material design and preparation. Semiconductors and conjugated polymers are the two popular components that have been widely investigated in the two systems for several decades. However, their low conductivity and charge mobility, and poor interfaces with conductive substrates lead to electrochromic and photovoltaic devices with some common drawbacks in practical applications, such as unsatisfactory performance, reliability, and stability, high production cost, and lack of flexibility under bending or mechanical deformation.

Nanocarbon hybrids show promise in addressing these problems. The family of nanocarbons has been drawing in more and more members recently, such as nanotubes, graphene, nanoscrolls, onions and nanofibers. These nanocarbons materials commonly have large surface areas and are electrically conducting and liable to be surface-functionalized, which lead them to serve as ideal templates or scaffolds for anchoring other functional components. Their unique structural, mechanical, optical and electronic properties have inspired numerous attempts in not only employing them as scaffolds for novel and more conductive hybrid structures with enhanced photon/electron transport and conversion performance, but also in designing and fabricating robust architectures for flexible, foldable and even wearable electronic devices.

17.2 Nanocarbon Hybrids for electrochromic materials and devices

Electrochromic materials have been widely used in developing functional optical devices such as information displays, light shutters, smart windows, variable-reflectance minors, and variable-emittance thermal radiators [1, 2]. There are a large number of materials, especially including metal oxides and conjugated polymers, which exhibit electrochromism, and a number of reviews on various categories of electrochromic materials and their applications have been published [3–8]. Essentially, the materials with good electrochromism should at least satisfy the following requirements: (1) high-efficiency optical modulation or coloration; (2) fast transport for ions and electrons; (3) good reversibility and stability.

17.2.1 Intrinsic electrochromism of nanocarbons

In the last two decades, nanocarbons have become increasingly popular due to their outstanding structural and physicochemical properties [9]. However, whether nanocarbons themselves can be used as intrinsically electrochromic components has been rarely explored. Recently, a pioneering study conducted by Yangai *et al.* demonstrated that intrinsically electrochromic color changes can be achieved with metallic single-walled carbon nanotubes (SWCNTs) in response to applied voltage [9]. As-prepared SWCNTs samples are in mixed chirality states, and such a situation induces several optical absorption bands covering the whole range of the visible spectrum [10]. This is the main reason that had prevented the discovery of intrinsically electrochromic color change in carbon nanotubes before. Recent achievement in separation of single-chirality SWCNTs enabled the study in detail [11]. Yangai *et al.* demonstrated very clear, stable and reversible color changes in metallic SWCNTs with diameters of 0.84, 1.0, and 1.4 nm, exhibiting yellow, magenta, and blue-green colors, respectively (Fig. 17.1). The color changes are reversible and repetitive, and a relatively good coloration efficiency $(1.9 \pm 0.2) \times 10^2$ cm^2C^{-1} is achieved. This efficiency is even better than metal-oxide compounds and Prussian blue. It is believed that the color changes are caused by filling or depleting Van Hove singularities of metallic SWCNTs, which modifies the optical transitions between states in SWCNTs [9]. Yangai *et al.*'s work opens up a new vista of both fundamental study and potential applications of "electrochromic carbon electrodes". However, many unknowns still remain in this field. For example, to the best of our knowledge, there are few reports dedicated to intrinsic electrochromic performances of other nanocarbon materials. It is very important for researchers to continue the exploration of the possibility of other nanocarbon materials such as graphene-based electrochromic components. The intrinsic electrochromic mechanism of nanocarbon materials is also worthy of further understanding.

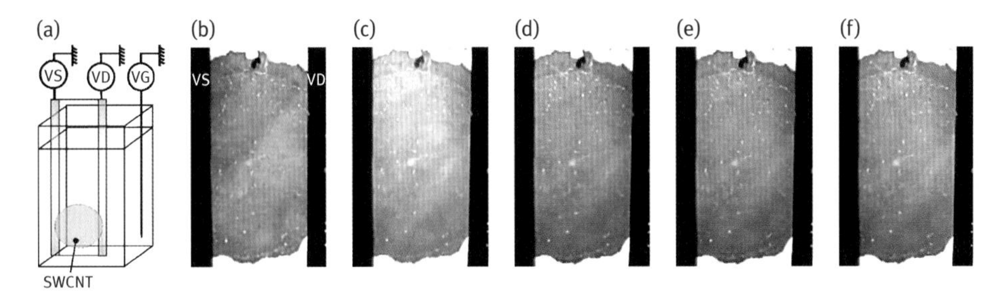

Fig. 17.1: (a) Schematic illustration of an SWCNT film and the source VS, drain VD, and gate VG voltages. Photographs of a film of metallic SWCNTs with diameters of 1.4 nm when (b) VS = VD = VG = 0 V, (c) VS = VD (= 0 V) < VG < 5 V, (d) VS (= 0 V) < VD ~ VG < 5 V, (e) VS (= 0 V) < VG < VD < 5 V, and (f) VS (= 0 V) < VG < VD ~ 5 V. Reprinted with permission from [9]. Copyright 2012 WILEY-VCH Verlag GmbH & Co. KGaA, Weinheim.

17.2.2 Synthesis and electrochromic properties of nanocarbon–metal oxide hybrids

Transition metal oxides are a major class of electrochromic materials. The oxides of the following transition metals are electrochromic: cerium, chromium, cobalt, copper, iridium, iron, manganese, molybdenum, nickel, niobium, palladium, praseodymium, rhodium, ruthenium, tantalum, titanium, tungsten and vanadium [1, 2]. Among these transition metal oxides, the oxides of tungsten, molybdenum, iridium and nickel show the most intense electrochromic color changes. However, the coloration efficiencies of inorganic oxide materials are typically rather low, in the range of tens cm^2/C, and their coloration-bleaching kinetic is also limited [12–14]. Conductive nanocarbon components have proven useful to facilitate the transition metal oxides with better electrochromic processes, which appear crucial to greatly enhancing the performances of electrochromic materials.

R. M. Osuna *et al.* showed that the electrochromic property of an inorganic oxide, WO_3, was considerably improved when SWCNT films were used instead of ITO as the transparent conductive layer. The SWCNT layer affords unprecedented electrochemical stability for WO_3 accompanied by enhanced electrochromism (high contrast, faster switching response, and better cycling stability) over other nanostructured crystalline WO_3 films. In this case, SWCNT films could effectively play the role of conducting and porous scaffold for facilitated proton intercalation/deintercalation, also imparting good adhesion and chemical and mechanical stability to WO_3 [15]. The enhanced electrochromic performance was also observed in multi-walled carbon nanotubes (MWCNTs)-WO_3 composite. It was found that the MWCNTs helped improve the intercalation/deintercalation kinetics due to the fact that the MWCNTs were anchored in WO_3 in a way that easily favored intercalation/deintercalation kinetics *e.g.* by forming suitable conduits [16]. Graphene, the newest member of the nanocarbon family, has become a new star material in recent years. Recent research has been launched into the development of graphene/metal oxide composites for electrochromic applications. Cai *et al.* developed an efficient route to a porous NiO/reduced graphene oxide (rGO) hybrid film. The porous hybrid film exhibits a noticeable electrochromism with reversible color changes from transparent to dark brown, and shows high coloration efficiency (76 cm^2C^{-1}), fast switching speed (7.2 s and 6.7 s) and better cycling performance compared with the porous NiO thin film. The enhancement of electrochromic performances is attributed to the reinforcement of the electrochemical activity of the rGO sheets and the greater amount of open space in the porous hybrid film which allows the electrolyte to penetrate and shorten the proton diffusion paths within the bulk of NiO [17].

Nanocarbon hybrids have recently been introduced as a new class of multifunctional composite materials [18]. In these hybrids, the nanocarbon is coated by a polymer or by the inorganic material in the form of a thin amorphous, polycrystalline or single-crystalline film. The close proximity and similar size domain/volume fraction of the two phases within a nanocarbon hybrid introduce the interface as a powerful new parameter. Interfacial processes such as charge and energy transfer create synergistic effects that improve the properties of the individual components and even create new properties [19]. We recently developed a simple dry wrapping method to fabricate a special class of nanocarbon hybrid, WO_3/carbon nanotube (CNT) coaxial cable structure (Fig. 17.2), in which WO_3 layers act as an electrochromic component while aligned

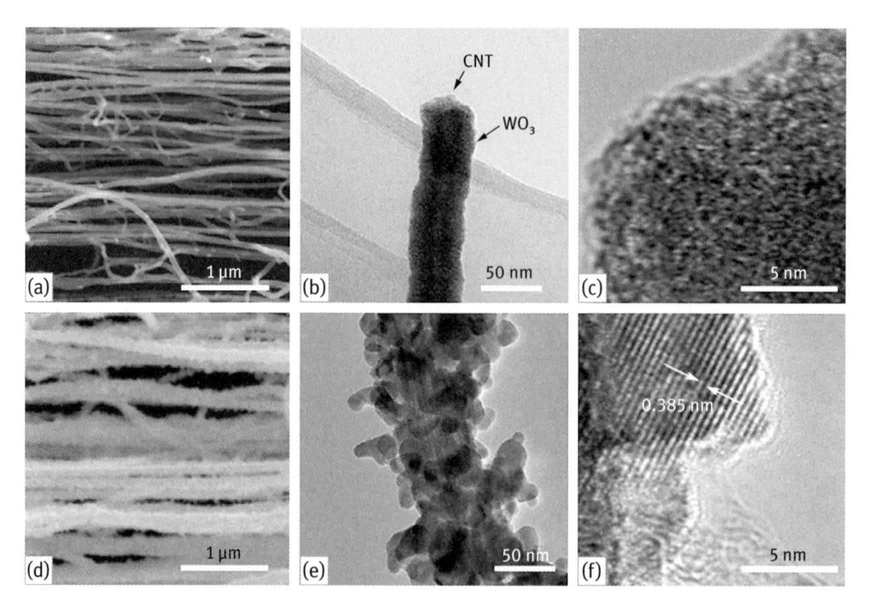

Fig. 17.2: Structural and morphological characterization of WO_3/CNT sheets prepared at 300 °C and 400 °C; (a), (d) SEM images showing the morphology of WO_3/CNT sheets; (b), (e) TEM images showing the WO_3 wrapping on the CNT surface; (c), (f) HRTEM images showing the lattice images of WO_3. Reprinted with permission from [20]. Copyright 2012, The Royal Society of Chemistry.

CNTs at the core provide mechanical support and an anisotropic, continuous electron transport pathway [20]. Both amorphous and crystalline forms of WO_3 can be rationally obtained. Interestingly, the resultant cable material exhibits an obvious gradient electrochromic phenomenon (Fig. 17.3). To the best of our knowledge, the aligned WO_3/CNT provides the first example of a nanosized coaxial cable that can be used as an obvious gradient electrochromic material, where a gradual color evolution from yellowish to blue along the CNT axial direction is observed although the biases are instantly applied. The construction of such WO_3/CNT coaxial 1D hybrid nanostructures offers not only an integrity of high conductivity, structural durability and flexibility, but also efficient charge transfer wired by CNT cores rather than by additional conductive substrates like ITO or other transparent conductive oxides (TCO).

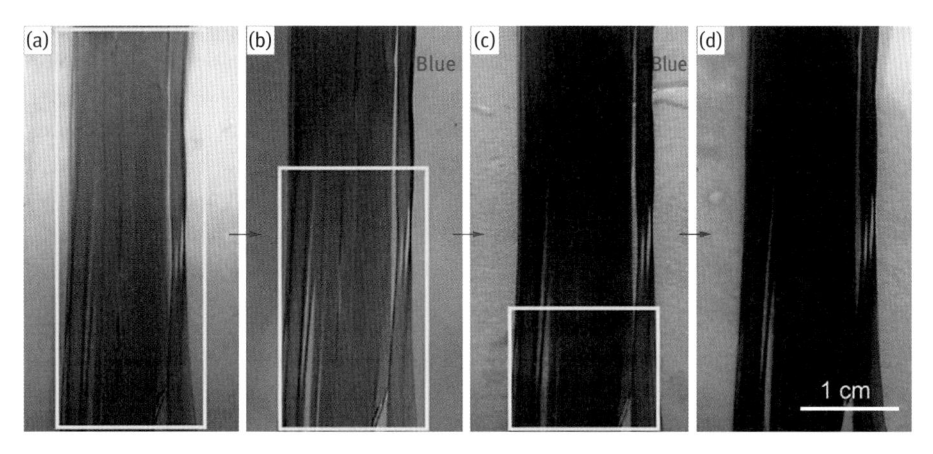

Fig. 17.3: Chromatic transition of crystalline WO_3/CNT sheets. (a) Original state appear yellow; colored states after applying –1.0 V for (b) 5 s, (c) 15 s and (d) 30 s. Reprinted with permission from [20]. Copyright 2012, The Royal Society of Chemistry.

17.2.3 Electrochromic properties of nanocarbon–polymer hybrids

Conjugated polymers such as polythiophene (PTP), polypyrroles (PPy) and polyanillines (PANI) are the most widely used organic electrochromic materials. Since electrochromism of most conjugated polymers involves oxidative doping, the stability of conjugated polymers still remains the main obstacle for their practical application. For electrochromic conjugated polymers, the most active research areas lie in design and synthesis of novel polymers to address the stability, while there are still considerable efforts spent on further enhancing coloration efficiency, switching speed, and processability of the polymers [14]. In the past decade, there have been continuous attempts to combine conjugated polymers with nanocarbon materials to form nanocarbon–conjugated polymer hybrid electrochromic materials. The strong π-π interactions between nanocarbon materials and conjugated polymers lead to synergistic effects in improving electrochromic performances.

Bhandari *et al.* reported PEDOT/functionalized MWCNTs composite thin films, which show lower charge transfer resistance and higher coloration efficiency (414 $cm^2\ C^{-1}$) than PEDOT (201 $cm^2\ C^{-1}$) films [21, 22]. Xiong *et al.* demonstrated that covalently bonded PANI-SWCNT hybrids can be readily synthesized *via* copolymerization of aniline with SWCNT-*p*-phenylenediamine in aqueous solutions containing poly(styrene sulfonate). This new class of water-processable hybrid materials exhibits significantly enhanced electrochromic contrast and switching kinetics [23]. On a similar note, graphene has also been incorporated into conjugated polymers to form electrochromic hybrid composites. Sheng *et al.* prepared graphene/PANI multilayer films by the combination of LBL assembly and chemical reduction. The electrochromic device based on graphene/PANI multilayer films showed improved electrochemical stability compared with a similar device with a conventional ITO electrode [24]. Specifically, another typical form of macroscopic CNT assembly, CNT fiber, was also used as a mechanical support for electrochromic conjugated polymers. Peng *et al.* reported the facile synthesis of CNT/polydiacetylene nanocomposite fibers that rapidly and reversibly respond to electrical current, with the resulting color change being readily observable with the naked eye (Fig. 17.4). The electrochromic nanocomposite fibers could have various applications in sensing [25]. We also recently sandwiched another macroscopic CNT assembly, a CNT sheet within polyurethane (PU) films to

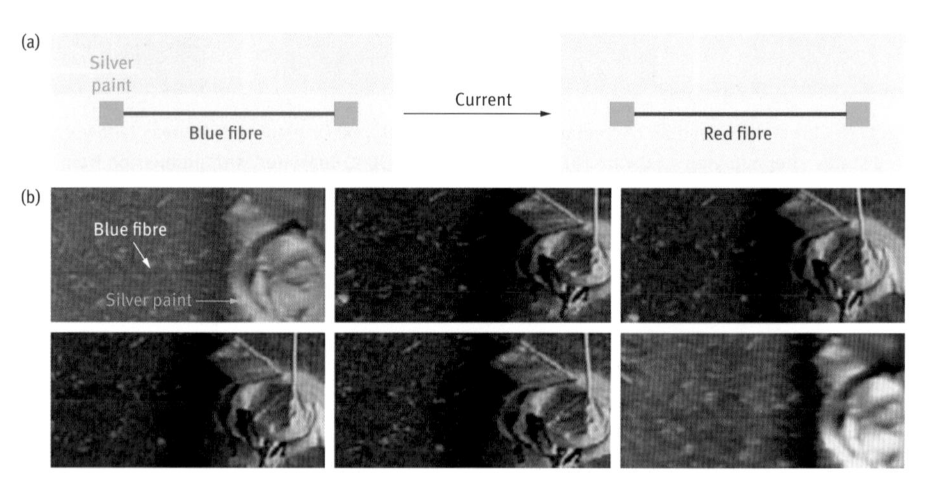

Fig. 17.4: (a) Schematic showing that the color of the CNT/PDA fiber changes from blue to red when a current is passed through it in a two-probe experiment. (b) Passing a d.c. current of 10 mA through a CNT/PDA fiber (diameter, 11 μm) causes a blue fiber (top left) to become red after 1 s (top middle). The current is turned off after 3 s (bottom left), and the fiber becomes blue again after a further 2 s (bottom right). Silver paint was used to hold the CNT/PDA fibers in place and to connect to the external current sources. All experiments were performed at room temperature. Reprinted with permission from [25]. Copyright 2009, Nature Publishing Group.

fabricate a composite film with switchable transparency. The introduction of CNTs not only makes the composite film electrically conductive, but also induces a rapid crystal melting of soft segments in the PU. As a result, the film can be switched from opaque to transparent in just several seconds after turning on voltage, and reversed back to opaque after turning off voltage [26]. On the whole, the improved coloration efficiency of the nanocarbon–polymer hybrids is attributed to the reduced charge transfer resistance brought by the bridging effect of the conductive nanoparticles such as CNT and graphene between crystalline domains of conjugated polymers, while the fast switching speed is due to the shorter diffusion length brought by nanocarbons [14].

17.3 Nanocarbon hybrids for photovoltaic applications

The increasing global demand for energy is becoming an impelling force for us to exploit renewable solar energy for electricity or fuels (see Chapter 18). Therefore, considerable efforts have been devoted to developing low-cost, environmentally clean, and high-efficiency solar conversion devices such as photoelectrochemical cells (PECs) and organic photovoltaics (OPVs). The potential contribution of nanocarbons such as fullerene, CNT and graphene to high-efficiency light energy harvesting devices are reviewed.

17.3.1 Working mechanisms of PECs and OPVs

Dye-sensitized solar cells (DSSC) and solar water splitting cells can be classified as PECs due to their similar cell configurations. Typically, a PEC is composed of an anode, a cathode, and electrolyte. However, their working mechanisms are different. The principle of operation and energy level scheme of DSSC is schematically shown in Fig. 17.5(a) [27]. The heart of DSSC is a mesoporous semiconductor electrode (TiO_2, ZnO, or Nb_2O_5) that has absorbed dye molecules. The dye molecules are excited upon light illumination and inject electrons to the conduction band of the oxide, leaving the dye molecules oxidized. The oxidized dyes are then reduced to their original state by electron donation from the electrolyte such as the iodide/triiodide redox couple in organic solution. The electrons in the oxide layer transport through the external load to the counter electrode at which the triiodide is regenerated in turn by the reduction of iodide. During the charge transfer cycle, the transport of electrons across the semiconductor electrode plays a very important role in the performance of the DSSC.

While for a solar water splitting cell, light is directly absorbed by the semiconductor electrode (anode or cathode). The separation of electron-hole pairs is achieved in the built-in electric field near the semiconductor surface. The electric field is formed due to the charge transfer between the semiconductor electrode and the electrolyte as schematically shown in Fig. 17.5(b) [28]. Take an n-type semiconductor electrode for example

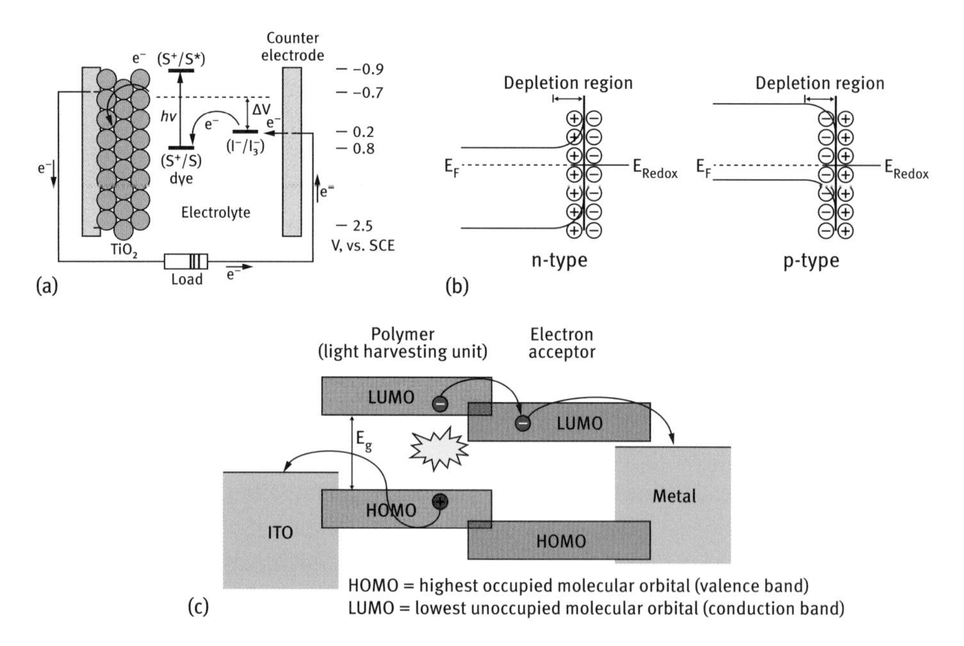

Fig. 17.5: Scheme of basic processes occurring in DSSCs (a) and organic solar cells (c). (b) Band bending for an n-type semiconductor and a p-type semiconductor in equilibrium with an electrolyte.

[29], the photogenerated electrons in the oxide coatings are transported through the external circuit, followed by reacting with protons to form hydrogen at the the counter electrode; the holes migrate to the electrode surface, being trapped by hydroxide ions to yield oxygen. This kind of electrode is called a photoanode. The the electrical flow direction is opposite to that of the n-type electrode in the p-type semiconductor. This means that hydrogen and oxygen are generated at the semiconductive electrode and the counter electrode, respectively; the electrode is called a photocathode.

OPVs commonly involve no electrolyte reactions, and work following a relatively simple operating mechanism. Light is harvested by donor–acceptor bilayers in such devices. As shown in Fig. 17.5(c), excitons (bonded electron–hole pairs) are initially generated upon light illumination on the conjugated polymer. They then diffuse to the interface of the donor and the acceptor, followed by dissociation to produce free electrons and holes. Finally, free carriers are either transported through segregated phases and finally collected by electrodes or lost *via* recombination during transportation.

17.3.2 Nanocarbon hybrids for PECs

The employment of a mesoporous semiconductor electrode dramatically increases the contact area between the semiconductor and the absorbed photoactive components and electrolyte, and eventually improves the conversion efficiency of the PECs. However, it also increases the semiconductor boundaries in the photoelectrodes. As shown in Fig. 17.6(a), the forward charge transport cycle in a PEC is largely limited by the slow transport of electrons across the particle network [30], which has been regarded as a major hurdle in attaining higher photoconversion efficiency. This is due to the fact that electrons have to encounter lots of grain boundaries before they are collected at the FTO electrode. It has been demonstrated that electron transport in such illuminated nanoparticle networks proceeds by a trap-limited diffusion process, in which photogenerated electrons repeatedly interact with a distribution of traps as they undertake a random walk through the particles as schematically shown in Fig. 17.6(b) [31, 32]. Such a random transit behavior of the photogenerated electrons increases the probability of their recombination with the oxidized dye molecules.

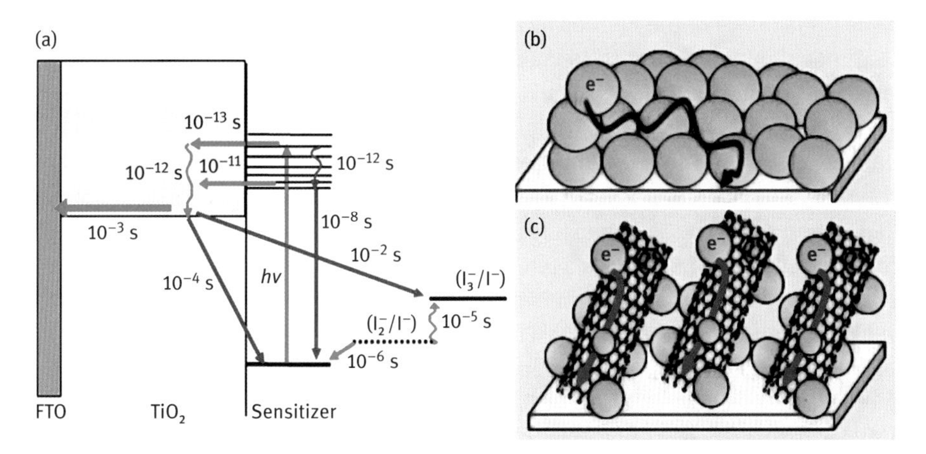

Fig. 17.6: (a) Overview of processes and typical time constants under working conditions (1 sun) in a Ru-dye-sensitized solar cell with iodide/triiodide electrolyte. Recombination processes are indicated by red arrows. Reprinted with permission from [30]. Copyright 2010, American Chemical Society. Electron transport across nanostructured semiconductor films: (b) in the absence and (c) in the presence of a nanotube support architecture. Reprinted with permission from [38]. Copyright 2007, American Chemical Society.

In order to inhibit the grain boundary recombination, an efficient pathway for the photogenerated charges is needed. CNTs, one-dimensional nanostructures with a surface area as high as $1600\ m^2\ g^{-1}$ [33], are characterized by a ballistic charge transport along the axis (see Chapter 1) [34]. They show strong electron-accepting ability, and their band gaps are tunable over a wide range by employing different radii and chirality [35]. Graphene is a two-dimensional material and has a remarkably high electron mobility ($15\,000\ cm^2\ V_{S^{-1}}^{-1}$) (see Chapter 2) [36]. The theoretical surface area of the exfoliated graphene is ~ $2600\ m^2\ g^{-1}$ [37]. Therefore, it is desirable to use these nanocarbons to make up the recombination deficiency by integrating their unique properties with the semiconductors to form nanocarbon hybrids. CNT/semiconductor hybrids were prepared by embedding semiconductor nanoparticles on the surface of CNTs [18]. P. Kamat *et al.* proposed a novel hybrid structure in which a CNT networks was used as support to anchor light-harvesting semiconductor particles (TiO_2), achieving a CNT/TiO_2 hybrid (Fig. 17.6(c)) [38]. Enhancement in the photoconversion efficiency by a factor of two was achieved by the nanohybrid compared to the pure TiO_2 nanoparticle electrode in the TiO_2-based PECs. Semiconductor quantum dots (QDs) such as CdS [39, 40], PdS [41], and CdSe [42], have also been coated on CNTs for harvesting visible light. Fast electron transfer from the quantum dots to CNTs was observed, which renders the CNT/QDs hybrids promising in optoelectronic applications [43].

In order to further improve the efficiency, two significant challenges still remain in architecting nanostructures for the nanocarbon hybrids. One is to achieve a good interface between the nanotubes and the semiconductor; the other is to interconnect the CNTs inside of the hybrids as an efficient conducting network. Both of them are of fundamental importance to make full use of the unique features of the nanotubes and the semiconductors. A good interface can be satisfied by constructing the CNT/semiconductor hybrid with a core/shell structure, which ensures not only a large contact area between the two components but also an efficient radial charge transfer path for the photogenerated electrons. D. Xiao *et al.* reported the synthesis of single-walled carbon nanotube/TiO_2 nanocrystal (SWCNT/TiO_2) core/shell hybrid using a genetically engineered M13 virus as a template, and they showed that small fractions of nanotubes improve the power conversion efficiency up to 10.6 % when the hybrid was used as photoanodes in DSSCs [44]. They found that the performance of the DSSCs, especially the short-circuit current density (I_{sc}), is closely related with the electronic type of SWCNT as shown in Fig. 17.7(a). The I_{sc} for the hybrids using semiconducting nanotubes (s-CNT) increases by 27 %, but I_{sc} for the hybrids using metallic nanotubes (m-CNT) decreases by 20 % compared to the TiO_2-only DSSCs. All fill factors are ~ 0.7, and all open-circuit voltages are ~ 780 mV. They attributed the difference to fact that, due to the non-continuous band structure of s-CNTs, the s-CNT/TiO_2 hybrid has longer electron diffusion length than the m-CNT/TiO_2 hybrid and the TiO_2 electrode as shown in Fig. 17.7(b). These different electron diffusion lengths result in different electron collection efficiencies for the devices, and account for the different power conversion efficiencies. The M13 virus as a template renders the uniform oxide coating around the CNTs. However the CNTs in the photoanode are still randomly distributed. Moreover, the chemical treatment on the CNTs may alter the electronic properties of the CNTs and induce lots of defects which have been regarded as the recombination sites. Therefore, a structure with the aligned CNTs directly contacting with the collecting substrates could be more effective. X. Quan and co-workers developed a CNT/TiO_2 heterojunction array in which the semiconductor is coated around the vertically aligned CNTs [45]. Photoelectrochemical investigations certified that the hybrid could minimize recombination of photoinduced electrons and holes and had a more effective photoconversion capability than the aligned TiO_2 nanotubes on the titanium substrate. Liu and co-workers further assembled a DSSC using the vertically aligned CNT/TiO_2 hybrid ar-

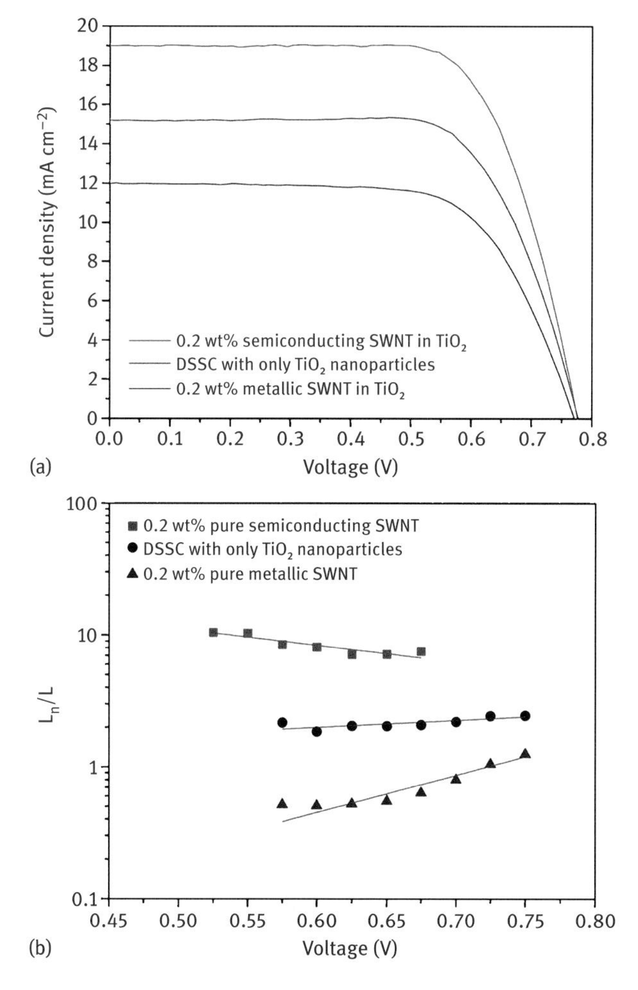

Fig. 17.7: Current–voltage curves (a) and calculated electron diffusion lengths (b) from three DSSCs: with TiO_2 nanoparticles only, with 0.2 wt% pure semiconducting SWCNT, and with 0.2 wt% pure metallic SWCNT. L is the film thickness and Ln is the electron diffusion length. A virus-to-SWCNT ratio of 1 : 5 was used for all devices. Reprinted with permission from [44]. Copyright 2011, Macmillan Publishers.

ray as schematically shown in Fig. 17.8(a) [46]. CNT arrays were grown on a transparent fluorine-doped tin oxide (FTO) coated quartz slides using a nanometer thick Ni layer as catalyst. TiO_2 coatings were deposited on the CNT array *via* MOCVD using titanium isopropoxide vapor under 150 mTorr at 550 °C. Figure 17.8(b) and (c) shows the nanostructures before and after oxide deposition, respectively. Using the hybrid absorbed with dye as a photoanode, the solar cell shows an encouraging overall conversion efficiency of ~ 1.09 % and a rather high open-circuit voltage of ~ 0.64 V. Further improvement in the conversion efficiency is limited by the low surface area of the oxides coatings as the CNTs in their work have very large diameters, and the CNT arrays show very strong light shielding. Very recently, Peng's group reported a DSSC using a CNT fiber absorbed with dye as the photoanode. The photoexited dye molecules directly inject electrons into the CNT fiber. The fiber-based solar cell shows an encouraging conversion efficiency of 2.6 % [47]. We demonstrated an aligned nanocarbon hybrid

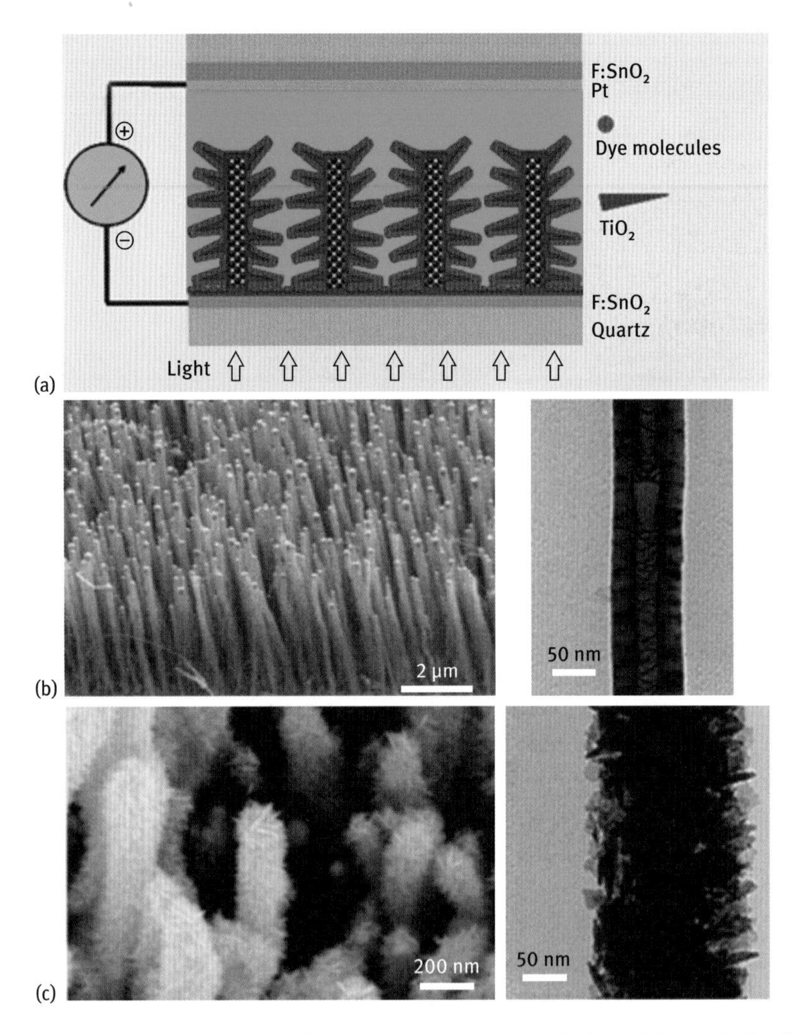

Fig. 17.8: (a) Schematic of CNT/TiO_2 arrays for DSSCs. Morphology of the vertically aligned CNT array (b) and the CNT/TiO_2 array (c). Reprinted with permission from [46]. Copyright 2009, American Chemical Society.

film in which continuous TiO_2 is conformally coated on a free-standing CNT sheet [48]. The CNTs are well-aligned and interconnected inside of the conformal oxide coating, achieving efficient conducting paths, as shown in Fig. 17.9(a). This architecture can effectively decrease the contact resistance of the nanotubes and simultaneously enhance the mechanical flexibility of the heterostructure film (Fig. 17.9(b), (c)), which is different to that of the CNT/TiO_2 hybrids synthesized by wet chemistry approaches as shown in Fig. 17.9(a) [18, 49, 50]. When applied as photoelectrodes, the hybrid film is free from high-cost transparent conducting electrodes due to the preformed nanotube conducting paths (Fig. 17.9(f)). Figure 17.9(d) shows the photograph of an as-prepared

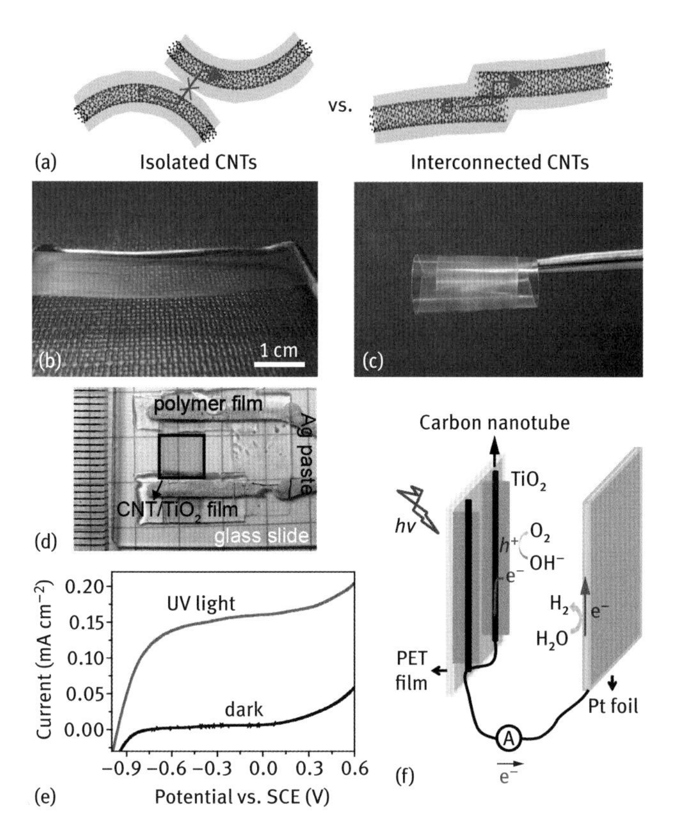

Fig. 17.9: (a) Schematic illustrating that charges transfer more smoothly between CNTs that are interconnected inside of oxide coatings (right) than those isolated by oxide coatings (left). (b) A freestanding CNT/TiO_2 film. (c) Photograph of a CNT/TiO2 film attached on a PET film. (d) Photograph of a CNT/TiO_2 photoelectrode. (e) Current–potential curves of a CNT/TiO_2 film deposited for 12 min under dark and UV illumination. (f) Schematic illustrating the structure of a water-splitting solar cell containing a CNT/TiO_2 film in which the CNTs serve as the charge transport paths and current collector. Reprinted with permission from [48]. Copyright 2012, WILEY-VCH Verlag GmbH & Co. KGaA, Weinheim.

CNT/TiO_2 electrode. A freestanding CNT/TiO_2 film was transformed onto a glass slide or a PET film, followed by connecting both ends of the film using silver paste. The silver paste was then sealed with polyurethane film, leaving the sample an opening for light illumination. This electrode geometry is in contrast to the TiO_2 or CNT/TiO_2 composite electrodes reported previously where the photoactive materials are commonly coated onto conducting substrates such as FTO and Ti foils [38, 51]. A three-electrode system was used for the photoelectrochemical measurement, where Pt foil and saturated calomel electrode serve as the counter electrode and the reference electrode, respectively. Figure 17.9(e) plots the current densities from a CNT/TiO_2 film with optimized deposition time as a function of applied potential under dark and UV illumination (300–400 nm, 1 mW/cm^2). The electrode generates a very low cathodic current density ($j < 7 \times 10^{-6}$ A/cm^2) with the potential swept from −0.75 to 0 V under dark, indicative of the formation of a depletion layer in the electrode–electrolyte interface [52, 53]. Upon light illumination, a large anodic photocurrent is observed. This strongly demonstrates that photogenerated electrons in the TiO_2 layer transport through CNTs to the external circuit. This unique feature of the hybrid film boosted the incident photon-to-electron conversion efficiency to be 32 % at 320 nm, outperforming TiO_2 nanoparticle electrodes fabricated on TOC substrates [38]. Moreover, the hybrid film is robust, and it can generate stable photocurrent even after being bent hundreds of times.

The scenario of the graphene-based hybrids in photovoltaics is consistent with that of the CNT-based hybrid as shown in Fig. 17.10. Photoexcitation of the semiconductor particles in the hybrid leads to charge transfer to the graphene sheet. On one hand, the electrons from large band gap semiconductors could reduce graphene oxide into reduced graphene oxide (rGO) [54]. On the other hand, the rGO can provide more efficient transport paths for the electrons to the charge collecting substrates. An rGO/TiO_2 composite was prepared by mixing rGO with TiO_2 nanoparticles in solution. When the composite was used as photoanode in solar water splitting cells, nearly 90 % enhancement in the photocurrent was observed compared with the TiO_2-based photoanodes [55]. N. Yang [56] and S. Sun [57] individually introduced this concept into DSSCs using the rGO/TiO_2 composite. The DSSCs show improved conversion efficiency compared with the solar cells using the nanocrystalline TiO_2 photoanodes.

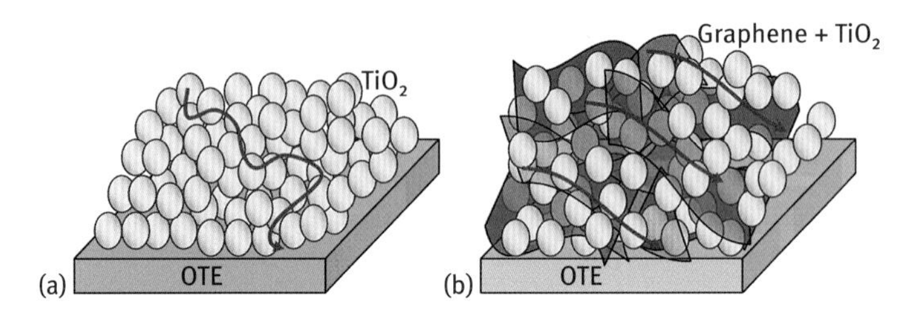

Fig. 17.10: Electron transport in pristine TiO_2 (left) and rGO–TiO_2 nanocomposite (right) films cast on an optically transparent electrode (OTE). Reprinted with permission from [55]. Copyright 2012, American Chemical Society.

17.3.3 Nanocarbon hybrids for OPVs

Apart from the aforementioned nanocarbon/inorganic oxide hybrids, nanocarbon/polymer hybrids have shown promising applications in OPVs. It is reported that the major factors that limit the conversion efficiency of OPVs include high-exciton binding energy (0.2–0.6 eV), short-exciton diffusing length (10–20 nm), and mismatch of the band position of the donor and acceptor [58]. Fullerene shows strong electron-accepting ability, and its derivatives have been widely used as acceptors for the bulk heterojunction structure [59]. The hybrid of poly(3-hexylthiophene) (P3HT) and 1-(3-methoxycarbonyl) propyl-1-phenyl[6, 6] C_{61} ($PC_{61}BM$) is a very popular active material for solar cells and efficiency is reported in the range of 5 %–6 % [60, 61]. By using the hybrids composed of suitable polymers and fullerene derivatives, the efficiency of the solar cells reported in literature works has been approaching 10 % [62]. Some novel hybrids are PTB7/$PC_{71}BM$ [62, 63], and PCDTBT/$PC_{71}BM$ [64]. Nevertheless, the efficiency of the solar cells based on fullerene derivatives is limited by the low charge mobility and aggregation of fullerenes [65]. Therefore, other hybrids containing CNTs or graphene are then proposed to fill these gaps, considering the high charge mobility of these nanocarbons. Inceptive work on CNT/polymer hybrids showed that the SWCNTs help separation of the exciton generated in the poly (3-octylthiophene). However, the conversion efficiency of the solar cells is only 0.04 % [66]. The low efficiency could be partially attributed to the limited interface between CNTs and polymers due to the poor dispersion of CNTs. In order to get good hybrid interfaces, Nicholas's group advanced the utilization of photoactive polymers to disperse CNTs, and individual CNT/polymer hybrids with polymer molecular wrapping around CNTs [67]. His group further showed that the photoexcitation of P3HT produces an ultrafast (~ 430 fs) charge transfer to the SWCNTs and P3HT/SWCNT hybrids incorporating only small fractions (1 %) of nanotubes individually embedded in the P3HT matrix allowing photon-to-charge conversion with efficiencies comparable to those for conventional P3HT/fullerene blends [68]. Further characterization of the CNT/P3OT hybrid by ultrafast transient absorption demonstrated that the life time of the photogenerated polarons is strongly dependent on the type of CNTs and the percolation pathways for both the separated electrons and holes. Indeed, it is reported that the semiconducting SWCNTs can extend the lifetime of the separated charges, while metallic ones act as recombination sites possible by electron transfer from the CNTs to the polymer [69, 70]. J. Holt *et al.* reported that the elimination of metallic SWCNTs can improve long-lived mobile carriers available for photocurrent by a factor of three [71]. The highest conversion efficiency achieved using SWCNT/P3HT hybrid is 0.72% by S. Ren *et al.* [72]. Further improvement relies on the sorting out or synthesis of SWCNTs with suitable band gaps, individual dispersion of SWCNTs, and good interface between the polymers and SWCNTs. Alternatively, functionalized graphene shows better solubility in polymers than CNTs, and thus this could improve the interface between the two components. Chen *et al.* [73] reported the fabrication of solar cells using a graphene/P3OT hybrid, and the optimized conversion efficiency is 1.4 %, higher than the solar cells using the SWCNT/P3OT hybrid.

17.4 Summary

The utility concept of nanocarbon hybrids in electrochromic and photovoltaic applications is the integration of the unique properties of nanocarbons to compensate for the deficiencies of the active components. As shown by the aforementioned works, synergetic effects have been observed in the nanocarbon hybrids by the remarkable enhancements of the electrochromic response and the photoconversion efficiency. However, the performance of the nanocarbon hybrid-based devices is still inferior to the state-of-the-art devices based on traditional materials. For example, the highest conversion efficiency of the DSSCs using CNT/polymer hybrids is 10.6 % [44], lower than 11.4 % [74] and 11.1 % [75] for the TiO_2-based DSSCs. The following aspects, we think, should be taken into consideration when designing and fabricating the nanocarbon hybrids to make full use of each component.

Energy Level Match Previous works have demonstrated that the hybrids containing only some types of semiconducting CNTs show positive effect on enhancing the conversion efficiency of the solar cells while the hybrids containing metallic CNTs tend to lower the efficiency [44, 71]. For the hybrids where nanocarbons mainly serve as fast conducting channels, energy barriers should be avoided between the components; metallic nanotubes could be favorably selected. Therefore, single-electronic-type CNTs with desirable electrical properties are needed in order to experimentally figure out a proper combination in the nanocarbon hybrids.

Good interface Provided that the energy levels of the components have been well matched, charge transfer between the guest materials and the nanocarbons becomes a 'core' issue for the hybrids' application in related areas. Thus, well-bonded interfaces are of great importance. Currently, the nanocarbons such as fullerenes, CNTs, and graphene are commonly chemically modified in order to disperse well in solutions and further get a uniform coating by guest materials. However, the dispersion process can induce lots of surface defects in the nanocarbons which increase the resistance of the charge transfer by recombination or charge scattering. Physical vapor deposition of target materials on nanocarbons could be a non-destructive method for the preparation of the nanocarbon hybrids. Moreover, the structure of the nanocarbons can be well preserved after the deposition. Recent works also showed that the good interface can be readily achieved not only for the nanocarbon hybrids but also for other novel hybrid materials using the method, and further performance improvement has been observed compared with the hybrids prepared by traditional chemical methods.

Acknowledgments

We are grateful for the financial support from the National Natural Science Foundation of China (No. 10834004, 21273269, 51372266, 51102274, and 61274130), and the National Basic Research Program of China (2011CB932600) by Ministry of Science and Technology of China.

Bibliography

[1] Granqvist, C. G., *Handbook of Inorganic Electrochromic Materials*. Elsevier Science: 1995.

[2] Monk, P.; Mortimer, R.; Rosseinsky, D., *Electrochromism and Electrochromic Devices*. Cambridge University Press: 2007.

[3] Rauh, R. D., Electrochromic windows: An Overview. *Electrochim Acta* **1999**, 44, 3165–3176.

[4] Rosseinsky, D. R.; Mortimer, R. J., Electrochromic systems and the prospects for devices. *Adv. Mater.* **2001**, 13, 783–793.

[5] Granqvist, C. G., Electrochromic tungsten oxide films: Review of progress 1993–1998. *Sol Energ Mater Sol C* **2000**, 60, 201–262.

[6] Granqvist, C., Progress in electrochromics: Tungsten oxide revisited. *Electrochim Acta* **1999**, 44, 3005–3015.

[7] Mortimer, R. J., Organic electrochromic materials. *Electrochim Acta* **1999**, 44, 2971–2981.

[8] Mortimer, R. J., Electrochromic materials. *Chem. Soc. Rev.* **1997**, 26.

[9] Yanagi, K.; Moriya, R.; Yomogida, Y.; Takenobu, T.; Naitoh, Y.; Ishida, T.; Kataura, H.; Matsuda, K.; Maniwa, Y., Electrochromic carbon electrodes: Controllable visible color changes in metallic single-wall carbon nanotubes. *Adv. Mater* **2011**, 23, 2811–2814.

[10] Kataura, H.; Kumazawa, Y.; Maniwa, Y.; Umezu, I.; Suzuki, S.; Ohtsuka, Y.; Achiba, Y., Optical properties of single-wall carbon nanotubes. *Synthetic Met* **1999**, 103, 2555–2558.

[11] Hersam, M. C., Progress towards monodisperse single-walled carbon nanotubes. *Nature Nanotechnology* **2008**, 3, 387–394.

[12] Oi, T., Electrochromic materials. *Annu. Rev. Mater. Sci.* **1986**, 16, 185–201.

[13] Amb, C. M.; Dyer, A. L.; Reynolds, J. R., Navigating the color palette of solution-processable electrochromic polymers. *Chem. Mater* **2010**, 23, 397–415.

[14] Thakur, V. K.; Ding, G.; Ma, J.; Lee, P. S.; Lu, X., Hybrid materials and polymer electrolytes for electrochromic device applications. *Adv. Mater* **2012**.

[15] Malaveì Osuna, R.; Hernaìndez, V. c.; Loìpez Navarrete, J. T.; Kauppinen, E. I.; Ruiz, V., Ultrafast and high-contrast electrochromism on bendable transparent carbon nanotube electrodes. *J. Phys. Chem. Lett.* **2010**, 1, 1367–1371.

[16] Kadam, P. M.; Tarwal, N. L.; Mali, S. S.; Deshmukh, H. P.; Patil, P. S., Enhanced electrochromic performance of F-Mwcnt-Wo3 composite. *Electrochim Acta* **2011**, 58, 556–561.

[17] Cai, G.; Tu, J.; Zhang, J.; Mai, Y.; Lu, Y.; Gu, C.; Wang, X., An efficient route to porous Nio/Rgo hybrid film with highly improved electrochromic properties. *Nanoscale* **2012**.

[18] Eder, D., Carbon nanotube-inorganic hybrids. *Chem. Rev.* **2010**, 110, 1348–1385.

[19] Vilatela, J. J.; Eder, D., Nanocarbon composites and hybrids in sustainability: A Review. *ChemSusChem* **2012**, 5, 456–478.

[20] Yao, Z.; Di, J.; Yong, Z.; Zhao, Z. G.; Li, Q., Aligned coaxial tungsten oxide/carbon nanotube sheet: A flexible and gradient electrochromic film. *Chem. Commun.* **2012**, 48, 8252–8254.

[21] Bhandari, S.; Deepa, M.; Srivastava, A. K.; Lal, C.; Kant, R., Poly(3,4-Ethylenedioxythiophene) (Pedot)-Coated Mwcnts tethered to conducting substrates: Facile electrochemistry and enhanced coloring efficiency. *Macromol Rapid Commun* **2009**, 30, 138–138.

[22] Bhandari, S.; Deepa, M.; Srivastava, A. K.; Joshi, A. G.; Kant, R., Poly (3, 4-Ethylene-dioxythiophene)– multiwalled carbon nanotube composite films: structure-directed amplified electrochromic response and improved redox activity. *J. Phys. Chem. B* **2009**, 113, 9416–9428.

[23] Xiong, S.; Wei, J.; Jia, P.; Yang, L.; Ma, J.; Lu, X., Water-processable polyaniline with covalently bonded single-walled carbon nanotubes: Enhanced electrochromic properties and impedance analysis. *Acs. Appl. Mater. Interfaces* **2011**, 3, 782–788.

[24] Sheng, K.; Bai, H.; Sun, Y.; Li, C.; Shi, G., Layer-by-layer assembly of graphene/polyaniline multilayer films and their application for electrochromic devices. *Polymer* **2011**.

[25] Peng, H.; Sun, X.; Cai, F.; Chen, X.; Zhu, Y.; Liao, G.; Chen, D.; Li, Q.; Lu, Y.; Zhu, Y., Electrochromatic carbon nanotube/polydiacetylene nanocomposite fibres. *Nature nanotechnology* **2009**, 4, 738–741.
[26] Meng, F.; Zhang, X.; Xu, G.; Yong, Z.; Chen, H.; Chen, M.; Li, Q.; Zhu, Y., Carbon nanotube composite films with switchable transparency. *Acs. Appl. Mater Interfaces* **2011**, 3, 658–661.
[27] Grätzel, M., Perspectives for dye-sensitized nanocrystalline solar cells. *Progress in photovoltaics: research and applications* **2000**, 8, 171–185.
[28] Bard, A. J., Photoelectrochemistry. *Science* **1980**, 207, 139–144.
[29] Bott, A. W., Electrochemistry of semiconductors. *Curr. Sep.* **1998**, 17, 87–92.
[30] Hagfeldt, A.; Boschloo, G.; Sun, L.; Kloo, L.; Pettersson, H., Dye-sensitized solar cells. *Chem. Rev.* **2010**, 110, 6595–6663.
[31] Oekermann, T.; Zhang, D.; Yoshida, T.; Minoura, H., Electron transport and back reaction in nanocrystalline Tio2 films prepared by hydrothermal crystallization. *J. Phys. Chem. B* **2004**, 108, 2227–2235.
[32] Fisher, A. C.; Peter, L. M.; Ponomarev, E. A.; Walker, A. B.; Wijayantha, K. G. U., Intensity dependence of the back reaction and transport of electrons in dye-sensitized nanocrystalline Tio2 solar cells. *J. Phys. Chem. B* **2000**, 104, 949–958.
[33] Baughman, R. H.; Cui, C.; Zakhidov, A. A.; Iqbal, Z.; Barisci, J. N.; Spinks, G. M.; Wallace, G. G.; Mazzoldi, A.; De Rossi, D.; Rinzler, A. G., Carbon nanotube actuators. *Science* **1999**, 284, 1340–1344.
[34] Javey, A.; Guo, J.; Wang, Q.; Lundstrom, M.; Dai, H. J., Ballistic carbon nanotube field-effect transistors. *Nature* **2003**, 424, 654–657.
[35] Dresselhaus, M.; Dresselhaus, G.; Saito, R., Physics of carbon nanotubes. *Carbon* **1995**, 33, 883–891.
[36] Geim, A. K.; Novoselov, K. S., The rise of graphene. *Nat. Mater* **2007**, 6, 183–191.
[37] Stankovich, S.; Dikin, D. A.; Dommett, G. H. B.; Kohlhaas, K. M.; Zimney, E. J.; Stach, E. A.; Piner, R. D.; Nguyen, S. B. T.; Ruoff, R. S., Graphene-based composite materials. *Nature* **2006**, 442, 282–286.
[38] Kongkanand, A.; Martínez Domínguez, R.; Kamat, P. V., Single wall carbon nanotube scaffolds for photoelectrochemical solar cells. Capture and transport of photogenerated electrons. *Nano Lett.* **2007**, 7, 676–680.
[39] Robel, I.; Bunker, B. A.; Kamat, P. V., Single-walled carbon nanotube–Cds nanocomposites as light-harvesting assemblies: Photoinduced charge-transfer interactions. *Adv. Mater* **2005**, 17, 2458–2463.
[40] Li, X.; Jia, Y.; Cao, A., Tailored single-walled carbon nanotube–Cds nanoparticle hybrids for tunable optoelectronic devices. *ACS Nano* **2009**, 4, 506–512.
[41] Ka, I.; Le Borgne, V.; Ma, D.; El Khakani, M. A., Pulsed laser ablation based direct synthesis of single-wall carbon nanotube/Pbs quantum dot nanohybrids exhibiting strong, spectrally wide and fast photoresponse. *Adv. Mater.* **2012**, n/a-n/a.
[42] Jeong, S.; Shim, H. C.; Kim, S.; Han, C.-S., Efficient Electron Transfer in Functional Assemblies of Pyridine-Modified Nqds on SWNTs. *ACS Nano* **2009**, 4, 324–330.
[43] Peng, X.; Chen, J.; Misewich, J. A.; Wong, S. S., Carbon nanotube–nanocrystal heterostructures. *Chem. Soc. Rev.* **2009**, 38, 1076–1098.
[44] Dang, X.; Yi, H.; Ham, M.-H.; Qi, J.; Yun, D. S.; Ladewski, R.; Strano, M. S.; Hammond, P. T.; Belcher, A. M., Virus-templated self-assembled single-walled carbon nanotubes for highly efficient electron collection in photovoltaic devices. *Nat. Nano.* **2011**, 6, 377–384.
[45] Yu, H.; Quan, X.; Chen, S.; Zhao, H., TiO2–multiwalled carbon nanotube heterojunction arrays and their charge separation capability. *J. Phys. Chem. C* **2007**, 111, 12987–12991.

[46] Liu, J.; Kuo, Y.-T.; Klabunde, K. J.; Rochford, C.; Wu, J.; Li, J., Novel dye-sensitized solar cell architecture using Tio2-coated vertically aligned carbon nanofiber arrays. *Acs. Appl. Mater. Interfaces* **2009**, 1, 1645–1649.

[47] Chen, T.; Wang, S.; Yang, Z.; Feng, Q.; Sun, X.; Li, L.; Wang, Z.-S.; Peng, H., Flexible, light-weight, ultrastrong, and semiconductive carbon nanotube fibers for a highly efficient solar cell. *Angewandte Chemie International Edition* **2011**, 50, 1815–1819.

[48] Di, J.; Yong, Z.; Yao, Z.; Liu, X.; Shen, X.; Sun, B.; Zhao, Z.; He, H.; Li, Q., Robust and aligned carbon nanotube/titania core/shell films for flexible Tco-free photoelectrodes. *Small* **2012**.

[49] Woan, K.; Pyrgiotakis, G.; Sigmund, W., Photocatalytic carbon-nanotube–Tio2 composites. *Adv. Mater.* **2009**, 21, 2233–2239.

[50] Eder, D.; Windle, A. H., Carbon-inorganic hybrid materials: The carbon-nanotube/Tio_2 interface. *Adv. Mater.* **2008**, 20, 1787–1793.

[51] Feng, X.; Shankar, K.; Varghese, O. K.; Paulose, M.; Latempa, T. J.; Grimes, C. A., Vertically aligned single crystal Tio2 nanowire arrays grown directly on transparent conducting oxide coated glass: synthesis details and applications. *Nano Lett.* **2008**, 8, 3781–3786.

[52] Walter, M. G.; Warren, E. L.; McKone, J. R.; Boettcher, S. W.; Mi, Q.; Santori, E. A.; Lewis, N. S., Solar water splitting cells. *Chem. Rev.* **2010**, 110, 6446–6473.

[53] Hagfeldt, A.; Grätzel, M., Light-induced redox reactions in nanocrystalline systems. *Chem. Rev.* **1995**, 95, 49–68.

[54] Kamat, P. V., Graphene-based nanoarchitectures. anchoring semiconductor and metal nanoparticles on a two-dimensional carbon support. *J. Phys. Chem. Lett* **2009**, 1, 520–527.

[55] Ng, Y. H.; Lightcap, I. V.; Goodwin, K.; Matsumura, M.; Kamat, P. V., To what extent do graphene scaffolds improve the photovoltaic and photocatalytic response of Tio2 nanostructured films? *J. Phys. Chem. Lett* **2010**, 1, 2222–2227.

[56] Yang, N.; Zhai, J.; Wang, D.; Chen, Y.; Jiang, L., Two-dimensional graphene bridges enhanced photoinduced charge transport in dye-sensitized solar cells. *ACS Nano* **2010**, 4, 887–894.

[57] Sun, S.; Gao, L.; Liu, Y., Enhanced dye-sensitized solar cell using graphene-Tio photoanode prepared by heterogeneous coagulation. *Appl. Phys. Lett.* **2010**, 96, 083113.

[58] Thompson, B. C.; Fréchet, J. M. J., Polymer–fullerene composite solar cells. *Angewandte Chemie International Edition* **2007**, 47, 58–77.

[59] Brabec, C. J.; Sariciftci, N. S.; Hummelen, J. C., Plastic solar cells. *Adv. Funct. Mater.* **2001**, 11, 15–26.

[60] Dang, M. T.; Hirsch, L.; Wantz, G., P3ht: Pcbm, best seller in polymer photovoltaic research. *Adv. Mater.* **2011**, 23, 3597–3602.

[61] Dennler, G.; Scharber, M. C.; Brabec, C. J., Polymer-fullerene bulk-heterojunction solar cells. *Adv. Mater.* **2009**, 21, 1323–1338.

[62] He, Z.; Zhong, C.; Su, S.; Xu, M.; Wu, H.; Cao, Y., Enhanced power-conversion efficiency in polymer solar cells using an inverted device structure. *Nat Photon* **2012**, 6, 593–597.

[63] Liang, Y.; Xu, Z.; Xia, J.; Tsai, S.-T.; Wu, Y.; Li, G.; Ray, C.; Yu, L., For the bright future – bulk heterojunction polymer solar cells with power conversion efficiency of 7.4 %. *Adv. Mater.* **2010**, 22, E135-E138.

[64] He, Z.; Zhong, C.; Huang, X.; Wong, W.-Y.; Wu, H.; Chen, L.; Su, S.; Cao, Y., Simultaneous enhancement of open-circuit voltage, short-circuit current density, and fill factor in polymer solar cells. *Adv. Mater.* **2011**, 23, 4636–4643.

[65] Thompson, B. C.; Fréchet, J. M. J., Polymer–fullerene composite solar cells. *Angewandte Chemie International Edition* **2008**, 47, 58–77.

[66] Kymakis, E.; Amaratunga, G., Single-wall carbon nanotube/conjugated polymer photovoltaic devices. *Appl. Phys. Lett.* **2002**, 80, 112–114.

[67] Nish, A.; Hwang, J.-Y.; Doig, J.; Nicholas, R. J., highly selective dispersion of single-walled carbon nanotubes using aromatic polymers. *Nat. Nano.* **2007**, 2, 640–646.

[68] Stranks, S. D.; Weisspfennig, C.; Parkinson, P.; Johnston, M. B.; Herz, L. M.; Nicholas, R. J., Ultrafast charge separation at a polymer–single-walled carbon nanotube molecular junction. *Nano Lett.* **2011**, 11, 66.

[69] Cataldo, S.; Salice, P.; Menna, E.; Pignataro, B., Carbon nanotubes and organic solar cells. *Energy & Environmental Science* **2012**, 5, 5919–5940.

[70] Li, G.; Liu, L., Carbon nanotubes for organic solar cells. *Nanotechnology Magazine, IEEE* **2011**, 5, 18–24.

[71] Holt, J. M.; Ferguson, A. J.; Kopidakis, N.; Larsen, B. A.; Bult, J.; Rumbles, G.; Blackburn, J. L., Prolonging charge separation in P3ht–SWNT composites using highly enriched semiconducting nanotubes. *Nano Lett.* **2010**, 10, 4627–4633.

[72] Ren, S.; Bernardi, M.; Lunt, R. R.; Bulovic, V.; Grossman, J. C.; Gradečak, S., Toward efficient carbon nanotube/P3ht solar cells: Active layer morphology, electrical, and optical properties. *Nano Lett.* **2011**, 11, 5316–5321.

[73] Liu, Z.; Liu, Q.; Huang, Y.; Ma, Y.; Yin, S.; Zhang, X.; Sun, W.; Chen, Y., Organic photovoltaic devices based on a novel acceptor material: Graphene. *Adv. Mater.* **2008**, 20, 3924–3930.

[74] Han, L.; Islam, A.; Chen, H.; Malapaka, C.; Chiranjeevi, B.; Zhang, S.; Yang, X.; Yanagida, M., High-efficiency dye-sensitized solar cell with a novel co-adsorbent. *Energy & Environmental Science* **2012**, 5, 6057–6060.

[75] Chiba, Y.; Islam, A.; Watanabe, Y.; Komiya, R.; Koide, N.; Han, L., Dye-sensitized solar cells with conversion efficiency of 11.1 %. *Jpn. J. Appl. Phys.* **2006**, 45, 638.

Rubén D. Costa and Dirk M. Guldi

18 Carbon nanomaterials as integrative components in dye-sensitized solar cells

18.1 Today's dye-sensitized solar cells. Definition and potential

One of the major challenges that humanity faces is the development of renewable energy resources to secure our current lifestyle. In fact, linking problems such as pollution and global warming to the use of fossil fuels as an energy source is eye opening [1–3]. In this context, different renewable energy sources such as windmill-powered plants, hydropower, biomass production, geothermal power, and solar energy are currently under scrutiny. Among the aforementioned, solar energy bears one of the greatest potentials for securing the current energy supply. This notion is based on the amount of energy that reaches the earth's surface at zero costs and at continuous supply. Notably, one hour of solar irradiation covers more than the world's current energy demand for a complete year. As such, photovoltaic devices/solar cells are considered as a superior means of energy harvesting [2, 4, 5].

Solar cells are separated into three different categories, namely first, second, and third generation. First generation solar cells, which currently lead the market, are either based on single-crystalline (sc-Si) or on multicrystalline (poly-Si) silicon technology as well as on GaAs and InGaP semiconductors. Although Si, GaAs, and InGaP technologies are most mature, especially in the context of their high efficiencies (*i.e.*, > 15 %), the cost factor of materials processing is a major hurdle. The production costs of such modules have undergone an enormous increase in recent years, owing to the needs of high purity silicon materials. By means of introducing low-cost, thin-film technologies, second generation solar cells have emerged. Leading examples of second generation solar cells are chemical vapor deposited (CVD) amorphous silicon (a-Si) and polycrystalline CdS, InSe, $CuInSe_2$, and CdTe devices [6]. These solar cells feature lower production costs but also lower efficiencies when compared to first generation solar cells. With the advent of exploring novel materials and inexpensive fabrication approaches, solvent deposition techniques have been developed to process inorganic and organic materials. Particularly important are dye-sensitized solar cells (DSSC) and organic photovoltaics (OPV) in the form of either bilayer or bulk heterojunction (BHJ) architectures. The latter two have led to efficiencies beyond 10 % at very low production costs [7–10]. Finally, third generation solar cells should be considered, which are based on a multilayer architecture of amorphous silicon and gallium arsenide that assists in bypassing the Shockley–Queisser limit found in first generation solar cells [11, 12].

During recent decades, research efforts have focused on the development of DSSCs. Their advantages include use of cheap materials, ease of device fabrication, high stability, and reasonably high efficiencies of currently around 11–13 % [9, 10, 13–

16]. Indeed, the results in development of DSSCs as technically and economically credible solar energy-conversion systems are overwhelming. G24 Innovation Ltd., for example, uses a low-cost, roll-to-roll process to fabricate flexible DSSCs modules, which feature 0.5W of power under sunlight illumination.[17]. Furthermore, first demonstrators displayed at the 1st International Photovoltaic Power Generation Expo in 2008 showed 2.1 m x 0.8 m DSSC modules with conversion efficiencies of around 3 %. More relevant are recent breakthroughs announced by, for example, Sony and Apple, which are gambling on implementing DSSCs in our daily life like decoration, iPads, etc. [18].

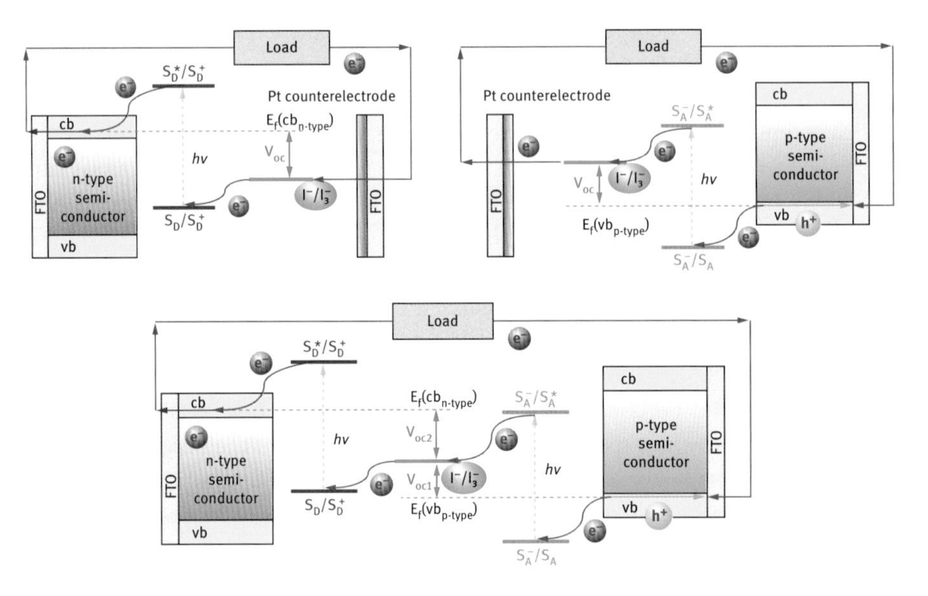

Fig. 18.1: *Modus operandi* of n-type (top left), p-type (top right), and tandem DSSCs (bottom). S_A and S_D are acceptor and donor dyes, respectively. E_f, cb, and vb are Fermi level, conduction band, and valance band, respectively. e^- and h^+ refer to electron and hole current, respectively.

To detail DSSC technologies, Fig. 18.1 illustrates the *modus operandi* of DSSCs. Initially, light is absorbed by a dye, which is anchored to the surface of either n- or p-type semiconductor mesoporous electrodes. Importantly, the possibility of integrating both types of electrodes into single DSSCs has evoked the potential of developing tandem DSSCs, which feature better overall device performances compared to just n- or p-type based DSSCs [19–26]. Briefly, n-type DSSCs, such as TiO_2 or ZnO mesoporous films, are deposited on top of indium-tin oxide (ITO) or fluorine-doped tin oxide (FTO) substrates and constitute the photoanodes. Here, charge separation takes place at the dye/electrode interface by means of electron injection from the photoexcited dye into the conduction band (cb) of the semiconductor [27, 28]. A different mechanism governs p-type DSSCs, which are mainly based on NiO electrodes on ITO and/or FTO substrates

[29–32]. In this type of DSSCs, once the dye is photoexcited, charge separation drives electrons from the valence band (vb) of the semiconductor to the photoexcited dye. Common to both types of DSSCs is the regeneration of the oxidized or reduced dye by a redox mediating electrolyte. The latter is mainly in the form of a liquid and/or a solid. Platinum films deposited onto ITO or FTO are the most utilized counter-electrodes and are required to close the electronic circuit.

18.2 Major challenges in improving the performance of DSSCs

Despite the plethora of outstanding contributions to the field of DSSCs and giant efforts to implement DSSCs as common technology, different challenges remain to date. These are divided into five main groups described as follows.

(i) The optimization of electron transport within the semiconductor electrode [13, 33]. The semiconductor electrode is the cornerstone of the DSSCs, owing to its main role in the dye adsorption, charge collection, and diffusion processes. The most widely used electrodes are based either on TiO_2, ZnO as photoanodes or on NiO as photocathodes. In this context, the major bottleneck is the transport of injected charges across the electrode network. Notably, charge transport competes with charge recombination and charge transfer processes that involve either the attached dye or the electrolyte, respectively. Recently, implementing carbon nanomaterials into mesoporous semiconductor electrodes has emerged as a highly viable strategy to overcome this particular problem, as will be discussed in Section 18.3.

(ii) The optimization of solid-state electrolytes [34–36]. Electrolyte solutions are unstable over time, owing to solvent leakages, evaporation processes under indoor and outdoor operation conditions (*i.e.*, 50–80 °C and more), and sealant corrosion. In addition, industrial large-scale production by, for example, roll-to-roll process is very inefficient when liquid electrolytes are used. Promising solutions involve the fabrication of solid-state DSSCs (ssDSSCs) using organic ionic plastic crystal electrolytes [37], organic hole conductors [38–44], polymer gel electrolytes [45, 46], light-trapping schemes adding nematic liquid crystals [47], physically cross-linked gelators [48, 49], and ionic liquid (IL) based electrolytes [45, 50–54]. Still, the rather poor pore filling of mesoporous semiconductor electrodes with solid conductive materials, the low charge mobility, and the moderate charge diffusivity are major obstacles for highly-efficient DSSCs. The implementation of carbon nanomaterials in solid-state electrolytes based on ionic liquids provide yet another promising strategy to circumvent the aforementioned problems and is described in Section 18.4.

(iii) The use of iodine-free redox couples [55]. To date, the most widely used and efficient electrolyte is the I^-/I_3^- redox couple. But this electrolyte has severe limitations, namely relatively high Nernst potential, absorption features in the visible region, rapid mass loss of I_2, enhanced charge recombination process due to I_3^- ion pair formation in the proximity of the electrodes, and unwanted corrosion that limits the DSSC stability. Practical alternatives are metal complexes, inorganic materials, and organic compounds. Cobalt complexes stand out among them all, owing to their excellent DSSC performance [56]. Recent examples, in which carbon nanomaterials function as electrolytes without the needs of I_2 that are utilized in solid-state DSSCs, are dealt with in Section 18.4.

(iv) The development of broad band sensitizers [27, 57–59]. Currently, the most utilized dyes are based on ruthenium(II) polypyridine complexes. These materials are expensive and environmentally unfriendly. The synthesis and implementation of organic dyes such as perylenediimides, squarines, thiophenes, porphyrins, and phthalocyanines, featuring absorption cross-sections that span from the ultraviolet (UV) to the near infrared (IR) regions of the solar spectrum is, therefore, at the focal point of investigations. Implicit in the design of new DSSC dyes is the modification/functionalization of existing dyes, the development of new linkers, and the synthesis of push-pull chromophores [27, 57–59]. Notably, porphyrins are the most promising dyes, owing to recent breakthroughs in device efficiencies with values as high as 13 % [9]. In addition, porphyrins exhibit an excellent ability to exfoliate graphene from bulk graphite. Thanks to this fact, two major findings have been recently established. On one hand, the combination of both materials, that is, porphyrins/porphycenes and graphene, enhanced the overall performance of DSSCs. Secondly, two well-established methods to provide large amounts of nanographene hybrids in solution have allowed its implementation into solar cells by using solution-based techniques. Both aspects are discussed in Sections 18.5 and 18.6, respectively.

(v) The search of platinum-free counter-electrodes [13, 60, 61]. Platinum (Pt) is the superior catalyst of DSSCs to date. Pt also presents several drawbacks including high price, rarity, and susceptibility to corrosion by iodide electrolyte. To this end, several sorts of low-cost materials – carbon nanomaterials [62–68], metal oxide [69], and transition metal carbides and sulfides [70–72] – have been probed as alternatives. The myriad of publications in the field of carbon nanomaterial-based counter-electrodes has been unsurpassable in the last five years. Instead of including a separate section, we refer to two excellent reviews that we recommend to the interested reader [60, 61].

As aforementioned, the introduction of carbon nanomaterials is an effective strategy to take on some of the contemporary challenges in the field of DSSCs. In particular, enhanced charge injection and charge transport processes in carbon nanomaterial-doped electrodes, efficient carbon nanomaterial-based, iodine-free, quasi-solid state electrolytes, and the use of novel nanographene hybrids as dyes are some of the most stunning milestones. All of these milestones are considered as solid proof for the excellent prospect of carbon nanomaterials in DSSCs. The major goal of this chapter is to provide an overview of recent advances of carbon nanomaterials as integrative components in DSSCs.

18.3 Carbon nanomaterials as integrative materials in semiconducting electrodes

As discussed in the previous section, one of the major DSSC bottlenecks is the charge transport across the electrode network. Recently, different groups have demonstrated that implementation of nanocarbons in the form of (i) interlayers on the bottom and/or on top of the mesoporous film, and (ii) dopants inside the electrode network, is a very powerful strategy to overcome this issue. In this section, the most relevant aspects are outlined.

18.3.1 Interlayers made out of carbon nanomaterials

The FTO/electrode/electrolyte and electrode/dye/electrolyte are the critical interfaces, where charge recombination processes takes place. Among all of them, the predominant place for charge recombination in DSSCs is the FTO/electrode/electrolyte interface [73]. To this end, the most common strategy to retard the interfacial charge recombination is the preparation of a thin film of TiO_2 *via* hydrolyzing aqueous solutions of $TiCl_4$ [10].

As a replacement for the thin film of TiO_2, the first example of carbon nanomaterial interlayer was reported by Kim *et al.* in 2009 [74]. In this work, GO was mixed with TiO_2 nanoparticles (Degussa P25). Importantly, the authors discovered that ultraviolet light irradiation assists in reducing GO to graphene. Exact conditions were a 450 W xenon arc lamp and irradiation for about 2 hours. The resulting paste was used as a buffer layer before a second TiO_2 layer was deposited on top by doctor-blading. Impedance spectroscopy was used to demonstrate that the interfacial layer effectively reduces the charge recombination processes leading to better fill factors (FF) and higher open-circuit voltages (V_{oc}) and, in turn, to an efficiency improvement of around 10 %.

Following this earlier work, Chen *et al.* have recently reported an approach in which graphene sheets with sizes ranging from 0.5 to 3 μm have been used to functionalize the surface of the FTO substrate [75]. They used the spin-coating technique for film forming and demonstrated by impedance spectroscopy that the presence of graphene between FTO and TiO_2 is critical in terms of collecting the injected charges. Moreover, the graphene sheets block the reversed electron flow, that is, to the TiO_2 electrode, owing to the low work function of graphene (−4.4 eV) when compared to TiO_2 (−4.2 eV). This led to an impressive improvement of the overall DSSC efficiencies from 7.54 to 8.13 % in devices featuring thin TiO_2 and graphene interlayers, respectively.

Recently, a similar sort of interlayer, which was deposited onto the electrode/dye/electrolyte interface, has been used to enhance photocurrent and fill factor in TiO_2-based DSSCs [76]. Song *et al.* proposed, for example, the use of GO to form a GO-TiO_2 Schottky junction that blocks recombination processes. In addition, the electron transport through the TiO_2 network and the adsorption of the dye by means of π-π interactions were enhanced. The layer was prepared by spray-coating of GO aqueous solutions at 120 °C to form a thin layer of GO on top of the TiO_2 electrode. A head-to-head comparison by impedance spectroscopy of a control device and a modified device in terms of charge transfer and charge recombination led to the conclusion that the overall device resistances were reduced and the electron lifetimes were increased. As a consequence, fill factors and photocurrents were improved leading to an enhancement of around 10 % in the efficiency.

18.3.2 Implementation of carbon nanomaterials into electrode networks

One of the first attempts was the preparation of SWCNT-doped TiO_2 electrodes by dispersing SWCNTs on top of carbon fiber electrodes *via* electrophoresis followed by a deposition of bare TiO_2 nanoparticles [77, 78]. The authors demonstrated mutual interactions between both materials in this new type of SWCNT-doped TiO_2-based electrodes. Overall, these devices doubled the efficiency when compared to reference devices without SWCNTs. However, once the electrodes were modified with $Ru^{II}(bpy)_2(dcbpy)$ as photosensitizer, the amplification in the overall efficiency was only marginal. Responsible for this trend is the reduction of the V_{oc}, owing to the low Fermi level of SWCNTs [78]. Nevertheless, the authors demonstrated that doping with SWCNTs enhanced the charge separation in the excited state and the transport of charges to the FTO substrate. Interestingly, steady-state and time-resolved measurements revealed that the charge injection process was not affected by the presence of SWCNTs. An additional asset is that these electrodes have also been utilized to drive the photoelectrolysis of water with moderate success [77].

The next step was the use of MWCNTs in ZnO and TiO_2 electrodes [79–83]. MWCNT containing photoanodes were fabricated either by a modified acid-catalyzed sol-gel process [82] or by step-wise mixing of oxidized MWCNT dispersions in metal-oxide particle pastes [83]. A common trend seen in all of these investigations is that the doping with MWCNTs enhanced the electrode roughness and, in turn, larger quantities of adsorbed dye allowing higher photocurrents (Fig. 18.2). In addition, charge recombination of electron-hole pairs was strongly reduced leading to enhanced overall device performance.

Graphene has also been used, owing to its high surface area providing extended contacts with TiO_2 or NiO [84–91]. Intermolecular forces such as physisorption, elec-

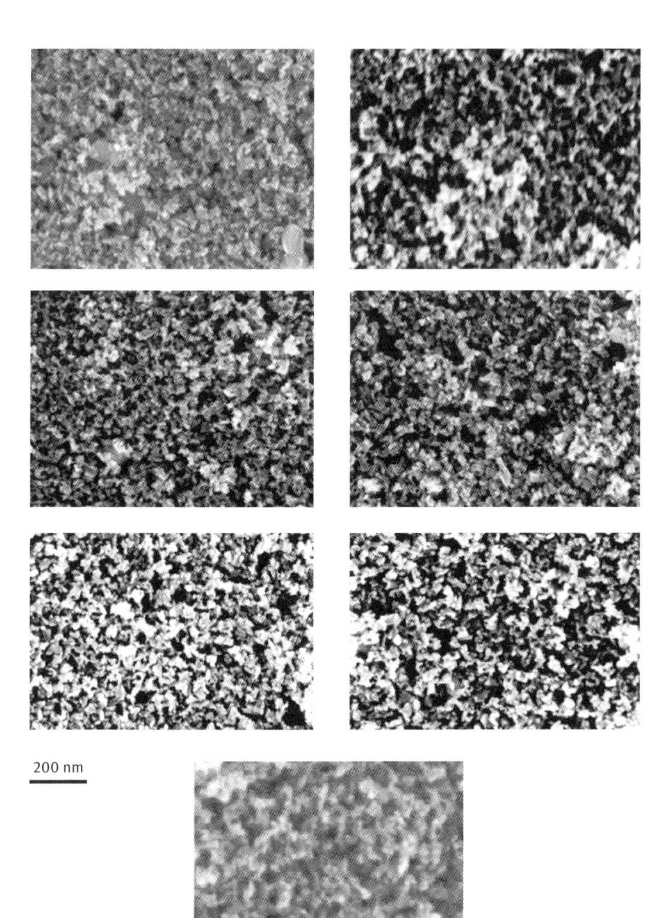

Fig. 18.2: Upper part – SEM images of pristine (left) and oxidized (right) nanocarbon/TiO_2 films. The TiO_2 films are doped with 0.5 wt% SWCNHs (top), 0.5 wt% graphene (middle), and 0.1 wt% SWCNTs (bottom). Lower part – SEM image of pristine TiO_2.

trostatic binding, and charge transfer interactions are at work. In general, two different procedures are known to prepare this new type of electrodes. Firstly, GO and the nanoparticles suspensions were mixed and the new composite was further treated under ultraviolet irradiation to induce reduction of GO [84, 85]. Secondly, graphene suspensions were prepared. This was realized by chemical exfoliation process under oxidative or reductive conditions [86, 87, 91] or by ultrasonication with Nafion as stabilizer [88]. Subsequently, the prepared graphene suspensions were mixed with TiO_2 or NiO leading to homogenous and robust composites that were used to prepare graphene-doped electrodes.

Similar to what has been seen with CNT-doped electrodes, the presence of graphene resulted in enhanced charge transport and inhibited charge recombination. Additionally, as a consequence of doping rougher surfaces evolved, which, in turn, increased light scattering and amount of dye adsorbed onto the surface (Fig. 18.2).

Importantly, all of the investigations agree on the fact that high graphene loading provokes aggregation and, as such, cracks start to appear throughout films. With the aforementioned advantages in hand, photocurrents were increased by 40–60 %, while a slight loss of V_{oc} is seen in TiO_2 based devices [84–90]. Interestingly, Li *et al.* showed that V_{oc} increases upon graphene doping from 73 to 105 mV in NiO based devices [91]. This fact was ascribed to the reduction of hole density, since graphene acts as a hole trap and, in addition, as a hole transport material. Independently of the type of electrode, the overall device efficiency was found to be greatly enhanced. An important fact is that the use of graphene to dope DSSCs led to a more notable enhancement of the overall device performance than in devices that are doped with SWCNTs – Fig. 18.3.

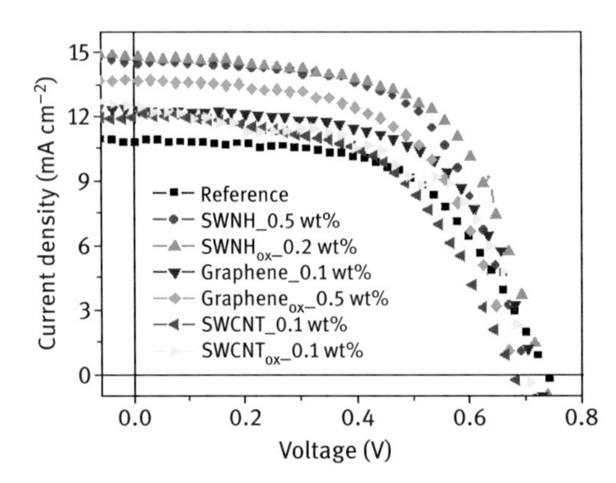

Fig. 18.3: Current versus applied voltage characteristics for DSSCs prepared with different carbon nanomaterials doped TiO_2 photoelectrodes.

The use of exfoliated graphene is common, although intrinsically difficult to achieve. Notably, ultrasonication or oxidation/reduction-assisted chemical exfoliation procedures destroy the excellent features of graphene by introducing defects in the graphene lattice [92]. Tang *et al.* have doubtlessly demonstrated that the impact of oxidation during the chemical exfoliation process of graphite is crucial on the device performance [87]. In addition, the resulting graphene suspensions are unstable, which hampers the processability of graphene in liquid media. To circumvent these aspects, synthetic graphene analogous, which combine the ease of managing and the high conductivity, have recently been explored [93]. To this end, hexa-tert-butyl-substituted hexa-peri-hexabenzocoronene (HBC), which is considered the smallest model system for graphene (42 sp^2 carbons), was employed to dope TiO_2 based DSSCs (Fig. 18.4). The most valuable finding is that, similar to graphene-doped devices, the roughness, the amount of dye, and the light scattering were enhanced in comparison to reference devices that implemented pristine TiO_2 electrodes. As a matter of fact, higher photocurrents and higher device efficiencies evolved. A direct comparison between HBC-doped and graphene-doped devices indicates that the main difference rests on the impact that these materials exert on the electronic properties of the electrodes.

Due to a better delocalization of charges within graphene relative to HBC, the charge transport and the charge recombination processes are profoundly impacted by doping with graphene rather than with HBC. Nevertheless, these results suggest that the delocalized structure of HBC is insufficient to mimic the characteristics of exfoliated graphene. Therefore, establishing a comparison between size and solubility remains as a future challenge.

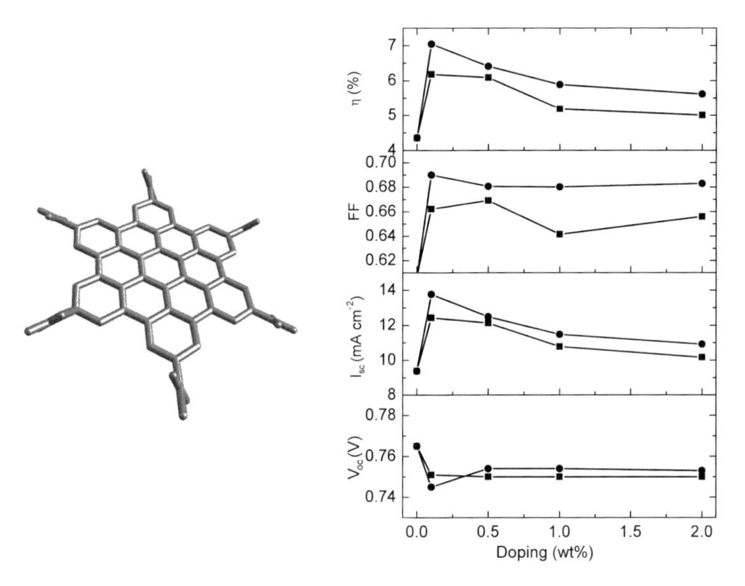

Fig. 18.4: Left part – chemical structure of HBC. Right part – device performance upon doping with different concentrations of HBC (squares) and graphene (circles).

Finally, the last carbon nanomaterials studied as DSSC dopant were SWCNHs [94]. Here, the preparation of the composites was realized by drop-wise addition of small amounts of either pristine or oxidized SWCNHs, which were suspended into the TiO_2 paste by means of ultrasonication. The results were similar to those found for other carbon nanomaterials. In particular, the major impact of electrode doping relates to enhanced roughness, quantity of dye adsorbed, light scattering, and electrical resistance features (Fig. 18.2). Additionally, transient absorption spectroscopy showed that charge recombination was reduced. Again, the photocurrent was found to increase with a simultaneous decrease of V_{oc} leading, nevertheless, to enhanced efficiencies. Intriguingly, the authors complemented their investigation by comparative assays with doped electrodes, that is, the use of SWCNHs, pristine SWCNTs, oxidized SWCNTs, and graphene. The results indicated that SWCNH- and graphene-doping resulted in devices that performed very similarly, while they outperformed those prepared with SWCNTs (Fig. 18.3). In short, the prospect of SWCNHs as integrative materials in TiO_2 electrodes is, indeed, very promising.

18.4 Carbon nanomaterials for solid-state electrolytes

As mentioned in Section 18.2, iodine-free solid-state electrolytes that feature good contact with the mesoporous electrode, high ionic mobility, and good diffusion, are one of the most challenging objectives in ssDSSCs. In recent years, the use of carbon nanomaterials has been explored to tackle the above highlighted drawbacks.

18.4.1 Fullerene-based solid-state electrolytes

One of the most popular approaches towards solid-state electrolytes is the use of inorganic p-type semiconductors – CuI, CuSCN, *etc.* [95, 96] – and/or organic hole transporting materials – thiophenes, polypyrroles, polyanilines, *etc.* [97, 98] Very recently, Jung *et al.* have reported a family of fullerene derivatives as primary components for iodine-free solid-state electrolytes (Fig. 18.5) [99]. Specifically, the electrolyte consisted of fullerene derivatives that were dissolved in ethylene carbonate (EC)/propylene carbonate (PC) [EC/PC=4 : 1 (v/v)] and that were deposited on top of TiO_2 electrodes. The latter were dried at 55 °C to remove any solvent residues. The resulting electrolytes lack absorption of visible light and are thermodynamically capable of regenerating the dye. Without any further additives, the devices gave rise to remarkable efficiencies reaching values between 1 and 2 %. In a parallel experiment, iodine (I_2) was employed as an additive. In that particular case, the efficiency increases up to 3 %. However, the cell stability was compromised and the electrolyte revealed distinct absorption features in the visible region.

Other examples have shown that pristine fullerenes serve as electron shuttles in liquid-based electrolytes [100, 101]. In this context, clusters of pristine fullerenes were deposited on top of TiO_2 electrodes that were modified with ruthenium (II) complex, porphyrins, and fluorescein. All of these examples featured overall enhancements of

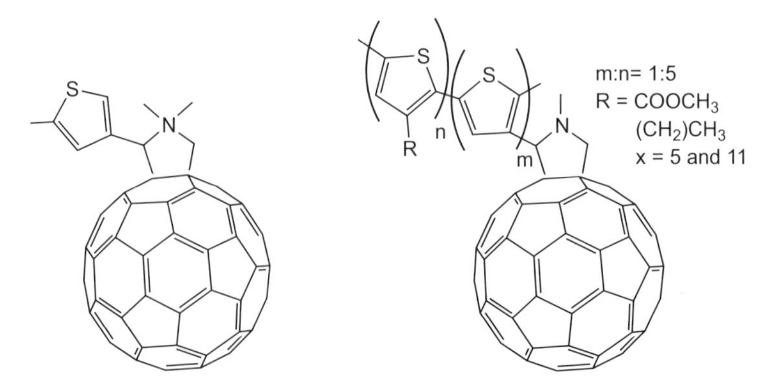

Fig. 18.5: Chemical structures of the family of fullerenes serving as hole transport materials in solid-state electrolytes.

the device photocurrents. However, this trend was only observed when the I^-/I_3^- based liquid electrolyte was present and used as a secondary electron shuttle.

18.4.2 CNTs-based solid-state electrolytes

Two early reports documented the benefits in terms of device performance when carbon nanotubes, carbon fibers, carbon nanoparticles and TiO_2 nanoparticles are individually incorporated into an ionic liquid electrolyte. The latter is based on a mixture of 1-ethyl-3-methylimidazolium bis(trifluoromethylsulfonyl)imide (EMIm-TFSI), 1-ethyl-3-methylimidazolium iodide (EMIm-I), lithium iodide (LiI), and iodine (I_2) [102]. The new electrolytes exhibited higher viscosities and better conductivities leading, in turn, to better device performances. Upon doping, as a consequence of higher FF and V_{oc} values, an enhanced efficiency evolved, which featured 5 % higher values with respect to reference devices. Following this pioneering work, Ikeda *et al.* prepared a new type of solid-state electrolyte based on clay-like conductive materials containing SWCNTs and imidazolium iodide. The plastic, solid-state device exhibited excellent efficiencies of 1.63 and 2.3 % under 1 and 0.23 sun illumination, respectively [103].

This area gained a new momentum in 2010 when Akhatar *et al.* explored a hybrid system consisting of a mixture of MWCNTs and polyethylene oxide (PEO), in which LiI and I_2 were dispersed by means of sonication and stirring [104]. They noted that doping induced higher roughness and better ionic conductivity. As such, interfacial contacts were improved, especially with the photoanode, and more favorable I_3^- diffusion was realized. Indeed, the incident photo-to-current conversion efficiency (IPCE) improved significantly from about 27 % in a control device to about 47 % when 1 % MWCNT/PEO doping was performed.

Next, Wang and Chang *et al.* proposed two interesting strategies to improve the overall dispersion features of CNTs within solid-state electrolytes [105, 106]. They proceeded *via* using either poly(oxyethylene)-segmented amides, imides polymer derivatives (POEM), or 1-(2-acryloyloxy-ethyl)-2-methyl-benzoimidazol-1-ium iodide (AMImI) to break-off MWCNT aggregates (Fig. 18.6). In both examples, the solid-state electrolyte was prepared in the same manner. In brief, all components, namely the I^-/I_3^- couple together with different additives, were dissolved in acetonitrile. Then, the dispersed MWCNTs were added to the acetonitrile solution, followed by mixing that was augmented by ultrasonication. In the last step, acetonitrile was removed under vacuum. In any of the cases, similar benefits emerged from doping with the MWCNT-based hybrids. The remarkable enhancement of the I_3^- diffusion in the solid-state stands out among the many benefits. This was corroborated by impedance spectroscopy and linear sweep voltammetry. Overall, a clear correlation between the photocurrent and the diffusion was established, while V_{oc}s and FFs remain nearly unchanged. The efficiencies were 6.86 % (POEM) and 3.6 % (AMImI) relative to the reference devices, which feature efficiencies of 4.63 and 1.15 %, respectively.

Oxidized MWCNTs have also been tested in conjunction with solid-state electrolytes [107]. Compared to pristine MWCNTs, the oxidized MWCNTs have a better miscibility with the ionic liquids used in the electrolyte. Overall, a much improved gel-forming ability resulted. The latter was clearly reflected in the device performance. In particular, devices with oxidized MWCNTs outperformed those with pristine MWCNTs and the reference devices in terms of photocurrents, V_{oc}s, and efficiencies. Importantly, the device stability was also greatly enhanced when oxidized MWCNTs were implemented – 100 days with a loss of overall efficiency by less than 10 %. The authors ascribed the drop in efficiency to phase separation and subsequent leakage of ionic liquids.

Another very interesting example was presented by Huang *et al.* in 2011 [108]. They used 15-crown-5-functionalized MWCNTs as dopant material (Fig. 18.6). The main incentive for their work is based on the binding Li^+ ions that are present in the electrolyte by means of crown ether complexation. As a matter of fact, Li^+ is expected to attract I^- ions, which would render any movement difficult within the electrolyte

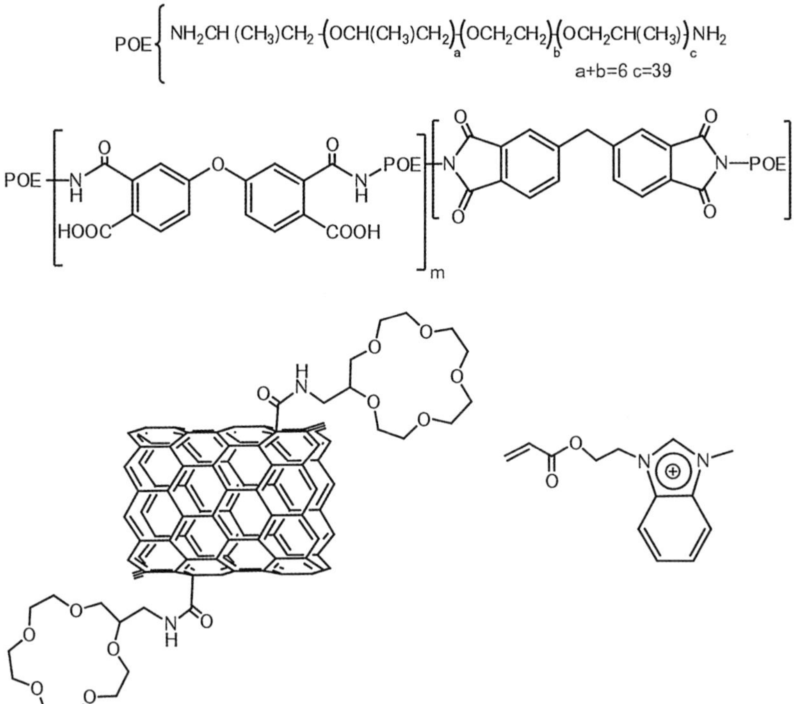

Fig. 18.6: Upper part – chemical structures of poly(oxyethylene)-segmented (POE) and amides and imides polymer derivatives (POEM). Lower left part – chemical structure of 15-crown-5-functionalized MWCNT. Lower right part – chemical structure of 1-(2-acryloyloxy-ethyl)-2-methyl-benzoimidazol-1-ium iodide (AMImI).

limiting the ionic diffusion. Similar to their preceding investigations, they prepared the electrolyte mixture in MeOH by ultrasonication and, finally, removed the solvent under vacuum. As proof of concept, they demonstrated a better diffusion by impedance spectroscopy. The latter led to moderately efficient (2.11 %) and highly stable (1200 hours at 100 mW cm^{-2}) devices.

All of the aforementioned examples focused on the use of I^-/I_3^- as redox couple. Most notably, the most appealing aim is to develop iodine-free electrolytes. Lee *et al.*, for example, have utilized SWCNTs, MWCNTs, and carbon black materials that were mixed with 1-ethyl-3-methylimidazolium iodide (EMII) and crystal growth inhibitors with excellent success [109, 110]. In initial work, a high efficiency of 3.5 % was achieved in ssDSSCs with hybrid SWCNT-EMII under 1 sun illumination [109]. By using impedance spectroscopy and laser-induced photovoltage transient spectroscopy techniques, the authors demonstrated that the addition of SWCNTs reduces the interfacial charge recombination and, simultaneously, improves the charge transport of EMII-based electrolytes. Additionally, SWCNTs act as filler for the physical gelation of the electrolyte and assist in the catalysis of I_3^- reduction formed close to the interface with the electrode. In follow-up work, they reported on an interesting comparison between carbon black, MWCNTs, and SWCNTs as dopants [110]. Common to all of them are similar effects – *vide supra*. However, devices with SWCNTs outperformed those with MWCNTs and carbon black. In particular, efficiencies of 4.01, 3.53, and 3.09 % were reported for devices doped with SWCNTs, MWCNTs, and carbon black, respectively. Additionally, all of the devices that were prepared with the aforementioned electrolytes showed remarkable stabilities with values of more than 1000 hours under room temperature and 1 sun illumination.

18.4.3 Graphene-based solid-state electrolytes

To date, only two investigations have focused on the implementation of graphene in iodine-free solid-state electrolytes. On one hand, Ahmad *et al.* prepared a series of quasi-solid electrolytes by mixing pre-exfoliated graphene into 1-methy-3-propylimidazolium iodide (PMII) [111]. For comparison, they also prepared corresponding electrolytes containing SWCNTs and a mixture of graphene and SWCNTs. In general, all of these electrolytes showed excellent thermal stability until 300°C and improved ionic conductivities relative to the reference electrolyte, that is, a PMII ionic liquid. By direct comparison with control devices, significant enhancement of the device efficiency of around 13 and 15 times were obtained with graphene and graphene-SWCNTs electrolytes, respectively. In both scenarios, a mechanism has been proposed that implies larger charge transfer surfaces that facilitate charge transfer processes from the counter-electrode to the PMII ionic liquid and that reduce the charge diffusion length. Worthwhile to mention is the catalytic reduction of I_3^- noted in these experiments.

On the other hand, Jung *et al.* developed a new strategy to implement graphene into electrolytes [112]. They utilized 1-octyl-2,3-dimethylimidazolium iodide (ODI) as both a reducing agent and a stabilizer in the preparation of graphene suspensions. More specifically, positively charged ODI was mixed in water with GO *via* electrostatic interactions reducing GO and acting as stabilized for the latter. The partial generation of graphene during the reduction procedure was demonstrated by Raman spectroscopy. The performance of the devices prepared with ODI-graphene hybrid was much superior to those with just ODI. High photocurrents, large open-circuit voltages, and high efficiencies were noted. The increase in photocurrent was attributed to enhanced ionic conductivity as well as efficient I_3^- reduction. Although there is neither experimental confirmation nor theoretical modeling, the authors postulated that graphene is mainly responsible for the reduction mechanism by means of delocalizing the positive charges throughout graphene.

18.5 Versatility of carbon nanomaterials-based hybrids as novel type of dyes

18.5.1 Fullerene-based dyes

Among all the carbon nanomaterials, fullerenes are by far the most studied systems in terms of chemical modification. It is safe to say that this field of research has been one of the most active for more than 20 years [113, 114]. Therefore, it is not surprising to find several examples of well-known dyes that have been functionalized with fullerene derivatives and have been tested in DSSCs. The first report in this regard was given in 2007 by Kim *et al.* [115]. They described a route to attach C_{60} to N3 dye (cis-bis(4,4′-dicarboxy-2,2′-bipyridine)dithiocyanato ruthenium(II)) *via* diaminohydorcarbon linkers with different alkyl chains (Fig. 18.7). While the photocurrents were almost equal for all devices, V_{oc}s increased up to 0.70 V with the functionalized ruthenium(II) complex from 0.68 V with pristine N3. In terms of efficiency, the values were 4.0 and 4.5 % in the absence and presence of C_{60}, respectively. The authors suggest that the linker used to connect C_{60} with N3 changes the basicity of the TiO_2 layers and, in turn, shifts the Fermi level negatively. Additionally, C_{60} creates a hydrophobic layer around the TiO_2 separating I_3^- from the TiO_2 surface. This might affect the back electron transfer process from the TiO_2 to the I_3^- anions, leading to an increase of V_{oc}.

As a complement, Shiga *et al.* reported on porphyrins that link to the TiO_2 nanoparticles by using a fullerene bearing carboxylic acid groups (Fig. 18.7) [116]. Although the authors measured very low efficiencies of around 0.53 %, the success of their approach was demonstrated by the incident photo-to-current efficiency spectra, in which the porphyrin features, that is, the Soret band, are discernible. In addition, they reported extraordinary stabilities with values up to 3000 hours.

Quite differently, Pleux *et al.* tested a series of three different organic dyads comprising a perylene monoimide (PMI) dye linked to a naphthalene diimide (NDI) or C_{60} for application in NiO-based DSSCs (Fig. 18.7) [117]. They corroborated a cascade electron flow from the valance band of NiO to PMI and, finally, to C_{60}. Transient absorption measurements in the nanosecond time regime revealed that the presence of C_{60} extends the charge-separated state lifetime compared to just PMI. This fact enhanced the device efficiencies up to values of 0.04 and 0.06 % when $Co^{II/III}$ and I^-/I_3^- electrolytes were utilized, respectively. More striking than the efficiencies is the remarkable incident photon-to-current efficiency spectrum, which features values of around 57 % associated to photocurrent densities of 1.88 mA/cm^2.

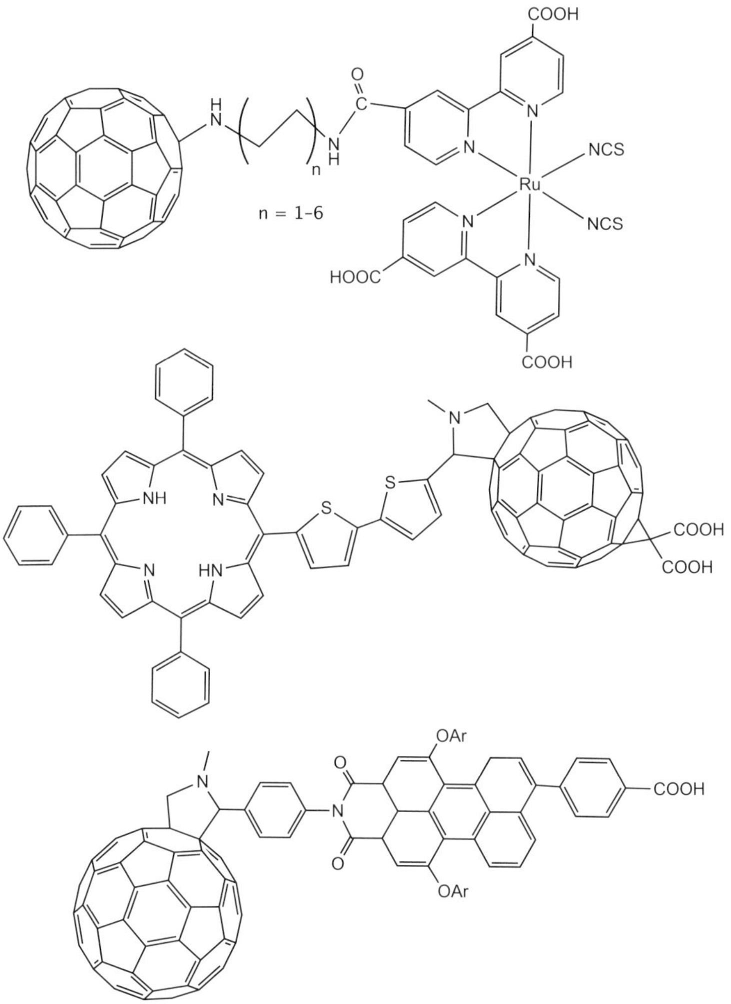

Fig. 18.7: Chemical structures of fullerene derivatives applied in DSSCs.

18.5.2 Graphene-based dyes

In 2010, Yan *et al.* were among the first to use graphene as sensitizer in DSSCs [118]. To this end, they synthetized solution-processable, graphene quantum dots with well-defined molecular size by shielding the graphene skeleton with 1,3,5-trialkyl-substituted phenyl substituents. Notable features of this novel material involve high solubility and absorption extinction coefficients of around 1.0×10^5 M^{-1} cm^{-1} at 591 nm. Upon implementing this material in TiO_2-based devices by just standard immersion of the electrode into toluene/ethanol solutions at 50 ºC for 96 hours, photocurrent densities of 200 μA/cm^2 and V_{oc}s of 0.48 V came along with a FF of 0.58. Unfortunately, the low photocurrents are attributed to weak interactions between the graphene material and the TiO_2 electrodes. The latter were barely colored reaching optical densities of 0.2 at 591 nm. Nevertheless, the authors conclude that the use of linkers, such as alkyl chains endowed with carboxylic groups to protect the graphene core and to connect to the TiO_2, might enhance the device performance.

Another way to explore the development of graphene-based dyes is the direct exfoliation of graphene from graphite using a photoactive compound with the necessary linker for the electrodes as stabilizer. This idea has recently been realized by using four different porphyrins bearing 4,4-dicarboxybuta-1,3-dienyl linkers (Fig. 18.8) [119]. The preparation of the nanographene hybrids consists of several consecutive steps, namely (i) ultrasonication of solutions of the porphyrins with natural graphite, (ii) centrifugation of the suspension to eliminate the remaining graphite particles, (iii) ultrasonication of the supernatant with newly added natural graphite, and (iv) repetition of the steps. Such a treatment resulted in the direct exfoliation of graphite and the concomitant immobilization of the porphyrins onto the basal plane of graphene. Importantly, the presence of graphene was probed by Raman, AFM, TEM, and femtosecond transient absorption spectroscopy. The most striking observation is the chemical doping of graphene by shifting electron density from the porphyrins to nanographene. SEM images of TiO_2 electrodes revealed that, even when the porphyrin interacts with the basal plane of graphene, the linker is operative to attach to the electrodes (Fig. 18.8). The authors demonstrated that devices that were fabricated using this hybrid, outperform those devices in which only the porphyrins are present. Under optimum photosensitizer uptake, the enhancement amounts to a 50 % increase of the device performance (Fig. 18.9). They postulate that the graphene flakes prevent aggregation benefiting the charge regeneration process of the porphyrin moiety, on one hand, and act as a Schottky barrier reducing the charge recombination processes, on the other hand.

In another work, the same authors demonstrated the versatility of graphene as donor material when preparing a similar nanographene hybrid with porphycenes, a well-known electron acceptor [120]. The preparation was followed along the same lines that were established for the porphyrins. By using Raman, TEM, and AFM, the nature of the graphene moiety was clearly demonstrated. Contrary to the electron ac-

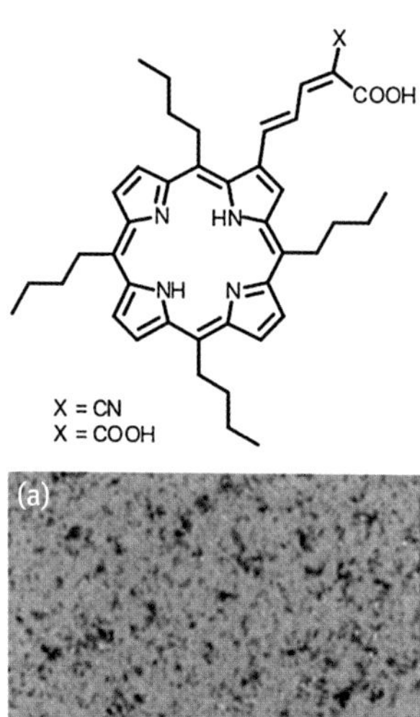

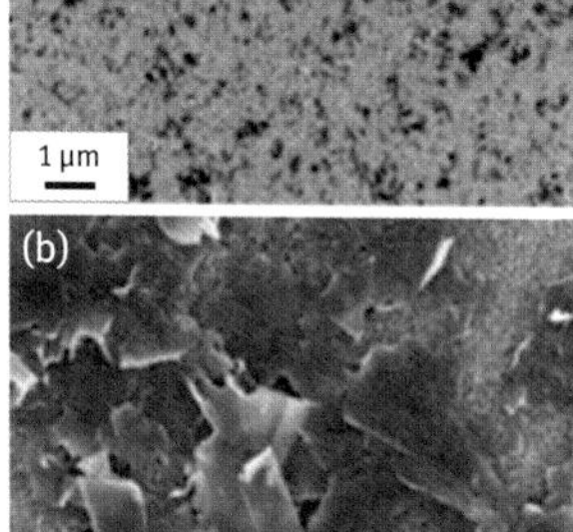

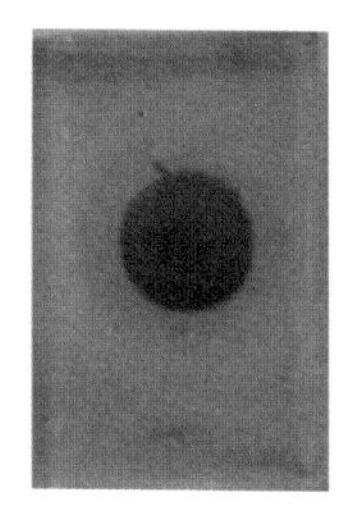

Fig. 18.8: Upper part – chemical structures of a porphyrin derivative used to prepare graphene hybrids. Lower part – SEM images (left) at high (a) and low (b) magnification and photograph (right) of TiO_2 electrodes soaked with nanographene hybrid for 120 hours.

ceptor nature of graphene, which has been seen when porphyrins or phthalocyanines are used, transient absorption spectroscopy with nanographene/porphycene settled for the first time the electron donor features of graphene. This fact was further explored in devices based on ZnO electrodes. In particular, devices with the new hybrid outperformed devices with only porphycenes (Fig. 18.9). As such, benefits stem from an enhancement of the charge regeneration process, Schottky barriers preventing charge recombination process, and, finally, the chemical doping of graphene by means of electron transfer to the porphycenes. The latter allows a *modus operandi* that implies a cascade of electron transfer from graphene to porphycenes and to ZnO. This was further corroborated by employing NiO electrodes, which is an electron donor material rather than an electron acceptor material. Here, the presence of graphene did not exert any enhancement of the device performance when compared to devices with only porphycenes. As a matter of fact, graphene acts as NiO, that is, as an electron donor material and, therefore, competes with NiO in the overall reduction of the porphycene.

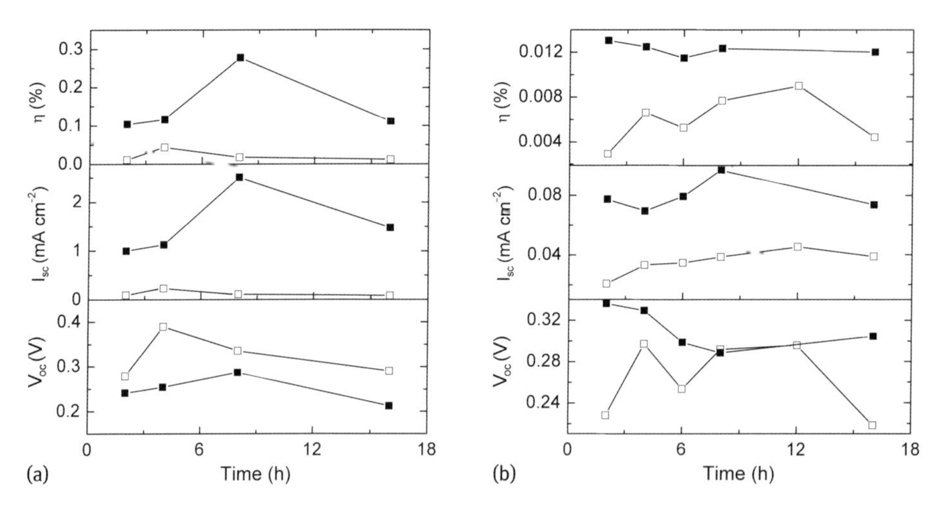

Fig. 18.9: Left part – device performance as a function of soaking time for TiO_2-based devices using nanographene/porphyrin hybrids (closed) and porphyrin dyes (open). Right part – device performance as a function of soaking time for ZnO-based devices using nanographene/porphycene hybrids (closed) and porphycenes (open).

18.6 Photoelectrodes prepared by nanographene hybrids

As mentioned in Section 18.5.2, a major challenge in the chemistry and technology of graphene is the preparation of solution-processable materials, on one hand, and the proper control over their properties through chemical functionalization, on the other hand. The use of phthalocyanines, porphyrins, and porphycenes as stabilizers to prepare highly-stable exfoliated graphene has been established as a powerful strategy to tackle these issues. As aforementioned, this approach allows characterizing and handling large amounts of nanographene in solution and, importantly, introducing novel functionalities such as electron transfer that bring out key benefits for solar cell devices. Thanks to this fact, it is currently possible to assess the full potential of nanographene in the area of photovoltaic applications. In order to provide additional support for this notion, the most recent examples of nanographene hybrid-based devices are described in this section.

18.6.1 Preparation of photoelectrodes by using noncovalently functionalized graphene

Two leading examples, that is, the preparation of high-quality graphene flakes by processing THF solutions of a zinc-based phathalocyanine (ZnPc) linked to an n-type oligo-para-phenylene-vinylene (oPPV) backbone of different chain lengths (Fig. 18.10), have been recently reported [121, 122]. More specifically, the authors established the influence of the size of the ZnPc-oPPV oligomer on the graphite exfoliation, the stabilization of the suspensions, the charge transfer features of the resulting nanographene hybrids, and, finally, the corresponding behavior as photoelectrode materials in photovoltaic cells.

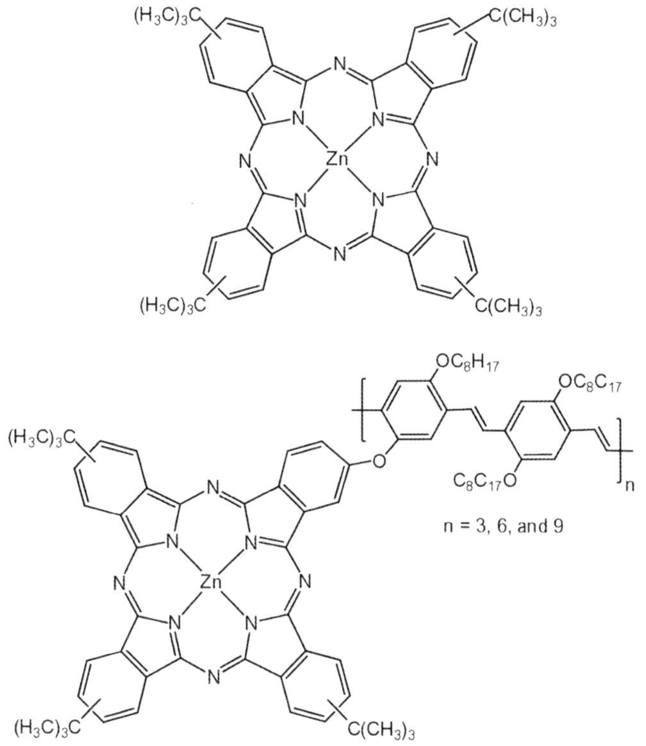

Fig. 18.10: Structures of ZnPc and ZnPc-oPPV used to prepare nanographene hybrids.

The preparation of the nanographene hybrids differed slightly from that mentioned above. In particular, assistance of a very long ultrasonication step (1 hour) and a centrifugation step at 500 rpm (20 G) for 90 minutes were required. The resulting suspensions were stable over several months. Quite likely, the p-type character and the length of the oPPV backbone are key aspects as means to facilitate interactions with graphene layers by π-π interactions (*i.e.*, exfoliation) and strong electron coupling (*i.e.*, charge transfer from the phthalocyanine unit to graphene). The former was elucidated by Raman, AFM, and TEM assays, which confirmed that longer oPPV backbones are better suited for the exfoliation of graphene. In detail, the authors achieved single and/or few layer graphene with an average height distribution as small as 2 nm and average flake size in the range from 1 to 3 μm^2. The latter was derived from steady-state and time-resolved spectroscopy. In particular, the longest oPPV backbone featured the strongest ZnPc fluorescence quenching in very good agreement with the exfoliation. This fact went hand-in-hand with transient absorption experiments, in which the dynamics of charge separation (a few picoseconds) and charge recombination (a few hundred of picoseconds) were established.

In the final step, photovoltaic devices were prepared by using the filtering deposition technique. The latter consists of transfer deposition of nanographene hybrid suspensions from a 0.20 μm pore-sized PTFE filter onto ITO substrates. Importantly, the film thickness was hereby adjusted by the number of filtering steps. In the devices, the I^-/I_3^- redox couple and a platinum electrode were used as redox electrolyte and counter electrode, respectively. In accordance with the features of the nanographene hybrids, only those prepared with the longest oPPV backbone showed instantaneous, stable, and reproducible photocurrents during several "on-off" cycles of illumination and current versus applied voltage characterizations. Notably, monochromatic IPCE values of up to 1% were noted.

18.6.2 Preparation of photoelectrodes by preparing nanographene-based building blocks *via* electrostatic interactions

In a first approach, suspensions of graphene were realized in aqueous media by using trimethyl-(2-oxo-2-pyren-1-yl-ethyl)-ammonium bromide as stabilizer [123]. In this context, natural graphite was ultrasonicated in a saturated aqueous solution of the pyrene derivative for 30 minutes, followed by stirring for two days, and centrifugation for 30 minutes at 10.000 G. The obtained dispersions were stable for several weeks. Again, TEM, AFM, Raman, and steady-state and time-resolved spectroscopy experiments confirmed the excellent grade of monolayer and few-layer graphene flakes formation in water. In the next step, the positively-charge functionalized nanographene was brought together with a negatively-charged zinc-based porphyrin (Fig. 18.11). Importantly, absorption and emission titration assays confirmed the electronic communication between both moieties.

Here, a different technique was utilized to fabricate photoelectrodes, that is, layer-by-layer (LbL). Briefly, ITO substrates, which were negatively functionalized by using potassium hydroxide, were immersed into aqueous suspensions of positively-charged graphene and, subsequently, into another solution of negatively functionalized porphyrins. The repetition of these steps led to sandwich architectures, in which the graphene flakes are preserved and the absorption features of the porphyrins increase upon deposition steps. In this particular case, sodium ascorbate/Na_3PO_4 solution as electrolyte and platinum as counter-electrode were used to finalize the devices. The most valuable finding is a reproducible and stable photocurrent response of around 360 nA during several "on-off" cycles of illumination.

In a corresponding investigation, the application of a new nanographene hybrid that consists of negatively-charged GO and a positively-charged porphycene

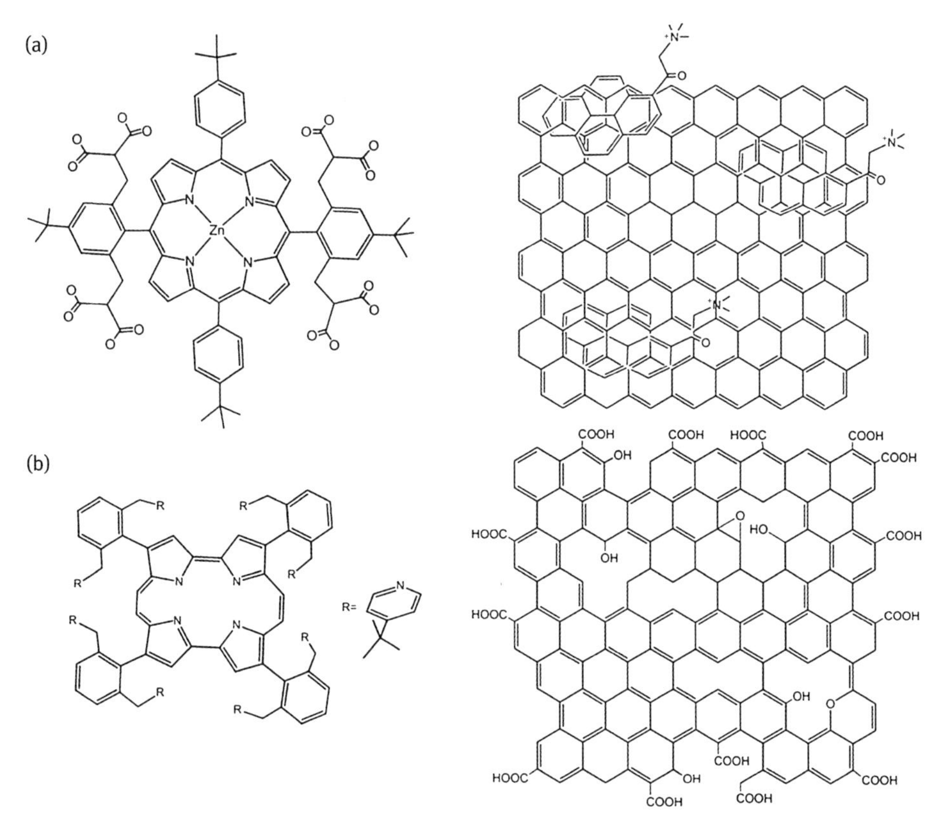

Fig. 18.11: Upper part – chemical structures of negatively-charged porphyrin and positively-charged graphene. Lower part – chemical structures of positively-charged porphycene and negatively-charged graphene oxide.

(Fig. 18.11) was probed [124]. The two major findings of this study are the synthesis of the most charged porphycene to date, on one hand, and its use for the first time in photovoltaic devices, on the other hand. Once again, absorption and emission titration experiments confirmed the mutual communication between GO and the porphycene in solution. Owing to the fact that GO and the porphycene are highly charged, similar experiments in the solid state led to the same conclusion. For photoelectrode fabrication, the LbL technique was again utilized – *vide supra*. In this particular case, two photoactive ITO electrodes modified by the sequential LbL deposition of the porphycene and GO, that is, ITO/PDDA/PSS/$(\text{porphycene/GO})_x$ and ITO/PDDA/PSS/$(\text{porphycene/PSS})_x$ used as reference electrodes, were prepared. Here, PDDA is poly-(diallyl-dimethylammonium), PSS is poly-(sodium-4-styrene-sulfonate), and x varies from 1 to 14. To complete the devices, polysulfide electrolytes – equimolar 3 M Na_2S/S/NaOH – and Cu_2S-based counter electrodes were utilized. These devices showed an enhancement of the overall device performance upon increasing the porphycene/GO sandwich layers reaching a plateau from 10 layers on. The best performing devices featured photocurrent density values of around 120 $\mu A/cm^2$ and V_{oc} values of 0.15 V. Importantly, in the absence of GO, namely in ITO/PDDA/PSS/(porphycene/PSS)$_{10}$ devices, no photoresponse was observed. As a matter of fact, charge separation at the interface, charge transport through the sandwich layers, and charge collection at the ITO electrodes were blocked. The direct comparison of the aforementioned devices clearly confirmed the benefits that stem from collecting and transporting electrons through the different GO layers.

18.7 Summary

No doubt, the usage of carbon nanomaterials in every part of DSSCs has established appreciable breakthroughs. Along these lines, we have systematically surveyed recent advances in the field. Key points are the electronic transport, the collection of photogenerated charges at the electrodes, the enhancement of the ionic diffusion and the reduction rate process of the electrolyte in the solid-state, and the design of new dyes. All the aforementioned benefits are meant to improve the overall device performance. In addition, carbon-based transparent conducting electrodes as viable alternatives to traditional ITO and FTO electrodes are currently finding application as flexible and versatile electrodes for DSSCs [60, 61].

DSSCs have recently undergone a renaissance thanks to the implementation of carbon nanomaterials. Indeed, the development of efficient electrodes, counterelectrodes, and solid-state electrolytes are poised to provide highly stable and efficient DSSCs. Importantly, such DSSCs are fabricated by using well-established printing techniques owing to the full compatibility of carbon nanomaterials with all sorts of substrates, that is, glass, metal, plastic, *etc.* It is, indeed, expected that printable DSSCs will, in the not so distant future, dominate the market. Hence, basic needs for high performance and low-cost production have emerged as important milestones. This applies not only for DSSCs but also in other photovoltaic devices, lighting, and battery technologies.

Although the prospect of carbon nanomaterials, with graphene in particular, seems to be unsurpassable, there are still remaining challenges. A minor concern is connected with the fact that all of the improvements have been achieved in different device architectures. This, in turn, generates the needs to combine all of these strategies in a single device and to confirm the overall versatility of carbon nanomaterials. Major concerns involve the exploration of ways to master them for solution process techniques, on one hand, and the proper control over their features through chemical functionalization, on the other hand. These issues constitute, especially for graphene, a giant challenge. Notably, the synthesis of soluble graphene-type molecules with properties similar to large area graphene sheets is still in its infancy. Still, a few prominent examples have demonstrated possible strategies for the future. On one hand, introduction of alkyl-substituents might be used to provide soluble graphene hybrids with a large aromatic core. Nevertheless, a clear relationship between materials properties and device performance has to be established. On the other hand, exfoliated graphene by means of ultrasonication process in combination with stabilizers like porphyrins, phthalocyanines, and porphycenes has recently appeared as a promising method to handle graphene for solution-based process techniques. There are, however, open questions regarding the control over the exfoliation in terms of flake sizes, the relationship between size and stability in solution as well as those with the device performance, *etc.*

In any case, it is safe to conclude that solving all the aforementioned concerns is just a matter of time. Once they have been dealt with, the industrial production of carbon-based printable DSSCs as well as the common use of graphene hybrids in electronic devices will become a reality.

Acknowledgments

For financial support the DFG (Cluster of Excellence – Engineering of Advanced Materials), ICMM, and GSMS are gratefully acknowledged. R.D.C. acknowledges the Humboldt Foundation for support.

Bibliography

[1] Hall, P. J. (2008) *Energy Policy*, **36**,4363.
[2] Armaroli, N., Balzani, V. (2007) *Angew. Chem., Int. Ed.*, **46**, 52.
[3] Dorian, J. P., Franssen, H. T., Simbeck, D. R. (2006) *Energy Policy*, **34**, 1984.
[4] Morton, O. (2006) *Nature*, **443**, 19.
[5] Service, R. F. (2005) *Science*, **309**, 548.
[6] Contreras, M. A., Egaas, B., Ramanathan, K., Hiltner, J., Swartzlander, A., Hasoon, F., Noufi, R. (1999) *Progress in Photovoltaics* **7**.
[7] http://www.konarka.com/index.php/site/pressreleasedetail/konarka_technologies_advances_award_winning_power_plastic_solar_cell_effici.
[8] http://www.heliatek.com/newscenter/latest_news/heliatek-erzielt-mit-107-effizienz-neuen-weltrekord-fur-seine-organische-tandemzelle/?lang=en.
[9] Yella, A., Lee, H.-W., Tsao, H. N., Yi, C., Chandiran, A. K., Nazeeruddin, M. K., Diau, E. W.-G., Yeh, C. Y., Zakeer uddin, S. M., Grätzel, M. (2011) *Science*, **334**, 629.
[10] Ito, S., Murakami, T. N., Comte, P., Liska, P., Grätzel, C., Nazeeruddin, M. K., Grätzel, M. (2008) *Thin Solid Films*, **516**, 4613.
[11] Shockley, W., Queisser, H. J. (1961) *J. Appl. Phys.*, **32**, 510.
[12] Green, M. A., *Third Generation Photovoltaics, Vol. 12*, Springer-Verlag, Berlin, **2003**.
[13] Hagfeldt, A., Boschloo, G., Sun, L., Kloo, L., Pettersson, H. (2010) *Chem. Rev.*, **110**, 6595.
[14] Grätzel, M. (2001) *Nature*, **414**, 338.
[15] Grätzel, M. (2001) *Pure. Appl. Chem.*, **73**, 459.
[16] Grätzel, M. (2009) *Acc. Chem. Res.*, **42**, 1788.

[17] http://www.g24i.com/.

[18] http://www.materialsviews.com/graetzel-dsscs-used-in-new-ipad-keyboard/; http://www.sony.net/SonyInfo/csr/SonyEnvironment/technology/solar_cells.html.

[19] Guldin, S., Hüttner, S., Tiwana, P., Orilall, M. C., Ülgüt, B., Stefik, M., Docampo, P., Kolle, M., Divitini, G., Ducati, C., Redfern, S. A. T., Snaith, H. J., Wiesner, U., Eder, D., Steiner, U., (2011) *Energy Environ. Sci.*, **4**, 225.

[20] Dürr, M., Bamedi, A., Yasuda, A., Nelles, G. (2004) *Appl. Phys. Lett.*, **84**, 3397.

[21] Uzaki, K., Pandey, S. S., Hayase, S. (2010) *J. Photochem. Photobiol. A*, **216**, 104.

[22] Usagawa, J., Kaya, M., Ogomi, Y., Pandey, S. S., Hayase, S. (2011) *J. Photon. Energy*, **1**, 011110.

[23] Andrew, N., Michael, F., Robert, K., Yi-Bing, C., Udo, B. (2008) *Nanotechnology*, **19**, 295304.

[24] Nattestad, A., Mozer, A. J., Fischer, M. K. R., Cheng, Y.-B., Mishra, A., Bäuerle, P., Bach, U. (2010) *Nat. Mater.*, **9**, 31.

[25] He, J., Lindström, H., Hagfeldt, A., Lindquist, S.-E. (2000) *Sol. Energ. Mat. Sol. Cells*, **62**, 265.

[26] Odobel, F., Le Pleux, L. c., Pellegrin, Y., Blart, E. (2010) *Acc. Chem. Res.*, **43**, 1063.

[27] Barnes, P. R. F., Miettunen, K., Li, X., Anderson, A. Y., Bessho, T., Gratzel, M., O'reagan, B. (2013) *Adv. Mater.*, **25**, 1881.

[28] Listorti, A., O'Regan, B., Durrant, J. R. (2011) *Chem. Mater.*, **23**, 3381.

[29] Gibson, E. A., Le Pleux, L., Fortage, J., Pellegrin, Y., Blart, E., Odobel, F., Hagfeldt, A., Boschloo, G. (2012) *Langmuir*, **28**, 6485.

[30] Qin, P., Wiberg, J., Gibson, E. A., Linder, M., Li, L., Brinck, T., Hagfeldt, A., Albinsson, B., Sun, L. (2010) *J. Phys. Chem. C*, **114**, 4738.

[31] Gibson, E. A., Smeigh, A. L., Le Pleux, L., Fortage, J., Boschloo, G., Blart, E., Pellegrin, Y., Odobel, F., Hagfeldt, A., Hammarström, L. (2009) *Angew. Chem., Int. Ed.*, **48**, 4402.

[32] Le Pleux, L., Smeigh, A. L., Gibson, E., Pellegrin, Y., Blart, E., Boschloo, G., Hagfeldt, A., Hammarstrom, L., Odobel, F. (2011) *Ener. Environ. Sci.*, **4**, 2075.

[33] Nazeeruddin, M. K., Baranoff, E., Graetzel, M. (2011) *Solar Energy*, **85**,1172.

[34] Li, D., Qin, D., Deng, M., Luo, Y., Meng, Q. (2009) *Ener. Environ. Sci.*, **2**, 283.

[35] Zakeruddin, S. M., Grätzel, M. (2009) *Adv. Func. Mater*, **19**, 2187.

[36] de Freitas, J. N., Nogueira, A. F., De Paoli, M.-A. (2009) *J. Mater. Chem.*, **19**, 5279.

[37] Li, Q., Zhao, J., Sun, B., Lin, B. i., Qiu, L., Zhang, Y., Chen, X., Lu, J., Yan, F. (2012) *Adv. Mater.*, **24**, 945.

[38] Docampo, P., Guldin, S., Stefik, M., Tiwana, P., Orilall, M. C., Huttner, S., Sai, H., Wiesner, U., Steiner, U., Snaith, H. J. (2010) *Adv. Funct. Mater.*, **20**, 1787.

[39] Saito, Y., Azechi, T., Kitamura, T., Hasegawa, Y., Wada, Y., Yanagida, S. (2004) *Coord. Chem. Rev.*, **248**, 1469.

[40] Karthikeyan, C. S., Wietasch, H., Thelakkat, M. (2007) *Adv. Mater.*, **19**, 1091.

[41] Leijtens, T., Ding, I.-K., Giovenzana, T., Bloking, J. T., McGehee, M. D., Sellinger, A. (2012) *Nano Lett.*, **6**, 1455.

[42] Koh, J. K., Kim, J., Kim, B., Kim, J. H., Kim, E. (2011) *Adv. Mater.*, **23**, 1641.

[43] Liu, X. Z., Zhang, W., Uchida, S., Cai, L. P., Liu, B., Ramakrishna, S. (2010) *Adv. Energy Mater.*, **22**, E150.

[44] Ferrere, S., Gregg, B. A. (2001) *J. Phys. Chem. B*, **105**, 7602.

[45] Zhao, J., Shen, X., Yan, F., Qiu, L., Lee, S., Sun, B. (2011) *J. Mater. Chem.*, **21**, 7326.

[46] Wu, J. H., Hao, S. C., Lan, Z., Lin, J. M., Huang, M. L., Huang, Y. F., Li, P. J., Yin, S., Sato, T. (2008) *J. Am. Chem. Soc.*, **130**, 11568.

[47] Wang, M., Pan, X., Fang, X., Guo, L., Liu, W., Zhang, C., Huang, Y., Hu, L., S., D. (2010) *Adv. Mater.*, **22**, 5526.

[48] Wang, P., Zakkeeruddin, S. M., Exnar, I., Grätzel, M. (2002) *Chem. Commun.*, 2972

[49] Zhou, Y., Xiang, W., Chen, S., Fang, S., Zhou, X., Zhou, J., Lin, Y. (2009) *Chem. Commun.*, 3895

[50] Wang, P., Zakeeruddin, S. M., Humphry-Bakerm, R., Gratzel, M. (2004) *Chem. Mater.*, **16**, 2694.

[51] Yamanaka, N., Kawano, R., Kubo, W., Masaki, N., Kitamura, T., Wada, Y., Watanabe, M., Yanagida, S. (2007) *J. Phys. Chem. B*, **111**, 4763.

[52] Wang, H., Zhang, X., Gong, F., Zhou, G., Wang, Z.-S. (2012) *Adv. Mater.*, **24**, 121.

[53] Zhao, Y., Zhai, J., He, J. L., Chen, X., Chen, L., Zhang, L. B., Tian, Y. X., Jiang, L., Zhu, D. B. (2008) *Chem. Mater.*, **20**, 6022.

[54] Yamanaka, N., Kawano, R., Kubo, W., Kitamura, T., Wada, Y., Watanabe, M., Yanagida, S. (2005) *Chem. Commun.*, 740.

[55] Tian, H., Sun, L. (2011) *J. Mater. Chem.*, **21**, 10592.

[56] Yum, J.-H., Baranoff, E., Kessler, F., Moehl, T., Ahmad, S., Bessho, T., Marchioro, A., Ghadiri, E., Moser, J.-E., Yi, C., Nazeeruddin, M. K., Grätzel, M. (2012) *Nat. Commun.*, **17**, 631.

[57] Martínez-Díaz, M. V., de la Torre, G., Torres, T. (2010) *Chem. Commun.*, **46**, 7090.

[58] Yen, Y.-S., Chou, H.-H., Chen, Y.-C., Hsu, C.-Y., Lin, J. T. (2012) *J. Mater. Chem.*, **22**, 8734.

[59] Griffith, M. J., Sunahara, K., Wagner, P., Wagner, K., Wallace, G.G., Officer, D. L., Furube, A., Katoh, R., Mori, S., Mozer, A. J. (2012) *Chem. Commun.*, **48**, 4145.

[60] Brennan, L. J., Byrne, M. T., Bari, M., Gunko, Y. K. (2011) *Adv. Ener. Mater.*, **1**, 472.

[61] Wang, H., Hu, Y. H. (2012) *Ener. Environ. Sci.*, **5**, 8182.

[62] Ramasamy, E., Lee, W. J., Lee, D. Y., Song, J. S. (2007) *Appl. Phys. Lett.*, **90**, 173103.

[63] Lee, W. J., Ramasamya, E., Lee, D. Y., Song, J. S. (2008) *Sol. Ener. Mater. Sol. Cells*, **92**, 814.

[64] Hsieh, C. T., Yang, B. H., Lin, J. Y. (2011) *Carbon*, **49**, 3092.

[65] Zhang, D. W., Li, X. D., Li, H. B., Chen, S., Sun, Z., Yin, X. J., Huanga, S. M. (2011) *Carbon*, **49**, 5382.

[66] Kavan, L., Yum, J.-H., Grätzel, M. (2011) *Nano Lett.*, **11**, 5501.

[67] Kavan, L., Yum, J.-H., Nazeeruddin, M. K., Grätzel, M. (2011) *ACS Nano*, **5**, 9171.

[68] Zhang, D. W., Li, X. D., Li, H. B., Chen, S., Sun, Z., Yin, X. J., Huang, S. M. (2011) *Carbon*, **49**, 5382.

[69] Wu, M., Lin, X., Hagfeldt, A., Ma, T. (2011) *Chem. Commun.*, **47**, 4535.

[70] Wu, M., Lin, X., Hagfeldt, A., Ma, T. (2011) *Angew. Chem., Int. Ed.*, **50**, 3520.

[71] Jang, J. S., Ham, D. J., Ramasamy, E., Lee, J. W., Lee, J. S. (2010) *Chem. Commun.*, **46**, 8600.

[72] Liu, C. J., Tai, S.-Y., Chou, H.-H., Yu, Y.-C., Chang, K. D., Wang, S., Shih-Sen, F., Lin, J. Y., Lin, T. W. (2012) *J. Mater. Chem.*, **22**, 21057.

[73] Wang, X., Zhi, L. J., Müllen, K. (2008) *Nano Lett.*, **8**, 323.

[74] Kim, S. R., Parvez, M. K., Chhowalla, M. (2009) *Chem. Phys. Lett.*, **483**, 124.

[75] Chen, T., Hu, W., Song, G., Guai, G. H., Li, C. (2012) *Adv. Func. Mater.*, **22**, 5245.

[76] Song, J., Yin, Z., Yang, Z., Amaladass, P., Wu, S., Ye, J., Zhao, Y., Deng, W.-Q., Zhang, H., Liu, X.-W. (2011) *Chem. Eur. J.*, **17**, 10832.

[77] Kongkanand, A., Dominguez, R. M., Kamat, P. V. (2007) *Nano Lett.*, **7**, 676.

[78] Brown, P., Takechi, K., Kamat, P. V. (2008) *J. Phys. Chem. C*, **112**, 4776.

[79] Sawatsuk T., Chindaduang A., Sae-kung C., Pratontep S., G., T. (2009) *Diam. Rel. Mater.*, **18**, 524.

[80] Yu J., Fan J., B., C. (2011) *J. Pow. Sour.*, **196**, 7891.

[81] Lin W.J., Hsu C.T., Tsai Y.C. (2011) *J. Colloid. Interface Sci.*, **358**, 562.

[82] Yen, C. Y., Lin, Y. F., Liao, S. H., Weng, C. C., Huang, C. C., Hsiao, Y. H., M. Ma, C. C., Chang, M. C., Shao, H., Tsai, M. C., Hsieh, C. K., Tsai, C. H., Weng, F. B. (2008) *Nanotechnology*, **19**, 045604

[83] Chang, W.-C., Cheng, Y.-Y., Yu, W.-C., Yao, Y.-C., Lee, C.-H., Ko, H.-H. (2012) *Nanoscale Research Letters*, **7**, 166.

[84] Kim, H.-I., Moon, G.-H., Monllor-Satoca, D., Park, Y., Choi, W. (2012) *J. Phys. Chem. C*, **116**, 1535.

[85] Williams, G., Seger, B., Kamat, P. V. (2008) *ACS Nano*, **2**, 1487.

[86] Yang, N., Zhai, J., Wang, D., Chen, Y., Jiang, L. (2010) *ACS Nano*, **4**, 887.

[87] Tang, Y. B., Lee, C. S., Xu, J., Liu, Z. T., Chen, Z. H., He, Z., Cao, Y. L., Yuan, G., Song, H., Chen, L., Luo, L., Cheng, H. M., Zhang, W. J., Bello, I., Lee, S. T. (2010) *ACS Nano*, **4**, 3482.

[88] Sun, S., Gao, L., Liu, Y. (2010) *App. Phys. Lett.*, **96**, 083113.

[89] Du, A., Ng, Y. H., Bell, N. J., Zhu, Z., Amal, R., Smith, S. C. (2011) *J. Phys. Chem. Lett.*, **2**, 894.

[90] Madhavan, A. A., Kalluri, S., Chacko, D. K., Arun, T. A., Nagarajan, S., Subramanian, K. R. V., Nair, A. S., Nair, S. V., Balakrishna, M. S. (2012) *RSC Advances*, **2**, 13032.

[91] Yang, H., Guai, G. H., Guo, C., Song, Q., Jiang, S. P., Wang, Y., Zhang, W., Li, C. M. (2011) *J. Phys. Chem. C*, **115**, 12209.

[92] Gómez-Navarro, C., Weitz, R. T., Bittner, A. M., Scolari, M., Mews, A., Burghard, M., Kern, K. (2007) *Nano Lett.*, **7**, 3499.

[93] Costa, R. D., Ruland, A., Englert, J. M., Hirsch, A. (2013) *J. Am. Chem. Soc., (submitted).*

[94] Costa, R. D., Feihl, S., Kahnt, A., Gambhir, S., Officer, D. L., Lucio, M. I., Herrero, M. A., Vázquez, E., Syrgiannis, Z., Prato, M., Guldi, D. M. (2013) *Adv. Mater.*, **25**, 6513.

[95] Wedemeyer, H., Michels, J., Chmielowski, R., Bourdais, S., Muto, T., Sugiura, M., Dennler, G., Bachmann, J. (2013) *Ener. Environ. Sci.*, **6**, 67.

[96] Tennakone, K., Perera, V. P. S., Kottegoda, I. R. M., Kumura, G. (1999) *J. Phys. D: Appl. Phys.*, **32**, 374.

[97] Xia, J., Masaki, N., Lira-Cantu, M., Kim, Y., Jiang, K., Yanagida, S. (2008) *J. Am. Chem. Soc.*, **130**, 1258.

[98] Koh, J. K., Kim, J., Kim, B., Kim, J. H., Kim, E. (2011) *Adv. Mater.*, **23**, 1641.

[99] Jung, K.-S., Lee, S.-B., Kim, Y.-K., Seo, M.-H., Hwang, W.-P., Jang, Y.-W., Park, H.-W., Jin, B.-R., Kim, M.-R., Lee, J.-K. (2012) *J. Photochem. Photobiol. A Chem.*, **231**, 64.

[100] Kamat, P. V., Haria, M., Hotchandani, S. (2004) *J. Phys. Chem. B*, **108**, 5166.

[101] Hasobe, T., Hattori, S., Kamat, P. V., Urano, Y., Umezawa, N., Nagano, T., Fukuzumi, S. (2008) *Thin Solid Films*, **516**, 1204.

[102] Usui, H., Matsui, H., Tanabe, N., Yanagida, S. (2004) *J. Photochem. Photobiol. Chem. A*, **164**, 97.

[103] Ikeda, N., Miyasaka, T. (2007) *Chem. Lett.*, **3**, 466.

[104] Akhatar, M. S., Park, J. G., Lee, H. C., Lee, S. K., Yang, O. B. (2010) *Electrochemica Acta*, **55**, 2418.

[105] Wang, Y. C., Huang, K. C., Dong, R. X., Liu, C. T., Wang, C. C., Ho, K. C., Lin, J. J. (2012) *J. Mater. Chem.*, **22**, 6982.

[106] Chang, Y. H., Lin, P. Y., Huang, S. R., Liu, K. Y., Lin, K. F. (2012) *J. Mater. Chem.*, **22**, 15592.

[107] Zhang, Y., Zhao, J., Sun, B., Chen, X., Li, Q., Qiu, L., Yan, F. (2012) *Electrochemica Acta*, **61**, 185.

[108] Huang, K. C., Chang, Y. H., Chen, C. Y., Liu, C. Y., Lin, L. Y., Vittal, R., Wu, C. G., Lin, K. F., Ho, K. C. (2011) *J. Mater. Chem.*, **21**, 18467.

[109] Lee, C. P., Lin, L. Y., Chen, P. Y., Vittal, R., Ho, K. C. (2010) *J. Mater. Chem.*, **20**, 3619.

[110] Lee, C. P., Yeh, M. H., Vittal, R., Ho, K. C. (2011) *J. Mater. Chem.*, **21**, 15471.

[111] Ahmad, I., Khan, U., Gun'ko, Y. K. (2011) *J. Mater. Chem.*, **21**, 16990.

[112] Jung, M. H., Kang, M. G., Chu, M. J. (2012) *J. Mater. Chem.*, **22**, 16477.

[113] Guldi, D. M., Illescas, B. M., Atienza, C. M., Wielopolski, M., Martín, N. (2009) *Chem. Rev.*, **38**, 1587.

[114] Babu, S. S., Möhwald, H., Nakanishi, T. (2010) *Chem. Soc. Rev.*, **39**, 4021.

[115] Lim, M. K., Jang, S. R., Vittal, R., Lee, J., Kim, K. J. (2007) *J. Photochem. Photobiol. A Chem.*, **190**, 128.

[116] Shiga, T., Motohiro, T. (2008) *Thin Solid Films*, **516**, 1204.
[117] Pleux, L. L., Smeigh, A. L., Gibson, E., Pellegrin, Y., Blart, E., Boschloo, G., Hagfeldt, A., Hammarstrom, L., Odobel, F. (2011) *Ener. Environ. Sciv.*, **4**, 2075.
[118] Yan, X., Cui, X., Li, B., Li, L.-S. (2010) *Nano Lett.*, **10**, 1869.
[119] Kiessling, D., Costa, R. D., Katsukis, G., Malig, J., Lodermeyer, F., Feihl, S., Roth, A., Wibmer, L., Kehrer, M., Volland, M., Wagner, P., Wallace, G. G., Officer, D. L., Guldi, D. M. (2013) *Chem. Sci.*, **4**, 3085.
[120] Costa, R. D., Malig, J., Brenner, W., Jux, N., Guldi, D. M. (2013) *Adv. Mater.*, **25**, 2600.
[121] Brinkhaus, L., Katsukis, G., Malig, J., Costa, R. D., Garcia-Iglesias, M., Vázquez, P., Torres, T., Guldi, D. M. (2013) *Small*, **9**, 2348.
[122] Malig, J., Jux, N., Kiessling, D., Cid, J.-J., Vázquez, P., Torres, T., Guldi, D. M. (2011) *Angew. Chem., Int. Ed.*, **50**, 3561.
[123] Malig, J., Romero-Nieto, C., Jux, N., Guldi, D. M. (2011) *Adv. Mater.*, **24**, 800.
[124] Brenner, W., Malig, J., Costa, R. D., Jux, N., Guldi, D. M. (2013) *Adv. Mater.*, **25**, 2314.

Ljubisa R. Radovic

19 Importance of edge atoms

19.1 Introduction

The issues of grain boundaries and surface (re)construction in solids, either crystalline or amorphous, are a very important part of materials science and engineering. In sp^2 hybridized carbon materials, the atoms at graphene edges have been of interest for many decades and in several different contexts; more recently, the interest in graphene nanoribbon (GNRs) materials has brought these issues into sharp focus again. As the crystallite dimensions of any carbon material decrease, the importance of 'external' edge atoms increases. This applies particularly to 'nanocarbons' where either the size or the structure is controlled at the nanometer scale [1]. The often ubiquitous presence of defects in carbon materials invariably results also in the exposure of 'internal' edge atoms to the environment. Thus, as shown in Fig. 19.1 and analyzed in some detail in Section 19.3, the presence of a single vacancy in the graphene basal plane reveals the existence of three zigzag edge sites and four potentially unpaired electrons (three of σ and one of π type). And, as reviewed recently by Biró and Lambin [2] for graphene grown by chemical vapor deposition (CVD) on a Cu substrate, its number of grain boundaries depends on the "number of nucleation centers from which the growth starts" and this in turn is very much dependent on the conditions of graphene growth.

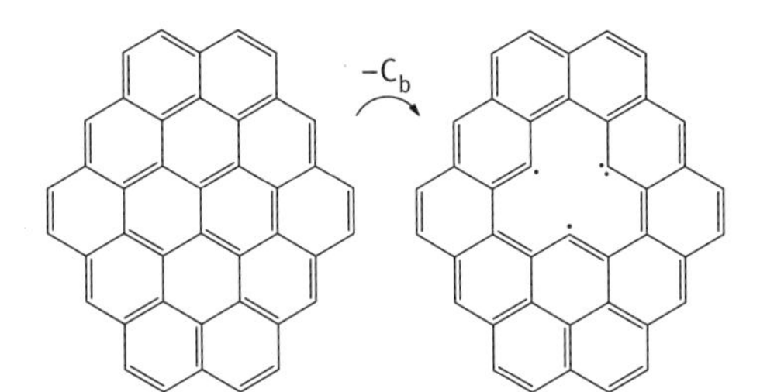

Fig. 19.1: Creation of three internal edge sites, or a (mono)vacancy, by removal of one basal-plane carbon atom in graphene.

The confluence of progress in nanoscience and the chemistry and physics of carbon surfaces can be gauged by analyzing representative publications whose titles contain the terms 'graphene' and 'edge*'. As summarized in Fig. 19.2, in the records of the Science Citation Index (thomsonreuters.com) there are close to 700 such papers as of December 2013, the vast majority published in the last five years (primarily in 'physics' journals), including several reviews [3–7]; this is in contrast with less than 200 papers dealing with 'graphite' and 'edge*' that span over at least half a century.

It is important to emphasize at the outset that the distinction between 'chemistry' and 'physics', while largely artificial almost a century after Dirac proclaimed

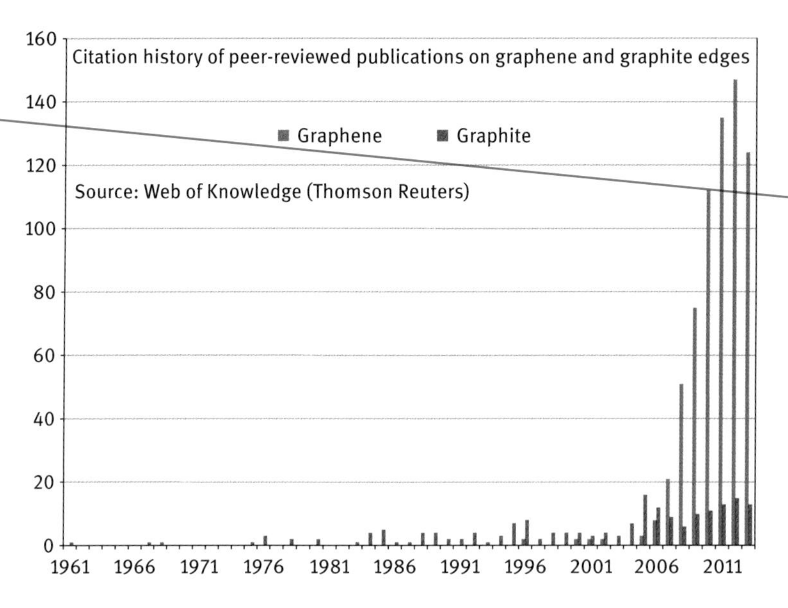

Fig. 19.2: Web of Knowledge (Science Citation Index) record of peer-reviewed publications whose title contains 'edge*' and 'graphene*' or 'graphite' in their title.

that the "underlying physical laws necessary for the mathematical theory of a large part of physics and the whole of chemistry are ... completely known" [8], remains disturbingly and unnecessarily sharp in the published literature. This will be illustrated in the following sections. Just one example is offered here, dealing with the basic issue of graphene edge identification. While there is no question that the scanning tunneling microscopy (STM) study of Tian *et al.* [9] provided "unprecedented details of the atomic structure and roughness of zigzag edges", the fact that the "graphene grains as-grown on Cu substrates [were] annealed at ~ 400 °C for 48 h before STM is performed at room temperature" would puzzle a chemistry-savvy investigator. True annealing, including removal of any surface oxygen chemisorbed at room temperature during the transfer and handling of samples (if air-exposed) [10–12] requires a much higher temperature; and from the perspective of a materials scientist, it would be important to know what STM can reveal about zigzag *vs.* armchair edge sites not just with respect to recent reports on exfoliated graphite (or graphene) [13, 14], but following up on the incisive pioneering work of Hennig [15] and Thomas [16–19]. These authors also do not seem to recognize that a direct observation of graphene edges had been carried out in an elegant early study by Kelemen and Mims [20], in which low energy electron diffraction (LEED) was used "to monitor changes in surface structure" of graphite (whose surface was annealed "by heating to 800 °C in vacuum") and Auger electron spectroscopy (AES) "to monitor elemental composition". Here then – because graphene edges have been investigated for many decades in the context of their chemical reactivity, while the emphasis over the last decade has been on their profound role in the physical characterization of graphene – it will be important to distinguish between "wheel reinventions" and genuinely novel issues and solutions called for by the nanoscale character of recently discovered sp^2 hybridized carbon materials such as nanotubes (CNTs) and GNRs.

19.2 External edges

Two basic issues can be distinguished here: (i) the physico-chemical nature of zigzag and armchair sites at the edges of fused-benzene-ring structures, including their differences and similarities with respect to polycyclic aromatic hydrocarbon (PAH) molecules, and (ii) the reconstruction of these edges to potentially more stable structures.

A convenient point of departure is the analysis of phenalenyl, the prototypical and extensively studied π radical, especially so because two very recent reports [21, 22] give it an unnecessarily confusing theoretical treatment. Hydrogen abstraction is a well understood process in the context of soot growth [23] and stabilization of a carbene-type edge site is of interest in the analogous context of nascent site deactivation [24]. Gorjizadeh *et al.* [22] have confirmed that "$C_{13}H_9$ become[s] ferromagnetic with high-spin state" when one H atom is removed not from a zigzag site, but from the zigzag-cusp graphene edge [24]; they also confirmed that spin contamination complicates the use and reliability of unrestricted Hartree–Fock methodology [25]. In their analysis of the ability of graphene oxide to "stabilize graphene radicals (GR)", Hou *et al.* [21] found that the "GRs were air-stable over a long period, both in the solution state and in freeze-dried powders" and thus suggested that "they are graphene-based phenalenyl-like radicals". However, their justification for this suggestion is based on false premises, stemming primarily from a superficial treatment of π and σ radicals, about which there exists, of course, a voluminous literature not only in the context of PAH chemistry and soot growth, but also dealing specifically with electron spin resonance of carbon materials [26–28]. The "carbene-like [radicals] with the triplet ground state" [21] are not π radicals: they are a result of free edge stabilization by virtue of σ-π coupling [29]. This misinterpretation also invalidates the argument that "the GRs were not oxidized by oxygen in air, which is known to readily occur for σ and carbene-like π radicals" [21]. Both references cited in its support are inappropriate, one dealing with interaction of hot air-dried onion with DPPH and peroxide radicals and the other being a classic paper by the group of Michael Dewar that shows the applicability of the semi-empirical PMO method to conventional hydrocarbon radicals; furthermore, the room-temperature air stability of free-carbon-site graphene-based materials arguably has a solid experimental foundation [11, 12, 29].

So, the true relevance of the phenalenyl radical to graphene edge chemistry and physics resides in the fact that, as graphene layers grow by condensation/dehydrogenation processes, and rather than trapping π radicals as a consequence of edge saturation of graphenes with odd number of H atoms (or forming unadulterated σ radicals by complete dehydrogenation), an outcome that is much more consistent with a large body of experimental and theoretical results is the stabilization of free edge sites. This is discussed in detail below.

As mentioned in Section 19.1, we are witnesses to an exponential rise in the relevant literature and therefore this review cannot purport to be comprehensive. Instead, I shall claim that a combination of a historical perspective and a thematic approach [30] should offer a representative state of knowledge in this field. And here the concept of an *active site* is the most convenient and the most appropriate point of departure: although it has a long history in the field of heterogeneous catalysis [31], the progress in its qualitative and quantitative understanding has been incremental, culminating recently in a Nobel Prize (www.nobelprize.org/nobel_prizes/chemistry/laureates/2007/ertl-lecture.html). Similar progress has occurred in the graphene/carbon field, the pioneering studies having been carried out half a century ago [32, 33]. The quantitative development of the same concept has been hampered, however, by the need to distinguish between an active and a *reactive site*: the latter not only participates in chemisorption but is also being consumed during reaction, while the former may host either a (temporary) spectator or a true reaction intermediate [34–38]. It is somewhat surprising that such a high level of understanding of graphene edges, rescued from decades of experimental and theoretical studies of 'traditional' and 'modern' graphene-based materials [1], does not yet find echo in the avalanche of studies of novel carbon materials (nanocarbons), mainly CNTs and GNRs. Representative examples from the recent literature are analyzed below. The physical basis for these studies has been established in a chemistry-oriented pioneering study by Thomas and Hughes [16] using optical microscopy. Their analysis of the orientation of etch pits due to gasification of graphite single crystals with O_2 revealed a consistent though surprisingly small difference between zigzag and armchair edges: the former receded *ca.* 20 % faster than the latter. The authors themselves rationalized their surprise with the following insightful [29] argument: "the electronic configuration of the exposed carbon atoms at these two faces is thought to be quite different [since] some of the C-C links at the [armchair] faces would acquire a partial triple-bond character, whereas the carbon atoms at the [zigzag] faces would tend to be in the divalent state, s^2p^2" [16].

Among the first studies in the new wave of interest in graphene edges is that of Fujita and co-workers [39, 40]. They emphasized the obvious fact that "if materials of a nanometer scale have edges, quite a high proportion of atoms are on the surface" and used tight binding band calculations to show that "the presence of edges gives rise to the critical difference in the electronic state from that of bulk graphite". In particular, they noted the "presence of the almost flat bands" and thus the "existence of local ferromagnetic structure" in ribbons dominated by H-terminated zigzag edges; but they displayed a lack of familiarity with well-established (albeit chemistry-oriented) literature on high-surface-area carbon materials when they "*propose*[d] that porous carbons *may have* many edge sites on the periphery of their micrographite constituents"

[39] (italics added here). This well-known experimental fact had been known for several decades. And Stein [41] had already clarified the essential electronic difference between armchair and zigzag edge sites in (H-terminated) PAH molecules: "[f]or the armchair-edge structure, electron densities are rather evenly distributed over the structure, while for the zigzag-edge series, densities are concentrated at the edges".

The consequences of the fact that a GNR has "the nonbonding molecular orbitals localized mainly along the zigzag edges" and that these "have partly flat bands and cause a sharp peak in the density of states at the Fermi energy" [42] have been explored by Harigaya and co-workers [42, 43]. They addressed "the qualitatively important problem of the competition between the spin and charge orderings in this system due to the on-site and nearest-neighbor Coulomb interactions" [42]. Using the "extended Hubbard model within the unrestricted Hartree-Fock approximation", they "demonstrate[d] that the nearest-neighbor Coulomb interaction stabilizes a charge-polarized (CP) state with a finite electric dipole moment in zigzag ribbons and it competes with the [spin-polarized] state" [42]; the experimentally observable implications of such competition, especially from a chemical point of view, have yet to be clarified.

The issue of the chemical nature of graphene edges, and thus of (re)active sites, was addressed head-on by Radovic and Bockrath [29]. Density functional theory was used to minimize problems that truly *ab initio* approaches such as Hartree–Fock are known to have, but careful comparison with experimental results in several seemingly unrelated fields was also performed. The key argument of relevance here is that, rather than being H-terminated, the sites that are most relevant for the chemical (re)activity of sp^2 hybridized carbon materials are of carbene- and carbyne-type, as illustrated below for the zigzag (a) and armchair edges (b).

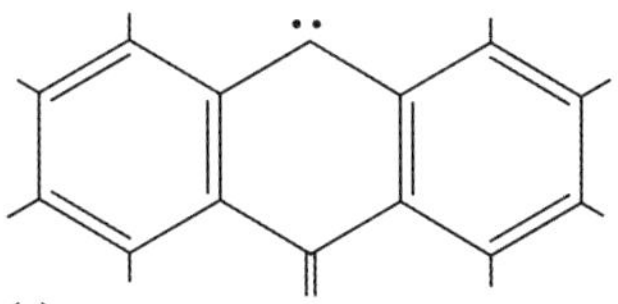

(a) 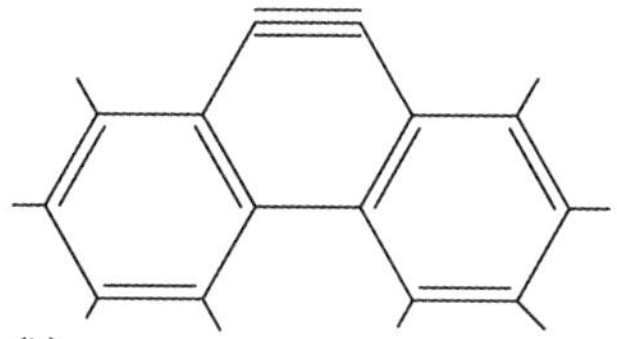 (b)

In other words, rather than being unadulterated σ-radical sites with unpaired electrons, and rather than being "unconventional radical sites involving stabilization by delocalization" [41], their relative stability is the consequence of localization of itinerant π electrons through σ-π coupling. This postulate is consistent with the properties and/or behavior of a wide variety of carbon materials, including elemental analysis (*i.e.*, H/C ratio), thermoelectric power and nascent site deactivation. It also offers a straightforward explanation for ferromagnetism in some impurity-free carbons because triplet, rather than singlet, is the ground state of carbene-site-containing graphene. In addition, the luminescence behavior of some carbon materials can be explained using the same concept, as presented by Pan and co-workers [44, 45] and reviewed recently [46]. As summarized in Fig. 19.3 – whose qualitative essence is

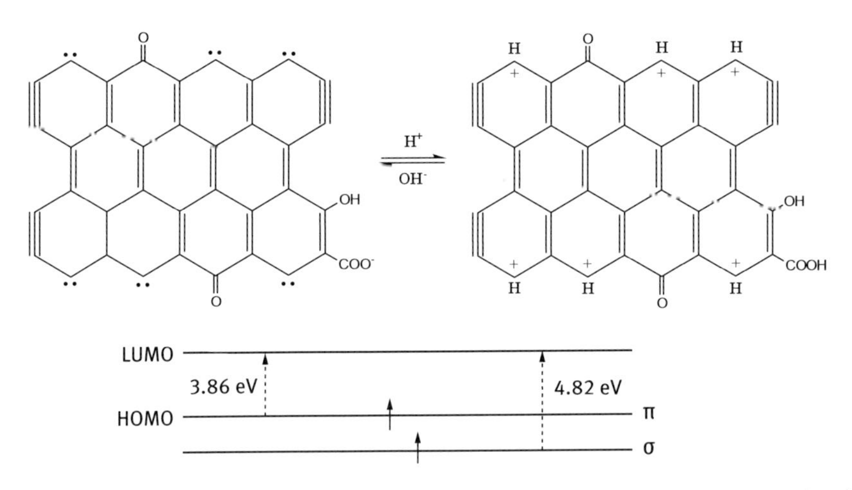

Fig. 19.3: Schematic representation of the role of carbene-type graphene edge sites in the photoluminescence effect of graphene quantum dots. (Adapted from [44].)

more important than its quantitative details, because the energies of the indicated transitions are dependent on graphene geometry and charge distribution – these investigators proposed that the origin of photoluminescence excitation (PLE) is "reversible protonation of carbene-like zigzag sites" [44], because the presence of such sites is consistent with the requirements of relatively large HOMO-LUMO gaps, Lewis base character and singly occupied orbitals [45]. Similar arguments regarding the origin of luminescence of graphene quantum dots were invoked by Gupta *et al.* [47] and Shen *et al.* [48].

It should be mentioned that a different mechanism of stabilization of edge states in graphene has been proposed by Sasaki *et al.* [49] who insist in analyzing graphene structures terminated with H-stabilized edges, as if they were identical to PAH molecules. The conclusion of these authors is that the unusual "decrease [in] the energy eigenvalue of the edge state", to a value "*below* the Fermi energy by about 20 meV", is a consequence of the "next nearest-neighbor (NNN) hopping process", but they avoid addressing the many other relevant issues summarized above.

It is also of interest to examine what earlier reviews have to say about these key issues regarding the structural characterization of graphene edges. Enoki *et al.* [3] emphasize the importance of edges in graphene, relative to fullerenes and carbon nanotubes, and assume that these are terminated as PAH molecules are; they also analyze dehydrogenation, monohydrogenation and dihydrogenation of the zigzag edges in the context of graphene magnetism (see a more extensive discussion of this issue below). The authoritative review of Dresselhaus and co-workers [5] emphasizes the various edge characterization techniques and the practical importance of "obtaining atomically sharp edges in narrow graphene nanoribbons of controlled width". These authors also mention theoretical predictions that "zigzag edges are metastable and a planar reconstruction into pentagon-heptagon configurations spontaneously takes place at room temperature" (see Section 19.4 for a more detailed discussion of this issue). They do point out that "[i]n practice, hydrogenated or oxygenated group passivation during heat treatment is commonly used to stabilize the edges in air", and conclude that the "detailed chemistry and carrier transport associated with these functionalized edges seem very exciting and further work is necessary along this direction". It is thus puzzling to note that Acik and Chabal [6] boldly state that "[e]ach carbon atom of the zigzag edge has an unpaired electron, which is active to combine with other reactants" whereas "the carbon atoms of the armchair edge side are more stable … because of a triple covalent bond between the two open edge carbon atoms of each hexagonal ring". Their only support for this assertion is the paper by Xu and Ye [50], who in turn make the exact same statement in their Results and Discussion section without any evidence or discussion of the relevant literature. Furthermore, these authors inappropriately analyzed 1,8-naphthalenediyl and 2,3-naphthalyne as prototypical graphenes containing zigzag and armchair sites, when in reality these sites are subsets of an armchair-cusp edge [24]. Similarly, Zhang *et al.* [7] do not seem to appreciate the importance of decades of experimental studies and instead emphasize the fact that the "atomic structures and thermal stabilities of the pristine graphene edges have … been intensively investigated theoretically". They say nothing definitive about zigzag edges, except that "[i]n contrast to the bare edge, the formation energy difference between passivated AC and ZZ edges is significantly lowered". Like Acik and Chabal [6], they state, without any evidence or discussion, that the "C-C bond length between two neighbouring edge atoms of the AC edge is reduced to 1.23 Å, forming a C≡C triple bond, and the pairing of the π electrons [sic] of the AC edge lowers the formation energy"; the latter part of this assertion reveals a superficial treatment of a key electron redistribution process (describing rehybridization of σ electrons to form a second π bond in terms of simple pairing of π electrons) and is a typical example of the communication gap that still exists between researchers with expertise in graphene 'physics' *vs*. graphene 'chemistry'.

The role of edge sites in the quantum Hall effect of graphene has also been of interest [51–54]. The profound difference between zigzag and armchair edge sites is manifested here as well. It is instructive to compare such analyses with that of the conventional Hall effect, characteristic of three-dimensional graphitic carbon materials [55, 56] and closely related to their thermoelectric power [29]. The essence of the issue being the origin of graphene's transverse voltage when placed in a magnetic field, it is closely related to the band model of graphite [57] and the nature of charge carriers whose investigations have a long, distinguished and too often ignored history: "a variable energy gap between [the conduction and valence] bands, and the presence of surface traps" [55]. These surface traps "are the carbon valencies which are not engaged in cross-bonds and which are present mostly on the peripheries of crystallites: [e]lectrons from the π band jump into these σ-orbitals, forming spin pairs and thus quenching the chemical activity of these free valencies" [55]. In modern terminology, as summarized above, the stabilization of zigzag sites results in the formation of carbene-type edges and armchair sites stabilize to form carbyne-type edges [29]. Mrozowski's insightful arguments apply also to internal edges (see Section 19.3 below): "three free valencies per vacancy are available as traps" and "an average from two to three excess holes in the band [is expected to] correspond to each vacant carbon site" [55]. In the study of Brey *et al.* [51] these issues are not even mentioned. The authors note that "transport in the quantum Hall regime is typically dominated by edge states", but they do not attempt to characterize these states in any essential detail. Instead, their conclusions are framed rather vaguely (at least from a chemical perspective) in terms of Landau level bands. In the same vein, in a study bearing the almost identical title, Castro Neto *et al.* [52] concluded that, "although the surface modes are localized by disorder in the absence of a magnetic field, they become delocalized by the edge modes that drag the surface modes in their motion via electron-electron interactions". In a study of bilayer graphene, Mazo *et al.* [54] do not acknowledge any relationship to the band structure of graphene-based carbon materials, but they do emphasize that a "basic explanation for the existence of Hall currents in clean systems involves edge states"; they studied ribbons with zigzag edges and concluded that the observed complicated edge-state dispersions "yield a variety of possible crossings and anticrossings, particularly when spin is included as a degree of freedom and the effects of Zeeman coupling are considered".

The effect of edges on conductivity was studied by Banerjee *et al.* [58]. Echoing the findings of both Fujita *et al.* [40] and Stein and Brown [59], they offered the following explanation: "Recently, detailed theoretical calculations have shown that zigzag edges have edge states localized at the edges having a higher electron density. Hence these edges will have higher electrical conductance".

In 2007, there were several reports on chemical aspects of graphene edges. Gunlycke *et al.* [60] analyzed the effects of zigzag surface functionalities on the electronic structure of graphene 'nanostrips'. While failing to cite a single study in which the nature of such functional groups has been firmly established (see, for example, several reviews in *Chemistry and Physics of Carbon*), these authors concluded that, "[u]nlike the zigzag-edge graphene nanostrips terminated with hydrogen atoms or hydroxyl groups, the nanostrips terminated with oxygen atoms or imine groups are spin unpolarized in equilibrium"; among the potential implications of this difference, the authors speculated that the former "could represent a good starting point for spintronics applications" while the latter "might find use as metallic or quasimetallic interconnects". The general applicability of such a conclusion is doubtful in light of the results summarized in Fig. 19.4: spin polarization, and its multiplicity (M) is a very sensitive function of the details of graphene geometry and electron density distribution [24, 29, 61], and here exactly the opposite situation is shown, even though the interatomic distances and angles (obtained using B3LYP/6-31G(d) model chemistry as implemented in the Gaussian03 software) are essentially the same as those reported by Gunlycke *et al.* [60]. The obvious difference, and indeed one that has been ex-

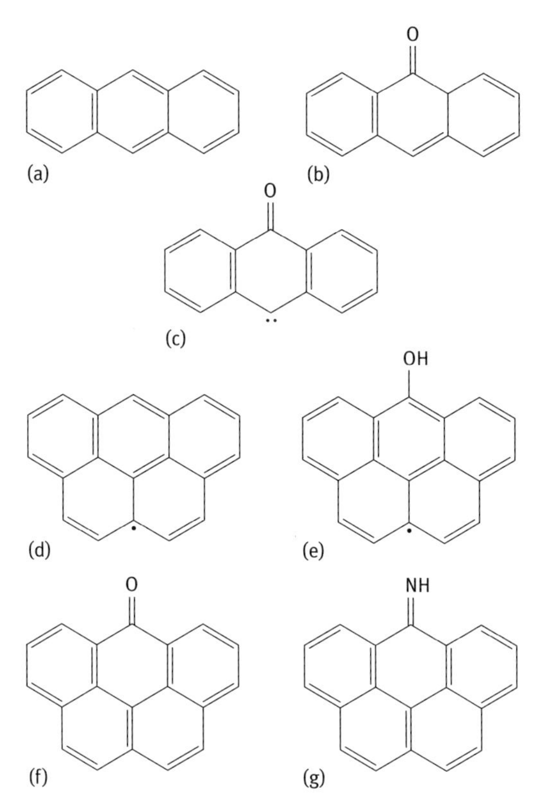

Fig. 19.4: Illustration of the effect of the number of fused benzene rings and edge functionality on the electron spin pairing processes (M = spin multiplicity) in graphene: (a) $C_{14}H_{10}$ (M = 1); (b) $C_{14}H_9O$ (M = 2); (c) $C_{14}H_8O$ (M = 3); (d) $C_{19}H_{11}$ (M = 2); (e) $C_{19}H_{10}OH$ (M = 2); (f) $C_{19}H_{10}O$ (M = 1); (g) $C_{19}H_{10}NH$ (M = 1).

tensively investigated in mechanistic studies of soot formation [62] and hydrocarbon carbonization [26], is that the spin pairing process is very much dependent on the number of fused benzene rings in graphene, which in turn is a delicate balance between edge dehydrogenation and ring condensation reactions. In the reports of Hod and co-workers [63, 64], the authors claim that theirs is the first study of "the effect of edge oxidation on the relative stability, the electronic properties, and the half-metallic nature of zigzag graphene nanoribbons" [62]. Their results highlight the gap that exists between the physics- and chemistry-oriented approaches to what should be essentially the same issues; periodic boundary conditions are typically used in the former, whereas prototypical clusters (or molecules) are more informative in the latter, especially when edge geometry and charge distribution are of primary interest. Indeed, based on their own results, these authors note that "careful consideration of finite-size and edge effects should be applied when designing new nanoelectronic devices based on graphene nanoribbons" [64]. Thus, for example, they conclude, rather counter intuitively (at least based on experience with PAH molecules), that the "structure of the oxidized ribbons is found to be stabilized with respect to the fully hydrogenated counterparts", the corresponding ground-state band gaps being 0.03 and 1.05 eV [63]; in contrast, the HOMO-LUMO gaps for the analogous prototypical graphene clusters

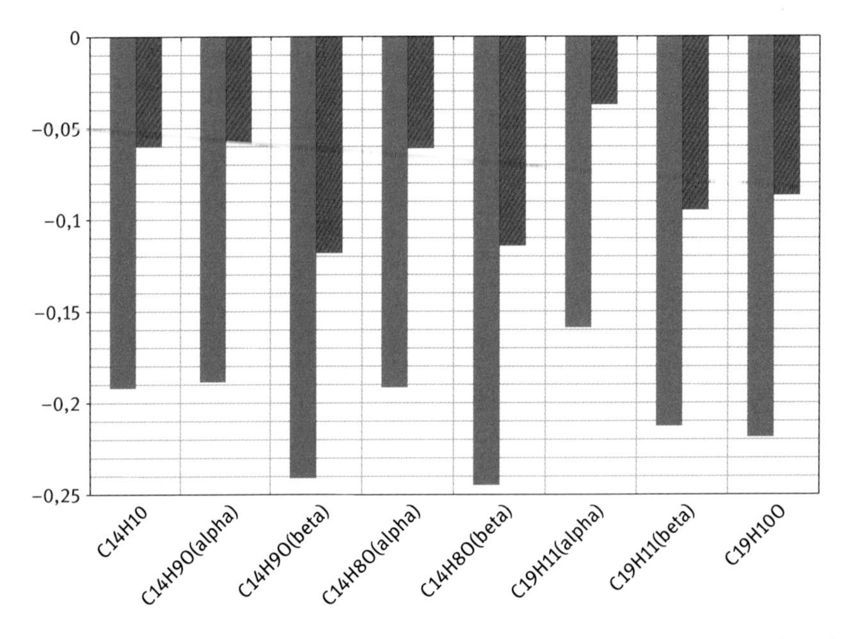

Fig. 19.5: HOMO and LUMO levels (in hartrees) of the PAH molecules and prototypical graphene clusters shown in Fig. 19.4.

shown in Fig. 19.5 are seen to depend on the details of electron density distribution and spin pairing; thus, for example, the doublet $C_{14}H_9O$ and the triplet $C_{14}H_8O$ are predicted (using DFT at the B3LYP/6-31G(d) level) to have smaller gaps, by 1.7 and 1.5 eV, than the singlet $C_{14}H_{10}$, but the HOMO-LUMO gap of the singlet $C_{19}H_{10}O$ is larger, by 1.9 eV, than that of its parent doublet $C_{19}H_{11}$. A similarly ambitious claim is offered by Jiang *et al.* [65] who addressed the reactivity of zigzag edge sites by "examining their reaction energetics with common radicals from first principles". These authors argue that both "H-free carbene-like zigzag edges and H-free dangling s-bond zigzag edges [are] unlikely to survive under ambient conditions (room temperature in air) due to their high chemical reactivity", and thus analyze H-saturated graphene edges. They explore the implications of such analysis for experimental studies of carbon materials (including carbon blacks, carbon fibers and glassy carbon), but they ignore the experimental fact that the H/C atomic ratios in many sp^2 hybridized carbons are much lower than what would result from complete H-saturation of their edges [29]. They reported a good correlation between the interaction of such zigzag edge with H, OH, CH_3 and halogen radicals and the analogous interaction in ethanol, without clarifying how that makes the important distinction from armchair edges or why such a comparative analysis is more appropriate than, say, a comparison with the abundant literature on the reactivity of polycyclic aromatic hydrocarbons [41, 66].

The issue of zigzag graphene edge saturation was explicitly addressed by Kudin [67], albeit by considering not the more reasonable aromatic H saturation but the conversion to a hydroaromatic or sp^3 hybridized structure. In a very much related study, Wassmann *et al.* [68] did not offer any comments regarding Kudin's results; instead they analyzed H-terminated GNR edges as a function of hydrogen content and thus included several structures with mono- and di-hydrogenated edge sites (including a few with carbyne-type armchair sites but, alas, none with carbene-type zigzag sites). They confirmed that the "electronic structure of the edges can be interpreted in terms of aromaticity and according to Clar's rule, well known in organic chemistry since the 1960s" and rationalized their results "by means of simple concepts of organic chemistry", but did not acknowledge the fact that the H/C ratio in many practical graphene applications is significantly lower than that available in PAH molecules. In a study that does cite both papers discussed above but makes no substantial comments about the key issues of interest here, Jiang *et al.* [69] explored "the issue of magnetic ground states" and concluded that a polyacene with two zigzag C-C chains across the GNR "can be considered the narrowest [zigzag-edge] GNR with an antiferromagnetic insulating ground state and a metastable, metallic ferromagnetic state". The role of edges in determining the magnetic properties of carbon materials has been the single most popular topic for researchers of graphene edges [22, 29, 70–101]. And this report by Jiang *et al.* [69] is indeed prototypical. Too many studies are based on the assumption that all graphene edges are terminated in the same fashion as PAH molecules; as summarized here and discussed in detail elsewhere [29], a triplet ground state of carbene-like free zigzag sites is a graphene structure that is much more consistent with a wide variety of properties of sp^2 hybridized carbon materials, including ferromagnetic conducting rather than antiferromagnetic insulating behavior.

The chemistry of edge termination was confirmed in many recent studies to depend on the experimental conditions used in the preparation of graphene-based materials. Thus, Zhang *et al.* [102] demonstrated that PAH-like GNRs are obtained (*i.e.*, with H-termination and no rehybridization or reconstruction of edges) in a hydrogen plasma. Another chemistry-oriented contribution is that of Kubo and co-workers [103], who synthesized and characterized quarteranthene with the aim to "[elucidate] the characteristics of the edge state of graphene nanoribbons at the molecular level". Following a surprisingly large number of graphene investigators, their premise is that PAHs "are possible candidates for this purpose, since they are structural components of GNRs with distinct edge structures and have been estimated to possess a spin state similar to [that] of GNRs". They discovered the singlet biradical (or diradical) character of the anthene series, which confers them antiferromagnetic properties; furthermore, the experimental finding that the singlet-triplet transition energy in 2,6,13,17-tetra-tert-butyl-4,15-dimesitylquarteranthene is very small (< 1 kcal/mol) suggested that such a quarteranthene derivative "is easily activated to a triplet state (a ferromagnetically correlated state) and that the population of the triplet species is ~ 50 % at room temperature". They also explored the chemical reactivity of zigzag edges by noting that these "are well known to be thermodynamically unstable"; intriguingly, however, both theoretical studies [104, 105] that they cite in support of the latter statement analyze H-free edges, rather than H-terminated PAH molecules. Finally, they speculated that the zigzag graphene edges "can be easily transformed into the more stable armchair edges by attacks by oxygen and various small molecules", but the only edge sites involved in the putative formation of "a peroxide bond connecting the neighboring molecules" were, again intriguingly, not the previously identified high-spin-density zigzag carbon atoms.

A related ambitious attempt to use PAH molecules to explore the nature of graphene edges [106] focused on the electron density distribution in terms of isodensity maps. The authors argued that a novelty of their theoretical treatment is that the "occurrence of enhanced electron density at the edges of a variety of PAHs has been shown to be a general edge effect" but, like too many graphene investigators, they do not seem to be familiar with the decades-old relevant literature dealing with graphene-based materials such as soot or coal-derived chars; so much so, that their earliest reference for the well-established fact [33] that "the edges of graphene and GNRs are appreciably more reactive than the bulk" is from three years ago. They did explore the possibility of having "different types of structures and sizes whether these have singlet or triplet ground state multiplicity" and confirmed the transition from singlet to triplet ground state as the number of benzene rings increases in linear polyacenes, as well as the preservation of enhanced negative charge at the edge carbon atoms with increasing PAH size. An intriguing result of their analysis is that the armchair edges "would be more reactive towards electrophiles than the zig-zag edges"; even though there is a long history of similar predictions, as well as some experimental results, the authors did not venture to make the relevant comparisons. In sharp contrast, the contemporaneous experimental and theoretical study by Ohba and Kanoh [106] exploited the reactivity differences between edge and basal-plane sites and showed "the first evidence of edge effects of nanographenes on gas adsorption abilities". Their intuitively appealing results are illustrated in Fig. 19.6. "Molecules with a strong dipole or quadrupole moment, such as H_2O or CO_2, are therefore selectively adsorbed on the edge sites, whereas dispersion-interaction-dominant molecules such as Ar, CH_4, and N_2 are selectively adsorbed on the basal planes". And more specifically, they found that "[t]his preference difference leads to extremely high selectivity of CO_2 adsorption over N_2 adsorption on the edge sites, with the selectivity being over 30 below 0.02 MPa".

Unraveling the role of edges in the electrochemical behavior of graphene also suffers from unappreciated similarities (and differences?) with the other graphene-based carbon materials. In a typical such study, Banerjee *et al.* point out that "[e]dge sites have already been demonstrated to possess enhanced electron transport rates and reactivity in studies of CNT ends" [108], as if this had not been well known and extensively investigated for many years and for all graphene-based materials in the context of their widespread use in batteries and fuel cells [109]. In their attempt to "demonstrate a graphene edge embedded nanopore (GEEN) structure to isolate graphene edge electrochemical activity from basal plane activity", they claimed to have carried out "the first known study of electrochemical current exchange at the graphene (potentially as thin as single layer) edge in an ionic solution". Apart from the dubious ar-

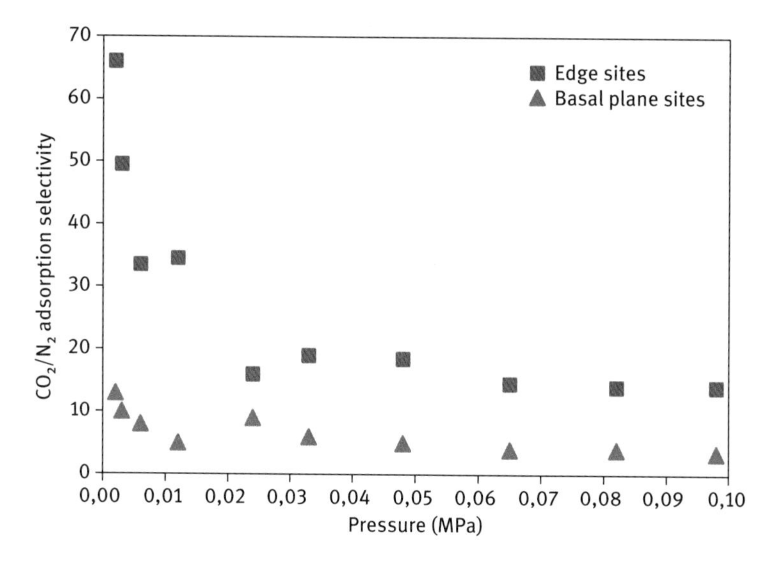

Fig. 19.6: Comparison of relative affinities of edge *vs.* basal-plane sites for adsorption of CO_2 and N_2 on graphene. (Adapted from [107].)

gument that one can define a 'nanopore' in the presence of single-layer graphene, it remains to be confirmed, and explained in much more specific terms, that "the high currents [are a consequence of] a combination of the nanopore edge structures produced by electron beam sculpting along with the convergent diffusion mechanisms due to nanosized electrodes" [108].

19.3 Internal edges

Vacancies in graphene have become again a very popular research subject. A crucial issue is their fate under thermal stress and/or in a reactive atmosphere. In particular, it is of interest to assess the similarities and differences in the behavior of these 'internal' *vs.* 'external' (*i.e.*, edge) active sites. The logical point of departure in such analysis is the consideration of two obvious (and too often neglected) facts: (i) removal of a basal-plane carbon atom produces four, and not three, potentially unpaired electrons (see Fig. 19.1); and (ii) each one of the free sites thus created is essentially a zigzag site. Here I summarize the results of a computational quantum chemical analysis (using the commercially available Gaussian software) that focuses on the key fundamental issue: the stabilization of vacancies under the influence of potentially competing electronic and geometric driving forces, *i.e.*, the electron pairing process and the preservation of planarity. Optimized geometries of representative graphene 'molecules' (or clusters) were determined using density functional theory at the B3LYP/6-31G(d) level, widely acknowledged to represent a judicious compromise between computational expediency and physical reality, especially with regard to electron correlation effects. Special attention was devoted to the dependence of ground-state thermodynamic stability on the realistic alternatives for the spin pairing processes.

In a pioneering, authoritative and yet essentially ignored study, Ubbelohde [110] analyzed what today many researchers would call "graphene nanoribbons" and proposed that this "network must bulge around the 'hole' or 'claw', owing to the repulsions between the [oxygen-containing] groups bonded to C_a and C_b [atoms]". A test of this assumption is reported and the fate of the four potentially unpaired electrons is carefully analyzed; thus, for example, vacancy reconstruction, much like edge reconstruction [24], may lead to the formation of multiplets with unpaired electrons, which is in turn a straightforward explanation [29] for the existence of ferromagnetism in impurity-free sp^2 hybridized carbons.

Figures 19.7 and 19.8 illustrate an essential finding: both singlets deviate from planarity, the curved one (Fig. 19.7(b)) being more stable than the twisted one (Fig. 19.7(c))

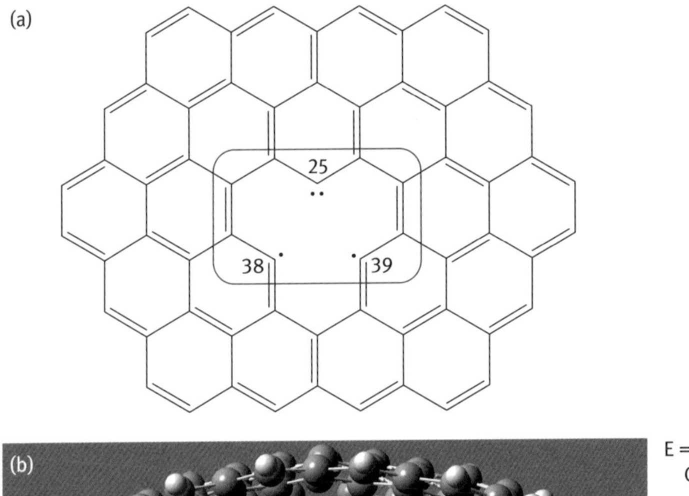

E = −2489.12490 hartrees
C38–C39 = 0.169 nm
C25–C38 = 0.258
C25–C39 = 0.258

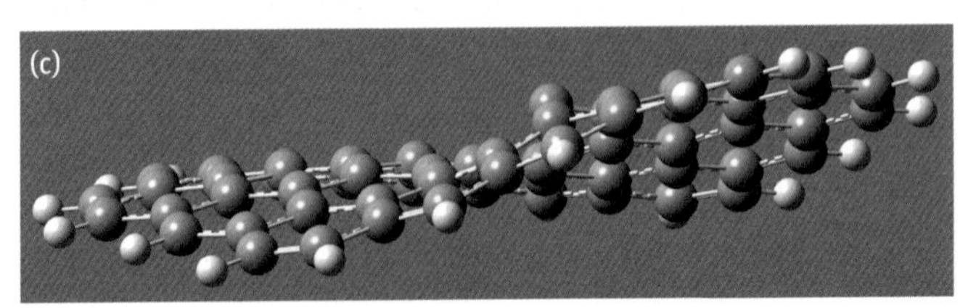

E = −2489.12140 hartrees
C38–C39 = 0.259 nm
C25–C38 = 0.172
C25–C39 = 0.256

Fig. 19.7: Illustration of the fate of an internal edge (vacancy): (a) schematic representation of graphene cluster $C_{65}H_{20}$; (b) optimized geometry of $C_{65}H_{20}$ singlet when C38-C39 bond is formed; (c) optimized geometry of $C_{65}H_{20}$ singlet when C25-C38 bond is formed.

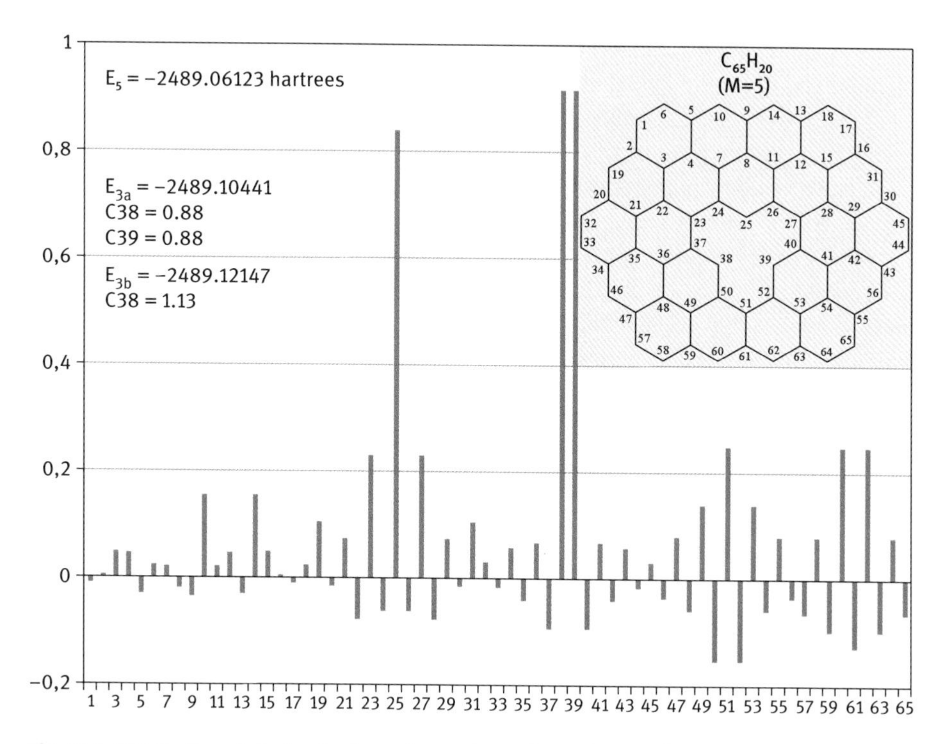

Fig. 19.8: Distribution of electronic charge in graphene cluster $C_{65}H_{20}$. (The details for the quintet are shown and only the essential information for the two triplets.)

by 2.2 kcal/mol. The triplet and the quintet (see Fig. 19.8) both preserve planarity but at a cost: 2.2 or 12.9 kcal/mol for the triplets, and 40 kcal/mol for the quintet (C38–C39 = 0.252 nm, C25–C38 = 0.252 nm, C25–C39 = 0.252 nm). The electron density distributions (Fig. 19.8) suggest that carbene site stabilization at monovacancies is not as thermodynamically favorable as it is for the graphene edge zigzag sites. In the quintet ground state, contrary to the intuitive expectation illustrated in Fig. 19.1(a), the spin density is larger at C38 and C39 than at C25, and neither is as high (< 1.0) as is typical [3, 4] at ('external') zigzag edges (*ca.* 1.3–1.4). Similarly, the electron density distribution in the several possible triplet states is atypical of zigzag sites: 1.13 at C38 and < 1.0 at both C38 and C39.

In a reactive atmosphere (*e.g.*, room-temperature air or H atoms), the fate of a vacancy is a delicate balance between stabilization by reconstruction and reaction (*e.g.*, with the ubiquitous O_2 and/or H radicals, if available). The reconstruction process (see Section 19.4) is analogous to, but different from, the formation of a Thrower–Stone–Wales (TSW) defect [111, 112]: As illustrated in Fig. 19.9, the energetically favorable diffusion (migration) of a free zigzag carbon atom results in complete valence saturation and formation of a pentagon-pentagon pair. This in turn results in significant graphene buckling and thus emphasizes the "chemical" in addition to the physical ori-

Fig. 19.9: Optimized geometry of $C_{41}H_{16}$ singlet (see also Fig. 19.1).

gin of this intriguing phenomenon [113]. At the other extreme, Fig. 19.10 illustrates the fate of a vacancy in the presence of quinone oxygen at graphene edges and partial hydrogen saturation of the carbon atoms: the residual free carbon site in this graphene, with its formally two unpaired electrons, is much more carbene-like (spin density = 1.34) than in the $C_{65}H_{20}$ triplets or quintet (see Fig. 19.8).

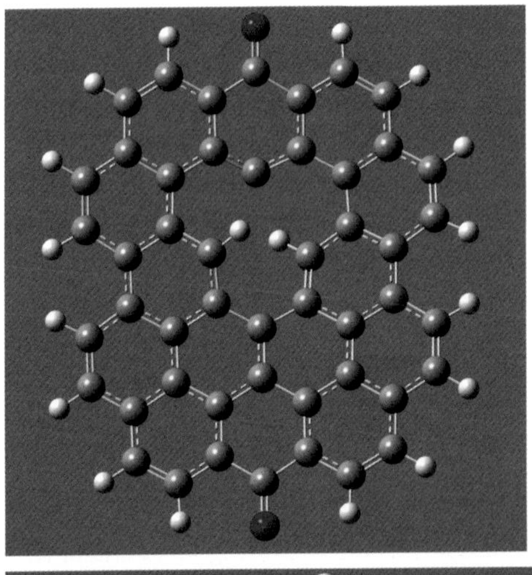

Fig. 19.10: Illustration of the fate of a vacancy in a reactive atmosphere: front and side view of optimized geometry of graphene $C_{49}H_{18}O_2$ (M = 3).

19.4 Edge reconstruction

Surface reconstruction processes have been studied extensively in both materials science [114] and heterogeneous catalysis [115]; this has not been the case, however, in the study of graphene-based materials. Under what circumstances do the hexagonal ring edges reconstruct to form pentagons and/or heptagons has been a topic of relatively recent interest, however, no doubt triggered by the stability of the latter rings in fullerenes. The pioneering insights in this regard have been provided by Thrower [116] and Stone and Wales [117], and there is now an abundant literature on TSW defects, mostly located in the graphene basal plane. No such studies seem to have been carried out for sp^2 hybridized carbon materials until the discovery of carbon nanotubes and the successful isolation of graphene. (An early attempt to understand the surface reconstruction in diamond, including the effect of oxygen, was carried out by Whitten *et al.* [118].) What follows is a summary of representative studies that deal with (or avoid!?) these issues.

In an early study, Ajayan *et al.* [119] showed that an "obvious way in which the defects in an irradiated [single-wall] nanotube can be reconstituted and the energy lowered is by dangling bond saturation and Stone-Wales type transformation". The latter process was found to account for the disappearance of octagons and larger rings, thus leading to a structure "mainly constituted of five-, six- and seven-membered rings". Their tight-binding molecular dynamics simulation of the former process in a monovacancy – in essence a nascent site deactivation process [24] – predicted the intuitively obvious formation of a pentagon, and this in turn offers a straightforward path for the closure of a growing nanotube.

In contrast to such intralayer reconstruction of broken C-C bonds, Rotkin and Gogotsi [120] argued that interlayer edge reconstruction is important as well. They examined the (external) edges of nonplanar graphitic materials, as well as of graphite, whose surface reconstruction has indeed been much less studied than that of, say, silicon. Using TEM analysis, they found what might seem to be a surprising abundance of evidence for the existence of what they called graphite polyhedral crystals (GPC), which "bridge a gap between the nanotubes and planar graphite" and whose "typical feature ... is a presence of continuous nano-arches on the pyramidal edge planes". They argued that such edge reconstruction, involving two or more graphene layers, occurs "in a very broad range of growth temperatures and environments", and concluded that "nanotube-like zipping of graphite faces describes the dominant mechanism of edge termination of graphite". The feasibility of this curling mechanism was supported by Ivanovskaya *et al.* [121], whose simulations showed that "free graphene sheet edges can curl back on themselves, reconstructing as nanotubes" and that this "results in lower formation energies than any other nonfunctionalized edge structure reported to date in the literature"; the latter is rationalized by the assumption that the dangling bonds are replaced "with sp^3-like hybridized carbon atoms". Hurt and co-workers [122] carried out an analysis of a similar process, prompted by the experimental observation that the initially hydrophilic graphene-edge-rich carbon materials (*e.g.*, platelet-type vapor-grown carbon fibers) become surprisingly hydrophobic even though they retain their exposed graphene edge surfaces after high-temperature annealing. Using a theoretical analysis based on the elastic constant of single graphene layers and the energies of unreconstructed free radical edge sites, they calculated the interlayer reconstruction energies for several cases; these are reproduced in Fig. 19.11 and suggest, in agreement with intuition, a monotonic increase from a hairpin arch (0.37 eV) to adjacent bonding(1.3 eV). This led them to conclude that the "high hydrophobicity is the result of extensive reconstruction into hairpin arches that cap nearly all active sites and expose uniform basal planes to the surrounding fluid". Of course, graphite being the ultimate lubricant and the lubrication mechanism (as well as that of pencil writing!) being the sliding of graphene layers, such edge termination must be a reversible process and the interlayer bonding must be quite weak. Furthermore, whether many of the other properties of graphene-based materials (*e.g.*, porosity, thermoelectric power, ferromagnetism or photoluminescence) are consistent with interlayer edge stabilization, rather than intralayer stabilization [24, 29], remains to be demonstrated.

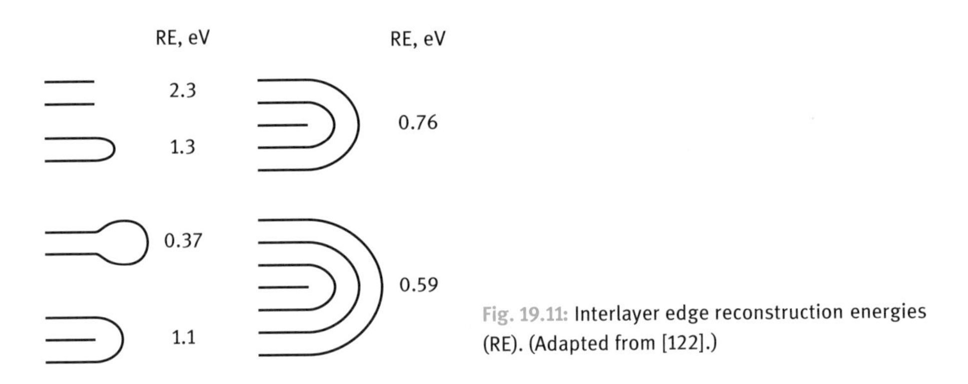

Fig. 19.11: Interlayer edge reconstruction energies (RE). (Adapted from [122].)

The first detailed study of intralayer edge reconstruction was performed by Koskinen and co-workers [123–127]. They used density functional theory to analyze the transformation of two adjacent zigzag-edge hexagons into a heptagon-pentagon pair [123] and "unexpectedly found that the zigzag edge is metastable and a planar reconstruction spontaneously takes place at room temperature". They correctly assumed that "the armchair has low energy due to triple bonds" [29, 128], but their statement that the "zigzag does not have such triple bonds and ends up with strong and expensive dangling bonds" is unsubstantiated, excessively simplistic and ignores the relevant literature.

In a follow-up study, Malola *et al.* [126] used density-functional tight-binding theory to explore the "structural, chemical and dynamical trends" in a graphene sample formed by "merging two graphene edges with the same chirality but different orientation and random translation along the boundary". One consequence of this defect (grain boundary) construction process is the postulated appearance of "dangling bonds" (DB) and these authors calculated their average density per unit length along a grain boundary as a function of the chiral angle (zero corresponding to a zigzag edge and 30° to an armchair edge). They reported that "54 % of the samples have one DB every 7.5–20 Å, and 16 % of the samples have one DB every 20–50 Å". The authors interpreted these findings to be in agreement with the results of Cervenka *et al.* [129], who argued that the ferromagnetic behavior of highly oriented pyrolytic graphite (HOPG), observed by magnetic force microscopy at room temperature, is due to the possibility that "some of the step edges are created on HOPG surfaces at places where bulk grain boundaries cross the surface". They thus concluded that their proposed grain boundary construction process is objective and may help "to design graphene structures for given functions". It should be recognized, however, that the central issue of dangling bond formation and survival has been discussed at length much before the resurgence of interest in graphene, and that much of the recent graphene literature seems to be oblivious to these relevant and often important contributions and insights. Since the early days of the very extensive ESR studies of sp^2 hybridized carbons [26, 29, 130–133], the key issue has been graphene's electron pairing propensity and σ-π interaction; indeed, carbons are well known not only to be paramagnetic but some are also ferromagnetic even in the absence of transition-metal impurities. This has recently been shown to depend very much on the details of graphene geometry and (edge) surface chemistry [29]; thus the propagation of defective graphene structures is expected to depend on the effectiveness of 'healing' of vacancies and the formation of carbene-type zigzag edge sites.

Chia and Crespi [134] recently analyzed the stabilization of zigzag GNR edges and proposed a method that exploits "steric crowding effects along the one-dimensional edge of a two-dimensional system [as] a general way to control edge reconstructions across a range of emerging single-layer systems". Somewhat surprisingly, in view of the ubiquitous gas-phase formation of soot and its PAH precursors, they argued that "a careful computational study of hydrogen-terminated ribbons demonstrated that the less interesting sp^3 termination ... is thermodynamically favored across a wide range of conditions" and that "the sp^2 geometry is favored only at extremely low hydrogen partial pressures". The studies on which these statements seem to be based [68, 135] do invoke "simple concepts from organic chemistry" [68] but their simplistic treatment considers only sp^2 and sp^3 hybridization and thus these authors too neglect the suggestion made by none other than Coulson, more than half a century ago, that free zigzag edges undergo s^2p^2 hybridization [29, 128]. And in their analysis [135] of edge interactions with molecules such as O_2, H_2O and CO_2, an issue with an exceedingly abundant literature for many graphene-based carbon materials, these authors cite only one pre-twenty-first-century study, and this one not devoted to carbon materials at all, but to GaAs!?

Dubois and co-workers [136] rightly point out how important the edges are, especially for GNRs: "In contrast to carbon nanotubes, GNRs exhibit a high degree of edge chemical reactivity, which, for instance, prevents the existence of truly metallic nanoribbons"; and "the discrepancy between the theoretical electronic confinement gap and the experimentally measured transport gap has been attributed to localized states induced by edge disorder". They too invoke charge transport rationalizations based on conventional organic chemistry (Clar´s sextet theory) and compare the effects of edge mono- *vs.* dihydrogenation. Thus, the presence of monohydrogenated pentagons or heptagons imbedded in an armchair edge "can be seen as the superposition of two mirroring Kekulé structures that partially destroy the benzenoid character of the aGNRs"; and "[o]n the contrary, the dihydrogenation of [these] defects restores the complete benzenoid character of the ribbon". This study highlights the fact that, apart from spin pairing by virtue of dehydrogenation and triplet carbene stabilization processes, "the controlled deposition of monatomic hydrogen is a key ingredient to explore a metal/insulator transition in these materials". As a final example, Zeng *et al.* [137] optimistically argued that "the ability of tuning the electronic transport of [zigzag graphene nanoribbons] could be improved through edge reconstruction activated by energetic particle radiation". Clearly, there are many possibilities for controlling and taking advantage of internal graphene edge reactivity in pursuit of novel physico-chemical applications.

19.5 Summary

The decades-old interest in chemical reactivity of graphene edges in sp^2 hybridized carbon materials ranging from soot to graphite has been augmented recently by a closer scrutiny of the physico-chemical properties of novel graphene-based materials such as nanotubes and nanoribbons. The increasingly voluminous literature on the distribution of electron density at external and internal edge sites is now scattered over an almost overwhelming number of publications in a wide variety of disciplines, mostly dominated by physical rather than chemical principles. As the 'dust' eventually settles, it will be quite a challenge to reconcile many of the essential facts and, especially so, of the often diverging points of view. But the path forward appears to be clear. At external edges it is necessary to unravel where PAH molecules end and graphenes begin, and to confirm experimentally the existence of, and the spin pairing effects in, free zigzag sites. And at internal edges (vacancies) it remains to quantify the outcome of the competition between geometric and electronic effects, the former being reflected in the tendency to preserve planarity and the latter in bonding rearrangements that reduce the number of unpaired electrons in a graphene layer.

Acknowledgment

Financial support from CONICYT-Chile, projects PFB-27 and Fondecyt 1120609, is gratefully acknowledged.

Bibliography

[1] M. Inagaki, L.R. Radovic, *Carbon* 40 (2002) 2279.
[2] L.P. Biro, P. Lambin, *New J. Phys.* 15 (2013) 035024.
[3] T. Enoki, Y. Kobayashi, K.I. Fukui, *International Reviews in Physical Chemistry* 26 (2007) 609.
[4] V.P. Gusynin, V.A. Miransky, S.G. Sharapov, I.A. Shovkovy, *Low Temp. Phys.* 34 (2008) 778.
[5] X.T. Jia, J. Campos-Delgado, M. Terrones, V. Meunier, M.S. Dresselhaus, *Nanoscale* 3 (2011) 86.
[6] M. Acik, Y.J. Chabal, *Jpn. J. Appl. Phys.* 50 (2011) 070101.
[7] X.Y. Zhang, J. Xin, F. Ding, *Nanoscale* 5 (2013) 2556.
[8] P.A.M. Dirac, *Proc. Roy. Soc. A* 123 (1929) 714.
[9] J.F. Tian, H.L. Cao, W. Wu, Q.K. Yu, Y.P. Chen, *Nano Lett.* 11 (2011) 3663.
[10] A.A. Lizzio, A. Piotrowski, L.R. Radovic, *Fuel* 67 (1988) 1691.
[11] J.A. Menéndez, J. Phillips, B. Xia, L.R. Radovic, *Langmuir* 12 (1996) 4404.
[12] J.A. Menéndez, B. Xia, J. Phillips, L.R. Radovic, *Langmuir* 13 (1997) 3414.
[13] A.K. Geim, K.S. Novoselov, *Nat. Mater.* 6 (2007) 183.
[14] S. Neubeck, Y.M. You, Z.H. Ni, P. Blake, Z.X. Shen, A.K. Geim, K.S. Novoselov, *Appl. Phys. Lett.* 97 (2010) 053110.
[15] G.R. Hennig, in: P.L. Walker, Jr. (Ed.), *Chemistry and Physics of Carbon*, Marcel Dekker, New York, 1966, p. 1.
[16] J.M. Thomas, E.E.G. Hughes, *Carbon* 1 (1964) 209.
[17] J.M. Thomas, K.M. Jones, *J. Nucl. Mater.* 11 (1964) 236.
[18] J.M. Thomas, C. Roscoe, in: P.L. Walker Jr. (Ed.), *Chemistry and Physics of Carbon*, Vol. 3, Marcel Dekker, New York, 1968.
[19] J.M. Thomas, in: P.L. Walker Jr. (Ed.), *Chemistry and Physics of Carbon*, Vol. 1, Marcel Dekker, New York, 1965, p. 121.
[20] S.R. Kelemen, C.A. Mims, *Surf. Sci.* 136 (1984) L35.
[21] X.L. Hou, J.L. Li, S.C. Drew, B. Tang, L. Sun, X.G. Wang, *J. Phys. Chem.* C 117 (2013) 6788.
[22] N. Gorjizadeh, N. Ota, Y. Kawazoe, *Chem. Phys.* 415 (2013) 64.
[23] M.J. Rossi, *Int. J. Chem. Kinet.* 40 (2008) 395.
[24] L.R. Radovic, A.F. Silva-Villalobos, A.B. Silva-Tapia, F. Vallejos-Burgos, *Carbon* 49 (2011) 3471.
[25] A. Montoya, T. Truong, A. Sarofim, *J. Phys. Chem.* A 104 (2000) 6108.
[26] I.C. Lewis, L.S. Singer, in: P.L. Walker Jr, P.A. Thrower (Eds.), *Chemistry and Physics of Carbon*, Vol. 17, Marcel Dekker, New York, 1981, p. 1.
[27] L.S. Singer, I.C. Lewis, *Applied Spectroscopy* 36 (1982) 52.
[28] J.W. Feng, S.K. Zheng, G.E. Maciel, *Energy Fuels* 18 (2004) 560.
[29] L.R. Radovic, B. Bockrath, *J. Amer. Chem. Soc.* 127 (2005) 5917.
[30] G. Holton, *Thematic Origins of Scientific Thought (Kepler to Einstein)*, Revised Edition, Harvard University Press, Cambridge, MA, 1988.
[31] M. Boudart, *Am. Scientist* 57 (1969) 97.
[32] N.R. Laine, F.J. Vastola, P.L. Walker Jr, *Proc. 5th Conf. Carbon*, Vol. 2 (1961) 211.
[33] N.R. Laine, F.J. Vastola, P.L. Walker Jr, *J. Phys. Chem.* 67 (1963) 2030.

[34] P.L. Walker Jr., R.L. Taylor, J.M. Ranish, *Carbon* 29 (1991) 411.

[35] L.R. Radovic, A.A. Lizzio, H. Jiang, in: J. Lahaye, P. Ehrburger (Eds.), *Fundamental Issues in Control of Carbon Gasification Reactivity*, Kluwer Academic Publishers, Dordrecht (The Netherlands), 1991, p. 235.

[36] O.W. Fritz, K.J. Hüttinger, *Carbon* 31 (1993) 923.

[37] K.J. Huttinger, J.S. Nill, *Carbon* 28 (1990) 457.

[38] J.M. Calo, P.J. Hall, in: J. Lahaye, P. Ehrburger (Eds.), *Fundamental Issues in Control of Carbon Gasification Reactivity*, Kluwer Academic Publishers, Dordrecht (The Netherlands), 1991, p. 329.

[39] K. Nakada, M. Fujita, G. Dresselhaus, M.S. Dresselhaus, *Phys. Rev.* B 54 (1996) 17954.

[40] M. Fujita, K. Wakabayashi, K. Nakada, K. Kusakabe, *J. Phys. Soc. Jpn.* 65 (1996) 1920.

[41] S.E. Stein, *Acc. Chem. Res.* 24 (1991) 350.

[42] A. Yamashiro, Y. Shimoi, K. Harigaya, K. Wakabayashi, *Phys Rev B* 68 (2003) 193410.

[43] A. Yamashiro, Y. Shimoi, K. Harigaya, K. Wakabayashi, *Physica E: Low-dimensional Systems and Nanostructures* 22 (2004) 688.

[44] D.Y. Pan, J.C. Zhang, Z. Li, M.H. Wu, *Adv. Mater.* 22 (2010) 734.

[45] D.Y. Pan, J.C. Zhang, Z. Li, C. Wu, X.M. Yan, M.H. Wu, *Chem. Commun.* 46 (2010) 3681.

[46] L.L. Li, G.H. Wu, G.H. Yang, J. Peng, J.W. Zhao, J.J. Zhu, *Nanoscale* 5 (2013) 4015.

[47] V. Gupta, N. Chaudhary, R. Srivastava, G.D. Sharma, R. Bhardwaj, S. Chand, J. *Am. Chem. Soc.* 133 (2011) 9960.

[48] J.H. Shen, Y.H. Zhu, C. Chen, X.L. Yang, C.Z. Li, *Chem. Commun.* 47 (2011) 2580.

[49] K. Sasaki, S. Murakami, R. Saito, *Appl. Phys. Lett.* 88 (2006) 113110.

[50] K. Xu, P.D. Ye, *J. Phys. Chem.* C 114 (2010) 10505.

[51] L. Brey, H.A. Fertig, *Physical Review B (Condensed Matter and Materials Physics)* 73 (2006) 195408.

[52] A.H. Castro Neto, F. Guinea, N.M.R. Peres, *Phys. Rev. B* 73 (2006) 205408.

[53] V.R. Gusynin, V.A. Miransky, S.G. Sharapov, I.A. Shovkovy, *Phys. Rev. B* 77 (2008) 778.

[54] V. Mazo, E. Shimshoni, H.A. Fertig, *Phys. Rev. B* 84 (2011) 045405.

[55] S. Mrozowski, A. Chaberski, *Phys. Rev.* 104 (1956) 74.

[56] S. Mrozowski, A. Chaberski, *Phys. Rev.* 94 (1954) 1427.

[57] S. Mrozowski, *Carbon* 9 (1971) 97.

[58] S. Banerjee et al., *Appl. Phys. Lett.* 88 (2006) 602111.

[59] S.E. Stein, R.L. Brown, in: J.F. Liebman, A. Greenberg (Eds.), *Molecular Structure and Energetics*, Vol. 2, VCH Publishers, New York, 1987, p. 37.

[60] D. Gunlycke, J. Li, J.W. Mintmire, C.T. White, *Appl. Phys. Lett.* 91 (2007) 3.

[61] L.R. Radovic, *J. Am. Chem. Soc.* 131 (2009) 17166.

[62] M. Frenklach, *Phys. Chem. Chem. Phys.* 4 (2002) 2028.

[63] O. Hod, V. Barone, J.E. Peralta, G.E. Scuseria, *Nano Lett.* 7 (2007) 2295.

[64] O. Hod, J.E. Peralta, G.E. Scuseria, *Phys. Rev. B* 76 (2007) 4.

[65] D.E. Jiang, B.G. Sumpter, S. Dai, *J. Chem. Phys.* 126 (2007) 124701.

[66] S.E. Stein, R.L. Brown, *J. Amer. Chem. Soc.* 109 (1987) 3721.

[67] K.N. Kudin, *ACS Nano* 2 (2008) 516.

[68] T. Wassmann, A.P. Seitsonen, A.M. Saitta, M. Lazzeri, F. Mauri, *Phys. Rev. Lett.* 101 (2008) 096402.

[69] D.E. Jiang, X.Q. Chen, W.D. Luo, W.A. Shelton, *Chem. Phys. Lett.* 483 (2009) 120.

[70] S. Bhowmick, V.B. Shenoy, *J. Chem. Phys.* 128 (2008) 244717.

[71] T. Enoki, K. Takai, *Solid State Commun.* 149 (2008) 1144.

[72] K.I. Sasaki, R. Saito, *J. Phys. Soc. Jpn.* 77 (2008) 054703.

[73] O.V. Yazyev, M.I. Katsnelson, *Phys. Rev. Lett.* 1 (2008) 047209.

[74] K.S. Yi, D. Kim, K.S. Park, J.J. Quinn, *Physica E* 40 (2008) 1715.
[75] V.P. Gusynin, V.A. Miransky, S.G. Sharapov, I.A. Shovkovy, C.M. Wyenberg, *Phys. Rev. B* 79 (2009) 115431.
[76] P. Lu, Z.H. Zhang, W.L. Guo, *Phys. Lett. A* 373 (2009) 3354.
[77] M.R. Philpott, Y. Kawazoe, *J. Chem. Phys.* 131 (2009) 214706.
[78] S. Dutta, S.K. Pati, *Carbon* 48 (2010) 4409.
[79] K. Furukawa, H. Yoshioka, Y. Mochizuki, *J. Phys. Soc. Jpn.* 79 (2010) 084708.
[80] A.D. Hernandez-Nieves, B. Partoens, F.M. Peeters, *Phys. Rev. B* 82 (2010) 165412.
[81] V.L.J. Joly, M. Kiguchi, S.J. Hao, K. Takai, T. Enoki, R. Sumii, K. Amemiya, H. Muramatsu, T. Hayashi, Y.A. Kim, M. Endo, J. Campos-Delgado, F. Lopez-Urias, A. Botello-Mendez, H. Terrones, M. Terrones, M.S. Dresselhaus, *Phys. Rev. B* 81 (2010) 245428.
[82] H.Z. Pan, M. Xu, L. Chen, Y.Y. Sun, Y.L. Wang, *Acta Phys. Sin.* 59 (2010) 6443.
[83] M.R. Philpott, S. Vukovic, Y. Kawazoe, W.A. Lester, *J. Chem. Phys.* 133 (2010) 044708.
[84] M.J. Schmidt, D. Loss, *Phys. Rev. B* 82 (2010) 085422.
[85] L.L. Song, X.H. Zheng, R.L. Wang, Z. Zeng, *J. Phys. Chem. C* 114 (2010) 12145.
[86] W.Z. Wu, Z.H. Zhang, P. Lu, W.L. Guo, *Phys. Rev. B* 82 (2010) 085425.
[87] J. Jung, *Phys. Rev. B* 83 (2011) 165415.
[88] M. Kiguchi, K. Takai, V.L.J. Joly, T. Enoki, R. Sumii, K. Amemiya, *Phys. Rev. B* 84 (2011) 045421.
[89] J. Kunstmann, C. Ozdogan, A. Quandt, H. Fehske, *Phys. Rev. B* 83 (2011) 045414.
[90] D.J. Luitz, F.F. Assaad, M.J. Schmidt, *Phys. Rev. B* 83 (2011) 195432.
[91] M.R. Philpott, Y. Kawazoe, *J. Chem. Phys.* 134 (2011) 124706.
[92] L.L. Sun, P. Wei, J.H. Wei, S. Sanvito, S.M. Hou, *J. Phys.-Condens. Matter* 23 (2011) 425301.
[93] O. Voznyy, A.D. Guclu, P. Potasz, P. Hawrylak, *Phys. Rev. B* 83 (2011) 165417.
[94] O.V. Yazyev, R.B. Capaz, S.G. Louie, *Phys. Rev. B* 84 (2011) 115406.
[95] H.Q. Zhou, H.C. Yang, C.Y. Qiu, Z. Liu, F. Yu, L.J. Hu, X.X. Xia, H.F. Yang, C.Z. Gu, L.F. Sun, *Chin. Phys. B* 20 (2011) 026803.
[96] T. Enoki, Fuller. Nanotub. *Carbon Nanostruct.* 20 (2012) 310.
[97] T. Enoki, K. Takai, M. Kiguchi, *Bull. Chem. Soc. Jpn.* 85 (2012) 249.
[98] W. Fa, J. Zhou, *Phys. Lett. A* 377 (2012) 112.
[99] T. Hu, J. Zhou, J.M. Dong, Y. Kawazoe, *Phys. Rev. B* 86 (2012) 125420.
[100] L.F. Huang, G.R. Zhang, X.H. Zheng, P.L. Gong, T.F. Cao, Z. Zeng, *J. Phys.-Condens. Matter* 25 (2013) 055304.
[101] A.T. Lee, K.J. Chang, *Phys. Rev. B* 87 (2013) 085435.
[102] X.W. Zhang, O.V. Yazyev, J.J. Feng, L.M. Xie, C.G. Tao, Y.C. Chen, L.Y. Jiao, Z. Pedramrazi, A. Zettl, S.G. Louie, H.J. Dai, M.F. Crommie, *ACS Nano* 7 (2013) 198.
[103] A. Konishi, Y. Hirao, K. Matsumoto, H. Kurata, R. Kishi, Y. Shigeta, M. Nakano, K. Tokunaga, K. Kamada, T. Kubo, *J. Am. Chem. Soc.* 135 (2013) 1430.
[104] Y.H. Lee, S.G. Kim, D. Tomanek, *Phys. Rev. Lett.* 78 (1997) 2393.
[105] T. Kawai, Y. Miyamoto, O. Sugino, Y. Koga, *Phys. Rev. B* 62 (2000) R16349.
[106] P.C. Mishra, A. Yadav, *Chem. Phys.* 402 (2012) 56.
[107] T. Ohba, H. Kanoh, *J. Phys. Chem. Lett.* 3 (2012) 511.
[108] S. Banerjee, J. Shim, J. Rivera, X.Z. Jin, D. Estrada, V. Solovyeva, X. You, J. Pak, E. Pop, N. Aluru, R. Bashir, *ACS Nano* 7 (2013) 834.
[109] F. Béguin, E. Frackowiak (Ed.), *Carbons for Electrochemical Energy Storage and Conversion Systems*. CRC Press, 2009.
[110] A.R. Ubbelohde, *Nature* 180 (1957) 380.
[111] L. Chen, J.F. Li, D.Y. Li, M.Z. Wei, X.L. Wang, *Solid State Commun.* 152 (2012) 1985.
[112] T. Fujimori, L.R. Radovic, A.B. Silva-Tapia, M. Endo, K. Kaneko, *Carbon* 50 (2012) 3274.
[113] H.-S. Shen, Y.-M. Xu, C.-L. Zhang, *Appl. Phys. Lett.* 102 (2013) 131905.

[114] C.B. Duke, *Chem. Rev.* 96 (1996) 1237.

[115] G.A. Somorjai, *Annu. Rev. Phys. Chem.* 45 (1994) 721.

[116] P.A. Thrower, in: P.L. Walker Jr (Ed.), *Chem. Phys. Carbon*, Vol 5, Marcel Dekker, 1969, p. 217.

[117] A.J. Stone, D.J. Wales, *Chem. Phys. Lett.* 128 (1986) 501.

[118] J.L. Whitten, P. Cremaschi, R.E. Thomas, R.A. Rudder, R.J. Markunas, *Appl. Surf. Sci.* 75 (1994) 45.

[119] P.M. Ajayan, V. Ravikumar, J.C. Charlier, *Phys. Rev. Lett.* 81 (1998) 1437.

[120] S.V. Rotkin, Y. Gogotsi, *Mater. Res. Innov.* 5 (2002) 191.

[121] V.V. Ivanovskaya, A. Zobelli, P. Wagner, M.I. Heggie, P.R. Briddon, M.J. Rayson, C.P. Ewels, *Phys. Rev. Lett.* 107 (2011) 065502.

[122] K. Jian, A. Yan, I. Kulaots, G.P. Crawford, R.H. Hurt, *Carbon* 44 (2006) 2102.

[123] P. Koskinen, S. Malola, H. Hakkinen, *Phys. Rev. Lett.* 101 (2008) 115502.

[124] P. Koskinen, S. Malola, H. Hakkinen, *Phys. Rev. B* 80 (2009) 073401.

[125] S. Malola, H. Hakkinen, P. Koskinen, *Eur. Phys. J. D* 52 (2009) 71.

[126] S. Malola, H. Hakkinen, P. Koskinen, *Phys. Rev.* B 81 (2010) 165447.

[127] P. Rakyta, A. Kormanyos, J. Cserti, P. Koskinen, *Phys. Rev. B* 81 (2010) 115411.

[128] C.A. Coulson, *The electronic structure of the boundary atoms of a graphite layer*, Fourth Conference on Carbon, Pergamon Press, University of Buffalo, Buffalo, NY, 1960, p. 215.

[129] J. Cervenka, M.I. Katsnelson, C.F.J. Flipse, *Nat. Phys.* 5 (2009) 840.

[130] R.C. Wayne, *J. Chem. Phys.* 39 (1963) 1337.

[131] L. Singer, I. Lewis, *Carbon* 2 (1964) 115.

[132] I. Lewis, L. Singer, *Carbon* 7 (1969) 93.

[133] S. Mrozowski, *Carbon* 26 (1988) 521.

[134] C.I. Chia, V.H. Crespi, *Phys. Rev. Lett.* 109 (2012) 076802.

[135] A.P. Seitsonen, A.M. Saitta, T. Wassmann, M. Lazzeri, F. Mauri, *Phys. Rev. B* 82 (2010) 115425.

[136] S.M.M. Dubois, A. Lopez-Bezanilla, A. Cresti, F. Triozon, B. Biel, J.C. Charlier, S. Roche, *ACS Nano* 4 (2010) 1971.

[137] H. Zeng, J. Zhao, D.H. Xu, J.W. Wei, H.F. Zhang, *Eur. Phys. J.* B 86 (2013) 80.

Index

π-π interactions, 403, 493
π-π stacking, 130
in situ hybridization, 323
1-(3-methoxycarbonyl) propyl-1-phenyl[6, 6] C_{61} ($PC_{61}BM$), 469
1-octyl-2,3-dimethylimidazolium iodide (ODI), 488

supercapacitors
– electrical double-layer capacitor (EDLC), 215
– redox capacitor, 215

acid treatment, 398, 413
acid/base properties, 407
acidity, 284
activated carbon, 215
adhesive, 234
adsorbent, 281
adsorption, 299
aerospace composites, 234
AgCl, 218
agglomeration, 321, 394
Al_2O_3, 238
annealing, 504
anode, 299, 300, 303
anti-flammability, 93
arc discharge, 12, 173, 312
aromatic compounds, 411
aromatic linker, 131
aromatic linking agents
– benzyl alcohol, 64
– coronene tetracarboxylic acid, 183
– perelene, 183
– phthalocyanine, 63
– porphyrin, 49, 59, 63, 184
– pyrene, 57, 59, 63
– pyrene derivatives, 183
– triphenylphosphine, 64
aromaticity, 47
artificial muscle, 248
aspect ratio, 232, 401
assembly, 18
atomic force microscopy (AFM), 34

B_2O_3, 405
ballistic electronic transport, 72
ballistic transport, 8
band gap, 511
Batteries
– Li-ion battery, 210
benzyl alcohol, 408
binding sites, 281
Bingel–Hirsch cyclopropanation, 51
biocompatibility, 82, 98
biomass, 202, 209
biomolecules
– antibody, 49
– DNA, 15, 49, 183
– enzyme, 49
– glucose oxidase, 58
– hemoglobin, 58
– myoglobin, 58
– ovalbumin, 207
– protein, 49, 59
biosensor, 57, 98, 182
– glucose, 58
boron doped graphene, 173
boron nitride, 174
– nanotubes, 94

capillary effect, 413
carbene, 50
carboboronitrides, 174
carbohydrates, 202
– cellulose, 202, 209
– chitosane, 208
– glucosamine, 208
– glucose, 202, 207
– hexose, 203
– starch, 202
– xylose, 202
carbon black, 357, 361, 369, 396, 404
carbon fibers, 434
carbon nanofibers (CNF), 205, 301, 360, 372, 395, 407, 433
carbon nanohorns, 435, 444, 483
carbon nitrides
– melon, 218
carbon onion, 404
carbon support, 377
catalyst deactivation, 419
catalyst support, 369, 370, 372, 396, 412, 417

cathode, 306, 309
CdS, 439
CdSe, 62, 188
cell proliferation, 98
cerium, 457
charge mobility, 319, 431
charge recombination, 446
charge separation, 194, 430, 476
charge transfer, 187, 188, 322, 377, 410
charge transport, 445, 522
chemical exfoliation, 320
chemical vapor deposition (CVD), 13, 31, 172, 237, 307, 379, 397, 503
chemisorption, 273
chirality, 4, 73
– armchair, 44
– chiral, 44
– zigzag, 44
chromium, 457
CNT array, 302, 306, 397
CNT fibers
– direct spinning, 241
– forest drawing, 241
– wet spinning, 241
CNT-CNF hybrids, 301, 311
CNT-graphene hybrids, 178, 301, 307, 311, 487
CNT-inorganic hybrids, 303, 304, 306
CNT-metal hybrids, 185, 304, 370, 413, 416, 419
CNT-metal oxide hybrids, 151, 305, 309, 313, 413, 480
CNT-polymer hybrids, 370
CNT-sulfur hybrids, 307
CNTs
– chirality, 359, 435
– defects, 371, 405, 435
– doping, 397, 407, 437
– functionalization, 370, 413, 415, 437, 486
– properties, 297
– structure, 297, 359, 394, 397, 435
– surface area, 297
CO_2 sequestration, 206
Co_3O_4, 309, 324, 327, 328
coal, 202
cobalt, 457
colloidal nanoparticles, 415
coloration, 456
composite fillers
– carbon black, 228
– carbon fiber, 228
– CNT fiber, 240
– glass fiber, 230, 236
– graphene, 231
– graphite platelets, 231
confinement effect, 95, 400, 413, 419
conjugated polymers, 459
copper, 457
core-shell structures, 206
corrosion, 309, 313
coulombic efficiency, 190, 192, 302, 304, 307, 325
covalent functionalization, 16, 46
– acid oxidation, 48, 81, 184
– amidation, 48, 62, 82, 184
– arylation, 184
– carbene addition, 81
– carbodiimide coupling, 127
– cycloaddition, 50, 54, 80
– esterification, 48, 62, 82
– fullerenes, 80
– graphene, 82
– halogenation, 49, 81
– nucleophilic substitution, 59
– polymer grafting, 80
– radical addition, 52
covalent hybridization
– azide, 128
– nitrene, 128
crack propagation, 89
Cu_2O, 305, 438
current density, 182, 311
curvature, 46, 72, 75, 76
cyclability, 301
cyclic voltammetry (CV), 313, 383
cytotoxicity, 102

dangling bonds, 371, 512, 519
decoration with metal NPs
– covalent, 62
defects, 5, 46
– buckling, 517
– grain boundary, 521
– non-sp^2 carbon, 78
– structural, 33, 76, 517
– topological, 32, 76
– vacancy, 78, 503, 509
degree of graphitization, 371
deintercalation, 457
dendrimer, 51, 64, 80, 136

density functional theory (DFT), 410, 441, 507, 520
diamond-like carbon (DLC), 92, 358
diamondoid, 74
diazonium, 45, 53
dielectrophoresis, 16
Diels–Alder reaction, 52, 80, 203
diffusion rate, 304
dispersion, 74, 81, 86, 182, 274, 443
donor–acceptor bilayers, 462
doping, 6, 47
- boron, 75, 90
- carbon, 218
- endohedral, 74
- exohedral, 74
- heterodoping, 76
- nitrogen, 75, 89, 91, 208, 216, 218, 376, 379, 383, 400
- phosphor, 75
- silicon, 91
- substitutional, 75
- sulfur, 75
drug delivery, 98
dyads, 217
dye adsorption, 480
dye-sensitized solar cell, 461

edge atoms, 10
electric double layer capacitor (EDLC), 320, 361
electrical conductivity, 72, 74, 95, 180, 232, 236, 240, 301, 313
- doping, 92
electrical double-layer capacitor (EDLC), 311
electrical properties, 322, 510
electroactive materials, 298, 360
electrocatalysis, 357
- aluminum, 363
- aromatics, 362
- chlorine, 362
- hydrogen peroxide, 365
- metal refining, 363
electrocatalytic activity, 186
electrochemical energy storage, 297, 314
electrochemical stability, 186, 370, 372
electrochromism, 455
electrode materials
- $LiCoO_2$, 210, 306
- $LiMe_2O_4$, 306
- $LiMePO_4$, 213, 306
electroless deposition, 141
electrolysis, 52
electromagnetic absorption, 234
electron
- beam irradiation, 78, 278
- density, 400, 507, 517
- diffraction, 32
- energy loss spectroscopy (EELS), 32, 404
- hole recombination, 431, 432, 441
- irradiation, 437
- mobility, 25
- transfer, 58, 175, 194, 288, 377, 416, 431, 432, 439, 446, 491
- transport, 480
electronic conductivity, 368, 444
electronic properties, 8
encapsulated model, 327
encapsulation, 299
- fullerenes, 74
- nanoconfinement, 445
- transition metals, 74
endohedral filling, 60
- biomolecule, 60
- fullerene, 60
- inorganic salt, 61
energy density, 307, 308, 327, 335, 421
enzymatic biodegradation, 99
epitaxial growth, 31
epoxy, 86, 233, 236, 245, 280
exfoliation, 26, 29, 182, 482

Fe_2O_3, 204, 365, 438
Fe_3O_4, 186, 327
FeMgAl, 307
Fermi energy, 508
ferromagnetism, 75, 506
fiber-reinforced polymer composites (FRPC), 229, 235
field-emission device, 176
fill factors (FF), 479
flocculation, 232
fluorescence quenching, 57, 184, 494
fluorine doped tin oxide (FTO), 476
Fourier transform infrared (FTIR) spectroscopy, 34
fracture toughness, 236
fullerenes, 175, 177, 184, 358, 411, 434, 484, 488, 519

functional groups, 28, 34, 274, 323, 398, 400
– amine, 415
– carboxylic acid, 407
– quinone, 402, 403
– thiol, 415

gas adsorption, 514
– SO_2, 288
– CO_2, 514
– H_2, 276
– H_2O, 285, 287
– H_2S, 286, 288
– N_2, 514
– NH_4, 282
– NO_2, 284, 288
gas diffusion layer (GDL), 368
gas permeability, 102
gauge factor, 233
glass fiber, 102
glucose, 365
gradient electrochromic, 459
graphene
– catalyst, 31
– characterization, 32
– electronic configuration, 506
– functionalization, 182
– nanoribbons, 437
– properties, 25, 171, 319
– radicals, 505
– reactivity, 507, 513
– synthesis, 26, 172
graphene analogues, 26, 174
graphene edge atoms, 28, 76, 194, 394, 398, 411, 503
– external, 505
– internal, 515
graphene edges, 323
graphene effects on
– morphology, 194
– particle size, 194
graphene materials
– few-layer graphene, 34, 172, 178, 182, 395
– graphene oxide (GO), 28, 174, 231, 273, 320, 323, 408, 413, 494
– nanoribbons (GNRs), 503
– nanosheets, 301, 321
– quantum dots, 508
– reduced graphene oxide (rGO), 173, 175, 181, 231, 320
– single layer graphene, 26, 29, 30, 319
– single-layer, 515
– single-layer graphene, 172, 394
graphene nanoribbons, 76, 91
graphene nanosheets, 182
graphene oxide
– reduction, 138
graphene paper, 183, 302
graphene-fullerene hybrids, 177
graphene-metal hybrid, 418
graphene-metal oxide hybrids, 187, 190, 321, 323, 480
graphene-porphyrine hybrids, 490
graphene-semiconductor hybrids, 187
graphite, 25, 303, 365

Hall–Heroult process, 363
hardness, 92
heat transfer, 234, 244
heat-sink effect, 147, 323
Heck coupling, 53
helicity, 44
heterogeneous catalysis, 395
– alcohol dehydration, 407
– alcohol synthesis, 400
– ammonia decomposition, 400, 421
– Fischer Tropsch synthesis, 419
– hydrogenation of cinnamaldehyde, 416
– oxidative dehydrogenation, 402
– reduction of nitroarenes, 411
– selective oxidation reactions, 408, 410
HOPG, 172
Hummers method, 28, 276, 408
hybridization, 44
– *ex situ*, 127, 415
– *in situ*, 134, 412
– atomic layer deposition (ALD), 150
– chemical vapor deposition (CVD), 149
– covalent, 127
– electrochemical, 142
– electroless deposition, 141
– electrophoresis, 132
– electropolymerization, 144
– graphitic carbon nitride, 153
– hydrophobic, 129
– noncovalent, 129
– physical vapor deposition (PVD), 148
– polymerization, 135
– single-walled carbon nanohorns, 152

- sol-gel, 146
- with supercritical solvent, 139
- within CNT, 139
- zeolite, 146
hydrogen peroxide (H_2O_2), 365
hydrophilicity, 74
hydrothermal carbonization (HTC), 201, 207
hydrothermal synthesis, 174, 191

immobilization, 396
impregnation, 412
impurities, 398
indium-doped tin oxide (ITO), 25
indium-tin oxide (ITO), 476
Infrared spectroscopy (FTIR), 11
infusion techniques, 236
interactions
- π-π bonding, 370
- π-stacking, 55, 57, 63
- ∂–∂ stacking, 183
- electrostatic, 54, 63, 327
- hydrogen bond, 63, 179
- van der Waals, 54, 63
intercalation, 299, 361, 457
- acids, 74
- alkali atoms, 74
- alkali metals, 27, 299
- halides, 28, 74
- oxides, 74
interface, 16
- metal oxide-nanocarbon, 322
- nanocarbon-nanocarbon, 231
- polymer-nanocarbon, 231
- semiconductor-nanocarbon, 431, 443
- semiconductor-semiconductor, 445
interfacial shear strength (IFSS), 231, 237
interlaminar shear strength (ILSS, 236
ion bombardment, 89
ionic liquid, 133, 415, 485
iridium, 457
IrO_2, 313

Joule heating, 234

laminate, 228
Laser Ablation, 12
layer-by-layer (LbL), 494
layer-by-layer (LbL) deposition, 332
layered silicates, 88
leaching, 416
$LiFePO_4$, 335
light emitting devices, 55
lightning strike protection, 236
linking agents
- alkyammonium ions, 86
- amino acids, 85
$LiPF_6$, 301
lithiation, 304
lithium air battery, 308
lithium ion battery (LIB), 189, 298, 320
lithium phosphate olivine, 306
lithium storage capacity, 190, 192, 300, 301, 303, 304, 307, 309, 321
lithium sulfur battery, 307
low energy electron diffraction (LEED), 504
luminescence, 507

magnetic properties, 513
Mars–van Krevelen mechanism, 10
mechanical properties, 7, 26, 179, 232, 244
- diamond-like carbon (DLC), 92
- doping, 89
- graphene nanoribbon, 91
- single layer graphene, 91
- SWCNTs, 89
mesoporous carbon, 204, 209, 308, 331, 366, 373, 396, 406, 433, 445
mesoporous metal oxides, 330, 476
metal nanoparticles, 206
- Ag, 206
- alloy, 304, 417, 421
- Au, 186, 206, 416
- Co, 383
- electrochemical synthesis, 143
- Fe, 381
- Ni, 384
- non covalent, 63
- Pd, 185, 206, 415, 417, 418
- photoreduction, 140
- Pt, 186, 370
- Rh, 417
- Ru, 400, 417, 419, 421
- Si, 211, 304
- Sn, 212, 304
metal organic frameworks (MOFs), 273
metal sulfides, 288
metal-organic-frameworks (MOFs), 192
metal-support interaction, 372, 377, 416, 422

metallic single-walled carbon nanotubes, 456
MgO, 415
micromechanical cleavage, 26, 172
micropores, 286
microreactor, 396
microspheres
– carbon, 204
– metal oxides, 213
microwave, 81, 234
microwave synthesis, 194
MIL (Material of Institut Lavoisier), 275
MnO_2, 309, 332
Mo_2C, 413
moisture, 103, 282
molecular dynamic simulation, 89, 91
Molten salt route, 13
molybdenum, 457
MoO_2, 305
MoS_2, 174, 191

Nafion, 373, 374, 481
nanocarbon
– classification, 394
nanocarbon composites, 26
– fiber route, 240
– filler route, 179, 229
– hierarchical route, 235
– integration route, 179, 228
nanocarbon effects on
– adsorption properties, 445
– band gap, 440
– morphology, 322, 323, 443
– particle size, 323
nanocarbon hybrids
– benefits, 443
– definition, 430
– overview, 432
– synthesis, 412
nanocarbon-metal nanoparticle hybrid, 140, 143
nanocarbon-metal oxide hybrid, 139, 141, 146
nanocarbon-metal thin film hybrid, 142
nanocarbon-polymer hybrid, 85, 135, 144
nanodiamond, 74, 179, 404, 411
nanofiller, 178
– aggregation, 85
– exfoliation, 84
– general, 84
– nanoclay, 84, 86, 87
nanoindenter, 89
nanoreactor, 60, 273
nanotubes
– carbon, 206
nanowires, 323
Nernst potential, 477
NiO, 212, 313, 476, 482, 489
nitrene, 51
nitrobenzene, 362
nitrogen-doped carbon, 206, 379
nitrogen-doped CNTs, 309, 410, 415, 418
nitrogen-doped graphene, 75, 173, 187, 377, 400, 409
non-covalent functionalization
– ð stacking , 183
– surfactants, 183
noncovalent functionalization, 17, 54
– π-π-stacking, 82
– aromatic molecules, 59, 131, 183
– biomolecules, 58, 82, 183
– polymer, 55
– surfactants, 55, 129
noncovalent hybridization
– electrostatic, 132
nonlinear optical (NLO), 184
nucleation, 322, 412
Nylon 6, 85

oligo(phenylenevinylene) amine (OPV), 184
optical properties, 97
optical transmittance, 319
optoelectronic materials, 455
optoelectronics, 56, 81
organic dyes
– methyl orange (MO), 220
organic liquid electrolyte, 309
organic photovoltaics, 461
oxidation, 16
oxidative dehydrogenation
– alcohols, 407
– alkanes, 405
– ethylbenzene, 402, 422
– kinetics, 404
oxygen reduction reaction (ORR), 357, 372, 374, 376, 383

P_2O_5, 405
p-p stacking, 17
p-phenylenediamine, 362
panel repair, 234

particle size effect, 421
percolation, 232
percolation threshold, 95, 97, 232
perlocation, 181
permittivity, 96
peroxide, 52, 53
photo-to-current conversion efficiency (IPCE), 485
photoanode, 365, 439, 445, 446, 477, 480
photocatalysis
– degradation of organic compounds, 187, 194, 218, 430, 432, 443
– overall water splitting, 443
– plasmon assisted, 218
– sacrificial water splitting, 187, 218, 365, 447
photocatalytic activity, 194
photocathode, 477
photoconversion efficiency, 365
photoelectrochemical cells, 461
photoelectrochemistry, 221
photoelectrolysis, 365, 442, 448
photoexcitation, 431
photoluminescence, 11, 438
photolysis, 52
photosensitization, 218, 432
photosensitizer, 480
photovoltaic devices, 55
– DSSC, 439, 446, 447
photovoltaics, 455
– bulk heterojunction cells (BHJ), 475
– dye-sensitized solar cell (DSSC), 248
– dye-sensitized solar cells (DSSC), 475
– hybrid cells, 188
– organic cells (OPV), 177, 475
piezoresistance, 233, 248
plastic deformation, 89
poly (3-methylthiophene (P3MT), 383
poly(1-phenyl-1-butyne) (PPB), 80
poly(3-hexylthiophene), 469
poly(acrylonitriles) (PAN), 182
poly(bisphenol A carbonate), 92
poly(diallyldimethylammonium PDDA), 332
poly(ethylene glycol), 82
poly(methyl methacrylate) (PMMA), 80, 87, 179, 181
poly(p-phenylene benzobisoxazole), 87
poly(p-phenylene terephthalamide) (PPTA), 242
poly(p-phenylene-2,6-benzobisoxazole) (PBO), 242
poly(phenylene vinylene -co -2,5 - dioctoxy-m-phenylene vinylene) (PMPV), 96
poly(sodium 4-styrenesulfonate), 183
poly-(diallyl-dimethylammonium) (PDDA), 495
poly-(sodium-4-styrenesulfonate) (PSS), 495
polyacrylonitrile (PAN), 360, 385, 403
polyaniline, 87
polyaniline (PANI), 179, 180, 182, 370, 383, 403, 484
polycaprolactone, 93
polycarbonate, 80
polycyclic aromatic hydrocarbon (PAH, 505
polyester, 80, 86
polyetherimide, 103
polyethylene (PE), 81, 93
polyethylene glycol (PEG), 48, 307
polyethylene oxide (PEO), 80, 485
polyethylene terephthalate, 97
polyethylenimine (PEI), 103
polyimide, 81, 103
polylactic acid, 98
polymer
– crystallization, 94
– electrolyte fuel cell (PEFC), 357, 368
– grafting, 48, 54
– nanocomposites, 178
– poly(aminobenzene sulfonic acid) (PABS), 48
– poly(diallyldimethylammonium chloride) (PDDA), 65
– poly(phenylenevinylene) (PPV), 56
– poly(sodium 4-styrenesulfonate) (PSS), 65
– polyacrylonitriles (PAN), 215
– polyfurane, 203
– polystyrene (PS), 54
– polystyrene-sulfonate, 54
– polythiophene (PT), 56
– polyvinyl alcohol (PVA), 206
– polyvinylpyridine (PVP), 54, 215
polymer synthesis
– *in situ* polymerization, 86
– melt blending, 88
– solvent methods, 87
polynorbornene, 103
polypropylene (PP), 93
polypyrrole (PPy), 370, 383, 484
polystyrene (PS), 80, 81, 92, 103, 181
polysulfide, 307
polyurethane, 80, 103
polyvinyl alcohol (PVA), 179

pore structure, 395, 416
porosity, 194, 276
porphyrin, 478
power conversion efficiency (PCE), 177, 476, 488
power density, 310, 327, 371
proton exchange membrane (PEM), 186, 365
pseudocapacitor, 310
Purification, 14
pyrolysis, 384

quantum dots (QDs), 62, 138, 187, 438, 439
quantum Hall effect, 319, 509

Raman spectroscopy, 10, 26, 33, 172, 178, 186
reduced graphene oxide (rGO), 188, 468
reinforcement, 230
resin transfer molding, 236
rotating disk electrode (RDE), 381
rubber, 228
rule of mixtures, 230
RuO_2, 313, 314, 323, 326

scanning tunneling microscopy (STM), 504
Schottky junction, 444, 480
selective deposition, 413
selectivity, 395, 406, 410
self-assembly, 333
sensor
– chemical, 245
– mechanical, 248
separation, 15
separation of SWCNTs, 45
shape memory polymers, 182
shear mixing, 97
Shockley–Queisser limit, 475
SiC, 31
SiO_2, 62, 211
SnO_2, 190, 213, 324, 334
solar water splitting cells, 461
solid state electrolyte, 309
solid-electrolyte interface (SEI), 300, 304
solid-state electrolytes, 477, 484
soot, 396
specific capacitance, 182, 311, 312, 324, 326, 332
specific capacity, 321
specific surface area, 179, 275, 276, 310, 321, 369, 394, 395
spintronics, 510
stiffness, 25, 73, 319, 395
stress transfer, 244
structural health monitoring (SHM), 233
sulfur, 307
supercapacitors, 215, 310, 320
superconductivity, 27, 75
superoxygen ions, 289
superparamagnetism, 64
supramolecular, 55
surface chemistry, 396, 397
surface energy, 29
surface reactions, 361
surface reconstruction, 503, 519
surfactant, 29, 129
– cetyltrimethylammonium bromide (CTAB), 55
– octadecylamine (ODA), 45, 184
– sodium dodecylbenzene sulfonate (SDBS), 55
– sodium dodecylsulfate (SDS), 55
– Triton X, 55
Suzuki reaction, 53

templates
– anodic alumina membranes, 205
– metal nanowires, 205, 207
– SBA 15, 204
– SiO_2, 204
tensile strength, 8, 90, 91, 360
thermal conductivity, 9, 26, 72, 93, 182, 238, 319, 396
– doping, 90
Thermal expansion, 10
Thermal stability, 10
thermoplast, 88
thermoplastics, 229, 234
thermoset, 86, 229
Thrower–Stone–Wales (TSW), 517
Thrower–Stone–Wales defects (TSW), 76
TiO_2, 62, 64, 187, 218, 365, 415, 430, 432, 433, 439, 440, 479, 485
– band gap, 440
– carbon doping, 432, 441, 442
– morphology, 444
tip *vs.* sidewall, 47
tips vs walls, 413
tissue regeneration, 98
Tollens reaction, 206
Toxic Industrial Compounds, TICs, 273
transition metal carbides, 413
transition metal oxides, 313, 457

transparent conducting films, 176, 248
transparent conducting oxide (TCO), 439
transparent conductive films, 56
triple phase boundary, 374

ultrasonication, 29, 88, 97, 483

Van Hove singularities, 456
Volmer/Heyrovsky mechanism, 362
volume fraction, 233, 430

WO_3, 365
work function, 479
WS_2, 174, 192

X-ray photo-emission spectroscopy (XPS), 34
X-ray photoelectron spectroscopy (XPS), 11, 384
X-ray photoemission spectroscopy (XPS), 400

Young's modulus, 7, 73, 85, 91, 179, 230, 319

Zeeman coupling, 510
zeolite, 194, 275, 403
zinc-based phathalocyanine (ZnPc), 492
ZnO, 186, 438, 491
$Zr(SO_4)_2$, 415